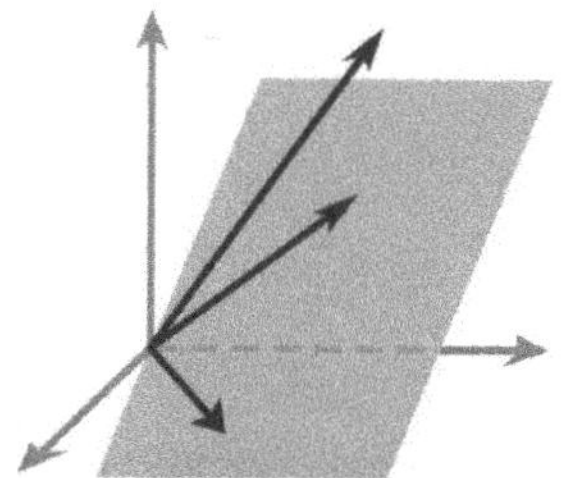

Student Solutions Manual

to accompany

Elementary Linear Algebra

11th Edition

and

Elementary Linear Algebra: Applications Version

11th Edition

Howard Anton
Professor Emeritus, Drexel University

Chris Rorres
University of Pennsylvania

Prepared by

Przemyslaw Bogacki
Old Dominion University

Cover Design/Illustration Norm Christiansen

ISBN: 978-1-118-46442-7

To order books or for customer service please, call 1-800-CALL WILEY (225-5945).

Printed in the United States of America

V10013116_081519

Table of Contents

CHAPTER 1: SYSTEMS OF LINEAR EQUATIONS AND MATRICES

1.1 Introduction to Systems of Linear Equations

1. **(a)** This is a linear equation in x_1, x_2, and x_3.

(b) This is not a linear equation in x_1, x_2, and x_3 because of the term x_1x_3.

(c) We can rewrite this equation in the form $x_1 + 7x_2 - 3x_3 = 0$ therefore it is a linear equation in x_1, x_2, and x_3.

(d) This is not a linear equation in x_1, x_2, and x_3 because of the term x_1^{-2}.

(e) This is not a linear equation in x_1, x_2, and x_3 because of the term $x_1^{3/5}$.

(f) This is a linear equation in x_1, x_2, and x_3.

3. **(a)**
$$\begin{aligned} a_{11}x_1 + a_{12}x_2 &= b_1 \\ a_{21}x_1 + a_{22}x_2 &= b_2 \end{aligned}$$

(b)
$$\begin{aligned} a_{11}x_1 + a_{12}x_2 + a_{13}x_3 &= b_1 \\ a_{21}x_1 + a_{22}x_2 + a_{23}x_3 &= b_2 \\ a_{31}x_1 + a_{32}x_2 + a_{33}x_3 &= b_3 \end{aligned}$$

(c)
$$\begin{aligned} a_{11}x_1 + a_{12}x_2 + a_{13}x_3 + a_{14}x_4 &= b_1 \\ a_{21}x_1 + a_{22}x_2 + a_{23}x_3 + a_{24}x_4 &= b_2 \end{aligned}$$

5. **(a)**
$$\begin{aligned} 2x_1 \qquad\quad &= 0 \\ 3x_1 - 4x_2 &= 0 \\ x_2 &= 1 \end{aligned}$$

(b)
$$\begin{aligned} 3x_1 \qquad\quad - 2x_3 &= 5 \\ 7x_1 + x_2 + 4x_3 &= -3 \\ -2x_2 + x_3 &= 7 \end{aligned}$$

7. **(a)**
$$\begin{bmatrix} -2 & 6 \\ 3 & 8 \\ 9 & -3 \end{bmatrix}$$

(b)
$$\begin{bmatrix} 6 & -1 & 3 & 4 \\ 0 & 5 & -1 & 1 \end{bmatrix}$$

(c)
$$\begin{bmatrix} 0 & 2 & 0 & -3 & 1 & 0 \\ -3 & -1 & 1 & 0 & 0 & -1 \\ 6 & 2 & -1 & 2 & -3 & 6 \end{bmatrix}$$

9. The values in (a), (d), and (e) satisfy all three equations – these 3-tuples are solutions of the system. The 3-tuples in (b) and (c) are not solutions of the system.

11. **(a)** We can eliminate x from the second equation by adding -2 times the first equation to the second. This yields the system

$$\begin{aligned} 3x - 2y &= 4 \\ 0 &= 1 \end{aligned}$$

The second equation is contradictory, so the original system has no solutions. The lines represented by the equations in that system have no points of intersection (the lines are parallel and distinct).

(b) We can eliminate x from the second equation by adding -2 times the first equation to the second. This yields the system

$$\begin{array}{rcrcr} 2x & - & 4y & = & 1 \\ & & 0 & = & 0 \end{array}$$

The second equation does not impose any restriction on x and y therefore we can omit it. The lines represented by the original system have infinitely many points of intersection. Solving the first equation for x we obtain $x = \frac{1}{2} + 2y$. This allows us to represent the solution using parametric equations

$$x = \frac{1}{2} + 2t, \quad y = t$$

where the parameter t is an arbitrary real number.

(c) We can eliminate x from the second equation by adding -1 times the first equation to the second. This yields the system

$$\begin{array}{rcrcr} x & - & 2y & = & 0 \\ & - & 2y & = & 8 \end{array}$$

From the second equation we obtain $y = -4$. Substituting -4 for y into the first equation results in $x = -8$. Therefore, the original system has the unique solution

$$x = -8, \quad y = -4$$

The represented by the equations in that system have one point of intersection: $(-8, -4)$.

13. (a) Solving the equation for x we obtain $x = \frac{3}{7} + \frac{5}{7}y$ therefore the solution set of the original equation can be described by the parametric equations

$$x = \frac{3}{7} + \frac{5}{7}t, \quad y = t$$

where the parameter t is an arbitrary real number.

(b) Solving the equation for x_1 we obtain $x_1 = \frac{7}{3} + \frac{5}{3}x_2 - \frac{4}{3}x_3$ therefore the solution set of the original equation can be described by the parametric equations

$$x_1 = \frac{7}{3} + \frac{5}{3}r - \frac{4}{3}s, \quad x_2 = r, \quad x_3 = s$$

where the parameters r and s are arbitrary real numbers.

(c) Solving the equation for x_1 we obtain $x_1 = -\frac{1}{8} + \frac{1}{4}x_2 - \frac{5}{8}x_3 + \frac{3}{4}x_4$ therefore the solution set of the original equation can be described by the parametric equations

$$x_1 = -\frac{1}{8} + \frac{1}{4}r - \frac{5}{8}s + \frac{3}{4}t, \quad x_2 = r, \quad x_3 = s, \quad x_4 = t$$

where the parameters r, s, and t are arbitrary real numbers.

(d) Solving the equation for v we obtain $v = \frac{8}{3}w - \frac{2}{3}x + \frac{1}{3}y - \frac{4}{3}z$ therefore the solution set of the original equation can be described by the parametric equations

$$v = \frac{8}{3}t_1 - \frac{2}{3}t_2 + \frac{1}{3}t_3 - \frac{4}{3}t_4, \quad w = t_1, \quad x = t_2, \quad y = t_3, \quad z = t_4$$

where the parameters t_1, t_2, t_3, and t_4 are arbitrary real numbers.

15. (a) We can eliminate x from the second equation by adding -3 times the first equation to the second. This yields the system

$$\begin{aligned} 2x - 3y &= 1 \\ 0 &= 0 \end{aligned}$$

The second equation does not impose any restriction on x and y therefore we can omit it. Solving the first equation for x we obtain $x = \frac{1}{2} + \frac{3}{2}y$. This allows us to represent the solution using parametric equations

$$x = \frac{1}{2} + \frac{3}{2}t, \quad y = t$$

where the parameter t is an arbitrary real number.

(b) We can see that the second and the third equation are multiples of the first: adding -3 times the first equation to the second, then adding the first equation to the third yields the system

$$\begin{aligned} x_1 + 3x_2 - x_3 &= -4 \\ 0 &= 0 \\ 0 &= 0 \end{aligned}$$

The last two equations do not impose any restriction on the unknowns therefore we can omit them. Solving the first equation for x_1 we obtain $x_1 = -4 - 3x_2 + x_3$. This allows us to represent the solution using parametric equations

$$x_1 = -4 - 3r + s, \quad x_2 = r, \quad x_3 = s$$

where the parameters r and s are arbitrary real numbers.

17. (a) Add 2 times the second row to the first to obtain $\begin{bmatrix} 1 & -7 & 8 & 8 \\ 2 & -3 & 3 & 2 \\ 0 & 2 & -3 & 1 \end{bmatrix}$.

(b) Add the third row to the first to obtain $\begin{bmatrix} 1 & 3 & -8 & 3 \\ 2 & -9 & 3 & 2 \\ 1 & 4 & -3 & 3 \end{bmatrix}$

(another solution: interchange the first row and the third row to obtain $\begin{bmatrix} 1 & 4 & -3 & 3 \\ 2 & -9 & 3 & 2 \\ 0 & -1 & -5 & 0 \end{bmatrix}$).

19. (a) Add -4 times the first row to the second to obtain $\begin{bmatrix} 1 & k & -4 \\ 0 & 8-4k & 18 \end{bmatrix}$ which corresponds to the system

$$\begin{aligned} x + \qquad ky &= -4 \\ (8-4k)y &= 18 \end{aligned}$$

If $k = 2$ then the second equation becomes $0 = 18$, which is contradictory thus the system becomes inconsistent.

If $k \neq 2$ then we can solve the second equation for y and proceed to substitute this value into the first equation and solve for x.

Consequently, for all values of $k \neq 2$ the given augmented matrix corresponds to a consistent linear system.

(b) Add -4 times the first row to the second to obtain $\begin{bmatrix} 1 & k & -1 \\ 0 & 8-4k & 0 \end{bmatrix}$ which corresponds to the system

$$\begin{aligned} x + \qquad ky &= -1 \\ (8-4k)y &= 0 \end{aligned}$$

If $k = 2$ then the second equation becomes $0 = 0$, which does not impose any restriction on x and y therefore we can omit it and proceed to determine the solution set using the first equation. There are infinitely many solutions in this set.

If $k \neq 2$ then the second equation yields $y = 0$ and the first equation becomes $x = -1$.

Consequently, for all values of k the given augmented matrix corresponds to a consistent linear system.

21. Substituting the coordinates of the first point into the equation of the curve we obtain

$$y_1 = ax_1^2 + bx_1 + c$$

Repeating this for the other two points and rearranging the three equations yields

$$\begin{aligned} x_1^2a + x_1b + c &= y_1 \\ x_2^2a + x_2b + c &= y_2 \\ x_3^2a + x_3b + c &= y_3 \end{aligned}$$

This is a linear system in the unknowns a, b, and c. Its augmented matrix is $\begin{bmatrix} x_1^2 & x_1 & 1 & y_1 \\ x_2^2 & x_2 & 1 & y_2 \\ x_3^2 & x_3 & 1 & y_3 \end{bmatrix}$.

23. Solving the first equation for x_1 we obtain $x_1 = c - kx_2$ therefore the solution set of the original equation can be described by the parametric equations

$$x_1 = c - kt, \qquad x_2 = t$$

where the parameter t is an arbitrary real number.

Substituting these into the second equation yields

$$c - kt + lt = d$$

which can be rewritten as

$$c - kt = d - lt$$

This equation must hold true for all real values t, which requires that the coefficients associated with the same power of t on both sides must be equal. Consequently, $c = d$ and $k = l$.

25.
$$\begin{array}{rcrcrcr} 2x & + & 3y & + & z & = & 7 \\ 2x & + & y & + & 3z & = & 9 \\ 4x & + & 2y & + & 5z & = & 16 \end{array}$$

27.
$$\begin{array}{rcrcrcr} x & + & y & + & z & = & 12 \\ 2x & + & y & + & 2z & = & 5 \\ -x & & & + & z & = & 1 \end{array}$$

True-False Exercises

(a) True. $(0,0, \ldots, 0)$ is a solution.

(b) False. Only multiplication by a **nonzero** constant is a valid elementary row operation.

(c) True. If $k = 6$ then the system has infinitely many solutions; otherwise the system is inconsistent.

(d) True. According to the definition, $a_1x_1 + a_2x_2 + \cdots + a_nx_n = b$ is a linear equation if the a's are not all zero. Let us assume $a_j \neq 0$. The values of all x's except for x_j can be set to be arbitrary parameters, and the equation can be used to express x_j in terms of those parameters.

(e) False. E.g. if the equations are all homogeneous then the system must be consistent. (See True-False Exercise (a) above.)

(f) False. If $c \neq 0$ then the new system has the same solution set as the original one.

(g) True. Adding -1 times one row to another amounts to the same thing as subtracting one row from another.

(h) False. The second row corresponds to the equation $0 = -1$, which is contradictory.

1.2 Gaussian Elimination

1. **(a)** This matrix has properties 1-4. It is in reduced row echelon form, therefore it is also in row echelon form.

(b) This matrix has properties 1-4. It is in reduced row echelon form, therefore it is also in row echelon form.

(c) This matrix has properties 1-4. It is in reduced row echelon form, therefore it is also in row echelon form.

(d) This matrix has properties 1-4. It is in reduced row echelon form, therefore it is also in row echelon form.

(e) This matrix has properties 1-4. It is in reduced row echelon form, therefore it is also in row echelon form.

(f) This matrix has properties 1-4. It is in reduced row echelon form, therefore it is also in row echelon form.

(g) This matrix has properties 1-3 but does not have property 4: the second column contains a leading 1 and a nonzero number (−7) above it. The matrix is in row echelon form but not reduced row echelon form.

3. **(a)** The linear system

$$\begin{array}{rcrcrcl} x & - & 3y & + & 4z & = & 7 \\ & & y & + & 2z & = & 2 \\ & & & & z & = & 5 \end{array} \quad \text{can be rewritten as} \quad \begin{array}{rcl} x & = & 7+3y-4z \\ y & = & 2-2z \\ z & = & 5 \end{array}$$

and solved by back-substitution:

$$z = 5$$
$$y = 2 - 2(5) = -8$$
$$x = 7 + 3(-8) - 4(5) = -37$$

therefore the original linear system has a unique solution: $x = -37,\ y = -8,\ z = 5$.

(b) The linear system

$$\begin{array}{rcrcrcrcl} w & & & + & 8y & - & 5z & = & 6 \\ & & x & + & 4y & - & 9z & = & 3 \\ & & & & y & + & z & = & 2 \end{array} \quad \text{can be rewritten as} \quad \begin{array}{rcl} w & = & 6-8y+5z \\ x & = & 3-4y+9z \\ y & = & 2-z \end{array}$$

Let $z = t$. Then

$$y = 2 - t$$
$$x = 3 - 4(2 - t) + 9t = -5 + 13t$$
$$w = 6 - 8(2 - t) + 5t = -10 + 13t$$

therefore the original linear system has infinitely many solutions:

$$w = -10 + 13t, x = -5 + 13t, y = 2 - t, z = t$$

where t is an arbitrary value.

(c) The linear system

$$\begin{array}{rcrcrcrcrcl} x_1 & + & 7x_2 & - & 2x_3 & & & - & 8x_5 & = & -3 \\ & & & & x_3 & + & x_4 & + & 6x_5 & = & 5 \\ & & & & & & x_4 & + & 3x_5 & = & 9 \\ & & & & & & & & 0 & = & 0 \end{array}$$

can be rewritten: $x_1 = -3 - 7x_2 + 2x_3 + 8x_5,\ x_3 = 5 - x_4 - 6x_5,\ x_4 = 9 - 3x_5$.

Let $x_2 = s$ and $x_5 = t$. Then

$$x_4 = 9 - 3t$$
$$x_3 = 5 - (9 - 3t) - 6t = -4 - 3t$$
$$x_1 = -3 - 7s + 2(-4 - 3t) + 8t = -11 - 7s + 2t$$

therefore the original linear system has infinitely many solutions:

$$x_1 = -11 - 7s + 2t,\ x_2 = s,\ x_3 = -4 - 3t,\ x_4 = 9 - 3t,\ x_5 = t$$

where s and t are arbitrary values.

(d) The system is inconsistent since the third row of the augmented matrix corresponds to the equation

$$0x + 0y + 0z = 1.$$

5.

$$\begin{bmatrix} 1 & 1 & 2 & 8 \\ -1 & -2 & 3 & 1 \\ 3 & -7 & 4 & 10 \end{bmatrix} \longleftarrow \text{The augmented matrix for the system.}$$

$$\begin{bmatrix} 1 & 1 & 2 & 8 \\ 0 & -1 & 5 & 9 \\ 3 & -7 & 4 & 10 \end{bmatrix} \longleftarrow \text{The first row was added to the second row.}$$

$$\begin{bmatrix} 1 & 1 & 2 & 8 \\ 0 & -1 & 5 & 9 \\ 0 & -10 & -2 & -14 \end{bmatrix} \longleftarrow -3 \text{ times the first row was added to the third row.}$$

$$\begin{bmatrix} 1 & 1 & 2 & 8 \\ 0 & 1 & -5 & -9 \\ 0 & -10 & -2 & -14 \end{bmatrix} \longleftarrow \text{The second row was multiplied by } -1.$$

$$\begin{bmatrix} 1 & 1 & 2 & 8 \\ 0 & 1 & -5 & -9 \\ 0 & 0 & -52 & -104 \end{bmatrix} \longleftarrow 10 \text{ times the second row was added to the third row.}$$

$$\begin{bmatrix} 1 & 1 & 2 & 8 \\ 0 & 1 & -5 & -9 \\ 0 & 0 & 1 & 2 \end{bmatrix} \longleftarrow \text{The third row was multiplied by } -\tfrac{1}{52}.$$

The system of equations corresponding to this augmented matrix in row echelon form is

$$\begin{aligned} x_1 + x_2 + 2x_3 &= 8 \\ x_2 - 5x_3 &= -9 \\ x_3 &= 2 \end{aligned} \qquad \text{and can be rewritten as} \qquad \begin{aligned} x_1 &= 8 - x_2 - 2x_3 \\ x_2 &= -9 + 5x_3 \\ x_3 &= 2 \end{aligned}$$

Back-substitution yields

$$\begin{aligned} x_3 &= 2 \\ x_2 &= -9 + 5(2) = 1 \\ x_1 &= 8 - 1 - 2(2) = 3 \end{aligned}$$

The linear system has a unique solution: $x_1 = 3,\ x_2 = 1,\ x_3 = 2.$

7.

$$\begin{bmatrix} 1 & -1 & 2 & -1 & -1 \\ 2 & 1 & -2 & -2 & -2 \\ -1 & 2 & -4 & 1 & 1 \\ 3 & 0 & 0 & -3 & -3 \end{bmatrix} \longleftarrow \text{The augmented matrix for the system.}$$

$$\begin{bmatrix} 1 & -1 & 2 & -1 & -1 \\ 0 & 3 & -6 & 0 & 0 \\ -1 & 2 & -4 & 1 & 1 \\ 3 & 0 & 0 & -3 & -3 \end{bmatrix} \longleftarrow -2 \text{ times the first row was added to the second row.}$$

$$\begin{bmatrix} 1 & -1 & 2 & -1 & -1 \\ 0 & 3 & -6 & 0 & 0 \\ 0 & 1 & -2 & 0 & 0 \\ 3 & 0 & 0 & -3 & -3 \end{bmatrix} \longleftarrow \text{The first row was added to the third row.}$$

$$\begin{bmatrix} 1 & -1 & 2 & -1 & -1 \\ 0 & 3 & -6 & 0 & 0 \\ 0 & 1 & -2 & 0 & 0 \\ 0 & 3 & -6 & 0 & 0 \end{bmatrix}$$ ⟵ -3 times the first row was added to the fourth row.

$$\begin{bmatrix} 1 & -1 & 2 & -1 & -1 \\ 0 & 1 & -2 & 0 & 0 \\ 0 & 1 & -2 & 0 & 0 \\ 0 & 3 & -6 & 0 & 0 \end{bmatrix}$$ ⟵ The second row was multiplied by $\frac{1}{3}$.

$$\begin{bmatrix} 1 & -1 & 2 & -1 & -1 \\ 0 & 1 & -2 & 0 & 0 \\ 0 & 0 & 0 & 0 & 0 \\ 0 & 3 & -6 & 0 & 0 \end{bmatrix}$$ ⟵ -1 times the second row was added to the third row.

$$\begin{bmatrix} 1 & -1 & 2 & -1 & -1 \\ 0 & 1 & -2 & 0 & 0 \\ 0 & 0 & 0 & 0 & 0 \\ 0 & 0 & 0 & 0 & 0 \end{bmatrix}$$ ⟵ -3 times the second row was added to the fourth row.

The system of equations corresponding to this augmented matrix in row echelon form is

$$\begin{aligned} x - y + 2z - w &= -1 \\ y - 2z &= 0 \\ 0 &= 0 \\ 0 &= 0 \end{aligned}$$

Solve the equations for the leading variables

$$\begin{aligned} x &= -1 + y - 2z + w \\ y &= 2z \end{aligned}$$

then substitute the second equation into the first

$$\begin{aligned} x &= -1 + 2z - 2z + w = -1 + w \\ y &= 2z \end{aligned}$$

If we assign z and w the arbitrary values s and t, respectively, the general solution is given by the formulas

$$x = -1 + t, \qquad y = 2s, \qquad z = s, \qquad w = t$$

9.

$$\begin{bmatrix} 1 & 1 & 2 & 8 \\ -1 & -2 & 3 & 1 \\ 3 & -7 & 4 & 10 \end{bmatrix}$$ ⟵ The augmented matrix for the system.

$$\begin{bmatrix} 1 & 1 & 2 & 8 \\ 0 & -1 & 5 & 9 \\ 3 & -7 & 4 & 10 \end{bmatrix}$$ ⟵ The first row was added to the second row.

$$\begin{bmatrix} 1 & 1 & 2 & 8 \\ 0 & -1 & 5 & 9 \\ 0 & -10 & -2 & -14 \end{bmatrix}$$ ⟵ -3 times the first row was added to the third row.

$$\begin{bmatrix} 1 & 1 & 2 & 8 \\ 0 & 1 & -5 & -9 \\ 0 & -10 & -2 & -14 \end{bmatrix}$$ ⟵ The second row was multiplied by -1.

$$\begin{bmatrix} 1 & 1 & 2 & 8 \\ 0 & 1 & -5 & -9 \\ 0 & 0 & -52 & -104 \end{bmatrix}$$ ⟵ 10 times the second row was added to the third row.

$$\begin{bmatrix} 1 & 1 & 2 & 8 \\ 0 & 1 & -5 & -9 \\ 0 & 0 & 1 & 2 \end{bmatrix}$$ ⟵ The third row was multiplied by $-\frac{1}{52}$.

$$\begin{bmatrix} 1 & 1 & 2 & 8 \\ 0 & 1 & 0 & 1 \\ 0 & 0 & 1 & 2 \end{bmatrix}$$ ⟵ 5 times the third row was added to the second row.

$$\begin{bmatrix} 1 & 1 & 0 & 4 \\ 0 & 1 & 0 & 1 \\ 0 & 0 & 1 & 2 \end{bmatrix}$$ ⟵ -2 times the third row was added to the first row.

$$\begin{bmatrix} 1 & 0 & 0 & 3 \\ 0 & 1 & 0 & 1 \\ 0 & 0 & 1 & 2 \end{bmatrix}$$ ⟵ -1 times the second row was added to the first row.

The linear system has a unique solution: $x_1 = 3,\ x_2 = 1,\ x_3 = 2$.

11. $$\begin{bmatrix} 1 & -1 & 2 & -1 & -1 \\ 2 & 1 & -2 & -2 & -2 \\ -1 & 2 & -4 & 1 & 1 \\ 3 & 0 & 0 & -3 & -3 \end{bmatrix}$$ ⟵ The augmented matrix for the system.

$$\begin{bmatrix} 1 & -1 & 2 & -1 & -1 \\ 0 & 3 & -6 & 0 & 0 \\ -1 & 2 & -4 & 1 & 1 \\ 3 & 0 & 0 & -3 & -3 \end{bmatrix}$$ ⟵ -2 times the first row was added to the second row.

$$\begin{bmatrix} 1 & -1 & 2 & -1 & -1 \\ 0 & 3 & -6 & 0 & 0 \\ 0 & 1 & -2 & 0 & 0 \\ 3 & 0 & 0 & -3 & -3 \end{bmatrix}$$ ⟵ the first row was added to the third row.

$$\begin{bmatrix} 1 & -1 & 2 & -1 & -1 \\ 0 & 3 & -6 & 0 & 0 \\ 0 & 1 & -2 & 0 & 0 \\ 0 & 3 & -6 & 0 & 0 \end{bmatrix}$$ ⟵ -3 times the first row was added to the fourth row.

$$\begin{bmatrix} 1 & -1 & 2 & -1 & -1 \\ 0 & 1 & -2 & 0 & 0 \\ 0 & 1 & -2 & 0 & 0 \\ 0 & 3 & -6 & 0 & 0 \end{bmatrix}$$ ⟵ The second row was multiplied by $\frac{1}{3}$.

$$\begin{bmatrix} 1 & -1 & 2 & -1 & -1 \\ 0 & 1 & -2 & 0 & 0 \\ 0 & 0 & 0 & 0 & 0 \\ 0 & 3 & -6 & 0 & 0 \end{bmatrix}$$ ⟵ -1 times the second row was added to the third row.

$$\begin{bmatrix} 1 & -1 & 2 & -1 & -1 \\ 0 & 1 & -2 & 0 & 0 \\ 0 & 0 & 0 & 0 & 0 \\ 0 & 0 & 0 & 0 & 0 \end{bmatrix} \longleftarrow \text{-3 times the second row was added to the fourth row.}$$

$$\begin{bmatrix} 1 & 0 & 0 & -1 & -1 \\ 0 & 1 & -2 & 0 & 0 \\ 0 & 0 & 0 & 0 & 0 \\ 0 & 0 & 0 & 0 & 0 \end{bmatrix} \longleftarrow \text{the second row was added to the first row.}$$

The system of equations corresponding to this augmented matrix in row echelon form is

$$\begin{array}{rcrcrcrcr} x & & & & & - & w & = & -1 \\ & & y & - & 2z & & & = & 0 \\ & & & & & & 0 & = & 0 \\ & & & & & & 0 & = & 0 \end{array}$$

Solve the equations for the leading variables

$$x = -1 + w$$
$$y = 2z$$

If we assign z and w the arbitrary values s and t, respectively, the general solution is given by the formulas

$$x = -1 + t, \qquad y = 2s, \qquad z = s, \qquad w = t$$

13. Since the number of unknowns (4) exceeds the number of equations (3), it follows from Theorem 1.2.2 that this system has infinitely many solutions. Those include the trivial solution and infinitely many nontrivial solutions.

15. We present two different solutions.
Solution I uses Gauss-Jordan elimination

$$\begin{bmatrix} 2 & 1 & 3 & 0 \\ 1 & 2 & 0 & 0 \\ 0 & 1 & 1 & 0 \end{bmatrix} \longleftarrow \text{The augmented matrix for the system.}$$

$$\begin{bmatrix} 1 & \frac{1}{2} & \frac{3}{2} & 0 \\ 1 & 2 & 0 & 0 \\ 0 & 1 & 1 & 0 \end{bmatrix} \longleftarrow \text{The first row was multiplied by $\tfrac{1}{2}$.}$$

$$\begin{bmatrix} 1 & \frac{1}{2} & \frac{3}{2} & 0 \\ 0 & \frac{3}{2} & -\frac{3}{2} & 0 \\ 0 & 1 & 1 & 0 \end{bmatrix} \longleftarrow \text{-1 times the first row was added to the second row.}$$

$$\begin{bmatrix} 1 & \frac{1}{2} & \frac{3}{2} & 0 \\ 0 & 1 & -1 & 0 \\ 0 & 1 & 1 & 0 \end{bmatrix} \longleftarrow \text{The second row was multiplied by $\tfrac{2}{3}$.}$$

$$\begin{bmatrix} 1 & \frac{1}{2} & \frac{3}{2} & 0 \\ 0 & 1 & -1 & 0 \\ 0 & 0 & 2 & 0 \end{bmatrix}$$ ⟵ -1 times the second row was added to the third row.

$$\begin{bmatrix} 1 & \frac{1}{2} & \frac{3}{2} & 0 \\ 0 & 1 & -1 & 0 \\ 0 & 0 & 1 & 0 \end{bmatrix}$$ ⟵ The third row was multiplied by $\frac{1}{2}$.

$$\begin{bmatrix} 1 & \frac{1}{2} & 0 & 0 \\ 0 & 1 & 0 & 0 \\ 0 & 0 & 1 & 0 \end{bmatrix}$$ ⟵ The third row was added to the second row and $-\frac{3}{2}$ times the third row was added to the first row

$$\begin{bmatrix} 1 & 0 & 0 & 0 \\ 0 & 1 & 0 & 0 \\ 0 & 0 & 1 & 0 \end{bmatrix}$$ ⟵ $-\frac{1}{2}$ times the second row was added to the first row.

Unique solution: $x_1 = 0,\ x_2 = 0,\ x_3 = 0.$

Solution II. This time, we shall choose the order of the elementary row operations differently in order to avoid introducing fractions into the computation. (Since every matrix has a unique reduced row echelon form, the exact sequence of elementary row operations being used does not matter – see part 1 of the discussion "Some Facts About Echelon Forms" on p. 21)

$$\begin{bmatrix} 2 & 1 & 3 & 0 \\ 1 & 2 & 0 & 0 \\ 0 & 1 & 1 & 0 \end{bmatrix}$$ ⟵ The augmented matrix for the system.

$$\begin{bmatrix} 1 & 2 & 0 & 0 \\ 2 & 1 & 3 & 0 \\ 0 & 1 & 1 & 0 \end{bmatrix}$$ ⟵ The first and second rows were interchanged (to avoid introducing fractions into the first row).

$$\begin{bmatrix} 1 & 2 & 0 & 0 \\ 0 & -3 & 3 & 0 \\ 0 & 1 & 1 & 0 \end{bmatrix}$$ ⟵ -2 times the first row was added to the second row.

$$\begin{bmatrix} 1 & 2 & 0 & 0 \\ 0 & 1 & -1 & 0 \\ 0 & 1 & 1 & 0 \end{bmatrix}$$ ⟵ The second row was multiplied by $-\frac{1}{3}$.

$$\begin{bmatrix} 1 & 2 & 0 & 0 \\ 0 & 1 & -1 & 0 \\ 0 & 0 & 2 & 0 \end{bmatrix}$$ ⟵ -1 times the second row was added to the third row.

$$\begin{bmatrix} 1 & 2 & 0 & 0 \\ 0 & 1 & -1 & 0 \\ 0 & 0 & 1 & 0 \end{bmatrix}$$ ⟵ The third row was multiplied by $\frac{1}{2}$.

$$\begin{bmatrix} 1 & 2 & 0 & 0 \\ 0 & 1 & 0 & 0 \\ 0 & 0 & 1 & 0 \end{bmatrix}$$ ⟵ The third row was added to the second row.

$$\begin{bmatrix} 1 & 0 & 0 & 0 \\ 0 & 1 & 0 & 0 \\ 0 & 0 & 1 & 0 \end{bmatrix} \longleftarrow \text{ -2 times the second row was added to the first row.}$$

Unique solution: $x_1 = 0,\ x_2 = 0,\ x_3 = 0.$

17. $$\begin{bmatrix} 3 & 1 & 1 & 1 & 0 \\ 5 & -1 & 1 & -1 & 0 \end{bmatrix} \longleftarrow \text{The augmented matrix for the system.}$$

$$\begin{bmatrix} 1 & \frac{1}{3} & \frac{1}{3} & \frac{1}{3} & 0 \\ 5 & -1 & 1 & -1 & 0 \end{bmatrix} \longleftarrow \text{The first row was multiplied by } \tfrac{1}{3}.$$

$$\begin{bmatrix} 1 & \frac{1}{3} & \frac{1}{3} & \frac{1}{3} & 0 \\ 0 & -\frac{8}{3} & -\frac{2}{3} & -\frac{8}{3} & 0 \end{bmatrix} \longleftarrow \text{ -5 times the first row was added to the second row.}$$

$$\begin{bmatrix} 1 & \frac{1}{3} & \frac{1}{3} & \frac{1}{3} & 0 \\ 0 & 1 & \frac{1}{4} & 1 & 0 \end{bmatrix} \longleftarrow \text{The second row was multiplied by } -\tfrac{3}{8}.$$

$$\begin{bmatrix} 1 & 0 & \frac{1}{4} & 0 & 0 \\ 0 & 1 & \frac{1}{4} & 1 & 0 \end{bmatrix} \longleftarrow -\tfrac{1}{3} \text{ times the second row was added to the first row.}$$

If we assign x_3 and x_4 the arbitrary values s and t, respectively, the general solution is given by the formulas

$$x_1 = -\frac{1}{4}s,\quad x_2 = -\frac{1}{4}s - t,\quad x_3 = s,\quad x_4 = t.$$

(Note that fractions in the solution could be avoided if we assigned $x_3 = 4s$ instead, which along with $x_4 = t$ would yield $x_1 = -s,\ x_2 = -s - t,\ x_3 = 4s,\ x_4 = t$.)

19. $$\begin{bmatrix} 0 & 2 & 2 & 4 & 0 \\ 1 & 0 & -1 & -3 & 0 \\ 2 & 3 & 1 & 1 & 0 \\ -2 & 1 & 3 & -2 & 0 \end{bmatrix} \longleftarrow \text{The augmented matrix for the system.}$$

$$\begin{bmatrix} 1 & 0 & -1 & -3 & 0 \\ 0 & 2 & 2 & 4 & 0 \\ 2 & 3 & 1 & 1 & 0 \\ -2 & 1 & 3 & -2 & 0 \end{bmatrix} \longleftarrow \text{The first and second rows were interchanged.}$$

$$\begin{bmatrix} 1 & 0 & -1 & -3 & 0 \\ 0 & 2 & 2 & 4 & 0 \\ 0 & 3 & 3 & 7 & 0 \\ 0 & 1 & 1 & -8 & 0 \end{bmatrix} \longleftarrow \text{ -2 times the first row was added to the third row and 2 times the first row was added to the fourth row.}$$

$$\begin{bmatrix} 1 & 0 & -1 & -3 & 0 \\ 0 & 1 & 1 & 2 & 0 \\ 0 & 3 & 3 & 7 & 0 \\ 0 & 1 & 1 & -8 & 0 \end{bmatrix} \longleftarrow \text{The second row was multiplied by } \tfrac{1}{2}.$$

$$\begin{bmatrix} 1 & 0 & -1 & -3 & 0 \\ 0 & 1 & 1 & 2 & 0 \\ 0 & 0 & 0 & 1 & 0 \\ 0 & 0 & 0 & -10 & 0 \end{bmatrix}$$ ← -3 times the second row was added to the third and -1 times the second row was added to the fourth row.

$$\begin{bmatrix} 1 & 0 & -1 & -3 & 0 \\ 0 & 1 & 1 & 2 & 0 \\ 0 & 0 & 0 & 1 & 0 \\ 0 & 0 & 0 & 0 & 0 \end{bmatrix}$$ ← 10 times the third row was added to the fourth row.

$$\begin{bmatrix} 1 & 0 & -1 & 0 & 0 \\ 0 & 1 & 1 & 0 & 0 \\ 0 & 0 & 0 & 1 & 0 \\ 0 & 0 & 0 & 0 & 0 \end{bmatrix}$$ ← -2 times the third row was added to the second and 3 times the third row was added to the first row.

If we assign y an arbitrary value t the general solution is given by the formulas

$$w = t, \quad x = -t, \quad y = t, \quad z = 0.$$

21.

$$\begin{bmatrix} 2 & -1 & 3 & 4 & 9 \\ 1 & 0 & -2 & 7 & 11 \\ 3 & -3 & 1 & 5 & 8 \\ 2 & 1 & 4 & 4 & 10 \end{bmatrix}$$ ← The augmented matrix for the system.

$$\begin{bmatrix} 1 & 0 & -2 & 7 & 11 \\ 2 & -1 & 3 & 4 & 9 \\ 3 & -3 & 1 & 5 & 8 \\ 2 & 1 & 4 & 4 & 10 \end{bmatrix}$$ ← The first and second rows were interchanged (to avoid introducing fractions into the first row).

$$\begin{bmatrix} 1 & 0 & -2 & 7 & 11 \\ 0 & -1 & 7 & -10 & -13 \\ 0 & -3 & 7 & -16 & -25 \\ 0 & 1 & 8 & -10 & -12 \end{bmatrix}$$ ← -2 times the first row was added to the second row, -3 times the first row was added to the third row, and -2 times the first row was added to the fourth.

$$\begin{bmatrix} 1 & 0 & -2 & 7 & 11 \\ 0 & 1 & -7 & 10 & 13 \\ 0 & -3 & 7 & -16 & -25 \\ 0 & 1 & 8 & -10 & -12 \end{bmatrix}$$ ← The second row was multiplied by -1.

$$\begin{bmatrix} 1 & 0 & -2 & 7 & 11 \\ 0 & 1 & -7 & 10 & 13 \\ 0 & 0 & -14 & 14 & 14 \\ 0 & 0 & 15 & -20 & -25 \end{bmatrix}$$ ← 3 times the second row was added to the third row and -1 times the second row was added to the fourth row.

$$\begin{bmatrix} 1 & 0 & -2 & 7 & 11 \\ 0 & 1 & -7 & 10 & 13 \\ 0 & 0 & 1 & -1 & -1 \\ 0 & 0 & 15 & -20 & -25 \end{bmatrix}$$ ← The third row was multiplied by $-\frac{1}{14}$.

$$\begin{bmatrix} 1 & 0 & -2 & 7 & 11 \\ 0 & 1 & -7 & 10 & 13 \\ 0 & 0 & 1 & -1 & -1 \\ 0 & 0 & 0 & -5 & -10 \end{bmatrix}$$ ← -15 times the third row was added to the fourth row.

$$\begin{bmatrix} 1 & 0 & -2 & 7 & 11 \\ 0 & 1 & -7 & 10 & 13 \\ 0 & 0 & 1 & -1 & -1 \\ 0 & 0 & 0 & 1 & 2 \end{bmatrix}$$ ← The fourth row was multiplied by $-\frac{1}{5}$.

$$\begin{bmatrix} 1 & 0 & -2 & 0 & -3 \\ 0 & 1 & -7 & 0 & -7 \\ 0 & 0 & 1 & 0 & 1 \\ 0 & 0 & 0 & 1 & 2 \end{bmatrix}$$ ← The fourth row was added to the third row, -10 times the fourth row was added to the second, and -7 times the fourth row was added to the first.

$$\begin{bmatrix} 1 & 0 & 0 & 0 & -1 \\ 0 & 1 & 0 & 0 & 0 \\ 0 & 0 & 1 & 0 & 1 \\ 0 & 0 & 0 & 1 & 2 \end{bmatrix}$$ ← 7 times the third row was added to the second row, and 2 times the third row was added to the first row.

Unique solution: $I_1 = -1,\ I_2 = 0,\ I_3 = 1,\ I_4 = 2.$

23. **(a)** The system is consistent; it has a unique solution (back-substitution can be used to solve for all three unknowns).

(b) The system is consistent; it has infinitely many solutions (the third unknown can be assigned an arbitrary value t, then back-substitution can be used to solve for the first two unknowns).

(c) The system is inconsistent since the third equation $0 = 1$ is contradictory.

(d) There is insufficient information to decide whether the system is consistent as illustrated by these examples:

- For $\begin{bmatrix} 1 & * & * & * \\ 0 & 0 & 0 & 0 \\ 0 & 0 & 1 & * \end{bmatrix}$ the system is consistent with infinitely many solutions.
- For $\begin{bmatrix} 1 & * & * & * \\ 0 & 0 & 1 & 0 \\ 0 & 0 & 1 & 1 \end{bmatrix}$ the system is inconsistent (the matrix can be reduced to $\begin{bmatrix} 1 & * & * & * \\ 0 & 0 & 1 & 0 \\ 0 & 0 & 0 & 1 \end{bmatrix}$).

25. $$\begin{bmatrix} 1 & 2 & -3 & 4 \\ 3 & -1 & 5 & 2 \\ 4 & 1 & a^2-14 & a+2 \end{bmatrix}$$ ← The augmented matrix for the system.

$$\begin{bmatrix} 1 & 2 & -3 & 4 \\ 0 & -7 & 14 & -10 \\ 0 & -7 & a^2-2 & a-14 \end{bmatrix}$$ ← -3 times the first row was added to the second row and -4 times the first row was added to the third row.

$$\begin{bmatrix} 1 & 2 & -3 & 4 \\ 0 & -7 & 14 & -10 \\ 0 & 0 & a^2-16 & a-4 \end{bmatrix}$$ ← -1 times the second row was added to the third row.

$$\begin{bmatrix} 1 & 2 & -3 & 4 \\ 0 & 1 & -2 & \frac{10}{7} \\ 0 & 0 & a^2-16 & a-4 \end{bmatrix}$$ ← The second row was multiplied by $-\frac{1}{7}$.

The system has no solutions when $a = -4$ (since the third row of our last matrix would then correspond to a contradictory equation $0 = -8$).

The system has infinitely many solutions when $a = 4$ (since the third row of our last matrix would then correspond to the equation $0 = 0$).

For all remaining values of a (i.e., $a \neq -4$ and $a \neq 4$) the system has exactly one solution.

27. $\begin{bmatrix} 1 & 3 & -1 & a \\ 1 & 1 & 2 & b \\ 0 & 2 & -3 & c \end{bmatrix}$ ⟵ The augmented matrix for the system.

$\begin{bmatrix} 1 & 3 & -1 & a \\ 0 & -2 & 3 & -a+b \\ 0 & 2 & -3 & c \end{bmatrix}$ ⟵ -1 times the first row was added to the second row.

$\begin{bmatrix} 1 & 3 & -1 & a \\ 0 & -2 & 3 & -a+b \\ 0 & 0 & 0 & -a+b+c \end{bmatrix}$ ⟵ The second row was added to the third row.

$\begin{bmatrix} 1 & 3 & -1 & a \\ 0 & 1 & -\frac{3}{2} & \frac{a}{2}-\frac{b}{2} \\ 0 & 0 & 0 & -a+b+c \end{bmatrix}$ ⟵ The second row was multiplied by $-\frac{1}{2}$.

If $-a+b+c=0$ then the linear system is consistent. Otherwise (if $-a+b+c \neq 0$) it is inconsistent.

29. $\begin{bmatrix} 2 & 1 & a \\ 3 & 6 & b \end{bmatrix}$ ⟵ The augmented matrix for the system.

$\begin{bmatrix} 1 & \frac{1}{2} & \frac{1}{2}a \\ 3 & 6 & b \end{bmatrix}$ ⟵ The first row was multiplied by $\frac{1}{2}$.

$\begin{bmatrix} 1 & \frac{1}{2} & \frac{1}{2}a \\ 0 & \frac{9}{2} & -\frac{3}{2}a+b \end{bmatrix}$ ⟵ -3 times the first row was added to the second row.

$\begin{bmatrix} 1 & \frac{1}{2} & \frac{1}{2}a \\ 0 & 1 & -\frac{1}{3}a+\frac{2}{9}b \end{bmatrix}$ ⟵ The third row was multiplied by $\frac{2}{9}$.

$\begin{bmatrix} 1 & 0 & \frac{2}{3}a-\frac{1}{9}b \\ 0 & 1 & -\frac{1}{3}a+\frac{2}{9}b \end{bmatrix}$ ⟵ $-\frac{1}{2}$ times the second row was added to the first row.

The system has exactly one solution: $x = \frac{2}{3}a - \frac{1}{9}b$ and $y = -\frac{1}{3}a + \frac{2}{9}b$.

31. Adding -2 times the first row to the second yields a matrix in row echelon form $\begin{bmatrix} 1 & 3 \\ 0 & 1 \end{bmatrix}$.

Adding -3 times its second row to the first results in $\begin{bmatrix} 1 & 0 \\ 0 & 1 \end{bmatrix}$, which is also in row echelon form.

33. We begin by substituting $x = \sin\alpha$, $y = \cos\beta$, and $z = \tan\gamma$ so that the system becomes

$$\begin{array}{rcrcrcr} x & + & 2y & + & 3z & = & 0 \\ 2x & + & 5y & + & 3z & = & 0 \\ -x & - & 5y & + & 5z & = & 0 \end{array}$$

$\begin{bmatrix} 1 & 2 & 3 & 0 \\ 2 & 5 & 3 & 0 \\ -1 & -5 & 5 & 0 \end{bmatrix}$ ⟵ The augmented matrix for the system.

$\begin{bmatrix} 1 & 2 & 3 & 0 \\ 0 & 1 & -3 & 0 \\ 0 & -3 & 8 & 0 \end{bmatrix}$ ⟵ -2 times the first row was added to the second row and the first row was added to the third row.

$\begin{bmatrix} 1 & 2 & 3 & 0 \\ 0 & 1 & -3 & 0 \\ 0 & 0 & -1 & 0 \end{bmatrix}$ ⟵ 3 times the second row was added to the third row.

$\begin{bmatrix} 1 & 2 & 3 & 0 \\ 0 & 1 & -3 & 0 \\ 0 & 0 & 1 & 0 \end{bmatrix}$ ⟵ The third row was multiplied by -1.

$\begin{bmatrix} 1 & 2 & 0 & 0 \\ 0 & 1 & 0 & 0 \\ 0 & 0 & 1 & 0 \end{bmatrix}$ ⟵ 3 times the third row was added to the second row and -3 times the third row was added to the first row.

$\begin{bmatrix} 1 & 0 & 0 & 0 \\ 0 & 1 & 0 & 0 \\ 0 & 0 & 1 & 0 \end{bmatrix}$ ⟵ -2 times the second row was added to the first row.

This system has exactly one solution $x = 0, \ y = 0, \ z = 0$.

On the interval $0 \le \alpha \le 2\pi$, the equation $\sin\alpha = 0$ has three solutions: $\alpha = 0, \alpha = \pi$, and $\alpha = 2\pi$.

On the interval $0 \le \beta \le 2\pi$, the equation $\cos\beta = 0$ has two solutions: $\beta = \frac{\pi}{2}$ and $\beta = \frac{3\pi}{2}$.

On the interval $0 \le \gamma \le 2\pi$, the equation $\tan\gamma = 0$ has three solutions: $\gamma = 0, \gamma = \pi$, and $\gamma = 2\pi$.

Overall, $3 \cdot 2 \cdot 3 = 18$ solutions (α, β, γ) can be obtained by combining the values of α, β, and γ listed above: $\left(0, \frac{\pi}{2}, 0\right), \left(\pi, \frac{\pi}{2}, 0\right)$, etc.

35. We begin by substituting $X = x^2$, $Y = y^2$, and $Z = z^2$ so that the system becomes

$$\begin{array}{rcrcrcr} X & + & Y & + & Z & = & 6 \\ X & - & Y & + & 2Z & = & 2 \\ 2X & + & Y & - & Z & = & 3 \end{array}$$

$\begin{bmatrix} 1 & 1 & 1 & 6 \\ 1 & -1 & 2 & 2 \\ 2 & 1 & -1 & 3 \end{bmatrix}$ ⟵ The augmented matrix for the system.

$\begin{bmatrix} 1 & 1 & 1 & 6 \\ 0 & -2 & 1 & -4 \\ 0 & -1 & -3 & -9 \end{bmatrix}$ ⟵ -1 times the first row was added to the second row and -2 times the first row was added to the third row.

$\begin{bmatrix} 1 & 1 & 1 & 6 \\ 0 & -1 & -3 & -9 \\ 0 & -2 & 1 & -4 \end{bmatrix}$ ⟵ The second and third rows were interchanged (to avoid introducing fractions into the second row).

$$\begin{bmatrix} 1 & 1 & 1 & 6 \\ 0 & 1 & 3 & 9 \\ 0 & -2 & 1 & -4 \end{bmatrix}$$ ← The second row was multiplied by -1.

$$\begin{bmatrix} 1 & 1 & 1 & 6 \\ 0 & 1 & 3 & 9 \\ 0 & 0 & 7 & 14 \end{bmatrix}$$ ← 2 times the second row was added to the third row.

$$\begin{bmatrix} 1 & 1 & 1 & 6 \\ 0 & 1 & 3 & 9 \\ 0 & 0 & 1 & 2 \end{bmatrix}$$ ← The third row was multiplied by $\frac{1}{7}$.

$$\begin{bmatrix} 1 & 1 & 0 & 4 \\ 0 & 1 & 0 & 3 \\ 0 & 0 & 1 & 2 \end{bmatrix}$$ ← -3 times the third row was added to the second row and -1 times the third row was added to the first row.

$$\begin{bmatrix} 1 & 0 & 0 & 1 \\ 0 & 1 & 0 & 3 \\ 0 & 0 & 1 & 2 \end{bmatrix}$$ ← -1 times the second row was added to the first row.

We obtain

$$\begin{aligned} X = 1 &\Rightarrow x = \pm 1 \\ Y = 3 &\Rightarrow y = \pm\sqrt{3} \\ Z = 2 &\Rightarrow z = \pm\sqrt{2} \end{aligned}$$

37. Each point on the curve yields an equation, therefore we have a system of four equations

equation corresponding to (1,7): $a + b + c + d = 7$

equation corresponding to $(3,-11)$: $27a + 9b + 3c + d = -11$

equation corresponding to $(4,-14)$: $64a + 16b + 4c + d = -14$

equation corresponding to (0,10): $d = 10$

$$\begin{bmatrix} 1 & 1 & 1 & 1 & 7 \\ 27 & 9 & 3 & 1 & -11 \\ 64 & 16 & 4 & 1 & -14 \\ 0 & 0 & 0 & 1 & 10 \end{bmatrix}$$ ← The augmented matrix for the system.

$$\begin{bmatrix} 1 & 1 & 1 & 1 & 7 \\ 0 & -18 & -24 & -26 & -200 \\ 0 & -48 & -60 & -63 & -462 \\ 0 & 0 & 0 & 1 & 10 \end{bmatrix}$$ ← -27 times the first row was added to the second row and -64 times the first row was added to the third.

$$\begin{bmatrix} 1 & 1 & 1 & 1 & 7 \\ 0 & 1 & \frac{4}{3} & \frac{13}{9} & \frac{100}{9} \\ 0 & -48 & -60 & -63 & -462 \\ 0 & 0 & 0 & 1 & 10 \end{bmatrix}$$ ← The second row was multiplied by $-\frac{1}{18}$.

$$\begin{bmatrix} 1 & 1 & 1 & 1 & 7 \\ 0 & 1 & \frac{4}{3} & \frac{13}{9} & \frac{100}{9} \\ 0 & 0 & 4 & \frac{19}{3} & \frac{214}{3} \\ 0 & 0 & 0 & 1 & 10 \end{bmatrix}$$ ← 48 times the second row was added to the third row.

$$\begin{bmatrix} 1 & 1 & 1 & 1 & 7 \\ 0 & 1 & \frac{4}{3} & \frac{13}{9} & \frac{100}{9} \\ 0 & 0 & 1 & \frac{19}{12} & \frac{107}{6} \\ 0 & 0 & 0 & 1 & 10 \end{bmatrix}$$ ⟵ The third row was multiplied by $\frac{1}{4}$.

$$\begin{bmatrix} 1 & 1 & 1 & 0 & -3 \\ 0 & 1 & \frac{4}{3} & 0 & -\frac{10}{3} \\ 0 & 0 & 1 & 0 & 2 \\ 0 & 0 & 0 & 1 & 10 \end{bmatrix}$$ ⟵ $-\frac{19}{12}$ times the fourth row was added to the third row, $-\frac{13}{9}$ times the fourth row was added to the second row, and -1 times the fourth row was added to the first.

$$\begin{bmatrix} 1 & 1 & 0 & 0 & -5 \\ 0 & 1 & 0 & 0 & -6 \\ 0 & 0 & 1 & 0 & 2 \\ 0 & 0 & 0 & 1 & 10 \end{bmatrix}$$ ⟵ $-\frac{4}{3}$ times the third row was added to the second row and -1 times the third row was added to the first row.

$$\begin{bmatrix} 1 & 0 & 0 & 0 & 1 \\ 0 & 1 & 0 & 0 & -6 \\ 0 & 0 & 1 & 0 & 2 \\ 0 & 0 & 0 & 1 & 10 \end{bmatrix}$$ ⟵ -1 times the second row was added to the first row.

The linear system has a unique solution: $a = 1$, $b = -6$, $c = 2$, $d = 10$. These are the coefficient values required for the curve $y = ax^3 + bx^2 + cx + d$ to pass through the four given points.

39. Since the homogeneous system has only the trivial solution, its augmented matrix must be possible to reduce via a sequence of elementary row operations to the reduced row echelon form $\begin{bmatrix} 1 & 0 & 0 & 0 \\ 0 & 1 & 0 & 0 \\ 0 & 0 & 1 & 0 \end{bmatrix}$.

Applying the **same** sequence of elementary row operations to the augmented matrix of the nonhomogeneous system yields the reduced row echelon form $\begin{bmatrix} 1 & 0 & 0 & r \\ 0 & 1 & 0 & s \\ 0 & 0 & 1 & t \end{bmatrix}$ where r, s, and t are some real numbers. Therefore, the nonhomogeneous system has one solution.

41. **(a)** There are eight possible reduced row echelon forms:

$$\begin{bmatrix} 1 & 0 & 0 \\ 0 & 1 & 0 \\ 0 & 0 & 1 \end{bmatrix}, \begin{bmatrix} 1 & 0 & r \\ 0 & 1 & s \\ 0 & 0 & 0 \end{bmatrix}, \begin{bmatrix} 1 & r & 0 \\ 0 & 0 & 1 \\ 0 & 0 & 0 \end{bmatrix}, \begin{bmatrix} 1 & r & s \\ 0 & 0 & 0 \\ 0 & 0 & 0 \end{bmatrix}, \begin{bmatrix} 0 & 1 & 0 \\ 0 & 0 & 1 \\ 0 & 0 & 0 \end{bmatrix}, \begin{bmatrix} 0 & 1 & r \\ 0 & 0 & 0 \\ 0 & 0 & 0 \end{bmatrix}, \begin{bmatrix} 0 & 0 & 1 \\ 0 & 0 & 0 \\ 0 & 0 & 0 \end{bmatrix}, \text{and} \begin{bmatrix} 0 & 0 & 0 \\ 0 & 0 & 0 \\ 0 & 0 & 0 \end{bmatrix}$$

where r and s can be any real numbers.

(b) There are sixteen possible reduced row echelon forms:

$$\begin{bmatrix}1&0&0&0\\0&1&0&0\\0&0&1&0\\0&0&0&1\end{bmatrix},\begin{bmatrix}1&0&0&r\\0&1&0&s\\0&0&1&t\\0&0&0&0\end{bmatrix},\begin{bmatrix}1&0&r&0\\0&1&s&0\\0&0&0&1\\0&0&0&0\end{bmatrix},\begin{bmatrix}1&0&r&t\\0&1&s&u\\0&0&0&0\\0&0&0&0\end{bmatrix},\begin{bmatrix}1&r&0&0\\0&0&1&0\\0&0&0&1\\0&0&0&0\end{bmatrix},\begin{bmatrix}1&r&0&s\\0&0&1&t\\0&0&0&0\\0&0&0&0\end{bmatrix},$$

$$\begin{bmatrix}1&r&s&0\\0&0&0&1\\0&0&0&0\\0&0&0&0\end{bmatrix},\begin{bmatrix}1&r&s&t\\0&0&0&0\\0&0&0&0\\0&0&0&0\end{bmatrix},\begin{bmatrix}0&1&0&0\\0&0&1&0\\0&0&0&1\\0&0&0&0\end{bmatrix},\begin{bmatrix}0&1&0&r\\0&0&1&s\\0&0&0&0\\0&0&0&0\end{bmatrix},\begin{bmatrix}0&1&r&0\\0&0&0&1\\0&0&0&0\\0&0&0&0\end{bmatrix},\begin{bmatrix}0&1&r&s\\0&0&0&0\\0&0&0&0\\0&0&0&0\end{bmatrix},$$

$$\begin{bmatrix}0&0&1&0\\0&0&0&1\\0&0&0&0\\0&0&0&0\end{bmatrix},\begin{bmatrix}0&0&1&r\\0&0&0&0\\0&0&0&0\\0&0&0&0\end{bmatrix},\begin{bmatrix}0&0&0&1\\0&0&0&0\\0&0&0&0\\0&0&0&0\end{bmatrix},\text{ and }\begin{bmatrix}0&0&0&0\\0&0&0&0\\0&0&0&0\\0&0&0&0\end{bmatrix}.$$

where r, s, t, and u can be any real numbers.

43. (a) We consider two possible cases: (i) $a = 0$, and (ii) $a \neq 0$.

(i) If $a = 0$ then the assumption $ad - bc \neq 0$ implies that $b \neq 0$ and $c \neq 0$. Gauss-Jordan elimination yields

$\begin{bmatrix}0&b\\c&d\end{bmatrix}$ ⟵ We assumed $a = 0$

$\begin{bmatrix}c&d\\0&b\end{bmatrix}$ ⟵ The rows were interchanged.

$\begin{bmatrix}1&\frac{d}{c}\\0&1\end{bmatrix}$ ⟵ The first row was multiplied by $\frac{1}{c}$ and the second row was multiplied by $\frac{1}{b}$. (Note that $b, c \neq 0$.)

$\begin{bmatrix}1&0\\0&1\end{bmatrix}$ ⟵ $-\frac{d}{c}$ times the second row was added to the first row.

(ii) If $a \neq 0$ then we perform Gauss-Jordan elimination as follows:

$\begin{bmatrix}a&b\\c&d\end{bmatrix}$

$\begin{bmatrix}1&\frac{b}{a}\\c&d\end{bmatrix}$ ⟵ The first row was multiplied by $\frac{1}{a}$.

$\begin{bmatrix}1&\frac{b}{a}\\0&\frac{ad-bc}{a}\end{bmatrix}$ ⟵ $-c$ times the first row was added to the second row.

$\begin{bmatrix}1&\frac{b}{a}\\0&1\end{bmatrix}$ ⟵ The second row was multiplied by $\frac{a}{ad-bc}$. (Note that both a and $ad - bc$ are nonzero.)

$\begin{bmatrix}1&0\\0&1\end{bmatrix}$ ⟵ $-\frac{b}{a}$ times the second row was added to the first row.

In both cases ($a = 0$ as well as $a \neq 0$) we established that the reduced row echelon form of $\begin{bmatrix} a & b \\ c & d \end{bmatrix}$ is $\begin{bmatrix} 1 & 0 \\ 0 & 1 \end{bmatrix}$ provided that $ad - bc \neq 0$.

(b) Applying the **same** elementary row operation steps as in part (a) the augmented matrix $\begin{bmatrix} a & b & k \\ c & d & l \end{bmatrix}$ will be transformed to a matrix in reduced row echelon form $\begin{bmatrix} 1 & 0 & p \\ 0 & 1 & q \end{bmatrix}$ where p and q are some real numbers. We conclude that the given linear system has exactly one solution: $x = p,\ y = q$.

True-False Exercises

(a) True. A matrix in reduced row echelon form has all properties required for the row echelon form.

(b) False. For instance, interchanging the rows of $\begin{bmatrix} 1 & 0 \\ 0 & 1 \end{bmatrix}$ yields a matrix that is not in row echelon form.

(c) False. See Exercise 31.

(d) True. In a reduced row echelon form, the number of nonzero rows equals to the number of leading 1's. The result follows from Theorem 1.2.1.

(e) True. This is implied by the third property of a row echelon form (see p. 11).

(f) False. Nonzero entries are permitted above the leading 1's in a row echelon form.

(g) True. In a reduced row echelon form, the number of nonzero rows equals to the number of leading 1's. From Theorem 1.2.1 we conclude that the system has $n - n = 0$ free variables, i.e. it has only the trivial solution.

(h) False. The row of zeros imposes no restriction on the unknowns and can be omitted. Whether the system has infinitely many, one, or no solution(s) depends *solely* on the nonzero rows of the reduced row echelon form.

(i) False. For example, the following system is clearly inconsistent:

$$\begin{aligned} x + y + z &= 1 \\ x + y + z &= 2 \end{aligned}$$

1.3 Matrices and Matrix Operations

1. **(a)** Undefined (the number of columns in B does not match the number of rows in A)

   (b) Defined; 4×4 matrix

   (c) Defined; 4×2 matrix

   (d) Defined; 5×2 matrix

   (e) Defined; 4×5 matrix

(f) Defined; 5×5 matrix

3. (a) $\begin{bmatrix} 1+6 & 5+1 & 2+3 \\ -1+(-1) & 0+1 & 1+2 \\ 3+4 & 2+1 & 4+3 \end{bmatrix} = \begin{bmatrix} 7 & 6 & 5 \\ -2 & 1 & 3 \\ 7 & 3 & 7 \end{bmatrix}$

(b) $\begin{bmatrix} 1-6 & 5-1 & 2-3 \\ -1-(-1) & 0-1 & 1-2 \\ 3-4 & 2-1 & 4-3 \end{bmatrix} = \begin{bmatrix} -5 & 4 & -1 \\ 0 & -1 & -1 \\ -1 & 1 & 1 \end{bmatrix}$

(c) $\begin{bmatrix} 5\cdot 3 & 5\cdot 0 \\ 5\cdot(-1) & 5\cdot 2 \\ 5\cdot 1 & 5\cdot 1 \end{bmatrix} = \begin{bmatrix} 15 & 0 \\ -5 & 10 \\ 5 & 5 \end{bmatrix}$

(d) $\begin{bmatrix} -7\cdot 1 & -7\cdot 4 & -7\cdot 2 \\ -7\cdot 3 & -7\cdot 1 & -7\cdot 5 \end{bmatrix} = \begin{bmatrix} -7 & -28 & -14 \\ -21 & -7 & -35 \end{bmatrix}$

(e) Undefined (a 2×3 matrix C cannot be subtracted from a 2×2 matrix $2B$)

(f) $\begin{bmatrix} 4\cdot 6 & 4\cdot 1 & 4\cdot 3 \\ 4\cdot(-1) & 4\cdot 1 & 4\cdot 2 \\ 4\cdot 4 & 4\cdot 1 & 4\cdot 3 \end{bmatrix} - \begin{bmatrix} 2\cdot 1 & 2\cdot 5 & 2\cdot 2 \\ 2\cdot(-1) & 2\cdot 0 & 2\cdot 1 \\ 2\cdot 3 & 2\cdot 2 & 2\cdot 4 \end{bmatrix} = \begin{bmatrix} 24-2 & 4-10 & 12-4 \\ -4-(-2) & 4-0 & 8-2 \\ 16-6 & 4-4 & 12-8 \end{bmatrix}$
$= \begin{bmatrix} 22 & -6 & 8 \\ -2 & 4 & 6 \\ 10 & 0 & 4 \end{bmatrix}$

(g) $-3\left(\begin{bmatrix} 1 & 5 & 2 \\ -1 & 0 & 1 \\ 3 & 2 & 4 \end{bmatrix} + \begin{bmatrix} 2\cdot 6 & 2\cdot 1 & 2\cdot 3 \\ 2\cdot(-1) & 2\cdot 1 & 2\cdot 2 \\ 2\cdot 4 & 2\cdot 1 & 2\cdot 3 \end{bmatrix}\right) = -3\begin{bmatrix} 1+12 & 5+2 & 2+6 \\ -1+(-2) & 0+2 & 1+4 \\ 3+8 & 2+2 & 4+6 \end{bmatrix}$
$= \begin{bmatrix} -3\cdot 13 & -3\cdot 7 & -3\cdot 8 \\ -3\cdot(-3) & -3\cdot 2 & -3\cdot 5 \\ -3\cdot 11 & -3\cdot 4 & -3\cdot 10 \end{bmatrix} = \begin{bmatrix} -39 & -21 & -24 \\ 9 & -6 & -15 \\ -33 & -12 & -30 \end{bmatrix}$

(h) $\begin{bmatrix} 3-3 & 0-0 \\ -1-(-1) & 2-2 \\ 1-1 & 1-1 \end{bmatrix} = \begin{bmatrix} 0 & 0 \\ 0 & 0 \\ 0 & 0 \end{bmatrix}$

(i) $1 + 0 + 4 = 5$

(j) $\operatorname{tr}\left(\begin{bmatrix} 1 & 5 & 2 \\ -1 & 0 & 1 \\ 3 & 2 & 4 \end{bmatrix} - \begin{bmatrix} 3\cdot 6 & 3\cdot 1 & 3\cdot 3 \\ 3\cdot(-1) & 3\cdot 1 & 3\cdot 2 \\ 3\cdot 4 & 3\cdot 1 & 3\cdot 3 \end{bmatrix}\right) = \operatorname{tr}\left(\begin{bmatrix} 1-18 & 5-3 & 2-9 \\ -1-(-3) & 0-3 & 1-6 \\ 3-12 & 2-3 & 4-9 \end{bmatrix}\right)$
$= \operatorname{tr}\left(\begin{bmatrix} -17 & 2 & -7 \\ 2 & -3 & -5 \\ -9 & -1 & -5 \end{bmatrix}\right) = -17 - 3 - 5 = -25$

(k) $4\operatorname{tr}\left(\begin{bmatrix} 7\cdot 4 & 7\cdot(-1) \\ 7\cdot 0 & 7\cdot 2 \end{bmatrix}\right) = 4\operatorname{tr}\left(\begin{bmatrix} 28 & -7 \\ 0 & 14 \end{bmatrix}\right) = 4(28+14) = 4\cdot 42 = 168$

(l) Undefined (trace is only defined for square matrices)

5. (a) $\begin{bmatrix} (3\cdot 4)+(0\cdot 0) & -(3\cdot 1)+(0\cdot 2) \\ -(1\cdot 4)+(2\cdot 0) & (1\cdot 1)+(2\cdot 2) \\ (1\cdot 4)+(1\cdot 0) & -(1\cdot 1)+(1\cdot 2) \end{bmatrix} = \begin{bmatrix} 12 & -3 \\ -4 & 5 \\ 4 & 1 \end{bmatrix}$

(b) Undefined (the number of columns of B does not match the number of rows in A)

(c) $\begin{bmatrix} 3\cdot 6 & 3\cdot 1 & 3\cdot 3 \\ 3\cdot(-1) & 3\cdot 1 & 3\cdot 2 \\ 3\cdot 4 & 3\cdot 1 & 3\cdot 3 \end{bmatrix}\begin{bmatrix} 1 & 5 & 2 \\ -1 & 0 & 1 \\ 3 & 2 & 4 \end{bmatrix}$

$$= \begin{bmatrix} (18\cdot 1)-(3\cdot 1)+(9\cdot 3) & (18\cdot 5)+(3\cdot 0)+(9\cdot 2) & (18\cdot 2)+(3\cdot 1)+(9\cdot 4) \\ -(3\cdot 1)-(3\cdot 1)+(6\cdot 3) & -(3\cdot 5)+(3\cdot 0)+(6\cdot 2) & -(3\cdot 2)+(3\cdot 1)+(6\cdot 4) \\ (12\cdot 1)-(3\cdot 1)+(9\cdot 3) & (12\cdot 5)+(3\cdot 0)+(9\cdot 2) & (12\cdot 2)+(3\cdot 1)+(9\cdot 4) \end{bmatrix}$$

$$= \begin{bmatrix} 42 & 108 & 75 \\ 12 & -3 & 21 \\ 36 & 78 & 63 \end{bmatrix}$$

(d) $\begin{bmatrix} (3\cdot 4)+(0\cdot 0) & -(3\cdot 1)+(0\cdot 2) \\ -(1\cdot 4)+(2\cdot 0) & (1\cdot 1)+(2\cdot 2) \\ (1\cdot 4)+(1\cdot 0) & -(1\cdot 1)+(1\cdot 2) \end{bmatrix}\begin{bmatrix} 1 & 4 & 2 \\ 3 & 1 & 5 \end{bmatrix} = \begin{bmatrix} 12 & -3 \\ -4 & 5 \\ 4 & 1 \end{bmatrix}\begin{bmatrix} 1 & 4 & 2 \\ 3 & 1 & 5 \end{bmatrix}$

$$= \begin{bmatrix} (12\cdot 1)-(3\cdot 3) & (12\cdot 4)-(3\cdot 1) & (12\cdot 2)-(3\cdot 5) \\ -(4\cdot 1)+(5\cdot 3) & -(4\cdot 4)+(5\cdot 1) & -(4\cdot 2)+(5\cdot 5) \\ (4\cdot 1)+(1\cdot 3) & (4\cdot 4)+(1\cdot 1) & (4\cdot 2)+(1\cdot 5) \end{bmatrix} = \begin{bmatrix} 3 & 45 & 9 \\ 11 & -11 & 17 \\ 7 & 17 & 13 \end{bmatrix}$$

(e) $\begin{bmatrix} 3 & 0 \\ -1 & 2 \\ 1 & 1 \end{bmatrix}\begin{bmatrix} (4\cdot 1)-(1\cdot 3) & (4\cdot 4)-(1\cdot 1) & (4\cdot 2)-(1\cdot 5) \\ (0\cdot 1)+(2\cdot 3) & (0\cdot 4)+(2\cdot 1) & (0\cdot 2)+(2\cdot 5) \end{bmatrix} = \begin{bmatrix} 3 & 0 \\ -1 & 2 \\ 1 & 1 \end{bmatrix}\begin{bmatrix} 1 & 15 & 3 \\ 6 & 2 & 10 \end{bmatrix}$

$$= \begin{bmatrix} (3\cdot 1)+(0\cdot 6) & (3\cdot 15)+(0\cdot 2) & (3\cdot 3)+(0\cdot 10) \\ -(1\cdot 1)+(2\cdot 6) & -(1\cdot 15)+(2\cdot 2) & -(1\cdot 3)+(2\cdot 10) \\ (1\cdot 1)+(1\cdot 6) & (1\cdot 15)+(1\cdot 2) & (1\cdot 3)+(1\cdot 10) \end{bmatrix} = \begin{bmatrix} 3 & 45 & 9 \\ 11 & -11 & 17 \\ 7 & 17 & 13 \end{bmatrix}$$

(f) $\begin{bmatrix} 1 & 4 & 2 \\ 3 & 1 & 5 \end{bmatrix}\begin{bmatrix} 1 & 3 \\ 4 & 1 \\ 2 & 5 \end{bmatrix} = \begin{bmatrix} (1\cdot 1)+(4\cdot 4)+(2\cdot 2) & (1\cdot 3)+(4\cdot 1)+(2\cdot 5) \\ (3\cdot 1)+(1\cdot 4)+(5\cdot 2) & (3\cdot 3)+(1\cdot 1)+(5\cdot 5) \end{bmatrix} = \begin{bmatrix} 21 & 17 \\ 17 & 35 \end{bmatrix}$

(g) $\left(\begin{bmatrix} (1\cdot 3)-(5\cdot 1)+(2\cdot 1) & (1\cdot 0)+(5\cdot 2)+(2\cdot 1) \\ -(1\cdot 3)-(0\cdot 1)+(1\cdot 1) & -(1\cdot 0)+(0\cdot 2)+(1\cdot 1) \\ (3\cdot 3)-(2\cdot 1)+(4\cdot 1) & (3\cdot 0)+(2\cdot 2)+(4\cdot 1) \end{bmatrix}\right)^T = \begin{bmatrix} 0 & -2 & 11 \\ 12 & 1 & 8 \end{bmatrix}$

(h) $\left(\begin{bmatrix} 1 & 3 \\ 4 & 1 \\ 2 & 5 \end{bmatrix}\begin{bmatrix} 4 & -1 \\ 0 & 2 \end{bmatrix}\right)\begin{bmatrix} 3 & -1 & 1 \\ 0 & 2 & 1 \end{bmatrix} = \begin{bmatrix} (1\cdot 4)+(3\cdot 0) & -(1\cdot 1)+(3\cdot 2) \\ (4\cdot 4)+(1\cdot 0) & -(4\cdot 1)+(1\cdot 2) \\ (2\cdot 4)+(5\cdot 0) & -(2\cdot 1)+(5\cdot 2) \end{bmatrix}\begin{bmatrix} 3 & -1 & 1 \\ 0 & 2 & 1 \end{bmatrix}$

$$= \begin{bmatrix} 4 & 5 \\ 16 & -2 \\ 8 & 8 \end{bmatrix}\begin{bmatrix} 3 & -1 & 1 \\ 0 & 2 & 1 \end{bmatrix} = \begin{bmatrix} (4\cdot 3)+(5\cdot 0) & -(4\cdot 1)+(5\cdot 2) & (4\cdot 1)+(5\cdot 1) \\ (16\cdot 3)-(2\cdot 0) & -(16\cdot 1)-(2\cdot 2) & (16\cdot 1)-(2\cdot 1) \\ (8\cdot 3)+(8\cdot 0) & -(8\cdot 1)+(8\cdot 2) & (8\cdot 1)+(8\cdot 1) \end{bmatrix}$$

$$= \begin{bmatrix} 12 & 6 & 9 \\ 48 & -20 & 14 \\ 24 & 8 & 16 \end{bmatrix}$$

(i) $\operatorname{tr}\left(\begin{bmatrix} 1 & 5 & 2 \\ -1 & 0 & 1 \\ 3 & 2 & 4 \end{bmatrix}\begin{bmatrix} 1 & -1 & 3 \\ 5 & 0 & 2 \\ 2 & 1 & 4 \end{bmatrix}\right)$

$$= \operatorname{tr}\left(\begin{bmatrix} (1\cdot 1)+(5\cdot 5)+(2\cdot 2) & -(1\cdot 1)+(5\cdot 0)+(2\cdot 1) & (1\cdot 3)+(5\cdot 2)+(2\cdot 4) \\ -(1\cdot 1)+(0\cdot 5)+(1\cdot 2) & (1\cdot 1)+(0\cdot 0)+(1\cdot 1) & -(1\cdot 3)+(0\cdot 2)+(1\cdot 4) \\ (3\cdot 1)+(2\cdot 5)+(4\cdot 2) & -(3\cdot 1)+(2\cdot 0)+(4\cdot 1) & (3\cdot 3)+(2\cdot 2)+(4\cdot 4) \end{bmatrix}\right)$$

$$= \operatorname{tr}\left(\begin{bmatrix} 30 & 1 & 21 \\ 1 & 2 & 1 \\ 21 & 1 & 29 \end{bmatrix}\right) = 30+2+29 = 61$$

(j) $\text{tr}\left(4\begin{bmatrix}6 & -1 & 4\\1 & 1 & 1\\3 & 2 & 3\end{bmatrix} - \begin{bmatrix}1 & 5 & 2\\-1 & 0 & 1\\3 & 2 & 4\end{bmatrix}\right) = \text{tr}\left(\begin{bmatrix}4\cdot 6-1 & 4\cdot(-1)-5 & 4\cdot 4-2\\4\cdot 1-(-1) & 4\cdot 1-0 & 4\cdot 1-1\\4\cdot 3-3 & 4\cdot 2-2 & 4\cdot 3-4\end{bmatrix}\right)$

$= \text{tr}\left(\begin{bmatrix}23 & -9 & 14\\5 & 4 & 3\\9 & 6 & 8\end{bmatrix}\right) = 23+4+8 = 35$

(k) $\text{tr}\left(\begin{bmatrix}1 & 3\\4 & 1\\2 & 5\end{bmatrix}\begin{bmatrix}3 & -1 & 1\\0 & 2 & 1\end{bmatrix} + 2\begin{bmatrix}6 & -1 & 4\\1 & 1 & 1\\3 & 2 & 3\end{bmatrix}\right)$

$\text{tr}\left(\begin{bmatrix}(1\cdot 3)+(3\cdot 0) & -(1\cdot 1)+(3\cdot 2) & (1\cdot 1)+(3\cdot 1)\\(4\cdot 3)+(1\cdot 0) & -(4\cdot 1)+(1\cdot 2) & (4\cdot 1)+(1\cdot 1)\\(2\cdot 3)+(5\cdot 0) & -(2\cdot 1)+(5\cdot 2) & (2\cdot 1)+(5\cdot 1)\end{bmatrix} + \begin{bmatrix}2\cdot 6 & 2\cdot(-1) & 2\cdot 4\\2\cdot 1 & 2\cdot 1 & 2\cdot 1\\2\cdot 3 & 2\cdot 2 & 2\cdot 3\end{bmatrix}\right)$

$\text{tr}\left(\begin{bmatrix}3 & 5 & 4\\12 & -2 & 5\\6 & 8 & 7\end{bmatrix} + \begin{bmatrix}12 & -2 & 8\\2 & 2 & 2\\6 & 4 & 6\end{bmatrix}\right) = \text{tr}\left(\begin{bmatrix}15 & 3 & 12\\14 & 0 & 7\\12 & 12 & 13\end{bmatrix}\right) = 15+0+13 = 28$

(l) $\text{tr}\left(\left(\begin{bmatrix}6 & 1 & 3\\-1 & 1 & 2\\4 & 1 & 3\end{bmatrix}\begin{bmatrix}1 & 3\\4 & 1\\2 & 5\end{bmatrix}\right)^{\text{T}}\begin{bmatrix}3 & 0\\-1 & 2\\1 & 1\end{bmatrix}\right)$

$= \text{tr}\left(\left(\begin{bmatrix}(6\cdot 1)+(1\cdot 4)+(3\cdot 2) & (6\cdot 3)+(1\cdot 1)+(3\cdot 5)\\-(1\cdot 1)+(1\cdot 4)+(2\cdot 2) & -(1\cdot 3)+(1\cdot 1)+(2\cdot 5)\\(4\cdot 1)+(1\cdot 4)+(3\cdot 2) & (4\cdot 3)+(1\cdot 1)+(3\cdot 5)\end{bmatrix}\right)^{T}\begin{bmatrix}3 & 0\\-1 & 2\\1 & 1\end{bmatrix}\right)$

$= \text{tr}\left(\left(\begin{bmatrix}16 & 34\\7 & 8\\14 & 28\end{bmatrix}\right)^{\text{T}}\begin{bmatrix}3 & 0\\-1 & 2\\1 & 1\end{bmatrix}\right) = \text{tr}\left(\begin{bmatrix}16 & 7 & 14\\34 & 8 & 28\end{bmatrix}\begin{bmatrix}3 & 0\\-1 & 2\\1 & 1\end{bmatrix}\right)$

$= \text{tr}\left(\begin{bmatrix}(16\cdot 3)-(7\cdot 1)+(14\cdot 1) & (16\cdot 0)+(7\cdot 2)+(14\cdot 1)\\(34\cdot 3)-(8\cdot 1)+(28\cdot 1) & (34\cdot 0)+(8\cdot 2)+(28\cdot 1)\end{bmatrix}\right)$

$= \text{tr}\left(\begin{bmatrix}55 & 28\\122 & 44\end{bmatrix}\right) = 55+44 = 99$

7. (a) first row of AB = [first row of A] B $= \begin{bmatrix}3 & -2 & 7\end{bmatrix}\begin{bmatrix}6 & -2 & 4\\0 & 1 & 3\\7 & 7 & 5\end{bmatrix}$

$= [(3\cdot 6)-(2\cdot 0)+(7\cdot 7) \quad -(3\cdot 2)-(2\cdot 1)+(7\cdot 7) \quad (3\cdot 4)-(2\cdot 3)+(7\cdot 5)]$
$= [67 \quad 41 \quad 41]$

(b) third row of AB = [third row of A] B $= \begin{bmatrix}0 & 4 & 9\end{bmatrix}\begin{bmatrix}6 & -2 & 4\\0 & 1 & 3\\7 & 7 & 5\end{bmatrix}$

$= [(0\cdot 6)+(4\cdot 0)+(9\cdot 7) \quad -(0\cdot 2)+(4\cdot 1)+(9\cdot 7) \quad (0\cdot 4)+(4\cdot 3)+(9\cdot 5)]$
$= [63 \quad 67 \quad 57]$

(c) second column of AB = A [second column of B]

$= \begin{bmatrix}3 & -2 & 7\\6 & 5 & 4\\0 & 4 & 9\end{bmatrix}\begin{bmatrix}-2\\1\\7\end{bmatrix} = \begin{bmatrix}-(3\cdot 2)-(2\cdot 1)+(7\cdot 7)\\-(6\cdot 2)+(5\cdot 1)+(4\cdot 7)\\-(0\cdot 2)+(4\cdot 1)+(9\cdot 7)\end{bmatrix} = \begin{bmatrix}41\\21\\67\end{bmatrix}$

(d) first column of BA = B [first column of A]

$$= \begin{bmatrix} 6 & -2 & 4 \\ 0 & 1 & 3 \\ 7 & 7 & 5 \end{bmatrix} \begin{bmatrix} 3 \\ 6 \\ 0 \end{bmatrix} = \begin{bmatrix} (6\cdot 3)-(2\cdot 6)+(4\cdot 0) \\ (0\cdot 3)+(1\cdot 6)+(3\cdot 0) \\ (7\cdot 3)+(7\cdot 6)+(5\cdot 0) \end{bmatrix} = \begin{bmatrix} 6 \\ 6 \\ 63 \end{bmatrix}$$

(e) third row of AA = [third row of A] A = $[0 \;\; 4 \;\; 9]\begin{bmatrix} 3 & -2 & 7 \\ 6 & 5 & 4 \\ 0 & 4 & 9 \end{bmatrix}$

$$= [(0\cdot 3)+(4\cdot 6)+(9\cdot 0) \quad -(0\cdot 2)+(4\cdot 5)+(9\cdot 4) \quad (0\cdot 7)+(4\cdot 4)+(9\cdot 9)]$$
$$= [24 \;\; 56 \;\; 97]$$

(f) third column of AA = A [third column of A]

$$= \begin{bmatrix} 3 & -2 & 7 \\ 6 & 5 & 4 \\ 0 & 4 & 9 \end{bmatrix} \begin{bmatrix} 7 \\ 4 \\ 9 \end{bmatrix} = \begin{bmatrix} (3\cdot 7)-(2\cdot 4)+(7\cdot 9) \\ (6\cdot 7)+(5\cdot 4)+(4\cdot 9) \\ (0\cdot 7)+(4\cdot 4)+(9\cdot 9) \end{bmatrix} = \begin{bmatrix} 76 \\ 98 \\ 97 \end{bmatrix}$$

9. **(a)** first column of $AA = 3\begin{bmatrix} 3 \\ 6 \\ 0 \end{bmatrix} + 6\begin{bmatrix} -2 \\ 5 \\ 4 \end{bmatrix} + 0\begin{bmatrix} 7 \\ 4 \\ 9 \end{bmatrix} = \begin{bmatrix} -3 \\ 48 \\ 24 \end{bmatrix}$

second column of $AA = -2\begin{bmatrix} 3 \\ 6 \\ 0 \end{bmatrix} + 5\begin{bmatrix} -2 \\ 5 \\ 4 \end{bmatrix} + 4\begin{bmatrix} 7 \\ 4 \\ 9 \end{bmatrix} = \begin{bmatrix} 12 \\ 29 \\ 56 \end{bmatrix}$

third column of $AA = 7\begin{bmatrix} 3 \\ 6 \\ 0 \end{bmatrix} + 4\begin{bmatrix} -2 \\ 5 \\ 4 \end{bmatrix} + 9\begin{bmatrix} 7 \\ 4 \\ 9 \end{bmatrix} = \begin{bmatrix} 76 \\ 98 \\ 97 \end{bmatrix}$

(b) first column of $BB = 6\begin{bmatrix} 6 \\ 0 \\ 7 \end{bmatrix} + 0\begin{bmatrix} -2 \\ 1 \\ 7 \end{bmatrix} + 7\begin{bmatrix} 4 \\ 3 \\ 5 \end{bmatrix} = \begin{bmatrix} 64 \\ 21 \\ 77 \end{bmatrix}$

second column of $BB = -2\begin{bmatrix} 6 \\ 0 \\ 7 \end{bmatrix} + 1\begin{bmatrix} -2 \\ 1 \\ 7 \end{bmatrix} + 7\begin{bmatrix} 4 \\ 3 \\ 5 \end{bmatrix} = \begin{bmatrix} 14 \\ 22 \\ 28 \end{bmatrix}$

third column of $BB = 4\begin{bmatrix} 6 \\ 0 \\ 7 \end{bmatrix} + 3\begin{bmatrix} -2 \\ 1 \\ 7 \end{bmatrix} + 5\begin{bmatrix} 4 \\ 3 \\ 5 \end{bmatrix} = \begin{bmatrix} 38 \\ 18 \\ 74 \end{bmatrix}$

11. **(a)** $A = \begin{bmatrix} 2 & -3 & 5 \\ 9 & -1 & 1 \\ 1 & 5 & 4 \end{bmatrix}$, $\mathbf{x} = \begin{bmatrix} x_1 \\ x_2 \\ x_3 \end{bmatrix}$, $\mathbf{b} = \begin{bmatrix} 7 \\ -1 \\ 0 \end{bmatrix}$; the matrix equation: $\begin{bmatrix} 2 & -3 & 5 \\ 9 & -1 & 1 \\ 1 & 5 & 4 \end{bmatrix}\begin{bmatrix} x_1 \\ x_2 \\ x_3 \end{bmatrix} = \begin{bmatrix} 7 \\ -1 \\ 0 \end{bmatrix}$

(b) $A = \begin{bmatrix} 4 & 0 & -3 & 1 \\ 5 & 1 & 0 & -8 \\ 2 & -5 & 9 & -1 \\ 0 & 3 & -1 & 7 \end{bmatrix}$, $\mathbf{x} = \begin{bmatrix} x_1 \\ x_2 \\ x_3 \\ x_4 \end{bmatrix}$, $\mathbf{b} = \begin{bmatrix} 1 \\ 3 \\ 0 \\ 2 \end{bmatrix}$; the matrix equation: $\begin{bmatrix} 4 & 0 & -3 & 1 \\ 5 & 1 & 0 & -8 \\ 2 & -5 & 9 & -1 \\ 0 & 3 & -1 & 7 \end{bmatrix}\begin{bmatrix} x_1 \\ x_2 \\ x_3 \\ x_4 \end{bmatrix} = \begin{bmatrix} 1 \\ 3 \\ 0 \\ 2 \end{bmatrix}$

13. **(a)**
$$\begin{aligned} 5x_1 + 6x_2 - 7x_3 &= 2 \\ -x_1 - 2x_2 + 3x_3 &= 0 \\ 4x_2 - x_3 &= 3 \end{aligned}$$

(b)
$$\begin{aligned} x + y + z &= 2 \\ 2x + 3y &= 2 \\ 5x - 3y - 6z &= -9 \end{aligned}$$

15. $[k \;\; 1 \;\; 1]\begin{bmatrix} 1 & 1 & 0 \\ 1 & 0 & 2 \\ 0 & 2 & -3 \end{bmatrix}\begin{bmatrix} k \\ 1 \\ 1 \end{bmatrix} = [k \;\; 1 \;\; 1]\begin{bmatrix} k+1 \\ k+2 \\ -1 \end{bmatrix} = k^2 + k + k + 2 - 1 = k^2 + 2k + 1 = (k+1)^2$

The only value of k that satisfies the equation is $k = -1$.

17. $\begin{bmatrix}4\\2\end{bmatrix}\begin{bmatrix}0 & 1 & 2\end{bmatrix} + \begin{bmatrix}-3\\-1\end{bmatrix}\begin{bmatrix}-2 & 3 & 1\end{bmatrix} = \begin{bmatrix}0 & 4 & 8\\0 & 2 & 4\end{bmatrix} + \begin{bmatrix}6 & -9 & -3\\2 & -3 & -1\end{bmatrix} = \begin{bmatrix}6 & -5 & 5\\2 & -1 & 3\end{bmatrix}$

19. $\begin{bmatrix}1\\4\end{bmatrix}\begin{bmatrix}1 & 2\end{bmatrix} + \begin{bmatrix}2\\5\end{bmatrix}\begin{bmatrix}3 & 4\end{bmatrix} + \begin{bmatrix}3\\6\end{bmatrix}\begin{bmatrix}5 & 6\end{bmatrix} = \begin{bmatrix}1 & 2\\4 & 8\end{bmatrix} + \begin{bmatrix}6 & 8\\15 & 20\end{bmatrix} + \begin{bmatrix}15 & 18\\30 & 36\end{bmatrix} = \begin{bmatrix}22 & 28\\49 & 64\end{bmatrix}$

21. $\begin{bmatrix}x_1\\x_2\\x_3\\x_4\\x_5\\x_6\end{bmatrix} = \begin{bmatrix}-3r-4s-2t\\r\\-2s\\s\\t\\\frac{1}{3}\end{bmatrix} = \begin{bmatrix}0\\0\\0\\0\\0\\\frac{1}{3}\end{bmatrix} + \begin{bmatrix}-3r\\r\\0\\0\\0\\0\end{bmatrix} + \begin{bmatrix}-4s\\0\\-2s\\s\\0\\0\end{bmatrix} + \begin{bmatrix}-2t\\0\\0\\0\\t\\0\end{bmatrix} = \begin{bmatrix}0\\0\\0\\0\\0\\\frac{1}{3}\end{bmatrix} + r\begin{bmatrix}-3\\1\\0\\0\\0\\0\end{bmatrix} + s\begin{bmatrix}-4\\0\\-2\\1\\0\\0\end{bmatrix} + t\begin{bmatrix}-2\\0\\0\\0\\1\\0\end{bmatrix}$

23. The given matrix equation is equivalent to the linear system

$$\begin{aligned} a &= 4 \\ 3 &= d - 2c \\ -1 &= d + 2c \\ a + b &= -2 \end{aligned}$$

After subtracting first equation from the fourth, adding the second to the third, and back-substituting, we obtain the solution: $a = 4$, $b = -6$, $c = -1$, and $d = 1$.

25. **(a)** If the ith row vector of A is $[0 \ \cdots \ 0]$ then it follows from Formula (9) in Section 1.3 that

ith row vector of $AB = [0 \ \cdots \ 0]\,B = [0 \ \cdots \ 0]$

(b) If the jth column vector of B is $\begin{bmatrix}0\\\vdots\\0\end{bmatrix}$ then it follows from Formula (8) in Section 1.3 that

the jth column vector of $AB = A\begin{bmatrix}0\\\vdots\\0\end{bmatrix} = \begin{bmatrix}0\\\vdots\\0\end{bmatrix}$

27. Setting the left hand side $A\begin{bmatrix}x\\y\\z\end{bmatrix} = \begin{bmatrix}a_{11} & a_{12} & a_{13}\\a_{21} & a_{22} & a_{23}\\a_{31} & a_{32} & a_{33}\end{bmatrix}\begin{bmatrix}x\\y\\z\end{bmatrix} = \begin{bmatrix}a_{11}x + a_{12}y + a_{13}z\\a_{21}x + a_{22}y + a_{23}z\\a_{31}x + a_{32}y + a_{33}z\end{bmatrix}$ equal to $\begin{bmatrix}x+y\\x-y\\0\end{bmatrix}$ yields

$$\begin{aligned} a_{11}x + a_{12}y + a_{13}z &= x + y \\ a_{21}x + a_{22}y + a_{23}z &= x - y \\ a_{31}x + a_{32}y + a_{33}z &= 0 \end{aligned}$$

Assuming the entries of A are real numbers that do not depend on x, y, and z, this requires that the coefficients corresponding to the same variable on both sides of each equation must match.

Therefore, the only matrix satisfying the given condition is $A = \begin{bmatrix}1 & 1 & 0\\1 & -1 & 0\\0 & 0 & 0\end{bmatrix}$.

29. **(a)** $\begin{bmatrix}1 & 1\\1 & 1\end{bmatrix}$ and $\begin{bmatrix}-1 & -1\\-1 & -1\end{bmatrix}$

(b) Four square roots can be found: $\begin{bmatrix}\sqrt{5} & 0\\0 & 3\end{bmatrix}$, $\begin{bmatrix}-\sqrt{5} & 0\\0 & 3\end{bmatrix}$, $\begin{bmatrix}\sqrt{5} & 0\\0 & -3\end{bmatrix}$, and $\begin{bmatrix}-\sqrt{5} & 0\\0 & -3\end{bmatrix}$.

33. The given matrix product represents $\begin{bmatrix} \text{the total cost of items purchased in January} \\ \text{the total cost of items purchased in February} \\ \text{the total cost of items purchased in March} \\ \text{the total cost of items purchased in April} \end{bmatrix}$.

True-False Exercises

(a) True. The main diagonal is only defined for square matrices.

(b) False. An $m \times n$ matrix has m row vectors and n column vectors.

(c) False. E.g., if $A = \begin{bmatrix} 1 & 0 \\ 0 & 0 \end{bmatrix}$ and $B = \begin{bmatrix} 0 & 0 \\ 1 & 0 \end{bmatrix}$ then $AB = \begin{bmatrix} 0 & 0 \\ 0 & 0 \end{bmatrix}$ does not equal $BA = B$.

(d) False. The ith row vector of AB can be computed by multiplying the ith row vector of A by B.

(e) True. Using Formula (14), $((A^T)^T)_{ij} = (A^T)_{ji} = (A)_{ij}$.

(f) False. E.g., if $A = \begin{bmatrix} 1 & 0 \\ 0 & 0 \end{bmatrix}$ and $B = \begin{bmatrix} 0 & 0 \\ 0 & 1 \end{bmatrix}$ then the trace of $AB = \begin{bmatrix} 0 & 0 \\ 0 & 0 \end{bmatrix}$ is 0, which does not equal $\text{tr}(A)\text{tr}(B) = 1$.

(g) False. E.g., if $A = \begin{bmatrix} 1 & 0 \\ 0 & 0 \end{bmatrix}$ and $B = \begin{bmatrix} 0 & 0 \\ 1 & 0 \end{bmatrix}$ then $(AB)^T = \begin{bmatrix} 0 & 0 \\ 0 & 0 \end{bmatrix}$ does not equal $A^T B^T = \begin{bmatrix} 0 & 1 \\ 0 & 0 \end{bmatrix}$.

(h) True. The main diagonal entries in a square matrix A are the same as those in A^T.

(i) True. Since A^T is a 4×6 matrix, it follows from $B^T A^T$ being a 2×6 matrix that B^T must be a 2×4 matrix. Consequently, B is a 4×2 matrix.

(j) True.

$$\text{tr}\left(c\begin{bmatrix} a_{11} & \cdots & a_{1n} \\ \vdots & \ddots & \vdots \\ a_{n1} & \cdots & a_{nn} \end{bmatrix}\right) = \text{tr}\left(\begin{bmatrix} ca_{11} & \cdots & ca_{1n} \\ \vdots & \ddots & \vdots \\ ca_{n1} & \cdots & ca_{nn} \end{bmatrix}\right)$$

$$= ca_{11} + \cdots + ca_{nn} = c(a_{11} + \cdots + a_{nn}) = c\,\text{tr}\left(\begin{bmatrix} a_{11} & \cdots & a_{1n} \\ \vdots & \ddots & \vdots \\ a_{n1} & \cdots & a_{nn} \end{bmatrix}\right)$$

(k) True. The equality of the matrices $A - C$ and $B - C$ implies that $a_{ij} - c_{ij} = b_{ij} - c_{ij}$ for all i and j. Adding c_{ij} to both sides yields $a_{ij} = b_{ij}$ for all i and j. Consequently, the matrices A and B are equal.

(l) False. E.g., if $A = \begin{bmatrix} 1 & 0 \\ 0 & 0 \end{bmatrix}$ and $B = C = \begin{bmatrix} 0 & 0 \\ 1 & 0 \end{bmatrix}$ then $AC = BC = \begin{bmatrix} 0 & 0 \\ 0 & 0 \end{bmatrix}$ even though $A \neq B$.

(m) True. If A is a $p \times q$ matrix and B is an $r \times s$ matrix then AB being defined requires $q = r$ and BA being defined requires $s = p$. For the $p \times p$ matrix AB to be possible to add to the $q \times q$ matrix BA, we must have $p = q$.

(n) True. If the jth column vector of B is $\begin{bmatrix} 0 \\ \vdots \\ 0 \end{bmatrix}$ then it follows from Formula (8) in Section 1.3 that

the jth column vector of $AB = A\begin{bmatrix} 0 \\ \vdots \\ 0 \end{bmatrix} = \begin{bmatrix} 0 \\ \vdots \\ 0 \end{bmatrix}$.

(o) False. E.g., if $A = \begin{bmatrix} 1 & 1 \\ 1 & 1 \end{bmatrix}$ and $B = \begin{bmatrix} 1 & 0 \\ 1 & 0 \end{bmatrix}$ then $BA = A$ does not have a column of zeros even though B does.

1.4 Inverses; Algebraic Properties of Matrices

1. **(a)** $A + (B + C) = (A + B) + C = \begin{bmatrix} 7 & 2 \\ 0 & -2 \end{bmatrix}$ **(b)** $A(BC) = (AB)C = \begin{bmatrix} -34 & -21 \\ 52 & 28 \end{bmatrix}$

(c) $A(B + C) = AB + AC = \begin{bmatrix} 14 & 15 \\ 0 & -18 \end{bmatrix}$ **(d)** $(a + b)C = aC + bC = \begin{bmatrix} -12 & -3 \\ 9 & 6 \end{bmatrix}$

3. **(a)** $(A^T)^T = A = \begin{bmatrix} 3 & -1 \\ 2 & 4 \end{bmatrix}$ **(b)** $(AB)^T = B^T A^T = \begin{bmatrix} -1 & 4 \\ 10 & -12 \end{bmatrix}$

5. The determinant of A, $\det(A) = (2)(4) - (-3)(4) = 20$, is nonzero. Therefore A is invertible and its inverse is $A^{-1} = \frac{1}{20}\begin{bmatrix} 4 & 3 \\ -4 & 2 \end{bmatrix} = \begin{bmatrix} \frac{1}{5} & \frac{3}{20} \\ -\frac{1}{5} & \frac{1}{10} \end{bmatrix}$.

7. The determinant of C, $\det(C) = (2)(3) - (0)(0) = 6$, is nonzero. Therefore C is invertible and its inverse is $C^{-1} = \frac{1}{6}\begin{bmatrix} 3 & 0 \\ 0 & 2 \end{bmatrix} = \begin{bmatrix} \frac{1}{2} & 0 \\ 0 & \frac{1}{3} \end{bmatrix}$.

9. The determinant of $A = \begin{bmatrix} \frac{1}{2}(e^x + e^{-x}) & \frac{1}{2}(e^x - e^{-x}) \\ \frac{1}{2}(e^x - e^{-x}) & \frac{1}{2}(e^x + e^{-x}) \end{bmatrix}$,

$\det(A) = \frac{1}{4}(e^x + e^{-x})^2 - \frac{1}{4}(e^x - e^{-x})^2 = \frac{1}{4}(e^{2x} + 2 + e^{-2x}) - \frac{1}{4}(e^{2x} - 2 + e^{-2x}) = \frac{1}{4}(2 + 2) = 1$ is nonzero. Therefore A is invertible and its inverse is $A^{-1} = \begin{bmatrix} \frac{1}{2}(e^x + e^{-x}) & -\frac{1}{2}(e^x - e^{-x}) \\ -\frac{1}{2}(e^x - e^{-x}) & \frac{1}{2}(e^x + e^{-x}) \end{bmatrix}$.

11. $A^T = \begin{bmatrix} 2 & 4 \\ -3 & 4 \end{bmatrix}$; $(A^T)^{-1} = \frac{1}{(2)(4)-(4)(-3)}\begin{bmatrix} 4 & -4 \\ 3 & 2 \end{bmatrix} = \frac{1}{20}\begin{bmatrix} 4 & -4 \\ 3 & 2 \end{bmatrix} = \begin{bmatrix} \frac{1}{5} & -\frac{1}{5} \\ \frac{3}{20} & \frac{1}{10} \end{bmatrix}$

$A^{-1} = \frac{1}{(2)(4)-(-3)(4)}\begin{bmatrix} 4 & 3 \\ -4 & 2 \end{bmatrix} = \frac{1}{20}\begin{bmatrix} 4 & 3 \\ -4 & 2 \end{bmatrix} = \begin{bmatrix} \frac{1}{5} & \frac{3}{20} \\ -\frac{1}{5} & \frac{1}{10} \end{bmatrix}$; $(A^{-1})^T = \begin{bmatrix} \frac{1}{5} & -\frac{1}{5} \\ \frac{3}{20} & \frac{1}{10} \end{bmatrix}$

13. $ABC = \begin{bmatrix} -18 & -12 \\ 64 & 36 \end{bmatrix}$; $(ABC)^{-1} = \frac{1}{(-18)(36)-(-12)(64)}\begin{bmatrix} 36 & 12 \\ -64 & -18 \end{bmatrix} = \frac{1}{120}\begin{bmatrix} 36 & 12 \\ -64 & -18 \end{bmatrix} = \begin{bmatrix} \frac{3}{10} & \frac{1}{10} \\ -\frac{8}{15} & -\frac{3}{20} \end{bmatrix}$

$C^{-1}B^{-1}A^{-1} = \left(\frac{1}{6}\begin{bmatrix} 3 & 0 \\ 0 & 2 \end{bmatrix}\right)\begin{bmatrix} 2 & -1 \\ -5 & 3 \end{bmatrix}\left(\frac{1}{20}\begin{bmatrix} 4 & 3 \\ -4 & 2 \end{bmatrix}\right) = \begin{bmatrix} \frac{1}{2} & 0 \\ 0 & \frac{1}{3} \end{bmatrix}\begin{bmatrix} 2 & -1 \\ -5 & 3 \end{bmatrix}\begin{bmatrix} \frac{1}{5} & \frac{3}{20} \\ -\frac{1}{5} & \frac{1}{10} \end{bmatrix} = \begin{bmatrix} \frac{3}{10} & \frac{1}{10} \\ -\frac{8}{15} & -\frac{3}{20} \end{bmatrix}$

15. From part (a) of Theorem 1.4.7 it follows that the inverse of $(7A)^{-1}$ is $7A$.

Thus $7A = \frac{1}{(-3)(-2)-(7)(1)}\begin{bmatrix}-2 & -7\\ -1 & -3\end{bmatrix} = \frac{1}{-1}\begin{bmatrix}-2 & -7\\ -1 & -3\end{bmatrix} = \begin{bmatrix}2 & 7\\ 1 & 3\end{bmatrix}$. Consequently, $A = \frac{1}{7}\begin{bmatrix}2 & 7\\ 1 & 3\end{bmatrix} = \begin{bmatrix}\frac{2}{7} & 1\\ \frac{1}{7} & \frac{3}{7}\end{bmatrix}$.

17. From part (a) of Theorem 1.4.7 it follows that the inverse of $(I + 2A)^{-1}$ is $I + 2A$.

Thus $I + 2A = \frac{1}{(-1)(5)-(2)(4)}\begin{bmatrix}5 & -2\\ -4 & -1\end{bmatrix} = \frac{1}{-13}\begin{bmatrix}5 & -2\\ -4 & -1\end{bmatrix} = \begin{bmatrix}-\frac{5}{13} & \frac{2}{13}\\ \frac{4}{13} & \frac{1}{13}\end{bmatrix}$.

Consequently, $A = \frac{1}{2}\left(\begin{bmatrix}-\frac{5}{13} & \frac{2}{13}\\ \frac{4}{13} & \frac{1}{13}\end{bmatrix} - \begin{bmatrix}1 & 0\\ 0 & 1\end{bmatrix}\right) = \begin{bmatrix}-\frac{9}{13} & \frac{1}{13}\\ \frac{2}{13} & -\frac{6}{13}\end{bmatrix}$

19. **(a)** $A^3 = AAA = \begin{bmatrix}41 & 15\\ 30 & 11\end{bmatrix}$

(b) $(A^3)^{-1} = \frac{1}{(41)(11)-(15)(30)}\begin{bmatrix}11 & -15\\ -30 & 41\end{bmatrix} = \begin{bmatrix}11 & -15\\ -30 & 41\end{bmatrix}$

(c) $A^2 - 2A + I = \begin{bmatrix}3 & 1\\ 2 & 1\end{bmatrix}\begin{bmatrix}3 & 1\\ 2 & 1\end{bmatrix} - 2\begin{bmatrix}3 & 1\\ 2 & 1\end{bmatrix} + \begin{bmatrix}1 & 0\\ 0 & 1\end{bmatrix} = \begin{bmatrix}11 & 4\\ 8 & 3\end{bmatrix} - \begin{bmatrix}6 & 2\\ 4 & 2\end{bmatrix} + \begin{bmatrix}1 & 0\\ 0 & 1\end{bmatrix} = \begin{bmatrix}6 & 2\\ 4 & 2\end{bmatrix}$

21. **(a)** $A - 2I = \begin{bmatrix}1 & 1\\ 2 & -1\end{bmatrix}$ **(b)** $2A^2 - A + I = \begin{bmatrix}20 & 7\\ 14 & 6\end{bmatrix}$ **(c)** $A^3 - 2A + I = \begin{bmatrix}36 & 13\\ 26 & 10\end{bmatrix}$

23. $AB = \begin{bmatrix}a & b\\ c & d\end{bmatrix}\begin{bmatrix}0 & 1\\ 0 & 0\end{bmatrix} = \begin{bmatrix}0 & a\\ 0 & c\end{bmatrix}$; $BA = \begin{bmatrix}0 & 1\\ 0 & 0\end{bmatrix}\begin{bmatrix}a & b\\ c & d\end{bmatrix} = \begin{bmatrix}c & d\\ 0 & 0\end{bmatrix}$.

The matrices A and B commute if $\begin{bmatrix}0 & a\\ 0 & c\end{bmatrix} = \begin{bmatrix}c & d\\ 0 & 0\end{bmatrix}$, i.e.

$$\begin{aligned}0 &= c\\ a &= d\\ 0 &= 0\\ c &= 0\end{aligned}$$

Therefore, $\begin{bmatrix}a & b\\ c & d\end{bmatrix}$ and $\begin{bmatrix}0 & 1\\ 0 & 0\end{bmatrix}$ commute if $c = 0$ and $a = d$.

If we assign b and d the arbitrary values s and t, respectively, the general solution is given by the formulas

$$a = t,\quad b = s,\quad c = 0,\quad d = t$$

25. $x_1 = \frac{(5)(-1)-(-2)(3)}{(3)(5)-(-2)(4)} = \frac{1}{23}$, $x_2 = \frac{(3)(3)-(4)(-1)}{(3)(5)-(-2)(4)} = \frac{13}{23}$

27. $x_1 = \frac{(-3)(0)-(1)(-2)}{(6)(-3)-(1)(4)} = \frac{2}{-22} = -\frac{1}{11}$, $x_2 = \frac{(6)(-2)-(4)(0)}{(6)(-3)-(1)(4)} = \frac{-12}{-22} = \frac{6}{11}$

29. $p(A) = A^2 - 9I = \begin{bmatrix}2 & 4\\ 8 & -6\end{bmatrix}$,

$p_1(A) = A + 3I = \begin{bmatrix}6 & 1\\ 2 & 4\end{bmatrix}$, $p_2(A) = A - 3I = \begin{bmatrix}0 & 1\\ 2 & -2\end{bmatrix}$, $p_1(A)p_2(A) = \begin{bmatrix}2 & 4\\ 8 & -6\end{bmatrix}$

31. **(a)** If $A = \begin{bmatrix}1 & 0\\ 0 & 0\end{bmatrix}$ and $B = \begin{bmatrix}0 & 1\\ 0 & 0\end{bmatrix}$ then $(A + B)(A - B) = \begin{bmatrix}1 & 1\\ 0 & 0\end{bmatrix}\begin{bmatrix}1 & -1\\ 0 & 0\end{bmatrix} = \begin{bmatrix}1 & -1\\ 0 & 0\end{bmatrix}$ does not equal $A^2 - B^2 = \begin{bmatrix}1 & 0\\ 0 & 0\end{bmatrix} - \begin{bmatrix}0 & 0\\ 0 & 0\end{bmatrix} = \begin{bmatrix}1 & 0\\ 0 & 0\end{bmatrix}$.

(b) Using the properties in Theorem 1.4.1 we can write

$$(A+B)(A-B) = A(A-B) + B(A-B) = A^2 - AB + BA - B^2$$

(c) If the matrices A and B commute (i.e., $AB = BA$) then $(A+B)(A-B) = A^2 - B^2$.

33. **(a)** We can rewrite the equation

$$\begin{aligned} A^2 + 2A + I &= O \\ A^2 + 2A &= -I \\ -A^2 - 2A &= I \\ A(-A-2I) &= I \end{aligned}$$

which shows that A is invertible and $A^{-1} = -A - 2I$.

(b) Let $p(x) = c_n x^n + \cdots + c_2 x^2 + c_1 x + c_0$ with $c_0 \neq 0$. The equation $p(A) = O$ can be rewritten as

$$\begin{aligned} c_n A^n + \cdots + c_2 A^2 + c_1 A + c_0 I &= O \\ c_n A^n + \cdots + c_2 A^2 + c_1 A &= -c_0 I \\ -\frac{c_n}{c_0}A^n - \cdots - \frac{c_2}{c_0}A^2 - \frac{c_1}{c_0}A &= I \\ A\left(-\frac{c_n}{c_0}A^{n-1} - \cdots - \frac{c_2}{c_0}A - \frac{c_1}{c_0}I\right) &= I \end{aligned}$$

which shows that A is invertible and $A^{-1} = -\frac{c_n}{c_0}A^{n-1} - \cdots - \frac{c_2}{c_0}A - \frac{c_1}{c_0}I$.

35. If the ith row vector of A is $[0 \ \cdots \ 0]$ then it follows from Formula (9) in Section 1.3 that ith row vector of $AB = [0 \ \cdots \ 0]\,B = [0 \ \cdots \ 0]$.
Consequently no matrix B can be found to make the product $AB = I$ thus A does not have an inverse.

If the jth column vector of A is $\begin{bmatrix}0\\ \vdots\\ 0\end{bmatrix}$ then it follows from Formula (8) in Section 1.3 that

the jth column vector of $BA = B\begin{bmatrix}0\\ \vdots\\ 0\end{bmatrix} = \begin{bmatrix}0\\ \vdots\\ 0\end{bmatrix}$.

Consequently no matrix B can be found to make the product $BA = I$ thus A does not have an inverse.

37. Letting $X = \begin{bmatrix}x_{11} & x_{12} & x_{13}\\ x_{21} & x_{22} & x_{23}\\ x_{31} & x_{32} & x_{33}\end{bmatrix}$, the matrix equation $AX = I$ becomes

$$\begin{bmatrix}x_{11}+x_{31} & x_{12}+x_{32} & x_{13}+x_{33}\\ x_{11}+x_{21} & x_{12}+x_{22} & x_{13}+x_{23}\\ x_{21}+x_{31} & x_{22}+x_{32} & x_{23}+x_{33}\end{bmatrix} = \begin{bmatrix}1 & 0 & 0\\ 0 & 1 & 0\\ 0 & 0 & 1\end{bmatrix}$$

Setting the first columns on both sides equal yields the system

$$\begin{aligned} x_{11} + x_{31} &= 1 \\ x_{11} + x_{21} &= 0 \\ x_{21} + x_{31} &= 0 \end{aligned}$$

Subtracting the second and third equations from the first leads to $-2x_{21}=1$. Therefore $x_{21}=-\frac{1}{2}$ and (after substituting this into the remaining equations) $x_{11}=x_{31}=\frac{1}{2}$.

The second and the third columns can be treated in a similar manner to result in $X=\begin{bmatrix}\frac{1}{2} & \frac{1}{2} & -\frac{1}{2}\\ -\frac{1}{2} & \frac{1}{2} & \frac{1}{2}\\ \frac{1}{2} & -\frac{1}{2} & \frac{1}{2}\end{bmatrix}$. We conclude that A invertible and its inverse is $A^{-1}=\begin{bmatrix}\frac{1}{2} & \frac{1}{2} & -\frac{1}{2}\\ -\frac{1}{2} & \frac{1}{2} & \frac{1}{2}\\ \frac{1}{2} & -\frac{1}{2} & \frac{1}{2}\end{bmatrix}$.

39. $(AB)^{-1}(AC^{-1})(D^{-1}C^{-1})^{-1}D^{-1}$

$= (B^{-1}A^{-1})(AC^{-1})((C^{-1})^{-1}(D^{-1})^{-1})D^{-1}$ ⟵ Theorem 1.4.6

$= (B^{-1}A^{-1})(AC^{-1})(CD)D^{-1}$ ⟵ Theorem 1.4.7(a)

$= B^{-1}(A^{-1}A)(C^{-1}C)(DD^{-1})$ ⟵ Theorem 1.4.1(c)

$= B^{-1}III$ ⟵ Formula (1) in Section 1.4

$= B^{-1}$ ⟵ Property $AI = IA = A$ on p. 43

41. If $R=[r_1 \ \cdots \ r_n]$ and $C=\begin{bmatrix}c_1\\ \vdots\\ c_n\end{bmatrix}$ then $CR=\begin{bmatrix}c_1r_1 & \cdots & c_1r_n\\ \vdots & \ddots & \vdots\\ c_nr_1 & \cdots & c_nr_n\end{bmatrix}$ and $RC=[r_1c_1+\cdots+r_nc_n]=[\mathrm{tr}(CR)]$.

43. **(a)** Assuming A is invertible, we can multiply (on the left) each side of the equation by A^{-1}:

$AB = AC$

$A^{-1}(AB) = A^{-1}(AC)$ ⟵ Multiply (on the left) each side by A^{-1}

$(A^{-1}A)B = (A^{-1}A)C$ ⟵ Theorem 1.4.1(c)

$IB = IC$ ⟵ Formula (1) in Section 1.4

$B = C$ ⟵ Property $AI = IA = A$ on p. 43

(b) If A is not an invertible matrix then $AB = AC$ does not generally imply $B = C$ as evidenced by Example 3.

45. **(a)** $A(A^{-1}+B^{-1})B(A+B)^{-1}$

$= (AA^{-1}B + AB^{-1}B)(A+B)^{-1}$ ⟵ Theorem 1.4.1(d) and (e)

$= (IB + AI)(A+B)^{-1}$ ⟵ Formula (1) in Section 1.4

$= (B + A)(A+B)^{-1}$ ⟵ Property $AI = IA = A$ on p. 43

$= (A + B)(A+B)^{-1}$ ⟵ Theorem 1.4.1(a)

$= I$ ⟵ Formula (1) in Section 1.4

(b) We can multiply each side of the equality from part (a) on the left by A^{-1}, then on the right by A to obtain

$$(A^{-1}+B^{-1})B(A+B)^{-1}A = I$$

which shows that if A, B, and $A+B$ are invertible then so is $A^{-1}+B^{-1}$.
Furthermore, $(A^{-1}+B^{-1})^{-1} = B(A+B)^{-1}A$.

47. Applying Theorem 1.4.1(d) and (g), property $AI = IA = A$, and the assumption $A^k = O$ we can write

$$\begin{aligned}&(I-A)\left(I+A+A^2+\cdots+A^{k-2}+A^{k-1}\right)\\ &= I-A+A-A^2+A^2-A^3+\cdots+A^{k-2}-A^{k-1}+A^{k-1}-A^k\\ &= I-A^k\\ &= I-O\\ &= I\end{aligned}$$

True-False Exercises

(a) False. A and B are inverses of one another if and only if $AB = BA = I$.

(b) False. $(A+B)^2 = (A+B)(A+B) = A^2+AB+BA+B^2$ does not generally equal $A^2+2AB+B^2$ since AB may not equal BA.

(c) False. $(A-B)(A+B) = A^2+AB-BA-B^2$ does not generally equal A^2-B^2 since AB may not equal BA.

(d) False. $(AB)^{-1} = B^{-1}A^{-1}$ does not generally equal $A^{-1}B^{-1}$.

(e) False. $(AB)^T = B^TA^T$ does not generally equal A^TB^T.

(f) True. This follows from Theorem 1.4.5.

(g) True. This follows from Theorem 1.4.8.

(h) True. This follows from Theorem 1.4.9. (The inverse of A^T is the transpose of A^{-1}.)

(i) False. $p(I) = (a_0+a_1+a_2+\cdots+a_m)I$.

(j) True.
If the ith row vector of A is $[0 \;\; \cdots \;\; 0]$ then it follows from Formula (9) in Section 1.3 that ith row vector of $AB = [0 \;\; \cdots \;\; 0]\,B = [0 \;\; \cdots \;\; 0]$.
Consequently no matrix B can be found to make the product $AB = I$ thus A does not have an inverse.

If the jth column vector of A is $\begin{bmatrix}0\\ \vdots\\ 0\end{bmatrix}$ then it follows from Formula (8) in Section 1.3 that

the jth column vector of $BA = B\begin{bmatrix}0\\ \vdots\\ 0\end{bmatrix} = \begin{bmatrix}0\\ \vdots\\ 0\end{bmatrix}$.

Consequently no matrix B can be found to make the product $BA = I$ thus A does not have an inverse.

(k) False. E.g. I and $-I$ are both invertible but $I+(-I) = O$ is not.

1.5 Elementary Matrices and a Method for Finding A^{-1}

1. **(a)** Elementary matrix (corresponds to adding -5 times the first row to the second row)

(b) Not an elementary matrix

(c) Not an elementary matrix

(d) Not an elementary matrix

3. **(a)** Add 3 times the second row to the first row: $\begin{bmatrix} 1 & 3 \\ 0 & 1 \end{bmatrix}$

(b) Multiply the first row by $-\frac{1}{7}$: $\begin{bmatrix} -\frac{1}{7} & 0 & 0 \\ 0 & 1 & 0 \\ 0 & 0 & 1 \end{bmatrix}$

(c) Add 5 times the first row to the third row: $\begin{bmatrix} 1 & 0 & 0 \\ 0 & 1 & 0 \\ 5 & 0 & 1 \end{bmatrix}$

(d) Interchange the first and third rows: $\begin{bmatrix} 0 & 0 & 1 & 0 \\ 0 & 1 & 0 & 0 \\ 1 & 0 & 0 & 0 \\ 0 & 0 & 0 & 1 \end{bmatrix}$

5. **(a)** Interchange the first and second rows: $EA = \begin{bmatrix} 3 & -6 & -6 & -6 \\ -1 & -2 & 5 & -1 \end{bmatrix}$

(b) Add -3 times the second row to the third row: $EA = \begin{bmatrix} 2 & -1 & 0 & -4 & -4 \\ 1 & -3 & -1 & 5 & 3 \\ -1 & 9 & 4 & -12 & -10 \end{bmatrix}$

(c) Add 4 times the third row to the first row: $EA = \begin{bmatrix} 13 & 28 \\ 2 & 5 \\ 3 & 6 \end{bmatrix}$

7. **(a)** $\begin{bmatrix} 0 & 0 & 1 \\ 0 & 1 & 0 \\ 1 & 0 & 0 \end{bmatrix}$ (B was obtained from A by interchanging the first row and the third row)

(b) $\begin{bmatrix} 0 & 0 & 1 \\ 0 & 1 & 0 \\ 1 & 0 & 0 \end{bmatrix}$ (A was obtained from B by interchanging the first row and the third row)

(c) $\begin{bmatrix} 1 & 0 & 0 \\ 0 & 1 & 0 \\ -2 & 0 & 1 \end{bmatrix}$ (C was obtained from A by adding -2 times the first row to the third row)

(d) $\begin{bmatrix} 1 & 0 & 0 \\ 0 & 1 & 0 \\ 2 & 0 & 1 \end{bmatrix}$ (A was obtained from C by adding 2 times the first row to the third row)

9. **(a)** (Method I: using Theorem 1.4.5)

The determinant of A, $\det(A) = (1)(7) - (4)(2) = -1$, is nonzero. Therefore A is invertible and its inverse is $A^{-1} = \frac{1}{-1}\begin{bmatrix} 7 & -4 \\ -2 & 1 \end{bmatrix} = \begin{bmatrix} -7 & 4 \\ 2 & -1 \end{bmatrix}$.

(Method II: using the inversion algorithm)

$\left[\begin{array}{cc|cc} 1 & 4 & 1 & 0 \\ 2 & 7 & 0 & 1 \end{array}\right]$ ⟵ The identity matrix was adjoined to the given matrix.

$\left[\begin{array}{cc|cc} 1 & 4 & 1 & 0 \\ 0 & -1 & -2 & 1 \end{array}\right]$ ⟵ -2 times the first row was added to the second row.

$\left[\begin{array}{cc|cc} 1 & 4 & 1 & 0 \\ 0 & 1 & 2 & -1 \end{array}\right]$ ⟵ The second row was multiplied by -1.

$\left[\begin{array}{cc|cc} 1 & 0 & -7 & 4 \\ 0 & 1 & 2 & -1 \end{array}\right]$ ⟵ -4 times the second row was added to the first row.

The inverse is $\begin{bmatrix} -7 & 4 \\ 2 & -1 \end{bmatrix}$.

(b) (Method I: using Theorem 1.4.5)
The determinant of A, $\det(A) = (2)(8) - (-4)(-4) = 0$. Therefore A is not invertible.
(Method II: using the inversion algorithm)

$\left[\begin{array}{cc|cc} 2 & -4 & 1 & 0 \\ -4 & 8 & 0 & 1 \end{array}\right]$ ⟵ The identity matrix was adjoined to the given matrix.

$\left[\begin{array}{cc|cc} 2 & -4 & 1 & 0 \\ 0 & 0 & 2 & 1 \end{array}\right]$ ⟵ 2 times the first row was added to the second row.

A row of zeros was obtained on the left side, therefore A is not invertible.

11. (a) $\left[\begin{array}{ccc|ccc} 1 & 2 & 3 & 1 & 0 & 0 \\ 2 & 5 & 3 & 0 & 1 & 0 \\ 1 & 0 & 8 & 0 & 0 & 1 \end{array}\right]$ ⟵ The identity matrix was adjoined to the given matrix.

$\left[\begin{array}{ccc|ccc} 1 & 2 & 3 & 1 & 0 & 0 \\ 0 & 1 & -3 & -2 & 1 & 0 \\ 0 & -2 & 5 & -1 & 0 & 1 \end{array}\right]$ ⟵ -2 times the first row was added to the second row and -1 times the first row was added to the third row.

$\left[\begin{array}{ccc|ccc} 1 & 2 & 3 & 1 & 0 & 0 \\ 0 & 1 & -3 & -2 & 1 & 0 \\ 0 & 0 & -1 & -5 & 2 & 1 \end{array}\right]$ ⟵ 2 times the second row was added to the third row.

$\left[\begin{array}{ccc|ccc} 1 & 2 & 3 & 1 & 0 & 0 \\ 0 & 1 & -3 & -2 & 1 & 0 \\ 0 & 0 & 1 & 5 & -2 & -1 \end{array}\right]$ ⟵ The third row was multiplied by -1.

$\left[\begin{array}{ccc|ccc} 1 & 2 & 0 & -14 & 6 & 3 \\ 0 & 1 & 0 & 13 & -5 & -3 \\ 0 & 0 & 1 & 5 & -2 & -1 \end{array}\right]$ ⟵ 3 times the third row was added to the second row and -3 times the third row was added to the first row.

$\left[\begin{array}{ccc|ccc} 1 & 0 & 0 & -40 & 16 & 9 \\ 0 & 1 & 0 & 13 & -5 & -3 \\ 0 & 0 & 1 & 5 & -2 & -1 \end{array}\right]$ ⟵ -2 times the second row was added to the first row.

The inverse is $\begin{bmatrix} -40 & 16 & 9 \\ 13 & -5 & -3 \\ 5 & -2 & -1 \end{bmatrix}$.

(b) $\left[\begin{array}{ccc|ccc} -1 & 3 & -4 & 1 & 0 & 0 \\ 2 & 4 & 1 & 0 & 1 & 0 \\ -4 & 2 & -9 & 0 & 0 & 1 \end{array}\right]$ ⟵ The identity matrix was adjoined to the given matrix.

$\left[\begin{array}{ccc|ccc} 1 & -3 & 4 & -1 & 0 & 0 \\ 2 & 4 & 1 & 0 & 1 & 0 \\ -4 & 2 & -9 & 0 & 0 & 1 \end{array}\right]$ ⟵ The first row was multiplied by -1.

$\left[\begin{array}{ccc|ccc} 1 & -3 & 4 & -1 & 0 & 0 \\ 0 & 10 & -7 & 2 & 1 & 0 \\ 0 & -10 & 7 & -4 & 0 & 1 \end{array}\right]$ ⟵ -2 times the first row was added to the second row and 4 times the first row was added to the third row.

$\left[\begin{array}{ccc|ccc} 1 & -3 & 4 & -1 & 0 & 0 \\ 0 & 10 & -7 & 2 & 1 & 0 \\ 0 & 0 & 0 & -2 & 1 & 1 \end{array}\right]$ ⟵ The second row was added to the third row.

A row of zeros was obtained on the left side, therefore the matrix is not invertible.

13. $\left[\begin{array}{ccc|ccc} 1 & 0 & 1 & 1 & 0 & 0 \\ 0 & 1 & 1 & 0 & 1 & 0 \\ 1 & 1 & 0 & 0 & 0 & 1 \end{array}\right]$ ⟵ The identity matrix was adjoined to the given matrix.

$\left[\begin{array}{ccc|ccc} 1 & 0 & 1 & 1 & 0 & 0 \\ 0 & 1 & 1 & 0 & 1 & 0 \\ 0 & 1 & -1 & -1 & 0 & 1 \end{array}\right]$ ⟵ -1 times the first row was added to the third row.

$\left[\begin{array}{ccc|ccc} 1 & 0 & 1 & 1 & 0 & 0 \\ 0 & 1 & 1 & 0 & 1 & 0 \\ 0 & 0 & -2 & -1 & -1 & 1 \end{array}\right]$ ⟵ -1 times the second row was added to the third row.

$\left[\begin{array}{ccc|ccc} 1 & 0 & 1 & 1 & 0 & 0 \\ 0 & 1 & 1 & 0 & 1 & 0 \\ 0 & 0 & 1 & \frac{1}{2} & \frac{1}{2} & -\frac{1}{2} \end{array}\right]$ ⟵ The third row was multiplied by $-\frac{1}{2}$.

$\left[\begin{array}{ccc|ccc} 1 & 0 & 0 & \frac{1}{2} & -\frac{1}{2} & \frac{1}{2} \\ 0 & 1 & 0 & -\frac{1}{2} & \frac{1}{2} & \frac{1}{2} \\ 0 & 0 & 1 & \frac{1}{2} & \frac{1}{2} & -\frac{1}{2} \end{array}\right]$ ⟵ -1 times the third row was added to the second and -1 times the third row was added to the first row

The inverse is $\begin{bmatrix} \frac{1}{2} & -\frac{1}{2} & \frac{1}{2} \\ -\frac{1}{2} & \frac{1}{2} & \frac{1}{2} \\ \frac{1}{2} & \frac{1}{2} & -\frac{1}{2} \end{bmatrix}$.

15. $\left[\begin{array}{ccc|ccc} 2 & 6 & 6 & 1 & 0 & 0 \\ 2 & 7 & 6 & 0 & 1 & 0 \\ 2 & 7 & 7 & 0 & 0 & 1 \end{array}\right]$ ⟵ The identity matrix was adjoined to the given matrix.

$\left[\begin{array}{ccc|ccc} 2 & 6 & 6 & 1 & 0 & 0 \\ 0 & 1 & 0 & -1 & 1 & 0 \\ 0 & 1 & 1 & -1 & 0 & 1 \end{array}\right]$ ⟵ -1 times the first row was added to the second and -1 times the first row was added to the third row

$\left[\begin{array}{ccc|ccc} 2 & 6 & 6 & 1 & 0 & 0 \\ 0 & 1 & 0 & -1 & 1 & 0 \\ 0 & 0 & 1 & 0 & -1 & 1 \end{array}\right]$ ⟵ -1 times the second row was added to the third row.

$\left[\begin{array}{ccc|ccc} 2 & 6 & 0 & 1 & 6 & -6 \\ 0 & 1 & 0 & -1 & 1 & 0 \\ 0 & 0 & 1 & 0 & -1 & 1 \end{array}\right]$ ⟵ -6 times the third row was added to the first row

$\left[\begin{array}{ccc|ccc} 2 & 0 & 0 & 7 & 0 & -6 \\ 0 & 1 & 0 & -1 & 1 & 0 \\ 0 & 0 & 1 & 0 & -1 & 1 \end{array}\right]$ ⟵ -6 times the second row was added to the first row

$\left[\begin{array}{ccc|ccc} 1 & 0 & 0 & \frac{7}{2} & 0 & -3 \\ 0 & 1 & 0 & -1 & 1 & 0 \\ 0 & 0 & 1 & 0 & -1 & 1 \end{array}\right]$ ⟵ The first row was multiplied by $\frac{1}{2}$.

The inverse is $\begin{bmatrix} \frac{7}{2} & 0 & -3 \\ -1 & 1 & 0 \\ 0 & -1 & 1 \end{bmatrix}$.

17. $\left[\begin{array}{cccc|cccc} 2 & -4 & 0 & 0 & 1 & 0 & 0 & 0 \\ 1 & 2 & 12 & 0 & 0 & 1 & 0 & 0 \\ 0 & 0 & 2 & 0 & 0 & 0 & 1 & 0 \\ 0 & -1 & -4 & -5 & 0 & 0 & 0 & 1 \end{array}\right]$ ⟵ The identity matrix was adjoined to the given matrix.

$\left[\begin{array}{cccc|cccc} 1 & 2 & 12 & 0 & 0 & 1 & 0 & 0 \\ 2 & -4 & 0 & 0 & 1 & 0 & 0 & 0 \\ 0 & 0 & 2 & 0 & 0 & 0 & 1 & 0 \\ 0 & -1 & -4 & -5 & 0 & 0 & 0 & 1 \end{array}\right]$ ⟵ The first and second rows were interchanged.

$\left[\begin{array}{cccc|cccc} 1 & 2 & 12 & 0 & 0 & 1 & 0 & 0 \\ 0 & -8 & -24 & 0 & 1 & -2 & 0 & 0 \\ 0 & 0 & 2 & 0 & 0 & 0 & 1 & 0 \\ 0 & -1 & -4 & -5 & 0 & 0 & 0 & 1 \end{array}\right]$ ⟵ -2 times the first row was added to the second.

$\left[\begin{array}{cccc|cccc} 1 & 2 & 12 & 0 & 0 & 1 & 0 & 0 \\ 0 & -1 & -4 & -5 & 0 & 0 & 0 & 1 \\ 0 & 0 & 2 & 0 & 0 & 0 & 1 & 0 \\ 0 & -8 & -24 & 0 & 1 & -2 & 0 & 0 \end{array}\right]$ ⟵ The second and fourth rows were interchanged.

$\left[\begin{array}{cccc|cccc} 1 & 2 & 12 & 0 & 0 & 1 & 0 & 0 \\ 0 & 1 & 4 & 5 & 0 & 0 & 0 & -1 \\ 0 & 0 & 2 & 0 & 0 & 0 & 1 & 0 \\ 0 & -8 & -24 & 0 & 1 & -2 & 0 & 0 \end{array}\right]$ ⟵ The second row was multiplied by -1.

$\left[\begin{array}{cccc|cccc} 1 & 2 & 12 & 0 & 0 & 1 & 0 & 0 \\ 0 & 1 & 4 & 5 & 0 & 0 & 0 & -1 \\ 0 & 0 & 2 & 0 & 0 & 0 & 1 & 0 \\ 0 & 0 & 8 & 40 & 1 & -2 & 0 & -8 \end{array}\right]$ ⟵ 8 times the second row was added to the fourth.

$$\left[\begin{array}{cccc|cccc} 1 & 2 & 12 & 0 & 0 & 1 & 0 & 0 \\ 0 & 1 & 4 & 5 & 0 & 0 & 0 & -1 \\ 0 & 0 & 1 & 0 & 0 & 0 & \frac{1}{2} & 0 \\ 0 & 0 & 8 & 40 & 1 & -2 & 0 & -8 \end{array}\right]$$ ← The third row was multiplied by $\frac{1}{2}$.

$$\left[\begin{array}{cccc|cccc} 1 & 2 & 12 & 0 & 0 & 1 & 0 & 0 \\ 0 & 1 & 4 & 5 & 0 & 0 & 0 & -1 \\ 0 & 0 & 1 & 0 & 0 & 0 & \frac{1}{2} & 0 \\ 0 & 0 & 0 & 40 & 1 & -2 & -4 & -8 \end{array}\right]$$ ← -8 times the third row was added to the fourth row.

$$\left[\begin{array}{cccc|cccc} 1 & 2 & 12 & 0 & 0 & 1 & 0 & 0 \\ 0 & 1 & 4 & 5 & 0 & 0 & 0 & -1 \\ 0 & 0 & 1 & 0 & 0 & 0 & \frac{1}{2} & 0 \\ 0 & 0 & 0 & 1 & \frac{1}{40} & -\frac{1}{20} & -\frac{1}{10} & -\frac{1}{5} \end{array}\right]$$ ← The fourth row was multiplied by $\frac{1}{40}$.

$$\left[\begin{array}{cccc|cccc} 1 & 2 & 12 & 0 & 0 & 1 & 0 & 0 \\ 0 & 1 & 4 & 0 & -\frac{1}{8} & \frac{1}{4} & \frac{1}{2} & 0 \\ 0 & 0 & 1 & 0 & 0 & 0 & \frac{1}{2} & 0 \\ 0 & 0 & 0 & 1 & \frac{1}{40} & -\frac{1}{20} & -\frac{1}{10} & -\frac{1}{5} \end{array}\right]$$ ← -5 times the fourth row was added to the second row.

$$\left[\begin{array}{cccc|cccc} 1 & 2 & 0 & 0 & 0 & 1 & -6 & 0 \\ 0 & 1 & 0 & 0 & -\frac{1}{8} & \frac{1}{4} & -\frac{3}{2} & 0 \\ 0 & 0 & 1 & 0 & 0 & 0 & \frac{1}{2} & 0 \\ 0 & 0 & 0 & 1 & \frac{1}{40} & -\frac{1}{20} & -\frac{1}{10} & -\frac{1}{5} \end{array}\right]$$ ← -4 times the third row was added to the second row and -12 times the third row was added to the first row.

$$\left[\begin{array}{cccc|cccc} 1 & 0 & 0 & 0 & \frac{1}{4} & \frac{1}{2} & -3 & 0 \\ 0 & 1 & 0 & 0 & -\frac{1}{8} & \frac{1}{4} & -\frac{3}{2} & 0 \\ 0 & 0 & 1 & 0 & 0 & 0 & \frac{1}{2} & 0 \\ 0 & 0 & 0 & 1 & \frac{1}{40} & -\frac{1}{20} & -\frac{1}{10} & -\frac{1}{5} \end{array}\right]$$ ← -2 times the second row was added to the first row.

The inverse is $\begin{bmatrix} \frac{1}{4} & \frac{1}{2} & -3 & 0 \\ -\frac{1}{8} & \frac{1}{4} & -\frac{3}{2} & 0 \\ 0 & 0 & \frac{1}{2} & 0 \\ \frac{1}{40} & -\frac{1}{20} & -\frac{1}{10} & -\frac{1}{5} \end{bmatrix}$.

19. (a) $\left[\begin{array}{cccc|cccc} k_1 & 0 & 0 & 0 & 1 & 0 & 0 & 0 \\ 0 & k_2 & 0 & 0 & 0 & 1 & 0 & 0 \\ 0 & 0 & k_3 & 0 & 0 & 0 & 1 & 0 \\ 0 & 0 & 0 & k_4 & 0 & 0 & 0 & 1 \end{array}\right]$ ⟵ The identity matrix was adjoined to the given matrix.

$\left[\begin{array}{cccc|cccc} 1 & 0 & 0 & 0 & \frac{1}{k_1} & 0 & 0 & 0 \\ 0 & 1 & 0 & 0 & 0 & \frac{1}{k_2} & 0 & 0 \\ 0 & 0 & 1 & 0 & 0 & 0 & \frac{1}{k_3} & 0 \\ 0 & 0 & 0 & 1 & 0 & 0 & 0 & \frac{1}{k_4} \end{array}\right]$ ⟵ The first row was multiplied by $1/k_1$, the second row was multiplied by $1/k_2$, the third row was multiplied by $1/k_3$, and the fourth row was multiplied by $1/k_4$.

The inverse is $\begin{bmatrix} \frac{1}{k_1} & 0 & 0 & 0 \\ 0 & \frac{1}{k_2} & 0 & 0 \\ 0 & 0 & \frac{1}{k_3} & 0 \\ 0 & 0 & 0 & \frac{1}{k_4} \end{bmatrix}$.

(b) $\left[\begin{array}{cccc|cccc} k & 1 & 0 & 0 & 1 & 0 & 0 & 0 \\ 0 & 1 & 0 & 0 & 0 & 1 & 0 & 0 \\ 0 & 0 & k & 1 & 0 & 0 & 1 & 0 \\ 0 & 0 & 0 & 1 & 0 & 0 & 0 & 1 \end{array}\right]$ ⟵ The identity matrix was adjoined to the given matrix.

$\left[\begin{array}{cccc|cccc} 1 & \frac{1}{k} & 0 & 0 & \frac{1}{k} & 0 & 0 & 0 \\ 0 & 1 & 0 & 0 & 0 & 1 & 0 & 0 \\ 0 & 0 & 1 & \frac{1}{k} & 0 & 0 & \frac{1}{k} & 0 \\ 0 & 0 & 0 & 1 & 0 & 0 & 0 & 1 \end{array}\right]$ ⟵ First row and third row were both multiplied by $1/k$.

$\left[\begin{array}{cccc|cccc} 1 & 0 & 0 & 0 & \frac{1}{k} & -\frac{1}{k} & 0 & 0 \\ 0 & 1 & 0 & 0 & 0 & 1 & 0 & 0 \\ 0 & 0 & 1 & 0 & 0 & 0 & \frac{1}{k} & -\frac{1}{k} \\ 0 & 0 & 0 & 1 & 0 & 0 & 0 & 1 \end{array}\right]$ ⟵ $-\frac{1}{k}$ times the fourth row was added to the third row and $-\frac{1}{k}$ times the second row was added to the first row.

The inverse is $\begin{bmatrix} \frac{1}{k} & -\frac{1}{k} & 0 & 0 \\ 0 & 1 & 0 & 0 \\ 0 & 0 & \frac{1}{k} & -\frac{1}{k} \\ 0 & 0 & 0 & 1 \end{bmatrix}$.

21. It follows from parts (a) and (d) of Theorem 1.5.3 that a square matrix is invertible if and only if its reduced row echelon form is identity.

$$\begin{bmatrix} c & c & c \\ 1 & c & c \\ 1 & 1 & c \end{bmatrix}$$

$$\begin{bmatrix} 1 & 1 & c \\ 1 & c & c \\ c & c & c \end{bmatrix} \longleftarrow \text{The first and third rows were interchanged.}$$

$$\begin{bmatrix} 1 & 1 & c \\ 0 & -1+c & 0 \\ 0 & 0 & c-c^2 \end{bmatrix} \longleftarrow \begin{array}{l} -1 \text{ times the first row was added to the second row and} \\ -c \text{ times the first row was added to the third row.} \end{array}$$

If $c - c^2 = c(1-c) = 0$ or $-1 + c = 0$, i.e. if $c = 0$ or $c = 1$ the last matrix contains at least one row of zeros, therefore it cannot be reduced to I by elementary row operations.

Otherwise (if $c \neq 0$ and $c \neq 1$), multiplying the second row by $\frac{1}{-1+c}$ and multiplying the third row by $\frac{1}{c-c^2}$ would result in a row echelon form with 1's on the main diagonal. Subsequent elementary row operations would then lead to the identity matrix.

We conclude that for any value of c other than 0 and 1 the matrix is invertible.

23. We perform a sequence of elementary row operations to reduce the given matrix to the identity matrix. As we do so, we keep track of each corresponding elementary matrix:

$$A = \begin{bmatrix} -3 & 1 \\ 2 & 2 \end{bmatrix}$$

$$\begin{bmatrix} 1 & 5 \\ 2 & 2 \end{bmatrix} \longleftarrow \text{2 times the second row was added to the first.} \qquad E_1 = \begin{bmatrix} 1 & 2 \\ 0 & 1 \end{bmatrix}$$

$$\begin{bmatrix} 1 & 5 \\ 0 & -8 \end{bmatrix} \longleftarrow -2 \text{ times the first row was added to the second.} \qquad E_2 = \begin{bmatrix} 1 & 0 \\ -2 & 1 \end{bmatrix}$$

$$\begin{bmatrix} 1 & 5 \\ 0 & 1 \end{bmatrix} \longleftarrow \text{The second row was multiplied by } -\frac{1}{8}. \qquad E_3 = \begin{bmatrix} 1 & 0 \\ 0 & -\frac{1}{8} \end{bmatrix}$$

$$\begin{bmatrix} 1 & 0 \\ 0 & 1 \end{bmatrix} \longleftarrow -5 \text{ times the second row was added to the first.} \qquad E_4 = \begin{bmatrix} 1 & -5 \\ 0 & 1 \end{bmatrix}$$

Since $E_4E_3E_2E_1A = I$, then

$A = (E_4E_3E_2E_1)^{-1}I = E_1^{-1}E_2^{-1}E_3^{-1}E_4^{-1} = \begin{bmatrix}1 & -2\\0 & 1\end{bmatrix}\begin{bmatrix}1 & 0\\2 & 1\end{bmatrix}\begin{bmatrix}1 & 0\\0 & -8\end{bmatrix}\begin{bmatrix}1 & 5\\0 & 1\end{bmatrix}$ and

$A^{-1} = E_4E_3E_2E_1 = \begin{bmatrix}1 & -5\\0 & 1\end{bmatrix}\begin{bmatrix}1 & 0\\0 & -\frac{1}{8}\end{bmatrix}\begin{bmatrix}1 & 0\\-2 & 1\end{bmatrix}\begin{bmatrix}1 & 2\\0 & 1\end{bmatrix}$.

Note that this answer is not unique since a different sequence of elementary row operations (and the corresponding elementary matrices) could be used instead.

25. We perform a sequence of elementary row operations to reduce the given matrix to the identity matrix. As we do so, we keep track of each corresponding elementary matrix:

$A = \begin{bmatrix}1 & 0 & -2\\0 & 4 & 3\\0 & 0 & 1\end{bmatrix}$

$\begin{bmatrix}1 & 0 & -2\\0 & 1 & \frac{3}{4}\\0 & 0 & 1\end{bmatrix}$ ⟵ The second row was multiplied by $\frac{1}{4}$. $E_1 = \begin{bmatrix}1 & 0 & 0\\0 & \frac{1}{4} & 0\\0 & 0 & 1\end{bmatrix}$

$\begin{bmatrix}1 & 0 & -2\\0 & 1 & 0\\0 & 0 & 1\end{bmatrix}$ ⟵ $-\frac{3}{4}$ times the third row was added to the second. $E_2 = \begin{bmatrix}1 & 0 & 0\\0 & 1 & -\frac{3}{4}\\0 & 0 & 1\end{bmatrix}$

$\begin{bmatrix}1 & 0 & 0\\0 & 1 & 0\\0 & 0 & 1\end{bmatrix}$ ⟵ 2 times the third row was added to the first row. $E_3 = \begin{bmatrix}1 & 0 & 2\\0 & 1 & 0\\0 & 0 & 1\end{bmatrix}$

Since $E_3E_2E_1A = I$, we have $A = (E_3E_2E_1)^{-1}I = E_1^{-1}E_2^{-1}E_3^{-1} = \begin{bmatrix}1 & 0 & 0\\0 & 4 & 0\\0 & 0 & 1\end{bmatrix}\begin{bmatrix}1 & 0 & 0\\0 & 1 & \frac{3}{4}\\0 & 0 & 1\end{bmatrix}\begin{bmatrix}1 & 0 & -2\\0 & 1 & 0\\0 & 0 & 1\end{bmatrix}$

and $A^{-1} = E_3E_2E_1 = \begin{bmatrix}1 & 0 & 2\\0 & 1 & 0\\0 & 0 & 1\end{bmatrix}\begin{bmatrix}1 & 0 & 0\\0 & 1 & -\frac{3}{4}\\0 & 0 & 1\end{bmatrix}\begin{bmatrix}1 & 0 & 0\\0 & \frac{1}{4} & 0\\0 & 0 & 1\end{bmatrix}$.

Note that this answer is not unique since a different sequence of elementary row operations (and the corresponding elementary matrices) could be used instead.

27. Let us perform a sequence of elementary row operations to produce B from A. As we do so, we keep track of each corresponding elementary matrix:

$A = \begin{bmatrix}1 & 2 & 3\\1 & 4 & 1\\2 & 1 & 9\end{bmatrix}$

$\begin{bmatrix}1 & 2 & 3\\0 & 2 & -2\\2 & 1 & 9\end{bmatrix}$ ⟵ -1 times the first row was added to the second row. $E_1 = \begin{bmatrix}1 & 0 & 0\\-1 & 1 & 0\\0 & 0 & 1\end{bmatrix}$

$\begin{bmatrix}1 & 0 & 5\\0 & 2 & -2\\2 & 1 & 9\end{bmatrix}$ ⟵ -1 times the second row was added to the first row. $E_2 = \begin{bmatrix}1 & -1 & 0\\0 & 1 & 0\\0 & 0 & 1\end{bmatrix}$

$B = \begin{bmatrix} 1 & 0 & 5 \\ 0 & 2 & -2 \\ 1 & 1 & 4 \end{bmatrix}$ ⟵ -1 times the first row was added to the third row. $E_3 = \begin{bmatrix} 1 & 0 & 0 \\ 0 & 1 & 0 \\ -1 & 0 & 1 \end{bmatrix}$

Since $E_3E_2E_1A = B$, the equality $CA = B$ is satisfied by the matrix

$$C = E_3E_2E_1 = \begin{bmatrix} 1 & 0 & 0 \\ 0 & 1 & 0 \\ -1 & 0 & 1 \end{bmatrix} \begin{bmatrix} 1 & -1 & 0 \\ 0 & 1 & 0 \\ 0 & 0 & 1 \end{bmatrix} \begin{bmatrix} 1 & 0 & 0 \\ -1 & 1 & 0 \\ 0 & 0 & 1 \end{bmatrix} = \begin{bmatrix} 2 & -1 & 0 \\ -1 & 1 & 0 \\ -2 & 1 & 1 \end{bmatrix}.$$

Note that this answer is not unique since a different sequence of elementary row operations (and the corresponding elementary matrices) could be used instead.

29. $A = \begin{bmatrix} 1 & 0 & 0 \\ 0 & 1 & 0 \\ a & b & c \end{bmatrix}$ cannot result from interchanging two rows of I_3 (since that would create a nonzero entry above the main diagonal).

A can result from multiplying the third row of I_3 by a nonzero number c (in this case, $a = b = 0,\ c \neq 0$).

The other possibilities are that A can be obtained by adding a times the first row to the third ($b = 0, c = 1$) or by adding b times the second row to the third ($a = 0, c = 1$).

In all three cases, at least one entry in the third row must be zero.

True-False Exercises

(a) False. An elementary matrix results from performing a *single* elementary row operation on an identity matrix; a product of two elementary matrices would correspond to a sequence of two such operations instead, which generally is not equivalent to a single elementary operation.

(b) True. This follows from Theorem 1.5.2.

(c) True. If A and B are row equivalent then there exist elementary matrices $E_1, \dots, E_p$ such that $B = E_p \cdots E_1A$. Likewise, if B and C are row equivalent then there exist elementary matrices $E_1^*, \dots, E_q^*$ such that $C = E_q^* \cdots E_1^*B$. Combining the two equalities yields $C = E_q^* \cdots E_1^*E_p \cdots E_1A$ therefore A and C are row equivalent.

(d) True. A homogeneous system $A\mathbf{x} = \mathbf{0}$ has either one solution (the trivial solution) or infinitely many solutions. If A is not invertible, then by Theorem 1.5.3 the system cannot have just one solution. Consequently, it must have infinitely many solutions.

(e) True. If the matrix A is not invertible then by Theorem 1.5.3 its reduced row echelon form is not I_n. However, the matrix resulting from interchanging two rows of A (an elementary row operation) must have the same reduced row echelon form as A does, so by Theorem 1.5.3 that matrix is not invertible either.

(f) True. Adding a multiple of the first row of a matrix to its second row is an elementary row operation. Denoting by E be the corresponding elementary matrix we can write $(EA)^{-1} = A^{-1}E^{-1}$ so the resulting matrix EA is invertible if A is.

(g) False. For instance, $\begin{bmatrix}1 & 0\\0 & 1\end{bmatrix} = \begin{bmatrix}1/2 & 0\\0 & 1\end{bmatrix}\begin{bmatrix}2 & 0\\0 & 1\end{bmatrix} = \begin{bmatrix}1/3 & 0\\0 & 1\end{bmatrix}\begin{bmatrix}3 & 0\\0 & 1\end{bmatrix}$.

1.6 More on Linear Systems and Invertible Matrices

1. The given system can be written in matrix form as $A\mathbf{x} = \mathbf{b}$, where $A = \begin{bmatrix}1 & 1\\5 & 6\end{bmatrix}$, $\mathbf{x} = \begin{bmatrix}x_1\\x_2\end{bmatrix}$, and $\mathbf{b} = \begin{bmatrix}2\\9\end{bmatrix}$. We begin by inverting the coefficient matrix A

$$\left[\begin{array}{cc|cc}1 & 1 & 1 & 0\\5 & 6 & 0 & 1\end{array}\right]$$ ⟵ The identity matrix was adjoined to the coefficient matrix.

$$\left[\begin{array}{cc|cc}1 & 1 & 1 & 0\\0 & 1 & -5 & 1\end{array}\right]$$ ⟵ -5 times the first row was added to the second row.

$$\left[\begin{array}{cc|cc}1 & 0 & 6 & -1\\0 & 1 & -5 & 1\end{array}\right]$$ ⟵ -1 times the second row was added to the first row.

Since $A^{-1} = \begin{bmatrix}6 & -1\\-5 & 1\end{bmatrix}$, Theorem 1.6.2 states that the system has exactly one solution $\mathbf{x} = A^{-1}\mathbf{b}$: $\begin{bmatrix}x_1\\x_2\end{bmatrix} = \begin{bmatrix}6 & -1\\-5 & 1\end{bmatrix}\begin{bmatrix}2\\9\end{bmatrix} = \begin{bmatrix}3\\-1\end{bmatrix}$, i.e., $x_1 = 3,\ x_2 = -1$.

3. The given system can be written in matrix form as $A\mathbf{x} = \mathbf{b}$, where $A = \begin{bmatrix}1 & 3 & 1\\2 & 2 & 1\\2 & 3 & 1\end{bmatrix}$, $\mathbf{x} = \begin{bmatrix}x_1\\x_2\\x_3\end{bmatrix}$, and $\mathbf{b} = \begin{bmatrix}4\\-1\\3\end{bmatrix}$. We begin by inverting the coefficient matrix A

$$\left[\begin{array}{ccc|ccc}1 & 3 & 1 & 1 & 0 & 0\\2 & 2 & 1 & 0 & 1 & 0\\2 & 3 & 1 & 0 & 0 & 1\end{array}\right]$$ ⟵ The identity matrix was adjoined to the coefficient matrix.

$$\left[\begin{array}{ccc|ccc}1 & 3 & 1 & 1 & 0 & 0\\0 & -4 & -1 & -2 & 1 & 0\\0 & -3 & -1 & -2 & 0 & 1\end{array}\right]$$ ⟵ -2 times the first row was added to the second and -2 times the first row was added to the third row.

$$\left[\begin{array}{ccc|ccc}1 & 3 & 1 & 1 & 0 & 0\\0 & -4 & -1 & -2 & 1 & 0\\0 & 1 & 0 & 0 & -1 & 1\end{array}\right]$$ ⟵ -1 times the second row was added to the third row.

$$\left[\begin{array}{ccc|ccc}1 & 3 & 1 & 1 & 0 & 0\\0 & 1 & 0 & 0 & -1 & 1\\0 & -4 & -1 & -2 & 1 & 0\end{array}\right]$$ ⟵ The second and third rows were interchanged.

$$\left[\begin{array}{ccc|ccc}1 & 3 & 1 & 1 & 0 & 0\\0 & 1 & 0 & 0 & -1 & 1\\0 & 0 & -1 & -2 & -3 & 4\end{array}\right]$$ ⟵ 4 times the second row was added to the third row.

$$\left[\begin{array}{ccc|ccc}1 & 3 & 1 & 1 & 0 & 0\\0 & 1 & 0 & 0 & -1 & 1\\0 & 0 & 1 & 2 & 3 & -4\end{array}\right]$$ ⟵ The third row was multiplied by -1.

$$\left[\begin{array}{ccc|ccc} 1 & 3 & 0 & -1 & -3 & 4 \\ 0 & 1 & 0 & 0 & -1 & 1 \\ 0 & 0 & 1 & 2 & 3 & -4 \end{array}\right] \longleftarrow \text{ -1 times the third row was added to the first row.}$$

$$\left[\begin{array}{ccc|ccc} 1 & 0 & 0 & -1 & 0 & 1 \\ 0 & 1 & 0 & 0 & -1 & 1 \\ 0 & 0 & 1 & 2 & 3 & -4 \end{array}\right] \longleftarrow \text{ -3 times the second row was added to the first row.}$$

Since $A^{-1} = \begin{bmatrix} -1 & 0 & 1 \\ 0 & -1 & 1 \\ 2 & 3 & -4 \end{bmatrix}$, Theorem 1.6.2 states that the system has exactly one solution

$\mathbf{x} = A^{-1}\mathbf{b}$: $\begin{bmatrix} x_1 \\ x_2 \\ x_3 \end{bmatrix} = \begin{bmatrix} -1 & 0 & 1 \\ 0 & -1 & 1 \\ 2 & 3 & -4 \end{bmatrix}\begin{bmatrix} 4 \\ -1 \\ 3 \end{bmatrix} = \begin{bmatrix} -1 \\ 4 \\ -7 \end{bmatrix}$, i.e., $x_1 = -1, x_2 = 4$, and $x_3 = -7$.

5. The given system can be written in matrix form as $A\mathbf{x} = \mathbf{b}$, where $A = \begin{bmatrix} 1 & 1 & 1 \\ 1 & 1 & -4 \\ -4 & 1 & 1 \end{bmatrix}$, $\mathbf{x} = \begin{bmatrix} x \\ y \\ z \end{bmatrix}$, and

$\mathbf{b} = \begin{bmatrix} 5 \\ 10 \\ 0 \end{bmatrix}$. We begin by inverting the coefficient matrix A

$$\left[\begin{array}{ccc|ccc} 1 & 1 & 1 & 1 & 0 & 0 \\ 1 & 1 & -4 & 0 & 1 & 0 \\ -4 & 1 & 1 & 0 & 0 & 1 \end{array}\right] \longleftarrow \text{ The identity matrix was adjoined to the coefficient matrix.}$$

$$\left[\begin{array}{ccc|ccc} 1 & 1 & 1 & 1 & 0 & 0 \\ 0 & 0 & -5 & -1 & 1 & 0 \\ 0 & 5 & 5 & 4 & 0 & 1 \end{array}\right] \longleftarrow \text{ -1 times the first row was added to the second row and 4 times the first row was added to the third row.}$$

$$\left[\begin{array}{ccc|ccc} 1 & 1 & 1 & 1 & 0 & 0 \\ 0 & 5 & 5 & 4 & 0 & 1 \\ 0 & 0 & -5 & -1 & 1 & 0 \end{array}\right] \longleftarrow \text{ The second and third rows were interchanged.}$$

$$\left[\begin{array}{ccc|ccc} 1 & 1 & 1 & 1 & 0 & 0 \\ 0 & 1 & 1 & \frac{4}{5} & 0 & \frac{1}{5} \\ 0 & 0 & 1 & \frac{1}{5} & -\frac{1}{5} & 0 \end{array}\right] \longleftarrow \text{ The second row was multiplied by $\tfrac{1}{5}$ and the third row was multiplied by $-\tfrac{1}{5}$.}$$

$$\left[\begin{array}{ccc|ccc} 1 & 1 & 0 & \frac{4}{5} & \frac{1}{5} & 0 \\ 0 & 1 & 0 & \frac{3}{5} & \frac{1}{5} & \frac{1}{5} \\ 0 & 0 & 1 & \frac{1}{5} & -\frac{1}{5} & 0 \end{array}\right] \longleftarrow \text{ -1 times the third row was added to the second row and to the first row.}$$

$$\left[\begin{array}{ccc|ccc} 1 & 0 & 0 & \frac{1}{5} & 0 & -\frac{1}{5} \\ 0 & 1 & 0 & \frac{3}{5} & \frac{1}{5} & \frac{1}{5} \\ 0 & 0 & 1 & \frac{1}{5} & -\frac{1}{5} & 0 \end{array}\right] \longleftarrow \text{ -1 times the second row was added to the first row.}$$

Since $A^{-1} = \begin{bmatrix} \frac{1}{5} & 0 & -\frac{1}{5} \\ \frac{3}{5} & \frac{1}{5} & \frac{1}{5} \\ \frac{1}{5} & -\frac{1}{5} & 0 \end{bmatrix}$, Theorem 1.6.2 states that the system has exactly one solution $\mathbf{x} = A^{-1}\mathbf{b}$:

$\begin{bmatrix} x \\ y \\ z \end{bmatrix} = \begin{bmatrix} \frac{1}{5} & 0 & -\frac{1}{5} \\ \frac{3}{5} & \frac{1}{5} & \frac{1}{5} \\ \frac{1}{5} & -\frac{1}{5} & 0 \end{bmatrix} \begin{bmatrix} 5 \\ 10 \\ 0 \end{bmatrix} = \begin{bmatrix} 1 \\ 5 \\ -1 \end{bmatrix}$, i.e., $x = 1, y = 5$, and $z = -1$.

7. The given system can be written in matrix form as $A\mathbf{x} = \mathbf{b}$, where $A = \begin{bmatrix} 3 & 5 \\ 1 & 2 \end{bmatrix}$, $\mathbf{x} = \begin{bmatrix} x_1 \\ x_2 \end{bmatrix}$, and $\mathbf{b} = \begin{bmatrix} b_1 \\ b_2 \end{bmatrix}$. We begin by inverting the coefficient matrix A

$\left[\begin{array}{cc|cc} 3 & 5 & 1 & 0 \\ 1 & 2 & 0 & 1 \end{array}\right]$ ⟵ The identity matrix was adjoined to the coefficient matrix.

$\left[\begin{array}{cc|cc} 1 & 2 & 0 & 1 \\ 3 & 5 & 1 & 0 \end{array}\right]$ ⟵ The first and second rows were interchanged.

$\left[\begin{array}{cc|cc} 1 & 2 & 0 & 1 \\ 0 & -1 & 1 & -3 \end{array}\right]$ ⟵ -3 times the first row was added to the second row.

$\left[\begin{array}{cc|cc} 1 & 2 & 0 & 1 \\ 0 & 1 & -1 & 3 \end{array}\right]$ ⟵ The second row was multiplied by -1.

$\left[\begin{array}{cc|cc} 1 & 0 & 2 & -5 \\ 0 & 1 & -1 & 3 \end{array}\right]$ ⟵ -2 times the second row was added to the first row.

Since $A^{-1} = \begin{bmatrix} 2 & -5 \\ -1 & 3 \end{bmatrix}$, Theorem 1.6.2 states that the system has exactly one solution $\mathbf{x} = A^{-1}\mathbf{b}$:

$\begin{bmatrix} x_1 \\ x_2 \end{bmatrix} = \begin{bmatrix} 2 & -5 \\ -1 & 3 \end{bmatrix} \begin{bmatrix} b_1 \\ b_2 \end{bmatrix} = \begin{bmatrix} 2b_1 - 5b_2 \\ -b_1 + 3b_2 \end{bmatrix}$, i.e., $x_1 = 2b_1 - 5b_2,\ x_2 = -b_1 + 3b_2$.

9. $\left[\begin{array}{cc|c|c} 1 & -5 & 1 & -2 \\ 3 & 2 & 4 & 5 \end{array}\right]$ ⟵ We augmented the coefficient matrix with two columns of constants on the right hand sides of the systems (i) and (ii) – refer to Example 2.

$\left[\begin{array}{cc|c|c} 1 & -5 & 1 & -2 \\ 0 & 17 & 1 & 11 \end{array}\right]$ ⟵ -3 times the first row was added to the second row.

$\left[\begin{array}{cc|c|c} 1 & -5 & 1 & -2 \\ 0 & 1 & \frac{1}{17} & \frac{11}{17} \end{array}\right]$ ⟵ The second row was multiplied by $\frac{1}{17}$.

$\left[\begin{array}{cc|c|c} 1 & 0 & \frac{22}{17} & \frac{21}{17} \\ 0 & 1 & \frac{1}{17} & \frac{11}{17} \end{array}\right]$ ⟵ 5 times the second row was added to the first row.

We conclude that the solutions of the two systems are:

(i) $x_1 = \frac{22}{17}, x_2 = \frac{1}{17}$ **(ii)** $x_1 = \frac{21}{17}, x_2 = \frac{11}{17}$

11. $\left[\begin{array}{cc|c|c|c|c} 4 & -7 & 0 & -4 & -1 & -5 \\ 1 & 2 & 1 & 6 & 3 & 1 \end{array}\right]$ ⟵ We augmented the coefficient matrix with four columns of constants on the right hand sides of the systems (i), (ii), (iii), and (iv) – refer to Example 2.

$\left[\begin{array}{cc|c|c|c|c} 1 & 2 & 1 & 6 & 3 & 1 \\ 4 & -7 & 0 & -4 & -1 & -5 \end{array}\right]$ ⟵ The first and second rows were interchanged.

$\left[\begin{array}{cc|c|c|c|c} 1 & 2 & 1 & 6 & 3 & 1 \\ 0 & -15 & -4 & -28 & -13 & -9 \end{array}\right]$ ⟵ -4 times the first row was added to the second row.

$\left[\begin{array}{cc|c|c|c|c} 1 & 2 & 1 & 6 & 3 & 1 \\ 0 & 1 & \frac{4}{15} & \frac{28}{15} & \frac{13}{15} & \frac{3}{5} \end{array}\right]$ ⟵ The second row was multiplied by $-\frac{1}{15}$.

$\left[\begin{array}{cc|c|c|c|c} 1 & 0 & \frac{7}{15} & \frac{34}{15} & \frac{19}{15} & -\frac{1}{5} \\ 0 & 1 & \frac{4}{15} & \frac{28}{15} & \frac{13}{15} & \frac{3}{5} \end{array}\right]$ ⟵ -2 times the second row was added to the first row.

We conclude that the solutions of the four systems are:

(i) $x_1 = \frac{7}{15}, x_2 = \frac{4}{15}$ **(ii)** $x_1 = \frac{34}{15}, x_2 = \frac{28}{15}$

(iii) $x_1 = \frac{19}{15}, x_2 = \frac{13}{15}$ **(iv)** $x_1 = -\frac{1}{5}, x_2 = \frac{3}{5}$

13. $\left[\begin{array}{cc|c} 1 & 3 & b_1 \\ -2 & 1 & b_2 \end{array}\right]$ ⟵ The augmented matrix for the system.

$\left[\begin{array}{cc|c} 1 & 3 & b_1 \\ 0 & 7 & 2b_1 + b_2 \end{array}\right]$ ⟵ 2 times the first row was added to the second row.

$\left[\begin{array}{cc|c} 1 & 3 & b_1 \\ 0 & 1 & \frac{2}{7}b_1 + \frac{1}{7}b_2 \end{array}\right]$ ⟵ The second row was multiplied by $\frac{1}{7}$.

The system is consistent for all values of b_1 and b_2.

15. $\left[\begin{array}{ccc|c} 1 & -2 & 5 & b_1 \\ 4 & -5 & 8 & b_2 \\ -3 & 3 & -3 & b_3 \end{array}\right]$ ⟵ The augmented matrix for the system.

$\left[\begin{array}{ccc|c} 1 & -2 & 5 & b_1 \\ 0 & 3 & -12 & -4b_1 + b_2 \\ 0 & -3 & 12 & 3b_1 + b_3 \end{array}\right]$ ⟵ -4 times the first row was added to the second row and 3 times the first row was added to the third row.

$$\left[\begin{array}{ccc|c} 1 & -2 & 5 & b_1 \\ 0 & 3 & -12 & -4b_1+b_2 \\ 0 & 0 & 0 & -b_1+b_2+b_3 \end{array}\right]$$ ⟵ The second row was added to the third row.

$$\left[\begin{array}{ccc|c} 1 & -2 & 5 & b_1 \\ 0 & 1 & -4 & -\frac{4}{3}b_1+\frac{1}{3}b_2 \\ 0 & 0 & 0 & -b_1+b_2+b_3 \end{array}\right]$$ ⟵ The second row was multiplied by $\frac{1}{3}$.

The system is consistent if and only if $-b_1+b_2+b_3=0$, i.e. $b_1=b_2+b_3$.

17.

$$\left[\begin{array}{cccc|c} 1 & -1 & 3 & 2 & b_1 \\ -2 & 1 & 5 & 1 & b_2 \\ -3 & 2 & 2 & -1 & b_3 \\ 4 & -3 & 1 & 3 & b_4 \end{array}\right]$$ ⟵ The augmented matrix for the system.

$$\left[\begin{array}{cccc|c} 1 & -1 & 3 & 2 & b_1 \\ 0 & -1 & 11 & 5 & 2b_1+b_2 \\ 0 & -1 & 11 & 5 & 3b_1+b_3 \\ 0 & 1 & -11 & -5 & -4b_1+b_4 \end{array}\right]$$ ⟵ 2 times the first row was added to the second row, 3 times the first row was added to the third row, and -4 times the first row was added to the fourth row.

$$\left[\begin{array}{cccc|c} 1 & -1 & 3 & 2 & b_1 \\ 0 & 1 & -11 & -5 & -2b_1-b_2 \\ 0 & -1 & 11 & 5 & 3b_1+b_3 \\ 0 & 1 & -11 & -5 & -4b_1+b_4 \end{array}\right]$$ ⟵ The second row was multiplied by -1.

$$\left[\begin{array}{cccc|c} 1 & -1 & 3 & 2 & b_1 \\ 0 & 1 & -11 & -5 & -2b_1-b_2 \\ 0 & 0 & 0 & 0 & b_1-b_2+b_3 \\ 0 & 0 & 0 & 0 & -2b_1+b_2+b_4 \end{array}\right]$$ ⟵ The second row was added to the third row and -1 times the second row was added to the fourth row.

The system is consistent for all values of b_1, b_2, b_3, and b_4 that satisfy the equations $b_1-b_2+b_3=0$ and $-2b_1+b_2+b_4=0$.

These equations form a linear system in the variables b_1, b_2, b_3, and b_4 whose augmented matrix $\begin{bmatrix} 1 & -1 & 1 & 0 & 0 \\ -2 & 1 & 0 & 1 & 0 \end{bmatrix}$ has the reduced row echelon form $\begin{bmatrix} 1 & 0 & -1 & -1 & 0 \\ 0 & 1 & -2 & -1 & 0 \end{bmatrix}$. Therefore the system is consistent if $b_1=b_3+b_4$ and $b_2=2b_3+b_4$.

19. $X=\begin{bmatrix} 1 & -1 & 1 \\ 2 & 3 & 0 \\ 0 & 2 & -1 \end{bmatrix}^{-1}\begin{bmatrix} 2 & -1 & 5 & 7 & 8 \\ 4 & 0 & -3 & 0 & 1 \\ 3 & 5 & -7 & 2 & 1 \end{bmatrix}$. Let us find $\begin{bmatrix} 1 & -1 & 1 \\ 2 & 3 & 0 \\ 0 & 2 & -1 \end{bmatrix}^{-1}$:

$$\left[\begin{array}{rrr|rrr} 1 & -1 & 1 & 1 & 0 & 0 \\ 2 & 3 & 0 & 0 & 1 & 0 \\ 0 & 2 & -1 & 0 & 0 & 1 \end{array}\right]$$ ⟵ The identity matrix was adjoined to the matrix.

$$\left[\begin{array}{rrr|rrr} 1 & -1 & 1 & 1 & 0 & 0 \\ 0 & 5 & -2 & -2 & 1 & 0 \\ 0 & 2 & -1 & 0 & 0 & 1 \end{array}\right]$$ ⟵ -2 times the first row was added to the second row.

$$\left[\begin{array}{rrr|rrr} 1 & -1 & 1 & 1 & 0 & 0 \\ 0 & 1 & 0 & -2 & 1 & -2 \\ 0 & 2 & -1 & 0 & 0 & 1 \end{array}\right]$$ ⟵ -2 times the third row was added to the second row.

$$\left[\begin{array}{rrr|rrr} 1 & -1 & 1 & 1 & 0 & 0 \\ 0 & 1 & 0 & -2 & 1 & -2 \\ 0 & 0 & -1 & 4 & -2 & 5 \end{array}\right]$$ ⟵ -2 times the second row was added to the third row.

$$\left[\begin{array}{rrr|rrr} 1 & -1 & 1 & 1 & 0 & 0 \\ 0 & 1 & 0 & -2 & 1 & -2 \\ 0 & 0 & 1 & -4 & 2 & -5 \end{array}\right]$$ ⟵ The third row was multiplied by -1.

$$\left[\begin{array}{rrr|rrr} 1 & -1 & 0 & 5 & -2 & 5 \\ 0 & 1 & 0 & -2 & 1 & -2 \\ 0 & 0 & 1 & -4 & 2 & -5 \end{array}\right]$$ ⟵ -1 times the third row was added to the first row.

$$\left[\begin{array}{rrr|rrr} 1 & 0 & 0 & 3 & -1 & 3 \\ 0 & 1 & 0 & -2 & 1 & -2 \\ 0 & 0 & 1 & -4 & 2 & -5 \end{array}\right]$$ ⟵ The second row was added to the first row.

Using $\begin{bmatrix} 1 & -1 & 1 \\ 2 & 3 & 0 \\ 0 & 2 & -1 \end{bmatrix}^{-1} = \begin{bmatrix} 3 & -1 & 3 \\ -2 & 1 & -2 \\ -4 & 2 & -5 \end{bmatrix}$ we obtain

$$X = \begin{bmatrix} 3 & -1 & 3 \\ -2 & 1 & -2 \\ -4 & 2 & -5 \end{bmatrix} \begin{bmatrix} 2 & -1 & 5 & 7 & 8 \\ 4 & 0 & -3 & 0 & 1 \\ 3 & 5 & -7 & 2 & 1 \end{bmatrix} = \begin{bmatrix} 11 & 12 & -3 & 27 & 26 \\ -6 & -8 & 1 & -18 & -17 \\ -15 & -21 & 9 & -38 & -35 \end{bmatrix}$$

True-False Exercises

(a) True. By Theorem 1.6.1, if a system of linear equation has more than one solution then it must have infinitely many.

(b) True. If A is a square matrix such that $A\mathbf{x} = \mathbf{b}$ has a unique solution then the reduced row echelon form of A must be I. Consequently, $A\mathbf{x} = \mathbf{c}$ must have a unique solution as well.

(c) True. Since B is a square matrix then by Theorem 1.6.3(b) $AB = I_n$ implies $B = A^{-1}$. Therefore, $BA = A^{-1}A = I_n$.

(d) True. Since A and B are row equivalent matrices, it must be possible to perform a sequence of elementary row operations on A resulting in B. Let E be the product of the corresponding elementary matrices, i.e., $EA = B$. Note that E must be an invertible matrix thus $A = E^{-1}B$.
Any solution of $A\mathbf{x} = \mathbf{0}$ is also a solution of $B\mathbf{x} = \mathbf{0}$ since $B\mathbf{x} = EA\mathbf{x} = E\mathbf{0} = \mathbf{0}$.
Likewise, any solution of $B\mathbf{x} = \mathbf{0}$ is also a solution of $A\mathbf{x} = \mathbf{0}$ since $A\mathbf{x} = E^{-1}B\mathbf{x} = E^{-1}\mathbf{0} = \mathbf{0}$.

(e) True. If $(S^{-1}AS)\mathbf{x} = \mathbf{b}$ then $SS^{-1}AS\mathbf{x} = A(S\mathbf{x}) = S\mathbf{b}$. Consequently, $\mathbf{y} = S\mathbf{x}$ is a solution of $A\mathbf{y} = S\mathbf{b}$.

(f) True. $A\mathbf{x} = 4\mathbf{x}$ is equivalent to $A\mathbf{x} = 4I_n\mathbf{x}$, which can be rewritten as $(A - 4I_n)\mathbf{x} = \mathbf{0}$. By Theorem 1.6.4, this homogeneous system has a unique solution (the trivial solution) if and only if its coefficient matrix $A - 4I_n$ is invertible.

(g) True. If AB were invertible, then by Theorem 1.6.5 both A and B would be invertible.

1.7 Diagonal, Triangular, and Symmetric Matrices

1. **(a)** The matrix is upper triangular. It is invertible (its diagonal entries are both nonzero).

(b) The matrix is lower triangular. It is not invertible (its diagonal entries are zero).

(c) This is a diagonal matrix, therefore it is also both upper and lower triangular. It is invertible (its diagonal entries are all nonzero).

(d) The matrix is upper triangular. It is not invertible (its diagonal entries include a zero).

3. $$\begin{bmatrix} 3 & 0 & 0 \\ 0 & -1 & 0 \\ 0 & 0 & 2 \end{bmatrix}\begin{bmatrix} 2 & 1 \\ -4 & 1 \\ 2 & 5 \end{bmatrix} = \begin{bmatrix} (3)(2) & (3)(1) \\ (-1)(-4) & (-1)(1) \\ (2)(2) & (2)(5) \end{bmatrix} = \begin{bmatrix} 6 & 3 \\ 4 & -1 \\ 4 & 10 \end{bmatrix}$$

5. $$\begin{bmatrix} 5 & 0 & 0 \\ 0 & 2 & 0 \\ 0 & 0 & -3 \end{bmatrix}\begin{bmatrix} -3 & 2 & 0 & 4 & -4 \\ 1 & -5 & 3 & 0 & 3 \\ -6 & 2 & 2 & 2 & 2 \end{bmatrix} = \begin{bmatrix} (5)(-3) & (5)(2) & (5)(0) & (5)(4) & (5)(-4) \\ (2)(1) & (2)(-5) & (2)(3) & (2)(0) & (2)(3) \\ (-3)(-6) & (-3)(2) & (-3)(2) & (-3)(2) & (-3)(2) \end{bmatrix}$$
$$= \begin{bmatrix} -15 & 10 & 0 & 20 & -20 \\ 2 & -10 & 6 & 0 & 6 \\ 18 & -6 & -6 & -6 & -6 \end{bmatrix}$$

7. $$A^2 = \begin{bmatrix} 1^2 & 0 \\ 0 & (-2)^2 \end{bmatrix} = \begin{bmatrix} 1 & 0 \\ 0 & 4 \end{bmatrix},\quad A^{-2} = \begin{bmatrix} 1^{-2} & 0 \\ 0 & (-2)^{-2} \end{bmatrix} = \begin{bmatrix} 1 & 0 \\ 0 & \frac{1}{4} \end{bmatrix},\quad A^{-k} = \begin{bmatrix} 1^{-k} & 0 \\ 0 & (-2)^{-k} \end{bmatrix} = \begin{bmatrix} 1 & 0 \\ 0 & \frac{1}{(-2)^k} \end{bmatrix}$$

9. $$A^2 = \begin{bmatrix} \left(\frac{1}{2}\right)^2 & 0 & 0 \\ 0 & \left(\frac{1}{3}\right)^2 & 0 \\ 0 & 0 & \left(\frac{1}{4}\right)^2 \end{bmatrix} = \begin{bmatrix} \frac{1}{4} & 0 & 0 \\ 0 & \frac{1}{9} & 0 \\ 0 & 0 & \frac{1}{16} \end{bmatrix},\quad A^{-2} = \begin{bmatrix} \left(\frac{1}{2}\right)^{-2} & 0 & 0 \\ 0 & \left(\frac{1}{3}\right)^{-2} & 0 \\ 0 & 0 & \left(\frac{1}{4}\right)^{-2} \end{bmatrix} = \begin{bmatrix} 4 & 0 & 0 \\ 0 & 9 & 0 \\ 0 & 0 & 16 \end{bmatrix},$$

$$A^{-k} = \begin{bmatrix} \left(\frac{1}{2}\right)^{-k} & 0 & 0 \\ 0 & \left(\frac{1}{3}\right)^{-k} & 0 \\ 0 & 0 & \left(\frac{1}{4}\right)^{-k} \end{bmatrix} = \begin{bmatrix} 2^k & 0 & 0 \\ 0 & 3^k & 0 \\ 0 & 0 & 4^k \end{bmatrix}$$

11. $$\begin{bmatrix} (1)(2)(0) & 0 & 0 \\ 0 & (0)(5)(2) & 0 \\ 0 & 0 & (3)(0)(1) \end{bmatrix} = \begin{bmatrix} 0 & 0 & 0 \\ 0 & 0 & 0 \\ 0 & 0 & 0 \end{bmatrix}$$

13. $$\begin{bmatrix} 1^{39} & 0 \\ 0 & (-1)^{39} \end{bmatrix} = \begin{bmatrix} 1 & 0 \\ 0 & -1 \end{bmatrix}$$

15. **(a)** $\begin{bmatrix} au & av \\ bw & bx \\ cy & cz \end{bmatrix}$ **(b)** $\begin{bmatrix} ra & sb & tc \\ ua & vb & wc \\ xa & yb & zc \end{bmatrix}$

17. **(a)** $\begin{bmatrix} 2 & -1 \\ -1 & 3 \end{bmatrix}$ **(b)** $\begin{bmatrix} 1 & 3 & 7 & 2 \\ 3 & 1 & -8 & -3 \\ 7 & -8 & 0 & 9 \\ 2 & -3 & 9 & 0 \end{bmatrix}$

19. From part (c) of Theorem 1.7.1, a triangular matrix is invertible if and only if its diagonal entries are all nonzero. Since this upper triangular matrix has a 0 on its diagonal, it is not invertible.

21. From part (c) of Theorem 1.7.1, a triangular matrix is invertible if and only if its diagonal entries are all nonzero. Since this lower triangular matrix has all four diagonal entries nonzero, it is invertible.

23. $AB = \begin{bmatrix} (3)(-1) & \times & \times \\ 0 & (1)(5) & \times \\ 0 & 0 & (-1)(6) \end{bmatrix}$. The diagonal entries of AB are: $-3, 5, -6$.

25. The matrix is symmetric if and only if $a + 5 = -3$. In order for A to be symmetric, we must have $a = -8$.

27. From part (c) of Theorem 1.7.1, a triangular matrix is invertible if and only if its diagonal entries are all nonzero. Therefore, the given upper triangular matrix is invertible for any real number x such that $x \neq 1$, $x \neq -2$, and $x \neq 4$.

29. By Theorem 1.7.1, A^{-1} is also an upper triangular or lower triangular invertible matrix. Its diagonal entries must all be nonzero - they are reciprocals of the corresponding diagonal entries of the matrix A.

31. $A = \begin{bmatrix} 1 & 0 & 0 \\ 0 & -1 & 0 \\ 0 & 0 & -1 \end{bmatrix}$

33. $$AB = \begin{bmatrix} (-1)(2)+(2)(0)+(5)(0) & (-1)(-8)+(2)(2)+(5)(0) & (-1)(0)+(2)(1)+(5)(3) \\ (0)(2)+(1)(0)+(3)(0) & (0)(-8)+(1)(2)+(3)(0) & (0)(0)+(1)(1)+(3)(3) \\ (0)(2)+(0)(0)+(-4)(0) & (0)(-8)+(0)(2)+(-4)(0) & (0)(0)+(0)(1)+(-4)(3) \end{bmatrix}$$

$= \begin{bmatrix} -2 & 12 & 17 \\ 0 & 2 & 10 \\ 0 & 0 & -12 \end{bmatrix}$. Since this is an upper triangular matrix, we have verified Theorem 1.7.1(b).

35. **(a)** $A^{-1} = \frac{1}{(2)(3)-(-1)(-1)}\begin{bmatrix} 3 & 1 \\ 1 & 2 \end{bmatrix} = \begin{bmatrix} \frac{3}{5} & \frac{1}{5} \\ \frac{1}{5} & \frac{2}{5} \end{bmatrix}$ is symmetric, therefore we verified Theorem 1.7.4.

(b) $\left[\begin{array}{rrr|rrr} 1 & -2 & 3 & 1 & 0 & 0 \\ -2 & 1 & -7 & 0 & 1 & 0 \\ 3 & -7 & 4 & 0 & 0 & 1 \end{array}\right]$ ← The identity matrix was adjoined to the matrix A.

$\left[\begin{array}{rrr|rrr} 1 & -2 & 3 & 1 & 0 & 0 \\ 0 & -3 & -1 & 2 & 1 & 0 \\ 0 & -1 & -5 & -3 & 0 & 1 \end{array}\right]$ ← 2 times the first row was added to the second row and -3 times the first row was added to the third row.

$$\left[\begin{array}{ccc|ccc} 1 & -2 & 3 & 1 & 0 & 0 \\ 0 & -1 & -5 & -3 & 0 & 1 \\ 0 & -3 & -1 & 2 & 1 & 0 \end{array}\right]$$ ← The second and third rows were interchanged.

$$\left[\begin{array}{ccc|ccc} 1 & -2 & 3 & 1 & 0 & 0 \\ 0 & 1 & 5 & 3 & 0 & -1 \\ 0 & -3 & -1 & 2 & 1 & 0 \end{array}\right]$$ ← The second row was multiplied by -1.

$$\left[\begin{array}{ccc|ccc} 1 & -2 & 3 & 1 & 0 & 0 \\ 0 & 1 & 5 & 3 & 0 & -1 \\ 0 & 0 & 14 & 11 & 1 & -3 \end{array}\right]$$ ← 3 times the second row was added to the third row.

$$\left[\begin{array}{ccc|ccc} 1 & -2 & 3 & 1 & 0 & 0 \\ 0 & 1 & 5 & 3 & 0 & -1 \\ 0 & 0 & 1 & \frac{11}{14} & \frac{1}{14} & -\frac{3}{14} \end{array}\right]$$ ← The third row was multiplied by $\frac{1}{14}$.

$$\left[\begin{array}{ccc|ccc} 1 & -2 & 3 & -\frac{19}{14} & -\frac{3}{14} & \frac{9}{14} \\ 0 & 1 & 0 & -\frac{13}{14} & -\frac{5}{14} & \frac{1}{14} \\ 0 & 0 & 1 & \frac{11}{14} & \frac{1}{14} & -\frac{3}{14} \end{array}\right]$$ ← -5 times the third row was added to the second row and -3 times the third row was added to the first row.

$$\left[\begin{array}{ccc|ccc} 1 & 0 & 3 & -\frac{45}{14} & -\frac{13}{14} & \frac{11}{14} \\ 0 & 1 & 0 & -\frac{13}{14} & -\frac{5}{14} & \frac{1}{14} \\ 0 & 0 & 1 & \frac{11}{14} & \frac{1}{14} & -\frac{3}{14} \end{array}\right]$$ ← 2 times the second row was added to the first row.

Since $A^{-1} = \begin{bmatrix} -\frac{45}{14} & -\frac{13}{14} & \frac{11}{14} \\ -\frac{13}{14} & -\frac{5}{14} & \frac{1}{14} \\ \frac{11}{14} & \frac{1}{14} & -\frac{3}{14} \end{bmatrix}$ is symmetric, we have verified Theorem 1.7.4

37. **(a)** $a_{ji} = j^2 + i^2 = i^2 + j^2 = a_{ij}$ for all i and j therefore A is symmetric.

(b) $a_{ji} = j^2 - i^2$ does not generally equal $a_{ij} = i^2 - j^2$ for $i \neq j$ therefore A is not symmetric (unless $n = 1$).

(c) $a_{ji} = 2j + 2i = 2i + 2j = a_{ij}$ for all i and j therefore A is symmetric.

(d) $a_{ji} = 2j^2 + 2i^3$ does not generally equal $a_{ij} = 2i^2 + 2j^3$ for $i \neq j$ therefore A is not symmetric (unless $n = 1$).

39. For a general upper triangular 2×2 matrix $A = \begin{bmatrix} a & b \\ 0 & c \end{bmatrix}$ we have

$$A^3 = \begin{bmatrix} a & b \\ 0 & c \end{bmatrix}\begin{bmatrix} a & b \\ 0 & c \end{bmatrix}\begin{bmatrix} a & b \\ 0 & c \end{bmatrix}$$
$$= \begin{bmatrix} a^2 & ab+bc \\ 0 & c^2 \end{bmatrix}\begin{bmatrix} a & b \\ 0 & c \end{bmatrix} = \begin{bmatrix} a^3 & a^2b + (ab+bc)c \\ 0 & c^3 \end{bmatrix} = \begin{bmatrix} a^3 & (a^2+ac+c^2)b \\ 0 & c^3 \end{bmatrix}$$

Setting $A^3 = \begin{bmatrix} 1 & 30 \\ 0 & -8 \end{bmatrix}$ we obtain the equations $a^3 = 1, (a^2 + ac + c^2)b = 30, c^3 = -8$.
The first and the third equations yield $a = 1, c = -2$.

Substituting these into the second equation leads to $(1-2+4)b = 30$, i.e., $b = 10$.

We conclude that the only upper triangular matrix A such that $A^3 = \begin{bmatrix} 1 & 30 \\ 0 & -8 \end{bmatrix}$ is $A = \begin{bmatrix} 1 & 10 \\ 0 & -2 \end{bmatrix}$.

41. **(a)** $\begin{bmatrix} 0 & 0 & 4 \\ 0 & 0 & 1 \\ -4 & -1 & 0 \end{bmatrix}$ **(b)** $\begin{bmatrix} 0 & 0 & -8 \\ 0 & 0 & -4 \\ 8 & 4 & 0 \end{bmatrix}$

43. No. If $AB = BA$, $A^T = -A$, and $B^T = -B$ then $(AB)^T = B^T A^T = (-B)(-A) = BA = AB$ which does not generally equal $-AB$. (The product of skew-symmetric matrices that commute is symmetric.)

45. **(a)** $(A^{-1})^T$

$= (A^T)^{-1}$ ⟵ Theorem 1.4.9(d)

$= (-A)^{-1}$ ⟵ The assumption: A is skew-symmetric

$= -A^{-1}$ ⟵ Theorem 1.4.7(c)

(b) $(A^T)^T$

$= A$ ⟵ Theorem 1.4.8(a)

$= -A^T$ ⟵ The assumption: A is skew-symmetric

$(A+B)^T$

$= A^T + B^T$ ⟵ Theorem 1.4.8(b)

$= -A - B$ ⟵ The assumption: A and B are skew-symmetric

$= -(A+B)$ ⟵ Theorem 1.4.1(h)

$(A-B)^T$

$= A^T - B^T$ ⟵ Theorem 1.4.8(c)

$= -A - (-B)$ ⟵ The assumption: A and B are skew-symmetric

$= -(A-B)$ ⟵ Theorem 1.4.1(i)

$(kA)^T$

$= kA^T$ ⟵ Theorem 1.4.8(d)

$= k(-A)$ ⟵ The assumption: A is skew-symmetric

$= -kA$ ⟵ Theorem 1.4.1(l)

47. $A^T = (A^TA)^T = A^T(A^T)^T = A^TA = A$ therefore A is symmetric; thus we have $A^2 = AA = A^TA = A$.

True-False Exercises

(a) True. Every diagonal matrix is symmetric: its transpose equals to the original matrix.

(b) False. The transpose of an upper triangular matrix is a *lower* triangular matrix.

(c) False. E.g., $\begin{bmatrix} 1 & 1 \\ 0 & 1 \end{bmatrix} + \begin{bmatrix} 1 & 0 \\ 1 & 1 \end{bmatrix} = \begin{bmatrix} 2 & 1 \\ 1 & 2 \end{bmatrix}$ is not a diagonal matrix.

(d) True. Mirror images of entries across the main diagonal must be equal - see the margin note next to Example 4.

(e) True. All entries below the main diagonal must be zero.

(f) False. By Theorem 1.7.1(d), the inverse of an invertible lower triangular matrix is a lower triangular matrix.

(g) False. A diagonal matrix is invertible if and only if all or its diagonal entries are nonzero (positive or negative).

(h) True. The entries above the main diagonal are zero.

(i) True. If A is upper triangular then A^T is lower triangular. However, if A is also symmetric then it follows that $A^T = A$ must be both upper triangular and lower triangular. This requires A to be a diagonal matrix.

(j) False. For instance, neither $A = \begin{bmatrix} 0 & 1 \\ 0 & 0 \end{bmatrix}$ nor $B = \begin{bmatrix} 0 & 0 \\ 1 & 0 \end{bmatrix}$ is symmetric even though $A + B = \begin{bmatrix} 0 & 1 \\ 1 & 0 \end{bmatrix}$ is.

(k) False. For instance, neither $A = \begin{bmatrix} 0 & 1 \\ -1 & 0 \end{bmatrix}$ nor $B = \begin{bmatrix} 0 & 0 \\ 1 & 0 \end{bmatrix}$ is upper triangular even though $A + B = \begin{bmatrix} 0 & 1 \\ 0 & 0 \end{bmatrix}$ is.

(l) False. For instance, $A = \begin{bmatrix} 0 & 0 \\ 1 & 0 \end{bmatrix}$ is not symmetric even though $A^2 = \begin{bmatrix} 0 & 0 \\ 0 & 0 \end{bmatrix}$ is.

(m) True. By Theorem 1.4.8(d), $(kA)^T = kA^T$. Since kA is symmetric, we also have $(kA)^T = kA$. For nonzero k the equality of the right hand sides $kA^T = kA$ implies $A^T = A$.

1.8 Matrix Transformations

1. **(a)** $T_A(\mathbf{x}) = A\mathbf{x}$ maps any vector $\mathbf{x}$ in R^2 into a vector $\mathbf{w} = A\mathbf{x}$ in R^3.
The domain of T_A is R^2; the codomain is R^3.

(b) $T_A(\mathbf{x}) = A\mathbf{x}$ maps any vector $\mathbf{x}$ in R^3 into a vector $\mathbf{w} = A\mathbf{x}$ in R^2.
The domain of T_A is R^3; the codomain is R^2.

(c) $T_A(\mathbf{x}) = A\mathbf{x}$ maps any vector $\mathbf{x}$ in R^3 into a vector $\mathbf{w} = A\mathbf{x}$ in R^3. The domain of T_A is R^3; the codomain is R^3.

(d) $T_A(\mathbf{x}) = A\mathbf{x}$ maps any vector $\mathbf{x}$ in R^6 into a vector $\mathbf{w} = A\mathbf{x}$ in $R^1 = R$. The domain of T_A is R^6; the codomain is R.

3. **(a)** The transformation maps any vector $\mathbf{x}$ in R^2 into a vector $\mathbf{w}$ in R^2. Its domain is R^2; the codomain is R^2.

(b) The transformation maps any vector $\mathbf{x}$ in R^2 into a vector $\mathbf{w}$ in R^3. Its domain is R^2; the codomain is R^3.

5. **(a)** The transformation maps any vector $\mathbf{x}$ in R^3 into a vector in R^2. Its domain is R^3; the codomain is R^2.

(b) The transformation maps any vector $\mathbf{x}$ in R^2 into a vector in R^3. Its domain is R^2; the codomain is R^3.

7. **(a)** The transformation maps any vector $\mathbf{x}$ in R^2 into a vector in R^2. Its domain is R^2; the codomain is R^2.

(b) The transformation maps any vector $\mathbf{x}$ in R^3 into a vector in R^2. Its domain is R^3; the codomain is R^2.

9. The transformation maps any vector $\mathbf{x}$ in R^2 into a vector in R^3. Its domain is R^2; the codomain is R^3.

11. **(a)** The given equations can be expressed in matrix form as $\begin{bmatrix} w_1 \\ w_2 \end{bmatrix} = \begin{bmatrix} 2 & -3 & 1 \\ 3 & 5 & -1 \end{bmatrix}\begin{bmatrix} x_1 \\ x_2 \\ x_3 \end{bmatrix}$

therefore the standard matrix for this transformation is $\begin{bmatrix} 2 & -3 & 1 \\ 3 & 5 & -1 \end{bmatrix}$

(b) The given equations can be expressed in matrix form as $\begin{bmatrix} w_1 \\ w_2 \\ w_3 \end{bmatrix} = \begin{bmatrix} 7 & 2 & -8 \\ 0 & -1 & 5 \\ 4 & 7 & -1 \end{bmatrix}\begin{bmatrix} x_1 \\ x_2 \\ x_3 \end{bmatrix}$

therefore the standard matrix for this transformation is $\begin{bmatrix} 7 & 2 & -8 \\ 0 & -1 & 5 \\ 4 & 7 & -1 \end{bmatrix}$.

13. **(a)** $T(x_1, x_2) = \begin{bmatrix} x_2 \\ -x_1 \\ x_1 + 3x_2 \\ x_1 - x_2 \end{bmatrix} = \begin{bmatrix} 0 & 1 \\ -1 & 0 \\ 1 & 3 \\ 1 & -1 \end{bmatrix}\begin{bmatrix} x_1 \\ x_2 \end{bmatrix}$; the standard matrix is $\begin{bmatrix} 0 & 1 \\ -1 & 0 \\ 1 & 3 \\ 1 & -1 \end{bmatrix}$

(b) $T(x_1, x_2, x_3, x_4) = \begin{bmatrix} 7x_1 + 2x_2 - x_3 + x_4 \\ x_2 + x_3 \\ -x_1 \end{bmatrix} = \begin{bmatrix} 7 & 2 & -1 & 1 \\ 0 & 1 & 1 & 0 \\ -1 & 0 & 0 & 0 \end{bmatrix}\begin{bmatrix} x_1 \\ x_2 \\ x_3 \\ x_4 \end{bmatrix}$;

the standard matrix is $\begin{bmatrix} 7 & 2 & -1 & 1 \\ 0 & 1 & 1 & 0 \\ -1 & 0 & 0 & 0 \end{bmatrix}$

(c) $T(x_1,x_2,x_3)=\begin{bmatrix}0\\0\\0\\0\\0\end{bmatrix}=\begin{bmatrix}0&0&0\\0&0&0\\0&0&0\\0&0&0\\0&0&0\end{bmatrix}\begin{bmatrix}x_1\\x_2\\x_3\end{bmatrix}$; the standard matrix is $\begin{bmatrix}0&0&0\\0&0&0\\0&0&0\\0&0&0\\0&0&0\end{bmatrix}$

(d) $T(x_1,x_2,x_3,x_4)=\begin{bmatrix}x_4\\x_1\\x_3\\x_2\\x_1-x_3\end{bmatrix}=\begin{bmatrix}0&0&0&1\\1&0&0&0\\0&0&1&0\\0&1&0&0\\1&0&-1&0\end{bmatrix}\begin{bmatrix}x_1\\x_2\\x_3\\x_4\end{bmatrix}$; the standard matrix is $\begin{bmatrix}0&0&0&1\\1&0&0&0\\0&0&1&0\\0&1&0&0\\1&0&-1&0\end{bmatrix}$

15. The given equations can be expressed in matrix form as $\begin{bmatrix}w_1\\w_2\\w_3\end{bmatrix}=\begin{bmatrix}3&5&-1\\4&-1&1\\3&2&-1\end{bmatrix}\begin{bmatrix}x_1\\x_2\\x_3\end{bmatrix}$ therefore the standard matrix for this operator is $\begin{bmatrix}3&5&-1\\4&-1&1\\3&2&-1\end{bmatrix}$.

By directly substituting $(-1,2,4)$ for (x_1,x_2,x_3) into the given equation we obtain

$$w_1=-(3)(1)+(5)(2)-(1)(4)=3$$
$$w_2=-(4)(1)-(1)(2)+(1)(4)=-2$$
$$w_3=-(3)(1)+(2)(2)-(1)(4)=-3$$

By matrix multiplication, $\begin{bmatrix}w_1\\w_2\\w_3\end{bmatrix}=\begin{bmatrix}3&5&-1\\4&-1&1\\3&2&-1\end{bmatrix}\begin{bmatrix}-1\\2\\4\end{bmatrix}=\begin{bmatrix}-(3)(1)+(5)(2)-(1)(4)\\-(4)(1)-(1)(2)+(1)(4)\\-(3)(1)+(2)(2)-(1)(4)\end{bmatrix}=\begin{bmatrix}3\\-2\\-3\end{bmatrix}$.

17. **(a)** $T(x_1,x_2)=\begin{bmatrix}-x_1+x_2\\x_2\end{bmatrix}=\begin{bmatrix}-1&1\\0&1\end{bmatrix}\begin{bmatrix}x_1\\x_2\end{bmatrix}$; the standard matrix is $\begin{bmatrix}-1&1\\0&1\end{bmatrix}$.

$T(\mathbf{x})=\begin{bmatrix}-1&1\\0&1\end{bmatrix}\begin{bmatrix}-1\\4\end{bmatrix}=\begin{bmatrix}(1)(1)+(1)(4)\\-(0)(1)+(1)(4)\end{bmatrix}=\begin{bmatrix}5\\4\end{bmatrix}$ matches $T(-1,4)=(1+4,4)=(5,4)$.

(b) $T(x_1,x_2,x_3)=\begin{bmatrix}2x_1-x_2+x_3\\x_2+x_3\\0\end{bmatrix}=\begin{bmatrix}2&-1&1\\0&1&1\\0&0&0\end{bmatrix}\begin{bmatrix}x_1\\x_2\\x_3\end{bmatrix}$; the standard matrix is $\begin{bmatrix}2&-1&1\\0&1&1\\0&0&0\end{bmatrix}$.

$T(\mathbf{x})=\begin{bmatrix}2&-1&1\\0&1&1\\0&0&0\end{bmatrix}\begin{bmatrix}2\\1\\-3\end{bmatrix}=\begin{bmatrix}(2)(2)-(1)(1)-(1)(3)\\(0)(2)+(1)(1)-(1)(3)\\(0)(2)+(0)(1)-(0)(3)\end{bmatrix}=\begin{bmatrix}0\\-2\\0\end{bmatrix}$

matches $T(2,1,-3)=(4-1-3,1-3,0)=(0,-2,0)$.

19. **(a)** $T_A(\mathbf{x})=A\mathbf{x}=\begin{bmatrix}1&2\\3&4\end{bmatrix}\begin{bmatrix}3\\-2\end{bmatrix}=\begin{bmatrix}-1\\1\end{bmatrix}$

(b) $T_A(\mathbf{x})=A\mathbf{x}=\begin{bmatrix}-1&2&0\\3&1&5\end{bmatrix}\begin{bmatrix}-1\\1\\3\end{bmatrix}=\begin{bmatrix}3\\13\end{bmatrix}$

21. **(a)** If $\mathbf{u}=(u_1,u_2)$ and $\mathbf{v}=(v_1,v_2)$ then

$$\begin{aligned}T(\mathbf{u}+\mathbf{v})&=T(u_1+v_1,u_2+v_2)\\&=\big(2(u_1+v_1)+(u_2+v_2),(u_1+v_1)-(u_2+v_2)\big)\\&=(2u_1+u_2,u_1-u_2)+(2v_1+v_2,v_1-v_2)\\&=T(\mathbf{u})+T(\mathbf{v})\end{aligned}$$

and $T(k\mathbf{u})=T(ku_1,ku_2)=(2ku_1+ku_2,ku_1-ku_2)=k(2u_1+u_2,u_1-u_2)=kT(\mathbf{u})$.

(b) If $\mathbf{u} = (u_1, u_2, u_3)$ and $\mathbf{v} = (v_1, v_2, v_3)$ then

$$\begin{aligned} T(\mathbf{u}+\mathbf{v}) &= T(u_1+v_1, u_2+v_2, u_3+v_3) \\ &= (u_1+v_1, u_3+v_3, u_1+v_1+u_2+v_2) \\ &= (u_1, u_3, u_1+u_2) + (v_1, v_3, v_1+v_2) \\ &= T(\mathbf{u}) + T(\mathbf{v}) \end{aligned}$$

and $T(k\mathbf{u}) = T(ku_1, ku_2, ku_3) = (ku_1, ku_3, ku_1 + ku_2) = k(u_1, u_3, u_1 + u_2) = kT(\mathbf{u})$.

23. **(a)** The homogeneity property fails to hold since $T(kx, ky) = ((kx)^2, ky) = (k^2x^2, ky)$ does not generally equal $kT(x, y) = k(x^2, y) = (kx^2, ky)$. (It can be shown that the additivity property fails to hold as well.)

(b) The homogeneity property fails to hold since $T(kx, ky, kz) = (kx, ky, kxkz) = (kx, ky, k^2xz)$ does not generally equal $kT(x, y, z) = k(x, y, xz) = (kx, ky, kxz)$. (It can be shown that the additivity property fails to hold as well.)

25. The homogeneity property fails to hold since for $b \neq 0$, $f(kx) = m(kx) + b$ does not generally equal $kf(x) = k(mx + b) = kmx + kb$. (It can be shown that the additivity property fails to hold as well.) On the other hand, both properties hold for $b = 0$: $f(x + y) = m(x + y) = mx + my = f(x) + f(y)$ and $f(kx) = m(kx) = k(mx) = kf(x)$.
Consequently, f is not a matrix transformation on R unless $b = 0$

27. By Formula (13), the standard matrix for T is $A = [\ T(\mathbf{e}_1) \ \mid T(\mathbf{e}_2) \ \mid \ T(\mathbf{e}_3)\ ]$. Therefore

$$A = \begin{bmatrix} 1 & 0 & 4 \\ 3 & 0 & -3 \\ 0 & 1 & -1 \end{bmatrix} \text{ and } T(\mathbf{x}) = A\mathbf{x} = \begin{bmatrix} (1)(2) + (0)(1) + (4)(0) \\ (3)(2) + (0)(1) - (3)(0) \\ (0)(2) + (1)(1) - (1)(0) \end{bmatrix} = \begin{bmatrix} 2 \\ 6 \\ 1 \end{bmatrix}.$$

29. By Formula (13), the standard matrix for T is $A = [\ T(\mathbf{e}_1) \ \mid T(\mathbf{e}_2)\ ]$. Therefore

$$A = \begin{bmatrix} a & c \\ b & d \end{bmatrix} \text{ and } T(1,1) = A\begin{bmatrix} 1 \\ 1 \end{bmatrix} = \begin{bmatrix} a+c \\ b+d \end{bmatrix}.$$

31. **(a)** $T_A(\mathbf{e}_1) = \begin{bmatrix} -1 \\ 2 \\ 4 \end{bmatrix}$, $T_A(\mathbf{e}_2) = \begin{bmatrix} 3 \\ 1 \\ 5 \end{bmatrix}$, $T_A(\mathbf{e}_3) = \begin{bmatrix} 0 \\ 2 \\ -3 \end{bmatrix}$.

(b) Since T_A is a matrix transformation,

$$T_A(\mathbf{e}_1 + \mathbf{e}_2 + \mathbf{e}_3) = T_A(\mathbf{e}_1) + T_A(\mathbf{e}_2) + T_A(\mathbf{e}_3) = \begin{bmatrix} -1 \\ 2 \\ 4 \end{bmatrix} + \begin{bmatrix} 3 \\ 1 \\ 5 \end{bmatrix} + \begin{bmatrix} 0 \\ 2 \\ -3 \end{bmatrix} = \begin{bmatrix} 2 \\ 5 \\ 6 \end{bmatrix}.$$

(c) Since T_A is a matrix transformation, $T_A(7e_3) = 7T_A(\mathbf{e}_3) = 7\begin{bmatrix} 0 \\ 2 \\ -3 \end{bmatrix} = \begin{bmatrix} 0 \\ 14 \\ -21 \end{bmatrix}$.

True-False Exercises

(a) False. The domain of T_A is R^3.

(b) False. The codomain of T_A is R^m.

(c) True. Since the statement requires the given equality to hold for <u>some</u> vector $\mathbf{x}$ in R^n, we can let $\mathbf{x} = \mathbf{0}$.

(d) False. (Refer to Theorem 1.8.3.)

(e) True. The columns of A are $T(\mathbf{e}_i) = \mathbf{0}$.

(f) False. The given equality must hold for every matrix transformation since it follows from the homogeneity property.

(g) False. The homogeneity property fails to hold since $T(k\mathbf{x}) = k\mathbf{x} + \mathbf{b}$ does not generally equal $kT(\mathbf{x}) = k(\mathbf{x} + \mathbf{b}) = k\mathbf{x} + k\mathbf{b}$.

1.9 Applications of Linear Systems

1. There are four nodes, which we denote by A, B, C, and D (see the figure on the left). We determine the unknown flow rates x_1, x_2, and x_3 assuming the counterclockwise direction (if any of these quantities are found to be negative then the flow direction along the corresponding branch will be reversed).

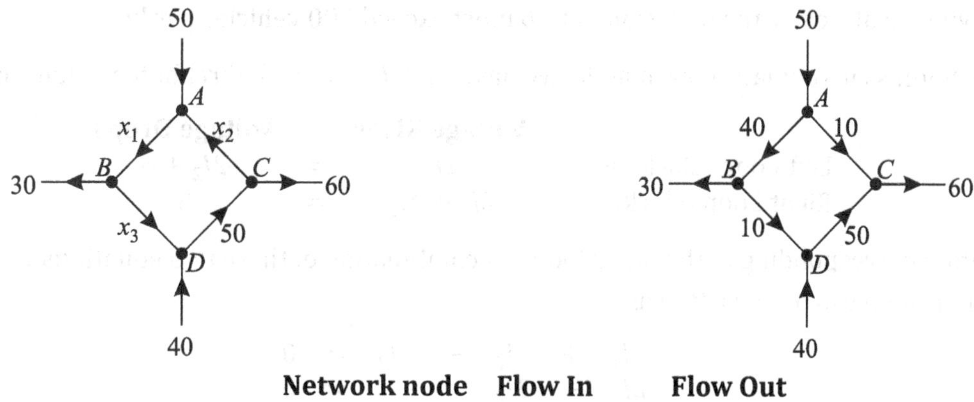

Network node	Flow In		Flow Out
A	$x_2 + 50$	=	x_1
B	x_1	=	$x_3 + 30$
C	50	=	$x_2 + 60$
D	$x_3 + 40$	=	50

This system can be rearranged as follows

$$\begin{array}{rcrcrcr} -x_1 & + & x_2 & & & = & -50 \\ x_1 & & & - & x_3 & = & 30 \\ & - & x_2 & & & = & 10 \\ & & & & x_3 & = & 10 \end{array}$$

By inspection, this system has a unique solution $x_1 = 40, x_2 = -10, x_3 = 10$. This yields the flow rates and directions shown in the figure on the right.

3. **(a)** There are four nodes – each of them corresponds to an equation.

Network node	Flow In		Flow Out
top left	$x_2 + 300$	=	$x_3 + 400$
top right (A)	$x_3 + 750$	=	$x_4 + 250$
bottom left	$x_1 + 100$	=	$x_2 + 400$
bottom right (B)	$x_4 + 200$	=	$x_1 + 300$

This system can be rearranged as follows

$$\begin{array}{rcrcrcrcr} & & x_2 & - & x_3 & & & = & 100 \\ & & & & x_3 & - & x_4 & = & -500 \\ x_1 & - & x_2 & & & & & = & 300 \\ -x_1 & & & & & + & x_4 & = & 100 \end{array}$$

(b) The augmented matrix of the linear system obtained in part (a) $\begin{bmatrix} 0 & 1 & -1 & 0 & 100 \\ 0 & 0 & 1 & -1 & -500 \\ 1 & -1 & 0 & 0 & 300 \\ -1 & 0 & 0 & 1 & 100 \end{bmatrix}$

has the reduced row echelon form $\begin{bmatrix} 1 & 0 & 0 & -1 & -100 \\ 0 & 1 & 0 & -1 & -400 \\ 0 & 0 & 1 & -1 & -500 \\ 0 & 0 & 0 & 0 & 0 \end{bmatrix}$. If we assign x_4 the arbitrary value s, the general solution is given by the formulas

$x_1 = -100 + s$, $x_2 = -400 + s$, $x_3 = -500 + s$, $x_4 = s$

(c) In order for all x_i values to remain positive, we must have $s > 500$. Therefore, to keep the traffic flowing on all roads, the flow from A to B must exceed 500 vehicles per hour.

5. From Kirchhoff's current law at each node, we have $I_1 + I_2 - I_3 = 0$. Kirchhoff's voltage law yields

	Voltage Rises		**Voltage Drops**
Left Loop (clockwise)	$2I_1$	=	$2I_2 + 6$
Right Loop (clockwise)	$2I_2 + 4I_3$	=	8

(An equation corresponding to the outer loop is a combination of these two equations.)
The linear system can be rewritten as

$$\begin{array}{rcrcrcr} I_1 & + & I_2 & - & I_3 & = & 0 \\ 2I_1 & - & 2I_2 & & & = & 6 \\ & & 2I_2 & + & 4I_3 & = & 8 \end{array}$$

Its augmented matrix has the reduced row echelon form $\begin{bmatrix} 1 & 0 & 0 & \frac{13}{5} \\ 0 & 1 & 0 & -\frac{2}{5} \\ 0 & 0 & 1 & \frac{11}{5} \end{bmatrix}$.

The solution is $I_1 = 2.6\text{A}$, $I_2 = -0.4\text{A}$, and $I_3 = 2.2\text{A}$.
Since I_2 is negative, this current is opposite to the direction shown in the diagram.

7. From Kirchhoff's current law, we have

	Current In		**Currrent Out**
Top Left Node	I_1	=	$I_2 + I_4$
Top Right Node	I_4	=	$I_3 + I_5$
Bottom Left Node	$I_2 + I_6$	=	I_1
Bottom Right Node	$I_3 + I_5$	=	I_6

Kirchhoff's voltage law yields

	Voltage Rises		**Voltage Drops**
Left Loop (clockwise)	10	=	$20I_1 + 20I_2$
Middle Loop (clockwise)	$20I_2$	=	$20I_3$
Right Loop (clockwise)	$20I_3 + 10$	=	$20I_5$

(Equations corresponding to the other loops are combinations of these three equations.)

The linear system can be rewritten as

$$\begin{array}{rcrcrcrcrcrcr}
I_1 & - & I_2 & & & - & I_4 & & & & & = & 0 \\
 & & & - & I_3 & + & I_4 & - & I_5 & & & = & 0 \\
-I_1 & + & I_2 & & & & & & & + & I_6 & = & 0 \\
 & & & & I_3 & & & + & I_5 & - & I_6 & = & 0 \\
-20I_1 & - & 20I_2 & & & & & & & & & = & -10 \\
 & & 20I_2 & - & 20I_3 & & & & & & & = & 0 \\
 & & & & 20I_3 & & & - & 20I_5 & & & = & -10
\end{array}$$

Its augmented matrix has the reduced row echelon form $\begin{bmatrix} 1 & 0 & 0 & 0 & 0 & 0 & \frac{1}{2} \\ 0 & 1 & 0 & 0 & 0 & 0 & 0 \\ 0 & 0 & 1 & 0 & 0 & 0 & 0 \\ 0 & 0 & 0 & 1 & 0 & 0 & \frac{1}{2} \\ 0 & 0 & 0 & 0 & 1 & 0 & \frac{1}{2} \\ 0 & 0 & 0 & 0 & 0 & 1 & \frac{1}{2} \\ 0 & 0 & 0 & 0 & 0 & 0 & 0 \end{bmatrix}$.

The solution is $I_1 = I_4 = I_5 = I_6 = 0.5\text{A}$, $I_2 = I_3 = 0\text{A}$.

9. We are looking for positive integers x_1, x_2, x_3, and x_4 such that

$$x_1(C_3H_8) + x_2(O_2) \rightarrow x_3(CO_2) + x_4(H_2O)$$

The number of atoms of carbon, hydrogen, and oxygen on both sides must equal:

	Left Side		**Right Side**
Carbon	$3x_1$	$=$	x_3
Hydrogen	$8x_1$	$=$	$2x_4$
Oxygen	$2x_2$	$=$	$2x_3 + x_4$

The linear system

$$\begin{array}{rcrcrcrcr}
3x_1 & & & - & x_3 & & & = & 0 \\
8x_1 & & & & & - & 2x_4 & = & 0 \\
 & & 2x_2 & - & 2x_3 & - & x_4 & = & 0
\end{array}$$

has the augmented matrix whose reduced row echelon form is $\begin{bmatrix} 1 & 0 & 0 & -\frac{1}{4} & 0 \\ 0 & 1 & 0 & -\frac{5}{4} & 0 \\ 0 & 0 & 1 & -\frac{3}{4} & 0 \end{bmatrix}$.

The general solution is $x_1 = \frac{1}{4}t,\ x_2 = \frac{5}{4}t,\ x_3 = \frac{3}{4}t,\ x_4 = t$ where t is arbitrary. The smallest positive integer values for the unknowns occur when $t = 4$, which yields the solution $x_1 = 1,\ x_2 = 5,\ x_3 = 3,\ x_4 = 4$. The balanced equation is

$$C_3H_8 + 5O_2 \rightarrow 3CO_2 + 4H_2O$$

11. We are looking for positive integers x_1, x_2, x_3, and x_4 such that

$$x_1(CH_3COF) + x_2(H_2O) \rightarrow x_3(CH_3COOH) + x_4(HF)$$

The number of atoms of carbon, hydrogen, oxygen, and fluorine on both sides must equal:

	Left Side		**Right Side**
Carbon	$2x_1$	$=$	$2x_3$
Hydrogen	$3x_1 + 2x_2$	$=$	$4x_3 + x_4$
Oxygen	$x_1 + x_2$	$=$	$2x_3$
Fluorine	x_1	$=$	x_4

The linear system

$$\begin{array}{rcrcrcrcl} 2x_1 & & & - & 2x_3 & & & = & 0 \\ 3x_1 & + & 2x_2 & - & 4x_3 & - & x_4 & = & 0 \\ x_1 & + & x_2 & - & 2x_3 & & & = & 0 \\ x_1 & & & & & - & x_4 & = & 0 \end{array}$$

has the augmented matrix whose reduced row echelon form is $\begin{bmatrix} 1 & 0 & 0 & -1 & 0 \\ 0 & 1 & 0 & -1 & 0 \\ 0 & 0 & 1 & -1 & 0 \\ 0 & 0 & 0 & 0 & 0 \end{bmatrix}$.

The general solution is $x_1 = t,\ x_2 = t,\ x_3 = t,\ x_4 = t$ where t is arbitrary. The smallest positive integer values for the unknowns occur when $t = 1$, which yields the solution $x_1 = 1,\ x_2 = 1\ x_3 = 1,$ $x_4 = 1$. The balanced equation is

$$CH_3COF + H_2O \rightarrow CH_3COOH + HF$$

13. We are looking for a polynomial of the form $p(x) = a_0 + a_1x + a_2x^2$ such that $p(1) = 1,\ p(2) = 2,$ and $p(3) = 5$. We obtain a linear system

$$\begin{array}{rcrcrcl} a_0 & + & a_1 & + & a_2 & = & 1 \\ a_0 & + & 2a_1 & + & 4a_2 & = & 2 \\ a_0 & + & 3a_1 & + & 9a_2 & = & 5 \end{array}$$

Its augmented matrix has the reduced row echelon form $\begin{bmatrix} 1 & 0 & 0 & 2 \\ 0 & 1 & 0 & -2 \\ 0 & 0 & 1 & 1 \end{bmatrix}$.

There is a unique solution $a_0 = 2,\ a_1 = -2,\ a_2 = 1$.
The quadratic polynomial is $p(x) = 2 - 2x + x^2$.

15. We are looking for a polynomial of the form $p(x) = a_0 + a_1x + a_2x^2 + a_3x^3$ such that $p(-1) = -1,$ $p(0) = 1,\ p(1) = 3$ and $p(4) = -1$. We obtain a linear system

$$\begin{array}{rcrcrcrcr} a_0 & - & a_1 & + & a_2 & - & a_3 & = & -1 \\ a_0 & & & & & & & = & 1 \\ a_0 & + & a_1 & + & a_2 & + & a_3 & = & 3 \\ a_0 & + & 4a_1 & + & 16a_2 & + & 64a_3 & = & -1 \end{array}$$

Its augmented matrix has the reduced row echelon form $\begin{bmatrix} 1 & 0 & 0 & 0 & 1 \\ 0 & 1 & 0 & 0 & \frac{13}{6} \\ 0 & 0 & 1 & 0 & 0 \\ 0 & 0 & 0 & 1 & -\frac{1}{6} \end{bmatrix}$.

There is a unique solution $a_0 = 1,\ a_1 = \frac{13}{6},\ a_2 = 0\,,\ a_3 = -\frac{1}{6}$.

The cubic polynomial is $p(x) = 1 + \frac{13}{6}x - \frac{1}{6}x^3$.

17. **(a)** We are looking for a polynomial of the form $p(x) = a_0 + a_1x + a_2x^2$ such that $p(0) = 1$ and $p(1) = 2$. We obtain a linear system

$$\begin{aligned} a_0 \qquad\qquad\quad &= 1 \\ a_0 + a_1 + a_2 &= 2 \end{aligned}$$

Its augmented matrix has the reduced row echelon form $\begin{bmatrix} 1 & 0 & 0 & 1 \\ 0 & 1 & 1 & 1 \end{bmatrix}$.

The general solution of the linear system is $a_0 = 1,\ a_1 = 1 - t,\ a_2 = t$ where t is arbitrary. Consequently, the family of all second-degree polynomials that pass through $(0,1)$ and $(1,2)$ can be represented by $p(x) = 1 + (1 - t)x + tx^2$ where t is an arbitrary real number.

(b)

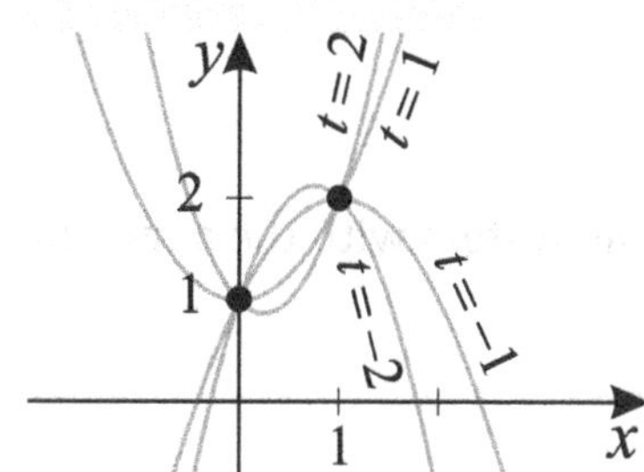

True-False Exercises

(a) False. In general, networks may or may not satisfy the property of flow conservation at each node (although the ones discussed in this section do).

(b) False. When a current passes through a resistor, there is a drop in the electrical potential in a circuit.

(c) True.

(d) False. A chemical equation is said to be balanced if *for each type of atom in the reaction*, the same number of atoms appears on each side of the equation.

(e) False. By Theorem 1.9.1, this is true if the points have distinct x-coordinates.

1.10 Leontief Input-Output Models

1. **(a)** $C = \begin{bmatrix} 0.50 & 0.25 \\ 0.25 & 0.10 \end{bmatrix}$

(b) The Leontief matrix is $I - C = \begin{bmatrix} 1 & 0 \\ 0 & 1 \end{bmatrix} - \begin{bmatrix} 0.50 & 0.25 \\ 0.25 & 0.10 \end{bmatrix} = \begin{bmatrix} 0.50 & -0.25 \\ -0.25 & 0.90 \end{bmatrix}$;

the outside demand vector is $\mathbf{d} = \begin{bmatrix} 7{,}000 \\ 14{,}000 \end{bmatrix}$.

The Leontief equation $(I - C)\mathbf{x} = \mathbf{d}$ leads to the linear system with the augmented matrix $\begin{bmatrix} 0.50 & -0.25 & 7{,}000 \\ -0.25 & 0.90 & 14{,}000 \end{bmatrix}$. Its reduced row echelon form is

$$\begin{bmatrix} 1 & 0 & \frac{784{,}000}{31} \\ 0 & 1 & \frac{700{,}000}{31} \end{bmatrix} \approx \begin{bmatrix} 1 & 0 & 25{,}290.32 \\ 0 & 1 & 22{,}580.65 \end{bmatrix}.$$

To meet the consumer demand, M must produce approximately \$25,290.32 worth of mechanical work and B must produce approximately \$22,580.65 worth of body work.

3. **(a)** $C = \begin{bmatrix} 0.10 & 0.60 & 0.40 \\ 0.30 & 0.20 & 0.30 \\ 0.40 & 0.10 & 0.20 \end{bmatrix}$

(b) The Leontief matrix is $I - C = \begin{bmatrix} 1 & 0 & 0 \\ 0 & 1 & 0 \\ 0 & 0 & 1 \end{bmatrix} - \begin{bmatrix} 0.10 & 0.60 & 0.40 \\ 0.30 & 0.20 & 0.30 \\ 0.40 & 0.10 & 0.20 \end{bmatrix} = \begin{bmatrix} 0.90 & -0.60 & -0.40 \\ -0.30 & 0.80 & -0.30 \\ -0.40 & -0.10 & 0.80 \end{bmatrix}$;

the outside demand vector is $\mathbf{d} = \begin{bmatrix} 1930 \\ 3860 \\ 5790 \end{bmatrix}$.

The Leontief equation $(I - C)\mathbf{x} = \mathbf{d}$ leads to the linear system with the augmented matrix $\begin{bmatrix} 0.90 & -0.60 & -0.40 & 1930 \\ -0.30 & 0.80 & -0.30 & 3860 \\ -0.40 & -0.10 & 0.80 & 5790 \end{bmatrix}$.

Its reduced row echelon form is $\begin{bmatrix} 1 & 0 & 0 & 31{,}500 \\ 0 & 1 & 0 & 26{,}500 \\ 0 & 0 & 1 & 26{,}300 \end{bmatrix}$.

The production vector that will meet the given demand is $\mathbf{x} = \begin{bmatrix} \$31{,}500 \\ \$26{,}500 \\ \$26{,}300 \end{bmatrix}$.

5. $I - C = \begin{bmatrix} 0.9 & -0.3 \\ -0.5 & 0.6 \end{bmatrix}$; $(I - C)^{-1} = \frac{100}{39}\begin{bmatrix} 0.6 & 0.3 \\ 0.5 & 0.9 \end{bmatrix} = \begin{bmatrix} \frac{20}{13} & \frac{10}{13} \\ \frac{50}{39} & \frac{30}{13} \end{bmatrix}$

$$\mathbf{x} = (I - C)^{-1}\mathbf{d} = \begin{bmatrix} \frac{20}{13} & \frac{10}{13} \\ \frac{50}{39} & \frac{30}{13} \end{bmatrix}\begin{bmatrix} 50 \\ 60 \end{bmatrix} = \begin{bmatrix} \frac{1600}{13} \\ \frac{7900}{39} \end{bmatrix} \approx \begin{bmatrix} 123.08 \\ 202.56 \end{bmatrix}$$

7. **(a)** The Leontief matrix is $I - C = \begin{bmatrix} \frac{1}{2} & 0 \\ 0 & 0 \end{bmatrix}$.

The Leontief equation $(I - C)\mathbf{x} = \begin{bmatrix} 2 \\ 0 \end{bmatrix}$ leads to the linear system with the augmented matrix

$\begin{bmatrix} \frac{1}{2} & 0 & 2 \\ 0 & 0 & 0 \end{bmatrix}$. Its reduced row echelon form is $\begin{bmatrix} 1 & 0 & 4 \\ 0 & 0 & 0 \end{bmatrix}$ therefore a production vector can be found (namely, $\begin{bmatrix} 4 \\ t \end{bmatrix}$ for an arbitrary nonnegative t) to meet the demand.

On the other hand, the Leontief equation $(I - C)\mathbf{x} = \begin{bmatrix} 2 \\ 1 \end{bmatrix}$ leads to the linear system with the augmented matrix $\begin{bmatrix} \frac{1}{2} & 0 & 2 \\ 0 & 0 & 1 \end{bmatrix}$. Its reduced row echelon form is $\begin{bmatrix} 1 & 0 & 0 \\ 0 & 0 & 1 \end{bmatrix}$; the system is inconsistent, therefore a production vector cannot be found to meet the demand.

(b) Mathematically, the linear system represented by $\begin{bmatrix} \frac{1}{2} & 0 \\ 0 & 0 \end{bmatrix}\begin{bmatrix} x_1 \\ x_2 \end{bmatrix} = \begin{bmatrix} d_1 \\ d_2 \end{bmatrix}$ can be rewritten as $\begin{bmatrix} \frac{1}{2}x_1 \\ 0 \end{bmatrix} = \begin{bmatrix} d_1 \\ d_2 \end{bmatrix}$.

Clearly, if $d_2 = 0$ the system has infinitely many solutions: $x_1 = 2d_1$; $x_2 = t$ where t is an arbitrary nonnegative number.
If $d_2 \neq 0$ the system is inconsistent. (Note that the Leontief matrix is not invertible.)

An economic explanation of the result in part (a) is that $\mathbf{c}_2 = \begin{bmatrix} 0 \\ 1 \end{bmatrix}$ therefore the second sector consumes all of its own output, making it impossible to meet any outside demand for its products.

9. From the assumption $c_{21}c_{12} < 1 - c_{11}$, it follows that the determinant of $\det(I - C) = \det\left(\begin{bmatrix} 1 - c_{11} & -c_{12} \\ -c_{21} & 1 \end{bmatrix}\right) = 1 - c_{11} - c_{12}c_{21}$ is nonzero. Consequently, the Leontief matrix is invertible; its inverse is $(I - C)^{-1} = \frac{1}{1-c_{11}-c_{12}c_{21}}\begin{bmatrix} 1 & c_{12} \\ c_{21} & 1 - c_{11} \end{bmatrix}$. Since the consumption matrix C has nonnegative entries and $1 - c_{11} > c_{21}c_{12} \geq 0$, we conclude that all entries of $(I - C)^{-1}$ are nonnegative as well. This economy is productive (see the discussion above Theorem 1.10.1) - the equation $\mathbf{x} - C\mathbf{x} = \mathbf{d}$ has a unique solution $\mathbf{x} = (I - C)^{-1}\mathbf{d}$ for every demand vector $\mathbf{d}$.

True-False Exercises

(a) False. Sectors that do *not* produce outputs are called open sectors.

(b) True.

(c) False. The ith row vector of a consumption matrix contains the monetary values required of the ith sector by the other sectors for each of them to produce one monetary unit of output.

(d) True. This follows from Theorem 1.10.1.

(e) True.

Chapter 1 Supplementary Exercises

1. The corresponding system of linear equations is

$$\begin{array}{rcrcrcrcr} 3x_1 & - & x_2 & & & + & 4x_4 & = & 1 \\ 2x_1 & & & + & 3x_3 & + & 3x_4 & = & -1 \end{array}$$

$\begin{bmatrix} 3 & -1 & 0 & 4 & 1 \\ 2 & 0 & 3 & 3 & -1 \end{bmatrix}$ ⟵ The original augmented matrix.

$\begin{bmatrix} 1 & -1 & -3 & 1 & 2 \\ 2 & 0 & 3 & 3 & -1 \end{bmatrix}$ ⟵ -1 times the second row was added to the first row.

$\begin{bmatrix} 1 & -1 & -3 & 1 & 2 \\ 0 & 2 & 9 & 1 & -5 \end{bmatrix}$ ⟵ -2 times the first row was added to the second row.

$\begin{bmatrix} 1 & -1 & -3 & 1 & 2 \\ 0 & 1 & \frac{9}{2} & \frac{1}{2} & -\frac{5}{2} \end{bmatrix}$ ⟵ The second row was multiplied by $\frac{1}{2}$.

This matrix is in row echelon form. It corresponds to the system of equations

$$\begin{array}{rcrcrcrcr} x_1 & - & x_2 & - & 3x_3 & + & x_4 & = & 2 \\ & & x_2 & + & \frac{9}{2}x_3 & + & \frac{1}{2}x_4 & = & -\frac{5}{2} \end{array}$$

Solve the equations for the leading variables

$$x_1 = x_2 + 3x_3 - x_4 + 2$$
$$x_2 = -\frac{9}{2}x_3 - \frac{1}{2}x_4 - \frac{5}{2}$$

then substitute the second equation into the first

$$x_1 = -\frac{3}{2}x_3 - \frac{3}{2}x_4 - \frac{1}{2}$$
$$x_2 = -\frac{9}{2}x_3 - \frac{1}{2}x_4 - \frac{5}{2}$$

If we assign x_3 and x_4 the arbitrary values s and t, respectively, the general solution is given by the formulas

$$x_1 = -\frac{3}{2}s - \frac{3}{2}t - \frac{1}{2}, \qquad x_2 = -\frac{9}{2}s - \frac{1}{2}t - \frac{5}{2}, \qquad x_3 = s, \qquad x_4 = t$$

3. The corresponding system of linear equations is

$$\begin{array}{rcrcrcr} 2x_1 & - & 4x_2 & + & x_3 & = & 6 \\ -4x_1 & & & + & 3x_3 & = & -1 \\ & & x_2 & - & x_3 & = & 3 \end{array}$$

$\begin{bmatrix} 2 & -4 & 1 & 6 \\ -4 & 0 & 3 & -1 \\ 0 & 1 & -1 & 3 \end{bmatrix}$ ⟵ The original augmented matrix.

$\begin{bmatrix} 1 & -2 & \frac{1}{2} & 3 \\ -4 & 0 & 3 & -1 \\ 0 & 1 & -1 & 3 \end{bmatrix}$ ⟵ The first row was multiplied by $\frac{1}{2}$.

$$\begin{bmatrix} 1 & -2 & \frac{1}{2} & 3 \\ 0 & -8 & 5 & 11 \\ 0 & 1 & -1 & 3 \end{bmatrix} \longleftarrow \text{4 times the first row was added to the second row.}$$

$$\begin{bmatrix} 1 & -2 & \frac{1}{2} & 3 \\ 0 & 1 & -1 & 3 \\ 0 & -8 & 5 & 11 \end{bmatrix} \longleftarrow \text{The second and third rows were interchanged.}$$

$$\begin{bmatrix} 1 & -2 & \frac{1}{2} & 3 \\ 0 & 1 & -1 & 3 \\ 0 & 0 & -3 & 35 \end{bmatrix} \longleftarrow \text{8 times the second row was added to the third row.}$$

$$\begin{bmatrix} 1 & -2 & \frac{1}{2} & 3 \\ 0 & 1 & -1 & 3 \\ 0 & 0 & 1 & -\frac{35}{3} \end{bmatrix} \longleftarrow \text{The third row was multiplied by } -\tfrac{1}{3}.$$

This matrix is in row echelon form. It corresponds to the system of equations

$$\begin{aligned} x_1 - 2x_2 + \frac{1}{2}x_3 &= 3 \\ x_2 - x_3 &= 3 \\ x_3 &= -\frac{35}{3} \end{aligned}$$

Solve the equations for the leading variables

$$\begin{aligned} x_1 &= 2x_2 - \frac{1}{2}x_3 + 3 \\ x_2 &= x_3 + 3 \\ x_3 &= -\frac{35}{3} \end{aligned}$$

then finish back-substituting to obtain the unique solution

$$x_1 = -\frac{17}{2}, \quad x_2 = -\frac{26}{3}, \quad x_3 = -\frac{35}{3}$$

5.

$$\begin{bmatrix} \frac{3}{5} & -\frac{4}{5} & x \\ \frac{4}{5} & \frac{3}{5} & y \end{bmatrix} \longleftarrow \text{The augmented matrix corresponding to the system.}$$

$$\begin{bmatrix} 1 & -\frac{4}{3} & \frac{5}{3}x \\ \frac{4}{5} & \frac{3}{5} & y \end{bmatrix} \longleftarrow \text{The first row was multiplied by } \tfrac{5}{3}.$$

$$\begin{bmatrix} 1 & -\frac{4}{3} & \frac{5}{3}x \\ 0 & \frac{5}{3} & -\frac{4}{3}x + y \end{bmatrix} \longleftarrow -\tfrac{4}{5} \text{ times the first row was added to the second row.}$$

$$\begin{bmatrix} 1 & -\frac{4}{3} & \frac{5}{3}x \\ 0 & 1 & -\frac{4}{5}x+\frac{3}{5}y \end{bmatrix}$$ ⟵ The second row was multiplied by $\frac{3}{5}$.

$$\begin{bmatrix} 1 & 0 & \frac{3}{5}x+\frac{4}{5}y \\ 0 & 1 & -\frac{4}{5}x+\frac{3}{5}y \end{bmatrix}$$ ⟵ $\frac{4}{3}$ times the second row was added to the first row.

The system has exactly one solution: $x' = \frac{3}{5}x + \frac{4}{5}y$ and $y' = -\frac{4}{5}x + \frac{3}{5}y$.

7.

$$\begin{bmatrix} 1 & 1 & 1 & 9 \\ 1 & 5 & 10 & 44 \end{bmatrix}$$ ⟵ The original augmented matrix.

$$\begin{bmatrix} 1 & 1 & 1 & 9 \\ 0 & 4 & 9 & 35 \end{bmatrix}$$ ⟵ -1 times the first row was added to the second row.

$$\begin{bmatrix} 1 & 1 & 1 & 9 \\ 0 & 1 & \frac{9}{4} & \frac{35}{4} \end{bmatrix}$$ ⟵ The second row was multiplied by $\frac{1}{4}$.

$$\begin{bmatrix} 1 & 0 & -\frac{5}{4} & \frac{1}{4} \\ 0 & 1 & \frac{9}{4} & \frac{35}{4} \end{bmatrix}$$ ⟵ -1 times the second row was added to the first row.

If we assign z an arbitrary value t, the general solution is given by the formulas

$$x = \frac{1}{4} + \frac{5}{4}t, \qquad y = \frac{35}{4} - \frac{9}{4}t, \qquad z = t$$

The positivity of the three variables requires that $\frac{1}{4} + \frac{5}{4}t > 0$, $\frac{35}{4} - \frac{9}{4}t > 0$, and $t > 0$. The first inequality can be rewritten as $t > -\frac{1}{4}$, while the second inequality is equivalent to $t < \frac{35}{9}$. All three unknowns are positive whenever $0 < t < \frac{35}{9}$. There are three integer values of $t = z$ in this interval: 1, 2, and 3. Of those, only $z = t = 3$ yields integer values for the remaining variables: $x = 4, y = 2$.

9.

$$\begin{bmatrix} a & 0 & b & 2 \\ a & a & 4 & 4 \\ 0 & a & 2 & b \end{bmatrix}$$ ⟵ The augmented matrix for the system.

$$\begin{bmatrix} a & 0 & b & 2 \\ 0 & a & 4-b & 2 \\ 0 & a & 2 & b \end{bmatrix}$$ ⟵ -1 times the first row was added to the second row.

$$\begin{bmatrix} a & 0 & b & 2 \\ 0 & a & 4-b & 2 \\ 0 & 0 & b-2 & b-2 \end{bmatrix}$$ ⟵ -1 times the second row was added to the third row.

(a) the system has a unique solution if $a \neq 0$ and $b \neq 2$ (multiplying the rows by $\frac{1}{a}, \frac{1}{a}$, and $\frac{1}{b-2}$, respectively, yields a row echelon form of the augmented matrix $\begin{bmatrix} 1 & 0 & \frac{b}{a} & \frac{2}{a} \\ 0 & 1 & \frac{4-b}{a} & \frac{2}{a} \\ 0 & 0 & 1 & 1 \end{bmatrix}$).

(b) the system has a one-parameter solution if $a \neq 0$ and $b = 2$ (multiplying the first two rows by $\frac{1}{a}$ yields a reduced row echelon form of the augmented matrix $\begin{bmatrix} 1 & 0 & \frac{2}{a} & \frac{2}{a} \\ 0 & 1 & \frac{2}{a} & \frac{2}{a} \\ 0 & 0 & 0 & 0 \end{bmatrix}$).

(c) the system has a two-parameter solution if $a = 0$ and $b = 2$ (the reduced row echelon form of the augmented matrix is $\begin{bmatrix} 0 & 0 & 1 & 1 \\ 0 & 0 & 0 & 0 \\ 0 & 0 & 0 & 0 \end{bmatrix}$).

(d) the system has no solution if $a = 0$ and $b \neq 2$ (the reduced row echelon form of the augmented matrix is $\begin{bmatrix} 0 & 0 & 1 & 0 \\ 0 & 0 & 0 & 1 \\ 0 & 0 & 0 & 0 \end{bmatrix}$).

11. For the product AKB to be defined, K must be a 2×2 matrix. Letting $K = \begin{bmatrix} a & b \\ c & d \end{bmatrix}$ we can write

$$ABC = \begin{bmatrix} 1 & 4 \\ -2 & 3 \\ 1 & -2 \end{bmatrix} \begin{bmatrix} a & b \\ c & d \end{bmatrix} \begin{bmatrix} 2 & 0 & 0 \\ 0 & 1 & -1 \end{bmatrix} = \begin{bmatrix} 1 & 4 \\ -2 & 3 \\ 1 & -2 \end{bmatrix} \begin{bmatrix} 2a & b & -b \\ 2c & d & -d \end{bmatrix} =$$

$$\begin{bmatrix} 2a + 8c & b + 4d & -b - 4d \\ -4a + 6c & -2b + 3d & 2b - 3d \\ 2a - 4c & b - 2d & -b + 2d \end{bmatrix}.$$

The matrix equation $AKB = C$ can be rewritten as a system of nine linear equations

$$\begin{array}{rrrrcr} 2a & & +\ 8c & & = & 8 \\ & b & & +\ 4d & = & 6 \\ & -\ b & & -\ 4d & = & -6 \\ -4a & & +\ 6c & & = & 6 \\ & -\ 2b & & +\ 3d & = & -1 \\ & 2b & & -\ 3d & = & 1 \\ 2a & & -\ 4c & & = & -4 \\ & b & & -\ 2d & = & 0 \\ & -\ b & & +\ 2d & = & 0 \end{array}$$

which has a unique solution $a = 0, b = 2, c = 1, d = 1$. (An easy way to solve this system is to first split it into two smaller systems. The system $2a + 8c = 8, -4a + 6c = 6, 2a - 4c = -4$ involves a and c only, whereas the remaining six equations involve just b and d.) We conclude that $K = \begin{bmatrix} 0 & 2 \\ 1 & 1 \end{bmatrix}$.

13. **(a)** X must be a 2×3 matrix. Letting $X = \begin{bmatrix} a & b & c \\ d & e & f \end{bmatrix}$ we can write

$$X \begin{bmatrix} -1 & 0 & 1 \\ 1 & 1 & 0 \\ 3 & 1 & -1 \end{bmatrix} = \begin{bmatrix} a & b & c \\ d & e & f \end{bmatrix} \begin{bmatrix} -1 & 0 & 1 \\ 1 & 1 & 0 \\ 3 & 1 & -1 \end{bmatrix} = \begin{bmatrix} -a + b + 3c & b + c & a - c \\ -d + e + 3f & e + f & d - f \end{bmatrix}$$

therefore the given matrix equation can be rewritten as a system of linear equations:

$$\begin{array}{rcrcrcrcrcrcrcr}
-a & + & b & + & 3c & & & & & & & = & 1\\
 & & b & + & c & & & & & & & = & 2\\
a & & & - & c & & & & & & & = & 0\\
 & & & & & - & d & + & e & + & 3f & = & -3\\
 & & & & & & & & e & + & f & = & 1\\
 & & & & & & d & & & - & f & = & 5
\end{array}$$

The augmented matrix of this system has the reduced row echelon form $\begin{bmatrix}1&0&0&0&0&0&-1\\0&1&0&0&0&0&3\\0&0&1&0&0&0&-1\\0&0&0&1&0&0&6\\0&0&0&0&1&0&0\\0&0&0&0&0&1&1\end{bmatrix}$ so the system has a unique solution $a=-1, b=3, c=-1, d=6, e=0, f=1$ and $X=\begin{bmatrix}-1&3&-1\\6&0&1\end{bmatrix}$.

(An alternative to dealing with this large system is to split it into two smaller systems instead: the first three equations involve a, b, and c only, whereas the remaining three equations involve just d, e, and f. Since the coefficient matrix for both systems is the same, we can follow the procedure of Example 2 in Section 1.6; the reduced row echelon form of the matrix $\left[\begin{array}{rrr|r|r}-1&1&3&1&-3\\0&1&1&2&1\\1&0&-1&0&5\end{array}\right]$ is $\left[\begin{array}{rrr|r|r}1&0&0&-1&6\\0&1&0&3&0\\0&0&1&-1&1\end{array}\right]$.)

Yet another way of solving this problem would be to determine the inverse $\begin{bmatrix}-1&0&1\\1&1&0\\3&1&-1\end{bmatrix}^{-1}=\begin{bmatrix}1&-1&-1\\-1&2&-1\\2&-1&1\end{bmatrix}$ using the method introduced in Section 1.5, then multiply both sides of the given matrix equation on the right by this inverse to determine X:

$$X=\begin{bmatrix}1&2&0\\-3&1&5\end{bmatrix}\begin{bmatrix}1&-1&-1\\-1&2&-1\\2&-1&1\end{bmatrix}=\begin{bmatrix}-1&3&-1\\6&0&1\end{bmatrix}$$

(b) X must be a 2×2 matrix. Letting $X=\begin{bmatrix}a&b\\c&d\end{bmatrix}$ we can write

$$X\begin{bmatrix}1&-1&2\\3&0&1\end{bmatrix}=\begin{bmatrix}a&b\\c&d\end{bmatrix}\begin{bmatrix}1&-1&2\\3&0&1\end{bmatrix}=\begin{bmatrix}a+3b&-a&2a+b\\c+3d&-c&2c+d\end{bmatrix}$$

therefore the given matrix equation can be rewritten as a system of linear equations:

$$\begin{array}{rcrcrcrcl}
a & + & 3b & & & & & = & -5\\
-a & & & & & & & = & -1\\
2a & + & b & & & & & = & 0\\
 & & & & c & + & 3d & = & 6\\
 & & & - & c & & & = & -3\\
 & & & & 2c & + & d & = & 7
\end{array}$$

The augmented matrix of this system has the reduced row echelon form $\begin{bmatrix}1&0&0&0&1\\0&1&0&0&-2\\0&0&1&0&3\\0&0&0&1&1\\0&0&0&0&0\\0&0&0&0&0\end{bmatrix}$ so the system has a unique solution $a=1, b=-2, c=3, d=1$. We conclude that $X=\begin{bmatrix}1&-2\\3&1\end{bmatrix}$.

(An alternative to dealing with this large system is to split it into two smaller systems instead: the first three equations involve a and b only, whereas the remaining three equations involve just c and d. Since the coefficient matrix for both systems is the same, we can follow the procedure of Example 2 in Section 1.6; the reduced row echelon form of the matrix $\left[\begin{array}{cc|c|c} 1 & 3 & -5 & 6 \\ -1 & 0 & -1 & -3 \\ 2 & 1 & 0 & 7 \end{array}\right]$ is $\left[\begin{array}{cc|c|c} 1 & 0 & 1 & 3 \\ 0 & 1 & -2 & 1 \\ 0 & 0 & 0 & 0 \end{array}\right]$.)

(c) X must be a 2×2 matrix. Letting $X = \begin{bmatrix} a & b \\ c & d \end{bmatrix}$ we can write

$$\begin{aligned} \begin{bmatrix} 3 & 1 \\ -1 & 2 \end{bmatrix} X - X \begin{bmatrix} 1 & 4 \\ 2 & 0 \end{bmatrix} &= \begin{bmatrix} 3 & 1 \\ -1 & 2 \end{bmatrix}\begin{bmatrix} a & b \\ c & d \end{bmatrix} - \begin{bmatrix} a & b \\ c & d \end{bmatrix}\begin{bmatrix} 1 & 4 \\ 2 & 0 \end{bmatrix} \\ &= \begin{bmatrix} 3a + c & 3b + d \\ -a + 2c & -b + 2d \end{bmatrix} - \begin{bmatrix} a + 2b & 4a \\ c + 2d & 4c \end{bmatrix} \\ &= \begin{bmatrix} 2a - 2b + c & -4a + 3b + d \\ -a + c - 2d & -b - 4c + 2d \end{bmatrix} \end{aligned}$$

therefore the given matrix equation can be rewritten as a system of linear equations:

$$\begin{array}{rcrcrcrcr} 2a & - & 2b & + & c & & & = & 2 \\ -4a & + & 3b & & & + & d & = & -2 \\ -a & & & + & c & - & 2d & = & 5 \\ & - & b & - & 4c & + & 2d & = & 4 \end{array}$$

The augmented matrix of this system has the reduced row echelon form $\begin{bmatrix} 1 & 0 & 0 & 0 & -\frac{113}{37} \\ 0 & 1 & 0 & 0 & -\frac{160}{37} \\ 0 & 0 & 1 & 0 & -\frac{20}{37} \\ 0 & 0 & 0 & 1 & -\frac{46}{37} \end{bmatrix}$

so the system has a unique solution $a = -\frac{113}{37}, b = -\frac{160}{37}, c = -\frac{20}{37}, d = -\frac{46}{37}$.

We conclude that $X = \begin{bmatrix} -\frac{113}{37} & -\frac{160}{37} \\ -\frac{20}{37} & -\frac{46}{37} \end{bmatrix}$.

15. We are looking for a polynomial of the form

$$p(x) = ax^2 + bx + c$$

such that $p(1) = 2,\ p(-1) = 6,$ and $p(2) = 3$. We obtain a linear system

$$\begin{array}{rcrcrcr} a & + & b & + & c & = & 2 \\ a & - & b & + & c & = & 6 \\ 4a & + & 2b & + & c & = & 3 \end{array}$$

Its augmented matrix has the reduced row echelon form $\begin{bmatrix} 1 & 0 & 0 & 1 \\ 0 & 1 & 0 & -2 \\ 0 & 0 & 1 & 3 \end{bmatrix}$.

There is a unique solution $a = 1,\ b = -2,\ c = 3$.

17. When multiplying the matrix J_n by itself, each entry in the product equals n. Therefore, $J_nJ_n = nJ_n$.

$(I - J_n)\left(I - \frac{1}{n-1}J_n\right)$

$$= I^2 - I\frac{1}{n-1}J_n - J_nI + J_n\frac{1}{n-1}J_n \quad \longleftarrow \quad \text{Theorem 1.4.1(f) and (g)}$$

$$= I - \frac{1}{n-1}J_n - J_n + J_n\frac{1}{n-1}J_n \quad \longleftarrow \quad \text{Property } AI = IA = A \text{ on p. 43}$$

$$= I - \frac{1}{n-1}J_n - J_n + \frac{1}{n-1}J_nJ_n \quad \longleftarrow \quad \text{Theorem 1.4.1(m)}$$

$$= I - \frac{1}{n-1}J_n - J_n + \frac{n}{n-1}J_n \quad \longleftarrow \quad J_nJ_n = nJ_n$$

$$= I + \left(\frac{-1}{n-1} - 1 + \frac{n}{n-1}\right)J_n \quad \longleftarrow \quad \text{Theorem 1.4.1(j) and (k)}$$

$$= I + \left(\frac{-1}{n-1} - \frac{n-1}{n-1} + \frac{n}{n-1}\right)J_n$$

$$= I$$

CHAPTER 2: DETERMINANTS

2.1 Determinants by Cofactor Expansion

1. $M_{11} = \begin{vmatrix} 1 & -2 & 3 \\ 6 & 7 & -1 \\ -3 & 1 & 4 \end{vmatrix} = \begin{vmatrix} 7 & -1 \\ 1 & 4 \end{vmatrix} = 29$ $\quad C_{11} = (-1)^{1+1}M_{11} = M_{11} = 29$

$M_{12} = \begin{vmatrix} 1 & -2 & 3 \\ 6 & 7 & -1 \\ -3 & 1 & 4 \end{vmatrix} = \begin{vmatrix} 6 & -1 \\ -3 & 4 \end{vmatrix} = 21$ $\quad C_{12} = (-1)^{1+2}M_{12} = -M_{12} = -21$

$M_{13} = \begin{vmatrix} 1 & -2 & 3 \\ 6 & 7 & -1 \\ -3 & 1 & 4 \end{vmatrix} = \begin{vmatrix} 6 & 7 \\ -3 & 1 \end{vmatrix} = 27$ $\quad C_{13} = (-1)^{1+3}M_{13} = M_{13} = 27$

$M_{21} = \begin{vmatrix} 1 & -2 & 3 \\ 6 & 7 & -1 \\ -3 & 1 & 4 \end{vmatrix} = \begin{vmatrix} -2 & 3 \\ 1 & 4 \end{vmatrix} = -11$ $\quad C_{21} = (-1)^{2+1}M_{21} = -M_{21} = 11$

$M_{22} = \begin{vmatrix} 1 & -2 & 3 \\ 6 & 7 & -1 \\ -3 & 1 & 4 \end{vmatrix} = \begin{vmatrix} 1 & 3 \\ -3 & 4 \end{vmatrix} = 13$ $\quad C_{22} = (-1)^{2+2}M_{22} = M_{22} = 13$

$M_{23} = \begin{vmatrix} 1 & -2 & 3 \\ 6 & 7 & -1 \\ -3 & 1 & 4 \end{vmatrix} = \begin{vmatrix} 1 & -2 \\ -3 & 1 \end{vmatrix} = -5$ $\quad C_{23} = (-1)^{2+3}M_{23} = -M_{23} = 5$

$M_{31} = \begin{vmatrix} 1 & -2 & 3 \\ 6 & 7 & -1 \\ -3 & 1 & 4 \end{vmatrix} = \begin{vmatrix} -2 & 3 \\ 7 & -1 \end{vmatrix} = -19$ $\quad C_{31} = (-1)^{3+1}M_{31} = M_{31} = -19$

$M_{32} = \begin{vmatrix} 1 & -2 & 3 \\ 6 & 7 & -1 \\ -3 & 1 & 4 \end{vmatrix} = \begin{vmatrix} 1 & 3 \\ 6 & -1 \end{vmatrix} = -19$ $\quad C_{32} = (-1)^{3+2}M_{32} = -M_{32} = 19$

$M_{33} = \begin{vmatrix} 1 & -2 & 3 \\ 6 & 7 & -1 \\ -3 & 1 & 4 \end{vmatrix} = \begin{vmatrix} 1 & -2 \\ 6 & 7 \end{vmatrix} = 19$ $\quad C_{33} = (-1)^{3+3}M_{33} = M_{33} = 19$

3. **(a)** $M_{13} = \begin{vmatrix} 0 & 0 & 3 \\ 4 & 1 & 14 \\ 4 & 1 & 2 \end{vmatrix} = 0\begin{vmatrix} 1 & 14 \\ 1 & 2 \end{vmatrix} - 0\begin{vmatrix} 4 & 14 \\ 4 & 2 \end{vmatrix} + 3\begin{vmatrix} 4 & 1 \\ 4 & 1 \end{vmatrix}$ ← cofactor expansion along the first row

$= 0 - 0 + 3(0) = 0$

$C_{13} = (-1)^{1+3}M_{13} = M_{13} = 0$

(b) $M_{23} = \begin{vmatrix} 4 & -1 & 6 \\ 4 & 1 & 14 \\ 4 & 1 & 2 \end{vmatrix} = 4\begin{vmatrix} 1 & 14 \\ 1 & 2 \end{vmatrix} - (-1)\begin{vmatrix} 4 & 14 \\ 4 & 2 \end{vmatrix} + 6\begin{vmatrix} 4 & 1 \\ 4 & 1 \end{vmatrix}$ ← cofactor expansion along the first row

$$= 4(-12) + 1(-48) + 6(0) = -96$$

$$C_{23} = (-1)^{2+3}M_{23} = -M_{23} = 96$$

(c) $M_{22} = \begin{vmatrix} 4 & 1 & 6 \\ 4 & 0 & 14 \\ 4 & 3 & 2 \end{vmatrix} = -4\begin{vmatrix} 1 & 6 \\ 3 & 2 \end{vmatrix} + 0\begin{vmatrix} 4 & 6 \\ 4 & 2 \end{vmatrix} - 14\begin{vmatrix} 4 & 1 \\ 4 & 3 \end{vmatrix}$ ← cofactor expansion along the second row

$$= -4(-16) + 0 - 14(8) = -48$$

$$C_{22} = (-1)^{2+2}M_{22} = M_{22} = -48$$

(d) $M_{21} = \begin{vmatrix} -1 & 1 & 6 \\ 1 & 0 & 14 \\ 1 & 3 & 2 \end{vmatrix} = -1\begin{vmatrix} 1 & 6 \\ 3 & 2 \end{vmatrix} + 0\begin{vmatrix} -1 & 6 \\ 1 & 2 \end{vmatrix} - 14\begin{vmatrix} -1 & 1 \\ 1 & 3 \end{vmatrix}$ ← cofactor expansion along the second row

$$= -1(-16) + 0 - 14(-4) = 72$$

$$C_{21} = (-1)^{2+1}M_{21} = -M_{21} = -72$$

5. $\begin{vmatrix} 3 & 5 \\ -2 & 4 \end{vmatrix} = (3)(4) - (5)(-2) = 12 + 10 = 22 \neq 0$. Inverse: $\frac{1}{22}\begin{bmatrix} 4 & -5 \\ 2 & 3 \end{bmatrix} = \begin{bmatrix} \frac{2}{11} & \frac{-5}{22} \\ \frac{1}{11} & \frac{3}{22} \end{bmatrix}$

7. $\begin{vmatrix} -5 & 7 \\ -7 & -2 \end{vmatrix} = (-5)(-2) - (7)(-7) = 10 + 49 = 59 \neq 0$. Inverse: $\frac{1}{59}\begin{bmatrix} -2 & -7 \\ 7 & -5 \end{bmatrix} = \begin{bmatrix} \frac{-2}{59} & \frac{-7}{59} \\ \frac{7}{59} & \frac{-5}{59} \end{bmatrix}$

9. $\begin{vmatrix} a-3 & 5 \\ -3 & a-2 \end{vmatrix} = \begin{vmatrix} a-3 & 5 \\ -3 & a-2 \end{vmatrix} = (a-3)(a-2) - 5(-3) = a^2 - 5a + 6 + 15 = a^2 - 5a + 21$

11. $\begin{vmatrix} -2 & 1 & 4 \\ 3 & 5 & -7 \\ 1 & 6 & 2 \end{vmatrix} = \begin{vmatrix} -2 & 1 & 4 \\ 3 & 5 & -7 \\ 1 & 6 & 2 \end{vmatrix}\begin{matrix} -2 & 1 \\ 3 & 5 \\ 1 & 6 \end{matrix} = [-20 - 7 + 72] - [20 + 84 + 6] = -65$

13. $\begin{vmatrix} 3 & 0 & 0 \\ 2 & -1 & 5 \\ 1 & 9 & -4 \end{vmatrix} = \begin{vmatrix} 3 & 0 & 0 \\ 2 & -1 & 5 \\ 1 & 9 & -4 \end{vmatrix}\begin{matrix} 3 & 0 \\ 2 & -1 \\ 1 & 9 \end{matrix} = [12 + 0 + 0] - [0 + 135 + 0] = -123$

15. $\det(A) = \begin{vmatrix} \lambda - 2 & 1 \\ -5 & \lambda + 4 \end{vmatrix} = (\lambda - 2)(\lambda + 4) - (1)(-5) = \lambda^2 + 2\lambda - 3 = (\lambda + 3)(\lambda - 1)$

The determinant is zero if $\lambda = -3$ or $\lambda = 1$.

17. $\det(A) = \begin{vmatrix} \lambda - 1 & 0 \\ 2 & \lambda + 1 \end{vmatrix} = (\lambda - 1)(\lambda + 1)$

The determinant is zero if $\lambda = 1$ or $\lambda = -1$.

19. **(a)** $3\begin{vmatrix} -1 & 5 \\ 9 & -4 \end{vmatrix} - 0 + 0 = 3(-41) = -123$

(b) $3\begin{vmatrix} -1 & 5 \\ 9 & -4 \end{vmatrix} - 2\begin{vmatrix} 0 & 0 \\ 9 & -4 \end{vmatrix} + 1\begin{vmatrix} 0 & 0 \\ -1 & 5 \end{vmatrix} = 3(-41) - 2(0) + 1(0) = -123$

(c) $-2\begin{vmatrix} 0 & 0 \\ 9 & -4 \end{vmatrix} + (-1)\begin{vmatrix} 3 & 0 \\ 1 & -4 \end{vmatrix} - 5\begin{vmatrix} 3 & 0 \\ 1 & 9 \end{vmatrix} = -2(0) - 1(-12) - 5(27) = -123$

(d) $-0 + (-1)\begin{vmatrix} 3 & 0 \\ 1 & -4 \end{vmatrix} - 9\begin{vmatrix} 3 & 0 \\ 2 & 5 \end{vmatrix} = -1(-12) - 9(15) = -123$

(e) $1\begin{vmatrix} 0 & 0 \\ -1 & 5 \end{vmatrix} - 9\begin{vmatrix} 3 & 0 \\ 2 & 5 \end{vmatrix} + (-4)\begin{vmatrix} 3 & 0 \\ 2 & -1 \end{vmatrix} = 1(0) - 9(15) - 4(-3) = -123$

(f) $0 - 5\begin{vmatrix} 3 & 0 \\ 1 & 9 \end{vmatrix} + (-4)\begin{vmatrix} 3 & 0 \\ 2 & -1 \end{vmatrix} = -5(27) - 4(-3) = -123$

21. Calculate the determinant by a cofactor expansion along the second column:

$$-0 + 5\begin{vmatrix} -3 & 7 \\ -1 & 5 \end{vmatrix} - 0 = 5(-8) = -40$$

23. Calculate the determinant by a cofactor expansion along the first column:

$$1\begin{vmatrix} k & k^2 \\ k & k^2 \end{vmatrix} - 1\begin{vmatrix} k & k^2 \\ k & k^2 \end{vmatrix} + 1\begin{vmatrix} k & k^2 \\ k & k^2 \end{vmatrix} = 1(0) - 1(0) + 1(0) = 0$$

25. Calculate the determinant by a cofactor expansion along the third column:

$$\det(A) = 0 - 0 + (-3)\begin{vmatrix} 3 & 3 & 5 \\ 2 & 2 & -2 \\ 2 & 10 & 2 \end{vmatrix} - 3\begin{vmatrix} 3 & 3 & 5 \\ 2 & 2 & -2 \\ 4 & 1 & 0 \end{vmatrix}$$

Calculate the determinants in the third and fourth terms by a cofactor expansion along the first row:

$$\begin{vmatrix} 3 & 3 & 5 \\ 2 & 2 & -2 \\ 2 & 10 & 2 \end{vmatrix} = 3\begin{vmatrix} 2 & -2 \\ 10 & 2 \end{vmatrix} - 3\begin{vmatrix} 2 & -2 \\ 2 & 2 \end{vmatrix} + 5\begin{vmatrix} 2 & 2 \\ 2 & 10 \end{vmatrix} = 3(24) - 3(8) + 5(16) = 128$$

$$\begin{vmatrix} 3 & 3 & 5 \\ 2 & 2 & -2 \\ 4 & 1 & 0 \end{vmatrix} = 3\begin{vmatrix} 2 & -2 \\ 1 & 0 \end{vmatrix} - 3\begin{vmatrix} 2 & -2 \\ 4 & 0 \end{vmatrix} + 5\begin{vmatrix} 2 & 2 \\ 4 & 1 \end{vmatrix} = 3(2) - 3(8) + 5(-6) = -48$$

Therefore $\det(A) = 0 - 0 - 3(128) - 3(-48) = -240$.

27. By Theorem 2.1.2, determinant of a diagonal matrix is the product of the entries on the main diagonal: $\det(A) = (1)(-1)(1) = -1$.

29. By Theorem 2.1.2, determinant of a lower triangular matrix is the product of the entries on the main diagonal: $\det(A) = (0)(2)(3)(8) = 0$.

31. By Theorem 2.1.2, determinant of an upper triangular matrix is the product of the entries on the main diagonal: $\det(A) = (1)(1)(2)(3) = 6$.

33. **(a)** $\begin{vmatrix} \sin\theta & \cos\theta \\ -\cos\theta & \sin\theta \end{vmatrix} = (\sin\theta)(\sin\theta) - (\cos\theta)(-\cos\theta) = \sin^2\theta + \cos^2\theta = 1$

(b) Calculate the determinant by a cofactor expansion along the third column:

$$0-0+1\begin{vmatrix}\sin\theta & \cos\theta\\ -\cos\theta & \sin\theta\end{vmatrix}=0-0+(1)(1)=1 \text{ (we used the result of part (a))}$$

35. The minor M_{11} in both determinants is $\begin{vmatrix}1 & f\\ 0 & 1\end{vmatrix}=1$. Expanding both determinants along the first row yields $d_1+\lambda=d_2$.

True-False Exercises

(a) False. The determinant is $ad-bc$.

(b) False. E.g., $\det(I_2)=\det(I_3)=1$.

(c) True. If $i+j$ is even then $(-1)^{i+j}=1$ therefore $C_{ij}=(-1)^{i+j}M_{ij}=M_{ij}$.

(d) True. Let $A=\begin{bmatrix}a & b & c\\ b & d & e\\ c & e & f\end{bmatrix}$.

Then $C_{12}=(-1)^{1+2}\begin{vmatrix}b & e\\ c & f\end{vmatrix}=-(bf-ec)$ and $C_{21}=(-1)^{2+1}\begin{vmatrix}b & c\\ e & f\end{vmatrix}=-(bf-ce)$ therefore $C_{12}=C_{21}$. In the same way, one can show $C_{13}=C_{31}$ and $C_{23}=C_{32}$.

(e) True. This follows from Theorem 2.1.1.

(f) True. In formulas (7) and (8), each cofactor C_{ij} is zero.

(g) False. The determinant of a lower triangular matrix is the *product* of the entries along the main diagonal.

(h) False. E.g. $\det(2I_2)=4\neq 2=2\det(I_2)$.

(i) False. E.g., $\det(I_2+I_2)=4\neq 2=\det(I_2)+\det(I_2)$.

(j) True. $\det\left(\begin{bmatrix}a & b\\ c & d\end{bmatrix}^2\right)=\begin{vmatrix}a^2+bc & ab+bd\\ ac+cd & bc+d^2\end{vmatrix}=(a^2+bc)(bc+d^2)-(ab+bd)(ac+cd)$
$=a^2bc+a^2d^2+b^2c^2+bcd^2-a^2bc-abcd-abcd-bcd^2=a^2d^2+b^2c^2-2abcd$.

$\begin{vmatrix}a & b\\ c & d\end{vmatrix}^2=(ad-bc)^2=a^2d^2-2adbc+b^2c^2$ therefore $\det\left(\begin{bmatrix}a & b\\ c & d\end{bmatrix}^2\right)=\left(\det\left(\begin{bmatrix}a & b\\ c & d\end{bmatrix}\right)\right)^2$.

2.2 Evaluating Determinants by Row Reduction

1. $\det(A)=\begin{vmatrix}-2 & 3\\ 1 & 4\end{vmatrix}=(-2)(4)-(3)(1)=-11$; $\det(A^T)=\begin{vmatrix}-2 & 1\\ 3 & 4\end{vmatrix}=(-2)(4)-(1)(3)=-11$

3. $\det(A)=\begin{vmatrix}2 & -1 & 3\\ 1 & 2 & 4\\ 5 & -3 & 6\end{vmatrix}=[24-20-9]-[30-24-6]=-5$;

$\det(A^T)=\begin{vmatrix}2 & 1 & 5\\ -1 & 2 & -3\\ 3 & 4 & 6\end{vmatrix}=[24-9-20]-[30-24-6]=-5$ (we used the arrow technique)

5. The third row of I_4 was multiplied by -5. By Theorem 2.2.4, the determinant equals -5.

7. The second and the third rows of I_4 were interchanged. By Theorem 2.2.4, the determinant equals -1.

9. $\begin{vmatrix} 3 & -6 & 9 \\ -2 & 7 & -2 \\ 0 & 1 & 5 \end{vmatrix} = 3\begin{vmatrix} 1 & -2 & 3 \\ -2 & 7 & -2 \\ 0 & 1 & 5 \end{vmatrix}$ ← A common factor of 3 from the first row was taken through the determinant sign.

$= 3\begin{vmatrix} 1 & -2 & 3 \\ 0 & 3 & 4 \\ 0 & 1 & 5 \end{vmatrix}$ ← 2 times the first row was added to the second row.

$= 3(-1)\begin{vmatrix} 1 & -2 & 3 \\ 0 & 1 & 5 \\ 0 & 3 & 4 \end{vmatrix}$ ← The second and third rows were interchanged.

$= (3)(-1)\begin{vmatrix} 1 & -2 & 3 \\ 0 & 1 & 5 \\ 0 & 0 & -11 \end{vmatrix}$ ← -3 times the second row was added to the third row.

$= (3)(-1)(-11)\begin{vmatrix} 1 & -2 & 3 \\ 0 & 1 & 5 \\ 0 & 0 & 1 \end{vmatrix}$ ← A common factor of -11 from the last row was taken through the determinant sign.

$= (3)(-1)(-11)(1) = 33$

Another way to evaluate the determinant would be to use cofactor expansion along the first column after the second step above:

$$\begin{vmatrix} 3 & -6 & 9 \\ -2 & 7 & -2 \\ 0 & 1 & 5 \end{vmatrix} = 3\begin{vmatrix} 1 & -2 & 3 \\ 0 & 3 & 4 \\ 0 & 1 & 5 \end{vmatrix} = 3\left[1\begin{vmatrix} 3 & 4 \\ 1 & 5 \end{vmatrix} - 0 + 0\right] = 3[(1)(11)] = 33.$$

11. $\begin{vmatrix} 2 & 1 & 3 & 1 \\ 1 & 0 & 1 & 1 \\ 0 & 2 & 1 & 0 \\ 0 & 1 & 2 & 3 \end{vmatrix} = (-1)\begin{vmatrix} 1 & 0 & 1 & 1 \\ 2 & 1 & 3 & 1 \\ 0 & 2 & 1 & 0 \\ 0 & 1 & 2 & 3 \end{vmatrix}$ ← The first and second rows were interchanged.

$= (-1)\begin{vmatrix} 1 & 0 & 1 & 1 \\ 0 & 1 & 1 & -1 \\ 0 & 2 & 1 & 0 \\ 0 & 1 & 2 & 3 \end{vmatrix}$ ← -2 times the first row was added to the second row.

$= (-1)\begin{vmatrix} 1 & 0 & 1 & 1 \\ 0 & 1 & 1 & -1 \\ 0 & 0 & -1 & 2 \\ 0 & 1 & 2 & 3 \end{vmatrix}$ ← -2 times the second row was added to the third row.

$= (-1)\begin{vmatrix} 1 & 0 & 1 & 1 \\ 0 & 1 & 1 & -1 \\ 0 & 0 & -1 & 2 \\ 0 & 0 & 1 & 4 \end{vmatrix}$ ← -1 times the second row was added to the fourth row.

$= (-1)(-1)\begin{vmatrix} 1 & 0 & 1 & 1 \\ 0 & 1 & 1 & -1 \\ 0 & 0 & 1 & -2 \\ 0 & 0 & 1 & 4 \end{vmatrix}$ ← A common factor of -1 from the third row was taken through the determinant sign.

$$= (-1)(-1)\begin{vmatrix} 1 & 0 & 1 & 1 \\ 0 & 1 & 1 & -1 \\ 0 & 0 & 1 & -2 \\ 0 & 0 & 0 & 6 \end{vmatrix}$$ ⟵ -1 times the third row was added to the fourth row.

$$= (-1)(-1)(6)\begin{vmatrix} 1 & 0 & 1 & 1 \\ 0 & 1 & 1 & -1 \\ 0 & 0 & 1 & -2 \\ 0 & 0 & 0 & 1 \end{vmatrix}$$ ⟵ A common factor of 6 from the third row was taken through the determinant sign.

$$= (-1)(-1)(6)(1) = 6$$

Another way to evaluate the determinant would be to use cofactor expansions along the first column after the fourth step above:

$$\begin{vmatrix} 2 & 1 & 3 & 1 \\ 1 & 0 & 1 & 1 \\ 0 & 2 & 1 & 0 \\ 0 & 1 & 2 & 3 \end{vmatrix} = (-1)\begin{vmatrix} 1 & 0 & 1 & 1 \\ 0 & 1 & 1 & -1 \\ 0 & 0 & -1 & 2 \\ 0 & 0 & 1 & 4 \end{vmatrix} = (-1)(1)\begin{vmatrix} 1 & 1 & -1 \\ 0 & -1 & 2 \\ 0 & 1 & 4 \end{vmatrix} = (-1)(1)(1)\begin{vmatrix} -1 & 2 \\ 1 & 4 \end{vmatrix}$$
$$= (-1)(1)(1)(-6) = 6\,.$$

13.
$$\begin{vmatrix} 1 & 3 & 1 & 5 & 3 \\ -2 & -7 & 0 & -4 & 2 \\ 0 & 0 & 1 & 0 & 1 \\ 0 & 0 & 2 & 1 & 1 \\ 0 & 0 & 0 & 1 & 1 \end{vmatrix}$$

$$= \begin{vmatrix} 1 & 3 & 1 & 5 & 3 \\ 0 & -1 & 2 & 6 & 8 \\ 0 & 0 & 1 & 0 & 1 \\ 0 & 0 & 2 & 1 & 1 \\ 0 & 0 & 0 & 1 & 1 \end{vmatrix}$$ ⟵ 2 times the first row was added to the second row.

$$= (-1)\begin{vmatrix} 1 & 3 & 1 & 5 & 3 \\ 0 & 1 & -2 & -6 & -8 \\ 0 & 0 & 1 & 0 & 1 \\ 0 & 0 & 2 & 1 & 1 \\ 0 & 0 & 0 & 1 & 1 \end{vmatrix}$$ ⟵ A common factor of -1 from the second row was taken through the determinant sign.

$$= (-1)\begin{vmatrix} 1 & 3 & 1 & 5 & 3 \\ 0 & 1 & -2 & -6 & -8 \\ 0 & 0 & 1 & 0 & 1 \\ 0 & 0 & 0 & 1 & -1 \\ 0 & 0 & 0 & 1 & 1 \end{vmatrix}$$ ⟵ -2 times the third row was added to the fourth row.

$$= (-1)\begin{vmatrix} 1 & 3 & 1 & 5 & 3 \\ 0 & 1 & -2 & -6 & -8 \\ 0 & 0 & 1 & 0 & 1 \\ 0 & 0 & 0 & 1 & -1 \\ 0 & 0 & 0 & 0 & 2 \end{vmatrix}$$ ⟵ -1 times the fourth row was added to the fifth row.

$$= (-1)(2)\begin{vmatrix} 1 & 3 & 1 & 5 & 3 \\ 0 & 1 & -2 & -6 & -8 \\ 0 & 0 & 1 & 0 & 1 \\ 0 & 0 & 0 & 1 & -1 \\ 0 & 0 & 0 & 0 & 1 \end{vmatrix}$$

← A common factor of 2 from the fifth row was taken through the determinant sign.

$$= (-1)(2)(1) = -2$$

Another way to evaluate the determinant would be to use cofactor expansions along the first column after the third step above:

$$\begin{vmatrix} 1 & 3 & 1 & 5 & 3 \\ -2 & -7 & 0 & -4 & 2 \\ 0 & 0 & 1 & 0 & 1 \\ 0 & 0 & 2 & 1 & 1 \\ 0 & 0 & 0 & 1 & 1 \end{vmatrix} = (-1)\begin{vmatrix} 1 & 3 & 1 & 5 & 3 \\ 0 & 1 & -2 & -6 & -8 \\ 0 & 0 & 1 & 0 & 1 \\ 0 & 0 & 0 & 1 & -1 \\ 0 & 0 & 0 & 1 & 1 \end{vmatrix} = (-1)(1)\begin{vmatrix} 1 & -2 & -6 & -8 \\ 0 & 1 & 0 & 1 \\ 0 & 0 & 1 & -1 \\ 0 & 0 & 1 & 1 \end{vmatrix}$$

$$= (-1)(1)(1)\begin{vmatrix} 1 & 0 & 1 \\ 0 & 1 & -1 \\ 0 & 1 & 1 \end{vmatrix} = (-1)(1)(1)(1)\begin{vmatrix} 1 & -1 \\ 1 & 1 \end{vmatrix} = (-1)(1)(1)(1)(2) = -2.$$

15.

$$\begin{vmatrix} d & e & f \\ g & h & i \\ a & b & c \end{vmatrix} = (-1)\begin{vmatrix} a & b & c \\ g & h & i \\ d & e & f \end{vmatrix}$$

← The first and third rows were interchanged.

$$= (-1)(-1)\begin{vmatrix} a & b & c \\ d & e & f \\ g & h & i \end{vmatrix}$$

← The second and third rows were interchanged.

$$= (-1)(-1)(-6) = -6$$

17.

$$\begin{vmatrix} 3a & 3b & 3c \\ -d & -e & -f \\ 4g & 4h & 4i \end{vmatrix} = 3\begin{vmatrix} a & b & c \\ -d & -e & -f \\ 4g & 4h & 4i \end{vmatrix}$$

← A common factor of 3 from the first row was taken through the determinant sign.

$$= 3(-1)\begin{vmatrix} a & b & c \\ d & e & f \\ 4g & 4h & 4i \end{vmatrix}$$

← A common factor of -1 from the second row was taken through the determinant sign.

$$= 3(-1)(4)\begin{vmatrix} a & b & c \\ d & e & f \\ g & h & i \end{vmatrix}$$

← A common factor of 4 from the third row was taken through the determinant sign.

$$= 3(-1)(4)(-6) = 72$$

19.

$$\begin{vmatrix} a+g & b+h & c+i \\ d & e & f \\ g & h & i \end{vmatrix} = \begin{vmatrix} a & b & c \\ d & e & f \\ g & h & i \end{vmatrix}$$

← -1 times the third row was added to the first row.

$$= -6$$

21.

$$\begin{vmatrix} -3a & -3b & -3c \\ d & e & f \\ g-4d & h-4e & i-4f \end{vmatrix}$$

$$= -3\begin{vmatrix} a & b & c \\ d & e & f \\ g-4d & h-4e & i-4f \end{vmatrix}$$ ← A common factor of −3 from the first row was taken through the determinant sign.

$$= -3\begin{vmatrix} a & b & c \\ d & e & f \\ g & h & i \end{vmatrix}$$ ← 4 times the second row was added to the last row.

$$= (-3)(-6) = 18$$

23.

$$\begin{vmatrix} 1 & 1 & 1 \\ a & b & c \\ a^2 & b^2 & c^2 \end{vmatrix} = \begin{vmatrix} 1 & 1 & 1 \\ 0 & b-a & c-a \\ a^2 & b^2 & c^2 \end{vmatrix}$$ ← $-a$ times the first row was added to the second row.

$$= \begin{vmatrix} 1 & 1 & 1 \\ 0 & b-a & c-a \\ 0 & b^2-a^2 & c^2-a^2 \end{vmatrix}$$ ← $-a^2$ times the first row was added to the third row.

$$= \begin{vmatrix} 1 & 1 & 1 \\ 0 & b-a & c-a \\ 0 & 0 & c^2-a^2-(c-a)(b+a) \end{vmatrix}$$ ← $-(b+a)$ times the second row was added to the third row.

$$= (1)(b-a)(c-a)(c+a-b-a)$$

$$= (b-a)(c-a)(c-b)$$

25.

$$\begin{vmatrix} a_1 & b_1 & a_1+b_1+c_1 \\ a_2 & b_2 & a_2+b_2+c_2 \\ a_3 & b_3 & a_3+b_3+c_3 \end{vmatrix}$$

$$= \begin{vmatrix} a_1 & b_1 & b_1+c_1 \\ a_2 & b_2 & b_2+c_2 \\ a_3 & b_3 & b_3+c_3 \end{vmatrix}$$ ← −1 times the first column was added to the third column.

$$= \begin{vmatrix} a_1 & b_1 & c_1 \\ a_2 & b_2 & c_2 \\ a_3 & b_3 & c_3 \end{vmatrix}$$ ← −1 times the second column was added to the third column.

27.

$$\begin{vmatrix} a_1+b_1 & a_1-b_1 & c_1 \\ a_2+b_2 & a_2-b_2 & c_2 \\ a_3+b_3 & a_3-b_3 & c_3 \end{vmatrix}$$

$$= \begin{vmatrix} a_1+b_1 & -2b_1 & c_1 \\ a_2+b_2 & -2b_2 & c_2 \\ a_3+b_3 & -2b_3 & c_3 \end{vmatrix}$$ ⟵ -1 times the first column was added to the second column.

$$= -2\begin{vmatrix} a_1+b_1 & b_1 & c_1 \\ a_2+b_2 & b_2 & c_2 \\ a_3+b_3 & b_3 & c_3 \end{vmatrix}$$ ⟵ A common factor of -2 from the second column was taken through the determinant sign.

$$= -2\begin{vmatrix} a_1 & b_1 & c_1 \\ a_2 & b_2 & c_2 \\ a_3 & b_3 & c_3 \end{vmatrix}$$ ⟵ -1 times the second column was added to the first column.

29. The second column vector is a scalar multiple of the fourth. By Theorem 2.2.5, the determinant is 0.

31. $\det(M) = \begin{vmatrix} 1 & 2 & 0 \\ 2 & 5 & 0 \\ -1 & 3 & 2 \end{vmatrix}\begin{vmatrix} 3 & 0 & 0 \\ 2 & 1 & 0 \\ -3 & 8 & -4 \end{vmatrix} = \left(0-0+2\begin{vmatrix} 1 & 2 \\ 2 & 5 \end{vmatrix}\right)\left(0-0+(-4)\begin{vmatrix} 3 & 0 \\ 2 & 1 \end{vmatrix}\right) = (2)(-12) = -24$

33. In order to reverse the order of rows in 2×2 and 3×3 matrix, the first and the last rows can be interchanged, so $\det(B) = -\det(A)$.

For 4×4 and 5×5 matrices, two such interchanges are needed: the first and last rows can be swapped, then the second and the penultimate one can follow.

Thus, $\det(B) = (-1)(-1)\det(A) = \det(A)$ in this case.

Generally, to rows in an $n\times n$ matrix can be reversed by

- interchanging row 1 with row n,
- interchanging row 2 with row $n-1$,
- $\vdots$
- interchanging row $\lfloor n/2 \rfloor$ with row $n - \lfloor n/2 \rfloor$

where $\lfloor x \rfloor$ is the greatest integer less than or equal to x (also known as the "floor" of x).

We conclude that $\det(B) = (-1)^{\lfloor n/2 \rfloor}\det(A)$.

True-False Exercises

(a) True. $\det(B) = (-1)(-1)\det(A) = \det(A)$.

(b) True. $\det(B) = (4)\left(\frac{3}{4}\right)\det(A) = 3\det(A)$.

(c) False. $\det(B) = \det(A)$.

(d) False. $\det(B) = n(n-1)\cdots 3\cdot 2\cdot 1\cdot \det(A) = (n!)\det(A)$.

(e) True. This follows from Theorem 2.2.5.

(f) True. Let B be obtained from A by adding the second row to the fourth row, so $\det(A) = \det(B)$. Since the fourth row and the sixth row of B are identical, by Theorem 2.2.5 $\det(B) = 0$.

2.3 Properties of Determinants; Cramer's Rule

1. $\det(2A) = \begin{vmatrix} -2 & 4 \\ 6 & 8 \end{vmatrix} = (-2)(8) - (4)(6) = -40$

$(2)^2 \det(A) = 4\begin{vmatrix} -1 & 2 \\ 3 & 4 \end{vmatrix} = 4((-1)(4) - (2)(3)) = (4)(-10) = -40$

3. We are using the arrow technique to evaluate both determinants.

$$\det(-2A) = \begin{vmatrix} -4 & 2 & -6 \\ -6 & -4 & -2 \\ -2 & -8 & -10 \end{vmatrix} = (-160 + 8 - 288) - (-48 - 64 + 120) = -448$$

$$(-2)^3 \det(A) = -8\begin{vmatrix} 2 & -1 & 3 \\ 3 & 2 & 1 \\ 1 & 4 & 5 \end{vmatrix} = (-8)((20 - 1 + 36) - (6 + 8 - 15)) = (-8)(56) = -448$$

5. We are using the arrow technique to evaluate the determinants in this problem.

$$\det(AB) = \begin{vmatrix} 9 & -1 & 8 \\ 31 & 1 & 17 \\ 10 & 0 & 2 \end{vmatrix} = (18 - 170 + 0) - (80 + 0 - 62) = -170;$$

$$\det(BA) = \begin{vmatrix} -1 & -3 & 6 \\ 17 & 11 & 4 \\ 10 & 5 & 2 \end{vmatrix} = (-22 - 120 + 510) - (660 - 20 - 102) = -170;$$

$$\det(A+B) = \begin{vmatrix} 3 & 0 & 3 \\ 10 & 5 & 2 \\ 5 & 0 & 3 \end{vmatrix} = (45 + 0 + 0) - (75 + 0 + 0) = -30;$$

$\det(A) = (16 + 0 + 0) - (0 + 0 + 6) = 10;$
$\det(B) = (1 - 10 + 0) - (15 + 0 - 7) = -17;$
$\det(A + B) \neq \det(A) + \det(B)$

7. $\det(A) = (-6 + 0 - 20) - (-10 + 0 - 15) = -1 \neq 0$ therefore A is invertible by Theorem 2.3.3

9. $\det(A) = (2)(1)(2) = 4 \neq 0$ therefore A is invertible by Theorem 2.3.3

11. $\det(A) = (24 - 24 - 16) - (24 - 16 - 24) = 0$ therefore A is not invertible by Theorem 2.3.3

13. $\det(A) = (2)(1)(6) = 12 \neq 0$ therefore A is invertible by Theorem 2.3.3

15. $\det(A) = (k-3)(k-2) - (-2)(-2) = k^2 - 5k + 2 = \left(k - \frac{5-\sqrt{17}}{2}\right)\left(k - \frac{5+\sqrt{17}}{2}\right)$. By Theorem 2.3.3, A is invertible if $k \neq \frac{5-\sqrt{17}}{2}$ and $k \neq \frac{5+\sqrt{17}}{2}$.

17. $\det(A) = (2 + 12k + 36) - (4k + 18 + 12) = 8 + 8k = 8(1 + k)$.
By Theorem 2.3.3, A is invertible if $k \neq -1$.

19. $\det(A) = (-6 + 0 - 20) - (-10 + 0 - 15) = -1 \neq 0$ therefore A is invertible by Theorem 2.3.3.

The cofactors of A are:

$C_{11} = \begin{vmatrix} -1 & 0 \\ 4 & 3 \end{vmatrix} = -3 \qquad C_{12} = -\begin{vmatrix} -1 & 0 \\ 2 & 3 \end{vmatrix} = 3 \qquad C_{13} = \begin{vmatrix} -1 & -1 \\ 2 & 4 \end{vmatrix} = -2$

$C_{21} = -\begin{vmatrix} 5 & 5 \\ 4 & 3 \end{vmatrix} = 5 \qquad C_{22} = \begin{vmatrix} 2 & 5 \\ 2 & 3 \end{vmatrix} = -4 \qquad C_{23} = -\begin{vmatrix} 2 & 5 \\ 2 & 4 \end{vmatrix} = 2$

$C_{31} = \begin{vmatrix} 5 & 5 \\ -1 & 0 \end{vmatrix} = 5 \qquad C_{32} = -\begin{vmatrix} 2 & 5 \\ -1 & 0 \end{vmatrix} = -5 \qquad C_{33} = \begin{vmatrix} 2 & 5 \\ -1 & -1 \end{vmatrix} = 3$

The matrix of cofactors is $\begin{bmatrix} -3 & 3 & -2 \\ 5 & -4 & 2 \\ 5 & -5 & 3 \end{bmatrix}$ and the adjoint matrix is $\text{adj}(A) = \begin{bmatrix} -3 & 5 & 5 \\ 3 & -4 & -5 \\ -2 & 2 & 3 \end{bmatrix}$.

From Theorem 2.3.6, we have $A^{-1} = \frac{1}{\det(A)}\text{adj}(A) = \frac{1}{-1}\begin{bmatrix} -3 & 5 & 5 \\ 3 & -4 & -5 \\ -2 & 2 & 3 \end{bmatrix} = \begin{bmatrix} 3 & -5 & -5 \\ -3 & 4 & 5 \\ 2 & -2 & -3 \end{bmatrix}$.

21. $\det(A) = (2)(1)(2) = 4 \neq 0$ therefore A is invertible by Theorem 2.3.3.

The cofactors of A are:

$C_{11} = \begin{vmatrix} 1 & -3 \\ 0 & 2 \end{vmatrix} = 2 \qquad C_{12} = -\begin{vmatrix} 0 & -3 \\ 0 & 2 \end{vmatrix} = 0 \qquad C_{13} = \begin{vmatrix} 0 & 1 \\ 0 & 0 \end{vmatrix} = 0$

$C_{21} = -\begin{vmatrix} -3 & 5 \\ 0 & 2 \end{vmatrix} = 6 \qquad C_{22} = \begin{vmatrix} 2 & 5 \\ 0 & 2 \end{vmatrix} = 4 \qquad C_{23} = -\begin{vmatrix} 2 & -3 \\ 0 & 0 \end{vmatrix} = 0$

$C_{31} = \begin{vmatrix} -3 & 5 \\ 1 & -3 \end{vmatrix} = 4 \qquad C_{32} = -\begin{vmatrix} 2 & 5 \\ 0 & -3 \end{vmatrix} = 6 \qquad C_{33} = \begin{vmatrix} 2 & -3 \\ 0 & 1 \end{vmatrix} = 2$

The matrix of cofactors is $\begin{bmatrix} 2 & 0 & 0 \\ 6 & 4 & 0 \\ 4 & 6 & 2 \end{bmatrix}$ and the adjoint matrix is $\text{adj}(A) = \begin{bmatrix} 2 & 6 & 4 \\ 0 & 4 & 6 \\ 0 & 0 & 2 \end{bmatrix}$.

From Theorem 2.3.6, we have $A^{-1} = \frac{1}{\det(A)}\text{adj}(A) = \frac{1}{4}\begin{bmatrix} 2 & 6 & 4 \\ 0 & 4 & 6 \\ 0 & 0 & 2 \end{bmatrix} = \begin{bmatrix} \frac{1}{2} & \frac{3}{2} & 1 \\ 0 & 1 & \frac{3}{2} \\ 0 & 0 & \frac{1}{2} \end{bmatrix}$.

23.

$$\begin{vmatrix} 1 & 3 & 1 & 1 \\ 2 & 5 & 2 & 2 \\ 1 & 3 & 8 & 9 \\ 1 & 3 & 2 & 2 \end{vmatrix} = \begin{vmatrix} 1 & 3 & 1 & 1 \\ 0 & -1 & 0 & 0 \\ 0 & 0 & 7 & 8 \\ 0 & 0 & 1 & 1 \end{vmatrix}$$

⟵ −2 times the first row was added to the second row; −1 times the first row was added to the third and fourth rows.

$$= -\begin{vmatrix} 1 & 3 & 1 & 1 \\ 0 & -1 & 0 & 0 \\ 0 & 0 & 1 & 1 \\ 0 & 0 & 7 & 8 \end{vmatrix}$$

⟵ The third row and the fourth row were interchanged.

$$= -\begin{vmatrix} 1 & 3 & 1 & 1 \\ 0 & -1 & 0 & 0 \\ 0 & 0 & 1 & 1 \\ 0 & 0 & 0 & 1 \end{vmatrix}$$

⟵ −7 times the third row was added to the fourth row

$$= -(1)(-1)(1)(1) = 1$$

The determinant of A is nonzero therefore by Theorem 2.3.3, A is invertible.

The cofactors of A are:

$$C_{11} = \begin{vmatrix} 5 & 2 & 2 \\ 3 & 8 & 9 \\ 3 & 2 & 2 \end{vmatrix} = (80 + 54 + 12) - (48 + 90 + 12) = -4$$

$$C_{12} = -\begin{vmatrix} 2 & 2 & 2 \\ 1 & 8 & 9 \\ 1 & 2 & 2 \end{vmatrix} = -[(32 + 18 + 4) - (16 + 36 + 4)] = 2$$

$$C_{13} = \begin{vmatrix} 2 & 5 & 2 \\ 1 & 3 & 9 \\ 1 & 3 & 2 \end{vmatrix} = (12 + 45 + 6) - (6 + 54 + 10) = -7$$

$$C_{14} = -\begin{vmatrix} 2 & 5 & 2 \\ 1 & 3 & 8 \\ 1 & 3 & 2 \end{vmatrix} = -[(12 + 40 + 6) - (6 + 48 + 10)] = 6$$

$$C_{21} = -\begin{vmatrix} 3 & 1 & 1 \\ 3 & 8 & 9 \\ 3 & 2 & 2 \end{vmatrix} = -[(48 + 27 + 6) - (24 + 54 + 6)] = 3$$

$$C_{22} = \begin{vmatrix} 1 & 1 & 1 \\ 1 & 8 & 9 \\ 1 & 2 & 2 \end{vmatrix} = (16 + 9 + 2) - (8 + 18 + 2) = -1$$

$$C_{23} = -\begin{vmatrix} 1 & 3 & 1 \\ 1 & 3 & 9 \\ 1 & 3 & 2 \end{vmatrix} = -[(6 + 27 + 3) - (3 + 27 + 6)] = 0$$

$$C_{24} = \begin{vmatrix} 1 & 3 & 1 \\ 1 & 3 & 8 \\ 1 & 3 & 2 \end{vmatrix} = (6 + 24 + 3) - (3 + 24 + 6) = 0$$

$$C_{31} = \begin{vmatrix} 3 & 1 & 1 \\ 5 & 2 & 2 \\ 3 & 2 & 2 \end{vmatrix} = (12 + 6 + 10) - (6 + 12 + 10) = 0$$

$$C_{32} = -\begin{vmatrix} 1 & 1 & 1 \\ 2 & 2 & 2 \\ 1 & 2 & 2 \end{vmatrix} = -[(4 + 2 + 4) - (2 + 4 + 4)] = 0$$

$$C_{33} = \begin{vmatrix} 1 & 3 & 1 \\ 2 & 5 & 2 \\ 1 & 3 & 2 \end{vmatrix} = (10 + 6 + 6) - (5 + 6 + 12) = -1$$

$$C_{34} = -\begin{vmatrix} 1 & 3 & 1 \\ 2 & 5 & 2 \\ 1 & 3 & 2 \end{vmatrix} = -[(10 + 6 + 6) - (5 + 6 + 12)] = 1$$

$$C_{41} = -\begin{vmatrix} 3 & 1 & 1 \\ 5 & 2 & 2 \\ 3 & 8 & 9 \end{vmatrix} = -[(54 + 6 + 40) - (6 + 48 + 45)] = -1$$

$$C_{42} = \begin{vmatrix} 1 & 1 & 1 \\ 2 & 2 & 2 \\ 1 & 8 & 9 \end{vmatrix} = (18 + 2 + 16) - (2 + 16 + 18) = 0$$

$$C_{43} = -\begin{vmatrix} 1 & 3 & 1 \\ 2 & 5 & 2 \\ 1 & 3 & 9 \end{vmatrix} = -[(45 + 6 + 6) - (5 + 6 + 54)] = 8$$

$$C_{44} = \begin{vmatrix} 1 & 3 & 1 \\ 2 & 5 & 2 \\ 1 & 3 & 8 \end{vmatrix} = (40 + 6 + 6) - (5 + 6 + 48) = -7$$

The matrix of cofactors is $\begin{bmatrix} -4 & 2 & -7 & 6 \\ 3 & -1 & 0 & 0 \\ 0 & 0 & -1 & 1 \\ -1 & 0 & 8 & -7 \end{bmatrix}$ and the adjoint matrix is

$$\text{adj}(A) = \begin{bmatrix} -4 & 3 & 0 & -1 \\ 2 & -1 & 0 & 0 \\ -7 & 0 & -1 & 8 \\ 6 & 0 & 1 & -7 \end{bmatrix}.$$

From Theorem 2.3.6, we have $A^{-1} = \frac{1}{\det(A)}\text{adj}(A) = \frac{1}{1}\begin{bmatrix} -4 & 3 & 0 & -1 \\ 2 & -1 & 0 & 0 \\ -7 & 0 & -1 & 8 \\ 6 & 0 & 1 & -7 \end{bmatrix} = \begin{bmatrix} -4 & 3 & 0 & -1 \\ 2 & -1 & 0 & 0 \\ -7 & 0 & -1 & 8 \\ 6 & 0 & 1 & -7 \end{bmatrix}.$

25. $\det(A) = \begin{vmatrix} 4 & 5 & 0 \\ 11 & 1 & 2 \\ 1 & 5 & 2 \end{vmatrix} = (8 + 10 + 0) - (0 + 40 + 110) = -132,$

$\det(A_1) = \begin{vmatrix} 2 & 5 & 0 \\ 3 & 1 & 2 \\ 1 & 5 & 2 \end{vmatrix} = (4 + 10 + 0) - (0 + 20 + 30) = -36,$

$\det(A_2) = \begin{vmatrix} 4 & 2 & 0 \\ 11 & 3 & 2 \\ 1 & 1 & 2 \end{vmatrix} = (24 + 4 + 0) - (0 + 8 + 44) = -24,$

$\det(A_3) = \begin{vmatrix} 4 & 5 & 2 \\ 11 & 1 & 3 \\ 1 & 5 & 1 \end{vmatrix} = (4 + 15 + 110) - (2 + 60 + 55) = 12;$

$x = \frac{\det(A_1)}{\det(A)} = \frac{-36}{-132} = \frac{3}{11},\ \ y = \frac{\det(A_2)}{\det(A)} = \frac{-24}{-132} = \frac{2}{11},\ \ z = \frac{\det(A_3)}{\det(A)} = \frac{12}{-132} = -\frac{1}{11}.$

27. $\det(A) = \begin{vmatrix} 1 & -3 & 1 \\ 2 & -1 & 0 \\ 4 & 0 & -3 \end{vmatrix} = (3 + 0 + 0) - (-4 + 0 + 18) = -11,$

$\det(A_1) = \begin{vmatrix} 4 & -3 & 1 \\ -2 & -1 & 0 \\ 0 & 0 & -3 \end{vmatrix} = -3\begin{vmatrix} 4 & -3 \\ -2 & -1 \end{vmatrix} = (-3)(-4 - 6) = 30,$

$\det(A_2) = \begin{vmatrix} 1 & 4 & 1 \\ 2 & -2 & 0 \\ 4 & 0 & -3 \end{vmatrix} = (6 + 0 + 0) - (-8 + 0 - 24) = 38,$

$\det(A_3) = \begin{vmatrix} 1 & -3 & 4 \\ 2 & -1 & -2 \\ 4 & 0 & 0 \end{vmatrix} = 4\begin{vmatrix} -3 & 4 \\ -1 & -2 \end{vmatrix} = (4)(6 + 4) = 40;$

$x_1 = \frac{\det(A_1)}{\det(A)} = \frac{30}{-11} = -\frac{30}{11},\ \ x_2 = \frac{\det(A_2)}{\det(A)} = \frac{38}{-11} = -\frac{38}{11},\ \ x_3 = \frac{\det(A_3)}{\det(A)} = \frac{40}{-11} = -\frac{40}{11}.$

29. $\det(A) = 0$ therefore Cramer's rule does not apply.

31. $\det(A) = \begin{vmatrix} 4 & 1 & 1 & 1 \\ 3 & 7 & -1 & 1 \\ 7 & 3 & -5 & 8 \\ 1 & 1 & 1 & 2 \end{vmatrix} = -424;\ \det(A_2) = \begin{vmatrix} 4 & 6 & 1 & 1 \\ 3 & 1 & -1 & 1 \\ 7 & -3 & -5 & 8 \\ 1 & 3 & 1 & 2 \end{vmatrix} = 0;\ y = \frac{\det(A_2)}{\det(A)} = \frac{0}{-424} = 0$

33. **(a)** $\det(3A) = 3^3\det(A) = (27)(-7) = -189$ (using Formula (1))

(b) $\det(A^{-1}) = \frac{1}{\det(A)} = \frac{1}{-7} = -\frac{1}{7}$ (using Theorem 2.3.5)

(c) $\det(2A^{-1}) = 2^3 \det(A^{-1}) = \frac{8}{\det(A)} = \frac{8}{-7} = -\frac{8}{7}$ (using Formula (1) and Theorem 2.3.5)

(d) $\det((2A)^{-1}) = \frac{1}{\det(2A)} = \frac{1}{2^3 \det(A)} = \frac{1}{(8)(-7)} = -\frac{1}{56}$ (using Theorem 2.3.5 and Formula (1))

(e) $\begin{vmatrix} a & g & d \\ b & h & e \\ c & i & f \end{vmatrix} = -\begin{vmatrix} a & d & g \\ b & e & h \\ c & f & i \end{vmatrix} = -\begin{vmatrix} a & b & c \\ d & e & f \\ g & h & i \end{vmatrix} = -(-7) = 7$ (in the first step we interchanged the last two columns applying Theorem 2.2.3(b); in the second step we transposed the matrix applying Theorem 2.2.2)

35. **(a)** $\det(3A) = 3^3 \det(A) = (27)(7) = 189$ (using Formula (1))

(b) $\det(A^{-1}) = \frac{1}{\det(A)} = \frac{1}{7}$ (using Theorem 2.3.5)

(c) $\det(2A^{-1}) = 2^3 \det(A^{-1}) = \frac{8}{\det(A)} = \frac{8}{7}$ (using Formula (1) and Theorem 2.3.5)

(d) $\det((2A)^{-1}) = \frac{1}{\det(2A)} = \frac{1}{2^3 \det(A)} = \frac{1}{(8)(7)} = \frac{1}{56}$ (using Theorem 2.3.5 and Formula (1))

True-False Exercises

(a) False. By Formula (1), $\det(2A) = 2^3 \det(A) = 8\det(A)$.

(b) False. E.g. $A = \begin{bmatrix} 1 & 0 \\ 0 & 0 \end{bmatrix}$ and $B = \begin{bmatrix} 0 & 0 \\ 0 & 1 \end{bmatrix}$ have $\det(A) = \det(B) = 0$ but $\det(A+B) = 1 \neq 2\det(A)$.

(c) True. By Theorems 2.3.4 and 2.3.5,
$\det(A^{-1}BA) = \det(A^{-1})\det(B)\det(A) = \frac{1}{\det(A)}\det(B)\det(A) = \det(B)$.

(d) False. A square matrix A is invertible if and only if $\det(A) \neq 0$.

(e) True. This follows from Definition 1.

(f) True. This is Formula (8).

(g) True. If $\det(A) \neq 0$ then by Theorem 2.3.8 $A\mathbf{x} = \mathbf{0}$ must have only the trivial solution, which contradicts our assumption. Consequently, $\det(A) = 0$.

(h) True. If the reduced row echelon form of A is I_n then by Theorem 2.3.8 $A\mathbf{x} = \mathbf{b}$ is consistent for every $\mathbf{b}$, which contradicts our assumption. Consequently, the reduced row echelon form of A cannot be I_n.

(i) True. Since the reduced row echelon form of E is I then by Theorem 2.3.8 $E\mathbf{x} = \mathbf{0}$ must have only the trivial solution.

(j) True. If A is invertible, so is A^{-1}. By Theorem 2.3.8, each system has only the trivial solution.

(k) True. From Theorem 2.3.6, $A^{-1} = \frac{1}{\det(A)}\operatorname{adj}(A)$ therefore $\operatorname{adj}(A) = \det(A)\,A^{-1}$. Consequently,
$\left(\frac{1}{\det(A)}A\right)\operatorname{adj}(A) = \left(\frac{1}{\det(A)}A\right)(\det(A)\,A^{-1}) = \frac{\det(A)}{\det(A)}(AA^{-1}) = I_n$ so $\left(\operatorname{adj}(A)\right)^{-1} = \frac{1}{\det(A)}A$.

(l) False. If the kth row of A contains only zeros then all cofactors C_{jk} where $j \neq i$ are zero (since each of them involves a determinant of a matrix with a zero row). This means the matrix of cofactors contains at least one zero row, therefore $\operatorname{adj}(A)$ has a *column* of zeros.

Chapter 2 Supplementary Exercises

1. **(a)** Cofactor expansion along the first row: $\begin{vmatrix}-4 & 2\\ 3 & 3\end{vmatrix} = (-4)(3) - (2)(3) = -12 - 6 = -18$

(b) $\begin{vmatrix}-4 & 2\\ 3 & 3\end{vmatrix} = -\begin{vmatrix}3 & 3\\ -4 & 2\end{vmatrix}$ ← The first and second rows were interchanged.

$= -(3)\begin{vmatrix}1 & 1\\ -4 & 2\end{vmatrix}$ ← A common factor of 3 from the first row was taken through the determinant sign.

$= -(3)\begin{vmatrix}1 & 1\\ 0 & 6\end{vmatrix}$ ← 4 times the first row was added to the second row

$= -(3)(1)(6) = -18$ ← Use Theorem 2.1.2.

3. **(a)** Cofactor expansion along the second row:

$$\begin{vmatrix}-1 & 5 & 2\\ 0 & 2 & -1\\ -3 & 1 & 1\end{vmatrix} = -0 + 2\begin{vmatrix}-1 & 2\\ -3 & 1\end{vmatrix} - (-1)\begin{vmatrix}-1 & 5\\ -3 & 1\end{vmatrix}$$

$$= 0 + 2[(-1)(1) - (2)(-3)] - (-1)[(-1)(1) - (5)(-3)]$$

$$= 0 + (2)(5) - (-1)(14) = 0 + 10 + 14 = 24$$

(b) $\begin{vmatrix}-1 & 5 & 2\\ 0 & 2 & -1\\ -3 & 1 & 1\end{vmatrix} = (-1)\begin{vmatrix}1 & -5 & -2\\ 0 & 2 & -1\\ -3 & 1 & 1\end{vmatrix}$ ← A common factor of -1 from the first row was taken through the determinant sign.

$= (-1)\begin{vmatrix}1 & -5 & -2\\ 0 & 2 & -1\\ 0 & -14 & -5\end{vmatrix}$ ← 3 times the first row was added to the third row.

$= (-1)\begin{vmatrix}1 & -5 & -2\\ 0 & 2 & -1\\ 0 & 0 & -12\end{vmatrix}$ ← 7 times the second row was added to the third.

$= (-1)(1)(2)(-12) = 24$ ← Use Theorem 2.1.2.

5. **(a)** Cofactor expansion along the first row:

$$\begin{vmatrix}3 & 0 & -1\\ 1 & 1 & 1\\ 0 & 4 & 2\end{vmatrix} = (3)\begin{vmatrix}1 & 1\\ 4 & 2\end{vmatrix} - 0 + (-1)\begin{vmatrix}1 & 1\\ 0 & 4\end{vmatrix}$$

$$= (3)[(1)(2) - (1)(4)] - 0 + (-1)[(1)(4) - (1)(0)]$$

$$= (3)(-2) - 0 + (-1)(4) = -6 + 0 - 4 = -10$$

(b) $\begin{vmatrix}3 & 0 & -1\\ 1 & 1 & 1\\ 0 & 4 & 2\end{vmatrix} = (-1)\begin{vmatrix}1 & 1 & 1\\ 3 & 0 & -1\\ 0 & 4 & 2\end{vmatrix}$ ← The first and second rows were interchanged.

$$= (-1)\begin{vmatrix} 1 & 1 & 1 \\ 0 & -3 & -4 \\ 0 & 4 & 2 \end{vmatrix}$$ ← -3 times the first row was added to the second.

$$= (-1)\begin{vmatrix} 1 & 1 & 1 \\ 0 & -3 & -4 \\ 0 & 1 & -2 \end{vmatrix}$$ ← The second row was added to the third row

$$= (-1)(-1)\begin{vmatrix} 1 & 1 & 1 \\ 0 & 1 & -2 \\ 0 & -3 & -4 \end{vmatrix}$$ ← The second and third rows were interchanged.

$$= (-1)(-1)\begin{vmatrix} 1 & 1 & 1 \\ 0 & 1 & -2 \\ 0 & 0 & -10 \end{vmatrix}$$ ← 3 times the second row was added to the third.

$$= (-1)(-1)(1)(1)(-10) = -10$$ ← Use Theorem 2.1.2.

7. **(a)** We perform cofactor expansions along the first row in the 4x4 determinant. In each of the 3x3 determinants, we expand along the second row:

$$\begin{vmatrix} 3 & 6 & 0 & 1 \\ -2 & 3 & 1 & 4 \\ 1 & 0 & -1 & 1 \\ -9 & 2 & -2 & 2 \end{vmatrix} = 3\begin{vmatrix} 3 & 1 & 4 \\ 0 & -1 & 1 \\ 2 & -2 & 2 \end{vmatrix} - 6\begin{vmatrix} -2 & 1 & 4 \\ 1 & -1 & 1 \\ -9 & -2 & 2 \end{vmatrix} + 0 - 1\begin{vmatrix} -2 & 3 & 1 \\ 1 & 0 & -1 \\ -9 & 2 & -2 \end{vmatrix}$$

$$=3\left(-0 + (-1)\begin{vmatrix} 3 & 4 \\ 2 & 2 \end{vmatrix} - 1\begin{vmatrix} 3 & 1 \\ 2 & -2 \end{vmatrix}\right) - 6\left(-1\begin{vmatrix} 1 & 4 \\ -2 & 2 \end{vmatrix} + (-1)\begin{vmatrix} -2 & 4 \\ -9 & 2 \end{vmatrix} - 1\begin{vmatrix} -2 & 1 \\ -9 & -2 \end{vmatrix}\right) + 0$$
$$-1\left(-1\begin{vmatrix} 3 & 1 \\ 2 & -2 \end{vmatrix} + 0 - (-1)\begin{vmatrix} -2 & 3 \\ -9 & 2 \end{vmatrix}\right)$$

$$=3(0 - 1(-2) - 1(-8)) - 6(-1(10) - 1(32) - 1(13)) + 0 - 1(-1(-8) + 0 + 1(23))$$
$$= 3(10) - 6(-55) + 0 - 1(31)$$
$$= 329$$

(b) $$\begin{vmatrix} 3 & 6 & 0 & 1 \\ -2 & 3 & 1 & 4 \\ 1 & 0 & -1 & 1 \\ -9 & 2 & -2 & 2 \end{vmatrix} = (-1)\begin{vmatrix} 1 & 0 & -1 & 1 \\ -2 & 3 & 1 & 4 \\ 3 & 6 & 0 & 1 \\ -9 & 2 & -2 & 2 \end{vmatrix}$$ ← The first and third rows were interchanged.

$$= (-1)\begin{vmatrix} 1 & 0 & -1 & 1 \\ 0 & 3 & -1 & 6 \\ 0 & 6 & 3 & -2 \\ 0 & 2 & -11 & 11 \end{vmatrix}$$ ← 2 times the first row was added to the second, -3 times the first row was added to the third and 9 times the first row was added to the fourth.

$$= (-1)\begin{vmatrix} 1 & 0 & -1 & 1 \\ 0 & 3 & -1 & 6 \\ 0 & 0 & 5 & -14 \\ 0 & 0 & -\frac{31}{3} & 7 \end{vmatrix}$$ ← -2 times the second row was added to the third and $-\frac{2}{3}$ times the second row was added to the fourth.

$$= (-1)\begin{vmatrix} 1 & 0 & -1 & 1 \\ 0 & 3 & -1 & 6 \\ 0 & 0 & 5 & -14 \\ 0 & 0 & 0 & -\frac{329}{15} \end{vmatrix}$$ ← $\frac{31}{15}$ times the third row was added to the fourth.

$$= (-1)(1)(3)(5)\left(-\tfrac{329}{15}\right) = 329$$ ← Use Theorem 2.1.2.

9. $\begin{vmatrix} -1 & 5 & 2 \\ 0 & 2 & -1 \\ -3 & 1 & 1 \end{vmatrix} = \begin{vmatrix} -1 & 5 & 2 \\ 0 & 2 & -1 \\ -3 & 1 & 1 \end{vmatrix}\begin{matrix} -1 & 5 \\ 0 & 2 \\ -3 & 1 \end{matrix} = [-2+15+0]-[-12+1+0] = 24$

$\begin{vmatrix} -1 & -2 & -3 \\ -4 & -5 & -6 \\ -7 & -8 & -9 \end{vmatrix} = \begin{vmatrix} -1 & -2 & -3 \\ -4 & -5 & -6 \\ -7 & -8 & -9 \end{vmatrix}\begin{matrix} -1 & -2 \\ -4 & -5 \\ -7 & -8 \end{matrix} = [-45-84-96]-[-105-48-72] = 0$

$\begin{vmatrix} 3 & 0 & -1 \\ 1 & 1 & 1 \\ 0 & 4 & 2 \end{vmatrix} = \begin{vmatrix} 3 & 0 & -1 \\ 1 & 1 & 1 \\ 0 & 4 & 2 \end{vmatrix}\begin{matrix} 3 & 0 \\ 1 & 1 \\ 0 & 4 \end{matrix} = [6+0-4]-[0+12+0] = -10$

$\begin{vmatrix} -5 & 1 & 4 \\ 3 & 0 & 2 \\ 1 & -2 & 2 \end{vmatrix} = \begin{vmatrix} -5 & 1 & 4 \\ 3 & 0 & 2 \\ 1 & -2 & 2 \end{vmatrix}\begin{matrix} -5 & 1 \\ 3 & 0 \\ 1 & -2 \end{matrix} = [0+2-24]-[0+20+6] = -48$

11. In Exercise 1: $\begin{vmatrix} -4 & 2 \\ 3 & 3 \end{vmatrix} = -18 \neq 0$ therefore the matrix is invertible.

In Exercise 2: $\begin{vmatrix} 7 & -1 \\ -2 & -6 \end{vmatrix} = -44 \neq 0$ therefore the matrix is invertible.

In Exercise 3: $\begin{vmatrix} -1 & 5 & 2 \\ 0 & 2 & -1 \\ -3 & 1 & 1 \end{vmatrix} = 24 \neq 0$ therefore the matrix is invertible.

In Exercise 4: $\begin{vmatrix} -1 & -2 & -3 \\ -4 & -5 & -6 \\ -7 & -8 & -9 \end{vmatrix} = 0$ therefore the matrix is not invertible.

13. $\begin{vmatrix} 5 & b-3 \\ b-2 & -3 \end{vmatrix} = (5)(-3)-(b-3)(b-2) = -15-b^2+2b+3b-6 = -b^2+5b-21$

15.

$$\begin{vmatrix} 0 & 0 & 0 & 0 & -3 \\ 0 & 0 & 0 & -4 & 0 \\ 0 & 0 & -1 & 0 & 0 \\ 0 & 2 & 0 & 0 & 0 \\ 5 & 0 & 0 & 0 & 0 \end{vmatrix}$$

$$= (-1)\begin{vmatrix} 5 & 0 & 0 & 0 & 0 \\ 0 & 0 & 0 & -4 & 0 \\ 0 & 0 & -1 & 0 & 0 \\ 0 & 2 & 0 & 0 & 0 \\ 0 & 0 & 0 & 0 & -3 \end{vmatrix}$$ ← The first row and the fifth row were interchanged.

$$= (-1)(-1)\begin{vmatrix} 5 & 0 & 0 & 0 & 0 \\ 0 & 2 & 0 & 0 & 0 \\ 0 & 0 & -1 & 0 & 0 \\ 0 & 0 & 0 & -4 & 0 \\ 0 & 0 & 0 & 0 & -3 \end{vmatrix}$$ ← The second row and the fourth row were interchanged.

$$= (-1)(-1)(5)(2)(-1)(-4)(-3) = -120$$

17. It was shown in the solution of Exercise 1 that $\begin{vmatrix} -4 & 2 \\ 3 & 3 \end{vmatrix} = -18$. The determinant is nonzero, therefore by Theorem 2.3.3, the matrix $A = \begin{bmatrix} -4 & 2 \\ 3 & 3 \end{bmatrix}$ is invertible.

The cofactors are:

$$C_{11} = 3 \qquad C_{12} = -3$$
$$C_{21} = -2 \qquad C_{22} = -4$$

The matrix of cofactors is $\begin{bmatrix} 3 & -3 \\ -2 & -4 \end{bmatrix}$ and the adjoint matrix is $\text{adj}(A) = \begin{bmatrix} 3 & -2 \\ -3 & -4 \end{bmatrix}$.

From Theorem 2.3.6, we have $A^{-1} = \frac{1}{\det(A)}\text{adj}(A) = \frac{1}{-18}\begin{bmatrix} 3 & -2 \\ -3 & -4 \end{bmatrix} = \begin{bmatrix} -\frac{1}{6} & \frac{1}{9} \\ \frac{1}{6} & \frac{2}{9} \end{bmatrix}$.

19. It was shown in the solution of Exercise 3 that $\begin{vmatrix} -1 & 5 & 2 \\ 0 & 2 & -1 \\ -3 & 1 & 1 \end{vmatrix} = 24$. The determinant is nonzero, therefore by Theorem 2.3.3, $A = \begin{bmatrix} -1 & 5 & 2 \\ 0 & 2 & -1 \\ -3 & 1 & 1 \end{bmatrix}$ is invertible.

The cofactors of A are:

$$C_{11} = \begin{vmatrix} 2 & -1 \\ 1 & 1 \end{vmatrix} = 3 \qquad C_{12} = -\begin{vmatrix} 0 & -1 \\ -3 & 1 \end{vmatrix} = 3 \qquad C_{13} = \begin{vmatrix} 0 & 2 \\ -3 & 1 \end{vmatrix} = 6$$
$$C_{21} = -\begin{vmatrix} 5 & 2 \\ 1 & 1 \end{vmatrix} = -3 \qquad C_{22} = \begin{vmatrix} -1 & 2 \\ -3 & 1 \end{vmatrix} = 5 \qquad C_{23} = -\begin{vmatrix} -1 & 5 \\ -3 & 1 \end{vmatrix} = -14$$
$$C_{31} = \begin{vmatrix} 5 & 2 \\ 2 & -1 \end{vmatrix} = -9 \qquad C_{32} = -\begin{vmatrix} -1 & 2 \\ 0 & -1 \end{vmatrix} = -1 \qquad C_{33} = \begin{vmatrix} -1 & 5 \\ 0 & 2 \end{vmatrix} = -2$$

The matrix of cofactors is $\begin{bmatrix} 3 & 3 & 6 \\ -3 & 5 & -14 \\ -9 & -1 & -2 \end{bmatrix}$ and the adjoint matrix is $\text{adj}(A) = \begin{bmatrix} 3 & -3 & -9 \\ 3 & 5 & -1 \\ 6 & -14 & -2 \end{bmatrix}$.

From Theorem 2.3.6, we have $A^{-1} = \frac{1}{\det(A)}\text{adj}(A) = \frac{1}{24}\begin{bmatrix} 3 & -3 & -9 \\ 3 & 5 & -1 \\ 6 & -14 & -2 \end{bmatrix} = \begin{bmatrix} \frac{1}{8} & -\frac{1}{8} & -\frac{3}{8} \\ \frac{1}{8} & \frac{5}{24} & -\frac{1}{24} \\ \frac{1}{4} & -\frac{7}{12} & -\frac{1}{12} \end{bmatrix}$.

21. It was shown in the solution of Exercise 5 that $\begin{vmatrix} 3 & 0 & -1 \\ 1 & 1 & 1 \\ 0 & 4 & 2 \end{vmatrix} = -10$. The determinant is nonzero, therefore by Theorem 2.3.3, $A = \begin{bmatrix} 3 & 0 & -1 \\ 1 & 1 & 1 \\ 0 & 4 & 2 \end{bmatrix}$ is invertible.

The cofactors of A are:

$$C_{11} = \begin{vmatrix} 1 & 1 \\ 4 & 2 \end{vmatrix} = -2 \qquad C_{12} = -\begin{vmatrix} 1 & 1 \\ 0 & 2 \end{vmatrix} = -2 \qquad C_{13} = \begin{vmatrix} 1 & 1 \\ 0 & 4 \end{vmatrix} = 4$$

$$C_{21} = -\begin{vmatrix} 0 & -1 \\ 4 & 2 \end{vmatrix} = -4 \qquad C_{22} = \begin{vmatrix} 3 & -1 \\ 0 & 2 \end{vmatrix} = 6 \qquad C_{23} = -\begin{vmatrix} 3 & 0 \\ 0 & 4 \end{vmatrix} = -12$$

$$C_{31} = \begin{vmatrix} 0 & -1 \\ 1 & 1 \end{vmatrix} = 1 \qquad C_{32} = -\begin{vmatrix} 3 & -1 \\ 1 & 1 \end{vmatrix} = -4 \qquad C_{33} = \begin{vmatrix} 3 & 0 \\ 1 & 1 \end{vmatrix} = 3$$

The matrix of cofactors is $\begin{bmatrix} -2 & -2 & 4 \\ -4 & 6 & -12 \\ 1 & -4 & 3 \end{bmatrix}$ and the adjoint matrix is $\operatorname{adj}(A) = \begin{bmatrix} -2 & -4 & 1 \\ -2 & 6 & -4 \\ 4 & -12 & 3 \end{bmatrix}$.

From Theorem 2.3.6, we have $A^{-1} = \frac{1}{\det(A)}\operatorname{adj}(A) = \frac{1}{-10}\begin{bmatrix} -2 & -4 & 1 \\ -2 & 6 & -4 \\ 4 & -12 & 3 \end{bmatrix} = \begin{bmatrix} \frac{1}{5} & \frac{2}{5} & -\frac{1}{10} \\ \frac{1}{5} & -\frac{3}{5} & \frac{2}{5} \\ -\frac{2}{5} & \frac{6}{5} & -\frac{3}{10} \end{bmatrix}$.

23. It was shown in the solution of Exercise 7 that $\begin{vmatrix} 3 & 6 & 0 & 1 \\ -2 & 3 & 1 & 4 \\ 1 & 0 & -1 & 1 \\ -9 & 2 & -2 & 2 \end{vmatrix} = 329$. The determinant of A is nonzero therefore by Theorem 2.3.3, $A = \begin{bmatrix} 3 & 6 & 0 & 1 \\ -2 & 3 & 1 & 4 \\ 1 & 0 & -1 & 1 \\ -9 & 2 & -2 & 2 \end{bmatrix}$ is invertible.

The cofactors of A are:

$$C_{11} = \begin{vmatrix} 3 & 1 & 4 \\ 0 & -1 & 1 \\ 2 & -2 & 2 \end{vmatrix} = (-6 + 2 + 0) - (-8 - 6 + 0) = 10$$

$$C_{12} = -\begin{vmatrix} -2 & 1 & 4 \\ 1 & -1 & 1 \\ -9 & -2 & 2 \end{vmatrix} = -[(4 - 9 - 8) - (36 + 4 + 2)] = 55$$

$$C_{13} = \begin{vmatrix} -2 & 3 & 4 \\ 1 & 0 & 1 \\ -9 & 2 & 2 \end{vmatrix} = (0 - 27 + 8) - (0 - 4 + 6) = -21$$

$$C_{14} = -\begin{vmatrix} -2 & 3 & 1 \\ 1 & 0 & -1 \\ -9 & 2 & -2 \end{vmatrix} = -[(0 + 27 + 2) - (0 + 4 - 6)] = -31$$

$$C_{21} = -\begin{vmatrix} 6 & 0 & 1 \\ 0 & -1 & 1 \\ 2 & -2 & 2 \end{vmatrix} = -[(-12 + 0 + 0) - (-2 - 12 + 0)] = -2$$

$$C_{22} = \begin{vmatrix} 3 & 0 & 1 \\ 1 & -1 & 1 \\ -9 & -2 & 2 \end{vmatrix} = (-6 + 0 - 2) - (9 - 6 + 0) = -11$$

$$C_{23} = -\begin{vmatrix} 3 & 6 & 1 \\ 1 & 0 & 1 \\ -9 & 2 & 2 \end{vmatrix} = -[(0-54+2)-(0+6+12)] = 70$$

$$C_{24} = \begin{vmatrix} 3 & 6 & 0 \\ 1 & 0 & -1 \\ -9 & 2 & -2 \end{vmatrix} = (0+54+0)-(0-6-12) = 72$$

$$C_{31} = \begin{vmatrix} 6 & 0 & 1 \\ 3 & 1 & 4 \\ 2 & -2 & 2 \end{vmatrix} = (12+0-6)-(2-48+0) = 52$$

$$C_{32} = -\begin{vmatrix} 3 & 0 & 1 \\ -2 & 1 & 4 \\ -9 & -2 & 2 \end{vmatrix} = -[(6+0+4)-(-9-24+0)] = -43$$

$$C_{33} = \begin{vmatrix} 3 & 6 & 1 \\ -2 & 3 & 4 \\ -9 & 2 & 2 \end{vmatrix} = (18-216-4)-(-27+24-24) = -175$$

$$C_{34} = -\begin{vmatrix} 3 & 6 & 0 \\ -2 & 3 & 1 \\ -9 & 2 & -2 \end{vmatrix} = -[(-18-54+0)-(0+6+24)] = 102$$

$$C_{41} = -\begin{vmatrix} 6 & 0 & 1 \\ 3 & 1 & 4 \\ 0 & -1 & 1 \end{vmatrix} = -[(6+0-3)-(0-24+0)] = -27$$

$$C_{42} = \begin{vmatrix} 3 & 0 & 1 \\ -2 & 1 & 4 \\ 1 & -1 & 1 \end{vmatrix} = (3+0+2)-(1-12+0) = 16$$

$$C_{43} = -\begin{vmatrix} 3 & 6 & 1 \\ -2 & 3 & 4 \\ 1 & 0 & 1 \end{vmatrix} = -[(9+24+0)-(3+0-12)] = -42$$

$$C_{44} = \begin{vmatrix} 3 & 6 & 0 \\ -2 & 3 & 1 \\ 1 & 0 & -1 \end{vmatrix} = (-9+6+0)-(0+0+12) = -15$$

The matrix of cofactors is $\begin{bmatrix} 10 & 55 & -21 & -31 \\ -2 & -11 & 70 & 72 \\ 52 & -43 & -175 & 102 \\ -27 & 16 & -42 & -15 \end{bmatrix}$ and $\operatorname{adj}(A) = \begin{bmatrix} 10 & -2 & 52 & -27 \\ 55 & -11 & -43 & 16 \\ -21 & 70 & -175 & -42 \\ -31 & 72 & 102 & -15 \end{bmatrix}$.

From Theorem 2.3.6, we have

$$A^{-1} = \frac{1}{\det(A)}\operatorname{adj}(A) = \frac{1}{329}\begin{bmatrix} 10 & -2 & 52 & -27 \\ 55 & -11 & -43 & 16 \\ -21 & 70 & -175 & -42 \\ -31 & 72 & 102 & -15 \end{bmatrix} = \begin{bmatrix} \frac{10}{329} & -\frac{2}{329} & \frac{52}{329} & -\frac{27}{329} \\ \frac{55}{329} & -\frac{11}{329} & -\frac{43}{329} & \frac{16}{329} \\ -\frac{3}{47} & \frac{10}{47} & -\frac{25}{47} & -\frac{6}{47} \\ -\frac{31}{329} & \frac{72}{329} & \frac{102}{329} & -\frac{15}{329} \end{bmatrix}.$$

25. $A = \begin{bmatrix} \frac{3}{5} & -\frac{4}{5} \\ \frac{4}{5} & \frac{3}{5} \end{bmatrix}$, $\det(A) = \left(\frac{3}{5}\right)\left(\frac{3}{5}\right) - \left(-\frac{4}{5}\right)\left(\frac{4}{5}\right) = \frac{9}{25} + \frac{16}{25} = 1$; $A_1 = \begin{bmatrix} x & -\frac{4}{5} \\ y & \frac{3}{5} \end{bmatrix}$, $A_2 = \begin{bmatrix} \frac{3}{5} & x \\ \frac{4}{5} & y \end{bmatrix}$;

$x' = \frac{\det(A_1)}{\det(A)} = \frac{3}{5}x + \frac{4}{5}y$, $y' = \frac{\det(A_2)}{\det(A)} = \frac{3}{5}y - \frac{4}{5}x$

27. The coefficient matrix of the given system is $A = \begin{bmatrix} 1 & 1 & \alpha \\ 1 & 1 & \beta \\ \alpha & \beta & 1 \end{bmatrix}$. Coefficient expansion along the first row yields

$$\det(A) = 1\begin{vmatrix} 1 & \beta \\ \beta & 1 \end{vmatrix} - 1\begin{vmatrix} 1 & \beta \\ \alpha & 1 \end{vmatrix} + \alpha\begin{vmatrix} 1 & 1 \\ \alpha & \beta \end{vmatrix}$$
$$= 1 - \beta^2 - (1 - \alpha\beta) + \alpha(\beta - \alpha) = -\alpha^2 + 2\alpha\beta - \beta^2 = -(\alpha - \beta)^2$$

By Theorem 2.3.8, the given system has a nontrivial solution if and only if $\det(A) = 0$, i.e., $\alpha = \beta$.

29. (a) We will justify the third equality, $a\cos\beta + b\cos\alpha = c$ by considering three cases:

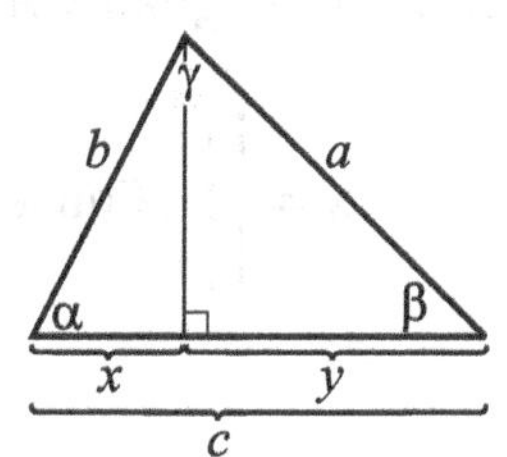

CASE I: $\alpha \le \frac{\pi}{2}$ and $\beta \le \frac{\pi}{2}$

Referring to the figure on the right side, we have
$x = b\cos\alpha$ and $y = a\cos\beta$.
Since $x + y = c$ we obtain , $a\cos\beta + b\cos\alpha = c$.

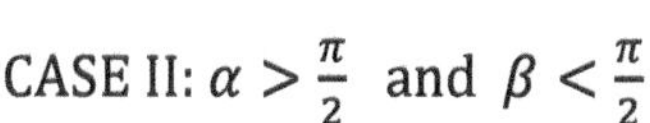

CASE II: $\alpha > \frac{\pi}{2}$ and $\beta < \frac{\pi}{2}$

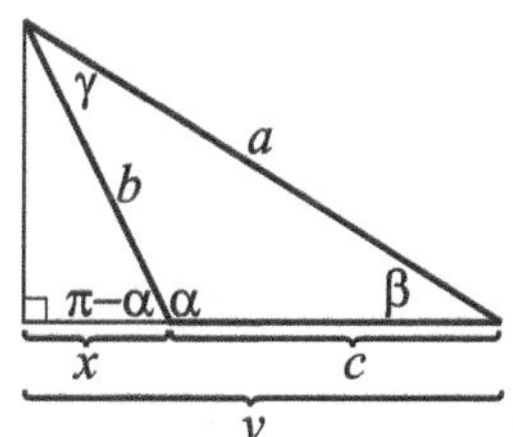

Referring to the picture on the right side, we can write
$x = b\cos(\pi - \alpha) = -b\cos\alpha$ and $y = a\cos\beta$
This time we can write $c = y - x = a\cos\beta - (-b\cos\alpha)$
therefore once again $a\cos\beta + b\cos\alpha = c$.

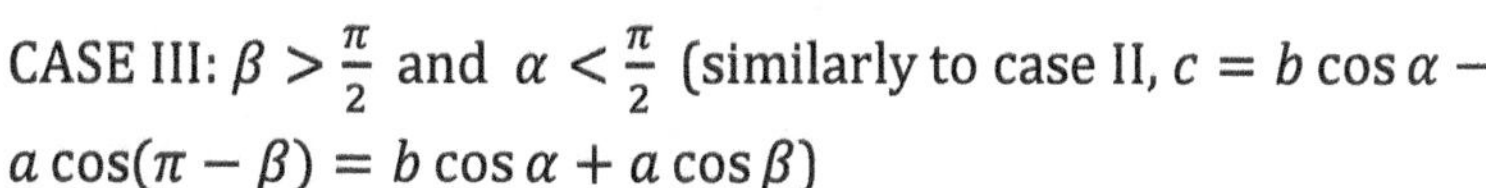
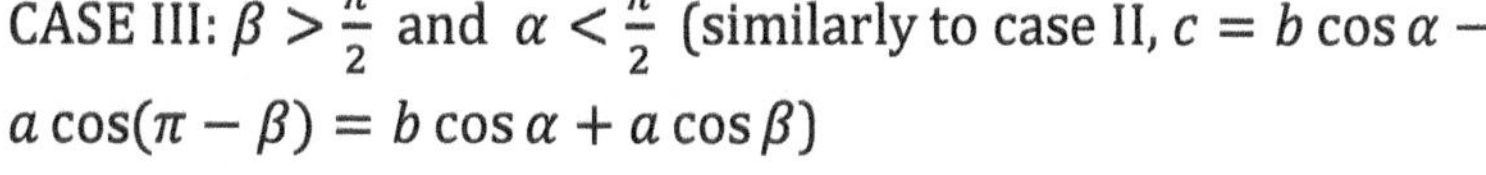

CASE III: $\beta > \frac{\pi}{2}$ and $\alpha < \frac{\pi}{2}$ (similarly to case II, $c = b\cos\alpha - a\cos(\pi - \beta) = b\cos\alpha + a\cos\beta$)

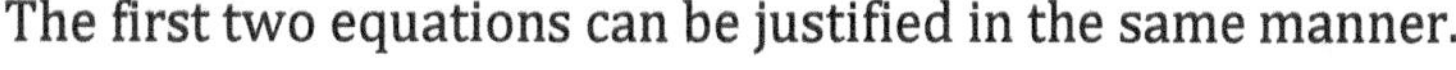

The first two equations can be justified in the same manner.

Denoting $X = \cos\alpha$, $Y = \cos\beta$, and $Z = \cos\gamma$ we can rewrite the linear system as

$$\begin{aligned} cY + bZ &= a \\ cX + aZ &= b \\ bX + aY &= c \end{aligned}$$

We have $\det(A) = \begin{vmatrix} 0 & c & b \\ c & 0 & a \\ b & a & 0 \end{vmatrix} = [0 + abc + abc] - [0 + 0 + 0] = 2abc$ and

$\det(A_1) = \begin{vmatrix} a & c & b \\ b & 0 & a \\ c & a & 0 \end{vmatrix} = [0 + ac^2 + ab^2] - [0 + a^3 + 0] = a(b^2 + c^2 - a^2)$ therefore by Cramer's rule

$$\cos\alpha = X = \frac{\det(A_1)}{\det(A)} = \frac{a(b^2+c^2-a^2)}{2abc} = \frac{b^2+c^2-a^2}{2bc}.$$

(b) Using the results obtained in part (a) along with

$\det(A_2) = \begin{vmatrix} 0 & a & b \\ c & b & a \\ b & c & 0 \end{vmatrix} = [0 + a^2b + bc^2] - [b^3 + 0 + 0] = b(a^2 + c^2 - b^2)$ and

$\det(A_3) = \begin{vmatrix} 0 & c & a \\ c & 0 & b \\ b & a & c \end{vmatrix} = [0 + b^2c + a^2c] - [0 + 0 + c^3] = c(a^2 + b^2 - c^2)$ therefore by

Cramer's rule

$\cos\beta = Y = \frac{\det(A_2)}{\det(A)} = \frac{a^2+c^2-b^2}{2ac}$ and $\cos\gamma = Z = \frac{\det(A_3)}{\det(A)} = \frac{a^2+b^2-c^2}{2ab}$.

31. From Theorem 2.3.6, $A^{-1} = \frac{1}{\det(A)}\text{adj}(A)$ therefore $\text{adj}(A) = \det(A)\, A^{-1}$. Consequently,

$$\left(\frac{1}{\det(A)}A\right)\text{adj}(A) = \left(\frac{1}{\det(A)}A\right)(\det(A)\ A^{-1}) = \frac{\det(A)}{\det(A)}(AA^{-1}) = I_n \text{ so } (\text{adj}(A))^{-1} = \frac{1}{\det(A)}A.$$

Using Theorem 2.3.5, we can also write $\text{adj}(A^{-1}) = \det(A^{-1})\,(A^{-1})^{-1} = \frac{1}{\det(A)}A$.

33. The equality $A\begin{bmatrix}1\\ \vdots\\ 1\end{bmatrix} = \begin{bmatrix}0\\ \vdots\\ 0\end{bmatrix}$ means that the homogeneous system $A\mathbf{x} = \mathbf{0}$ has a nontrivial solution

$\mathbf{x} = \begin{bmatrix}1\\ \vdots\\ 1\end{bmatrix}$. Consequently, it follows from Theorem 2.3.8 that $\det(A) = 0$.

CHAPTER 3: EUCLIDEAN VECTOR SPACES

3.1 Vectors in 2-Space, 3-Space, and n-Space

1. **(a)** $(4-1, 1-5) = (3, -4)$ **(b)** $(0-2, 0-3, 4-0) = (-2, -3, 4)$

3. **(a)** $\overrightarrow{P_1P_2} = (2-3, 8-5) = (-1, 3)$ **(b)** $\overrightarrow{P_1P_2} = (2-5, 4-(-2), 2-1) = (-3, 6, 1)$

5. **(a)** Denote the terminal point by $B(b_1, b_2)$. Since the vector $\overrightarrow{AB} = (b_1 - 1, b_2 - 1)$ is to be equivalent to the vector $\mathbf{u} = (1, 2)$, the coordinates of B must satisfy the equations

$$b_1 - 1 = 1 \quad \text{and} \quad b_2 - 1 = 2$$

therefore $b_1 = 2$ and $b_2 = 3$. The terminal point is $B(2, 3)$.

(b) Denote the initial point by $A(a_1, a_2, a_3)$. Since the vector $\overrightarrow{AB} = (-1 - a_1, -1 - a_2, 2 - a_3)$ is to be equivalent to the vector $\mathrm{u} = (1, 1, 3)$, the coordinates of A must satisfy the equations

$$-1 - a_1 = 1, \qquad -1 - a_2 = 1, \qquad \text{and} \qquad 2 - a_3 = 3$$

therefore $a_1 = -2$, $a_2 = -2$, and $a_3 = -1$. The initial point is $A(-2, -2, -1)$.

7. **(a)** For any positive real number k, the vector $\mathbf{u} = k\mathbf{v}$ has the same direction as $\mathbf{v}$. For example, letting $k = 1$, we have $\mathbf{u} = (4, -2, -1)$. If the terminal point is $Q(3, 0, -5)$ then the initial point has coordinates $(3-4, 0-(-2), -5-(-1))$, i.e., $(-1, 2, -4)$.

(b) For any negative real number k, the vector $\mathbf{u} = k\mathbf{v}$ is oppositely directed to $\mathbf{v}$. For example, letting $k = -1$, we have $\mathbf{u} = (-4, 2, 1)$. If the terminal point is $Q(3, 0, -5)$ then the initial point has coordinates $(3-(-4), 0-2, -5-1)$, i.e., $(7, -2, -6)$.

9. **(a)** $\mathbf{u} + \mathbf{w} = \big(4 + (-3), -1 + (-3)\big) = (1, -4)$

(b) $\mathbf{v} - 3\mathbf{u} = (0, 5) - (12, -3) = \big(0 - 12, 5 - (-3)\big) = (-12, 8)$

(c) $2(\mathbf{u} - 5\mathbf{w}) = 2[(4, -1) - (-15, -15)] = 2(19, 14) = (38, 28)$

(d) $3\mathbf{v} - 2(\mathbf{u} + 2\mathbf{w}) = (0, 15) - 2[(4, -1) + (-6, -6)] = (0, 15) - 2(-2, -7)$
$= (0, 15) - (-4, -14) = (4, 29)$

11. **(a)** $\mathbf{v} - \mathbf{w} = (4-5, 7-(-2), -3-8, 2-1) = (-1, 9, -11, 1)$

(b) $-\mathbf{u} + (\mathbf{v} - 4\mathbf{w}) = (3, -2, -1, 0) + [(4, 7, -3, 2) - (20, -8, 32, 4)]$
$= (3, -2, -1, 0) + (-16, 15, -35, -2) = (-13, 13, -36, -2)$

(c) $6(\mathbf{u} - 3\mathbf{v}) = 6[(-3, 2, 1, 0) - (12, 21, -9, 6)] = 6(-15, -19, 10, -6) = (-90, -114, 60, -36)$

(d) $(6\mathbf{v} - \mathbf{w}) - (4\mathbf{u} + \mathbf{v}) = [(24, 42, -18, 12) - (5, -2, 8, 1)] - [(-12, 8, 4, 0) + (4, 7, -3, 2)]$
$= (19, 44, -26, 11) - (-8, 15, 1, 2) = (27, 29, -27, 9)$

13. Solve the vector equation using the properties listed in Theorems 3.1.1 and 3.1.2:

$3\mathbf{u} + \mathbf{v} + (-2)\mathbf{w} = 3\mathbf{x} + 2\mathbf{w}$ [Part (c) of Theorem 3.1.2 and part (g) of Theorem 3.1.1]

$(3\mathbf{u} + \mathbf{v}) + (-4)\mathbf{w} = 3\mathbf{x} + 0\mathbf{w}$ [Add $-2\mathbf{w}$ to both sides, use parts (b) and (d) of Th. 3.1.1]

$(3\mathbf{u} + \mathbf{v}) + (-4)\mathbf{w} = 3\mathbf{x}$ [Use part (a) of Theorem 3.1.2]

$\frac{1}{3}[(3\mathbf{u} + \mathbf{v}) + (-4)\mathbf{w}] = \frac{1}{3}(3\mathbf{x})$ [Multiply both sides by $\frac{1}{3}$]

$\frac{1}{3}[(3\mathbf{u} + \mathbf{v}) + (-4)\mathbf{w}] = \mathbf{x}$ [Parts (g) and (h) of Theorem 3.1.1]

Therefore $\mathbf{x} = \frac{1}{3}[(-5,13,0,2) + (-20,8,-32,-4)] = \frac{1}{3}(-25,21,-32,-2) = (-\frac{25}{3},7,-\frac{32}{3},-\frac{2}{3})$.

15. Vectors **u** and **v** are parallel (collinear) if one of them is a scalar multiple of the other one, i.e. either $\mathbf{u} = a\mathbf{v}$ for some scalar a or $\mathbf{v} = b\mathbf{u}$ for some scalar b or both (the two conditions are not equivalent if one of the vectors is a zero vector, but the other one is not.)

(a) $\mathbf{v} = (4,2,0,6,10,2)$ does not equal $k\mathbf{u} = (-2k,k,0,3k,5k,k)$ for any scalar k; **v** is not parallel to **u**

(b) $\mathbf{v} = (4,-2,0,-6,-10,-2) = -2\mathbf{u}$; **v** is parallel to **u**

(c) $\mathbf{v} = (0,0,0,0,0,0) = 0\mathbf{u}$; **v** is parallel to **u**

17. The vector equation $a(1,-1,3,5) + b(2,1,0,-3) = (1,-4,9,18)$ is equivalent to the linear system

$$\begin{aligned} 1a &+ 2b &= 1 \\ -1a &+ 1b &= -4 \\ 3a &+ 0b &= 9 \\ 5a &- 3b &= 18 \end{aligned}$$

whose augmented matrix $\begin{bmatrix} 1 & 2 & 1 \\ -1 & 1 & -4 \\ 3 & 0 & 9 \\ 5 & -3 & 18 \end{bmatrix}$ has the reduced row echelon form $\begin{bmatrix} 1 & 0 & 3 \\ 0 & 1 & -1 \\ 0 & 0 & 0 \\ 0 & 0 & 0 \end{bmatrix}$.

Therefore, the unique solution is $a = 3$ and $b = -1$.

19. The vector equation $c_1(1,-1,0) + c_2(3,2,1) + c_3(0,1,4) = (-1,1,19)$ is equivalent to the linear system

$$\begin{aligned} 1c_1 &+ 3c_2 &+ 0c_3 &= -1 \\ -1c_1 &+ 2c_2 &+ 1c_3 &= 1 \\ 0c_1 &+ 1c_2 &+ 4c_3 &= 19 \end{aligned}$$

whose augmented matrix $\begin{bmatrix} 1 & 3 & 0 & -1 \\ -1 & 2 & 1 & 1 \\ 0 & 1 & 4 & 19 \end{bmatrix}$ has the reduced row echelon form $\begin{bmatrix} 1 & 0 & 0 & 2 \\ 0 & 1 & 0 & -1 \\ 0 & 0 & 1 & 5 \end{bmatrix}$.

Therefore, the unique solution is $c_1 = 2$, $c_2 = -1$, and $c_3 = 5$.

21. The vector equation $c_1(-2,9,6) + c_2(-3,2,1) + c_3(1,\ 7,5) = (0,5,4)$ is equivalent to the linear system

$$\begin{aligned} -2c_1 &- 3c_2 &+ 1c_3 &= 0 \\ 9c_1 &+ 2c_2 &+ 7c_3 &= 5 \\ 6c_1 &+ 1c_2 &+ 5c_3 &= 4 \end{aligned}$$

whose augmented matrix $\begin{bmatrix} -2 & -3 & 1 & 0 \\ 9 & 2 & 7 & 5 \\ 6 & 1 & 5 & 4 \end{bmatrix}$ has the reduced row echelon form $\begin{bmatrix} 1 & 0 & 1 & 0 \\ 0 & 1 & -1 & 0 \\ 0 & 0 & 0 & 1 \end{bmatrix}$.

The system has no solution.

23. **(a)** The midpoint of the segment is the terminal point of the vector
$\overrightarrow{OM} = \overrightarrow{OP} + \frac{1}{2}\overrightarrow{PQ} = (2,3,-2) + \frac{1}{2}(7-2,-4-3,1-(-2)) = \left(\frac{9}{2},-\frac{1}{2},-\frac{1}{2}\right)$
therefore the midpoint has coordinates $\left(\frac{9}{2},-\frac{1}{2},-\frac{1}{2}\right)$.

(b) The desired point is the terminal point of the vector
$\overrightarrow{ON} = \overrightarrow{OP} + \frac{3}{4}\overrightarrow{PQ} = (2,3,-2) + \frac{3}{4}(7-2,-4-3,1-(-2)) = \left(\frac{23}{4},-\frac{9}{4},\frac{1}{4}\right)$
therefore this point has coordinates $\left(\frac{23}{4},-\frac{9}{4},\frac{1}{4}\right)$.

25. **(a)** $\mathbf{u}+\mathbf{v}+\mathbf{w} = (5,-5)+(-10,2)+(3,8) = (-2,5)$

(b) $\mathbf{u}+\mathbf{v}+\mathbf{w} = (10,-7)+(-3,8)+(-4,-9) = (3,-8)$

27. The midpoint of the line segment connecting the points $P(x_1,y_1,z_1)$ and $Q(x_2,y_2,z_2)$ is

$$\left(\frac{x_1+x_2}{2},\frac{y_1+y_2}{2},\frac{z_1+z_2}{2}\right)$$

Therefore we have

$$\left(\frac{1+x_2}{2},\frac{3+y_2}{2},\frac{7+z_2}{2}\right) = (4,0,-6).$$

This vector equation is equivalent to a system of three linear equations in three unknowns that is easy to solve:

$$\begin{aligned} \frac{1+x_2}{2} &= 4 \quad \Leftrightarrow \quad x_2 = 7 \\ \frac{3+y_2}{2} &= 0 \quad \Leftrightarrow \quad y_2 = -3 \\ \frac{7+z_2}{2} &= -6 \quad \Leftrightarrow \quad z_2 = -19 \end{aligned}$$

We conclude that the point Q is $(7,-3,-19)$.

29. **(a)** We have $\mathbf{a}+\mathbf{d} = \mathbf{b}+\mathbf{e} = \mathbf{c}+\mathbf{f} = \mathbf{0}$ therefore $\mathbf{a}+\mathbf{b}+\mathbf{c}+\mathbf{d}+\mathbf{e}+\mathbf{f} = \mathbf{0}$.

(b) The sum is $\frac{1}{2}(\mathbf{0}) = \mathbf{0}$.

(c) From part (a), $\mathbf{b}+\mathbf{c}+\mathbf{d}+\mathbf{e}+\mathbf{f} = -\mathbf{a}$.

(d) From part (a), the sum of any five vectors remaining after one is removed equals to the negative of the removed vector.

True-False Exercises

(a) False. Equivalent vectors have the same length and direction - they may have different initial points.

(b) False. According to Definition 2, equivalent vectors must have the same number of components.

(c) False. **v** and $k\mathbf{v}$ are parallel for any k.

(d) True. This is a consequence of Theorem 3.1.1.

(e) True. This is a consequence of Theorem 3.1.1.

(f) False. At least one of the scalars must be nonzero for the vectors to be parallel.

(g) False. For nonzero vector **u**, the vectors **u** and $-\mathbf{u}$ are collinear and have the same length but are not equal.

(h) True.

(i) False. $(k+m)(\mathbf{u}+\mathbf{v}) = (k+m)\mathbf{u} + (k+m)\mathbf{v}$.

(j) True. $\mathbf{x} = \frac{5}{8}\mathbf{v} + \frac{1}{2}\mathbf{w}$.

(k) False. For instance, if $\mathbf{v}_2 = 2\mathbf{v}_1$ then $4\mathbf{v}_1 + 2\mathbf{v}_2 = 2\mathbf{v}_1 + 3\mathbf{v}_2$.

3.2 Norm, Dot Product, and Distance in R^n

1. **(a)** $\|\mathbf{v}\| = \sqrt{2^2+2^2+2^2} = \sqrt{12} = 2\sqrt{3}$;

$\frac{1}{\|\mathbf{v}\|}\mathbf{v} = \frac{1}{2\sqrt{3}}(2,2,2) = (\frac{1}{\sqrt{3}}, \frac{1}{\sqrt{3}}, \frac{1}{\sqrt{3}})$; $-\frac{1}{\|\mathbf{v}\|}\mathbf{v} = -\frac{1}{2\sqrt{3}}(2,2,2) = (-\frac{1}{\sqrt{3}}, -\frac{1}{\sqrt{3}}, -\frac{1}{\sqrt{3}})$

(b) $\|\mathbf{v}\| = \sqrt{1^2+0^2+2^2+1^2+3^2} = \sqrt{15}$;

$\frac{1}{\|\mathbf{v}\|}\mathbf{v} = \frac{1}{\sqrt{15}}(1,0,2,1,3) = (\frac{1}{\sqrt{15}}, 0, \frac{2}{\sqrt{15}}, \frac{1}{\sqrt{15}}, \frac{3}{\sqrt{15}})$;

$-\frac{1}{\|\mathbf{v}\|}\mathbf{v} = -\frac{1}{\sqrt{15}}(1,0,2,1,3) = (-\frac{1}{\sqrt{15}}, 0, -\frac{2}{\sqrt{15}}, -\frac{1}{\sqrt{15}}, -\frac{3}{\sqrt{15}})$

3. **(a)** $\mathbf{u}+\mathbf{v} = (3,-5,7)$; $\|\mathbf{u}+\mathbf{v}\| = \sqrt{3^2+(-5)^2+7^2} = \sqrt{83}$

(b) $\|\mathbf{u}\| + \|\mathbf{v}\| = \sqrt{2^2+(-2)^2+3^2} + \sqrt{1^2+(-3)^2+4^2} = \sqrt{17}+\sqrt{26}$

(c) $-2\mathbf{u}+2\mathbf{v} = (-4,4,-6)+(2,-6,8) = (-2,-2,2)$;

$\|-2\mathbf{u}+2\mathbf{v}\| = \sqrt{(-2)^2+(-2)^2+2^2} = \sqrt{12} = 2\sqrt{3}$

(d) $3\mathbf{u}-5\mathbf{v}+\mathbf{w} = (6,-6,9)-(5,-15,20)+(3,6,-4) = (4,15,-15)$;

$\|3\mathbf{u}-5\mathbf{v}+\mathbf{w}\| = \sqrt{4^2+15^2+(-15)^2} = \sqrt{466}$

5. **(a)** $3\mathbf{u}-5\mathbf{v}+\mathbf{w} = (-6,-3,12,15)-(15,5,-25,35)+(-6,2,1,1) = (-27,-6,38,-19)$;

$\|3\mathbf{u}-5\mathbf{v}+\mathbf{w}\| = \sqrt{(-27)^2+(-6)^2+38^2+(-19)^2} = \sqrt{2570}$

(b) $\|3\mathbf{u}\| - 5\|\mathbf{v}\| + \|\mathbf{w}\|$

$= \sqrt{(-6)^2+(-3)^2+12^2+15^2} - 5\sqrt{3^2+1^2+(-5)^2+7^2} + \sqrt{(-6)^2+2^2+1^2+1^2}$

$= \sqrt{414} - 5\sqrt{84} + \sqrt{42} = 3\sqrt{46} - 10\sqrt{21} + \sqrt{42}$

(c) $\|\mathbf{u}\| = \sqrt{(-2)^2+(-1)^2+4^2+5^2} = \sqrt{46}$;

$\|-\|\mathbf{u}\|\mathbf{v}\| = \|-\sqrt{46}\mathbf{v}\| = \sqrt{46}\sqrt{3^2+1^2+(-5)^2+7^2} = \sqrt{46}\sqrt{84} = 2\sqrt{966}$

$\|\mathbf{u}-\mathbf{v}\|\mathbf{w} = (-6\sqrt{114}, 2\sqrt{114}, \sqrt{114}, \sqrt{114})$; $\|\|\mathbf{u}-\mathbf{v}\|\mathbf{w}\| = \sqrt{4788} = 6\sqrt{133}$

7. $\|k\mathbf{v}\| = \sqrt{(-2k)^2 + (3k)^2 + 0^2 + (6k)^2} = \sqrt{49k^2} = 7\sqrt{k^2}$; this quantity equals 5 if $k = \frac{5}{7}$ or $k = -\frac{5}{7}$

9. **(a)** $\mathbf{u}\cdot\mathbf{v} = (3)(2) + (1)(2) + (4)(-4) = -8$
$\mathbf{u}\cdot\mathbf{u} = (3)(3) + (1)(1) + (4)(4) = 26$
$\mathbf{v}\cdot\mathbf{v} = (2)(2) + (2)(2) + (-4)(-4) = 24$

(b) $\mathbf{u}\cdot\mathbf{v} = (1)(2) + (1)(-2) + (4)(3) + (6)(-2) = 0$
$\mathbf{u}\cdot\mathbf{u} = (1)(1) + (1)(1) + (4)(4) + (6)(6) = 54$
$\mathbf{v}\cdot\mathbf{v} = (2)(2) + (-2)(-2) + (3)(3) + (-2)(-2) = 21$

11. **(a)** $d(\mathbf{u},\mathbf{v}) = \|\mathbf{u}-\mathbf{v}\| = \sqrt{(3-1)^2 + (3-0)^2 + (3-4)^2} = \sqrt{14}$

$\cos\theta = \frac{\mathbf{u}\cdot\mathbf{v}}{\|\mathbf{u}\|\|\mathbf{v}\|} = \frac{(3)(1)+(3)(0)+(3)(4)}{\sqrt{3^2+3^2+3^2}\sqrt{1^2+0^2+4^2}} = \frac{15}{\sqrt{27}\sqrt{17}} = \frac{5}{\sqrt{51}}$; the angle is acute since $\mathbf{u}\cdot\mathbf{v} > 0$

(b) $d(\mathbf{u},\mathbf{v}) = \|\mathbf{u}-\mathbf{v}\| = \sqrt{(0-(-3))^2 + (-2-2)^2 + (-1-4)^2 + (1-4)^2} = \sqrt{59}$

$\cos\theta = \frac{\mathbf{u}\cdot\mathbf{v}}{\|\mathbf{u}\|\|\mathbf{v}\|} = \frac{(0)(-3)+(-2)(2)+(-1)(4)+(1)(4)}{\sqrt{0^2+(-2)^2+(-1)^2+1^2}\sqrt{(-3)^2+2^2+4^2+4^2}} = \frac{-4}{\sqrt{6}\sqrt{45}}$; the angle is obtuse since $\mathbf{u}\cdot\mathbf{v} < 0$

13. The angle between the two vectors is 30°, so by Formula (1) we have $\mathbf{a}\cdot\mathbf{b} = \|\mathbf{a}\|\|\mathbf{b}\|\cos 30° = \frac{45\sqrt{3}}{2}$.

15. **(a)** $\mathbf{u}\cdot(\mathbf{v}\cdot\mathbf{w})$ does not make sense; $\mathbf{v}\cdot\mathbf{w}$ is a scalar, whereas the dot product is only defined for vectors

(b) $\mathbf{u}\cdot(\mathbf{v}+\mathbf{w})$ makes sense (the result is a scalar)

(c) $\|\mathbf{u}\cdot\mathbf{v}\|$ does not make sense; $\mathbf{u}\cdot\mathbf{v}$ is a scalar, whereas the norm is only defined for vectors

(d) $(\mathbf{u}\cdot\mathbf{v}) - \|\mathbf{u}\|$ makes sense (the result is a scalar)

17. **(a)** $|\mathbf{u}\cdot\mathbf{v}| = |(-3)(2) + (1)(-1) + (0)(3)| = 7$;
$\|\mathbf{u}\|\|\mathbf{v}\| = \sqrt{(-3)^2 + 1^2 + 0^2}\sqrt{2^2 + (-1)^2 + 3^2} = \sqrt{10}\sqrt{14}$
Since $|\mathbf{u}\cdot\mathbf{v}| = 7 = \sqrt{49} \le \sqrt{140} = \sqrt{10}\sqrt{14} = \|\mathbf{u}\|\|\mathbf{v}\|$, the Cauchy-Schwarz inequality holds.

(b) $|\mathbf{u}\cdot\mathbf{v}| = |(0)(1) + (2)(1) + (2)(1) + (1)(1)| = 5$
$\|\mathbf{u}\|\|\mathbf{v}\| = \sqrt{0^2 + 2^2 + 2^2 + 1^2}\sqrt{1^2 + 1^2 + 1^2 + 1^2} = \sqrt{9}\sqrt{4} = 6$
Since $|\mathbf{u}\cdot\mathbf{v}| = 5 \le 6 = \|\mathbf{u}\|\|\mathbf{v}\|$, the Cauchy-Schwarz inequality holds.

21. We have $\|\mathbf{i}\| = \|\mathbf{j}\| = \|\mathbf{k}\| = 1$. Therefore

$$\cos\alpha = \frac{\mathbf{v}\cdot\mathbf{i}}{\|\mathbf{v}\|\|\mathbf{i}\|} = \frac{(v_1)(1) + (v_2)(0) + (v_3)(0)}{\|\mathbf{v}\|} = \frac{v_1}{\|\mathbf{v}\|}$$
$$\cos\beta = \frac{\mathbf{v}\cdot\mathbf{j}}{\|\mathbf{v}\|\|\mathbf{j}\|} = \frac{(v_1)(0) + (v_2)(1) + (v_3)(0)}{\|\mathbf{v}\|} = \frac{v_2}{\|\mathbf{v}\|}$$
$$\cos\gamma = \frac{\mathbf{v}\cdot\mathbf{k}}{\|\mathbf{v}\|\|\mathbf{k}\|} = \frac{(v_1)(0) + (v_2)(0) + (v_3)(1)}{\|\mathbf{v}\|} = \frac{v_3}{\|\mathbf{v}\|}$$

23. Using the result of Exercise 21, and letting $\mathbf{v}_1 = (a_1, b_1, c_1)$ and $\mathbf{v}_2 = (a_2, b_2, c_2)$, we can have

$$\cos\alpha_1 \cos\alpha_2 + \cos\beta_1 \cos\beta_2 + \cos\gamma_1 \cos\gamma_2 = \frac{a_1}{\|\mathbf{v}_1\|}\frac{a_2}{\|\mathbf{v}_2\|} + \frac{b_1}{\|\mathbf{v}_1\|}\frac{b_2}{\|\mathbf{v}_2\|} + \frac{c_1}{\|\mathbf{v}_1\|}\frac{c_2}{\|\mathbf{v}_2\|} = \frac{\mathbf{v}_1 \cdot \mathbf{v}_2}{\|\mathbf{v}_1\|\|\mathbf{v}_2\|}$$

The left-hand side is zero if and only if the right-hand side is zero; this happens if and only if $\mathbf{v}_1$ and $\mathbf{v}_2$ are nonzero orthogonal vectors.

25. Align the edges of the box with the coordinate axes so that the diagonal becomes the vector $\mathbf{v} = (10,15,25)$. The length of this vector is $\|\mathbf{v}\| = \sqrt{10^2 + 15^2 + 25^2} = 5\sqrt{38}$ therefore

- the angle between $\mathbf{v}$ and the x-axis is $\cos^{-1}\left(\frac{\mathbf{v}\cdot\mathbf{i}}{\|\mathbf{v}\|\|\mathbf{i}\|}\right) = \cos^{-1}\left(\frac{2}{\sqrt{38}}\right) \approx 71°$,
- the angle between $\mathbf{v}$ and the y-axis is $\cos^{-1}\left(\frac{\mathbf{v}\cdot\mathbf{j}}{\|\mathbf{v}\|\|\mathbf{j}\|}\right) = \cos^{-1}\left(\frac{3}{\sqrt{38}}\right) \approx 61°$,
- the angle between $\mathbf{v}$ and the z-axis is $\cos^{-1}\left(\frac{\mathbf{v}\cdot\mathbf{k}}{\|\mathbf{v}\|\|\mathbf{k}\|}\right) = \cos^{-1}\left(\frac{5}{\sqrt{38}}\right) \approx 36°$. **29.** The scalar product of $\frac{m}{\|\mathbf{v}\|}\mathbf{v}$ has the same direction as $\mathbf{v}$ and its length is $\left\|\frac{m}{\|\mathbf{v}\|}\mathbf{v}\right\| = m$.

31. We are looking for the force $\mathbf{F}$ such that $\mathbf{F} + (10\cos 60°, 10\sin 60°) + (-8, 0) = (0, 0)$. This yields $\mathbf{F} = -(5, 5\sqrt{3}) - (-8, 0) = (3, -5\sqrt{3})$. The magnitude of $\mathbf{F}$ is $\sqrt{84}$ lb ≈ 9.17 lb; the vector forms the angle $\approx -70.9°$ with the positive x-axis.

True-False Exercises

(a) True. By Theorem 3.2.1(b), $\|2\mathbf{v}\| = |2|\|\mathbf{v}\| = 2\|\mathbf{v}\|$.

(b) True.

(c) False. Norm can be zero for the zero vector.

(d) True. The two vectors are $\frac{1}{\|\mathbf{v}\|}\mathbf{v}$ and $-\frac{1}{\|\mathbf{v}\|}\mathbf{v}$.

(e) True. This follows from Formula (13).

(f) False. The first expression does not make sense since the scalar $\mathbf{u}\cdot\mathbf{v}$ cannot be added to a vector.

(g) False. For example, let $\mathbf{u} = (1,0)$, $\mathbf{v} = (0,1)$, and $\mathbf{w} = (0,2)$. We have $\mathbf{v} \neq \mathbf{w}$ even though $\mathbf{u}\cdot\mathbf{v} = \mathbf{u}\cdot\mathbf{w}$.

(h) False. For example, for $\mathbf{u} = (1,1) \neq (0,0)$ and $\mathbf{v} = (1,-1) \neq (0,0)$ we have $\mathbf{u}\cdot\mathbf{v} = 0$.

(i) True. Cosine of such angle cannot be positive, therefore neither can $\mathbf{u}\cdot\mathbf{v}$.

(j) True. Applying triangle inequality twice, $\|\mathbf{u}+\mathbf{v}+\mathbf{w}\| \le \|\mathbf{u}+\mathbf{v}\| + \|\mathbf{w}\| \le \|\mathbf{u}\| + \|\mathbf{v}\| + \|\mathbf{w}\|$.

3.3 Orthogonality

1. **(a)** $\mathbf{u}\cdot\mathbf{v} = (6)(2) + (1)(0) + (4)(-3) = 0$ therefore $\mathbf{u}$ and $\mathbf{v}$ are orthogonal vectors

(b) $\mathbf{u}\cdot\mathbf{v} = (0)(1) + (0)(1) + (-1)(1) = -1 \neq 0$ therefore $\mathbf{u}$ and $\mathbf{v}$ are not orthogonal vectors

(c) $\mathbf{u}\cdot\mathbf{v} = (3)(-4) + (-2)(1) + (1)(-3) + (3)(7) = 4 \neq 0$ therefore $\mathbf{u}$ and $\mathbf{v}$ are not orthogonal vectors

(d) $\mathbf{u} \cdot \mathbf{v} = (5)(-4) + (-4)(1) + (0)(-3) + (3)(7) = -3 \neq 0$ therefore $\mathbf{u}$ and $\mathbf{v}$ are not orthogonal vectors

3. $-2(x-(-1)) + 1(y-3) - 1(z-(-2)) = 0$ can be rewritten as $-2(x+1) + (y-3) - (z+2) = 0$

5. $0(x-2) + 0(y-0) + 2(z-0) = 0$ can be rewritten as $2z = 0$

7. The plane $4x - y + 2z = 5$ has a normal vector $(4, -1, 2)$.
The plane $7x - 3y + 4z = 8$ has a normal vector $(7, -3, 4)$.
The two normal vectors are not parallel (neither of them can be expressed as a scalar multiple of the other one) therefore the planes are not parallel either.

9. Rewriting the first plane equation $2y = 8x - 4z + 5$ as $-8x + 2y + 4z = 5$ yields a normal vector $(-8, 2, 4)$.
Rewriting the second plane equation $x = \frac{1}{2}z + \frac{1}{4}y$ as $x - \frac{1}{4}y - \frac{1}{2}z = 0$ yields a normal vector $\left(1, -\frac{1}{4}, -\frac{1}{2}\right)$.
The two normal vectors are parallel: $(-8, 2, 4) = -8\left(1, -\frac{1}{4}, -\frac{1}{2}\right)$ therefore the planes are parallel as well.

11. Normal vectors of the two planes are not orthogonal:

$$(3, -1, 1) \cdot (1, 0, 2) = (3)(1) + (-1)(0) + (1)(2) = 5 \neq 0$$

therefore the given planes are not perpendicular.

13. **(a)** From Formula (12), $\|\text{proj}_{\mathbf{a}}\mathbf{u}\| = \frac{|\mathbf{u}\cdot\mathbf{a}|}{\|\mathbf{a}\|} = \frac{|(1)(-4)+(-2)(-3)|}{\sqrt{(-4)^2+(-3)^2}} = \frac{2}{\sqrt{25}} = \frac{2}{5}$

(b) From Formula (12), $\|\text{proj}_{\mathbf{a}}\mathbf{u}\| = \frac{|\mathbf{u}\cdot\mathbf{a}|}{\|\mathbf{a}\|} = \frac{|(3)(2)+(0)(3)+(4)(3)|}{\sqrt{2^2+3^2+3^2}} = \frac{18}{\sqrt{22}}$

15. $\mathbf{u} \cdot \mathbf{a} = (6)(3) + (2)(-9) = 0$, $\|\mathbf{a}\|^2 = (3)^2 + (-9)^2 = 90$,
the vector component of $\mathbf{u}$ along $\mathbf{a}$ is $\text{proj}_{\mathbf{a}}\mathbf{u} = \frac{\mathbf{u}\cdot\mathbf{a}}{\|\mathbf{a}\|^2}\mathbf{a} = \frac{0}{90}(3, -9) = (0, 0)$,
the vector component of $\mathbf{u}$ orthogonal to $\mathbf{a}$ is $\mathbf{u} - \text{proj}_{\mathbf{a}}\mathbf{u} = (6, 2) - (0, 0) = (6, 2)$

17. $\mathbf{u} \cdot \mathbf{a} = (3)(1) + (1)(0) + (-7)(5) = -32$, $\|\mathbf{a}\|^2 = 1^2 + 0^2 + 5^2 = 26$,
the vector component of $\mathbf{u}$ along $\mathbf{a}$ is
$\text{proj}_{\mathbf{a}}\mathbf{u} = \frac{\mathbf{u}\cdot\mathbf{a}}{\|\mathbf{a}\|^2}\mathbf{a} = \frac{-32}{26}(1, 0, 5) = \left(-\frac{32}{26}, 0, -\frac{160}{26}\right) = \left(-\frac{16}{13}, 0, -\frac{80}{13}\right)$,
the vector component of $\mathbf{u}$ orthogonal to $\mathbf{a}$ is
$\mathbf{u} - \text{proj}_{\mathbf{a}}\mathbf{u} = (3, 1, -7) - \left(-\frac{16}{13}, 0, -\frac{80}{13}\right) = \left(\frac{55}{13}, 1, -\frac{11}{13}\right)$

19. $\mathbf{u} \cdot \mathbf{a} = (2)(4) + (1)(-4) + (1)(2) + (2)(-2) = 2$, $\|\mathbf{a}\|^2 = 4^2 + (-4)^2 + 2^2 + (-2)^2 = 40$,
the vector component of $\mathbf{u}$ along $\mathbf{a}$ is $\text{proj}_{\mathbf{a}}\mathbf{u} = \frac{\mathbf{u}\cdot\mathbf{a}}{\|\mathbf{a}\|^2}\mathbf{a} = \frac{2}{40}(4, -4, 2, -2) = \left(\frac{1}{5}, -\frac{1}{5}, \frac{1}{10}, -\frac{1}{10}\right)$,
the vector component of $\mathbf{u}$ orthogonal to $\mathbf{a}$ is
$\mathbf{u} - \text{proj}_{\mathbf{a}}\mathbf{u} = (2, 1, 1, 2) - \left(\frac{1}{5}, -\frac{1}{5}, \frac{1}{10}, -\frac{1}{10}\right) = \left(\frac{9}{5}, \frac{6}{5}, \frac{9}{10}, \frac{21}{10}\right)$

21. From Theorem 3.3.4(a) the distance between the point and the line is $D = \frac{|(4)(-3)+(3)(1)+4|}{\sqrt{4^2+3^2}} = \frac{5}{\sqrt{25}} = 1$

23. From Theorem 3.3.4(a) the distance between the point and the line is $D = \frac{|(4)(2)+(1)(-5)-2|}{\sqrt{4^2+1^2}} = \frac{1}{\sqrt{17}}$ (the equation of the line had to be rewritten in the form $ax + by + c = 0$ as $4x + y - 2 = 0$)

25. From Theorem 3.3.4(b) the distance between the point and the plane is $D = \frac{|(1)(3)+(2)(1)+(-2)(-2)-4|}{\sqrt{1^2+2^2+(-2)^2}} = \frac{5}{\sqrt{9}} = \frac{5}{3}$ (the equation of the plane had to be rewritten in the form $ax + by + cz + d = 0$ as $x + 2y - 2z - 4 = 0$)

27. First, select an arbitrary point in the plane $2x - y - z = 5$ by setting $x = y = 0$; we obtain $P_0(0,0,-5)$. From Theorem 3.3.4(b) the distance between P_0 and the plane $-4x + 2y + 2z - 12 = 0$ is $D = \frac{|(-4)(0)+(2)(0)+(2)(-5)-12|}{\sqrt{(-4)^2+2^2+2^2}} = \frac{22}{\sqrt{24}} = \frac{11}{\sqrt{6}}$

29. In order for $\mathbf{w} = (a, b, c)$ to be orthogonal to both $(1,0,1)$ and $(0,1,1)$, we must have $a + c = 0$ and $b + c = 0$. These equations form a linear system whose augmented matrix $\begin{bmatrix} 1 & 0 & 1 & 0 \\ 0 & 1 & 1 & 0 \end{bmatrix}$ is already in reduced row echelon form. For arbitrary real number t, the solutions are $a = -t, b = -t, c = t$. Since $\mathbf{w}$ is also required to be a unit vector, we must have $\|\mathbf{w}\| = \sqrt{(-t)^2 + (-t)^2 + t^2} = \sqrt{3t^2} = 1$. This yields $t = \pm\frac{1}{\sqrt{3}}$, consequently there are two possible vectors that satisfy the given conditions: $\left(-\frac{1}{\sqrt{3}}, -\frac{1}{\sqrt{3}}, \frac{1}{\sqrt{3}}\right)$ and $\left(\frac{1}{\sqrt{3}}, \frac{1}{\sqrt{3}}, -\frac{1}{\sqrt{3}}\right)$.

31. $\overrightarrow{AB} = (-2 - 1, 0 - 1, 3 - 1) = (-3, -1, 2)$, $\overrightarrow{AC} = (-3 - 1, -1 - 1, 1 - 1) = (-4, -2, 0)$,
$\overrightarrow{BC} = (-3 - (-2), -1 - 0, 1 - 3) = (-1, -1, -2)$
$\overrightarrow{AB} \cdot \overrightarrow{BC} = (-3)(-1) + (-1)(-1) + (2)(-2) = 0$
therefore the points A, B, and C form the vertices of a right triangle

33. Assuming $\mathbf{v} \cdot \mathbf{w}_1 = \mathbf{v} \cdot \mathbf{w}_2 = 0$ and using Theorem 3.2.2, we have
$\mathbf{v} \cdot (k_1\mathbf{w}_1 + k_2\mathbf{w}_2) = \mathbf{v} \cdot (k_1\mathbf{w}_1) + \mathbf{v} \cdot (k_2\mathbf{w}_2) = k_1(\mathbf{v} \cdot \mathbf{w}_1) + k_2(\mathbf{v} \cdot \mathbf{w}_2) = (k_1)(0) + (k_2)(0) = 0$.

39. $W = \|\mathbf{F}\|\|\overrightarrow{PQ}\| \cos\theta = (500)(100)\cos\frac{\pi}{4} = \frac{50{,}000}{\sqrt{2}} \approx 35{,}355$ Nm.

True-False Exercises

(a) True. $(3, -1, 2) \cdot (0, 0, 0) = 0$.

(b) True. By Theorem 3.2.2(c) and Theorem 3.2.3(e), $(k\mathbf{u}) \cdot (m\mathbf{v}) = (km)(\mathbf{u} \cdot \mathbf{v}) = (km)(0) = 0$.

(c) True. This follows from Theorem 3.3.2.

(d) True. $\text{proj}_{\mathbf{a}}(\text{proj}_{\mathbf{b}}(\mathbf{u})) = \frac{\left(\frac{\mathbf{u}\cdot\mathbf{b}}{\|\mathbf{b}\|^2}\mathbf{b}\right)\cdot\mathbf{a}}{\|\mathbf{a}\|^2}\mathbf{a} = \frac{\frac{\mathbf{u}\cdot\mathbf{b}}{\|\mathbf{b}\|^2}(\mathbf{b}\cdot\mathbf{a})}{\|\mathbf{a}\|^2}\mathbf{a} = 0\mathbf{a} = \mathbf{0}$
($\text{proj}_{\mathbf{b}}(\mathbf{u})$ has the same direction as $\mathbf{b}$, so it is also orthogonal to $\mathbf{a}$).

(e) True. $\text{proj}_{\mathbf{a}}(\text{proj}_{\mathbf{a}}(\mathbf{u})) = \frac{\frac{\mathbf{u}\cdot\mathbf{a}}{\|\mathbf{a}\|^2}\mathbf{a}\cdot\mathbf{a}}{\|\mathbf{a}\|^2}\mathbf{a} = \frac{\frac{\mathbf{u}\cdot\mathbf{a}}{\|\mathbf{a}\|^2}\|\mathbf{a}\|^2}{\|\mathbf{a}\|^2}\mathbf{a} = \frac{\mathbf{u}\cdot\mathbf{a}}{\|\mathbf{a}\|^2}\mathbf{a} = \text{proj}_{\mathbf{a}}(\mathbf{u})$
($\text{proj}_{\mathbf{a}}(\mathbf{u}) = k\mathbf{a}$ for some scalar k and then $\text{proj}_{\mathbf{a}}(k\mathbf{a}) = k\mathbf{a}$).

(f) False. For instance, let $\mathbf{u}$ be a nonzero vector orthogonal to $\mathbf{a}$. Then $\text{proj}_{\mathbf{a}}(\mathbf{u}) = \text{proj}_{\mathbf{a}}(2\mathbf{u}) = \mathbf{0}$ even though $\mathbf{u} \neq 2\mathbf{u}$.

(g) False. By Theorem 3.2.5(a), $\|\mathbf{u}+\mathbf{v}\| \le \|\mathbf{u}\| + \|\mathbf{v}\|$. This becomes an equality only when **u** and **v** are collinear vectors in the same direction. (For instance, $\|(1,0)+(0,1)\| = \|(1,1)\| = \sqrt{2}$ does not equal $\|(1,0)\| + \|(0,1)\| = 1+1 = 2$.)

3.4 The Geometry of Linear Systems

1. The vector equation in Formula (5) can be expressed as $(x,y) = (-4,1) + t(0,-8)$.
This yields the parametric equations $x = -4,\ y = 1 - 8t$.

3. The vector equation in Formula (5) can be expressed as $(x,y,z) = t(-3,0,1)$.
This yields the parametric equations $x = -3t,\ y = 0,\ z = t$.

5. A point on the line: $(3,-6)$; a vector parallel to the line: $(-5,-1)$.

7. Rewriting the vector equation as $(x,y) = (4-6t, 6-6t)$ yields a point on the line: $(4,6)$ and a vector parallel to the line: $(-6,-6)$.

9. The vector equation in Formula (6) can be expressed as
$(x,y,z) = (-3,1,0) + t_1(0,-3,6) + t_2(-5,1,2)$.
This yields the parametric equations $x = -3 - 5t_2,\ y = 1 - 3t_1 + t_2,\ z = 6t_1 + 2t_2$.

11. The vector equation in Formula (6) can be expressed as
$(x,y,z) = (-1,1,4) + t_1(6,-1,0) + t_2(-1,3,1)$.
This yields the parametric equations $x = -1 + 6t_1 - t_2,\ y = 1 - t_1 + 3t_2,\ z = 4 + t_2$.

13. We find a nonzero vector orthogonal to **v**, e.g., $(3,2)$. The vector equation of the line passing through $(0,0)$ and parallel to $(3,2)$ can be expressed as $(x,y) = t(3,2)$. Parametric equations are $x = 3t$ and $y = 2t$.

15. We find two nonparallel nonzero vectors orthogonal to **v**, e.g., $(5,0,4)$ and $(0,1,0)$. The vector equation of the plane that contains the origin and these two vectors can be expressed as $(x,y,z) = t_1(5,0,4) + t_2(0,1,0)$. Parametric equations are $x = 5t_1,\ y = t_2$, and $z = 4t_1$.

17. The augmented matrix of the linear system $\begin{bmatrix} 1 & 1 & 1 & 0 \\ 2 & 2 & 2 & 0 \\ 3 & 3 & 3 & 0 \end{bmatrix}$ has the reduced row echelon form $\begin{bmatrix} 1 & 1 & 1 & 0 \\ 0 & 0 & 0 & 0 \\ 0 & 0 & 0 & 0 \end{bmatrix}$. A general solution of the system, $x_1 = -s - t,\ x_2 = s,\ x_3 = t$ expressed in vector form as $\mathbf{x} = (-s-t, s, t)$ is orthogonal to the rows of the coefficient matrix of the original system $\mathbf{r}_1 = (1,1,1)$, $\mathbf{r}_2 = (2,2,2)$, and $\mathbf{r}_3 = (3,3,3)$ since $\mathbf{r}_1 \cdot \mathbf{x} = (1)(-s-t) + (1)(s) + (1)(t) = 0$, $\mathbf{r}_2 \cdot \mathbf{x} = (2)(-s-t) + (2)(s) + (2)(t) = 0$, and $\mathbf{r}_3 \cdot \mathbf{x} = (3)(-s-t) + (3)(s) + (3)(t) = 0$.

19. The augmented matrix of the linear system $\begin{bmatrix} 1 & 5 & 1 & 2 & -1 & 0 \\ 1 & -2 & -1 & 3 & 2 & 0 \end{bmatrix}$ has the reduced row echelon form $\begin{bmatrix} 1 & 0 & -\frac{3}{7} & \frac{19}{7} & \frac{8}{7} & 0 \\ 0 & 1 & \frac{2}{7} & -\frac{1}{7} & -\frac{3}{7} & 0 \end{bmatrix}$. A general solution of the system,

$x_1 = \frac{3}{7}r - \frac{19}{7}s - \frac{8}{7}t$, $x_2 = -\frac{2}{7}r + \frac{1}{7}s + \frac{3}{7}t$, $x_3 = r, x_4 = s, x_5 = t$ expressed in vector form as $\mathbf{x} = \left(\frac{3}{7}r - \frac{19}{7}s - \frac{8}{7}t, -\frac{2}{7}r + \frac{1}{7}s + \frac{3}{7}t, r, s, t\right)$ is orthogonal to the rows of the coefficient matrix of the original system $\mathbf{r}_1 = (1,5,1,2,-1)$ and $\mathbf{r}_2 = (1,-2,-1,3,2)$ since

$\mathbf{r}_1 \cdot \mathbf{x} = (1)\left(\frac{3}{7}r - \frac{19}{7}s - \frac{8}{7}t\right) + (5)\left(-\frac{2}{7}r + \frac{1}{7}s + \frac{3}{7}t\right) + (1)(r) + (2)(s) + (-1)(t) = 0$ and

$\mathbf{r}_2 \cdot \mathbf{x} = (1)\left(\frac{3}{7}r - \frac{19}{7}s - \frac{8}{7}t\right) + (-2)\left(-\frac{2}{7}r + \frac{1}{7}s + \frac{3}{7}t\right) + (-1)(r) + (3)(s) + (2)(t) = 0$.

21. (a) The associated homogeneous system $x + y + z = 0$ has a general solution $x = -s - t,\ y = s,\ z = t$. The original nonhomogeneous system has a general solution $x = 1 - s - t,\ y = s,\ z = t$, which can be expressed in vector form as

$$(x,y,z) = (1 - s - t, s, t) = \underbrace{(1,0,0)}_{\substack{\text{particular}\\ \text{solution}\\ \text{of the}\\ \text{nonhomogeneous}\\ \text{system}}} + \underbrace{(-s - t, s, t)}_{\substack{\text{general}\\ \text{solution}\\ \text{of the}\\ \text{homogeneous}\\ \text{system}}}$$

(b) Geometrically, the points (x,y,z) corresponding to solutions of $x + y + z = 1$ form a plane passing through the point $(1,0,0)$ and parallel to the vectors $(-1,1,0)$ and $(-1,0,1)$.

23. (a) Theorem 3.4.3 yields the following homogeneous linear system that satisfies our requirements:

$$\begin{aligned} x + y + z &= 0 \\ -2x + 3y &= 0 \end{aligned}$$

(b) A straight line passing through the origin – this line is parallel to any vector that is orthogonal to both $\mathbf{a}$ and $\mathbf{b}$.

(c) The augmented matrix of the system obtained in part (a) has the reduced row echelon form $\begin{bmatrix} 1 & 0 & \frac{3}{5} & 0 \\ 0 & 1 & \frac{2}{5} & 0 \end{bmatrix}$. A general solution of the system is $x = -\frac{3}{5}t,\ y = -\frac{2}{5}t,\ z = t$. It can also be expressed in vector form as $\mathbf{u} = (x,y,z) = (-\frac{3}{5}t, -\frac{2}{5}t, t)$. To confirm that Theorem 3.4.3 holds, we verify that $\mathbf{u}$ is orthogonal to both $\mathbf{a}$ and $\mathbf{b}$:

$\mathbf{u} \cdot \mathbf{a} = \left(-\frac{3}{5}t\right)(1) + \left(-\frac{2}{5}t\right)(1) + (t)(1) = 0$, $\mathbf{u} \cdot \mathbf{b} = \left(-\frac{3}{5}t\right)(-2) + \left(-\frac{2}{5}t\right)(3) + (t)(0) = 0$.

25. (a) The augmented matrix of the homogeneous system has the reduced row echelon form $\begin{bmatrix} 1 & \frac{2}{3} & -\frac{1}{3} & 0 \\ 0 & 0 & 0 & 0 \\ 0 & 0 & 0 & 0 \end{bmatrix}$. A general solution of the system is $x_1 = -\frac{2}{3}s + \frac{1}{3}t,\ x_2 = s,\ x_3 = t$.

(b) Multiplying $\begin{bmatrix} 3 & 2 & -1 \\ 6 & 4 & -2 \\ -3 & -2 & 1 \end{bmatrix}\begin{bmatrix} 1 \\ 0 \\ 1 \end{bmatrix}$ yields $\begin{bmatrix} 2 \\ 4 \\ -2 \end{bmatrix}$ therefore $x_1 = 1,\ x_2 = 0,\ x_3 = 1$ is a solution of the nonhomogeneous system.

(c) The vector form of a general solution of the nonhomogeneous system is

$$(x_1, x_2, x_3) = \underbrace{(1,0,1)}_{\substack{\text{particular}\\ \text{solution}\\ \text{of the}\\ \text{nonhomogeneous}\\ \text{system}}} + \underbrace{(-\tfrac{2}{3}s + \tfrac{1}{3}t, s, t)}_{\substack{\text{general}\\ \text{solution}\\ \text{of the}\\ \text{homogeneous}\\ \text{system}}}$$

(d) The augmented matrix of the homogeneous system has the reduced row echelon form $\begin{bmatrix} 1 & \frac{2}{3} & -\frac{1}{3} & \frac{2}{3} \\ 0 & 0 & 0 & 0 \\ 0 & 0 & 0 & 0 \end{bmatrix}$. A general solution of the system is $x_1 = \frac{2}{3} - \frac{2}{3}p + \frac{1}{3}q,\ x_2 = p,\ x_3 = q.$

If we let $p = s$ and $q = t + 1$ then this agrees with the solution we obtained in part (c).

27. The augmented matrix of the nonhomogeneous system $\begin{bmatrix} 3 & 4 & 1 & 2 & 3 \\ 6 & 8 & 2 & 5 & 7 \\ 9 & 12 & 3 & 10 & 13 \end{bmatrix}$ has the reduced row echelon form $\begin{bmatrix} 1 & \frac{4}{3} & \frac{1}{3} & 0 & \frac{1}{3} \\ 0 & 0 & 0 & 1 & 1 \\ 0 & 0 & 0 & 0 & 0 \end{bmatrix}$. A general solution of this system

$x_1 = \frac{1}{3} - \frac{4}{3}r - \frac{1}{3}s,\ \ x_2 = r,\ \ x_3 = s,\ \ x_4 = 1$ can be expressed in vector form as

$$(x_1, x_2, x_3, x_4) = \underbrace{\left(\tfrac{1}{3}, 0, 0, 1\right)}_{\substack{\text{particular}\\ \text{solution}\\ \text{of the}\\ \text{nonhomogeneous}\\ \text{system}}} + \underbrace{(-\tfrac{4}{3}r - \tfrac{1}{3}s,\ r,\ s,\ 0)}_{\substack{\text{general}\\ \text{solution}\\ \text{of the}\\ \text{associated}\\ \text{homogeneous}\\ \text{system}}}$$

29. By Theorem 1.8.2, we can write $T(\mathbf{x}_0 + t\mathbf{v}) = T(\mathbf{x}_0) + T(t\mathbf{v}) = T(\mathbf{x}_0) + tT(\mathbf{v})$.
If $T(\mathbf{v}) = \mathbf{0}$ then the image of the entire line is a single point $T(\mathbf{x}_0)$.
Otherwise, the image is a line through $T(\mathbf{x}_0)$ that is parallel to $T(\mathbf{v})$.

True-False Exercises

(a) True. This follows from Definition 1.

(b) False. We need *two* vectors parallel to the plane that are not collinear.

(c) True. This follows from Theorem 3.4.1.

(d) True.
If $\mathbf{b} = \mathbf{0}$ then by Theorem 3.4.3, all solution vectors of $A\mathbf{x} = \mathbf{b}$ are orthogonal to the row vectors of A.
If all solution vectors of $A\mathbf{x} = \mathbf{b}$ are orthogonal to the row vectors of A, $\mathbf{r}_1, \dots, \mathbf{r}_m$ then the ith component of the product $A\mathbf{x}$ is $\mathbf{r}_i \cdot \mathbf{x} = 0$, so we must have $\mathbf{b} = \mathbf{0}$.

(e) False. By Theorem 3.4.4, the general solution of $A\mathbf{x} = \mathbf{b}$ can be obtained by adding any specific solution of $A\mathbf{x} = \mathbf{b}$ to the general solution of $A\mathbf{x} = \mathbf{0}$.

(f) True. Subtracting $A\mathbf{x}_1 = \mathbf{b}$ from $A\mathbf{x}_2 = \mathbf{b}$ yields $A\mathbf{x}_1 - A\mathbf{x}_2 = \mathbf{b} - \mathbf{b}$, i.e., $A(\mathbf{x}_1 - \mathbf{x}_2) = \mathbf{0}$.

3.5 Cross Product

1. (a) $\mathbf{v}\times\mathbf{w} = \left(\begin{vmatrix}2 & -3\\6 & 7\end{vmatrix}, -\begin{vmatrix}0 & -3\\2 & 7\end{vmatrix}, \begin{vmatrix}0 & 2\\2 & 6\end{vmatrix}\right) = (32,-6,-4)$

(b) $\mathbf{w}\times\mathbf{v} = \left(\begin{vmatrix}6 & 7\\2 & -3\end{vmatrix}, -\begin{vmatrix}2 & 7\\0 & -3\end{vmatrix}, \begin{vmatrix}2 & 6\\0 & 2\end{vmatrix}\right) = (-32,6,4)$

(c) $(\mathbf{u}+\mathbf{v})\times\mathbf{w} = (3,4,-4)\times(2,6,7) = \left(\begin{vmatrix}4 & -4\\6 & 7\end{vmatrix}, -\begin{vmatrix}3 & -4\\2 & 7\end{vmatrix}, \begin{vmatrix}3 & 4\\2 & 6\end{vmatrix}\right) = (52,-29,10)$

(d) Using the result of part (a),
$\mathbf{v}\cdot(\mathbf{v}\times\mathbf{w}) = (0,2,-3)\cdot(32,-6,-4) = (0)(32)+(2)(-6)+(-3)(-4) = 0$

(e) $\mathbf{v}\times\mathbf{v} = \left(\begin{vmatrix}2 & -3\\2 & -3\end{vmatrix}, -\begin{vmatrix}0 & -3\\0 & -3\end{vmatrix}, \begin{vmatrix}0 & 2\\0 & 2\end{vmatrix}\right) = (0,0,0)$

(f) $(\mathbf{u}-3\mathbf{w})\times(\mathbf{u}-3\mathbf{w}) = (-3,-16,-22)\times(-3,-16,-22)$
$= \left(\begin{vmatrix}-16 & -22\\-16 & -22\end{vmatrix}, -\begin{vmatrix}-3 & -22\\-3 & -22\end{vmatrix}, \begin{vmatrix}-3 & -16\\-3 & -16\end{vmatrix}\right) = (0,0,0)$

3. By Lagrange's identity (Theorem 3.5.1(c)) and Formula (18) in Section 3.2, we have
$\|\mathbf{u}\times\mathbf{w}\|^2 = \|\mathbf{u}\|^2\|\mathbf{w}\|^2 - (\mathbf{u}\cdot\mathbf{w})^2 = (\mathbf{u}\cdot\mathbf{u})(\mathbf{w}\cdot\mathbf{w}) - (\mathbf{u}\cdot\mathbf{w})^2$.

$\mathbf{u}\times\mathbf{w} = \left(\begin{vmatrix}2 & -1\\6 & 7\end{vmatrix}, -\begin{vmatrix}3 & -1\\2 & 7\end{vmatrix}, \begin{vmatrix}3 & 2\\2 & 6\end{vmatrix}\right) = (20,-23,14)$

$\|\mathbf{u}\times\mathbf{w}\|^2 = \left(\sqrt{20^2+(-23)^2+14^2}\right)^2 = 1125$

$(\mathbf{u}\cdot\mathbf{u})(\mathbf{w}\cdot\mathbf{w}) - (\mathbf{u}\cdot\mathbf{w})^2 = (3^2+2^2+(-1)^2)(2^2+6^2+7^2) - [(3)(2)+(2)(6)+(-1)(7)]^2$
$= (14)(89) - 11^2 = 1246 - 121 = 1125$

5. $\mathbf{v}\times\mathbf{w} = \left(\begin{vmatrix}2 & -3\\6 & 7\end{vmatrix}, -\begin{vmatrix}0 & -3\\2 & 7\end{vmatrix}, \begin{vmatrix}0 & 2\\2 & 6\end{vmatrix}\right) = (32,-6,-4)$
$\mathbf{u}\times(\mathbf{v}\times\mathbf{w}) = \left(\begin{vmatrix}2 & -1\\-6 & -4\end{vmatrix}, -\begin{vmatrix}3 & -1\\32 & -4\end{vmatrix}, \begin{vmatrix}3 & 2\\32 & -6\end{vmatrix}\right) = (-14,-20,-82)$
By Theorem 3.5.1(d),
$\mathbf{u}\times(\mathbf{v}\times\mathbf{w}) = (\mathbf{u}\cdot\mathbf{w})\mathbf{v} - (\mathbf{u}\cdot\mathbf{v})\mathbf{w}$
$= [(3)(2)+(2)(6)+(-1)(7)](0,2,-3) - [(3)(0)+(2)(2)+(-1)(-3)](2,6,7)$
$= 11(0,2,-3) - 7(2,6,7) = (0,22,-33) - (14,42,49) = (-14,-20,-82)$

7. $\mathbf{u}\times\mathbf{v} = \left(\begin{vmatrix}4 & 2\\1 & 5\end{vmatrix}, -\begin{vmatrix}-6 & 2\\3 & 5\end{vmatrix}, \begin{vmatrix}-6 & 4\\3 & 1\end{vmatrix}\right) = (18,36,-18)$ is orthogonal to both $\mathbf{u}$ and $\mathbf{v}$.

9. $\mathbf{u}\times\mathbf{v} = \left(\begin{vmatrix}-1 & 2\\3 & 1\end{vmatrix}, -\begin{vmatrix}1 & 2\\0 & 1\end{vmatrix}, \begin{vmatrix}1 & -1\\0 & 3\end{vmatrix}\right) = (-7,-1,3)$
The area of the parallelogram determined by both $\mathbf{u}$ and $\mathbf{v}$ is $\|\mathbf{u}\times\mathbf{v}\| = \sqrt{(-7)^2+(-1)^2+3} = \sqrt{59}$.

11. $\overrightarrow{P_1P_2} = (3,2) = \overrightarrow{P_4P_3}$, $\overrightarrow{P_1P_4} = (3,1) = \overrightarrow{P_2P_3}$

Viewing these as vectors in 3-space, we obtain

$$\begin{aligned}\overrightarrow{P_1P_2} \times \overrightarrow{P_1P_4} &= (3,2,0) \times (3,1,0) \\ &= \left(\begin{vmatrix} 2 & 0 \\ 1 & 0 \end{vmatrix}, -\begin{vmatrix} 3 & 0 \\ 3 & 0 \end{vmatrix}, \begin{vmatrix} 3 & 2 \\ 3 & 1 \end{vmatrix}\right) \\ &= (0,0,-3)\end{aligned}$$

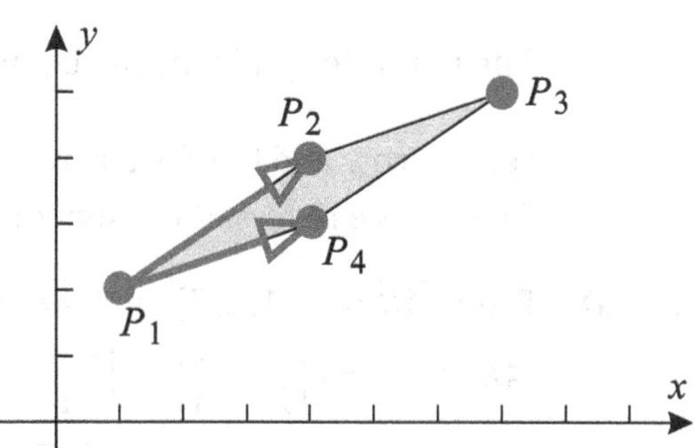

The area of the parallelogram is

$$\left\|\overrightarrow{P_1P_2} \times \overrightarrow{P_1P_4}\right\| = \sqrt{0^2 + 0^2 + (-3)^2} = 3.$$

13. We have $\overrightarrow{AB} = (1,4)$ and $\overrightarrow{AC} = (-3,2)$. Viewing these as vectors in 3-space, we obtain

$\overrightarrow{AB} \times \overrightarrow{AC} = (1,4,0) \times (-3,2,0) = \left(\begin{vmatrix} 4 & 0 \\ 2 & 0 \end{vmatrix}, -\begin{vmatrix} 1 & 0 \\ -3 & 0 \end{vmatrix}, \begin{vmatrix} 1 & 4 \\ -3 & 2 \end{vmatrix}\right) = (0,0,14).$

The area of the triangle is $\frac{1}{2}\left\|\overrightarrow{AB} \times \overrightarrow{AC}\right\| = \frac{1}{2}\sqrt{0^2 + 0^2 + 14^2} = 7.$

15. $\overrightarrow{P_1P_2} = (-1,-5,2)$, $\overrightarrow{P_1P_3} = (2,0,3)$

$\overrightarrow{P_1P_2} \times \overrightarrow{P_1P_3} = \left(\begin{vmatrix} -5 & 2 \\ 0 & 3 \end{vmatrix}, -\begin{vmatrix} -1 & 2 \\ 2 & 3 \end{vmatrix}, \begin{vmatrix} -1 & -5 \\ 2 & 0 \end{vmatrix}\right) = (-15,7,10).$

The area of the triangle is $\frac{1}{2}\left\|\overrightarrow{P_1P_2} \times \overrightarrow{P_1P_3}\right\| = \frac{1}{2}\sqrt{(-15)^2 + 7^2 + 10^2} = \frac{\sqrt{374}}{2}.$

17. From Theorem 3.5.4(b), the volume of the parallelepiped is equal to $\left|\det\begin{bmatrix} 2 & -6 & 2 \\ 0 & 4 & -2 \\ 2 & 2 & -4 \end{bmatrix}\right| = 16.$

19. $\begin{vmatrix} -1 & -2 & 1 \\ 3 & 0 & -2 \\ 5 & -4 & 0 \end{vmatrix} = 16 \neq 0$ therefore by Theorem 3.5.5 these vectors do not lie in the same plane when they have the same initial point.

21. From Formula (7), $\mathbf{u} \cdot (\mathbf{v} \times \mathbf{w}) = \begin{vmatrix} -2 & 0 & 6 \\ 1 & -3 & 1 \\ -5 & -1 & 1 \end{vmatrix} = -92.$

23. From Formula (7), $\mathbf{u} \cdot (\mathbf{v} \times \mathbf{w}) = \begin{vmatrix} a & 0 & 0 \\ 0 & b & 0 \\ 0 & 0 & c \end{vmatrix} = abc.$

25. **(a)** $\mathbf{u} \cdot (\mathbf{w} \times \mathbf{v}) = \begin{vmatrix} u_1 & u_2 & u_3 \\ w_1 & w_2 & w_3 \\ v_1 & v_2 & v_3 \end{vmatrix}$ can be obtained from $\mathbf{u} \cdot (\mathbf{v} \times \mathbf{w}) = \begin{vmatrix} u_1 & u_2 & u_3 \\ v_1 & v_2 & v_3 \\ w_1 & w_2 & w_3 \end{vmatrix}$ by interchanging the second row and the third row. This reverses the sign of the determinant, therefore $\mathbf{u} \cdot (\mathbf{w} \times \mathbf{v}) = -3$.

(b) By Theorem 3.2.2(a), $(\mathbf{v} \times \mathbf{w}) \cdot \mathbf{u} = \mathbf{u} \cdot (\mathbf{v} \times \mathbf{w}) = 3$.

(c) $\mathbf{w} \cdot (\mathbf{u} \times \mathbf{v}) = \begin{vmatrix} w_1 & w_2 & w_3 \\ u_1 & u_2 & u_3 \\ v_1 & v_2 & v_3 \end{vmatrix}$ can be obtained from $\begin{vmatrix} u_1 & u_2 & u_3 \\ w_1 & w_2 & w_3 \\ v_1 & v_2 & v_3 \end{vmatrix}$ by interchanging the first row and the second row.

The latter determinant can be obtained from $\mathbf{u}\cdot(\mathbf{v}\times\mathbf{w}) = \begin{vmatrix} u_1 & u_2 & u_3 \\ v_1 & v_2 & v_3 \\ w_1 & w_2 & w_3 \end{vmatrix}$ by interchanging the second row and the third row.

Overall, we reversed the sign of the determinant twice, therefore $\mathbf{w}\cdot(\mathbf{u}\times\mathbf{v}) = (-1)(-1)3 = 3$.

27. **(a)** From $\overrightarrow{AB} = (-1,2,2)$ and $\overrightarrow{AC} = (1,1,-1)$ we obtain

$\overrightarrow{AB}\times\overrightarrow{AC} = \left(\begin{vmatrix} 2 & 2 \\ 1 & -1 \end{vmatrix}, -\begin{vmatrix} -1 & 2 \\ 1 & -1 \end{vmatrix}, \begin{vmatrix} -1 & 2 \\ 1 & 1 \end{vmatrix}\right) = (-4,1,-3)$.

The area of the triangle is $\frac{1}{2}\|\overrightarrow{AB}\times\overrightarrow{AC}\| = \frac{1}{2}\sqrt{(-4)^2+1^2+(-3)^2} = \frac{\sqrt{26}}{2}$.

(b) Denoting the altitude from C to AB by h, we must have $\frac{1}{2}\|\overrightarrow{AB}\|h = \frac{\sqrt{26}}{2}$.

Since $\|\overrightarrow{AB}\| = \sqrt{(-1)^2+2^2+2^2} = 3$, we conclude that $h = \frac{\sqrt{26}}{3}$.

29. Using parts (a), (b), (c), and (f) of Theorem 3.5.2, we can write $(\mathbf{u}+\mathbf{v})\times(\mathbf{u}-\mathbf{v})$ $= (\mathbf{u}\times\mathbf{u}) - (\mathbf{u}\times\mathbf{v}) + (\mathbf{v}\times\mathbf{u}) + (\mathbf{v}\times\mathbf{v}) = \mathbf{0} - (-(\mathbf{v}\times\mathbf{u})) + (\mathbf{v}\times\mathbf{u}) + \mathbf{0} = 2(\mathbf{v}\times\mathbf{u})$.

31. **(a)** Taking $\mathbf{F} = 1000\left(-\frac{1}{\sqrt{2}}, 0, \frac{1}{\sqrt{2}}\right) = 500\sqrt{2}(-1,0,1)$ and $\mathbf{d} = \overrightarrow{PQ} = (0,2,-1)$ we obtain

$\mathbf{F}\times\mathbf{d} = 500\sqrt{2}[(-1,0,1)\times(0,2,-1)] = 500\sqrt{2}\left(\begin{vmatrix} 0 & 1 \\ 2 & -1 \end{vmatrix}, -\begin{vmatrix} -1 & 1 \\ 0 & -1 \end{vmatrix}, \begin{vmatrix} -1 & 0 \\ 0 & 2 \end{vmatrix}\right)$

$= 500\sqrt{2}(-2,-1,-2)$. $\|\mathbf{F}\times\mathbf{d}\| = 500\sqrt{2}\sqrt{(-2)^2+(-1)^2+(-2)^2} = 1500\sqrt{2}$ therefore the scalar moment of $\mathbf{F}$ about the point P is $1500\sqrt{2}$ Nm ≈ 2121.32 Nm.

(b) It was shown in the solution of part (a) that the vector moment of $\mathbf{F}$ about the point P is $\mathbf{F}\times\mathbf{d} = 500\sqrt{2}(-2,-1,-2)$ and its magnitude is $1500\sqrt{2}$. The direction angles are $\cos^{-1}\left(-\frac{1000\sqrt{2}}{1500\sqrt{2}}\right) \approx 132°$, $\cos^{-1}\left(-\frac{500\sqrt{2}}{1500\sqrt{2}}\right) \approx 109°$, and $\cos^{-1}\left(-\frac{1000\sqrt{2}}{1500\sqrt{2}}\right) \approx 132°$.

39. **(a)** The volume is $\frac{1}{6}|\overrightarrow{PQ}\cdot(\overrightarrow{PR}\times\overrightarrow{PS})| = \frac{1}{6}\left|\det\begin{bmatrix} 3 & -1 & -3 \\ 2 & -1 & 1 \\ 4 & -4 & 3 \end{bmatrix}\right| = \frac{1}{6}|17| = \frac{17}{6}$.

(b) The volume is $\frac{1}{6}|\overrightarrow{PQ}\cdot(\overrightarrow{PR}\times\overrightarrow{PS})| = \frac{1}{6}\left|\det\begin{bmatrix} 1 & 2 & -1 \\ 3 & 4 & 0 \\ -1 & -3 & 4 \end{bmatrix}\right| = \frac{1}{6}|-3| = \frac{1}{2}$.

True-False Exercises

(a) True. This follows from Formula (6): for nonzero vectors $\mathbf{u}$ and $\mathbf{v}$, $\|\mathbf{u}\times\mathbf{v}\| = \|\mathbf{u}\|\|\mathbf{v}\|\sin\theta$ is zero if and only if $\sin\theta = 0$ (i.e., the vectors are parallel).

(b) True. The cross product of two nonzero noncollinear vectors in a plane is a nonzero vector perpendicular to both vectors, and therefore to the entire plane.

(c) False. The scalar triple product is a scalar, rather than a vector.

(d) True. This follows from Theorem 3.5.3 and from the equality $\|\mathbf{u}\times\mathbf{v}\| = \|\mathbf{v}\times\mathbf{u}\|$.

(e) False. These two triple vector products are generally not the same, as evidenced by parts (d) and (e) of Theorem 3.5.1.

(f) False. For instance, let $\mathbf{u} = \mathbf{v} = \mathbf{i}$ and $\mathbf{w} = 2\mathbf{i}$. We have $\mathbf{u}\times\mathbf{v} = \mathbf{u}\times\mathbf{w} = \mathbf{0}$ even though $\mathbf{v} \neq \mathbf{w}$.

Chapter 3 Supplementary Exercises

1. **(a)** $3\mathbf{v} - 2\mathbf{u} = (9,-3,18) - (-4,0,8) = (13,-3,10)$

(b) $\mathbf{u}+\mathbf{v}+\mathbf{w} = (3,-6,5)$; $\|\mathbf{u}+\mathbf{v}+\mathbf{w}\| = \sqrt{3^2+(-6)^2+5^2} = \sqrt{70}$

(c) $-3\mathbf{u} - (\mathbf{v}+5\mathbf{w}) = (6,0,-12) - \big((3,-1,6)+(10,-25,-25)\big) = (-7,26,7)$

$d(-3\mathbf{u}, \mathbf{v}+5\mathbf{w}) = \|-3\mathbf{u} - (\mathbf{v}+5\mathbf{w})\| = \sqrt{(-7)^2+26^2+7^2} = \sqrt{774} = 3\sqrt{86}$

(d) $\mathbf{u}\cdot\mathbf{w} = (-2)(2)+(0)(-5)+(4)(-5) = -24$; $\|\mathbf{w}\|^2 = 2^2+(-5)^2+(-5)^2 = 54$;

$\text{proj}_{\mathbf{w}}\mathbf{u} = \frac{\mathbf{u}\cdot\mathbf{w}}{\|\mathbf{w}\|^2}\mathbf{w} = \frac{-24}{54}(2,-5,-5) = (-\frac{8}{9}, \frac{20}{9}, \frac{20}{9})$

(e) From Formula (7) in Section 3.5, $\mathbf{u}\cdot(\mathbf{v}\times\mathbf{w}) = \begin{vmatrix} -2 & 0 & 4 \\ 3 & -1 & 6 \\ 2 & -5 & -5 \end{vmatrix} = -122$

(f) $-5\mathbf{v}+\mathbf{w} = (-15,5,-30)+(2,-5,-5) = (-13,0,-35)$
$(\mathbf{u}\cdot\mathbf{v})\mathbf{w} = [(-2)(3)+(0)(-1)+(4)(6)]\mathbf{w} = 18\mathbf{w} = (36,-90,-90)$
$(-5\mathbf{v}+\mathbf{w})\times((\mathbf{u}\cdot\mathbf{v})\mathbf{w}) = \left(\begin{vmatrix} 0 & -35 \\ -90 & -90 \end{vmatrix}, -\begin{vmatrix} -13 & -35 \\ 36 & -90 \end{vmatrix}, \begin{vmatrix} -13 & 0 \\ 36 & -90 \end{vmatrix}\right)$
$= (-3150, -2430, 1170)$

3. **(a)** $3\mathbf{v} - 2\mathbf{u} = (-9,0,24,0) - (-4,12,4,2) = (-5,-12,20,-2)$

(b) $\mathbf{u}+\mathbf{v}+\mathbf{w} = (4,7,4,-5)$; $\|\mathbf{u}+\mathbf{v}+\mathbf{w}\| = \sqrt{4^2+7^2+4^2+(-5)^2} = \sqrt{106}$

(c) $-3\mathbf{u} - (\mathbf{v}+5\mathbf{w}) = (6,-18,-6,-3) - \big((-3,0,8,0)+(45,5,-30,-30)\big) = (-36,-23,16,27)$
$d(-3\mathbf{u}, \mathbf{v}+5\mathbf{w}) = \|-3\mathbf{u} - (\mathbf{v}+5\mathbf{w})\| = \sqrt{(-36)^2+(-23)^2+16^2+(-27)^2} = \sqrt{2810}$

(d) $\mathbf{u}\cdot\mathbf{w} = (-2)(9)+(6)(1)+(2)(-6)+(1)(-6) = -30$;
$\|\mathbf{w}\|^2 = 9^2+1^2+(-6)^2+(-6)^2 = 154$;

$\text{proj}_{\mathbf{w}}\mathbf{u} = \frac{\mathbf{u}\cdot\mathbf{w}}{\|\mathbf{w}\|^2}\mathbf{w} = \frac{-30}{154}(9,1,-6,-6) = \frac{-15}{77}(9,1,-6,-6) = \left(-\frac{135}{77}, -\frac{15}{77}, \frac{90}{77}, \frac{90}{77}\right)$

5. By Theorem 3.5.5, this set is the plane containing A, B, and C.

7. Denoting $S(-1,a,b)$ we have have $\overrightarrow{RS} = (-6, a-1, b-1)$. For this vector to be parallel to $\overrightarrow{PQ} = (3,1,-2)$ there must exist a scalar k such that $\overrightarrow{RS} = k\,\overrightarrow{PQ}$. The equality of the first components immediately leads to $k = -2$. Equating the remaining pairs of components yields the equations:

$$a-1 = (-2)(1),\quad b-1 = (-2)(-2)$$

therefore $a = -1$ and $b = 5$. We conclude that the point S has coordinates $(-1,-1,5)$.

9. $\overrightarrow{PQ} = (3,1,-2)$; $\overrightarrow{PR} = (2,2,-3)$; $\cos\theta = \frac{\overrightarrow{PQ}\cdot\overrightarrow{PR}}{\|\overrightarrow{PQ}\|\|\overrightarrow{PR}\|} = \frac{(3)(2)+(1)(2)+(-2)(-3)}{\sqrt{3^2+1^2+(-2)^2}\sqrt{2^2+2^2+(-3)^2}} = \frac{14}{\sqrt{14}\sqrt{17}} = \sqrt{\frac{14}{17}}$

11. From Theorem 3.3.4(b) the distance between the point and the plane is
$D = \frac{|(5)(-3)+(-3)(1)+(1)(3)+4|}{\sqrt{5^2+(-3)^2+1^2}} = \frac{11}{\sqrt{35}}$ (the equation of the plane had to be rewritten in the form $ax + by + cz + d = 0$ as $5x - 3y + z + 4 = 0$)

13. A vector equation of the plane that contains the point P and vectors $\overrightarrow{PQ} = (1,-2,-2)$ and $\overrightarrow{PR} = (5,-1,-5)$ can be expressed as $(x,y,z) = (-2,1,3) + t_1(1,-2,-2) + t_2(5,-1,-5)$. Parametric equations are $x = -2 + t_1 + 5t_2$, $y = 1 - 2t_1 - t_2$, and $z = 3 - 2t_1 - 5t_2$.

15. A vector equation of the line can be expressed as $(x,y) = (0,-3) + t(8,-1)$.
This yields parametric equations $x = 8t,\ y = -3 - t$.

17. Since the line has a slope 3, the vector $(1,3)$ is parallel to the line (any scalar multiple of (1,3) could be used instead).
Substituting an arbitrary number into the line equation for x, we can solve for y to obtain coordinates of a point on the line. For instance, letting $x = 0$ yields $y = -5$ resulting in the point $(0,-5)$.
A vector equation of the line can now be expressed as $(x,y) = (0,-5) + t(1,3)$.
This yields parametric equations $x = t,\ y = -5 + 3t$.

19. The given vector equation specifies a point on the plane, $(-1,5,6)$, as well as two vectors parallel to the plane. A normal vector can be obtained as a cross product of these two vectors:

$$(0,-1,3) \times (2,-1,0) = \left(\begin{vmatrix} -1 & 3 \\ -1 & 0 \end{vmatrix}, -\begin{vmatrix} 0 & 3 \\ 2 & 0 \end{vmatrix}, \begin{vmatrix} 0 & -1 \\ 2 & -1 \end{vmatrix} \right) = (3,6,2)$$

A point-normal equation for the plane can be written as $3(x+1) + 6(y-5) + 2(z-6) = 0$.

21. Begin by forming two vectors parallel to the plane: $\overrightarrow{PQ} = (-10,4,-1)$ and $\overrightarrow{PR} = (-9,6,-6)$. A normal vector can be obtained as a cross product of these two vectors:

$$\overrightarrow{PQ} \times \overrightarrow{PR} = \left(\begin{vmatrix} 4 & -1 \\ 6 & -6 \end{vmatrix}, -\begin{vmatrix} -10 & -1 \\ -9 & -6 \end{vmatrix}, \begin{vmatrix} -10 & 4 \\ -9 & 6 \end{vmatrix} \right) = (-18,-51,-24)$$

A point-normal equation for the plane can be written as $-18(x-9) - 51y - 24(z-4) = 0$.

25. The equation represents a plane through the origin perpendicular to the xy-plane. It intersects the xy-plane along the line $Ax + By = 0$.

CHAPTER 4: GENERAL VECTOR SPACES

4.1 Real Vector Spaces

1. **(a)** $\mathbf{u}+\mathbf{v}=(-1+3,2+4)=(2,6)$; $k\mathbf{u}=(0,3\cdot 2)=(0,6)$

(b) For any $\mathbf{u}=(u_1,u_2)$ and $\mathbf{v}=(v_1,v_2)$ in V, $\mathbf{u}+\mathbf{v}=(u_1+v_1,u_2+v_2)$ is an ordered pair of real numbers, therefore $\mathbf{u}+\mathbf{v}$ is in V. Consequently, V is closed under addition.

For any $\mathbf{u}=(u_1,u_2)$ in V and for any scalar k, $k\mathbf{u}=(0,ku_2)$ is an ordered pair of real numbers, therefore $k\mathbf{u}$ is in V. Consequently, V is closed under scalar multiplication.

(c) Axioms 1-5 hold for V because they are known to hold for R^2.

(d) Axiom 7: $k\big((u_1,u_2)+(v_1,v_2)\big)=k(u_1+v_1,u_2+v_2)=\big(0,k(u_2+v_2)\big)=(0,ku_2)+(0,kv_2)$ $=k(u_1,u_2)+k(v_1,v_2)$ for all real k, u_1,u_2,v_1, and v_2;

Axiom 8: $(k+m)(u_1,u_2)=(0,(k+m)u_2)=(0,ku_2+mu_2)=(0,ku_2)+(0,mu_2)$ $=k(u_1,u_2)+m(u_1,u_2)$ for all real k, m, u_1, and u_2;

Axiom 9: $k\big(m(u_1,u_2)\big)=k(0,mu_2)=(0,kmu_2)=(km)(u_1,u_2)$ for all real k, m, u_1, and u_2;

(e) Axiom 10 fails to hold: $1(u_1,u_2)=(0,u_2)$ does not generally equal (u_1,u_2). Consequently, V is not a vector space.

3. Let V denote the set of all real numbers.

Axiom 1: $x+y$ is in V for all real x and y;

Axiom 2: $x+y=y+x$ for all real x and y;

Axiom 3: $x+(y+z)=(x+y)+z$ for all real x, y, and z;

Axiom 4: taking $\mathbf{0}=0$, we have $0+x=x+0=x$ for all real x;

Axiom 5: for each $\mathbf{u}=x$, let $-\mathbf{u}=-x$; then $x+(-x)=(-x)+x=0$

Axiom 6: kx is in V for all real k and x;

Axiom 7: $k(x+y)=kx+ky$ for all real k, x, and y;

Axiom 8: $(k+m)x=kx+mx$ for all real k, m, and x;

Axiom 9: $k(mx)=(km)x$ for all real k, m, and x;

Axiom 10: $1x=x$ for all real x.

This is a vector space – all axioms hold.

5. Axiom 5 fails whenever $x\neq 0$ since it is then impossible to find (x',y') satisfying $x'\geq 0$ for which $(x,y)+(x',y')=(0,0)$. (The zero vector from axiom 4 must be $\mathbf{0}=(0,0)$.)

Axiom 6 fails whenever $k<0$ and $x\neq 0$.

This is not a vector space.

7. Axiom 8 fails to hold:
$(k+m)\mathbf{u} = ((k+m)^2x, (k+m)^2y, (k+m)^2z)$
$k\mathbf{u} + m\mathbf{u} = (k^2x, k^2y, k^2z) + (m^2x, m^2y, m^2z) = ((k^2+m^2)x, (k^2+m^2)y, (k^2+m^2)z)$
therefore in general $(k+m)\mathbf{u} \neq k\mathbf{u} + m\mathbf{u}$.

This is not a vector space.

9. Let V be the set of all 2×2 matrices of the form $\begin{bmatrix} a & 0 \\ 0 & b \end{bmatrix}$ (i.e., all diagonal 2×2 matrices)

Axiom 1: the sum of two diagonal 2×2 matrices is also a diagonal 2×2 matrix.
Axiom 2: follows from part (a) of Theorem 1.4.1.

Axiom 3: follows from part (b) of Theorem 1.4.1.

Axiom 4: taking $\mathbf{0} = \begin{bmatrix} 0 & 0 \\ 0 & 0 \end{bmatrix}$; follows from part (a) of Theorem 1.4.2.

Axiom 5: let the negative of $\begin{bmatrix} a & 0 \\ 0 & b \end{bmatrix}$ be $\begin{bmatrix} -a & 0 \\ 0 & -b \end{bmatrix}$;
follows from part (c) of Theorem 1.4.2 and Axiom 2.

Axiom 6: the scalar multiple of a diagonal 2×2 matrix is also a diagonal 2×2 matrix.

Axiom 7: follows from part (h) of Theorem 1.4.1.

Axiom 8: follows from part (j) of Theorem 1.4.1.

Axiom 9: follows from part (l) of Theorem 1.4.1.

Axiom 10: $1\begin{bmatrix} a & 0 \\ 0 & b \end{bmatrix} = \begin{bmatrix} a & 0 \\ 0 & b \end{bmatrix}$ for all real a and b.

This is a vector space – all axioms hold.

11. Let V denote the set of all pairs of real numbers of the form $(1, x)$.

Axiom 1: $(1,y) + (1,y') = (1, y+y')$ is in V for all real y and y';

Axiom 2: $(1,y) + (1,y') = (1,y+y') = (1,y'+y) = (1,y') + (1,y)$ for all real y and y';

Axiom 3: $(1,y) + \big((1,y') + (1,y'')\big) = (1,y) + (1,y'+y'') = (1,y+y'+y'') = (1,y+y') + (1,y'')$
$= \big((1,y) + (1,y')\big) + (1,y'')$ for all real y, y', and y'';

Axiom 4: taking $\mathbf{0} = (1,0)$, we have $(1,0) + (1,y) = (1,y)$ and $(1,y) + (1,0) = (1,y)$
for all real y;

Axiom 5: for each $\mathbf{u} = (1,y)$, let $-\mathbf{u} = (1,-y)$;
then $(1,y) + (1,-y) = (1,0)$ and $(1,-y) + (1,y) = (1,0)$;

Axiom 6: $k(1,y) = (1,ky)$ is in V for all real k and y;

Axiom 7: $k\big((1,y) + (1,y')\big) = k(1,y+y') = (1,ky+ky') = (1,ky) + (1,ky') = k(1,y) + k(1,y')$
for all real k, y, and y';

Axiom 8: $(k+m)(1,y)=(1,(k+m)y)=(1,ky+my)=(1,ky)+(1,my)=k(1,y)+m(1,y)$ for all real k, m, and y;

Axiom 9: $k(m(1,y))=k(1,my)=(1,kmy)=(km)(1,y)$ for all real k, m, and y;

Axiom 10: $1(1,y)=(1,y)$ for all real y.

This is a vector space – all axioms hold.

13. Axiom 3: follows from part (b) of Theorem 1.4.1 since

$$\mathbf{u}+(\mathbf{v}+\mathbf{w})=\begin{bmatrix}u_{11} & u_{12}\\ u_{21} & u_{22}\end{bmatrix}+\left(\begin{bmatrix}v_{11} & v_{12}\\ v_{21} & v_{22}\end{bmatrix}+\begin{bmatrix}w_{11} & w_{12}\\ w_{21} & w_{22}\end{bmatrix}\right)$$
$$=\left(\begin{bmatrix}u_{11} & u_{12}\\ u_{21} & u_{22}\end{bmatrix}+\begin{bmatrix}v_{11} & v_{12}\\ v_{21} & v_{22}\end{bmatrix}\right)+\begin{bmatrix}w_{11} & w_{12}\\ w_{21} & w_{22}\end{bmatrix}=(\mathbf{u}+\mathbf{v})+\mathbf{w}$$

Axiom 7: follows from part (h) of Theorem 1.4.1 since

$$k(\mathbf{u}+\mathbf{v})=k\left(\begin{bmatrix}u_{11} & u_{12}\\ u_{21} & u_{22}\end{bmatrix}+\begin{bmatrix}v_{11} & v_{12}\\ v_{21} & v_{22}\end{bmatrix}\right)=k\begin{bmatrix}u_{11} & u_{12}\\ u_{21} & u_{22}\end{bmatrix}+k\begin{bmatrix}v_{11} & v_{12}\\ v_{21} & v_{22}\end{bmatrix}=k\mathbf{u}+k\mathbf{v}$$

Axiom 8: follows from part (j) of Theorem 1.4.1 since

$$(k+m)\mathbf{u}=(k+m)\begin{bmatrix}u_{11} & u_{12}\\ u_{21} & u_{22}\end{bmatrix}=k\begin{bmatrix}u_{11} & u_{12}\\ u_{21} & u_{22}\end{bmatrix}+m\begin{bmatrix}u_{11} & u_{12}\\ u_{21} & u_{22}\end{bmatrix}=k\mathbf{u}+m\mathbf{u}$$

Axiom 9: follows from part (l) of Theorem 1.4.1 since

$$k(m\mathbf{u})=k\left(m\begin{bmatrix}u_{11} & u_{12}\\ u_{21} & u_{22}\end{bmatrix}\right)=(km)\begin{bmatrix}u_{11} & u_{12}\\ u_{21} & u_{22}\end{bmatrix}=(km)\mathbf{u}$$

15. Axiom 1: $(u_1,u_2)+(v_1,v_2)=(u_1+v_1,u_2+v_2)$ is in V

Axiom 2: $(u_1,u_2)+(v_1,v_2)=(u_1+v_1,u_2+v_2)=(v_1+u_1,v_2+u_2)=(v_1,v_2)+(u_1,u_2)$

Axiom 3: $(u_1,u_2)+((v_1,v_2)+(w_1,w_2))=(u_1,u_2)+(v_1+w_1,v_2+w_2)$
$=(u_1+v_1+w_1,u_2+v_2+w_2)=(u_1+v_1,u_2+v_2)+(w_1,w_2)$
$=((u_1,u_2)+(v_1,v_2))+(w_1,w_2)$

Axiom 4: taking $\mathbf{0}=(0,0)$, we have $(0,0)+(u_1,u_2)=(u_1,u_2)$ and $(u_1,u_2)+(0,0)=(u_1,u_2)$

Axiom 5: for each $\mathbf{u}=(u_1,u_2)$, let $-\mathbf{u}=(-u_1,-u_2)$;
then $(u_1,u_2)+(-u_1,-u_2)=(0,0)$ and $(-u_1,-u_2)+(u_1,u_2)=(0,0)$

Axiom 6: $k(u_1,u_2)=(ku_1,0)$ is in V

Axiom 7: $k((u_1,u_2)+(v_1,v_2))=k(u_1+v_1,u_2+v_2)=(ku_1+kv_1,0)$
$=(ku_1,0)+(kv_1,0)=k(u_1,u_2)+k(v_1,v_2)$

Axiom 8: $(k+m)(u_1,u_2)=((k+m)u_1,0)=(ku_1+mu_1,0)=(ku_1,0)+(mu_1,0)$
$=k(u_1,u_2)+m(u_1,u_2)$

Axiom 9: $k(m(u_1,u_2))=k(mu_1,0)=(kmu_1,0)=(km)(u_1,u_2)$

19. $\frac{1}{u}=u^{-1}$

21.

$\mathbf{u}+\mathbf{w}=\mathbf{v}+\mathbf{w}$	Hypothesis
$(\mathbf{u}+\mathbf{w})+(-\mathbf{w})=(\mathbf{v}+\mathbf{w})+(-\mathbf{w})$	Add $-\mathbf{w}$ to both sides
$\mathbf{u}+[\mathbf{w}+(-\mathbf{w})]=\mathbf{v}+[\mathbf{w}+(-\mathbf{w})]$	Axiom 3

$\mathbf{u}+\mathbf{0}=\mathbf{v}+\mathbf{0}$ Axiom 5

$\mathbf{u}=\mathbf{v}$ Axiom 4

True-False Exercises

(a) True. This is a part of Definition 1.

(b) False. Example 1 discusses a vector space containing only one vector.

(c) False. By part (d) of Theorem 4.1.1, if $k\mathbf{u}=\mathbf{0}$ then $k=0$ or $\mathbf{u}=\mathbf{0}$.

(d) False. Axiom 6 fails to hold if $k<0$. (Also, Axiom 4 fails to hold.)

(e) True. This follows from part (c) of Theorem 4.1.1.

(f) False. This function must have a value of zero at *every* point in $(-\infty,\infty)$.

4.2 Subspaces

1. **(a)** Let W be the set of all vectors of the form $(a,0,0)$, i.e. all vectors in R^3 with last two components equal to zero.
This set contains at least one vector, e.g. $(0,0,0)$.
Adding two vectors in W results in another vector in W: $(a,0,0)+(b,0,0)=(a+b,0,0)$ since the result has zeros as the last two components.
Likewise, a scalar multiple of a vector in W is also in W: $k(a,0,0)=(ka,0,0)$ - the result also has zeros as the last two components.
According to Theorem 4.2.1, W is a subspace of R^3.

(b) Let W be the set of all vectors of the form $(a,1,1)$, i.e. all vectors in R^3 with last two components equal to one. The set W is not closed under the operation of vector addition since $(a,1,1)+(b,1,1)=(a+b,2,2)$ does not have ones as its last two components thus it is outside W.
According to Theorem 4.2.1, W is not a subspace of R^3.

(c) Let W be the set of all vectors of the form (a,b,c), where $b=a+c$.
This set contains at least one vector, e.g. $(0,0,0)$. (The condition $b=a+c$ is satisfied when $a=b=c=0$.)
Adding two vectors in W results in another vector in W
$(a,a+c,c)+(a',a'+c',c')=(a+a',\ a+c+a'+c',\ c+c')$ since in this result, the second component is the sum of the first and the third: $a+c+a'+c'=(a+a')+(c+c')$.
Likewise, a scalar multiple of a vector in W is also in W: $k(a,a+c,c)=(ka,k(a+c),kc)$ since in this result, the second component is once again the sum of the first and the third: $k(a+c)=ka+kc$.
According to Theorem 4.2.1, W is a subspace of R^3.

(d) Let W be the set of all vectors of the form (a,b,c), where $b=a+c+1$. The set W is not closed under the operation of vector addition, since in the result of the following addition of two vectors from W

$(a, a+c+1, c)+(a', a'+c'+1, c')=(a+a',\ a+c+a'+c'+2,\ c+c')$
the second component does not equal to the sum of the first, the third, and 1:
$a+c+a'+c'+2 \neq (a+a')+(c+c')+1$. Consequently, this result is not a vector in W.
According to Theorem 4.2.1, W is not a subspace of R^3.

(e) Let W be the set of all vectors of the form $(a, b, 0)$, i.e. all vectors in R^3 with last component equal to zero.
This set contains at least one vector, e.g. $(0,0,0)$.
Adding two vectors in W results in another vector in W
$(a, b, 0)+(a', b', 0)=(a+a', b+b', 0)$ since the result has 0 as the last component.
Likewise, a scalar multiple of a vector in W is also in W: $k(a, b, 0)=(ka, kb, 0)$ - the result also has 0 as the last component.
According to Theorem 4.2.1, W is a subspace of R^3.

3. **(a)** Let W be the set of all polynomials $a_0+a_1x+a_2x^2+a_3x^3$ for which $a_0=0$.
This set contains at least one polynomial, $0+0x+0x^2+0x^3=0$.
Adding two polynomials in W results in another polynomial in W:
$(0+a_1x+a_2x^2+a_3x^3)+(0+b_1x+b_2x^2+b_3x^3)$
$=0+(a_1+b_1)x+(a_2+b_2)x^2+(a_3+b_3)x^3$.
Likewise, a scalar multiple of a polynomial in W is also in W:
$k(0+a_1x+a_2x^2+a_3x^3)=0+(ka_1)x+(ka_2)x^2+(ka_3)x^3$.
According to Theorem 4.2.1, W is a subspace of P_3.

(b) Let W be the set of all polynomials $a_0+a_1x+a_2x^2+a_3x^3$ for which $a_0+a_1+a_2+a_3=0$, i.e. all polynomials that can be expressed in the form $-a_1-a_2-a_3+a_1x+a_2x^2+a_3x^3$.
Adding two polynomials in W results in another polynomial in W
$(-a_1-a_2-a_3+a_1x+a_2x^2+a_3x^3)+(-b_1-b_2-b_3+b_1x+b_2x^2+b_3x^3)$
$=(-a_1-a_2-a_3-b_1-b_2-b_3)+(a_1+b_1)x+(a_2+b_2)x^2+(a_3+b_3)x^3$
since we have $(-a_1-a_2-a_3-b_1-b_2-b_3)+(a_1+b_1)+(a_2+b_2)+(a_3+b_3)=0$.
Likewise, a scalar multiple of a polynomial in W is also in W
$k(-a_1-a_2-a_3+a_1x+a_2x^2+a_3x^3)=-ka_1-ka_2-ka_3+ka_1x+ka_2x^2+ka_3x^3$
since it meets the condition $(-ka_1-ka_2-ka_3)+(ka_1)+(ka_2)+(ka_3)=0$.
According to Theorem 4.2.1, W is a subspace of P_3.

(c) Let W be the set of all polynomials $a_0+a_1x+a_2x^2+a_3x^3$ in which a_0, a_1, a_2, and a_3 are rational numbers. The set W is not closed under the operation of scalar multiplication, e.g., the scalar product of the polynomial x^3 in W by $k=\pi$ is πx^3, which is not in W.
According to Theorem 4.2.1, W is not a subspace of P_3.

(d) The set of all polynomials of degree ≤ 1 is a subset of P_3. It is also a vector space (called P_1) with same operations of addition and scalar multiplication as those defined in P_3. By Definition 1, we conclude that P_1 is a subspace of P_3.

5. **(a)** Let W be the set of all sequences in R^∞ of the form $(v, 0, v, 0, v, 0, \ldots)$.
This set contains at least one sequence, e.g. $(0,0,0,\ldots)$.
Adding two sequences in W results in another sequence in W:

$(v,0,v,0,v,0,\ldots)+(w,0,w,0,w,0,\ldots)=(v+w,0,v+w,0,v+w,0,\ldots)$.
Likewise, a scalar multiple of a vector in W is also in W:
$k(v,0,v,0,v,0,\ldots)=(kv,0,kv,0,kv,0,\ldots)$.
According to Theorem 4.2.1, W is a subspace of R^∞.

(b) Let W be the set of all sequences in R^∞ of the form $(v,1,v,1,v,1,\ldots)$.
This set is not closed under addition since
$(v,1,v,1,v,1,\ldots)+(w,1,w,1,w,1,\ldots)=(v+w,2,v+w,2,v+w,2,\ldots)$ is not in W.
We conclude that W is not a subspace of R^∞.

(c) Let W be the set of all sequences in R^∞ of the form $(v,2v,4v,8v,16v,\ldots)$.
This set contains at least one sequence, e.g. $(0,0,0,\ldots)$.
Adding two sequences in W results in another sequence in W:
$(v,2v,4v,8v,16v,\ldots)+(w,2w,4w,8w,16w,\ldots)$
$=(v+w,2(v+w),4(v+w),8(v+w),16(v+w),\ldots)$.
Likewise, a scalar multiple of a vector in W is also in W:
$k(v,2v,4v,8v,16v,\ldots)=(kv,2kv,4kv,8kv,16kv,\ldots)$.
According to Theorem 4.2.1, W is a subspace of R^∞.

(d) Let W be the set of all sequences in R^∞ whose components are 0 from some point on.
This set contains at least one sequence, e.g. $(0,0,0,\ldots)$.
Let a sequence $\mathbf{u}$ in W have 0 components starting from the ith element; also, let a sequence $\mathbf{v}$ in W have 0 components starting from the jth element. It follows that $\mathbf{u}+\mathbf{v}$ must have 0 component starting no later than from the position corresponding to max (i,j) - the larger of the two numbers. Therefore, $\mathbf{u}+\mathbf{v}$ is in W.
The scalar product $k\mathbf{u}$ must have 0 components starting no later than from the ith element, therefore $k\mathbf{u}$ is also in W.
According to Theorem 4.2.1, W is a subspace of R^∞.

7. (a) For $(2,2,2)$ to be a linear combination of the vectors $\mathbf{u}$ and $\mathbf{v}$, there must exist scalars a and b such that

$$a(0,-2,2)+b(1,3,-1)=(2,2,2)$$

Equating corresponding components on both sides yields the linear system

$$\begin{array}{rcrcr} 0a & + & 1b & = & 2 \\ -2a & + & 3b & = & 2 \\ 2a & - & 1b & = & 2 \end{array}$$

whose augmented matrix has the reduced row echelon form $\begin{bmatrix}1&0&2\\0&1&2\\0&0&0\end{bmatrix}$. The linear system is consistent, therefore $(2,2,2)$ is a linear combination of $\mathbf{u}$ and $\mathbf{v}$.

(b) For $(0,4,5)$ to be a linear combination of the vectors $\mathbf{u}$ and $\mathbf{v}$, there must exist scalars a and b such that

$$a(0,-2,2)+b(1,3,-1)=(0,4,5)$$

Equating corresponding components on both sides yields the linear system

$$\begin{array}{rcrcr} 0a & + & 1b & = & 0 \\ -2a & + & 3b & = & 4 \\ 2a & - & 1b & = & 5 \end{array}$$

whose augmented matrix has the reduced row echelon form $\begin{bmatrix} 1 & 0 & 0 \\ 0 & 1 & 0 \\ 0 & 0 & 1 \end{bmatrix}$. The last row corresponds to the equation $0 = 1$ which is contradictory. We conclude that $(0, 4, 5)$ is not a linear combination of $\mathbf{u}$ and $\mathbf{v}$.

(c) By inspection, the zero vector $(0, 0, 0)$ is a linear combination of $\mathbf{u}$ and $\mathbf{v}$ since

$$0(0, -2, 2) + 0(1, 3, -1) = (0, 0, 0)$$

9. **(a)** For $\begin{bmatrix} 6 & -8 \\ -1 & -8 \end{bmatrix}$ to be a linear combination of A, B, and C, there must exist scalars a, b, and c such that

$$a\begin{bmatrix} 4 & 0 \\ -2 & -2 \end{bmatrix} + b\begin{bmatrix} 1 & -1 \\ 2 & 3 \end{bmatrix} + c\begin{bmatrix} 0 & 2 \\ 1 & 4 \end{bmatrix} = \begin{bmatrix} 6 & -8 \\ -1 & -8 \end{bmatrix}$$

Equating corresponding entries on both sides yields the linear system

$$\begin{array}{rcrcrcr} 4a & + & 1b & + & 0c & = & 6 \\ 0a & - & 1b & + & 2c & = & -8 \\ -2a & + & 2b & + & 1c & = & -1 \\ -2a & + & 3b & + & 4c & = & -8 \end{array}$$

whose augmented matrix has the reduced row echelon form $\begin{bmatrix} 1 & 0 & 0 & 1 \\ 0 & 1 & 0 & 2 \\ 0 & 0 & 1 & -3 \\ 0 & 0 & 0 & 0 \end{bmatrix}$. The linear system is consistent, therefore $\begin{bmatrix} 6 & -8 \\ -1 & -8 \end{bmatrix}$ is a linear combination of A, B, and C.

(b) The zero matrix $\begin{bmatrix} 0 & 0 \\ 0 & 0 \end{bmatrix}$ is a linear combination of A, B, and C since $0A + 0B + 0C = \begin{bmatrix} 0 & 0 \\ 0 & 0 \end{bmatrix}$.

(c) For $\begin{bmatrix} -1 & 5 \\ 7 & 1 \end{bmatrix}$ to be a linear combination of A, B, and C, there must exist scalars a, b, and c such that

$$a\begin{bmatrix} 4 & 0 \\ -2 & -2 \end{bmatrix} + b\begin{bmatrix} 1 & -1 \\ 2 & 3 \end{bmatrix} + c\begin{bmatrix} 0 & 2 \\ 1 & 4 \end{bmatrix} = \begin{bmatrix} -1 & 5 \\ 7 & 1 \end{bmatrix}$$

Equating corresponding entries on both sides yields the linear system

$$\begin{array}{rcrcrcr} 4a & + & 1b & + & 0c & = & -1 \\ 0a & - & 1b & + & 2c & = & 5 \\ -2a & + & 2b & + & 1c & = & 7 \\ -2a & + & 3b & + & 4c & = & 1 \end{array}$$

whose augmented matrix has the reduced row echelon form $\begin{bmatrix} 1 & 0 & 0 & 0 \\ 0 & 1 & 0 & 0 \\ 0 & 0 & 1 & 0 \\ 0 & 0 & 0 & 1 \end{bmatrix}$. The last row corresponds to the equation $0 = 1$ which is contradictory. We conclude that $\begin{bmatrix} -1 & 5 \\ 7 & 1 \end{bmatrix}$ is not a linear combination of A, B, and C.

11. (a) The given vectors span R^3 if an arbitrary vector $\mathbf{b} = (b_1, b_2, b_3)$ can be expressed as a linear combination

$$(b_1, b_2, b_3) = k_1(2,2,2) + k_2(0,0,3) + k_3(0,1,1)$$

Equating corresponding components on both sides yields the linear system

$$\begin{aligned} 2k_1 + 0k_2 + 0k_3 &= b_1 \\ 2k_1 + 0k_2 + 1k_3 &= b_2 \\ 2k_1 + 3k_2 + 1k_3 &= b_3 \end{aligned}$$

By inspection, regardless of the right hand side values b_1, b_2, b_3, the first equation can be solved for k_1, then the second equation can be used to obtain k_3, and the third would yield k_2. We conclude that $\mathbf{v}_1$, $\mathbf{v}_2$, and $\mathbf{v}_3$ span R^3.

(b) The given vectors span R^3 if an arbitrary vector $\mathbf{b} = (b_1, b_2, b_3)$ can be expressed as a linear combination

$$(b_1, b_2, b_3) = k_1(2,-1,3) + k_2(4,1,2) + k_3(8,-1,8)$$

Equating corresponding components on both sides yields the linear system

$$\begin{aligned} 2k_1 + 4k_2 + 8k_3 &= b_1 \\ -1k_1 + 1k_2 - 1k_3 &= b_2 \\ 3k_1 + 2k_2 + 8k_3 &= b_3 \end{aligned}$$

The determinant of the coefficient matrix of this system is $\begin{vmatrix} 2 & 4 & 8 \\ -1 & 1 & -1 \\ 3 & 2 & 8 \end{vmatrix} = 0$, therefore by Theorem 2.3.8, the system cannot be consistent for all right hand side vectors $\mathbf{b}$. We conclude that $\mathbf{v}_1$, $\mathbf{v}_2$, and $\mathbf{v}_3$ do not span R^3.

13. The given polynomials span P_2 if an arbitrary polynomial in P_2, $\mathbf{p} = a_0 + a_1x + a_2x^2$ can be expressed as a linear combination

$$a_0 + a_1x + a_2x^2 = k_1(1 - x + 2x^2) + k_2(3 + x) + k_3(5 - x + 4x^2) + k_4(-2 - 2x + 2x^2)$$

Grouping the terms according to the powers of x yields

$$a_0 + a_1x + a_2x^2 = (k_1 + 3k_2 + 5k_3 - 2k_4) + (-k_1 + k_2 - k_3 - 2k_4)x + (2k_1 + 4k_3 + 2k_4)x^2$$

Since this equality must hold for every real value x, the coefficients associated with the like powers of x on both sides must match. This results in the linear system

$$\begin{aligned} 1k_1 + 3k_2 + 5k_3 - 2k_4 &= a_0 \\ -1k_1 + 1k_2 - 1k_3 - 2k_4 &= a_1 \\ 2k_1 + 0k_2 + 4k_3 + 2k_4 &= a_2 \end{aligned}$$

whose augmented matrix $\begin{bmatrix} 1 & 3 & 5 & -2 & a_0 \\ -1 & 1 & -1 & -2 & a_1 \\ 2 & 0 & 4 & 2 & a_2 \end{bmatrix}$ reduces to $\begin{bmatrix} 1 & 0 & 2 & 1 & \frac{1}{4}a_0 - \frac{3}{4}a_1 \\ 0 & 1 & 1 & -1 & \frac{1}{4}a_0 + \frac{1}{4}a_1 \\ 0 & 0 & 0 & 0 & -\frac{1}{2}a_0 + \frac{3}{2}a_1 + a_2 \end{bmatrix}$

therefore the system has no solution if $-\frac{1}{2}a_0 + \frac{3}{2}a_1 + a_2 \neq 0$.

Since polynomials $\mathbf{p} = a_0 + a_1x + a_2x^2$ for which $-\frac{1}{2}a_0 + \frac{3}{2}a_1 + a_2 \neq 0$ cannot be expressed as a linear combination of $\mathbf{p}_1, \mathbf{p}_2, \mathbf{p}_3$, and $\mathbf{p}_4$, we conclude that the polynomials $\mathbf{p}_1, \mathbf{p}_2, \mathbf{p}_3$, and $\mathbf{p}_4$ do not span P_2.

15. **(a)** The reduced row echelon form of the coefficient matrix A is $\begin{bmatrix} 1 & 0 & \frac{1}{2} \\ 0 & 1 & \frac{3}{2} \\ 0 & 0 & 0 \end{bmatrix}$ therefore the solutions are $x = -\frac{1}{2}t,\ y = -\frac{3}{2}t, z = t$. These are parametric equations of a line through the origin.

(b) The reduced row echelon form of the coefficient matrix A is $\begin{bmatrix} 1 & 0 & 0 \\ 0 & 1 & 0 \\ 0 & 0 & 1 \end{bmatrix}$ therefore the only solution is $x = y = z = 0$ - the origin.

(c) The reduced row echelon form of the coefficient matrix A is $\begin{bmatrix} 1 & -3 & 1 \\ 0 & 0 & 0 \\ 0 & 0 & 0 \end{bmatrix}$ which corresponds to an equation of a plane through the origin $x - 3y + z = 0$.

(d) The reduced row echelon form of the coefficient matrix A is $\begin{bmatrix} 1 & 0 & 3 \\ 0 & 1 & 2 \\ 0 & 0 & 0 \end{bmatrix}$ therefore the solutions are $x = -3t,\ y = -2t, z = t$. These are parametric equations of a line through the origin.

17. Let W denote the set of all continuous functions $f = f(x)$ on $[a,b]$ such that $\int_a^b f(x)dx = 0$.
This set contains at least one function $f(x) \equiv 0$.
Let us assume $\mathbf{f} = f(x)$ and $\mathbf{g} = g(x)$ are functions in W. From calculus,
$\int_a^b f(x) + g(x)dx = \int_a^b f(x)dx + \int_a^b g(x)dx = 0$ and $\int_a^b kf(x)dx = k\int_a^b f(x)dx = 0$ therefore both $\mathbf{f} + \mathbf{g}$ and $k\mathbf{f}$ are in W for any scalar k. According to Theorem 4.2.1, W is a subspace of $C[a,b]$.

19. **(a)** The vectors $T_A(1,2) = (-1,4)$ and $T_A(-1,1) = (-2,2)$ span R^2 if an arbitrary vector $\mathbf{b} = (b_1, b_2)$ can be expressed as a linear combination

$$(b_1, b_2) = k_1(-1,4) + k_2(-2,2)$$

Equating corresponding components on both sides yields the linear system

$$\begin{aligned} -1k_1 \ - \ 2k_2 &= b_1 \\ 4k_1 \ + \ 2k_2 &= b_2 \end{aligned}$$

The determinant of the coefficient matrix of this system is $\begin{vmatrix} -1 & -2 \\ 4 & 2 \end{vmatrix} = 6 \neq 0$, therefore by Theorem 2.3.8, the system is consistent for all right hand side vectors $\mathbf{b}$.
We conclude that $T_A(\mathbf{u}_1)$ and $T_A(\mathbf{u}_2)$ span R^2.

(b) The vectors $T_A(1,2) = (-1,2)$ and $T_A(-1,1) = (-2,4)$ span R^2 if an arbitrary vector $\mathbf{b} = (b_1, b_2)$ can be expressed as a linear combination

$$(b_1, b_2) = k_1(-1,2) + k_2(-2,4)$$

Equating corresponding components on both sides yields the linear system

$$\begin{aligned} -1k_1 &- 2k_2 &= b_1 \\ 2k_1 &+ 4k_2 &= b_2 \end{aligned}$$

The determinant of the coefficient matrix of this system is $\begin{vmatrix} -1 & -2 \\ 2 & 4 \end{vmatrix} = 0$, therefore by Theorem 2.3.8, the system cannot be consistent for all right hand side vectors $\mathbf{b}$.
We conclude that $T_A(\mathbf{u}_1)$ and $T_A(\mathbf{u}_2)$ do not span R^2.

21. Since $T_A: R^3 \to R^m$, it follows from Theorem 4.2.5 that the kernel of T_A must be a subspace of R^3. Hence, according to Table 1 the kernel can be one of the following four geometric objects:

- the origin,
- a line through the origin,
- a plane through the origin,
- R^3.

23. Let W be the set of all functions of the form $x(t) = c_1 \cos\omega t + c_2 \sin\omega t$ - W is a subset of $C^\infty(-\infty, \infty)$. This set contains at least one function $x(t) \equiv 0$.
A sum of two functions in W is also in W:
$(c_1 \cos\omega t + c_2 \sin\omega t) + (d_1 \cos\omega t + d_2 \sin\omega t) = (c_1 + d_1)\cos\omega t + (c_2 + d_2)\sin\omega t$.
A scalar product of a function in W by any scalar k is also a function in W:
$k(c_1 \cos\omega t + c_2 \sin\omega t) = (kc_1)\cos\omega t + (kc_2)\sin\omega t$.
According to Theorem 4.2.1, W is a subspace of $C^\infty(-\infty, \infty)$.

True-False Exercises

(a) True. This follows from Definition 1.

(b) True.

(c) False. The set of all nonnegative real numbers is a subset of the vector space R containing 0, but it is not closed under scalar multiplication.

(d) False. By Theorem 4.2.5, the kernel of $T_A: R^n \to R^m$ is a subspace of R^n.

(e) False. The solution set of a nonhomogeneous system is not closed under addition: $A\mathbf{x} = \mathbf{b}$ and $A\mathbf{y} = \mathbf{b}$ do not imply $A(\mathbf{x} + \mathbf{y}) = \mathbf{b}$.

(f) True. This follows from part (a) of Theorem 4.2.3.

(g) True. This follows from Theorem 4.2.2.

(h) False. Consider $W_1 = \text{span}\{(1,0)\}$ and $W_2 = \text{span}\{(0,1)\}$. The union of these sets is not closed under vector addition, e.g. $(1,0) + (0,1) = (1,1)$ is outside the union.

(i) False. For any nonzero vector $\mathbf{v}$ in a vector space V, both $\{\mathbf{v}\}$ and $\{2\mathbf{v}\}$ span the same subspace of V.

(j) True. This set contains at least one matrix (e.g., I_n). A sum of two upper triangular matrices is also upper triangular, therefore the set is closed under addition. A scalar multiple of an upper triangular matrix is also upper triangular, hence the set is closed under scalar multiplication.

(k) False. The constant polynomial $p(x) = 1$ cannot be represented as a linear combination of these, since at $x = 1$ all three are zero, whereas $p(1) = 1$.

4.3 Linear Independence

1. **(a)** Since $\mathbf{u}_2 = -5\mathbf{u}_1$, linear dependence follows from Definition 1.

(b) A set of 3 vectors in R^2 must be linearly dependent by Theorem 4.3.3.

(c) Since $\mathbf{p}_2 = 2\mathbf{p}_1$, linear dependence follows from Definition 1.

(d) Since $A = (-1)B$, linear dependence follows from Definition 1.

3. **(a)** The vector equation $a(3,8,7,-3) + b(1,5,3,-1) + c(2,-1,2,6) + d(4,2,6,4) = (0,0,0,0)$ can be rewritten as a homogeneous linear system by equating the corresponding components on both sides

$$\begin{array}{rcrcrcrcl} 3a & + & 1b & + & 2c & + & 4d & = & 0 \\ 8a & + & 5b & - & 1c & + & 2d & = & 0 \\ 7a & + & 3b & + & 2c & + & 6d & = & 0 \\ -3a & - & 1b & + & 6c & + & 4d & = & 0 \end{array}$$

The augmented matrix of this system has the reduced row echelon form $\begin{bmatrix} 1 & 0 & 0 & 1 & 0 \\ 0 & 1 & 0 & -1 & 0 \\ 0 & 0 & 1 & 1 & 0 \\ 0 & 0 & 0 & 0 & 0 \end{bmatrix}$

therefore a general solution of the system is $a = -t,\ b = t,\ c = -t,\ d = t$.

Since the system has nontrivial solutions, the given set of vectors is linearly dependent.

(b) The vector equation $a(3,0,-3,6) + b(0,2,3,1) + c(0,-2,-2,0) + d(-2,1,2,1) = (0,0,0,0)$ can be rewritten as a homogeneous linear system by equating the corresponding components on both sides

$$\begin{array}{rcrcrcrcl} 3a & + & 0b & + & 0c & - & 2d & = & 0 \\ 0a & + & 2b & - & 2c & + & 1d & = & 0 \\ -3a & + & 3b & - & 2c & + & 2d & = & 0 \\ 6a & + & 1b & + & 0c & + & 1d & = & 0 \end{array}$$

The augmented matrix of this system has the reduced row echelon form $\begin{bmatrix} 1 & 0 & 0 & 0 & 0 \\ 0 & 1 & 0 & 0 & 0 \\ 0 & 0 & 1 & 0 & 0 \\ 0 & 0 & 0 & 1 & 0 \end{bmatrix}$ therefore the system has only the trivial solution $a = b = c = d = 0$. We conclude that the given set of vectors is linearly independent.

5. **(a)** The matrix equation $a\begin{bmatrix} 1 & 0 \\ 1 & 2 \end{bmatrix} + b\begin{bmatrix} 1 & 2 \\ 2 & 1 \end{bmatrix} + c\begin{bmatrix} 0 & 1 \\ 2 & 1 \end{bmatrix} = \begin{bmatrix} 0 & 0 \\ 0 & 0 \end{bmatrix}$ can be rewritten as a homogeneous linear system

$$\begin{array}{rcrcrcr} 1a & + & 1b & + & 0c & = & 0 \\ 0a & + & 2b & + & 1c & = & 0 \\ 1a & + & 2b & + & 2c & = & 0 \\ 2a & + & 1b & + & 1c & = & 0 \end{array}$$

The augmented matrix of this system has the reduced row echelon form $\begin{bmatrix} 1 & 0 & 0 & 0 \\ 0 & 1 & 0 & 0 \\ 0 & 0 & 1 & 0 \\ 0 & 0 & 0 & 0 \end{bmatrix}$ therefore the system has only the trivial solution $a = b = c = 0$. We conclude that the given matrices are linearly independent.

(b) By inspection, the matrix equation $a\begin{bmatrix} 1 & 0 & 0 \\ 0 & 0 & 0 \end{bmatrix} + b\begin{bmatrix} 0 & 0 & 1 \\ 0 & 0 & 0 \end{bmatrix} + c\begin{bmatrix} 0 & 0 & 0 \\ 0 & 1 & 0 \end{bmatrix} = \begin{bmatrix} 0 & 0 & 0 \\ 0 & 0 & 0 \end{bmatrix}$ has only the trivial solution $a = b = c = 0$. We conclude that the given matrices are linearly independent.

7. Three vectors in R^3 lie in a plane if and only if they are linearly dependent when they have their initial points at the origin. (See the discussion following Example 6.)

(a) The vector equation $a(2,-2,0) + b(6,1,4) + c(2,0,-4) = (0,0,0)$ can be rewritten as a homogeneous linear system by equating the corresponding components on both sides

$$\begin{array}{rcrcrcr} 2a & + & 6b & + & 2c & = & 0 \\ -2a & + & 1b & + & 0c & = & 0 \\ 0a & + & 4b & - & 4c & = & 0 \end{array}$$

The augmented matrix of this system has the reduced row echelon form $\begin{bmatrix} 1 & 0 & 0 & 0 \\ 0 & 1 & 0 & 0 \\ 0 & 0 & 1 & 0 \end{bmatrix}$ therefore the system has only the trivial solution $a = b = c = 0$. We conclude that the given vectors are linearly independent, hence they do not lie in a plane.

(b) The vector equation $a(-6,7,2) + b(3,2,4) + c(4,-1,2) = (0,0,0)$ can be rewritten as a homogeneous linear system by equating the corresponding components on both sides

$$\begin{array}{rcrcrcr} -6a & + & 3b & + & 4c & = & 0 \\ 7a & + & 2b & - & 1c & = & 0 \\ 2a & + & 4b & + & 2c & = & 0 \end{array}$$

The augmented matrix of this system has the reduced row echelon form $\begin{bmatrix} 1 & 0 & -\frac{1}{3} & 0 \\ 0 & 1 & \frac{2}{3} & 0 \\ 0 & 0 & 0 & 0 \end{bmatrix}$

therefore a general solution of the system is $a = \frac{1}{3}t,\ b = -\frac{2}{3}t,\ c = t$.

Since the system has nontrivial solutions, the given vectors are linearly dependent, hence they lie in a plane.

9. **(a)** The vector equation $a(0,3,1,-1) + b(6,0,5,1) + c(4,-7,1,3) = (0,0,0,0)$ can be rewritten as a homogeneous linear system by equating the corresponding components on both sides

$$\begin{array}{rcrcrcl} 0a & + & 6b & + & 4c & = & 0 \\ 3a & + & 0b & - & 7c & = & 0 \\ 1a & + & 5b & + & 1c & = & 0 \\ -1a & + & 1b & + & 3c & = & 0 \end{array}$$

The augmented matrix of this system has the reduced row echelon form $\begin{bmatrix} 1 & 0 & -\frac{7}{3} & 0 \\ 0 & 1 & \frac{2}{3} & 0 \\ 0 & 0 & 0 & 0 \\ 0 & 0 & 0 & 0 \end{bmatrix}$

therefore a general solution of the system is $a = \frac{7}{3}t,\ b = -\frac{2}{3}t,\ c = t$.

Since the system has nontrivial solutions, the given set of vectors is linearly dependent.

(b) From part (a), we have $\frac{7}{3}t\mathbf{v}_1 - \frac{2}{3}t\mathbf{v}_2 + t\mathbf{v}_3 = 0$.

Letting $t = \frac{3}{7}$, we obtain $\mathbf{v}_1 = \frac{2}{7}\mathbf{v}_2 - \frac{3}{7}\mathbf{v}_3$.

Letting $t = -\frac{3}{2}$, we obtain $\mathbf{v}_2 = \frac{7}{2}\mathbf{v}_1 + \frac{3}{2}\mathbf{v}_3$.

Letting $t = 1$, we obtain $\mathbf{v}_3 = -\frac{7}{3}\mathbf{v}_1 + \frac{2}{3}\mathbf{v}_2$.

11. By inspection, when $\lambda = -\frac{1}{2}$, the vectors become linearly dependent (since they all become equal). We proceed to find the remaining values of λ.

The vector equation $a\left(\lambda, -\frac{1}{2}, -\frac{1}{2}\right) + b\left(-\frac{1}{2}, \lambda, -\frac{1}{2}\right) + c\left(-\frac{1}{2}, -\frac{1}{2}, \lambda\right) = (0,0,0)$ can be rewritten as a homogeneous linear system by equating the corresponding components on both sides

$$\begin{array}{rcrcrcl} \lambda a & - & \frac{1}{2}b & - & \frac{1}{2}c & = & 0 \\ -\frac{1}{2}a & + & \lambda b & - & \frac{1}{2}c & = & 0 \\ -\frac{1}{2}a & - & \frac{1}{2}b & + & \lambda c & = & 0 \end{array}$$

The determinant of the coefficient matrix is $\begin{vmatrix} \lambda & -\frac{1}{2} & -\frac{1}{2} \\ -\frac{1}{2} & \lambda & -\frac{1}{2} \\ -\frac{1}{2} & -\frac{1}{2} & \lambda \end{vmatrix} = \lambda^3 - \frac{3}{4}\lambda - \frac{1}{4}$. This determinant equals zero for all λ values for which the vectors are linearly dependent. Since we already know that $\lambda = -\frac{1}{2}$ is one of those values, we can divide $\lambda + \frac{1}{2}$ into $\lambda^3 - \frac{3}{4}\lambda - \frac{1}{4}$ to obtain
$\lambda^3 - \frac{3}{4}\lambda - \frac{1}{4} = \left(\lambda + \frac{1}{2}\right)\left(\lambda^2 - \frac{1}{2}\lambda - \frac{1}{2}\right) = \left(\lambda + \frac{1}{2}\right)\left(\lambda + \frac{1}{2}\right)(\lambda - 1)$.
We conclude that the vectors form a linearly dependent set for $\lambda = -\frac{1}{2}$ and for $\lambda = 1$.

13. **(a)** We calculate $T_A(1,2) = (-1,4)$ and $T_A(-1,1) = (-2,2)$. The vector equation

$$k_1(-1,4) + k_2(-2,2) = (0,0)$$

can be rewritten as a homogeneous linear system

$$\begin{aligned} -1k_1 - 2k_2 &= 0 \\ 4k_1 + 2k_2 &= 0 \end{aligned}$$

The determinant of the coefficient matrix of this system is $\begin{vmatrix} -1 & -1 \\ 4 & 1 \end{vmatrix} = 6 \neq 0$, therefore by Theorem 2.3.8, the system has only the trivial solution. We conclude that $T_A(\mathbf{u}_1)$ and $T_A(\mathbf{u}_2)$ form a linearly independent set.

(b) We calculate $T_A(1,2) = (-1,2)$ and $T_A(-1,1) = (-2,4)$. Since $(-2,4) = 2(-1,2)$, it follows by Definition 1 that $T_A(\mathbf{u}_1)$ and $T_A(\mathbf{u}_2)$ form a linearly dependent set.

15. Three vectors in R^3 lie in a plane if and only if they are linearly dependent when they have their initial points at the origin. (See the discussion following Example 6.)

(a) After the three vectors are moved so that their initial points are at the origin, the resulting vectors do not lie on the same plane. Hence these vectors are linearly independent.

(b) After the three vectors are moved so that their initial points are at the origin, the resulting vectors lie on the same plane. Hence these vectors are linearly dependent.

17. The Wronskian is $W(x) = \begin{vmatrix} x & \cos x \\ 1 & -\sin x \end{vmatrix} = -x\sin x - \cos x$. Since $W(x)$ is not identically 0 on $(-\infty, \infty)$ (e.g., $W(0) = -1 \neq 0$), the functions x and $\cos x$ are linearly independent.

19. **(a)** The Wronskian is $W(x) = \begin{vmatrix} 1 & x & e^x \\ 0 & 1 & e^x \\ 0 & 0 & e^x \end{vmatrix} = e^x$. Since $W(x)$ is not identically 0 on $(-\infty, \infty)$ (e.g., $W(0) = 1 \neq 0$), the functions 1, x and e^x are linearly independent.

(b) The Wronskian is $W(x) = \begin{vmatrix} 1 & x & x^2 \\ 0 & 1 & 2x \\ 0 & 0 & 2 \end{vmatrix} = 2$. Since $W(x)$ is not identically 0 on $(-\infty, \infty)$, the functions 1, x and x^2 are linearly independent.

21. $$W(x) = \begin{vmatrix} \sin x & \cos x & x\cos x \\ \cos x & -\sin x & \cos x - x\sin x \\ -\sin x & -\cos x & -2\sin x - x\cos x \end{vmatrix}$$ ← The Wronskian

$$= \begin{vmatrix} \sin x & \cos x & x\cos x \\ \cos x & -\sin x & \cos x - x\sin x \\ 0 & 0 & -2\sin x \end{vmatrix}$$ ← The first row was added to the third.

$$= -2\sin x \begin{vmatrix} \sin x & \cos x \\ \cos x & -\sin x \end{vmatrix}$$ ← Cofactor expansion along the third row

$$= -2\sin x\,(-\sin^2 x - \cos^2 x)$$

$$= (-2\sin x)(-1) = 2\sin x$$

Since $W(x)$ is not identically 0 on $(-\infty,\infty)$, $f_1(x)$, $f_2(x)$, and $f_3(x)$ are linearly independent.

True-False Exercises

(a) False. By part (b) of Theorem 4.3.2, a set containing a single *nonzero* vector is linearly independent.

(b) True. This follows directly from Definition 1.

(c) False. For instance $\{(1,1),(2,2)\}$ is a linearly dependent set that does not contain $(0,0)$.

(d) True. If $a\mathbf{v}_1 + b\mathbf{v}_2 + c\mathbf{v}_3 = \mathbf{0}$ has only one solution $a = b = c = 0$ then $a(k\mathbf{v}_1) + b(k\mathbf{v}_2) + c(k\mathbf{v}_3) = k(a\mathbf{v}_1 + b\mathbf{v}_2 + c\mathbf{v}_3)$ can only equal $\mathbf{0}$ when $a = b = c = 0$ as well.

(e) True. Since the vectors must be nonzero, $\{\mathbf{v}_1\}$ must be linearly independent.
Let us begin adding vectors to the set until the set $\{\mathbf{v}_1, \ldots, \mathbf{v}_k\}$ becomes linearly dependent, therefore, by construction, $\{\mathbf{v}_1, \ldots, \mathbf{v}_{k-1}\}$ is linearly independent. The equation $c_1\mathbf{v}_1 + \cdots + c_{k-1}\mathbf{v}_{k-1} + c_k\mathbf{v}_k = \mathbf{0}$ must have a solution with $c_k \neq 0$, therefore $\mathbf{v}_k = -\frac{c_1}{c_k}\mathbf{v}_1 - \cdots - \frac{c_{k-1}}{c_k}\mathbf{v}_{k-1}$. Let us assume there exists another representation $\mathbf{v}_k = d_1\mathbf{v}_1 + \cdots + d_{k-1}\mathbf{v}_{k-1}$. Subtracting both sides yields $\mathbf{0} = \left(d_1 + \frac{c_1}{c_k}\right)\mathbf{v}_1 + \cdots + \left(d_{k-1} + \frac{c_{k-1}}{c_k}\right)\mathbf{v}_{k-1}$. By linear independence of $\{\mathbf{v}_1, \ldots, \mathbf{v}_{k-1}\}$, we must have $d_1 = -\frac{c_1}{c_k}, \ldots, d_{k-1} = -\frac{c_{k-1}}{c_k}$, which shows that $\mathbf{v}_k$ is a *unique* linear combination of $\mathbf{v}_1, \ldots, \mathbf{v}_{k-1}$.

(f) False. The set $\left\{\begin{bmatrix} 1 & 1 \\ 0 & 0 \end{bmatrix}, \begin{bmatrix} 0 & 0 \\ 1 & 1 \end{bmatrix}, \begin{bmatrix} 1 & 0 \\ 1 & 0 \end{bmatrix}, \begin{bmatrix} 0 & 1 \\ 0 & 1 \end{bmatrix}, \begin{bmatrix} 1 & 0 \\ 0 & 1 \end{bmatrix}, \begin{bmatrix} 0 & 1 \\ 1 & 0 \end{bmatrix}\right\}$ is linearly dependent since $\begin{bmatrix} 1 & 1 \\ 0 & 0 \end{bmatrix} = (-1)\begin{bmatrix} 0 & 0 \\ 1 & 1 \end{bmatrix} + \begin{bmatrix} 1 & 0 \\ 1 & 0 \end{bmatrix} + \begin{bmatrix} 0 & 1 \\ 0 & 1 \end{bmatrix}$.

(g) True. Requiring that for all x values $a(x-1)(x+2) + bx(x+2) + cx(x-1) = 0$ holds true implies that the equality must be true for any specific x value. Setting $x = 0$ yields $a = 0$. Likewise, $x = 1$ implies $b = 0$, and $x = -2$ implies $c = 0$. Since $a = b = c = 0$ is required, we conclude that the three given polynomials are linearly independent.

(h) False. The functions f_1 and f_2 are linearly dependent if there exist scalars k_1 and k_2, not both equal 0, such that $k_1 f_1(x) + k_2 f_2(x) = 0$ for all real numbers x.

4.4 Coordinates and Basis

1. Vectors (2,1) and (3,0) are linearly independent if the vector equation

$$c_1(2,1) + c_2(3,0) = (0,0)$$

has only the trivial solution. For these vectors to span R^2, it must be possible to express every vector $\mathbf{b} = (b_1, b_2)$ in R^2 as

$$c_1(2,1) + c_2(3,0) = (b_1, b_2)$$

These two equations can be rewritten as linear systems

$$\begin{array}{rcrcl} 2c_1 & + & 3c_2 & = & 0 \\ c_1 & & & = & 0 \end{array} \quad \text{and} \quad \begin{array}{rcrcl} 2c_1 & + & 3c_2 & = & b_1 \\ c_1 & & & = & b_2 \end{array}$$

Since the coefficient matrix of both systems has determinant $\begin{vmatrix} 2 & 3 \\ 1 & 0 \end{vmatrix} = -3 \neq 0$, it follows from parts (b), (e), and (g) of Theorem 2.3.8 that the homogeneous system has only the trivial solution and the nonhomogeneous system is consistent for all real values b_1 and b_2. Therefore the vectors (2,1) and (3,0) are linearly independent and span R^2 so that they form a basis for R^2.

3. Polynomials $x^2 + 1, x^2 - 1$, and $2x - 1$ are linearly independent if the equation

$$c_1(x^2 + 1) + c_2(x^2 - 1) + c_3(2x - 1) = 0$$

has only the trivial solution. For these polynomials to span P_2, it must be possible to express every polynomial $a_0 + a_1x + a_2x^2$ as

$$c_1(x^2 + 1) + c_2(x^2 - 1) + c_3(2x - 1) = a_0 + a_1x + a_2x^2$$

Grouping the terms on the left hand side of both equations as $(c_1 - c_2 - c_3) + (2c_3)x + (c_1 + c_2)x^2$ these equations can be rewritten as linear systems

$$\begin{array}{rcrcrcl} 1c_1 & - & 1c_2 & - & 1c_3 & = & 0 \\ 0c_1 & + & 0c_2 & + & 2c_3 & = & 0 \\ 1c_1 & + & 1c_2 & + & 0c_3 & = & 0 \end{array} \quad \text{and} \quad \begin{array}{rcrcrcl} 1c_1 & - & 1c_2 & - & 1c_3 & = & a_0 \\ 0c_1 & + & 0c_2 & + & 2c_3 & = & a_1 \\ 1c_1 & + & 1c_2 & + & 0c_3 & = & a_2 \end{array}$$

Since the coefficient matrix of both systems has determinant $\begin{vmatrix} 1 & -1 & -1 \\ 0 & 0 & 2 \\ 1 & 1 & 0 \end{vmatrix} = -4 \neq 0$, it follows from parts (b), (e), and (g) of Theorem 2.3.8 that the homogeneous system has only the trivial solution and the nonhomogeneous system is consistent for all real values a_0, a_1, and a_2. Therefore the polynomials $x^2 + 1, x^2 - 1$, and $2x - 1$ are linearly independent and span P_2 so that they form a basis for P_2.

5. Matrices $\begin{bmatrix} 3 & 6 \\ 3 & -6 \end{bmatrix}, \begin{bmatrix} 0 & -1 \\ -1 & 0 \end{bmatrix}, \begin{bmatrix} 0 & -8 \\ -12 & -4 \end{bmatrix}$, and $\begin{bmatrix} 1 & 0 \\ -1 & 2 \end{bmatrix}$ are linearly independent if the equation

$$c_1\begin{bmatrix} 3 & 6 \\ 3 & -6 \end{bmatrix} + c_2\begin{bmatrix} 0 & -1 \\ -1 & 0 \end{bmatrix} + c_3\begin{bmatrix} 0 & -8 \\ -12 & -4 \end{bmatrix} + c_4\begin{bmatrix} 1 & 0 \\ -1 & 2 \end{bmatrix} = \begin{bmatrix} 0 & 0 \\ 0 & 0 \end{bmatrix}$$

has only the trivial solution. For these matrices to span M_{22}, it must be possible to express every matrix $\begin{bmatrix} a_{11} & a_{12} \\ a_{21} & a_{22} \end{bmatrix}$ as

$$c_1\begin{bmatrix} 3 & 6 \\ 3 & -6 \end{bmatrix} + c_2\begin{bmatrix} 0 & -1 \\ -1 & 0 \end{bmatrix} + c_3\begin{bmatrix} 0 & -8 \\ -12 & -4 \end{bmatrix} + c_4\begin{bmatrix} 1 & 0 \\ -1 & 2 \end{bmatrix} = \begin{bmatrix} a_{11} & a_{12} \\ a_{21} & a_{22} \end{bmatrix}$$

Equating corresponding entries on both sides yields linear systems

$$\begin{array}{rcrcrcrcl} 3c_1 & + & 0c_2 & + & 0c_3 & + & 1c_4 & = & 0 \\ 6c_1 & - & 1c_2 & - & 8c_3 & + & 0c_4 & = & 0 \\ 3c_1 & - & 1c_2 & - & 12c_3 & - & 1c_4 & = & 0 \\ -6c_1 & + & 0c_2 & - & 4c_3 & + & 2c_4 & = & 0 \end{array} \quad \text{and} \quad \begin{array}{rcrcrcrcl} 3c_1 & + & 0c_2 & + & 0c_3 & + & 1c_4 & = & a_{11} \\ 6c_1 & - & 1c_2 & - & 8c_3 & + & 0c_4 & = & a_{12} \\ 3c_1 & - & 1c_2 & - & 12c_3 & - & 1c_4 & = & a_{21} \\ -6c_1 & + & 0c_2 & - & 4c_3 & + & 2c_4 & = & a_{22} \end{array}$$

Since the coefficient matrix of both systems has determinant $\begin{vmatrix} 3 & 0 & 0 & 1 \\ 6 & -1 & -8 & 0 \\ 3 & -1 & -12 & -1 \\ -6 & 0 & -4 & 2 \end{vmatrix} = 48 \neq 0$, it follows from parts (b), (e), and (g) of Theorem 2.3.8 that the homogeneous system has only the trivial solution and the nonhomogeneous system is consistent for all real values a_{11}, a_{12}, a_{21}, and a_{22}. Therefore the matrices $\begin{bmatrix} 3 & 6 \\ 3 & -6 \end{bmatrix}, \begin{bmatrix} 0 & -1 \\ -1 & 0 \end{bmatrix}, \begin{bmatrix} 0 & -8 \\ -12 & -4 \end{bmatrix}$, and $\begin{bmatrix} 1 & 0 \\ -1 & 2 \end{bmatrix}$ are linearly independent and span M_{22} so that they form a basis for M_{22}.

7. **(a)** Vectors $(2,-3,1), (4,1,1)$, and $(0,-7,1)$ are linearly independent if the vector equation

$$c_1(2,-3,1) + c_2(4,1,1) + c_3(0,-7,1) = (0,0,0)$$

has only the trivial solution. This equation can be rewritten as a linear system

$$\begin{array}{rcrcrcl} 2c_1 & + & 4c_2 & + & 0c_3 & = & 0 \\ -3c_1 & + & 1c_2 & - & 7c_3 & = & 0 \\ 1c_1 & + & 1c_2 & + & 1c_3 & = & 0 \end{array}$$

Since the determinant of the coefficient matrix of this system is $\begin{vmatrix} 2 & 4 & 0 \\ -3 & 1 & -7 \\ 1 & 1 & 1 \end{vmatrix} = 0$, it follows from parts (b) and (g) of Theorem 2.3.8 that the homogeneous system has nontrivial solutions. Since the vectors $(2,-3,1), (4,1,1)$, and $(0,-7,1)$ are linearly dependent, they do not form a basis for R^3.

(b) Vectors $(1,6,4), (2,4,-1)$, and $(-1,2,5)$ are linearly independent if the vector equation

$$c_1(1,6,4) + c_2(2,4,-1) + c_3(-1,2,5) = (0,0,0)$$

has only the trivial solution. This equation can be rewritten as a linear system

$$\begin{array}{rcrcrcl} 1c_1 & + & 2c_2 & - & 1c_3 & = & 0 \\ 6c_1 & + & 4c_2 & + & 2c_3 & = & 0 \\ 4c_1 & - & 1c_2 & + & 5c_3 & = & 0 \end{array}$$

Since the determinant of the coefficient matrix of this system is $\begin{vmatrix} 1 & 2 & -1 \\ 6 & 4 & 2 \\ 4 & -1 & 5 \end{vmatrix} = 0$, it follows from parts (b) and (g) of Theorem 2.3.8 that the homogeneous system has nontrivial solutions.

Since the vectors $(1,6,4)$, $(2,4,-1)$, and $(-1,2,5)$ are linearly dependent, they do not form a basis for R^3.

9. Matrices $\begin{bmatrix}1 & 0\\1 & 1\end{bmatrix}, \begin{bmatrix}2 & -2\\3 & 2\end{bmatrix}, \begin{bmatrix}1 & -1\\1 & 0\end{bmatrix}$, and $\begin{bmatrix}0 & -1\\1 & 1\end{bmatrix}$ are linearly independent if the equation

$$c_1\begin{bmatrix}1 & 0\\1 & 1\end{bmatrix} + c_2\begin{bmatrix}2 & -2\\3 & 2\end{bmatrix} + c_3\begin{bmatrix}1 & -1\\1 & 0\end{bmatrix} + c_4\begin{bmatrix}0 & -1\\1 & 1\end{bmatrix} = \begin{bmatrix}0 & 0\\0 & 0\end{bmatrix}$$

has only the trivial solution. Equating corresponding entries on both sides yields a linear system

$$\begin{array}{rcrcrcrcl} 1c_1 & + & 2c_2 & + & 1c_3 & + & 0c_4 & = & 0\\ 0c_1 & - & 2c_2 & - & 1c_3 & - & 1c_4 & = & 0\\ 1c_1 & + & 3c_2 & + & 1c_3 & + & 1c_4 & = & 0\\ 1c_1 & + & 2c_2 & + & 0c_3 & + & 1c_4 & = & 0 \end{array}$$

Since the determinant of the coefficient matrix of this system is $\begin{vmatrix}1 & 2 & 1 & 0\\0 & -2 & -1 & -1\\1 & 3 & 1 & 1\\1 & 2 & 0 & 1\end{vmatrix} = 0$, it follows from parts (b) and (g) of Theorem 2.3.8 that the homogeneous system has nontrivial solutions. Since the matrices $\begin{bmatrix}1 & 0\\1 & 1\end{bmatrix}, \begin{bmatrix}2 & -2\\3 & 2\end{bmatrix}, \begin{bmatrix}1 & -1\\1 & 0\end{bmatrix}$, and $\begin{bmatrix}0 & -1\\1 & 1\end{bmatrix}$ are linearly dependent, we conclude that they do not form a basis for M_{22}.

11. **(a)** Expressing $\mathbf{w}$ as a linear combination of $\mathbf{u}_1$ and $\mathbf{u}_2$ we obtain

$$(1,1) = c_1(2,-4) + c_2(3,8)$$

Equating corresponding components on both sides yields the linear system

$$\begin{array}{rcrcl} 2c_1 & + & 3c_2 & = & 1\\ -4c_1 & + & 8c_2 & = & 1 \end{array}$$

whose augmented matrix has the reduced row echelon form $\begin{bmatrix}1 & 0 & \frac{5}{28}\\0 & 1 & \frac{3}{14}\end{bmatrix}$. The solution of the linear system is $c_1 = \frac{5}{28}$, $c_2 = \frac{3}{14}$, therefore the coordinate vector is $(\mathbf{w})_S = (\frac{5}{28}, \frac{3}{14})$.

(b) Expressing $\mathbf{w}$ as a linear combination of $\mathbf{u}_1$ and $\mathbf{u}_2$ we obtain

$$(a,b) = c_1(1,1) + c_2(0,2)$$

Equating corresponding components on both sides yields the linear system

$$\begin{array}{rcrcl} 1c_1 & + & 0c_2 & = & a\\ 1c_1 & + & 2c_2 & = & b \end{array}$$

whose augmented matrix has the reduced row echelon form $\begin{bmatrix}1 & 0 & a\\0 & 1 & \frac{b-a}{2}\end{bmatrix}$. The solution of the linear system is $c_1 = a$, $c_2 = \frac{b-a}{2}$, therefore the coordinate vector is $(\mathbf{w})_S = (a, \frac{b-a}{2})$.

13. **(a)** Expressing $\mathbf{v}$ as a linear combination of $\mathbf{v}_1, \mathbf{v}_2$, and $\mathbf{v}_3$ we obtain

$$(2,-1,3) = c_1(1,0,0) + c_2(2,2,0) + c_3(3,3,3)$$

Equating corresponding components on both sides yields the linear system

$$\begin{array}{rcrcrcr} c_1 & + & 2c_2 & + & 3c_3 & = & 2 \\ & & 2c_2 & + & 3c_3 & = & -1 \\ & & & & 3c_3 & = & 3 \end{array}$$

which can be solved by back-substitution to obtain $c_3 = 1, c_2 = -2$, and $c_1 = 3$. The coordinate vector is $(\mathbf{v})_S = (3, -2, 1)$.

(b) Expressing $\mathbf{v}$ as a linear combination of $\mathbf{v}_1, \mathbf{v}_2,$ and $\mathbf{v}_3$ we obtain

$$(5, -12, 3) = c_1(1, 2, 3) + c_2(-4, 5, 6) + c_3(7, -8, 9)$$

Equating corresponding components on both sides yields the linear system

$$\begin{array}{rcrcrcr} 1c_1 & - & 4c_2 & + & 7c_3 & = & 5 \\ 2c_1 & + & 5c_2 & - & 8c_3 & = & -12 \\ 3c_1 & + & 6c_2 & + & 9c_3 & = & 3 \end{array}$$

whose augmented matrix has the reduced row echelon form $\begin{bmatrix} 1 & 0 & 0 & -2 \\ 0 & 1 & 0 & 0 \\ 0 & 0 & 1 & 1 \end{bmatrix}$. The solution of the linear system is $c_1 = -2, c_2 = 0$, and $c_3 = 1$. The coordinate vector is $(\mathbf{v})_S = (-2, 0, 1)$.

15. Matrices (vectors in M_{22}) $A_1, A_2, A_3,$ and A_4 are linearly independent if the equation

$$k_1A_1 + k_2A_2 + k_3A_3 + k_4A_4 = \mathbf{0}$$

has only the trivial solution. For these matrices to span M_{22}, it must be possible to express every matrix $B = \begin{bmatrix} a & b \\ c & d \end{bmatrix}$ as

$$k_1A_1 + k_2A_2 + k_3A_3 + k_4A_4 = B$$

The left hand side of each of these equations is the matrix $\begin{bmatrix} k_1 & k_1 + k_2 \\ k_1 + k_2 + k_3 & k_1 + k_2 + k_3 + k_4 \end{bmatrix}$. Equating corresponding entries, these two equations can be rewritten as linear systems

$$\begin{array}{rcrcrcrcr} k_1 & & & & & & & = & 0 \\ k_1 & + & k_2 & & & & & = & 0 \\ k_1 & + & k_2 & + & k_3 & & & = & 0 \\ k_1 & + & k_2 & + & k_3 & + & k_4 & = & 0 \end{array} \quad \text{and} \quad \begin{array}{rcrcrcrcr} k_1 & & & & & & & = & a \\ k_1 & + & k_2 & & & & & = & b \\ k_1 & + & k_2 & + & k_3 & & & = & c \\ k_1 & + & k_2 & + & k_3 & + & k_4 & = & d \end{array}$$

Since the coefficient matrix of both systems has determinant $\begin{vmatrix} 1 & 0 & 0 & 0 \\ 1 & 1 & 0 & 0 \\ 1 & 1 & 1 & 0 \\ 1 & 1 & 1 & 1 \end{vmatrix} = 1 \neq 0$, it follows from parts (b), (e), and (g) of Theorem 2.3.8 that the homogeneous system has only the trivial solution and the nonhomogeneous system is consistent for all real values a, b, c and d. Therefore the matrices $A_1, A_2, A_3,$ and A_4 are linearly independent and span M_{22} so that they form a basis for M_{22}.

To express $A = \begin{bmatrix} 1 & 0 \\ 1 & 0 \end{bmatrix}$ as a linear combination of the matrices $A_1, A_2, A_3,$ and A_4, we form the nonhomogeneous system as above, with the appropriate right hand side values

$$\begin{array}{rcl} k_1 & & = 1 \\ k_1 + k_2 & & = 0 \\ k_1 + k_2 + k_3 & & = 1 \\ k_1 + k_2 + k_3 + k_4 & & = 0 \end{array}$$

which can be solved by forward-substitution to obtain $k_1 = 1,\ k_2 = -1,\ k_3 = 1,\ k_4 = -1$.
This allows us to express $A = 1A_1 - 1A_2 + 1A_3 - 1A_4$.
The coordinate vector is $(A)_S = (1, -1, 1, -1)$.

17. Vectors $\mathbf{p}_1$, $\mathbf{p}_2$, and $\mathbf{p}_3$ are linearly independent if the vector equation

$$c_1\mathbf{p}_1 + c_2\mathbf{p}_2 + c_3\mathbf{p}_3 = \mathbf{0}$$

has only the trivial solution. For these vectors to span P_2, it must be possible to express every vector $\mathbf{p} = a_0 + a_1x + a_2x^2$ in P_2 as

$$c_1\mathbf{p}_1 + c_2\mathbf{p}_2 + c_3\mathbf{p}_3 = \mathbf{p}$$

Grouping the terms on the left hand sides as $c_1(1 + x + x^2) + c_2(x + x^2) + c_3x^2 = c_1 + (c_1 + c_2)x + (c_1 + c_2 + c_3)x^2$ these two equations can be rewritten as linear systems

$$\begin{array}{rcl} c_1 & = & 0 \\ c_1 + c_2 & = & 0 \\ c_1 + c_2 + c_3 & = & 0 \end{array} \quad \text{and} \quad \begin{array}{rcl} c_1 & = & a_0 \\ c_1 + c_2 & = & a_1 \\ c_1 + c_2 + c_3 & = & a_2 \end{array}$$

Since the coefficient matrix of both systems has determinant $\begin{vmatrix} 1 & 0 & 0 \\ 1 & 1 & 0 \\ 1 & 1 & 1 \end{vmatrix} = 1 \neq 0$, it follows from parts (b), (e), and (g) of Theorem 2.3.8 that the homogeneous system has only the trivial solution and the nonhomogeneous system is consistent for all real values a_0, a_1, and a_2. Therefore the vectors $\mathbf{p}_1$, $\mathbf{p}_2$, and $\mathbf{p}_3$ are linearly independent and span P_2 so that they form a basis for P_2.
To express $\mathbf{p} = 7 - x + 2x^2$ as a linear combination of the vectors $\mathbf{p}_1$, $\mathbf{p}_2$, and $\mathbf{p}_3$, we form the nonhomogeneous system as above, with the appropriate right hand side values

$$\begin{array}{rcl} c_1 & = & 7 \\ c_1 + c_2 & = & -1 \\ c_1 + c_2 + c_3 & = & 2 \end{array}$$

which can be solved by forward-substitution to obtain $c_1 = 7,\ c_2 = -8,\ c_3 = 3$.
This allows us to express $\mathbf{p} = 7\mathbf{p}_1 - 8\mathbf{p}_2 + 3\mathbf{p}_3$. The coordinate vector is $(\mathbf{p})_S = (7, -8, 3)$.

19. **(a)** The third vector is a sum of the first two. This makes the set linearly dependent, hence it cannot be a basis for R^2.

(b) The two vectors generate a plane in R^3, but they do not span all of R^3. Consequently, the set is not a basis for R^3.

(c) For instance, the polynomial $\mathbf{p} = 1$ cannot be expressed as a linear combination of the given two polynomials. This means these two polynomials do not span P_2, hence they do not form a basis for P_2.

(d) For instance, the matrix $\begin{bmatrix} 0 & 1 \\ 0 & 0 \end{bmatrix}$ cannot be expressed as a linear combination of the given four matrices. This means these four matrices do not span M_{22}, hence they do not form a basis for M_{22}.

21. **(a)** We have $T_A(1,0,0) = (1,0,-1)$, $T_A(0,1,0) = (1,1,2)$, and $T_A(0,0,1) = (1,-3,0)$. The vector equation

$$k_1(1,0,-1) + k_2(1,1,2) + k_3(1,-3,0) = (0,0,0)$$

can be rewritten as a homogeneous linear system

$$\begin{array}{rcrcrcl} 1k_1 & + & 1k_2 & + & 1k_3 & = & 0 \\ 0k_1 & + & 1k_2 & - & 3k_3 & = & 0 \\ -1k_1 & + & 2k_2 & + & 0k_3 & = & 0 \end{array}$$

The determinant of the coefficient matrix of this system is $\det(A) = 10 \neq 0$, therefore by Theorem 2.3.8, the system has only the trivial solution. We conclude that the set $\{T_A(\mathbf{e}_1), T_A(\mathbf{e}_2), T_A(\mathbf{e}_3)\}$ is linearly independent.

(b) We have $T_A(1,0,0) = (1,0,-1)$, $T_A(0,1,0) = (1,1,2)$, and $T_A(0,0,1) = (2,1,1)$. By inspection,

$$(2,1,1) = (1,0,-1) + (1,1,2)$$

We conclude that the set $\{T_A(\mathbf{e}_1), T_A(\mathbf{e}_2), T_A(\mathbf{e}_3)\}$ is linearly dependent.

23. We have $\mathbf{u}_1 = (\cos 30°, \sin 30°) = \left(\frac{\sqrt{3}}{2}, \frac{1}{2}\right)$ and $\mathbf{u}_2 = (0,1)$.

(a) By inspection, we can express $\mathbf{w} = (\sqrt{3}, 1)$ as a linear combination of $\mathbf{u}_1$ and $\mathbf{u}_2$

$$(\sqrt{3}, 1) = 2\left(\frac{\sqrt{3}}{2}, \frac{1}{2}\right) + 0(0,1)$$

therefore the coordinate vector is $(\mathbf{w})_S = (2,0)$.

(b) Expressing $\mathbf{w} = (\sqrt{3}, 1)$ as a linear combination of $\mathbf{u}_1$ and $\mathbf{u}_2$ we obtain

$$(1,0) = c_1\left(\frac{\sqrt{3}}{2}, \frac{1}{2}\right) + c_2(0,1)$$

Equating corresponding components on both sides yields the linear system

$$\begin{array}{rcrcl} \frac{\sqrt{3}}{2}c_1 & & & = & 1 \\ \frac{1}{2}c_1 & + & c_2 & = & 0 \end{array}$$

The first equation yields $c_1 = \frac{2}{\sqrt{3}}$, then the second equation can be solved to obtain $c_2 = -\frac{1}{\sqrt{3}}$. The coordinate vector is $(\mathbf{w})_S = (\frac{2}{\sqrt{3}}, -\frac{1}{\sqrt{3}})$.

(c) By inspection, we can express $\mathbf{w} = (0,1)$ as a linear combination of $\mathbf{u}_1$ and $\mathbf{u}_2$

$$(0,1) = 0\left(\frac{\sqrt{3}}{2}, \frac{1}{2}\right) + 1(0,1)$$

therefore the coordinate vector is $(\mathbf{w})_S = (0, 1)$.

(d) Expressing $\mathbf{w} = (a, b)$ as a linear combination of $\mathbf{u}_1$ and $\mathbf{u}_2$ we obtain

$$(a, b) = c_1\left(\frac{\sqrt{3}}{2}, \frac{1}{2}\right) + c_2(0,1)$$

Equating corresponding components on both sides yields the linear system

$$\begin{array}{rcrcl} \frac{\sqrt{3}}{2}c_1 & & & = & a \\ \frac{1}{2}c_1 & + & c_2 & = & b \end{array}$$

The first equation yields $c_1 = \frac{2a}{\sqrt{3}}$, then the second equation can be solved to obtain $c_2 = b - \frac{a}{\sqrt{3}}$.

The coordinate vector is $(\mathbf{w})_S = \left(\frac{2a}{\sqrt{3}}, b - \frac{a}{\sqrt{3}}\right)$.

25. (a) Polynomials $1,\ 2t, -2 + 4t^2$, and $-12t + 8t^3$ are linearly independent if the equation

$$c_1(1) + c_2(2t) + c_3(-2 + 4t^2) + c_4(-12t + 8t^3) = 0$$

has only the trivial solution. For these polynomials to span P_3, it must be possible to express every polynomial $a_0 + a_1t + a_2t^2 + a_3t^3$ as

$$c_1(1) + c_2(2t) + c_3(-2 + 4t^2) + c_4(-12t + 8t^3) = a_0 + a_1t + a_2t^2 + a_3t^3$$

Grouping the terms on the left hand side of both equations as $(c_1 - 2c_3) + (2c_2 - 12c_4)t + 4c_3t^2 + 8c_4t^3$ these equations can be rewritten as linear systems

$$\begin{array}{rcrcrcrcl} 1c_1 & + & 0c_2 & - & 2c_3 & + & 0c_4 & = & 0 \\ 0c_1 & + & 2c_2 & + & 0c_3 & - & 12c_4 & = & 0 \\ 0c_1 & + & 0c_2 & + & 4c_3 & + & 0c_4 & = & 0 \\ 0c_1 & + & 0c_2 & + & 0c_3 & + & 8c_4 & = & 0 \end{array} \quad \text{and} \quad \begin{array}{rcrcrcrcl} 1c_1 & + & 0c_2 & - & 2c_3 & + & 0c_4 & = & a_0 \\ 0c_1 & + & 2c_2 & + & 0c_3 & - & 12c_4 & = & a_1 \\ 0c_1 & + & 0c_2 & + & 4c_3 & + & 0c_4 & = & a_2 \\ 0c_1 & + & 0c_2 & + & 0c_3 & + & 8c_4 & = & a_3 \end{array}$$

Since the coefficient matrix of both systems has determinant $\begin{vmatrix} 1 & 0 & -2 & 0 \\ 0 & 2 & 0 & -12 \\ 0 & 0 & 4 & 0 \\ 0 & 0 & 0 & 8 \end{vmatrix} = 64 \neq 0$, it follows from parts (b), (e), and (g) of Theorem 2.3.8 that the homogeneous system has only the trivial solution and the nonhomogeneous system is consistent for all real values a_0, a_1, a_2, and a_3. Therefore the polynomials $1,\ 2t, -2 + 4t^2$, and $-12t + 8t^3$ are linearly independent and span P_3 so that they form a basis for P_3.

(b) To express $\mathbf{p} = -1 - 4t + 8t^2 + 8t^3$ as a linear combination of the four vectors in B, we form the nonhomogeneous system as was done in part (a), with the appropriate right hand side values

$$\begin{array}{rcrcrcrcr} 1c_1 & + & 0c_2 & - & 2c_3 & + & 0c_4 & = & -1 \\ 0c_1 & + & 2c_2 & + & 0c_3 & - & 12c_4 & = & -4 \\ 0c_1 & + & 0c_2 & + & 4c_3 & + & 0c_4 & = & 8 \\ 0c_1 & + & 0c_2 & + & 0c_3 & + & 8c_4 & = & 8 \end{array}$$

Back-substitution yields $c_4 = 1, c_3 = 2, c_2 = 4$, and $c_1 = 3$.
The coordinate vector is $(\mathbf{p})_B = (3,4,2,1)$.

27. **(a)** $\mathbf{w} = 6(3,1,-4) - 1(2,5,6) + 4(1,4,8) = (20,17,2)$

(b) $\mathbf{q} = 3(x^2+1) + 0(x^2-1) + 4(2x-1) = 3x^2 + 8x - 1$

(c) $B = -8\begin{bmatrix} 3 & 6 \\ 3 & -6 \end{bmatrix} + 7\begin{bmatrix} 0 & -1 \\ -1 & 0 \end{bmatrix} + 6\begin{bmatrix} 0 & -8 \\ -12 & -4 \end{bmatrix} + 3\begin{bmatrix} 1 & 0 \\ -1 & 2 \end{bmatrix} = \begin{bmatrix} -21 & -103 \\ -106 & 30 \end{bmatrix}$

True-False Exercises

(a) False. The set must also be linearly independent.

(b) False. The subset must also span V.

(c) True. This follows from Theorem 4.4.1.

(d) True. For any vector $\mathbf{v} = (a_1, \dots, a_n)$ in R^n, we have $\mathbf{v} = a_1\mathbf{e}_1 + \dots + a_n\mathbf{e}_n$ therefore the coordinate vector of $\mathbf{v}$ with respect to the standard basis $S = \{\mathbf{e}_1, \dots, \mathbf{e}_n\}$ is $(\mathbf{v})_S = (a_1, \dots, a_n) = \mathbf{v}$.

(e) False. For instance, $\{1+t^4, t+t^4, t^2+t^4, t^3+t^4, t^4\}$ is a basis for P_4.

4.5 Dimension

1. The augmented matrix of the linear system $\begin{bmatrix} 1 & 1 & -1 & 0 \\ -2 & -1 & 2 & 0 \\ -1 & 0 & 1 & 0 \end{bmatrix}$ has the reduced row echelon form $\begin{bmatrix} 1 & 0 & -1 & 0 \\ 0 & 1 & 0 & 0 \\ 0 & 0 & 0 & 0 \end{bmatrix}$. The general solution is $x_1 = t,\ x_2 = 0,\ x_3 = t$. In vector form

$$(x_1, x_2, x_3) = (t, 0, t) = t(1,0,1)$$

therefore the solution space is spanned by a vector $\mathbf{v}_1 = (1,0,1)$. This vector is nonzero, therefore it forms a linearly independent set (Theorem 4.3.2(b)). We conclude that $\mathbf{v}_1$ forms a basis for the solution space and that the dimension of the solution space is 1.

3. The augmented matrix of the linear system $\begin{bmatrix} 2 & 1 & 3 & 0 \\ 1 & 0 & 5 & 0 \\ 0 & 1 & 1 & 0 \end{bmatrix}$ has the reduced row echelon form $\begin{bmatrix} 1 & 0 & 0 & 0 \\ 0 & 1 & 0 & 0 \\ 0 & 0 & 1 & 0 \end{bmatrix}$. The only solution is $x_1 = x_2 = x_3 = 0$.
The solution space has no basis - its dimension is 0.

5. The augmented matrix of the linear system $\begin{bmatrix} 1 & -3 & 1 & 0 \\ 2 & -6 & 2 & 0 \\ 3 & -9 & 3 & 0 \end{bmatrix}$ has the reduced row echelon form

$\begin{bmatrix} 1 & -3 & 1 & 0 \\ 0 & 0 & 0 & 0 \\ 0 & 0 & 0 & 0 \end{bmatrix}$. The general solution is $x_1 = 3s - t,\ x_2 = s,\ x_3 = t$. In vector form

$$(x_1, x_2, x_3) = (3s - t, s, t) = s(3,1,0) + t(-1,0,1)$$

therefore the solution space is spanned by vectors $\mathbf{v}_1 = (3,1,0)$ and $\mathbf{v}_2 = (-1,0,1)$. These vectors are linearly independent since neither of them is a scalar multiple of the other (Theorem 4.3.2(c)). We conclude that $\mathbf{v}_1$ and $\mathbf{v}_2$ form a basis for the solution space and that the dimension of the solution space is 2.

7. **(a)** If we let $y = s$ and $z = t$ be arbitrary values, we can solve the plane equation for x: $x = \frac{2}{3}s - \frac{5}{3}t$. Expressing the solution in vector form $(x, y, z) = \left(\frac{2}{3}s - \frac{5}{3}t, s, t\right) = s\left(\frac{2}{3}, 1,0\right) + t\left(-\frac{5}{3}, 0,1\right)$. By Theorem 4.3.2(c), $\left\{\left(\frac{2}{3}, 1,0\right), \left(-\frac{5}{3}, 0,1\right)\right\}$ is linearly independent since neither vector in the set is a scalar multiple of the other. A basis for the subspace is $\left\{\left(\frac{2}{3}, 1,0\right), \left(-\frac{5}{3}, 0,1\right)\right\}$. The dimension of the subspace is 2.

(b) If we let $y = s$ and $z = t$ be arbitrary values, we can solve the plane equation for x: $x = s$. Expressing the solution in vector form $(x, y, z) = (s, s, t) = s(1,1,0) + t(0,0,1)$. By Theorem 4.3.2(c), $\{(1,1,0), (0,0,1)\}$ is linearly independent since neither vector in the set is a scalar multiple of the other. A basis for the subspace is $\{(1,1,0), (0,0,1)\}$. The dimension of the subspace is 2.

(c) In vector form, $(x, y, z) = (2t, -t, 4t) = t(2, -1, 4)$. By Theorem 4.3.2(b), the vector $(2, -1,4)$ forms a linearly independent set since it is not the zero vector. A basis for the subspace is $\{(2, -1, 4)\}$. The dimension of the subspace is 1.

(d) The subspace contains all vectors $(a, a + c, c) = a(1, 1, 0) + c(0, 1, 1)$ thus we can express it as as span(S) where $S = \{(1, 1, 0), (0, 1, 1)\}$. By Theorem 4.3.2(c), S is linearly independent since neither vector in the set is a scalar multiple of the other. Consequently, S forms a basis for the given subspace. The dimension of the subspace is 2.

9. **(a)** Let W be the space of all diagonal $n \times n$ matrices. We can write

$$\begin{bmatrix} d_1 & 0 & \cdots & 0 \\ 0 & d_2 & \cdots & 0 \\ \vdots & \vdots & \ddots & \vdots \\ 0 & 0 & \cdots & d_n \end{bmatrix} = d_1 \underbrace{\begin{bmatrix} 1 & 0 & \cdots & 0 \\ 0 & 0 & \cdots & 0 \\ \vdots & \vdots & \ddots & \vdots \\ 0 & 0 & \cdots & 0 \end{bmatrix}}_{A_1} + d_2 \underbrace{\begin{bmatrix} 0 & 0 & \cdots & 0 \\ 0 & 1 & \cdots & 0 \\ \vdots & \vdots & \ddots & \vdots \\ 0 & 0 & \cdots & 0 \end{bmatrix}}_{A_2} + \cdots + d_n \underbrace{\begin{bmatrix} 0 & 0 & \cdots & 0 \\ 0 & 0 & \cdots & 0 \\ \vdots & \vdots & \ddots & \vdots \\ 0 & 0 & \cdots & 1 \end{bmatrix}}_{A_n}$$

The matrices $A_1, \ldots, A_n$ are linearly independent and they span W; hence, $A_1, \ldots, A_n$ form a basis for W. Consequently, the dimension of W is n.

(b) A basis for this space can be constructed by including the n matrices $A_1, \ldots, A_n$ from part (a), as well as $(n - 1) + (n - 2) + \cdots + 3 + 2 + 1 = \frac{n(n-1)}{2}$ matrices B_{ij} (for all $i < j$) where all entries

are 0 except for the (i,j) and (j,i) entries, which are both 1.
For instance, for $n = 3$, such a basis would be:

$$\underbrace{\begin{bmatrix}1&0&0\\0&0&0\\0&0&0\end{bmatrix}}_{A_1},\underbrace{\begin{bmatrix}0&0&0\\0&1&0\\0&0&0\end{bmatrix}}_{A_2},\underbrace{\begin{bmatrix}0&0&0\\0&0&0\\0&0&1\end{bmatrix}}_{A_3},\underbrace{\begin{bmatrix}0&1&0\\1&0&0\\0&0&0\end{bmatrix}}_{B_{12}},\underbrace{\begin{bmatrix}0&0&1\\0&0&0\\1&0&0\end{bmatrix}}_{B_{13}},\underbrace{\begin{bmatrix}0&0&0\\0&0&1\\0&1&0\end{bmatrix}}_{B_{23}}$$

The dimension is $n + \frac{n(n-1)}{2} = \frac{n(n+1)}{2}$.

(c) A basis for this space can be constructed by including the n matrices $A_1,\ldots,A_n$ from part (a), as well as $(n-1)+(n-2)+\cdots+3+2+1 = \frac{n(n-1)}{2}$ matrices C_{ij} (for all $i < j$) where all entries are 0 except for the (i,j) entry, which is 1.
For instance, for $n = 3$, such a basis would be:

$$\underbrace{\begin{bmatrix}1&0&0\\0&0&0\\0&0&0\end{bmatrix}}_{A_1},\underbrace{\begin{bmatrix}0&0&0\\0&1&0\\0&0&0\end{bmatrix}}_{A_2},\underbrace{\begin{bmatrix}0&0&0\\0&0&0\\0&0&1\end{bmatrix}}_{A_3},\underbrace{\begin{bmatrix}0&1&0\\0&0&0\\0&0&0\end{bmatrix}}_{C_{12}},\underbrace{\begin{bmatrix}0&0&1\\0&0&0\\0&0&0\end{bmatrix}}_{C_{13}},\underbrace{\begin{bmatrix}0&0&0\\0&0&1\\0&0&0\end{bmatrix}}_{C_{23}}$$

The dimension is $n + \frac{n(n-1)}{2} = \frac{n(n+1)}{2}$.

11. **(a)** W is the set of all polynomials $a_0 + a_1x + a_2x^2$ for which $a_0 + a_1 + a_2 = 0$, i.e. all polynomials that can be expressed in the form $-a_1 - a_2 + a_1x + a_2x^2$.
Adding two polynomials in W results in another polynomial in W
$(-a_1 - a_2 + a_1x + a_2x^2) + (-b_1 - b_2 + b_1x + b_2x^2)$
$= (-a_1 - a_2 - b_1 - b_2) + (a_1 + b_1)x + (a_2 + b_2)x^2$
since we have $(-a_1 - a_2 - b_1 - b_2) + (a_1 + b_1) + (a_2 + b_2) = 0$.
Likewise, a scalar multiple of a polynomial in W is also in W
$k(-a_1 - a_2 + a_1x + a_2x^2) = -ka_1 - ka_2 + ka_1x + ka_2x^2$
since it meets the condition $(-ka_1 - ka_2) + (ka_1) + (ka_2) = 0$.
According to Theorem 4.2.1, W is a subspace of P_2.

(c) From part (a), an arbitrary polynomial in W can be expressed in the form

$$-a_1 - a_2 + a_1x + a_2x^2 = a_1(-1 + x) + a_2(-1 + x^2)$$

therefore, the polynomials $-1 + x$ and $-1 + x^2$ span W. Also, $a_1(-1 + x) + a_2(-1 + x^2) = 0$ implies $a_1 = a_2 = 0$, so $-1 + x$ and $-1 + x^2$ are linearly independent, hence they form a basis for W. The dimension of W is 2.

13. The equation $k_1\mathbf{v}_1 + k_2\mathbf{v}_2 + k_3\mathbf{e}_1 + k_4\mathbf{e}_2 + k_5\mathbf{e}_3 + k_6\mathbf{e}_4 = \mathbf{0}$ can be rewritten as a linear system

$$\begin{array}{rcrcrcrcrcrcl} k_1 & - & 3k_2 & + & k_3 & & & & & & & = & 0\\ -4k_1 & + & 8k_2 & & & + & k_4 & & & & & = & 0\\ 2k_1 & - & 4k_2 & & & & & + & k_5 & & & = & 0\\ -3k_1 & + & 6k_2 & & & & & & & + & k_6 & = & 0 \end{array}$$

whose augmented matrix has the reduced row echelon form $\begin{bmatrix} 1 & 0 & -2 & 0 & 0 & -1 & 0 \\ 0 & 1 & -1 & 0 & 0 & -\frac{1}{3} & 0 \\ 0 & 0 & 0 & 1 & 0 & -\frac{4}{3} & 0 \\ 0 & 0 & 0 & 0 & 1 & \frac{2}{3} & 0 \end{bmatrix}$.

Based on the leading entries in the first, second, fourth, and fifth columns, the vector equation $k_1\mathbf{v}_1 + k_2\mathbf{v}_2 + k_4\mathbf{e}_2 + k_5\mathbf{e}_3 = \mathbf{0}$ has only the trivial solution (the corresponding augmented matrix has the reduced row echelon form $\begin{bmatrix} 1 & 0 & 0 & 0 & 0 \\ 0 & 1 & 0 & 0 & 0 \\ 0 & 0 & 1 & 0 & 0 \\ 0 & 0 & 0 & 1 & 0 \end{bmatrix}$). Therefore the vectors $\mathbf{v}_1$, $\mathbf{v}_2$, $\mathbf{e}_2$, and $\mathbf{e}_3$ are linearly independent. Since $\dim(R^4) = 4$, it follows by Theorem 4.5.4 that the vectors $\mathbf{v}_1$, $\mathbf{v}_2$, $\mathbf{e}_2$, and $\mathbf{e}_3$ form a basis for R^4. (The answer is not unique.)

15. The equation $k_1\mathbf{v}_1 + k_2\mathbf{v}_2 + k_3\mathbf{e}_1 + k_4\mathbf{e}_2 + k_5\mathbf{e}_3 = \mathbf{0}$ can be rewritten as a linear system

$$\begin{aligned} k_1 + k_3 &= 0 \\ -2k_1 + 5k_2 + k_4 &= 0 \\ 3k_1 - 3k_2 + k_5 &= 0 \end{aligned}$$

whose augmented matrix has the reduced row echelon form $\begin{bmatrix} 1 & 0 & 0 & \frac{1}{3} & \frac{5}{9} & 0 \\ 0 & 1 & 0 & \frac{1}{3} & \frac{2}{9} & 0 \\ 0 & 0 & 1 & -\frac{1}{3} & -\frac{5}{9} & 0 \end{bmatrix}$.

Based on the leading entries in the first three columns, the vector equation $k_1\mathbf{v}_1 + k_2\mathbf{v}_2 + k_3\mathbf{e}_1 = \mathbf{0}$ has only the trivial solution (the corresponding augmented matrix has the reduced row echelon form $\begin{bmatrix} 1 & 0 & 0 & 0 \\ 0 & 1 & 0 & 0 \\ 0 & 0 & 1 & 0 \end{bmatrix}$). Therefore the vectors $\mathbf{v}_1$, $\mathbf{v}_2$, and $\mathbf{e}_1$ are linearly independent. Since $\dim(R^3) = 3$, it follows by Theorem 4.5.4 that the vectors $\mathbf{v}_1$, $\mathbf{v}_2$, and $\mathbf{e}_1$ form a basis for R^3. (The answer is not unique.)

17. The equation $k_1\mathbf{v}_1 + k_2\mathbf{v}_2 + k_3\mathbf{v}_3 + k_4\mathbf{v}_4 = \mathbf{0}$ can be rewritten as a linear system

$$\begin{aligned} 1k_1 + 1k_2 + 2k_3 + 0k_4 &= 0 \\ 0k_1 + 0k_2 + 0k_3 + 0k_4 &= 0 \\ 0k_1 + 1k_2 + 1k_3 - 1k_4 &= 0 \end{aligned}$$

whose augmented matrix has the reduced row echelon form $\begin{bmatrix} 1 & 0 & 1 & 1 & 0 \\ 0 & 1 & 1 & -1 & 0 \\ 0 & 0 & 0 & 0 & 0 \end{bmatrix}$.

For arbitrary values of s and t, we have $k_1 = -s - t$, $k_2 = -s + t$, $x_3 = s$, $k_4 = t$.
Letting $s = 1$ and $t = 0$ allows us to express $\mathbf{v}_3$ as a linear combination of $\mathbf{v}_1$ and $\mathbf{v}_2$: $\mathbf{v}_3 = \mathbf{v}_1 + \mathbf{v}_2$.

Letting $s = 0$ and $t = 1$ allows us to express $\mathbf{v}_4$ as a linear combination of $\mathbf{v}_1$ and $\mathbf{v}_2$: $\mathbf{v}_4 = \mathbf{v}_1 - \mathbf{v}_2$. By part (b) of Theorem 4.5.3, $\text{span}\{\mathbf{v}_1, \mathbf{v}_2\} = \text{span}\{\mathbf{v}_1, \mathbf{v}_2, \mathbf{v}_3, \mathbf{v}_4\}$.

Based on the leading entries in the first two columns, the vector equation $k_1\mathbf{v}_1 + k_2\mathbf{v}_2 = \mathbf{0}$ has only the trivial solution (the corresponding augmented matrix has the reduced row echelon form $\begin{bmatrix} 1 & 0 & 0 \\ 0 & 1 & 0 \\ 0 & 0 & 0 \end{bmatrix}$). Therefore the vectors $\mathbf{v}_1$ and $\mathbf{v}_2$ are linearly independent. We conclude that the vectors $\mathbf{v}_1$ and $\mathbf{v}_2$ form a basis for $\text{span}\{\mathbf{v}_1, \mathbf{v}_2, \mathbf{v}_3, \mathbf{v}_4\}$. (The answer is not unique.)

19. The space of all vectors $\mathbf{x} = (x_1, x_2, x_3)$ for which $T_A(\mathbf{x}) = \mathbf{0}$ is the solution space of $A\mathbf{x} = \mathbf{0}$.

(a) The reduced row echelon form of A is $\begin{bmatrix} 1 & 0 & 1 \\ 0 & 1 & -1 \\ 0 & 0 & 0 \end{bmatrix}$ so $x_1 = -t, x_2 = t, x_3 = t$. In vector form, $(x_1, x_2, x_3) = (-t, t, t) = t(-1,1,1)$. Since $\{(-1,1,1)\}$ is a basis for the space, the dimension is 1.

(b) The reduced row echelon form of A is $\begin{bmatrix} 1 & 2 & 0 \\ 0 & 0 & 0 \\ 0 & 0 & 0 \end{bmatrix}$ so $x_1 = -2s, x_2 = s, x_3 = t$. In vector form, $(x_1, x_2, x_3) = (-2s, s, t) = s(-2,1,0) + t(0,0,1)$. Since $\{(-2,1,0), (0,0,1)\}$ is a basis for the space, the dimension is 2.

(c) The reduced row echelon form of A is $\begin{bmatrix} 1 & 0 & 0 \\ 0 & 1 & 1 \\ 0 & 0 & 0 \end{bmatrix}$ so $x_1 = 0, x_2 = -t, x_3 = t$. In vector form, $(x_1, x_2, x_3) = (0, -t, t) = t(0,-1,1)$. Since $\{(0,-1,1)\}$ is a basis for the space, the dimension is 1.

27. In parts (a) and (b), we will use the results of Exercises 18 and 19 by working with coordinate vectors with respect to the standard basis for P_2, $S = \{1, x, x^2\}$.

(a) Denote $\mathbf{v}_1 = -1 + x - 2x^2$, $\mathbf{v}_2 = 3 + 3x + 6x^2$, $\mathbf{v}_3 = 9$.
Then $(\mathbf{v}_1)_S = (-1,1,-2)$, $(\mathbf{v}_2)_S = (3,3,6)$, $(\mathbf{v}_3)_S = (9,0,0)$.
Setting $k_1(\mathbf{v}_1)_S + k_2(\mathbf{v}_2)_S + k_3(\mathbf{v}_3)_S = \mathbf{0}$ we obtain a linear system with augmented matrix $\begin{bmatrix} -1 & 3 & 9 & 0 \\ 1 & 3 & 0 & 0 \\ -2 & 6 & 0 & 0 \end{bmatrix}$ whose reduced row echelon form is $\begin{bmatrix} 1 & 0 & 0 & 0 \\ 0 & 1 & 0 & 0 \\ 0 & 0 & 1 & 0 \end{bmatrix}$. Since there is only the trivial solution, it follows that the three coordinate vectors are linearly independent, and, by the result of Exercise 22, so are the vectors $\mathbf{v}_1$, $\mathbf{v}_2$, and $\mathbf{v}_3$. Because the number of these vector matches $\dim(P_2) = 3$, from Theorem 4.5.4 the vectors $\mathbf{v}_1$, $\mathbf{v}_2$, and $\mathbf{v}_3$ form a basis for P_2.

(b) Denote $\mathbf{v}_1 = 1 + x$, $\mathbf{v}_2 = x^2$, $\mathbf{v}_3 = 2 + 2x + 3x^2$.
Then $(\mathbf{v}_1)_S = (1,1,0)$, $(\mathbf{v}_2)_S = (0,0,1)$, $(\mathbf{v}_3)_S = (2,2,3)$.
Setting $k_1(\mathbf{v}_1)_S + k_2(\mathbf{v}_2)_S + k_3(\mathbf{v}_3)_S = \mathbf{0}$ we obtain a linear system with augmented matrix $\begin{bmatrix} 1 & 0 & 2 & 0 \\ 1 & 0 & 2 & 0 \\ 0 & 1 & 3 & 0 \end{bmatrix}$ whose reduced row echelon form is $\begin{bmatrix} 1 & 0 & 2 & 0 \\ 0 & 1 & 3 & 0 \\ 0 & 0 & 0 & 0 \end{bmatrix}$.
This yields solutions $k_1 = -2t, k_2 = -3t, k_3 = t$. Taking $t = 1$, we can express $(\mathbf{v}_3)_S$ as a linear combination of $(\mathbf{v}_1)_S$ and $(\mathbf{v}_2)_S$: $(\mathbf{v}_3)_S = 2(\mathbf{v}_1)_S + 3(\mathbf{v}_2)_S$ - the same relationship holds true for the vectors themselves: $\mathbf{v}_3 = 2\mathbf{v}_1 + 3\mathbf{v}_2$. By part (b) of Theorem 4.5.3,

$\text{span}\{\mathbf{v}_1, \mathbf{v}_2\} = \text{span}\{\mathbf{v}_1, \mathbf{v}_2, \mathbf{v}_3\}$.

Based on the leading entries in the first two columns, the vector equation $k_1(\mathbf{v}_1)_S + k_2(\mathbf{v}_2)_S = \mathbf{0}$ has only the trivial solution (the corresponding augmented matrix $\begin{bmatrix} 1 & 0 & 0 \\ 1 & 0 & 0 \\ 0 & 1 & 0 \end{bmatrix}$ has the reduced row echelon form $\begin{bmatrix} 1 & 0 & 0 \\ 0 & 1 & 0 \\ 0 & 0 & 0 \end{bmatrix}$). Therefore the coordinate vectors $(\mathbf{v}_1)_S$ and $(\mathbf{v}_2)_S$ are linearly independent and, by the result of Exercise 18, so are the vectors $\mathbf{v}_1$ and $\mathbf{v}_2$.

We conclude that the vectors $\mathbf{v}_1$ and $\mathbf{v}_2$ form a basis for $\text{span}\{\mathbf{v}_1, \mathbf{v}_2, \mathbf{v}_3\}$.

(c) Clearly, $1 + x - 3x^2 = \frac{1}{2}(2 + 2x - 6x^2) = \frac{1}{3}(3 + 3x - 9x^2)$ therefore from Theorem 4.5.3(b), the subspace is spanned by $1 + x - 3x^2$. By Theorem 4.3.2(b), a set containing a single nonzero vector is linearly independent.

We conclude that $1 + x - 3x^2$ forms a basis for this subspace of P_2.

True-False Exercises

(a) True.

(b) True. For instance, $\mathbf{e}_1, \dots, \mathbf{e}_{17}$.

(c) False. This follows from Theorem 4.5.2(b).

(d) True. This follows from Theorem 4.5.4.

(e) True. This follows from Theorem 4.5.4.

(f) True. This follows from Theorem 4.5.5(a).

(g) True. This follows from Theorem 4.5.5(b).

(h) True. For instance, invertible matrices $\begin{bmatrix} 1 & 0 \\ 0 & 1 \end{bmatrix}, \begin{bmatrix} 1 & 0 \\ 0 & -1 \end{bmatrix}, \begin{bmatrix} 0 & 1 \\ 1 & 0 \end{bmatrix}, \begin{bmatrix} 0 & 1 \\ -1 & 0 \end{bmatrix}$ form a basis for M_{22}.

(i) True. The set has $n^2 + 1$ matrices, which exceeds $\dim(M_{nn}) = n^2$.

(j) False. This follows from Theorem 4.5.6(c).

(k) False. For instance, for any constant c, $\text{span}\{x - c, x^2 - c^2\}$ is a two-dimensional subspace of P_2 consisting of all polynomials in P_2 for which $p(c) = 0$. Clearly, there are infinitely many different subspaces of this type.

4.6 Change of Basis

1. **(a)** In this part, B' is the start basis and B is the end basis:

$$[\text{end basis} \mid \text{start basis}] = \left[\begin{array}{cc|cc} 2 & 4 & 1 & -1 \\ 2 & -1 & 3 & -1 \end{array}\right]$$

The reduced row echelon form of this matrix is

$$[I \mid \text{transition from start to end}] = \left[\begin{array}{cc|cc} 1 & 0 & \frac{13}{10} & -\frac{1}{2} \\ 0 & 1 & -\frac{2}{5} & 0 \end{array}\right]$$

The transition matrix is $P_{B'\to B} = \begin{bmatrix} \frac{13}{10} & -\frac{1}{2} \\ -\frac{2}{5} & 0 \end{bmatrix}$.

(b) In this part, B is the start basis and B' is the end basis:

$$[\text{end basis} \mid \text{start basis}] = \left[\begin{array}{cc|cc} 1 & -1 & 2 & 4 \\ 3 & -1 & 2 & -1 \end{array}\right]$$

The reduced row echelon form of this matrix is

$$[I \mid \text{transition from start to end}] = \left[\begin{array}{cc|cc} 1 & 0 & 0 & -\frac{5}{2} \\ 0 & 1 & -2 & -\frac{13}{2} \end{array}\right]$$

The transition matrix is $P_{B\to B'} = \begin{bmatrix} 0 & -\frac{5}{2} \\ -2 & -\frac{13}{2} \end{bmatrix}$.

(c) Expressing $\mathbf{w}$ as a linear combination of $\mathbf{u}_1$ and $\mathbf{u}_2$ we obtain

$$\begin{bmatrix} 3 \\ -5 \end{bmatrix} = c_1\begin{bmatrix} 2 \\ 2 \end{bmatrix} + c_2\begin{bmatrix} 4 \\ -1 \end{bmatrix}$$

Equating corresponding components on both sides yields the linear system

$$\begin{array}{rcrcr} 2c_1 & + & 4c_2 & = & 3 \\ 2c_1 & - & c_2 & = & -5 \end{array}$$

whose augmented matrix has the reduced row echelon form $\begin{bmatrix} 1 & 0 & -\frac{17}{10} \\ 0 & 1 & \frac{8}{5} \end{bmatrix}$. The solution of the linear system is $c_1 = -\frac{17}{10}$, $c_2 = \frac{8}{5}$, therefore the coordinate vector is $[\mathbf{w}]_B = \begin{bmatrix} -\frac{17}{10} \\ \frac{8}{5} \end{bmatrix}$.

Using Formula (12), $[\mathbf{w}]_{B'} = P_{B\to B'}[\mathbf{w}]_B = \begin{bmatrix} 0 & -\frac{5}{2} \\ -2 & -\frac{13}{2} \end{bmatrix}\begin{bmatrix} -\frac{17}{10} \\ \frac{8}{5} \end{bmatrix} = \begin{bmatrix} -4 \\ -7 \end{bmatrix}$.

(d) Expressing $\mathbf{w}$ as a linear combination of $\mathbf{u}'_1$ and $\mathbf{u}'_2$ we obtain

$$\begin{bmatrix} 3 \\ -5 \end{bmatrix} = c_1\begin{bmatrix} 1 \\ 3 \end{bmatrix} + c_2\begin{bmatrix} -1 \\ -1 \end{bmatrix}$$

Equating corresponding components on both sides yields the linear system

$$\begin{array}{rcrcr} c_1 & - & c_2 & = & 3 \\ 3c_1 & - & c_2 & = & -5 \end{array}$$

whose augmented matrix has the reduced row echelon form $\begin{bmatrix} 1 & 0 & -4 \\ 0 & 1 & -7 \end{bmatrix}$. The solution of the linear system is $c_1 = -4$, $c_2 = -7$, therefore the coordinate vector is $[\mathbf{w}]_{B'} = \begin{bmatrix} -4 \\ -7 \end{bmatrix}$. This matches the result obtained in part (c).

3. **(a)** In this part, B is the start basis and B' is the end basis:

$$[\text{end basis} \mid \text{start basis}] = \left[\begin{array}{ccc|ccc} 3 & 1 & -1 & 2 & 2 & 1 \\ 1 & 1 & 0 & 1 & -1 & 2 \\ -5 & -3 & 2 & 1 & 1 & 1 \end{array}\right]$$

The reduced row echelon form of this matrix is

$$[I \mid \text{transition from start to end}] = \left[\begin{array}{ccc|ccc} 1 & 0 & 0 & 3 & 2 & \frac{5}{2} \\ 0 & 1 & 0 & -2 & -3 & -\frac{1}{2} \\ 0 & 0 & 1 & 5 & 1 & 6 \end{array}\right]$$

The transition matrix is $P_{B \to B'} = \begin{bmatrix} 3 & 2 & \frac{5}{2} \\ -2 & -3 & -\frac{1}{2} \\ 5 & 1 & 6 \end{bmatrix}$.

(b) Expressing $\mathbf{w}$ as a linear combination of $\mathbf{u}_1$, $\mathbf{u}_2$, and $\mathbf{u}_3$ we obtain

$$\begin{bmatrix} -5 \\ 8 \\ -5 \end{bmatrix} = c_1 \begin{bmatrix} 2 \\ 1 \\ 1 \end{bmatrix} + c_2 \begin{bmatrix} 2 \\ -1 \\ 1 \end{bmatrix} + c_3 \begin{bmatrix} 1 \\ 2 \\ 1 \end{bmatrix}$$

Equating corresponding components on both sides yields the linear system

$$\begin{array}{rcrcrcr} 2c_1 & + & 2c_2 & + & c_3 & = & -5 \\ c_1 & - & c_2 & + & 2c_3 & = & 8 \\ c_1 & + & c_2 & + & c_3 & = & -5 \end{array}$$

whose augmented matrix has the reduced row echelon form $\begin{bmatrix} 1 & 0 & 0 & 9 \\ 0 & 1 & 0 & -9 \\ 0 & 0 & 1 & -5 \end{bmatrix}$. The solution of the linear system is $c_1 = 9$, $c_2 = -9$, $c_3 = -5$ therefore the coordinate vector is $[\mathbf{w}]_B = \begin{bmatrix} 9 \\ -9 \\ -5 \end{bmatrix}$.

Using Formula (12), $[\mathbf{w}]_{B'} = P_{B \to B'}[\mathbf{w}]_B = \begin{bmatrix} 3 & 2 & \frac{5}{2} \\ -2 & -3 & -\frac{1}{2} \\ 5 & 1 & 6 \end{bmatrix} \begin{bmatrix} 9 \\ -9 \\ -5 \end{bmatrix} = \begin{bmatrix} -\frac{7}{2} \\ \frac{23}{2} \\ 6 \end{bmatrix}$.

(c) Expressing $\mathbf{w}$ as a linear combination of $\mathbf{u}_1'$, $\mathbf{u}_2'$ and $\mathbf{u}_3'$ we obtain

$$\begin{bmatrix} -5 \\ 8 \\ -5 \end{bmatrix} = c_1 \begin{bmatrix} 3 \\ 1 \\ -5 \end{bmatrix} + c_2 \begin{bmatrix} 1 \\ 1 \\ -3 \end{bmatrix} + c_3 \begin{bmatrix} -1 \\ 0 \\ 2 \end{bmatrix}$$

Equating corresponding components on both sides yields the linear system

$$\begin{array}{rcrcrcr} 3c_1 & + & c_2 & - & c_3 & = & -5 \\ c_1 & + & c_2 & & & = & 8 \\ -5c_1 & - & 3c_2 & + & 2c_3 & = & -5 \end{array}$$

whose augmented matrix has the reduced row echelon form $\begin{bmatrix} 1 & 0 & 0 & -\frac{7}{2} \\ 0 & 1 & 0 & \frac{23}{2} \\ 0 & 0 & 1 & 6 \end{bmatrix}$.

The solution of the linear system is $c_1 = -\frac{7}{2}$, $c_2 = \frac{23}{2}$, $c_3 = 6$ therefore the coordinate vector is $[\mathbf{w}]_{B'} = \begin{bmatrix} -\frac{7}{2} \\ \frac{23}{2} \\ 6 \end{bmatrix}$, which matches the result we obtained in part (b).

5. **(a)** The set $\{\mathbf{f}_1, \mathbf{f}_2\}$ is linearly independent since neither vector is a scalar multiple of the other. Thus $\{\mathbf{f}_1, \mathbf{f}_2\}$ is a basis for V and $\dim(V) = 2$.
Likewise, the set $\{\mathbf{g}_1, \mathbf{g}_2\}$ of vectors in V is linearly independent since neither vector is a scalar multiple of the other. By Theorem 4.5.4, $\{\mathbf{g}_1, \mathbf{g}_2\}$ is a basis for V.

(b) Clearly, $[\mathbf{g}_1]_B = \begin{bmatrix} 2 \\ 1 \end{bmatrix}$ and $[\mathbf{g}_2]_B = \begin{bmatrix} 0 \\ 3 \end{bmatrix}$ hence $P_{B' \to B} = [[\mathbf{g}_1]_B \mid [\mathbf{g}_2]_B] = \begin{bmatrix} a_1 & b_1 \\ a_2 & b_2 \end{bmatrix} = \begin{bmatrix} 2 & 0 \\ 1 & 3 \end{bmatrix}$.

(c) We find the two columns of the transitions matrix $P_{B \to B'} = [[\mathbf{f}_1]_{B'} \mid [\mathbf{f}_2]_{B'}]$

$$\mathbf{f}_1 = a_1\mathbf{g}_1 + a_2\mathbf{g}_2 \qquad\qquad \mathbf{f}_2 = b_1\mathbf{g}_1 + b_2\mathbf{g}_2$$

$$\sin x = a_1(2\sin x + \cos x) + a_2(3\cos x) \qquad\qquad \cos x = b_1(2\sin x + \cos x) + b_2(3\cos x)$$

equate the coefficients corresponding to the same function on both sides of each equation

$$\begin{array}{rcrcr} 2a_1 & & & = & 1 \\ a_1 & + & 3a_2 & = & 0 \end{array} \qquad\qquad \begin{array}{rcrcr} 2b_1 & & & = & 0 \\ b_1 & + & 3b_2 & = & 1 \end{array}$$

reduced row echelon form of the augmented matrix of each system

$$\begin{bmatrix} 1 & 0 & \frac{1}{2} \\ 0 & 1 & -\frac{1}{6} \end{bmatrix} \qquad\qquad \begin{bmatrix} 1 & 0 & 0 \\ 0 & 1 & \frac{1}{3} \end{bmatrix}$$

We obtain the transition matrix $P_{B \to B'} = [[\mathbf{f}_1]_{B'} \mid [\mathbf{f}_2]_{B'}] = \begin{bmatrix} a_1 & b_1 \\ a_2 & b_2 \end{bmatrix} = \begin{bmatrix} \frac{1}{2} & 0 \\ -\frac{1}{6} & \frac{1}{3} \end{bmatrix}$.

(An alternate way to solve this part is to use Theorem 4.6.1 to yield

$$P_{B \to B'} = P_{B' \to B}^{-1} = \left(\begin{bmatrix} 2 & 0 \\ 1 & 3 \end{bmatrix} \right)^{-1} = \frac{1}{(2)(3)-(0)(1)} \begin{bmatrix} 3 & 0 \\ -1 & 2 \end{bmatrix} = \frac{1}{6} \begin{bmatrix} 3 & 0 \\ -1 & 2 \end{bmatrix} = \begin{bmatrix} \frac{1}{2} & 0 \\ -\frac{1}{6} & \frac{1}{3} \end{bmatrix}.)$$

(d) Clearly, the coordinate vector is $[\mathbf{h}]_B = \begin{bmatrix} 2 \\ -5 \end{bmatrix}$.

Using Formula (12), we obtain $[\mathbf{h}]_{B'} = P_{B \to B'}[\mathbf{h}]_B = \begin{bmatrix} \frac{1}{2} & 0 \\ -\frac{1}{6} & \frac{1}{3} \end{bmatrix} \begin{bmatrix} 2 \\ -5 \end{bmatrix} = \begin{bmatrix} 1 \\ -2 \end{bmatrix}$.

(e) By inspection, $2\sin x - 5\cos x = (2\sin x + \cos x) - 2(3\cos x)$, hence the coordinate vector is $[\mathbf{p}]_{B'} = \begin{bmatrix} 1 \\ -2 \end{bmatrix}$, which matches the result obtained in part (d).

7. **(a)** In this part, B_2 is the start basis and B_1 is the end basis:
$[\text{end basis} \mid \text{start basis}] = \left[\begin{array}{cc|cc} 1 & 2 & 1 & 1 \\ 2 & 3 & 3 & 4 \end{array}\right]$.

The reduced row echelon form of this matrix is
$[I \mid \text{transition from start to end}] = \left[\begin{array}{cc|cc} 1 & 0 & 3 & 5 \\ 0 & 1 & -1 & -2 \end{array}\right]$.

The transition matrix is $P_{B_2 \to B_1} = \begin{bmatrix} 3 & 5 \\ -1 & -2 \end{bmatrix}$.

(b) In this part, B_1 is the start basis and B_2 is the end basis:
$[\text{end basis} \mid \text{start basis}] = \left[\begin{array}{cc|cc} 1 & 1 & 1 & 2 \\ 3 & 4 & 2 & 3 \end{array}\right]$.

The reduced row echelon form of this matrix is
$[I \mid \text{transition from start to end}] = \left[\begin{array}{cc|cc} 1 & 0 & 2 & 5 \\ 0 & 1 & -1 & -3 \end{array}\right]$.

The transition matrix is $P_{B_1 \to B_2} = \begin{bmatrix} 2 & 5 \\ -1 & -3 \end{bmatrix}$.

(c) Since $\begin{bmatrix} 3 & 5 \\ -1 & -2 \end{bmatrix}\begin{bmatrix} 2 & 5 \\ -1 & -3 \end{bmatrix} = \begin{bmatrix} 1 & 0 \\ 0 & 1 \end{bmatrix}$ and $\begin{bmatrix} 2 & 5 \\ -1 & -3 \end{bmatrix}\begin{bmatrix} 3 & 5 \\ -1 & -2 \end{bmatrix} = \begin{bmatrix} 1 & 0 \\ 0 & 1 \end{bmatrix}$ it follows that $P_{B_2 \to B_1}$ and $P_{B_1 \to B_2}$ are inverses of one another.

(d) Expressing $\mathbf{w}$ as a linear combination of $\mathbf{u}_1$ and $\mathbf{u}_2$ we obtain

$$\begin{bmatrix} 0 \\ 1 \end{bmatrix} = c_1 \begin{bmatrix} 1 \\ 2 \end{bmatrix} + c_2 \begin{bmatrix} 2 \\ 3 \end{bmatrix}$$

Equating corresponding components on both sides yields the linear system

$$\begin{aligned} c_1 + 2c_2 &= 0 \\ 2c_1 + 3c_2 &= 1 \end{aligned}$$

whose augmented matrix has the reduced row echelon form $\begin{bmatrix} 1 & 0 & 2 \\ 0 & 1 & -1 \end{bmatrix}$. The solution of the linear system is $c_1 = 2$, $c_2 = -1$, therefore the coordinate vector is $[\mathbf{w}]_{B_1} = \begin{bmatrix} 2 \\ -1 \end{bmatrix}$.

From Formula (12), $[\mathbf{w}]_{B_2} = P_{B_1 \to B_2}[\mathbf{w}]_{B_1} = \begin{bmatrix} 2 & 5 \\ -1 & -3 \end{bmatrix}\begin{bmatrix} 2 \\ -1 \end{bmatrix} = \begin{bmatrix} -1 \\ 1 \end{bmatrix}$.

(e) Expressing $\mathbf{w}$ as a linear combination of $\mathbf{v}_1$ and $\mathbf{v}_2$ we obtain

$$\begin{bmatrix}2\\5\end{bmatrix} = c_1\begin{bmatrix}1\\3\end{bmatrix} + c_2\begin{bmatrix}1\\4\end{bmatrix}$$

Equating corresponding components on both sides yields the linear system

$$\begin{aligned} 1c_1 &+ 1c_2 &= 2 \\ 3c_1 &+ 4c_2 &= 5 \end{aligned}$$

whose augmented matrix has the reduced row echelon form $\begin{bmatrix}1 & 0 & 3\\0 & 1 & -1\end{bmatrix}$. The solution of the linear system is $c_1 = 3$, $c_2 = -1$, therefore the coordinate vector is $[\mathbf{w}]_{B_2} = \begin{bmatrix}3\\-1\end{bmatrix}$.

From Formula (12), $[\mathbf{w}]_{B_1} = P_{B_2\to B_1}[\mathbf{w}]_{B_2} = \begin{bmatrix}3 & 5\\-1 & -2\end{bmatrix}\begin{bmatrix}3\\-1\end{bmatrix} = \begin{bmatrix}4\\-1\end{bmatrix}$.

9. **(a)** By Theorem 4.6.2, $P_{B\to S} = \begin{bmatrix}1 & 2 & 3\\2 & 5 & 3\\1 & 0 & 8\end{bmatrix}$.

(b) In this part, S is the start basis and B is the end basis:

$$[\text{end basis} \mid \text{start basis}] = \left[\begin{array}{ccc|ccc}1 & 2 & 3 & 1 & 0 & 0\\2 & 5 & 3 & 0 & 1 & 0\\1 & 0 & 8 & 0 & 0 & 1\end{array}\right].$$

The reduced row echelon form of this matrix is

$$[I \mid \text{transition from start to end}] = \left[\begin{array}{ccc|ccc}1 & 0 & 0 & -40 & 16 & 9\\0 & 1 & 0 & 13 & -5 & -3\\0 & 0 & 1 & 5 & -2 & -1\end{array}\right].$$

The transition matrix is $P_{S\to B} = \begin{bmatrix}-40 & 16 & 9\\13 & -5 & -3\\5 & -2 & -1\end{bmatrix}$.

(c) Since $\begin{bmatrix}-40 & 16 & 9\\13 & -5 & -3\\5 & -2 & -1\end{bmatrix}\begin{bmatrix}1 & 2 & 3\\2 & 5 & 3\\1 & 0 & 8\end{bmatrix} = \begin{bmatrix}1 & 0 & 0\\0 & 1 & 0\\0 & 0 & 1\end{bmatrix}$ and $\begin{bmatrix}1 & 2 & 3\\2 & 5 & 3\\1 & 0 & 8\end{bmatrix}\begin{bmatrix}-40 & 16 & 9\\13 & -5 & -3\\5 & -2 & -1\end{bmatrix} = \begin{bmatrix}1 & 0 & 0\\0 & 1 & 0\\0 & 0 & 1\end{bmatrix}$ it follows that $P_{B\to S}$ and $P_{S\to B}$ are inverses of one another.

(d) Expressing $\mathbf{w}$ as a linear combination of $\mathbf{v}_1$, $\mathbf{v}_2$, and $\mathbf{v}_3$ we obtain

$$\begin{bmatrix}5\\-3\\1\end{bmatrix} = c_1\begin{bmatrix}1\\2\\1\end{bmatrix} + c_2\begin{bmatrix}2\\5\\0\end{bmatrix} + c_3\begin{bmatrix}3\\3\\8\end{bmatrix}$$

Equating corresponding components on both sides yields the linear system

$$\begin{aligned} c_1 &+ 2c_2 &+ 3c_3 &= 5 \\ 2c_1 &+ 5c_2 &+ 3c_3 &= -3 \\ c_1 & &+ 8c_3 &= 1 \end{aligned}$$

whose augmented matrix has the reduced row echelon form $\begin{bmatrix}1 & 0 & 0 & -239\\0 & 1 & 0 & 77\\0 & 0 & 1 & 30\end{bmatrix}$. The solution of the linear system is $c_1 = -239$, $c_2 = 77$, $c_3 = 30$ therefore the coordinate vector is $[\mathbf{w}]_B = \begin{bmatrix}-239\\77\\30\end{bmatrix}$. From Formula (12), $[\mathbf{w}]_S = P_{B\to S}[\mathbf{w}]_B = \begin{bmatrix}1 & 2 & 3\\2 & 5 & 3\\1 & 0 & 8\end{bmatrix}\begin{bmatrix}-239\\77\\30\end{bmatrix} = \begin{bmatrix}5\\-3\\1\end{bmatrix}$.

(e) By inspection, $[\mathbf{w}]_S = \begin{bmatrix} 3 \\ -5 \\ 0 \end{bmatrix}$.

From Formula (12), $[\mathbf{w}]_B = P_{S\to B}[\mathbf{w}]_S = \begin{bmatrix} -40 & 16 & 9 \\ 13 & -5 & -3 \\ 5 & -2 & -1 \end{bmatrix}\begin{bmatrix} 3 \\ -5 \\ 0 \end{bmatrix} = \begin{bmatrix} -200 \\ 64 \\ 25 \end{bmatrix}$.

11. **(a)** Clearly, $\mathbf{v}_1 = (\cos(2\theta), \sin(2\theta))$. Referring to the figure on the right, we see that the angle between the positive x-axis and $\mathbf{v}_2$ is $\frac{\pi}{2} - 2\left(\frac{\pi}{2} - \theta\right) = 2\theta - \frac{\pi}{2}$. Hence,

$$\mathbf{v}_2 = \left(\cos\left(2\theta - \frac{\pi}{2}\right), \sin\left(2\theta - \frac{\pi}{2}\right)\right) = (\sin(2\theta), -\cos(2\theta))$$

From Theorem 4.6.5, $P_{B\to S} = \begin{bmatrix} \cos(2\theta) & \sin(2\theta) \\ \sin(2\theta) & -\cos(2\theta) \end{bmatrix}$.

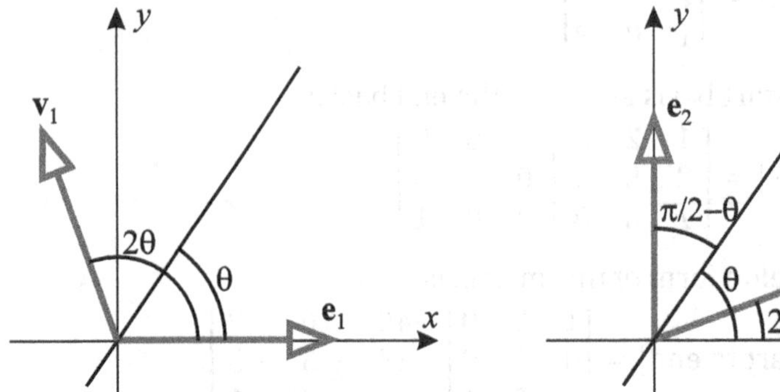

(b) Denoting $P = \begin{bmatrix} \cos(2\theta) & \sin(2\theta) \\ \sin(2\theta) & -\cos(2\theta) \end{bmatrix}$, it follows from Theorem 4.6.5 that $P_{S\to B} = P^{-1}$. In our case, $PP = I$ therefore $P = P^{-1}$. Furthermore, since P is symmetric, we also have $P_{S\to B} = P^T$.

13. Since for every vector $\mathbf{v}$ we have $[\mathbf{v}]_B = P[\mathbf{v}]_{B'}$ and $[\mathbf{v}]_C = Q[\mathbf{v}]_B$, it follows that $[\mathbf{v}]_C = QP[\mathbf{v}]_{B'}$ so that $P_{B'\to C} = QP$. From Theorem 4.6.1, $P_{C\to B'} = (QP)^{-1} = P^{-1}Q^{-1}$.

15. **(a)** By Theorem 4.6.2, P is the transition matrix from $B = \{(1,1,0), (1,0,2), (0,2,1)\}$ to S.

(b) By Theorem 4.6.1, $P^{-1} = \begin{bmatrix} \frac{4}{5} & \frac{1}{5} & -\frac{2}{5} \\ \frac{1}{5} & -\frac{1}{5} & \frac{2}{5} \\ -\frac{2}{5} & \frac{2}{5} & \frac{1}{5} \end{bmatrix}$ is the transition matrix from B to S, hence by Theorem 4.6.2, $B = \left\{\left(\frac{4}{5}, \frac{1}{5}, -\frac{2}{5}\right), \left(\frac{1}{5}, -\frac{1}{5}, \frac{2}{5}\right), \left(-\frac{2}{5}, \frac{2}{5}, \frac{1}{5}\right)\right\}$.

17. From $T(1,0) = (2,5)$, $T(0,1) = (3,-1)$, and Theorem 4.6.2 we obtain $P_{B\to S} = \begin{bmatrix} 2 & 3 \\ 5 & -1 \end{bmatrix}$.

19. By Formula (10), the transition matrix from the standard basis $S = \{\mathbf{e}_1, \ldots, \mathbf{e}_n\}$ to B is $P_{S\to B} = [[\mathbf{e}_1]_B | \ldots | [\mathbf{e}_n]_B] = [\mathbf{e}_1 | \ldots | \mathbf{e}_n] = I_n$ therefore B must be the standard basis.

True-False Exercises

(a) True. The matrix can be constructed according to Formula (10).

(b) True. This follows from Theorem 4.6.1.

(c) True.

(d) True.

(e) False. For instance, $B_1 = \{(0,2),(3,0)\}$ is a basis for R^2 made up of scalar multiples of vectors in the standard basis $B_2 = \{(1,0),(0,1)\}$. However, $P_{B_1 \to B_2} = \begin{bmatrix} 0 & 3 \\ 2 & 0 \end{bmatrix}$ (obtained by Theorem 4.6.2) is not a diagonal matrix.

(f) False. A must be invertible.

4.7 Row Space, Column Space, and Null Space

1. **(a)** $\begin{bmatrix} 2 & 3 \\ -1 & 4 \end{bmatrix}\begin{bmatrix} 1 \\ 2 \end{bmatrix} = 1\begin{bmatrix} 2 \\ -1 \end{bmatrix} + 2\begin{bmatrix} 3 \\ 4 \end{bmatrix}$

(b) $\begin{bmatrix} 4 & 0 & -1 \\ 3 & 6 & 2 \\ 0 & -1 & 4 \end{bmatrix}\begin{bmatrix} -2 \\ 3 \\ 5 \end{bmatrix} = -2\begin{bmatrix} 4 \\ 3 \\ 0 \end{bmatrix} + 3\begin{bmatrix} 0 \\ 6 \\ -1 \end{bmatrix} + 5\begin{bmatrix} -1 \\ 2 \\ 4 \end{bmatrix}$

3. **(a)** The reduced row echelon form of the augmented matrix of the system $A\mathbf{x} = \mathbf{b}$ is $\begin{bmatrix} 1 & 0 & 1 & 0 \\ 0 & 1 & 1 & 0 \\ 0 & 0 & 0 & 1 \end{bmatrix}$, thus $A\mathbf{x} = \mathbf{b}$ is inconsistent. By Theorem 4.7.1, $\mathbf{b}$ is not in the column space of A.

(b) The reduced row echelon form of the augmented matrix of the system $A\mathbf{x} = \mathbf{b}$ is $\begin{bmatrix} 1 & 0 & 0 & 1 \\ 0 & 1 & 0 & -3 \\ 0 & 0 & 1 & 1 \end{bmatrix}$, so the system has a unique solution $x_1 = 1, x_2 = -3, x_3 = 1$. By Theorem 4.7.1, $\mathbf{b}$ is in the column space of A. By Formula (2), we can write $\begin{bmatrix} 1 \\ 9 \\ 1 \end{bmatrix} - 3\begin{bmatrix} -1 \\ 3 \\ 1 \end{bmatrix} + \begin{bmatrix} 1 \\ 1 \\ 1 \end{bmatrix} = \begin{bmatrix} 5 \\ 1 \\ -1 \end{bmatrix}$.

5. **(a)** $\begin{bmatrix} x_1 \\ x_2 \\ x_3 \\ x_4 \end{bmatrix} = r\begin{bmatrix} 5 \\ 0 \\ 0 \\ 0 \end{bmatrix} + s\begin{bmatrix} -2 \\ 1 \\ 1 \\ 0 \end{bmatrix} + t\begin{bmatrix} 0 \\ 0 \\ 1 \\ 1 \end{bmatrix}$ **(b)** $\begin{bmatrix} x_1 \\ x_2 \\ x_3 \\ x_4 \end{bmatrix} = \begin{bmatrix} 3 \\ 0 \\ -1 \\ 5 \end{bmatrix} + r\begin{bmatrix} 5 \\ 0 \\ 0 \\ 0 \end{bmatrix} + s\begin{bmatrix} -2 \\ 1 \\ 1 \\ 0 \end{bmatrix} + t\begin{bmatrix} 0 \\ 0 \\ 1 \\ 1 \end{bmatrix}$

7. **(a)** The reduced row echelon form of the augmented matrix of the system $A\mathbf{x} = \mathbf{b}$ is $\begin{bmatrix} 1 & -3 & 1 \\ 0 & 0 & 0 \end{bmatrix}$.
The general solution of this system is $x_1 = 1 + 3t, x_2 = t$; in vector form,
$(x_1, x_2) = (1 + 3t, t) = (1,0) + t(3,1)$.
The vector form of the general solution of $A\mathbf{x} = \mathbf{0}$ is $(x_1, x_2) = t(3,1)$.

(b) The reduced row echelon form of the augmented matrix of the system $A\mathbf{x} = \mathbf{b}$ is $\begin{bmatrix} 1 & 0 & 1 & -2 \\ 0 & 1 & 1 & 7 \\ 0 & 0 & 0 & 0 \end{bmatrix}$. The general solution of this system is $x_1 = -2 - t, x_2 = 7 - t, x_3 = t$; in vector form, $(x_1, x_2, x_3) = (-2 - t, 7 - t, t) = (-2,7,0) + t(-1,-1,1)$.
The vector form of the general solution of $A\mathbf{x} = \mathbf{0}$ is $(x_1, x_2, x_3) = t(-1,-1,1)$.

9. **(a)** The reduced row echelon form of A is $\begin{bmatrix} 1 & 0 & -16 \\ 0 & 1 & -19 \\ 0 & 0 & 0 \end{bmatrix}$. The reduced row echelon form of the augmented matrix of the homogeneous system $A\mathbf{x} = \mathbf{0}$ would have an additional column of zeros appended to this matrix. The general solution of the system $x_1 = 16t,\ x_2 = 19t,\ x_3 = t$ can be written in the vector form $\begin{bmatrix} x_1 \\ x_2 \\ x_3 \end{bmatrix} = t\begin{bmatrix} 16 \\ 19 \\ 1 \end{bmatrix}$ therefore the vector $\begin{bmatrix} 16 \\ 19 \\ 1 \end{bmatrix}$ forms a basis for the null space of A.

A basis for the row space is formed by the nonzero rows of the reduced row echelon form of A: $[1 \quad 0 \quad -16]$ and $[0 \quad 1 \quad -19]$.

(b) The reduced row echelon form of A is $\begin{bmatrix} 1 & 0 & -\frac{1}{2} \\ 0 & 0 & 0 \\ 0 & 0 & 0 \end{bmatrix}$. The reduced row echelon form of the augmented matrix of the homogeneous system $A\mathbf{x} = \mathbf{0}$ would have an additional column of zeros appended to this matrix. The general solution of the system $x_1 = \frac{1}{2}t,\ x_2 = s,\ x_3 = t$ can be written in the vector form $\begin{bmatrix} x_1 \\ x_2 \\ x_3 \end{bmatrix} = s\begin{bmatrix} 0 \\ 1 \\ 0 \end{bmatrix} + t\begin{bmatrix} \frac{1}{2} \\ 0 \\ 1 \end{bmatrix}$ therefore the vectors $\begin{bmatrix} 0 \\ 1 \\ 0 \end{bmatrix}$ and $\begin{bmatrix} \frac{1}{2} \\ 0 \\ 1 \end{bmatrix}$ form a basis for the null space of A.

A basis for the row space is formed by the nonzero row of the reduced row echelon form of A: $\left[1 \quad 0 \quad -\frac{1}{2}\right]$.

11. We use Theorem 4.7.5 to obtain the following answers.

(a) Columns containing leading 1's form a basis for the column space: $\begin{bmatrix} 1 \\ 0 \\ 0 \end{bmatrix}, \begin{bmatrix} 2 \\ 1 \\ 0 \end{bmatrix}$.

Nonzero rows form a basis for the row space: $[1 \quad 0 \quad 2]$, $[0 \quad 0 \quad 1]$.

(b) Columns containing leading 1's form a basis for the column space: $\begin{bmatrix} 1 \\ 0 \\ 0 \\ 0 \end{bmatrix}, \begin{bmatrix} -3 \\ 1 \\ 0 \\ 0 \end{bmatrix}$.

Nonzero rows form a basis for the row space: $[1 \quad -3 \quad 0 \quad 0]$, $[0 \quad 1 \quad 0 \quad 0]$.

13. **(a)** The reduced row echelon form of A is $B = \begin{bmatrix} 1 & 0 & 11 & 0 & 3 \\ 0 & 1 & 3 & 0 & 0 \\ 0 & 0 & 0 & 1 & 0 \\ 0 & 0 & 0 & 0 & 0 \end{bmatrix}$.

By Theorems 4.7.4 and 4.7.5, the nonzero rows of B form a basis for the row space of A: $\mathbf{r}_1 = [1 \quad 0 \quad 11 \quad 0 \quad 3]$, $\mathbf{r}_2 = [0 \quad 1 \quad 3 \quad 0 \quad 0]$, and $\mathbf{r}_3 = [0 \quad 0 \quad 0 \quad 1 \quad 0]$.

By Theorem 4.7.5, columns of B containing leading 1's form a basis for the column space of B:

$\mathbf{c}_1' = \begin{bmatrix}1\\0\\0\\0\end{bmatrix}$, $\mathbf{c}_2' = \begin{bmatrix}0\\1\\0\\0\end{bmatrix}$, and $\mathbf{c}_4' = \begin{bmatrix}0\\0\\1\\0\end{bmatrix}$. By Theorem 4.7.6(b), a basis for the column space of A is formed by the corresponding columns of A: $\mathbf{c}_1 = \begin{bmatrix}1\\-2\\-1\\-3\end{bmatrix}$, $\mathbf{c}_2 = \begin{bmatrix}-2\\5\\3\\8\end{bmatrix}$, and $\mathbf{c}_4 = \begin{bmatrix}0\\0\\1\\1\end{bmatrix}$.

(b) We begin by transposing the matrix A.

We obtain $A^T = \begin{bmatrix}1 & -2 & -1 & -3\\-2 & 5 & 3 & 8\\5 & -7 & -2 & -9\\0 & 0 & 1 & 1\\3 & -6 & -3 & -9\end{bmatrix}$, whose reduced row echelon form is $C = \begin{bmatrix}1 & 0 & 0 & 0\\0 & 1 & 0 & 1\\0 & 0 & 1 & 1\\0 & 0 & 0 & 0\\0 & 0 & 0 & 0\end{bmatrix}$.

By Theorem 4.7.5, columns of C containing leading 1's form a basis for the column space of C:

$\mathbf{c}_1' = \begin{bmatrix}1\\0\\0\\0\\0\end{bmatrix}$, $\mathbf{c}_2' = \begin{bmatrix}0\\1\\0\\0\\0\end{bmatrix}$, and $\mathbf{c}_3' = \begin{bmatrix}0\\0\\1\\0\\0\end{bmatrix}$. By Theorem 4.7.6(b), a basis for the column space of A^T is formed by the corresponding columns of A^T: $\mathbf{c}_1 = \begin{bmatrix}1\\-2\\5\\0\\3\end{bmatrix}$, $\mathbf{c}_2 = \begin{bmatrix}-2\\5\\-7\\0\\-6\end{bmatrix}$, and $\mathbf{c}_3 = \begin{bmatrix}-1\\3\\-2\\1\\-3\end{bmatrix}$.

Since columns of A^T are rows of A, a basis for the row space of A is formed by $\mathbf{r}_1 = [1 \quad -2 \quad 5 \quad 0 \quad 3]$, $\mathbf{r}_2 = [-2 \quad 5 \quad -7 \quad 0 \quad -6]$, and $\mathbf{r}_3 = [-1 \quad 3 \quad -2 \quad 1 \quad -3]$.

15. We construct a matrix whose columns are the given vectors: $A = \begin{bmatrix}1 & 0 & -2 & 0\\1 & 0 & 0 & -3\\0 & 1 & 2 & 0\\0 & 1 & 2 & 3\end{bmatrix}$. The reduced row echelon form of A is $B = \begin{bmatrix}1 & 0 & 0 & 0\\0 & 1 & 0 & 0\\0 & 0 & 1 & 0\\0 & 0 & 0 & 1\end{bmatrix}$. By Theorem 4.7.5, the four columns of B form a basis for the column space of B. By Theorem 4.7.6(b), the four columns of A form a basis for the column space of A. We conclude that $\{(1,1,0,0), (0,0,1,1), (-2,0,2,2), (0,-3,0,3)\}$ is a basis for the subspace of R^4 spanned by these vectors.

17. Construct a matrix whose column vectors are the given vectors $\mathbf{v}_1$, $\mathbf{v}_2$, $\mathbf{v}_3$, $\mathbf{v}_4$, and $\mathbf{v}_5$:

$A = \begin{bmatrix}1 & -2 & 4 & 0 & -7\\-1 & 3 & -5 & 4 & 18\\5 & 1 & 9 & 2 & 2\\2 & 0 & 4 & -3 & -8\end{bmatrix}$. Since its reduced row echelon form

$$\begin{bmatrix} 1 & 0 & 2 & 0 & -1 \\ 0 & 1 & -1 & 0 & 3 \\ 0 & 0 & 0 & 1 & 2 \\ 0 & 0 & 0 & 0 & 0 \end{bmatrix}$$
$$\begin{matrix} \uparrow & \uparrow & \uparrow & \uparrow & \uparrow \\ \mathbf{w}_1 & \mathbf{w}_2 & \mathbf{w}_3 & \mathbf{w}_4 & \mathbf{w}_5 \end{matrix}$$

contains leading 1's in columns 1, 2, and 4, by Theorems 4.7.5 and 4.7.6(b), the vectors $\mathbf{v}_1$, $\mathbf{v}_2$ and $\mathbf{v}_4$ form a basis for the column space of A, and for $\text{span}\{\mathbf{v}_1, \mathbf{v}_2, \mathbf{v}_3, \mathbf{v}_4, \mathbf{v}_5\}$.

By inspection, the columns of the reduced row echelon form matrix satisfy $\mathbf{w}_3 = 2\mathbf{w}_1 - \mathbf{w}_2$ and $\mathbf{w}_5 = -\mathbf{w}_1 + 3\mathbf{w}_2 + 2\mathbf{w}_4$. Because elementary row operations preserve dependence relations between column vectors, we conclude that $\mathbf{v}_3 = 2\mathbf{v}_1 - \mathbf{v}_2$ and $\mathbf{v}_5 = -\mathbf{v}_1 + 3\mathbf{v}_2 + 2\mathbf{v}_4$.

19. We are employing the procedure developed in Example 9.

The reduced row echelon form of $A^T = \begin{bmatrix} 1 & 3 & -1 & 2 \\ 4 & -2 & 0 & 3 \\ 5 & 1 & -1 & 5 \\ 6 & 4 & -2 & 7 \\ 9 & -1 & -1 & 8 \end{bmatrix}$ is $\begin{bmatrix} 1 & 0 & -\frac{1}{7} & \frac{13}{14} \\ 0 & 1 & -\frac{2}{7} & \frac{5}{14} \\ 0 & 0 & 0 & 0 \\ 0 & 0 & 0 & 0 \\ 0 & 0 & 0 & 0 \end{bmatrix}$. Since the first two columns of the reduced row echelon form contain leading 1's, by Theorems 4.7.5 and 4.7.6(b) the first two columns of A^T form a basis for the column space of A^T. Consequently, the first two rows of A, $[1 \quad 4 \quad 5 \quad 6 \quad 9]$ and $[3 \quad -2 \quad 1 \quad 4 \quad -1]$, form a basis for the row space of A.

21. Since $T_A(\mathbf{x}) = A\mathbf{x}$, we are seeking the general solution of the linear system $A\mathbf{x} = \mathbf{b}$.

(a) The reduced row echelon form of the augmented matrix $\begin{bmatrix} 1 & 2 & 0 & 0 \\ 1 & -1 & 4 & 0 \end{bmatrix}$ is $\begin{bmatrix} 1 & 0 & \frac{8}{3} & 0 \\ 0 & 1 & -\frac{4}{3} & 0 \end{bmatrix}$.

The general solution is $x_1 = -\frac{8}{3}t, x_2 = \frac{4}{3}t, x_3 = t$. In vector form, $\mathbf{x} = t\left(-\frac{8}{3}, \frac{4}{3}, 1\right)$ where t is arbitrary.

(b) The reduced row echelon form of the augmented matrix $\begin{bmatrix} 1 & 2 & 0 & 1 \\ 1 & -1 & 4 & 3 \end{bmatrix}$ is $\begin{bmatrix} 1 & 0 & \frac{8}{3} & \frac{7}{3} \\ 0 & 1 & -\frac{4}{3} & -\frac{2}{3} \end{bmatrix}$.

The general solution is $x_1 = \frac{7}{3} - \frac{8}{3}t, x_2 = -\frac{2}{3} + \frac{4}{3}t, x_3 = t$.

In vector form, $\mathbf{x} = \left(\frac{7}{3}, -\frac{2}{3}, 0\right) + t\left(-\frac{8}{3}, \frac{4}{3}, 1\right)$ where t is arbitrary.

(c) The reduced row echelon form of the augmented matrix $\begin{bmatrix} 1 & 2 & 0 & -1 \\ 1 & -1 & 4 & 1 \end{bmatrix}$ is $\begin{bmatrix} 1 & 0 & \frac{8}{3} & \frac{1}{3} \\ 0 & 1 & -\frac{4}{3} & -\frac{2}{3} \end{bmatrix}$.

The general solution is $x_1 = \frac{1}{3} - \frac{8}{3}t, x_2 = -\frac{2}{3} + \frac{4}{3}t, x_3 = t$.

In vector form, $\mathbf{x} = \left(\frac{1}{3}, -\frac{2}{3}, 0\right) + t\left(-\frac{8}{3}, \frac{4}{3}, 1\right)$ where t is arbitrary.

23. **(a)** The reduced row echelon form of A is $B = \begin{bmatrix} 1 & 0 & 0 \\ 0 & 1 & 0 \\ 0 & 0 & 0 \end{bmatrix}$. The general solution $\mathbf{x} = (x, y, z)$ of $A\mathbf{x} = \mathbf{0}$ is $x = 0, y = 0, z = t$; in vector form, $\mathbf{x} = t(0,0,1)$. This shows that the null space of A consists of all points on the z-axis.
The column space of A, span$\{(1,0,0), (0,1,0)\}$ clearly consists of all points in the xy-plane.

(b) $\begin{bmatrix} 0 & 0 & 0 \\ 0 & 1 & 0 \\ 0 & 0 & 1 \end{bmatrix}$ is an example of such a matrix.

25. **(a)** By inspection, $\begin{bmatrix} 3 & -5 \\ 0 & 0 \end{bmatrix}$ has the desired null space. In general, this will hold true for all matrices of the form $\begin{bmatrix} 3a & -5a \\ 3b & -5b \end{bmatrix}$ where a and b are not both zero (if $a = b = 0$ then the null space is the entire plane).

(b) Only the zero vector forms the null space for both A and B (their determinants are nonzero, therefore in each case the corresponding homogeneous system has only the trivial solution).
The line $3x + y = 0$ forms the null space for C.
The entire plane forms the null space for D.

True-False Exercises

(a) True.

(b) False. The column space of A is the space spanned by all column vectors of A.

(c) False. Those column vectors form a basis for the column space of R.

(d) False. This would be true if A were in row echelon form.

(e) False. For instance $A = \begin{bmatrix} 1 & 0 \\ 2 & 0 \end{bmatrix}$ and $B = \begin{bmatrix} 1 & 0 \\ 3 & 0 \end{bmatrix}$ have the same row space, but different column spaces.

(f) True. This follows from Theorem 4.7.3.

(g) True. This follows from Theorem 4.7.4.

(h) False. Elementary row operations generally can change the column space of a matrix.

(i) True. This follows from Theorem 4.7.1.

(j) False. Let both A and B be $n \times n$ matrices. By Theorem 4.7.4, row operations do not change the row space of a matrix. An invertible matrix can be reduced to I thus its row space is always R^n. On the other hand, a singular matrix cannot be reduced to identity matrix - at least one row in its reduced row echelon form is made up of zeros. Consequently, its row space is spanned by fewer than n vectors, therefore the dimension of this space is less than n.

4.8 Rank, Nullity, and the Fundamental Matrix Spaces

1. **(a)** The reduced row echelon form of A is $\begin{bmatrix} 1 & 2 & -1 & 1 \\ 0 & 0 & 0 & 0 \\ 0 & 0 & 0 & 0 \\ 0 & 0 & 0 & 0 \end{bmatrix}$. We have

   - $\text{rank}(A) = 1$ (the number of leading 1's)
   - $\text{nullity}(A) = 3$ (by Theorem 4.8.2).

   (b) The reduced row echelon form of A is $\begin{bmatrix} 1 & -2 & 0 & -1 & 3 \\ 0 & 0 & 1 & 2 & -2 \\ 0 & 0 & 0 & 0 & 0 \end{bmatrix}$. We have

   - $\text{rank}(A) = 2$ (the number of leading 1's)
   - $\text{nullity}(A) = 3$ (by Theorem 4.8.2).

3. **(a)** $\text{rank}(A) = 3$; $\text{nullity}(A) = 0$

   (b) $\text{rank}(A) + \text{nullity}(A) = 3 + 0 = 3 = n \leftarrow$ number of columns of A

   (c) 3 leading variables; 0 parameters in the general solution (the solution is unique)

5. **(a)** $\text{rank}(A) = 1$; $\text{nullity}(A) = 2$

   (b) $\text{rank}(A) + \text{nullity}(A) = 1 + 2 = 3 = n \leftarrow$ number of columns of A

   (c) 1 leading variable; 2 parameters in the general solution

7. **(a)** If every column of the reduced row echelon form of a 4×4 matrix A contains a leading 1 then

   - the rank of A has its largest possible value: 4
   - the nullity of A has the smallest possible value: 0

   (b) If every row of the reduced row echelon form of a 3×5 matrix A contains a leading 1 then

   - the rank of A has its largest possible value: 3
   - the nullity of A has the smallest possible value: 2

   (c) If every column of the reduced row echelon form of a 5×3 matrix A contains a leading 1 then

   - the rank of A has its largest possible value: 3
   - the nullity of A has the smallest possible value: 0

9.

			(a)	(b)	(c)	(d)	(e)	(f)	(g)
	Size of A:	$m \times n$	3×3	3×3	3×3	5×9	5×9	4×4	6×2
	rank(A)	$= r$	3	2	1	2	2	0	2
	rank(A | **b**)	$= s$	3	3	1	2	3	0	2
(i)	dimension of the row space of A	$= r$	3	2	1	2	2	0	2
	dimension of the column space of A	$= r$	3	2	1	2	2	0	2
	dimension of the null space of A	$= n - r$	0	1	2	7	7	4	0
	dimension of the null space of A^T	$= m - r$	0	1	2	3	3	4	4
(ii)	is the system $A\mathbf{x} = \mathbf{b}$ consistent?	Is $r = s$?	Yes	No	Yes	Yes	No	Yes	Yes
(iii)	number of parameters in the general solution of $A\mathbf{x} = \mathbf{b}$	$= n - r$ if consistent	0	-	2	7	-	4	0

11. **(a)** Applying Formula (4) to both A and its transpose yields

$$2 + \text{nullity}(A) = 4 \text{ and } 2 + \text{nullity}(A^T) = 3$$

therefore

$$\text{nullity}(A) - \text{nullity}(A^T) = 1$$

(b) Applying Formula (4) to both A and its transpose yields

$$\text{rank}(A) + \text{nullity}(A) = n \text{ and } \text{rank}(A^T) + \text{nullity}(A^T) = m$$

By Theorem 4.8.4, $\text{rank}(A^T) = \text{rank}(A)$ therefore

$$\text{nullity}(A) - \text{nullity}(A^T) = n - m$$

13. $T(x_1, x_2, x_3, x_4, x_5) = \begin{bmatrix} x_1 + x_2 \\ x_2 + x_3 + x_4 \\ x_4 + x_5 \end{bmatrix} = \begin{bmatrix} 1 & 1 & 0 & 0 & 0 \\ 0 & 1 & 1 & 1 & 0 \\ 0 & 0 & 0 & 1 & 1 \end{bmatrix} \begin{bmatrix} x_1 \\ x_2 \\ x_3 \\ x_4 \\ x_5 \end{bmatrix}$; the standard matrix is

$A = \begin{bmatrix} 1 & 1 & 0 & 0 & 0 \\ 0 & 1 & 1 & 1 & 0 \\ 0 & 0 & 0 & 1 & 1 \end{bmatrix}$. Its reduced row echelon form is $\begin{bmatrix} 1 & 0 & -1 & 0 & 0 \\ 0 & 1 & 1 & 0 & -1 \\ 0 & 0 & 0 & 1 & 1 \end{bmatrix}$.

(a) $\text{rank}(A) = 3$ **(b)** $\text{nullity}(A) = 2$

15. By inspection, there must be leading 1's in the first column (because of the first row) and in the third column (because of the fourth row) regardless of the values of r and s, therefore the matrix cannot have rank 1.

It has rank 2 if $r = 2$ and $s = 1$, since there is no leading 1 in the second column in that case.

17. No, both row and column spaces of A must be planes through the origin since from $\text{nullity}(A) = 1$, it follows by Formula (4) that $\text{rank}(A) = 3 - 1 = 2$.

19. **(a)** 3; reduced row echelon form of A can contain at most 3 leading 1's when each of its rows is nonzero;

(b) 5; if A is the zero matrix, then the general solution of $A\mathbf{x} = \mathbf{0}$ has five parameters;

(c) 3; reduced row echelon form of A can contain at most 3 leading 1's when each of its columns has a leading 1;

(d) 3; if A is the zero matrix, then the general solution of $A\mathbf{x} = \mathbf{0}$ has three parameters;

21. **(a)** By Formula (4), nullity$(A) = 7 - 4 = 3$ thus the dimension of the solution space of $A\mathbf{x} = \mathbf{0}$ is 3.

(b) No, the column space of A is a subspace of R^5 of dimension 4, therefore there exist vectors $\mathbf{b}$ in R^5 that are outside this column space. For any such vector, the system $A\mathbf{x} = \mathbf{b}$ is inconsistent.

23. From the result of Exercise 22, the rank of the matrix being less than 2 implies that

$$\begin{vmatrix} x & y \\ 1 & x \end{vmatrix} = x^2 - y = 0, \quad \begin{vmatrix} x & z \\ 1 & y \end{vmatrix} = xy - z = 0, \quad \begin{vmatrix} y & z \\ x & y \end{vmatrix} = y^2 - xz = 0$$

therefore $y = x^2$ and $z = xy = x^3$. Letting $x = t$, we obtain $y = t^2$ and $z = t^3$.

25. The reduced row echelon form of A^T is $\begin{bmatrix} 1 & 0 & 10 & 0 \\ 0 & 1 & -5 & 0 \\ 0 & 0 & 0 & 1 \\ 0 & 0 & 0 & 0 \\ 0 & 0 & 0 & 0 \\ 0 & 0 & 0 & 0 \end{bmatrix}$. The general solution of $A^T\mathbf{x} = 0$ has components $x_1 = -10t, x_2 = 5t, x_3 = t, x_4 = 0$, so in vector form $\mathbf{x} = t(-10,5,1,0)$. Evaluating dot products of columns of A and $\mathbf{v} = (-10,5,1,0)$, which forms a basis for the null space of A^T we obtain

$\mathbf{c}_1 \cdot \mathbf{v} = (1,2,0,2) \cdot (-10,5,1,0) = (1)(-10) + (2)(5) + (0)(1) + (2)(0) = 0$

$\mathbf{c}_2 \cdot \mathbf{v} = (3,6,0,6) \cdot (-10,5,1,0) = (3)(-10) + (6)(5) + (0)(1) + (6)(0) = 0$

$\mathbf{c}_3 \cdot \mathbf{v} = (-2,-5,5,0) \cdot (-10,5,1,0) = (-2)(-10) + (-5)(5) + (5)(1) + (2)(0) = 0$

$\mathbf{c}_4 \cdot \mathbf{v} = (0,-2,10,8) \cdot (-10,5,1,0) = (0)(-10) + (-2)(5) + (10)(1) + (8)(0) = 0$

$\mathbf{c}_5 \cdot \mathbf{v} = (2,4,0,4) \cdot (-10,5,1,0) = (2)(-10) + (4)(5) + (0)(1) + (4)(0) = 0$

$\mathbf{c}_6 \cdot \mathbf{v} = (0,-3,15,18) \cdot (-10,5,1,0) = (0)(-10) + (-3)(5) + (15)(1) + (18)(0) = 0$

Since the column space of A is span$\{\mathbf{c}_1, \mathbf{c}_2, \mathbf{c}_3, \mathbf{c}_4, \mathbf{c}_5, \mathbf{c}_6\}$ and the null space of A^T is span$\{\mathbf{v}\}$, we conclude that the two spaces are orthogonal complements in R^4.

27. **(a)** $m = 3 > 2 = n$ so the system is overdetermined. The augmented matrix of the system is row equivalent to $\begin{bmatrix} 1 & 0 & b_1 + b_3 \\ 0 & 1 & b_3 \\ 0 & 0 & 3b_1 + b_2 + 2b_3 \end{bmatrix}$ hence the system is inconsistent for all b's that satisfy $3b_1 + b_2 + 2b_3 \neq 0$.

(b) $m = 2 < 3 = n$ so the system is underdetermined. The augmented matrix of the system is row equivalent to $\begin{bmatrix} 1 & 0 & 0 & \frac{1}{2}b_1 - \frac{1}{4}b_2 \\ 0 & 1 & -\frac{4}{3} & -\frac{1}{6}b_1 - \frac{1}{12}b_2 \end{bmatrix}$ hence the system has infinitely many solutions for all b's (no values of b's can make this system inconsistent).

(c) $m = 2 < 3 = n$ so the system is underdetermined. The augmented matrix of the system is row equivalent to $\begin{bmatrix} 1 & 0 & -\frac{3}{2} & -\frac{1}{2}b_1 - \frac{3}{2}b_2 \\ 0 & 1 & -\frac{1}{2} & -\frac{1}{2}b_1 - \frac{1}{2}b_2 \end{bmatrix}$ hence the system has infinitely many solutions for all b's (no values of b's can make this system inconsistent).

True-False Exercises

(a) False. For instance, in $\begin{bmatrix} 1 & 2 \\ 2 & 4 \end{bmatrix}$, neither row vectors nor column vectors are linearly independent.

(b) True. In an $m \times n$ matrix, if $m < n$ then by Theorem 4.5.2(a), the n columns in R^m must be linearly dependent. If $m > n$, then by the same theorem, the m rows in R^n must be linearly dependent. We conclude that $m = n$.

(c) False. The nullity in an $m \times n$ matrix is at most n.

(d) False. For instance, if the column contains all zeros, adding it to a matrix does not change the rank.

(e) True. In an $n \times n$ matrix A with linearly dependent rows, $\text{rank}(A) \leq n - 1$.
By Formula (4), $\text{nullity}(A) = n - \text{rank}(A) \geq 1$.

(f) False. By Theorem 4.8.7, the nullity must be nonzero.

(g) False. This follows from Theorem 4.8.1.

(h) False. By Theorem 4.8.4, $\text{rank}(A^T) = \text{rank}(A)$ for any matrix A.

(i) True. Since each of the two spaces has dimension 1, these dimensions would add up to 2 instead of 3 as required by Formula (4).

(j) False. For instance, if $n = 3$, $V = \text{span}\{\mathbf{i}, \mathbf{j}\}$ (the xy-plane), and $W = \text{span}\{\mathbf{i}\}$ (the x-axis) then $W^\perp = \text{span}\{\mathbf{j}, \mathbf{k}\}$ (the yz-plane) is not a subspace of $V^\perp = \text{span}\{\mathbf{k}\}$ (the z-axis).
(Note that it is true that $V^\perp$ is a subspace of $W^\perp$.)

4.9 Matrix Transformations from R^n to R^m

1. **(a)** $\begin{bmatrix} 1 & 0 \\ 0 & -1 \end{bmatrix}\begin{bmatrix} -1 \\ 2 \end{bmatrix} = \begin{bmatrix} -1 \\ -2 \end{bmatrix}$ **(b)** $\begin{bmatrix} -1 & 0 \\ 0 & 1 \end{bmatrix}\begin{bmatrix} -1 \\ 2 \end{bmatrix} = \begin{bmatrix} 1 \\ 2 \end{bmatrix}$ **(c)** $\begin{bmatrix} 0 & 1 \\ 1 & 0 \end{bmatrix}\begin{bmatrix} -1 \\ 2 \end{bmatrix} = \begin{bmatrix} 2 \\ -1 \end{bmatrix}$

3. **(a)** $\begin{bmatrix} 1 & 0 & 0 \\ 0 & 1 & 0 \\ 0 & 0 & -1 \end{bmatrix}\begin{bmatrix} 2 \\ -5 \\ 3 \end{bmatrix} = \begin{bmatrix} 2 \\ -5 \\ -3 \end{bmatrix}$ **(b)** $\begin{bmatrix} 1 & 0 & 0 \\ 0 & -1 & 0 \\ 0 & 0 & 1 \end{bmatrix}\begin{bmatrix} 2 \\ -5 \\ 3 \end{bmatrix} = \begin{bmatrix} 2 \\ 5 \\ 3 \end{bmatrix}$ **(c)** $\begin{bmatrix} -1 & 0 & 0 \\ 0 & 1 & 0 \\ 0 & 0 & 1 \end{bmatrix}\begin{bmatrix} 2 \\ -5 \\ 3 \end{bmatrix} = \begin{bmatrix} -2 \\ -5 \\ 3 \end{bmatrix}$

5. **(a)** $\begin{bmatrix} 1 & 0 \\ 0 & 0 \end{bmatrix}\begin{bmatrix} 2 \\ -5 \end{bmatrix} = \begin{bmatrix} 2 \\ 0 \end{bmatrix}$ **(b)** $\begin{bmatrix} 0 & 0 \\ 0 & 1 \end{bmatrix}\begin{bmatrix} 2 \\ -5 \end{bmatrix} = \begin{bmatrix} 0 \\ -5 \end{bmatrix}$

7. **(a)** $\begin{bmatrix} 1 & 0 & 0 \\ 0 & 1 & 0 \\ 0 & 0 & 0 \end{bmatrix}\begin{bmatrix} -2 \\ 1 \\ 3 \end{bmatrix} = \begin{bmatrix} -2 \\ 1 \\ 0 \end{bmatrix}$ **(b)** $\begin{bmatrix} 1 & 0 & 0 \\ 0 & 0 & 0 \\ 0 & 0 & 1 \end{bmatrix}\begin{bmatrix} -2 \\ 1 \\ 3 \end{bmatrix} = \begin{bmatrix} -2 \\ 0 \\ 3 \end{bmatrix}$ **(c)** $\begin{bmatrix} 0 & 0 & 0 \\ 0 & 1 & 0 \\ 0 & 0 & 1 \end{bmatrix}\begin{bmatrix} -2 \\ 1 \\ 3 \end{bmatrix} = \begin{bmatrix} 0 \\ 1 \\ 3 \end{bmatrix}$

9. **(a)** $\begin{bmatrix}\cos 30° & -\sin 30°\\ \sin 30° & \cos 30°\end{bmatrix}\begin{bmatrix}3\\-4\end{bmatrix}=\begin{bmatrix}\frac{\sqrt{3}}{2} & -\frac{1}{2}\\ \frac{1}{2} & \frac{\sqrt{3}}{2}\end{bmatrix}\begin{bmatrix}3\\-4\end{bmatrix}=\begin{bmatrix}\frac{3\sqrt{3}}{2}+2\\ \frac{3}{2}-2\sqrt{3}\end{bmatrix}\approx\begin{bmatrix}4.60\\-1.96\end{bmatrix}$

(b) $\begin{bmatrix}\cos(-60°) & -\sin(-60°)\\ \sin(-60°) & \cos(-60°)\end{bmatrix}\begin{bmatrix}3\\-4\end{bmatrix}=\begin{bmatrix}\frac{1}{2} & \frac{\sqrt{3}}{2}\\ -\frac{\sqrt{3}}{2} & \frac{1}{2}\end{bmatrix}\begin{bmatrix}3\\-4\end{bmatrix}=\begin{bmatrix}\frac{3}{2}-2\sqrt{3}\\ -\frac{3\sqrt{3}}{2}-2\end{bmatrix}\approx\begin{bmatrix}-1.96\\-4.60\end{bmatrix}$

(c) $\begin{bmatrix}\cos 45° & -\sin 45°\\ \sin 45° & \cos 45°\end{bmatrix}\begin{bmatrix}3\\-4\end{bmatrix}=\begin{bmatrix}\frac{\sqrt{2}}{2} & -\frac{\sqrt{2}}{2}\\ \frac{\sqrt{2}}{2} & \frac{\sqrt{2}}{2}\end{bmatrix}\begin{bmatrix}3\\-4\end{bmatrix}=\begin{bmatrix}\frac{7\sqrt{2}}{2}\\ -\frac{\sqrt{2}}{2}\end{bmatrix}\approx\begin{bmatrix}4.95\\-0.71\end{bmatrix}$

(d) $\begin{bmatrix}\cos 90° & -\sin 90°\\ \sin 90° & \cos 90°\end{bmatrix}\begin{bmatrix}3\\-4\end{bmatrix}=\begin{bmatrix}0 & -1\\ 1 & 0\end{bmatrix}\begin{bmatrix}3\\-4\end{bmatrix}=\begin{bmatrix}4\\3\end{bmatrix}$

11. **(a)** $\begin{bmatrix}1 & 0 & 0\\ 0 & \cos(-30°) & -\sin(-30°)\\ 0 & \sin(-30°) & \cos(-30°)\end{bmatrix}\begin{bmatrix}2\\-1\\2\end{bmatrix}=\begin{bmatrix}1 & 0 & 0\\ 0 & \frac{\sqrt{3}}{2} & \frac{1}{2}\\ 0 & -\frac{1}{2} & \frac{\sqrt{3}}{2}\end{bmatrix}\begin{bmatrix}2\\-1\\2\end{bmatrix}=\begin{bmatrix}2\\ 1-\frac{\sqrt{3}}{2}\\ \frac{1}{2}+\sqrt{3}\end{bmatrix}$

(b) $\begin{bmatrix}\cos 30° & 0 & \sin 30°\\ 0 & 1 & 0\\ -\sin 30° & 0 & \cos 30°\end{bmatrix}\begin{bmatrix}2\\-1\\2\end{bmatrix}=\begin{bmatrix}\frac{\sqrt{3}}{2} & 0 & \frac{1}{2}\\ 0 & 1 & 0\\ -\frac{1}{2} & 0 & \frac{\sqrt{3}}{2}\end{bmatrix}\begin{bmatrix}2\\-1\\2\end{bmatrix}=\begin{bmatrix}\sqrt{3}+1\\ -1\\ -1+\sqrt{3}\end{bmatrix}$

(c) $\begin{bmatrix}\cos(-45°) & 0 & \sin(-45°)\\ 0 & 1 & 0\\ -\sin(-45°) & 0 & \cos(-45°)\end{bmatrix}\begin{bmatrix}2\\-1\\2\end{bmatrix}=\begin{bmatrix}\frac{\sqrt{2}}{2} & 0 & -\frac{\sqrt{2}}{2}\\ 0 & 1 & 0\\ \frac{\sqrt{2}}{2} & 0 & \frac{\sqrt{2}}{2}\end{bmatrix}\begin{bmatrix}2\\-1\\2\end{bmatrix}=\begin{bmatrix}0\\ -1\\ 2\sqrt{2}\end{bmatrix}$

(d) $\begin{bmatrix}\cos 90° & -\sin 90° & 0\\ \sin 90° & \cos 90° & 0\\ 0 & 0 & 1\end{bmatrix}\begin{bmatrix}2\\-1\\2\end{bmatrix}=\begin{bmatrix}0 & -1 & 0\\ 1 & 0 & 0\\ 0 & 0 & 1\end{bmatrix}\begin{bmatrix}2\\-1\\2\end{bmatrix}=\begin{bmatrix}1\\2\\2\end{bmatrix}$

13. **(a)** $\begin{bmatrix}\frac{1}{2} & 0\\ 0 & \frac{1}{2}\end{bmatrix}\begin{bmatrix}-1\\2\end{bmatrix}=\begin{bmatrix}-\frac{1}{2}\\1\end{bmatrix}$ **(b)** $\begin{bmatrix}3 & 0\\ 0 & 3\end{bmatrix}\begin{bmatrix}-1\\2\end{bmatrix}=\begin{bmatrix}-3\\6\end{bmatrix}$

15. **(a)** $\begin{bmatrix}\frac{1}{4} & 0 & 0\\ 0 & \frac{1}{4} & 0\\ 0 & 0 & \frac{1}{4}\end{bmatrix}\begin{bmatrix}2\\-1\\3\end{bmatrix}=\begin{bmatrix}\frac{1}{2}\\ -\frac{1}{4}\\ \frac{3}{4}\end{bmatrix}$ **(b)** $\begin{bmatrix}2 & 0 & 0\\ 0 & 2 & 0\\ 0 & 0 & 2\end{bmatrix}\begin{bmatrix}2\\-1\\3\end{bmatrix}=\begin{bmatrix}4\\-2\\6\end{bmatrix}$

17. **(a)** $\begin{bmatrix}\frac{1}{2} & 0\\ 0 & 1\end{bmatrix}\begin{bmatrix}-1\\2\end{bmatrix}=\begin{bmatrix}-\frac{1}{2}\\2\end{bmatrix}$ **(b)** $\begin{bmatrix}1 & 0\\ 0 & \frac{1}{2}\end{bmatrix}\begin{bmatrix}-1\\2\end{bmatrix}=\begin{bmatrix}-1\\1\end{bmatrix}$

19. **(a)** $\begin{bmatrix}\frac{1}{\alpha} & 0\\ 0 & 1\end{bmatrix}\begin{bmatrix}a\\b\end{bmatrix}=\begin{bmatrix}\frac{a}{\alpha}\\b\end{bmatrix}$ **(b)** $\begin{bmatrix}1 & 0\\ 0 & \alpha\end{bmatrix}\begin{bmatrix}a\\b\end{bmatrix}=\begin{bmatrix}a\\ \alpha b\end{bmatrix}$

21. **(a)** the matrix A_1 corresponds to the contraction with factor $\frac{1}{2}$

(b) the matrix A_2 corresponds to the compression in the y-direction with factor $\frac{1}{2}$

(c) the matrix A_3 corresponds to the shear in the y-direction by a factor $\frac{1}{2}$

(d) the matrix A_4 corresponds to the shear in the y-direction by a factor $-\frac{1}{2}$

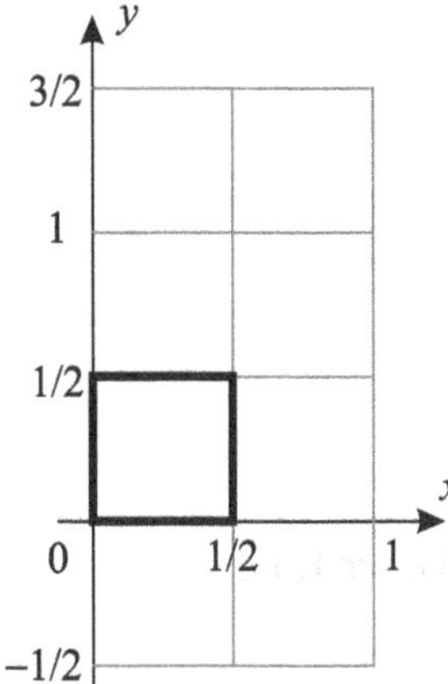

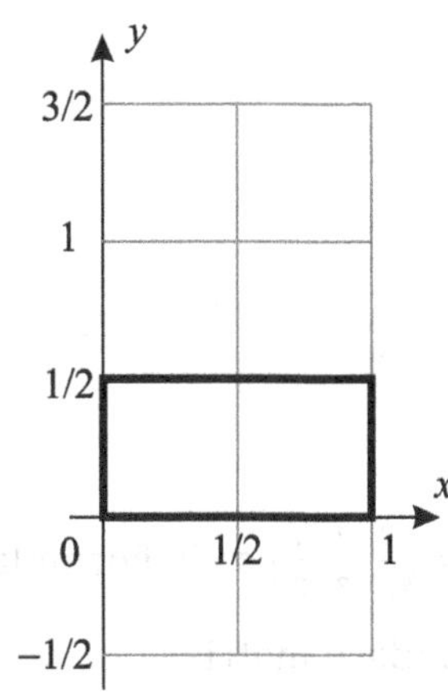

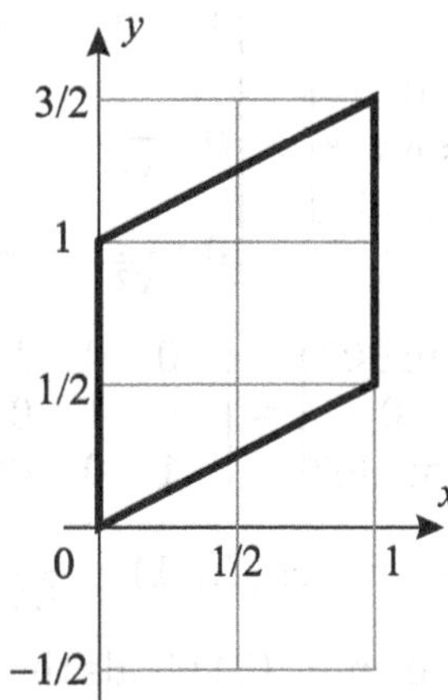

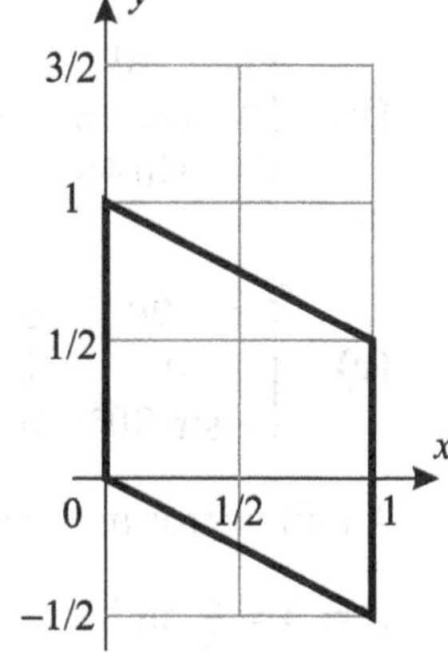

23. **(a)** $\begin{bmatrix} 2 & 0 \\ 0 & 2 \end{bmatrix}$ (dilation with factor 2)

(b) $\begin{bmatrix} 1 & 2 \\ 0 & 1 \end{bmatrix}$ (shear in the x-direction by a factor 2)

25. By Formula (4), the standard matrix for the projection is $P_{\pi/3} = \begin{bmatrix} \cos^2\frac{\pi}{3} & \sin\frac{\pi}{3}\cos\frac{\pi}{3} \\ \sin\frac{\pi}{3}\cos\frac{\pi}{3} & \sin^2\frac{\pi}{3} \end{bmatrix} = \begin{bmatrix} \frac{1}{4} & \frac{\sqrt{3}}{4} \\ \frac{\sqrt{3}}{4} & \frac{3}{4} \end{bmatrix}$.

From $\begin{bmatrix} \frac{1}{4} & \frac{\sqrt{3}}{4} \\ \frac{\sqrt{3}}{4} & \frac{3}{4} \end{bmatrix}\begin{bmatrix} 3 \\ 4 \end{bmatrix} = \begin{bmatrix} \frac{3}{4} + \sqrt{3} \\ \frac{3\sqrt{3}}{4} + 3 \end{bmatrix} \approx \begin{bmatrix} 2.48 \\ 4.30 \end{bmatrix}$ we obtain $P_{\pi/3}(3,4) \approx (2.48, 4.30)$.

27. By Formula (6), the standard matrix for the reflection $H_{\pi/3} = \begin{bmatrix} \cos\frac{2\pi}{3} & \sin\frac{2\pi}{3} \\ \sin\frac{2\pi}{3} & -\cos\frac{2\pi}{3} \end{bmatrix} = \begin{bmatrix} -\frac{1}{2} & \frac{\sqrt{3}}{2} \\ \frac{\sqrt{3}}{2} & \frac{1}{2} \end{bmatrix}$.

From $\begin{bmatrix} -\frac{1}{2} & \frac{\sqrt{3}}{2} \\ \frac{\sqrt{3}}{2} & \frac{1}{2} \end{bmatrix}\begin{bmatrix} 3 \\ 4 \end{bmatrix} = \begin{bmatrix} -\frac{3}{2} + 2\sqrt{3} \\ \frac{3\sqrt{3}}{2} + 2 \end{bmatrix} \approx \begin{bmatrix} 1.96 \\ 4.60 \end{bmatrix}$ we obtain $H_{\pi/3}(3,4) \approx (1.96, 4.60)$.

29. Reflection about the xy-plane: $T(1,2,3) = \begin{bmatrix} 1 & 0 & 0 \\ 0 & 1 & 0 \\ 0 & 0 & -1 \end{bmatrix}\begin{bmatrix} 1 \\ 2 \\ 3 \end{bmatrix} = \begin{bmatrix} 1 \\ 2 \\ -3 \end{bmatrix}$.

Reflection about the xz-plane: $T(1,2,3) = \begin{bmatrix} 1 & 0 & 0 \\ 0 & -1 & 0 \\ 0 & 0 & 1 \end{bmatrix}\begin{bmatrix} 1 \\ 2 \\ 3 \end{bmatrix} = \begin{bmatrix} 1 \\ -2 \\ 3 \end{bmatrix}$.

Reflection about the yz-plane: $T(1,2,3) = \begin{bmatrix} -1 & 0 & 0 \\ 0 & 1 & 0 \\ 0 & 0 & 1 \end{bmatrix}\begin{bmatrix} 1 \\ 2 \\ 3 \end{bmatrix} = \begin{bmatrix} -1 \\ 2 \\ 3 \end{bmatrix}$.

31. **(a)** $\begin{bmatrix} \cos 30° & -\sin 30° & 0 \\ \sin 30° & \cos 30° & 0 \\ 0 & 0 & 1 \end{bmatrix} = \begin{bmatrix} \frac{\sqrt{3}}{2} & -\frac{1}{2} & 0 \\ \frac{1}{2} & \frac{\sqrt{3}}{2} & 0 \\ 0 & 0 & 1 \end{bmatrix}$

(b) $\begin{bmatrix} 1 & 0 & 0 \\ 0 & \cos 45° & -\sin 45° \\ 0 & \sin 45° & \cos 45° \end{bmatrix} = \begin{bmatrix} 1 & 0 & 0 \\ 0 & \frac{1}{\sqrt{2}} & -\frac{1}{\sqrt{2}} \\ 0 & \frac{1}{\sqrt{2}} & \frac{1}{\sqrt{2}} \end{bmatrix}$

(c) $\begin{bmatrix} \cos 90° & 0 & \sin 90° \\ 0 & 1 & 0 \\ -\sin 90° & 0 & \cos 90° \end{bmatrix} = \begin{bmatrix} 0 & 0 & 1 \\ 0 & 1 & 0 \\ -1 & 0 & 0 \end{bmatrix}$

33. A unit vector in the direction of (2,2,1) is $\frac{1}{\|\mathbf{v}\|}\mathbf{v} = \left(\frac{2}{3}, \frac{2}{3}, \frac{1}{3}\right)$ so in Formula (3), we take $a = b = \frac{2}{3}$ and $c = \frac{1}{3}$. Formula (3) yields the standard matrix

$$\begin{bmatrix} \frac{4}{9}(1-\cos\pi)+\cos\pi & \frac{4}{9}(1-\cos\pi)-\frac{1}{3}\sin\pi & \frac{2}{9}(1-\cos\pi)+\frac{2}{3}\sin\pi \\ \frac{4}{9}(1-\cos\pi)+\frac{1}{3}\sin\pi & \frac{4}{9}(1-\cos\pi)+\cos\pi & \frac{2}{9}(1-\cos\pi)-\frac{2}{3}\sin\pi \\ \frac{2}{9}(1-\cos\pi)-\frac{2}{3}\sin\pi & \frac{2}{9}(1-\cos\pi)+\frac{2}{3}\sin\pi & \frac{1}{9}(1-\cos\pi)+\cos\pi \end{bmatrix} = \begin{bmatrix} -\frac{1}{9} & \frac{8}{9} & \frac{4}{9} \\ \frac{8}{9} & -\frac{1}{9} & \frac{4}{9} \\ \frac{4}{9} & \frac{4}{9} & -\frac{7}{9} \end{bmatrix}$$

35. For the rotation about the x-axis, we take $\mathbf{v} = (1,0,0)$, so in Formula (3) we have $a = 1, b = c = 0$:

$$\begin{bmatrix} (1)\left(1-\cos\frac{\pi}{2}\right)+\cos\frac{\pi}{2} & (0)\left(1-\cos\frac{\pi}{2}\right)-(0)\sin\frac{\pi}{2} & (0)\left(1-\cos\frac{\pi}{2}\right)+(0)\sin\frac{\pi}{2} \\ (0)\left(1-\cos\frac{\pi}{2}\right)+(0)\sin\frac{\pi}{2} & (0)\left(1-\cos\frac{\pi}{2}\right)+\cos\frac{\pi}{2} & (0)\left(1-\cos\frac{\pi}{2}\right)-(1)\sin\frac{\pi}{2} \\ (0)\left(1-\cos\frac{\pi}{2}\right)-(0)\sin\frac{\pi}{2} & (0)\left(1-\cos\frac{\pi}{2}\right)+(1)\sin\frac{\pi}{2} & (0)\left(1-\cos\frac{\pi}{2}\right)+\cos\frac{\pi}{2} \end{bmatrix} = \begin{bmatrix} 1 & 0 & 0 \\ 0 & 0 & -1 \\ 0 & 1 & 0 \end{bmatrix}.$$

For the rotation about the y-axis, we take $\mathbf{v} = (0,1,0)$, so in Formula (3) we have $b = 1, a = c = 0$:

$$\begin{bmatrix} (0)\left(1-\cos\frac{\pi}{2}\right)+\cos\frac{\pi}{2} & (0)\left(1-\cos\frac{\pi}{2}\right)-(0)\sin\frac{\pi}{2} & (0)\left(1-\cos\frac{\pi}{2}\right)+(1)\sin\frac{\pi}{2} \\ (0)\left(1-\cos\frac{\pi}{2}\right)+(0)\sin\frac{\pi}{2} & (1)\left(1-\cos\frac{\pi}{2}\right)+\cos\frac{\pi}{2} & (0)\left(1-\cos\frac{\pi}{2}\right)-(0)\sin\frac{\pi}{2} \\ (0)\left(1-\cos\frac{\pi}{2}\right)-(1)\sin\frac{\pi}{2} & (0)\left(1-\cos\frac{\pi}{2}\right)+(0)\sin\frac{\pi}{2} & (0)\left(1-\cos\frac{\pi}{2}\right)+\cos\frac{\pi}{2} \end{bmatrix} = \begin{bmatrix} 0 & 0 & 1 \\ 0 & 1 & 0 \\ -1 & 0 & 0 \end{bmatrix}.$$

For the rotation about the z-axis, we take $\mathbf{v} = (0,0,1)$, so in Formula (3) we have $a = b = 0, c = 1$:

$$\begin{bmatrix} (0)\left(1-\cos\frac{\pi}{2}\right)+\cos\frac{\pi}{2} & (0)\left(1-\cos\frac{\pi}{2}\right)-(1)\sin\frac{\pi}{2} & (0)\left(1-\cos\frac{\pi}{2}\right)+(0)\sin\frac{\pi}{2} \\ (0)\left(1-\cos\frac{\pi}{2}\right)+(1)\sin\frac{\pi}{2} & (0)\left(1-\cos\frac{\pi}{2}\right)+\cos\frac{\pi}{2} & (0)\left(1-\cos\frac{\pi}{2}\right)-(0)\sin\frac{\pi}{2} \\ (0)\left(1-\cos\frac{\pi}{2}\right)-(0)\sin\frac{\pi}{2} & (0)\left(1-\cos\frac{\pi}{2}\right)+(0)\sin\frac{\pi}{2} & (1)\left(1-\cos\frac{\pi}{2}\right)+\cos\frac{\pi}{2} \end{bmatrix} = \begin{bmatrix} 0 & -1 & 0 \\ 1 & 0 & 0 \\ 0 & 0 & 1 \end{bmatrix}.$$

37. Since $\cos^2\theta - \sin^2\theta = \cos(2\theta)$ and $2\sin\theta\cos\theta = \sin(2\theta)$, we have $A = \begin{bmatrix} \cos(2\theta) & -\sin(2\theta) \\ \sin(2\theta) & \cos(2\theta) \end{bmatrix}$. The geometric effect of multiplying A by $\mathbf{x}$ is to rotate the vector through the angle 2θ.

39. The terminal point of the vector is first rotated about the origin through the angle θ, then it is translated by the vector $\mathbf{x}_0$. No, this is not a matrix transformation, for instance it fails the additivity property: $T(\mathbf{u}+\mathbf{v}) = \mathbf{x}_0 + R_\theta(\mathbf{u}+\mathbf{v}) = \mathbf{x}_0 + R_\theta\mathbf{u} + R_\theta\mathbf{v} \neq \mathbf{x}_0 + R_\theta\mathbf{u} + \mathbf{x}_0 + R_\theta\mathbf{v} = T(\mathbf{u}) + T(\mathbf{v})$.

4.10 Properties of Matrix Transformations

1. **(a)** From Tables 1 and 3 in Section 4.9, $[T_1] = \begin{bmatrix} 0 & 1 \\ 1 & 0 \end{bmatrix}$ and $[T_2] = \begin{bmatrix} 1 & 0 \\ 0 & 0 \end{bmatrix}$;
$[T_1 \circ T_2] = [T_1][T_2] = \begin{bmatrix} 0 & 0 \\ 1 & 0 \end{bmatrix}$; $[T_2 \circ T_1] = [T_2][T_1] = \begin{bmatrix} 0 & 1 \\ 0 & 0 \end{bmatrix}$.
For these transformations, $T_1 \circ T_2 \neq T_2 \circ T_1$.

(b) From Table 1 in Section 4.9, $[T_1] = \begin{bmatrix} 1 & 0 \\ 0 & -1 \end{bmatrix}$ and $[T_2] = \begin{bmatrix} 0 & 1 \\ 1 & 0 \end{bmatrix}$;
$[T_1 \circ T_2] = [T_1][T_2] = \begin{bmatrix} 0 & 1 \\ -1 & 0 \end{bmatrix}$; $[T_2 \circ T_1] = [T_2][T_1] = \begin{bmatrix} 0 & -1 \\ 1 & 0 \end{bmatrix}$.
For these transformations, $T_1 \circ T_2 \neq T_2 \circ T_1$.

3. From Table 8 in Section 4.9, $[T_1] = \begin{bmatrix} k & 0 & 0 \\ 0 & k & 0 \\ 0 & 0 & k \end{bmatrix}$ and $[T_2] = \begin{bmatrix} 1/k & 0 & 0 \\ 0 & 1/k & 0 \\ 0 & 0 & 1/k \end{bmatrix}$;
$[T_1 \circ T_2] = [T_1][T_2] = \begin{bmatrix} 1 & 0 & 0 \\ 0 & 1 & 0 \\ 0 & 0 & 1 \end{bmatrix}$; $[T_2 \circ T_1] = [T_2][T_1] = \begin{bmatrix} 1 & 0 & 0 \\ 0 & 1 & 0 \\ 0 & 0 & 1 \end{bmatrix}$.
For these transformations, $T_1 \circ T_2 = T_2 \circ T_1$.

5. $[T_B \circ T_A] = [T_B][T_A] = BA = \begin{bmatrix} -10 & -7 \\ 5 & -10 \end{bmatrix}$; $[T_A \circ T_B] = [T_A][T_B] = AB = \begin{bmatrix} -8 & -3 \\ 13 & -12 \end{bmatrix}$

7. **(a)** We are looking for the standard matrix of $T = T_2 \circ T_1$ where T_1 is a rotation of $90°$ and T_2 is a reflection about the line $y = x$. From Tables 5 and 1 in Section 4.9,
$[T_1] = \begin{bmatrix} \cos 90° & -\sin 90° \\ \sin 90° & \cos 90° \end{bmatrix} = \begin{bmatrix} 0 & -1 \\ 1 & 0 \end{bmatrix}$, $[T_2] = \begin{bmatrix} 0 & 1 \\ 1 & 0 \end{bmatrix}$. Therefore, $[T] = [T_2][T_1] = \begin{bmatrix} 1 & 0 \\ 0 & -1 \end{bmatrix}$.

(b) We are looking for the standard matrix of $T = T_2 \circ T_1$ where T_1 is an orthogonal projection on the y-axis and T_2 is a contraction with factor $k = \frac{1}{2}$. From Tables 3 and 7 in Section 4.9,

$[T_1] = \begin{bmatrix} 0 & 0 \\ 0 & 1 \end{bmatrix}$, $[T_2] = \begin{bmatrix} \frac{1}{2} & 0 \\ 0 & \frac{1}{2} \end{bmatrix}$. Therefore, $[T] = [T_2][T_1] = \begin{bmatrix} 0 & 0 \\ 0 & \frac{1}{2} \end{bmatrix}$.

(c) We are looking for the standard matrix of $T = T_3 \circ T_2 \circ T_1$ where T_1 is a reflection about the x-axis, T_2 is a dilation with factor $k = 3$, and T_3 is a rotation of $60°$. From Tables 1, 7, and 5 in Section 4.9, $[T_1] = \begin{bmatrix} 1 & 0 \\ 0 & -1 \end{bmatrix}$, $[T_2] = \begin{bmatrix} 3 & 0 \\ 0 & 3 \end{bmatrix}$, and $[T_3] = \begin{bmatrix} \cos 60° & -\sin 60° \\ \sin 60° & \cos 60° \end{bmatrix} = \begin{bmatrix} \frac{1}{2} & -\frac{\sqrt{3}}{2} \\ \frac{\sqrt{3}}{2} & \frac{1}{2} \end{bmatrix}$.

Therefore, $[T] = [T_3][T_2][T_1] = \begin{bmatrix} \frac{3}{2} & \frac{3\sqrt{3}}{2} \\ \frac{3\sqrt{3}}{2} & -\frac{3}{2} \end{bmatrix}$.

9. **(a)** We are looking for the standard matrix of $T = T_2 \circ T_1$ where T_1 is a reflection about the yz-plane and T_2 is an orthogonal projection on the xz-plane. From Tables 2 and 4 in Section 4.9, $[T_1] = \begin{bmatrix} -1 & 0 & 0 \\ 0 & 1 & 0 \\ 0 & 0 & 1 \end{bmatrix}$ and $[T_2] = \begin{bmatrix} 1 & 0 & 0 \\ 0 & 0 & 0 \\ 0 & 0 & 1 \end{bmatrix}$. Therefore, $[T] = [T_2][T_1] = \begin{bmatrix} -1 & 0 & 0 \\ 0 & 0 & 0 \\ 0 & 0 & 1 \end{bmatrix}$.

(b) We are looking for the standard matrix of $T = T_2 \circ T_1$ where T_1 is a rotation of 45° about the y-axis and T_2 is a dilation with factor $k = \sqrt{2}$. From Tables 6 and 8 in Section 4.9,

$[T_1] = \begin{bmatrix} \cos 45° & 0 & \sin 45° \\ 0 & 1 & 0 \\ -\sin 45° & 0 & \cos 45° \end{bmatrix} = \begin{bmatrix} \frac{\sqrt{2}}{2} & 0 & \frac{\sqrt{2}}{2} \\ 0 & 1 & 0 \\ -\frac{\sqrt{2}}{2} & 0 & \frac{\sqrt{2}}{2} \end{bmatrix}$ and $[T_2] = \begin{bmatrix} \sqrt{2} & 0 & 0 \\ 0 & \sqrt{2} & 0 \\ 0 & 0 & \sqrt{2} \end{bmatrix}$. Therefore,

$[T] = [T_2][T_1] = \begin{bmatrix} 1 & 0 & 1 \\ 0 & \sqrt{2} & 0 \\ -1 & 0 & 1 \end{bmatrix}$.

(c) We are looking for the standard matrix of $T = T_2 \circ T_1$ where T_1 is an orthogonal projection on the xy-plane and T_2 is a reflection about the yz-plane. From Tables 4 and 2 in Section 4.9, $[T_1] = \begin{bmatrix} 1 & 0 & 0 \\ 0 & 1 & 0 \\ 0 & 0 & 0 \end{bmatrix}$, $[T_2] = \begin{bmatrix} -1 & 0 & 0 \\ 0 & 1 & 0 \\ 0 & 0 & 1 \end{bmatrix}$. Therefore, $[T] = [T_2][T_1] = \begin{bmatrix} -1 & 0 & 0 \\ 0 & 1 & 0 \\ 0 & 0 & 0 \end{bmatrix}$.

11. **(a)** In vector form, $T_1(x_1, x_2) = \begin{bmatrix} x_1 + x_2 \\ x_1 - x_2 \end{bmatrix} = \begin{bmatrix} 1 & 1 \\ 1 & -1 \end{bmatrix}\begin{bmatrix} x_1 \\ x_2 \end{bmatrix}$ so that $[T_1] = \begin{bmatrix} 1 & 1 \\ 1 & -1 \end{bmatrix}$.

Likewise, $T_2(x_1, x_2) = \begin{bmatrix} 3x_1 \\ 2x_1 + 4x_2 \end{bmatrix} = \begin{bmatrix} 3 & 0 \\ 2 & 4 \end{bmatrix}\begin{bmatrix} x_1 \\ x_2 \end{bmatrix}$ so that $[T_2] = \begin{bmatrix} 3 & 0 \\ 2 & 4 \end{bmatrix}$.

(b) $[T_2 \circ T_1] = [T_2][T_1] = \begin{bmatrix} 3 & 0 \\ 2 & 4 \end{bmatrix}\begin{bmatrix} 1 & 1 \\ 1 & -1 \end{bmatrix} = \begin{bmatrix} 3 & 3 \\ 6 & -2 \end{bmatrix}$

$[T_1 \circ T_2] = [T_1][T_2] = \begin{bmatrix} 1 & 1 \\ 1 & -1 \end{bmatrix}\begin{bmatrix} 3 & 0 \\ 2 & 4 \end{bmatrix} = \begin{bmatrix} 5 & 4 \\ 1 & -4 \end{bmatrix}$

(c) $T_1(T_2(x_1, x_2)) = (5x_1 + 4x_2,\ x_1 - 4x_2)$; $T_2(T_1(x_1, x_2)) = (3x_1 + 3x_2,\ 6x_1 - 2x_2)$

13. **(a)** Not one-to-one (maps distinct vectors with the same x components into the same vector).

(b) One-to-one (distinct vectors that are reflected have distinct images).

(c) One-to-one (distinct vectors that are reflected have distinct images).

(d) One-to-one (distinct vectors that are contracted have distinct images).

15. **(a)** The reflection about the x-axis in R^2 is its own inverse.

(b) The rotation through an angle of $-\pi/4$ in R^2 (i.e., the clockwise rotation through an angle $\pi/4$) is the desired inverse.

(c) The contraction by a factor of $\frac{1}{3}$ in R^2 is the desired inverse.

17. **(a)** $\begin{bmatrix} w_1 \\ w_2 \end{bmatrix} = \begin{bmatrix} 8x_1 + 4x_2 \\ 2x_1 + x_2 \end{bmatrix} = \begin{bmatrix} 8 & 4 \\ 2 & 1 \end{bmatrix}\begin{bmatrix} x_1 \\ x_2 \end{bmatrix}$; the standard matrix is $\begin{bmatrix} 8 & 4 \\ 2 & 1 \end{bmatrix}$; since $\begin{vmatrix} 8 & 4 \\ 2 & 1 \end{vmatrix} = 0$, it follows from parts (g) and (s) of Theorem 4.10.2 that the operator is not one-to-one

(b) $\begin{bmatrix} w_1 \\ w_2 \\ w_3 \end{bmatrix} = \begin{bmatrix} -x_1 + 3x_2 + 2x_3 \\ 2x_1 + 4x_3 \\ x_1 + 3x_2 + 6x_3 \end{bmatrix} = \begin{bmatrix} -1 & 3 & 2 \\ 2 & 0 & 4 \\ 1 & 3 & 6 \end{bmatrix} \begin{bmatrix} x_1 \\ x_2 \\ x_3 \end{bmatrix}$; the standard matrix is $\begin{bmatrix} -1 & 3 & 2 \\ 2 & 0 & 4 \\ 1 & 3 & 6 \end{bmatrix}$; since $\begin{vmatrix} -1 & 3 & 2 \\ 2 & 0 & 4 \\ 1 & 3 & 6 \end{vmatrix} = 0$, it follows from parts (g) and (s) of Theorem 4.10.2 that the operator is not one-to-one

19. (a) $\begin{bmatrix} w_1 \\ w_2 \end{bmatrix} = \begin{bmatrix} x_1 + 2x_2 \\ -x_1 + x_2 \end{bmatrix} = \begin{bmatrix} 1 & 2 \\ -1 & 1 \end{bmatrix} \begin{bmatrix} x_1 \\ x_2 \end{bmatrix}$; the standard matrix is $\begin{bmatrix} 1 & 2 \\ -1 & 1 \end{bmatrix}$; since $\begin{vmatrix} 1 & 2 \\ -1 & 1 \end{vmatrix} = 3 \neq 0$, it follows from parts (g) and (s) of Theorem 4.10.2 that the operator is one-to-one;

the standard matrix of T^{-1} is $\frac{1}{3}\begin{bmatrix} 1 & -2 \\ 1 & 1 \end{bmatrix} = \begin{bmatrix} \frac{1}{3} & -\frac{2}{3} \\ \frac{1}{3} & \frac{1}{3} \end{bmatrix}$; $T^{-1}(w_1, w_2) = \left(\frac{1}{3}w_1 - \frac{2}{3}w_2, \frac{1}{3}w_1 + \frac{1}{3}w_2\right)$

(b) $\begin{bmatrix} w_1 \\ w_2 \end{bmatrix} = \begin{bmatrix} 4x_1 - 6x_2 \\ -2x_1 + 3x_2 \end{bmatrix} = \begin{bmatrix} 4 & -6 \\ -2 & 3 \end{bmatrix} \begin{bmatrix} x_1 \\ x_2 \end{bmatrix}$; the standard matrix is $\begin{bmatrix} 4 & -6 \\ -2 & 3 \end{bmatrix}$; since $\begin{vmatrix} 4 & -6 \\ -2 & 3 \end{vmatrix} = 0$, it follows from parts (g) and (s) of Theorem 4.10.2 that the operator is not one-to-one

21. (a) Suppose $A\begin{bmatrix} x_1 \\ x_2 \end{bmatrix} = \begin{bmatrix} w_1 \\ w_2 \\ w_3 \end{bmatrix}$ and $A\begin{bmatrix} y_1 \\ y_2 \end{bmatrix} = \begin{bmatrix} w_1 \\ w_2 \\ w_3 \end{bmatrix}$. Subtracting both equations yields $A\begin{bmatrix} x_1 - y_1 \\ x_2 - y_2 \end{bmatrix} = \begin{bmatrix} 0 \\ 0 \\ 0 \end{bmatrix}$ therefore the transformation is one-to-one if and only if the nullity of A is 0 since that is equivalent to stating that $A\begin{bmatrix} x_1 \\ x_2 \end{bmatrix} = A\begin{bmatrix} y_1 \\ y_2 \end{bmatrix}$ implies $\begin{bmatrix} x_1 \\ x_2 \end{bmatrix} = \begin{bmatrix} y_1 \\ y_2 \end{bmatrix}$. The reduced row echelon form of A is $\begin{bmatrix} 1 & 0 \\ 0 & 1 \\ 0 & 0 \end{bmatrix}$, so we conclude that the nullity of A is 0, thus T_A is one-to-one.

(b) Proceeding as in part (a), we determine the reduced row echelon form of A to be $\begin{bmatrix} 1 & 0 & 4 \\ 0 & 1 & -\frac{1}{2} \end{bmatrix}$. Therefore A has nullity 1 and T_A is not one-to-one (e.g., $A\begin{bmatrix} 0 \\ 0 \\ 0 \end{bmatrix} = A\begin{bmatrix} -8 \\ 1 \\ 2 \end{bmatrix} = \begin{bmatrix} 0 \\ 0 \end{bmatrix}$).

23. (a) The range of $T(x) = Ax$ consists of all vectors (y_1, y_2, y_3) that are images of at least one vector (x_1, x_2, x_3) under this transformation $\begin{bmatrix} y_1 \\ y_2 \\ y_3 \end{bmatrix} = \begin{bmatrix} 1 & -1 & 3 \\ 5 & 6 & -4 \\ 7 & 4 & 2 \end{bmatrix} \begin{bmatrix} x_1 \\ x_2 \\ x_3 \end{bmatrix} = x_1\begin{bmatrix} 1 \\ 5 \\ 7 \end{bmatrix} + x_2\begin{bmatrix} -1 \\ 6 \\ 4 \end{bmatrix} + x_3\begin{bmatrix} 3 \\ -4 \\ 2 \end{bmatrix}$.

Since the reduced row echelon form of A is $\begin{bmatrix} 1 & 0 & \frac{14}{11} \\ 0 & 1 & -\frac{19}{11} \\ 0 & 0 & 0 \end{bmatrix}$, by Theorem 4.7.6 the vectors $\begin{bmatrix} 1 \\ 5 \\ 7 \end{bmatrix}$ and $\begin{bmatrix} -1 \\ 6 \\ 4 \end{bmatrix}$ are linearly independent, and they span $R(T)$. We conclude that $\begin{bmatrix} 1 \\ 5 \\ 7 \end{bmatrix}$ and $\begin{bmatrix} -1 \\ 6 \\ 4 \end{bmatrix}$ form a basis for $R(T)$.

(b) The kernel of T consists of all vectors (x_1, x_2, x_3) such that $\begin{bmatrix} 1 & -1 & 3 \\ 5 & 6 & -4 \\ 7 & 4 & 2 \end{bmatrix}\begin{bmatrix} x_1 \\ x_2 \\ x_3 \end{bmatrix} = \begin{bmatrix} 0 \\ 0 \\ 0 \end{bmatrix}$. Based on the reduced row echelon form of A obtained above, the general solution is $x_1 = -\frac{14}{11}t$, $x_2 = \frac{19}{11}t$, $x_3 = t$. Therefore a basis for $\ker(T)$ is formed by the vector $(-14, 19, 11)$.

(c) From part (a), $\text{rank}(T) = \dim(R(T)) = 2$. From part (b), $\text{nullity}(T) = \dim(\ker(T)) = 1$.

(d) Based on the reduced row echelon form of A obtained above, $\text{rank}(A) = 2$ and $\text{nullity}(A) = 1$.

25. The kernel of T_A consists of all vectors (x_1, x_2, x_3, x_4) such that $\begin{bmatrix} 1 & 2 & -1 & -2 \\ -3 & 1 & 3 & 4 \\ -3 & 8 & 4 & 2 \end{bmatrix}\begin{bmatrix} x_1 \\ x_2 \\ x_3 \\ x_4 \end{bmatrix} = \begin{bmatrix} 0 \\ 0 \\ 0 \end{bmatrix}$. Since the reduced row echelon form of A is $\begin{bmatrix} 1 & 0 & 0 & -\frac{10}{7} \\ 0 & 1 & 0 & -\frac{2}{7} \\ 0 & 0 & 1 & 0 \end{bmatrix}$, the general solution is $x_1 = \frac{10}{7}t$, $x_2 = \frac{2}{7}t$, $x_3 = 0$, $x_4 = t$. Therefore a basis for $\ker(T_A)$ is formed by the vector $(10, 2, 0, 7)$.

The range of $T_A(\mathbf{x}) = A\mathbf{x}$ consists of all vectors (y_1, y_2, y_3) that are images of at least one vector (x_1, x_2, x_3, x_4) under this transformation

$\begin{bmatrix} y_1 \\ y_2 \\ y_3 \end{bmatrix} = \begin{bmatrix} 1 & 2 & -1 & -2 \\ -3 & 1 & 3 & 4 \\ -3 & 8 & 4 & 2 \end{bmatrix}\begin{bmatrix} x_1 \\ x_2 \\ x_3 \\ x_4 \end{bmatrix} = x_1\begin{bmatrix} 1 \\ -3 \\ -3 \end{bmatrix} + x_2\begin{bmatrix} 2 \\ 1 \\ 8 \end{bmatrix} + x_3\begin{bmatrix} -1 \\ 3 \\ 4 \end{bmatrix} + x_4\begin{bmatrix} -2 \\ 4 \\ 2 \end{bmatrix}$. Based on the reduced row echelon form of A obtained above, by Theorem 4.7.6 the vectors $\begin{bmatrix} 1 \\ -3 \\ -3 \end{bmatrix}$, $\begin{bmatrix} 2 \\ 1 \\ 8 \end{bmatrix}$, and $\begin{bmatrix} -1 \\ 3 \\ 4 \end{bmatrix}$ are linearly independent, and they span $R(T_A)$. We conclude that $\begin{bmatrix} 1 \\ -3 \\ -3 \end{bmatrix}$, $\begin{bmatrix} 2 \\ 1 \\ 8 \end{bmatrix}$, and $\begin{bmatrix} -1 \\ 3 \\ 4 \end{bmatrix}$ form a basis for $R(T_A)$.

27. (a) By parts (g) and (s) of Theorem 4.10.2, the range of T cannot be R^n - it must be a proper subset of R^n instead.

For instance $A = \begin{bmatrix} 1 & 0 \\ 0 & 0 \end{bmatrix}$ is the standard matrix of the orthogonal projection onto the x-axis; the range of this transformation is the x-axis - a proper subset of R^2.

(b) By parts (g) and (r) of Theorem 4.10.2, the kernel of T must contain at least one nonzero vector $\mathbf{v}$. Consequently T maps infinitely many vectors (e.g., scalar multiples $k\mathbf{v}$) into $\mathbf{0}$.

29. (a) Yes. If $T_1: R^n \to R^m$ and $T_2: R^m \to R^k$ are both one-to-one then for any vectors $\mathbf{u}$ and $\mathbf{v}$ in R^n, $T_2(T_1(\mathbf{u})) = T_2(T_1(\mathbf{v}))$ must imply $T_1(\mathbf{u}) = T_1(\mathbf{v})$ (since T_2 is one-to-one), which further implies that $\mathbf{u} = \mathbf{v}$ (since T_1 is one-to-one), therefore the composition $T_2 \circ T_1: R^n \to R^k$ is also one-to-one.

(b) Yes. For instance, $T_1(x_1, x_2) = (x_1, x_2, 0)$ is one-to-one but $T_2(x_1, x_2, x_3) = (x_1, x_2)$ is not. However, the composition $T_2(T_1(x_1, x_2)) = T_2(x_1, x_2, 0) = (x_1, x_2)$ is obviously one-to-one.

However, if T_1 is not one-to-one, then the composition $T_2 \circ T_1$ is not one-to-one since there must exist two vectors $\mathbf{u} \neq \mathbf{v}$ such that $T_1(\mathbf{u}) = T_1(\mathbf{v})$ leading to $T_2(T_1(\mathbf{u})) = T_2(T_1(\mathbf{v}))$.

31. **(a)** From Table 1, the standard matrix of the reflection about the line $y = x$ is $A = \begin{bmatrix} 0 & 1 \\ 1 & 0 \end{bmatrix}$.

The inverse is $A^{-1} = \frac{1}{(0)(0)-(1)(1)} \begin{bmatrix} 0 & -1 \\ -1 & 0 \end{bmatrix} = \begin{bmatrix} 0 & 1 \\ 1 & 0 \end{bmatrix} = A$.

(b) By Table 9, the standard matrix of the compression in the x-direction with the factor k (such that $0 < k < 1$) is $A = \begin{bmatrix} k & 0 \\ 0 & 1 \end{bmatrix}$. The inverse is $A^{-1} = \frac{1}{(k)(1)-(0)(0)} \begin{bmatrix} 1 & 0 \\ 0 & k \end{bmatrix} = \begin{bmatrix} 1/k & 0 \\ 0 & 1 \end{bmatrix}$.

Since $\frac{1}{k} > 1$, this is a standard matrix of an expansion in the x-direction with the factor $1/k$. (Compressions in the y-direction can be treated analogously.)

True-False Exercises

(a) False. For instance, Example 1 shows two matrix operators on R^2 whose composition is not commutative.

(b) True. This is stated as Formula (4).

(c) True. This was established in Example 2.

(d) False. For instance, composition of any reflection operator with itself is the identity operator, which is not a reflection.

(e) True. This is stated in Formula (6).

(f) True. This follows from parts (b) and (d) of Theorem 4.10.1.

(g) True. This follows from parts (b) and (s) of Theorem 4.10.2.

4.11 Geometry of Matrix Operators on R^2

1. Coordinates (x, y) are being transformed to coordinates (x', y') according to the equation $\begin{bmatrix} x' \\ y' \end{bmatrix} = \begin{bmatrix} 5 & 2 \\ 2 & 1 \end{bmatrix} \begin{bmatrix} x \\ y \end{bmatrix}$. Thus $\begin{bmatrix} x \\ y \end{bmatrix} = \begin{bmatrix} 5 & 2 \\ 2 & 1 \end{bmatrix}^{-1} \begin{bmatrix} x' \\ y' \end{bmatrix} = \begin{bmatrix} 1 & -2 \\ -2 & 5 \end{bmatrix} \begin{bmatrix} x' \\ y' \end{bmatrix}$ so $x = x' - 2y'$ and $y = -2x' + 5y'$. Substituting these into $y = 4x$ yields $-2x' + 5y' = 4(x' - 2y')$ or equivalently $y' = \frac{6}{13}x'$.

3. From Table 1, the standard matrix for the shear in the x-direction by a factor 3 is $A = \begin{bmatrix} 1 & 3 \\ 0 & 1 \end{bmatrix}$. Coordinates (x, y) are being transformed to coordinates (x', y') according to the equation $\begin{bmatrix} x' \\ y' \end{bmatrix} = \begin{bmatrix} 1 & 3 \\ 0 & 1 \end{bmatrix} \begin{bmatrix} x \\ y \end{bmatrix}$. Thus $\begin{bmatrix} x \\ y \end{bmatrix} = \begin{bmatrix} 1 & 3 \\ 0 & 1 \end{bmatrix}^{-1} \begin{bmatrix} x' \\ y' \end{bmatrix} = \begin{bmatrix} 1 & -3 \\ 0 & 1 \end{bmatrix} \begin{bmatrix} x' \\ y' \end{bmatrix}$ so $x = x' - 3y'$ and $y = y'$. Substituting these into $y = 2x$ yields $y' = 2(x' - 3y')$ or equivalently $y' = \frac{2}{7}x'$.

5. Since $\begin{bmatrix}3 & -1\\1 & -2\end{bmatrix}\begin{bmatrix}0\\0\end{bmatrix} = \begin{bmatrix}0\\0\end{bmatrix}$, $\begin{bmatrix}3 & -1\\1 & -2\end{bmatrix}\begin{bmatrix}1\\0\end{bmatrix} = \begin{bmatrix}3\\1\end{bmatrix}$, $\begin{bmatrix}3 & -1\\1 & -2\end{bmatrix}\begin{bmatrix}0\\1\end{bmatrix} = \begin{bmatrix}-1\\-2\end{bmatrix}$, and $\begin{bmatrix}3 & -1\\1 & -2\end{bmatrix}\begin{bmatrix}1\\1\end{bmatrix} = \begin{bmatrix}2\\-1\end{bmatrix}$, the image of the unit square is a parallelogram with vertices $(0,0)$, $(3,1)$, $(-1,-2)$, and $(2,-1)$.

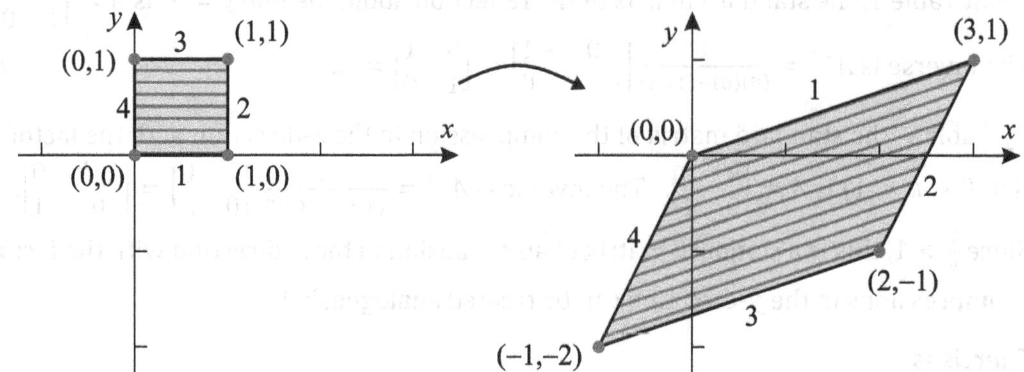

7. **(a)** We are looking for the standard matrix of $T = T_2 \circ T_1$ where T_1 is the compression by a factor of $\frac{1}{2}$ in the x-direction and T_2 is the expansion by a factor of 5 in the y-direction. From Table 1, $[T_1] = \begin{bmatrix}\frac{1}{2} & 0\\0 & 1\end{bmatrix}$ and $[T_2] = \begin{bmatrix}1 & 0\\0 & 5\end{bmatrix}$. Therefore, $[T] = [T_2][T_1] = \begin{bmatrix}\frac{1}{2} & 0\\0 & 5\end{bmatrix}$.

(b) We are looking for the standard matrix of $T = T_2 \circ T_1$ where T_1 is the expansion by a factor of 5 in the y-direction and T_2 is the shear by a factor of 2 in the y-direction. From Table 1, $[T_1] = \begin{bmatrix}1 & 0\\0 & 5\end{bmatrix}$ and $[T_2] = \begin{bmatrix}1 & 0\\2 & 1\end{bmatrix}$. Therefore, $[T] = [T_2][T_1] = \begin{bmatrix}1 & 0\\2 & 5\end{bmatrix}$.

(c) We are looking for the standard matrix of $T = T_2 \circ T_1$ where T_1 is the reflection about the line $y = x$ and T_2 is rotation through an angle of 180° about the origin . From Table 1, $[T_1] = \begin{bmatrix}0 & 1\\1 & 0\end{bmatrix}$ and $[T_2] = \begin{bmatrix}\cos 180° & -\sin 180°\\\sin 180° & \cos 180°\end{bmatrix} = \begin{bmatrix}-1 & 0\\0 & -1\end{bmatrix}$. Therefore, $[T] = [T_2][T_1] = \begin{bmatrix}0 & -1\\-1 & 0\end{bmatrix}$.

9. **(a)** From Table 1, T_1, a reflection about the x-axis has the standard matrix $[T_1] = \begin{bmatrix}1 & 0\\0 & -1\end{bmatrix}$ and T_2, a compression by a factor of $\frac{1}{3}$ in the x-direction has the standard matrix $[T_2] = \begin{bmatrix}\frac{1}{3} & 0\\0 & 1\end{bmatrix}$.

$[T_1 \circ T_2] = [T_1][T_2] = \begin{bmatrix}\frac{1}{3} & 0\\0 & -1\end{bmatrix}$; $[T_2 \circ T_1] = [T_2][T_1] = \begin{bmatrix}\frac{1}{3} & 0\\0 & -1\end{bmatrix}$.

Since $T_1 \circ T_2 = T_2 \circ T_1$, these operators commute.

(b) From Table 1, T_1, a reflection about the line $y = x$ has the standard matrix $[T_1] = \begin{bmatrix}0 & 1\\1 & 0\end{bmatrix}$ and T_2, an expansion by a factor of 2 in the x-direction has the standard matrix $[T_2] = \begin{bmatrix}2 & 0\\0 & 1\end{bmatrix}$.

$[T_1 \circ T_2] = [T_1][T_2] = \begin{bmatrix}0 & 1\\2 & 0\end{bmatrix}$; $[T_2 \circ T_1] = [T_2][T_1] = \begin{bmatrix}0 & 2\\1 & 0\end{bmatrix}$.

Since $T_1 \circ T_2 \neq T_2 \circ T_1$, these operators do not commute.

11. $A = \begin{bmatrix} 4 & 4 \\ 0 & -2 \end{bmatrix} \underset{\substack{\text{Multiply} \\ \text{the first} \\ \text{row by } \frac{1}{4}}}{\rightarrow} \begin{bmatrix} 1 & 1 \\ 0 & -2 \end{bmatrix} \underset{\substack{\text{Multiply} \\ \text{the second} \\ \text{row by } -\frac{1}{2}}}{\rightarrow} \begin{bmatrix} 1 & 1 \\ 0 & 1 \end{bmatrix} \underset{\substack{\text{Add} -1 \\ \text{times the} \\ \text{second} \\ \text{row to} \\ \text{the first} \\ \text{row}}}{\rightarrow} \begin{bmatrix} 1 & 0 \\ 0 & 1 \end{bmatrix}$

therefore $E_3E_2E_1A = I$ with $E_1 = \begin{bmatrix} \frac{1}{4} & 0 \\ 0 & 1 \end{bmatrix}$, $E_2 = \begin{bmatrix} 1 & 0 \\ 0 & -\frac{1}{2} \end{bmatrix}$, and $E_3 = \begin{bmatrix} 1 & -1 \\ 0 & 1 \end{bmatrix}$ so that

$A = E_1^{-1}E_2^{-1}E_3^{-1} = \begin{bmatrix} 4 & 0 \\ 0 & 1 \end{bmatrix}\begin{bmatrix} 1 & 0 \\ 0 & -2 \end{bmatrix}\begin{bmatrix} 1 & 1 \\ 0 & 1 \end{bmatrix}$.

Multiplication by A has the geometric effect of shearing by a factor of 1 in the x-direction, then reflection about the x-axis, then expanding by a factor of 2 in the y-direction, then expanding by a factor of 4 in the x-direction.

13. $A = \begin{bmatrix} 0 & -2 \\ 4 & 0 \end{bmatrix} \underset{\substack{\text{Interchange} \\ \text{the first row} \\ \text{and the} \\ \text{second row}}}{\rightarrow} \begin{bmatrix} 4 & 0 \\ 0 & -2 \end{bmatrix} \underset{\substack{\text{Multiply} \\ \text{the first} \\ \text{row by } \frac{1}{4}}}{\rightarrow} \begin{bmatrix} 1 & 0 \\ 0 & -2 \end{bmatrix} \underset{\substack{\text{Multiply} \\ \text{the second} \\ \text{row by } -\frac{1}{2}}}{\rightarrow} \begin{bmatrix} 1 & 0 \\ 0 & 1 \end{bmatrix}$

therefore $E_3E_2E_1A = I$ with $E_1 = \begin{bmatrix} 0 & 1 \\ 1 & 0 \end{bmatrix}$, $E_2 = \begin{bmatrix} \frac{1}{4} & 0 \\ 0 & 1 \end{bmatrix}$, and $E_3 = \begin{bmatrix} 1 & 0 \\ 0 & -\frac{1}{2} \end{bmatrix}$ so that

$A = E_1^{-1}E_2^{-1}E_3^{-1} = \begin{bmatrix} 0 & 1 \\ 1 & 0 \end{bmatrix}\begin{bmatrix} 4 & 0 \\ 0 & 1 \end{bmatrix}\begin{bmatrix} 1 & 0 \\ 0 & -2 \end{bmatrix}$.

Multiplication by A has the geometric effect of reflection about the x-axis, then expanding by a factor of 2 in the y-direction, then expanding by a factor of 4 in the x-direction, then reflection about the line $y = x$.

15. **(a)** The unit square is expanded in the x-direction by a factor of 3.

(b) The unit square is reflected about the x-axis and expanded in the y-direction by a factor of 5.

17. **(a)** Coordinates (a, b) are being transformed to coordinates (x, y) according to the equation $\begin{bmatrix} x \\ y \end{bmatrix} = \begin{bmatrix} 3 & 1 \\ 6 & 2 \end{bmatrix}\begin{bmatrix} a \\ b \end{bmatrix} = \begin{bmatrix} 3a + b \\ 6a + 2b \end{bmatrix}$. It follows that $x = 3a + b$ and $y = 6a + 2b$ satisfy the equation of the line $y = 2x$ (since $6a + 2b = 2(3a + b)$).

(b) $\begin{vmatrix} 3 & 1 \\ 6 & 2 \end{vmatrix} = 0$ so $\begin{bmatrix} 3 & 1 \\ 6 & 2 \end{bmatrix}$ is not invertible; Theorem 4.11.1 applies only to invertible matrices.

19. Coordinates (x, y) are being transformed to coordinates (x', y') according to the equation $\begin{bmatrix} x' \\ y' \end{bmatrix} = \begin{bmatrix} 3 & 2 \\ 1 & 1 \end{bmatrix}\begin{bmatrix} x \\ y \end{bmatrix}$. Thus $\begin{bmatrix} x \\ y \end{bmatrix} = \begin{bmatrix} 3 & 2 \\ 1 & 1 \end{bmatrix}^{-1}\begin{bmatrix} x' \\ y' \end{bmatrix} = \begin{bmatrix} 1 & -2 \\ -1 & 3 \end{bmatrix}\begin{bmatrix} x' \\ y' \end{bmatrix}$ so $x = x' - 2y'$ and $y = -x' + 3y'$.

Substituting these into $y = 3x + 1$ yields $-x' + 3y' = 3(x' - 2y') + 1$ or equivalently $y' = \frac{4}{9}x' + \frac{1}{9}$.

Substituting $x = x' - 2y'$ and $y = -x' + 3y'$ into $y = 3x - 2$ yields $-x' + 3y' = 3(x' - 2y') - 2$ or equivalently $y' = \frac{4}{9}x' - \frac{2}{9}$. Since both lines we obtained in the (x', y') coordinates have the same slope $(4/9)$, we conclude that the given parallel lines are mapped into parallel lines.

21. **(a)** We obtain $\begin{bmatrix}1 & -1\\1 & 1\end{bmatrix}\begin{bmatrix}0\\0\end{bmatrix}=\begin{bmatrix}0\\0\end{bmatrix}$, $\begin{bmatrix}1 & -1\\1 & 1\end{bmatrix}\begin{bmatrix}1\\0\end{bmatrix}=\begin{bmatrix}1\\1\end{bmatrix}$, and $\begin{bmatrix}1 & -1\\1 & 1\end{bmatrix}\begin{bmatrix}0.5\\1\end{bmatrix}=\begin{bmatrix}-0.5\\1.5\end{bmatrix}$.

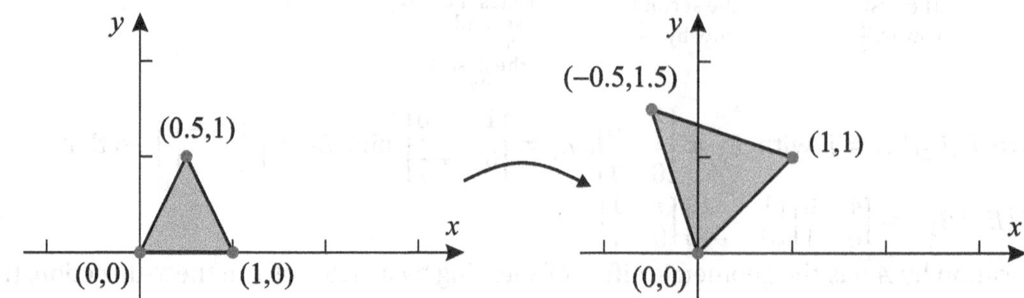

(b) $A=\begin{bmatrix}1 & -1\\1 & 1\end{bmatrix} \rightarrow \begin{bmatrix}1 & -1\\0 & 2\end{bmatrix} \rightarrow \begin{bmatrix}1 & -1\\0 & 1\end{bmatrix} \rightarrow \begin{bmatrix}1 & 0\\0 & 1\end{bmatrix}$

(Add −1 times the first row to the second row; Multiply the second row by $\frac{1}{2}$; Add the second row to the first row)

therefore $E_3E_2E_1A=I$ with $E_1=\begin{bmatrix}1 & 0\\-1 & 1\end{bmatrix}$, $E_2=\begin{bmatrix}1 & 0\\0 & \frac{1}{2}\end{bmatrix}$, and $E_3=\begin{bmatrix}1 & 1\\0 & 1\end{bmatrix}$ so that $A=E_1^{-1}E_2^{-1}E_3^{-1}=\begin{bmatrix}1 & 0\\1 & 1\end{bmatrix}\begin{bmatrix}1 & 0\\0 & 2\end{bmatrix}\begin{bmatrix}1 & -1\\0 & 1\end{bmatrix}$.

Shearing by a factor of −1 in the x-direction, then expanding by a factor of 2 in the y-direction, then shearing by a factor of 1 in the y-direction will produce the same image as in part (a).

23. We calculate the positions of corners of the image of the unit square in which the figure is inscribed:

$\begin{bmatrix}1 & \frac{1}{4}\\0 & 1\end{bmatrix}\begin{bmatrix}0\\0\end{bmatrix}=\begin{bmatrix}0\\0\end{bmatrix}$, $\begin{bmatrix}1 & \frac{1}{4}\\0 & 1\end{bmatrix}\begin{bmatrix}1\\0\end{bmatrix}=\begin{bmatrix}1\\0\end{bmatrix}$, $\begin{bmatrix}1 & \frac{1}{4}\\0 & 1\end{bmatrix}\begin{bmatrix}0\\1\end{bmatrix}=\begin{bmatrix}\frac{1}{4}\\1\end{bmatrix}$, and $\begin{bmatrix}1 & \frac{1}{4}\\0 & 1\end{bmatrix}\begin{bmatrix}1\\1\end{bmatrix}=\begin{bmatrix}\frac{5}{4}\\1\end{bmatrix}$.

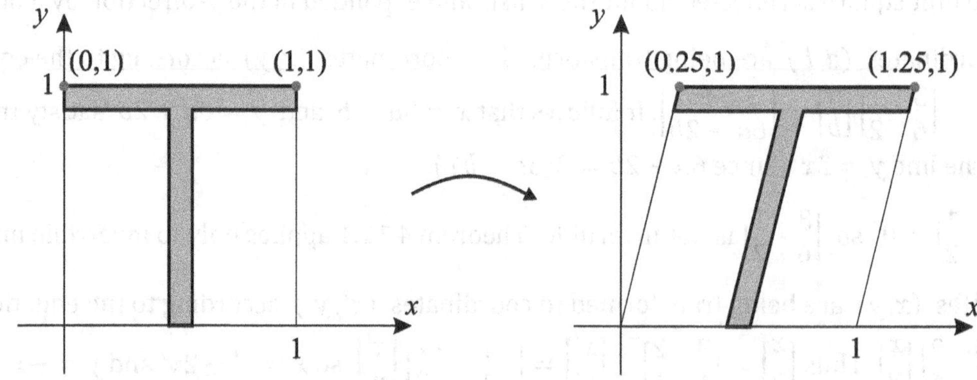

(The following calculations determine positions of the remaining endpoints of segments comprising the outline of the figure: $\begin{bmatrix}1 & \frac{1}{4}\\0 & 1\end{bmatrix}\begin{bmatrix}0.45\\0\end{bmatrix}=\begin{bmatrix}0.45\\0\end{bmatrix}$, $\begin{bmatrix}1 & \frac{1}{4}\\0 & 1\end{bmatrix}\begin{bmatrix}0.55\\0\end{bmatrix}=\begin{bmatrix}0.55\\0\end{bmatrix}$, $\begin{bmatrix}1 & \frac{1}{4}\\0 & 1\end{bmatrix}\begin{bmatrix}0\\0.95\end{bmatrix}=\begin{bmatrix}0.2375\\0.95\end{bmatrix}$, $\begin{bmatrix}1 & \frac{1}{4}\\0 & 1\end{bmatrix}\begin{bmatrix}0.45\\0.95\end{bmatrix}=\begin{bmatrix}0.6875\\0.95\end{bmatrix}$, $\begin{bmatrix}1 & \frac{1}{4}\\0 & 1\end{bmatrix}\begin{bmatrix}0.55\\0.95\end{bmatrix}=\begin{bmatrix}0.7875\\0.95\end{bmatrix}$, and $\begin{bmatrix}1 & \frac{1}{4}\\0 & 1\end{bmatrix}\begin{bmatrix}1\\0.95\end{bmatrix}=\begin{bmatrix}1.2375\\0.95\end{bmatrix}$.)

25. Since $\begin{bmatrix}2 & -1\\0 & 0\end{bmatrix}\begin{bmatrix}0\\0\end{bmatrix}=\begin{bmatrix}0\\0\end{bmatrix}$, $\begin{bmatrix}2 & -1\\0 & 0\end{bmatrix}\begin{bmatrix}1\\1\end{bmatrix}=\begin{bmatrix}1\\0\end{bmatrix}$, and $\begin{bmatrix}2 & -1\\0 & 0\end{bmatrix}\begin{bmatrix}2\\0\end{bmatrix}=\begin{bmatrix}2\\0\end{bmatrix}$, the image of the given triangle is the line segment from (0,0) to (2,0). Theorem 4.11.1 does not apply here because A is singular.

True-False Exercises

(a) False. The image is a parallelogram.

(b) True. This is a consequence of Theorem 4.11.2.

(c) True. This is the statement of part (a) of Theorem 4.11.1.

(d) True. Performing the same reflection twice amounts to no change (identity transformation).

(e) False. The matrix represents a composition of a reflection and a dilation.

(f) False. This matrix does not represent a shear in either x or y direction.

(g) True. This matrix represents an expansion by a factor of 3 in the y-direction.

Chapter 4 Supplementary Exercises

1. **(a)** $\mathbf{u}+\mathbf{v}=(3+1,-2+5,4-2)=(4,3,2)$; $k\mathbf{u}=(-1\cdot 3,0,0)=(-3,0,0)$

(b) For any $\mathbf{u}=(u_1,u_2,u_3)$ and $\mathbf{v}=(v_1,v_2,v_3)$ in V, $\mathbf{u}+\mathbf{v}=(u_1+v_1,u_2+v_2,u_3+v_3)$ is an ordered triple of real numbers, therefore $\mathbf{u}+\mathbf{v}$ is in V. Consequently, V is closed under addition.

For any $\mathbf{u}=(u_1,u_2,u_3)$ in V and for any scalar k, $k\mathbf{u}=(ku_1,0,0)$ is an ordered triple of real numbers, therefore $k\mathbf{u}$ is in V. Consequently, V is closed under scalar multiplication.

(c) Axioms 1-5 hold for V because they are known to hold for R^3.

(d) Axiom 7: $k\big((u_1,u_2,u_3)+(v_1,v_2,v_3)\big)=k(u_1+v_1,u_2+v_2,u_3+v_3)=(k(u_1+v_1),0,0)=k(u_1,u_2,u_3)+k(v_1,v_2,v_3)$ for all real k, u_1,u_2,u_3, v_1,v_2, and v_3.

Axiom 8: $(k+m)(u_1,u_2,u_3)=\big((k+m)u_1,0,0\big)=(ku_1+mu_1,0,0)=k(u_1,u_2,u_3)+m(u_1,u_2,u_3)$ for all real k, m, u_1, u_2, and u_3;

Axiom 9: $k\big(m(u_1,u_2,u_3)\big)=k(mu_1,0,0)=(kmu_1,0,0)=(km)(u_1,u_2,u_3)$ for all real k, m, u_1,u_2, and u_3;

(e) Axiom 10 fails to hold: $1(u_1,u_2,u_3)=(u_1,0,0)$ does not generally equal (u_1,u_2,u_3). Consequently, V is not a vector space.

3. $$A=\begin{bmatrix}1 & 1 & s\\1 & s & 1\\s & 1 & 1\end{bmatrix}$$ ← The coefficient matrix of the system

$$\begin{bmatrix}1 & 1 & s\\0 & s-1 & 1-s\\0 & 1-s & 1-s^2\end{bmatrix}$$ ← -1 times the first row was added to the second row and $-s$ times the first row was added to the third row.

$$\begin{bmatrix} 1 & 1 & s \\ 0 & s-1 & 1-s \\ 0 & 0 & 2-s-s^2 \end{bmatrix}$$ ← The second row was added to the third row.

After factoring $2-s-s^2=(2+s)(1-s)$, we conclude that

- the solution space is a plane through the origin if $s=1$ (the reduced row echelon form becomes $\begin{bmatrix} 1 & 1 & 1 \\ 0 & 0 & 0 \\ 0 & 0 & 0 \end{bmatrix}$, so nullity$(A)=2$),
- the solution space is a line through the origin if $s=-2$ (the reduced row echelon form becomes $\begin{bmatrix} 1 & 0 & -1 \\ 0 & 1 & -1 \\ 0 & 0 & 0 \end{bmatrix}$, so nullity$(A)=1$),
- the solution space is the origin if $s\neq-2$ and $s\neq1$ (the reduced row echelon form becomes $\begin{bmatrix} 1 & 0 & 0 \\ 0 & 1 & 0 \\ 0 & 0 & 1 \end{bmatrix}$, so nullity$(A)=0$),
- there are no values of s for which the solution space is R^3.

5. **(a)** Using trigonometric identities we can write

$$\mathbf{f}_1=\sin(x+\theta)=\sin x\cos\theta+\cos x\sin\theta=(\cos\theta)\mathbf{f}+(\sin\theta)\mathbf{g}$$
$$\mathbf{g}_1=\cos(x+\theta)=\cos x\cos\theta-\sin x\sin\theta=(-\sin\theta)\mathbf{f}+(\cos\theta)\mathbf{g}$$

which shows that $\mathbf{f}_1$ and $\mathbf{g}_1$ are both in $W=\text{span}\{\mathbf{f},\mathbf{g}\}$.

(b) The functions $\mathbf{f}_1=\sin(x+\theta)$ and $\mathbf{f}_2=\cos(x+\theta)$ are linearly independent since neither function is a scalar multiple of the other. By Theorem 4.5.4, these functions form a basis for W.

7. Denoting $B=[\ \mathbf{v}_1\ \mid\ \cdots\ \mid\ \mathbf{v}_n\ ]$ we can write $AB=[\ A\mathbf{v}_1\ \mid\ \cdots\ \mid\ A\mathbf{v}_n\ ]$.
By parts (g) and (h) of Theorem 4.8.7, the columns of AB are linearly independent if and only if $\det(AB)\neq0$. This implies that $\det(A)\neq0$, i.e., the matrix A must be invertible.

9. **(a)** The reduced row echelon form of $\begin{bmatrix} 1 & 0 & 1 \\ 0 & 1 & 0 \\ 1 & 0 & 1 \end{bmatrix}$ is $\begin{bmatrix} 1 & 0 & 1 \\ 0 & 1 & 0 \\ 0 & 0 & 0 \end{bmatrix}$, so the rank is 2 and the nullity is 1.

(b) The reduced row echelon form of $\begin{bmatrix} 1 & 0 & 1 & 0 \\ 0 & 1 & 0 & 1 \\ 1 & 0 & 1 & 0 \\ 0 & 1 & 0 & 1 \end{bmatrix}$ is $\begin{bmatrix} 1 & 0 & 1 & 0 \\ 0 & 1 & 0 & 1 \\ 0 & 0 & 0 & 0 \\ 0 & 0 & 0 & 0 \end{bmatrix}$, so the rank is 2 and the nullity is 2.

(c) For $n=1$, the rank is 1 and the nullity is 0.
For $n\geq2$, the reduced row echelon form will always have two nonzero rows; the rank is 2 and the nullity is $n-2$.

11. **(a)** Let W be the set of all polynomials p in P_n for which $p(-x)=p(x)$. In order for a polynomial $p(x)=a_0+a_1x+a_2x^2+\cdots+a_nx^n$ to be in W, we must have
$p(x)=a_0+a_1x+a_2x^2+\cdots+a_nx^n=a_0+a_1(-x)+a_2(-x)^2+\cdots+a_n(-x)^n=p(-x)$ which implies that for all x, $2a_1x+2a_3x^3+\cdots=0$ so $a_1=a_3=\cdots=0$.
Any polynomial of the form $p(x)=a_0+a_2x^2+a_4x^4+\cdots+a_{2\lfloor n/2\rfloor}x^{2\lfloor n/2\rfloor}$ satisfies

$p(-x) = p(x)$ (the notation $\lfloor t \rfloor$ represents the largest integer less than or equal to t).
This means $W = \text{span}\{1, x^2, x^4, \ldots, x^{2\lfloor n/2 \rfloor}\}$, so W is a subspace of P_n by Theorem 4.2.3(a).
The polynomials in $\{1, x^2, x^4, \ldots, x^{2\lfloor n/2 \rfloor}\}$ are linearly independent (since they form a subset of the standard basis for P_n), consequently they form a basis for W.

(b) Let W be the set of all polynomials p in P_n for which $p(0) = p(1)$.
In order for a polynomial $p(x) = a_0 + a_1x + a_2x^2 + \cdots + a_nx^n$ to be in W, we must have $p(0) = a_0 = a_0 + a_1 + a_2 + \cdots + a_n = p(1)$ which implies that $a_1 + a_2 + \cdots + a_n = 0$.
Therefore any polynomial in W can be expressed as
$p(x) = a_0 + a_1x + a_2x^2 + \cdots + a_{n-1}x^{n-1} + (-a_1 - a_2 - \cdots - a_{n-1})x^n$
$= a_0 + a_1(x - x^n) + a_2(x^2 - x^n) + \cdots + a_{n-1}(x^{n-1} - x^n)$.
This means $W = \text{span}\{1, x - x^n, x^2 - x^n, \ldots, x^{n-1} - x^n\}$, so W is a subspace of P_n by Theorem 4.2.3(a). Since $a_0 + a_1(x - x^n) + a_2(x^2 - x^n) + \cdots + a_{n-1}(x^{n-1} - x^n) = 0$ implies $a_0 = a_1 = a_2 = \cdots = a_{n-1} = 0$, it follows that $\{1, x - x^n, x^2 - x^n, \ldots, x^{n-1} - x^n\}$ is linearly independent, hence it is a basis for W.

13. (a) A general 3×3 symmetric matrix can be expressed as $\begin{bmatrix} a & b & c \\ b & d & e \\ c & e & f \end{bmatrix}$

$$= a\begin{bmatrix} 1 & 0 & 0 \\ 0 & 0 & 0 \\ 0 & 0 & 0 \end{bmatrix} + b\begin{bmatrix} 0 & 1 & 0 \\ 1 & 0 & 0 \\ 0 & 0 & 0 \end{bmatrix} + c\begin{bmatrix} 0 & 0 & 1 \\ 0 & 0 & 0 \\ 1 & 0 & 0 \end{bmatrix} + d\begin{bmatrix} 0 & 0 & 0 \\ 0 & 1 & 0 \\ 0 & 0 & 0 \end{bmatrix} + e\begin{bmatrix} 0 & 0 & 0 \\ 0 & 0 & 1 \\ 0 & 1 & 0 \end{bmatrix} + f\begin{bmatrix} 0 & 0 & 0 \\ 0 & 0 & 0 \\ 0 & 0 & 1 \end{bmatrix}.$$

Clearly the matrices $\begin{bmatrix} 1 & 0 & 0 \\ 0 & 0 & 0 \\ 0 & 0 & 0 \end{bmatrix}, \begin{bmatrix} 0 & 1 & 0 \\ 1 & 0 & 0 \\ 0 & 0 & 0 \end{bmatrix}, \begin{bmatrix} 0 & 0 & 1 \\ 0 & 0 & 0 \\ 1 & 0 & 0 \end{bmatrix}, \begin{bmatrix} 0 & 0 & 0 \\ 0 & 1 & 0 \\ 0 & 0 & 0 \end{bmatrix}, \begin{bmatrix} 0 & 0 & 0 \\ 0 & 0 & 1 \\ 0 & 1 & 0 \end{bmatrix}, \begin{bmatrix} 0 & 0 & 0 \\ 0 & 0 & 0 \\ 0 & 0 & 1 \end{bmatrix}$ span the space of all 3×3 symmetric matrices. Also, these matrices are linearly indpendent, since $\begin{bmatrix} a & b & c \\ b & d & e \\ c & e & f \end{bmatrix} = \begin{bmatrix} 0 & 0 & 0 \\ 0 & 0 & 0 \\ 0 & 0 & 0 \end{bmatrix}$ requires that all six coefficients in the linear combination above must be zero. We conclude that the matrices $\begin{bmatrix} 1 & 0 & 0 \\ 0 & 0 & 0 \\ 0 & 0 & 0 \end{bmatrix}, \begin{bmatrix} 0 & 1 & 0 \\ 1 & 0 & 0 \\ 0 & 0 & 0 \end{bmatrix}, \begin{bmatrix} 0 & 0 & 1 \\ 0 & 0 & 0 \\ 1 & 0 & 0 \end{bmatrix}, \begin{bmatrix} 0 & 0 & 0 \\ 0 & 1 & 0 \\ 0 & 0 & 0 \end{bmatrix}$, $\begin{bmatrix} 0 & 0 & 0 \\ 0 & 0 & 1 \\ 0 & 1 & 0 \end{bmatrix}$, and $\begin{bmatrix} 0 & 0 & 0 \\ 0 & 0 & 0 \\ 0 & 0 & 1 \end{bmatrix}$ form a basis for the space of all 3×3 symmetric matrices.

(b) A general 3×3 skew-symmetric matrix can be expressed as

$$\begin{bmatrix} 0 & a & b \\ -a & 0 & c \\ -b & -c & 0 \end{bmatrix} = a\begin{bmatrix} 0 & 1 & 0 \\ -1 & 0 & 0 \\ 0 & 0 & 0 \end{bmatrix} + b\begin{bmatrix} 0 & 0 & 1 \\ 0 & 0 & 0 \\ -1 & 0 & 0 \end{bmatrix} + c\begin{bmatrix} 0 & 0 & 0 \\ 0 & 0 & 1 \\ 0 & -1 & 0 \end{bmatrix}.$$

Clearly the matrices $\begin{bmatrix} 0 & 1 & 0 \\ -1 & 0 & 0 \\ 0 & 0 & 0 \end{bmatrix}, \begin{bmatrix} 0 & 0 & 1 \\ 0 & 0 & 0 \\ -1 & 0 & 0 \end{bmatrix}, \begin{bmatrix} 0 & 0 & 0 \\ 0 & 0 & 1 \\ 0 & -1 & 0 \end{bmatrix}$ span the space of all 3×3 skew-symmetric matrices. Also, these matrices are linearly indpendent, since

$$\begin{bmatrix} 0 & a & b \\ -a & 0 & c \\ -b & -c & 0 \end{bmatrix} = \begin{bmatrix} 0 & 0 & 0 \\ 0 & 0 & 0 \\ 0 & 0 & 0 \end{bmatrix}$$

requires that all three coefficients in the linear combination above must be zero. We conclude that the matrices $\begin{bmatrix} 0 & 1 & 0 \\ -1 & 0 & 0 \\ 0 & 0 & 0 \end{bmatrix}$, $\begin{bmatrix} 0 & 0 & 1 \\ 0 & 0 & 0 \\ -1 & 0 & 0 \end{bmatrix}$, and $\begin{bmatrix} 0 & 0 & 0 \\ 0 & 0 & 1 \\ 0 & -1 & 0 \end{bmatrix}$ form a basis for the space of all 3×3 skew-symmetric matrices.

15. All submatrices of size 3×3 or larger contain at least two rows that are scalar multiples of each other, so their determinants are 0. Therefore the rank cannot exceed 2. The possible values are:

- $\text{rank}(A) = 2$, e.g., if $a_{51} = a_{16} = 1$ regardless of the other values,
- $\text{rank}(A) = 1$, e.g., if $a_{16} = a_{26} = a_{36} = a_{46} = 0$ and $a_{56} = 1$ regardless of the other values, and
- $\text{rank}(A) = 0$ if all entries are 0.

17. The standard matrices for D_k, R_θ, and S_k are $\begin{bmatrix} k & 0 \\ 0 & k \end{bmatrix}$, $\begin{bmatrix} \cos\theta & -\sin\theta \\ \sin\theta & \cos\theta \end{bmatrix}$, and $\begin{bmatrix} 1 & k \\ 0 & 1 \end{bmatrix}$ (assuming a shear in the x-direction).

(a) $\begin{bmatrix} k & 0 \\ 0 & k \end{bmatrix}\begin{bmatrix} \cos\theta & -\sin\theta \\ \sin\theta & \cos\theta \end{bmatrix} = \begin{bmatrix} k\cos\theta & -k\sin\theta \\ k\sin\theta & k\cos\theta \end{bmatrix} = \begin{bmatrix} \cos\theta & -\sin\theta \\ \sin\theta & \cos\theta \end{bmatrix}\begin{bmatrix} k & 0 \\ 0 & k \end{bmatrix}$ therefore D_k and R_θ commute.

(b) $\begin{bmatrix} \cos\theta & -\sin\theta \\ \sin\theta & \cos\theta \end{bmatrix}\begin{bmatrix} 1 & k \\ 0 & 1 \end{bmatrix} = \begin{bmatrix} \cos\theta & k\cos\theta - \sin\theta \\ \sin\theta & k\sin\theta + \cos\theta \end{bmatrix}$ does not generally equal
$\begin{bmatrix} 1 & k \\ 0 & 1 \end{bmatrix}\begin{bmatrix} \cos\theta & -\sin\theta \\ \sin\theta & \cos\theta \end{bmatrix} = \begin{bmatrix} \cos\theta + k\sin\theta & -\sin\theta + k\cos\theta \\ \sin\theta & \cos\theta \end{bmatrix}$
therefore R_θ and S_k do not commute (same result is obtained if a shear in the y-direction is taken instead)

(c) $\begin{bmatrix} k & 0 \\ 0 & k \end{bmatrix}\begin{bmatrix} 1 & k \\ 0 & 1 \end{bmatrix} = \begin{bmatrix} k & k^2 \\ 0 & k \end{bmatrix} = \begin{bmatrix} 1 & k \\ 0 & 1 \end{bmatrix}\begin{bmatrix} k & 0 \\ 0 & k \end{bmatrix}$ therefore D_k and S_k commute (same result is obtained if a shear in the y-direction is taken instead)

CHAPTER 5: EIGENVALUES AND EIGENVECTORS

5.1 Eigenvalues and Eigenvectors

1. $A\mathbf{x} = \begin{bmatrix} 1 & 2 \\ 3 & 2 \end{bmatrix}\begin{bmatrix} 1 \\ -1 \end{bmatrix} = \begin{bmatrix} -1 \\ 1 \end{bmatrix} = -1\mathbf{x}$ therefore $\mathbf{x}$ is an eigenvector of A corresponding to the eigenvalue -1.

3. $A\mathbf{x} = \begin{bmatrix} 4 & 0 & 1 \\ 2 & 3 & 2 \\ 1 & 0 & 4 \end{bmatrix}\begin{bmatrix} 1 \\ 2 \\ 1 \end{bmatrix} = \begin{bmatrix} 5 \\ 10 \\ 5 \end{bmatrix} = 5\mathbf{x}$ therefore $\mathbf{x}$ is an eigenvector of A corresponding to the eigenvalue 5.

5. **(a)** $\det(\lambda I - A) = \begin{vmatrix} \lambda - 1 & -4 \\ -2 & \lambda - 3 \end{vmatrix} = (\lambda - 1)(\lambda - 3) - (-4)(-2) = \lambda^2 - 4\lambda - 5 = (\lambda - 5)(\lambda + 1)$.

The characteristic equation is $(\lambda - 5)(\lambda + 1) = 0$. The eigenvalues are $\lambda = 5$ and $\lambda = -1$.

The reduced row echelon form of $5I - A = \begin{bmatrix} 4 & -4 \\ -2 & 2 \end{bmatrix}$ is $\begin{bmatrix} 1 & -1 \\ 0 & 0 \end{bmatrix}$. The general solution of $(5I - A)\mathbf{x} = \mathbf{0}$ is $x_1 = t, x_2 = t$. In vector form, $\begin{bmatrix} x_1 \\ x_2 \end{bmatrix} = \begin{bmatrix} t \\ t \end{bmatrix} = t\begin{bmatrix} 1 \\ 1 \end{bmatrix}$.

A basis for the eigenspace corresponding to $\lambda = 5$ is $\{(1,1)\}$.

The reduced row echelon form of $-1I - A = \begin{bmatrix} -2 & -4 \\ -2 & -4 \end{bmatrix}$ is $\begin{bmatrix} 1 & 2 \\ 0 & 0 \end{bmatrix}$. The general solution of $(-1I - A)\mathbf{x} = \mathbf{0}$ is $x_1 = -2t, x_2 = t$. In vector form, $\begin{bmatrix} x_1 \\ x_2 \end{bmatrix} = \begin{bmatrix} -2t \\ t \end{bmatrix} = t\begin{bmatrix} -2 \\ 1 \end{bmatrix}$.

A basis for the eigenspace corresponding to $\lambda = -1$ is $\{(-2,1)\}$.

(b) $\det(\lambda I - A) = \begin{vmatrix} \lambda + 2 & 7 \\ -1 & \lambda - 2 \end{vmatrix} = (\lambda + 2)(\lambda - 2) - (7)(-1) = \lambda^2 + 3$.

The characteristic equation is $\lambda^2 + 3 = 0$. There are no real eigenvalues.

(c) $\det(\lambda I - A) = \begin{vmatrix} \lambda - 1 & 0 \\ 0 & \lambda - 1 \end{vmatrix} = (\lambda - 1)^2$.

The characteristic equation is $(\lambda - 1)^2 = 0$. The eigenvalue is $\lambda = 1$.

The matrix $I - A = \begin{bmatrix} 0 & 0 \\ 0 & 0 \end{bmatrix}$ is already in reduced row echelon form. The general solution of $(I - A)\mathbf{x} = \mathbf{0}$ is $x_1 = s, x_2 = t$. In vector form, $\begin{bmatrix} x_1 \\ x_2 \end{bmatrix} = \begin{bmatrix} s \\ t \end{bmatrix} = s\begin{bmatrix} 1 \\ 0 \end{bmatrix} + t\begin{bmatrix} 0 \\ 1 \end{bmatrix}$.

A basis for the eigenspace corresponding to $\lambda = 1$ is $\{(1,0), (0,1)\}$.

(d) $\det(\lambda I - A) = \begin{vmatrix} \lambda - 1 & 2 \\ 0 & \lambda - 1 \end{vmatrix} = (\lambda - 1)^2$.

The characteristic equation is $(\lambda - 1)^2 = 0$. The eigenvalue is $\lambda = 1$.

The reduced row echelon form of $I - A = \begin{bmatrix} 0 & 2 \\ 0 & 0 \end{bmatrix}$ is $\begin{bmatrix} 0 & 1 \\ 0 & 0 \end{bmatrix}$. The general solution of $(I - A)\mathbf{x} = \mathbf{0}$ is $x_1 = t, x_2 = 0$. In vector form, $\begin{bmatrix} x_1 \\ x_2 \end{bmatrix} = \begin{bmatrix} t \\ 0 \end{bmatrix} = t\begin{bmatrix} 1 \\ 0 \end{bmatrix}$.

A basis for the eigenspace corresponding to $\lambda = 1$ is $\{(1,0)\}$.

7. Cofactor expansion along the second column yields $\det(\lambda I - A) = \begin{vmatrix} \lambda-4 & 0 & -1 \\ 2 & \lambda-1 & 0 \\ 2 & 0 & \lambda-1 \end{vmatrix}$

$= (\lambda-1)\begin{vmatrix} \lambda-4 & -1 \\ 2 & \lambda-1 \end{vmatrix} = (\lambda-1)[(\lambda-4)(\lambda-1)-(-1)(2)] = (\lambda-1)(\lambda^2-5\lambda+6)$
$=(\lambda-1)(\lambda-2)(\lambda-3)$. The characteristic equation is $(\lambda-1)(\lambda-2)(\lambda-3)=0$.
The eigenvalues are $\lambda = 1$, $\lambda = 2$, and $\lambda = 3$.

The reduced row echelon form of $I - A = \begin{bmatrix} -3 & 0 & -1 \\ 2 & 0 & 0 \\ 2 & 0 & 0 \end{bmatrix}$ is $\begin{bmatrix} 1 & 0 & 0 \\ 0 & 0 & 1 \\ 0 & 0 & 0 \end{bmatrix}$. The general solution of $(I - A)\mathbf{x} = \mathbf{0}$ is $x_1 = 0, x_2 = t, x_3 = 0$. In vector form, $\begin{bmatrix} x_1 \\ x_2 \\ x_3 \end{bmatrix} = \begin{bmatrix} 0 \\ t \\ 0 \end{bmatrix} = t\begin{bmatrix} 0 \\ 1 \\ 0 \end{bmatrix}$. A basis for the eigenspace corresponding to $\lambda = 1$ is $\{(0,1,0)\}$.

The reduced row echelon form of $2I - A = \begin{bmatrix} -2 & 0 & -1 \\ 2 & 1 & 0 \\ 2 & 0 & 1 \end{bmatrix}$ is $\begin{bmatrix} 1 & 0 & \frac{1}{2} \\ 0 & 1 & -1 \\ 0 & 0 & 0 \end{bmatrix}$. The general solution of $(2I - A)\mathbf{x} = \mathbf{0}$ is $x_1 = -\frac{1}{2}t, x_2 = t, x_3 = t$. In vector form, $\begin{bmatrix} x_1 \\ x_2 \\ x_3 \end{bmatrix} = \begin{bmatrix} -\frac{1}{2}t \\ t \\ t \end{bmatrix} = t\begin{bmatrix} -\frac{1}{2} \\ 1 \\ 1 \end{bmatrix}$. A basis for the eigenspace corresponding to $\lambda = 2$ is $\{(-1,2,2)\}$ (scaled by a factor of 2 for convenience).

The reduced row echelon form of $3I - A = \begin{bmatrix} -1 & 0 & -1 \\ 2 & 2 & 0 \\ 2 & 0 & 2 \end{bmatrix}$ is $\begin{bmatrix} 1 & 0 & 1 \\ 0 & 1 & -1 \\ 0 & 0 & 0 \end{bmatrix}$. The general solution of $(3I - A)\mathbf{x} = \mathbf{0}$ is $x_1 = -t, x_2 = t, x_3 = t$. In vector form, $\begin{bmatrix} x_1 \\ x_2 \\ x_3 \end{bmatrix} = \begin{bmatrix} -t \\ t \\ t \end{bmatrix} = t\begin{bmatrix} -1 \\ 1 \\ 1 \end{bmatrix}$. A basis for the eigenspace corresponding to $\lambda = 3$ is $\{(-1,1,1)\}$.

9. Cofactor expansion along the second row yields $\det(\lambda I - A) = \begin{vmatrix} \lambda-6 & -3 & 8 \\ 0 & \lambda+2 & 0 \\ -1 & 0 & \lambda+3 \end{vmatrix}$

$= (\lambda+2)\begin{vmatrix} \lambda-6 & 8 \\ -1 & \lambda+3 \end{vmatrix} = (\lambda+2)[(\lambda-6)(\lambda+3)-(8)(-1)] = (\lambda+2)(\lambda^2-3\lambda-10)$
$=(\lambda+2)(\lambda+2)(\lambda-5)$. The characteristic equation is $(\lambda+2)^2(\lambda-5)=0$.
The eigenvalues are $\lambda = -2$ and $\lambda = 5$.

The reduced row echelon form of $-2I - A = \begin{bmatrix} -8 & -3 & 8 \\ 0 & 0 & 0 \\ -1 & 0 & 1 \end{bmatrix}$ is $\begin{bmatrix} 1 & 0 & -1 \\ 0 & 1 & 0 \\ 0 & 0 & 0 \end{bmatrix}$. The general solution of $(-2I - A)\mathbf{x} = \mathbf{0}$ is $x_1 = t, x_2 = 0, x_3 = t$. In vector form, $\begin{bmatrix} x_1 \\ x_2 \\ x_3 \end{bmatrix} = \begin{bmatrix} t \\ 0 \\ t \end{bmatrix} = t\begin{bmatrix} 1 \\ 0 \\ 1 \end{bmatrix}$. A basis for the eigenspace corresponding to $\lambda = -2$ is $\{(1,0,1)\}$.

The reduced row echelon form of $5I - A = \begin{bmatrix} -1 & -3 & 8 \\ 0 & 7 & 0 \\ -1 & 0 & 8 \end{bmatrix}$ is $\begin{bmatrix} 1 & 0 & -8 \\ 0 & 1 & 0 \\ 0 & 0 & 0 \end{bmatrix}$. The general solution of $(5I - A)\mathbf{x} = \mathbf{0}$ is $x_1 = 8t, x_2 = 0, x_3 = t$. In vector form, $\begin{bmatrix} x_1 \\ x_2 \\ x_3 \end{bmatrix} = \begin{bmatrix} 8t \\ 0 \\ t \end{bmatrix} = t\begin{bmatrix} 8 \\ 0 \\ 1 \end{bmatrix}$. A basis for the eigenspace corresponding to $\lambda = 5$ is $\{(8,0,1)\}$.

11. Cofactor expansion along the second column yields $\det(\lambda I - A) = \begin{vmatrix} \lambda-4 & 0 & 1 \\ 0 & \lambda-3 & 0 \\ -1 & 0 & \lambda-2 \end{vmatrix}$

$= (\lambda-3)\begin{vmatrix} \lambda-4 & 1 \\ -1 & \lambda-2 \end{vmatrix} = (\lambda-3)[(\lambda-4)(\lambda-2) - (1)(-1)] = (\lambda-3)(\lambda^2 - 6\lambda + 9) = (\lambda-3)^3.$

The characteristic equation is $(\lambda-3)^3 = 0$. The eigenvalue is $\lambda = 3$.

The reduced row echelon form of $3I - A = \begin{bmatrix} -1 & 0 & 1 \\ 0 & 0 & 0 \\ -1 & 0 & 1 \end{bmatrix}$ is $\begin{bmatrix} 1 & 0 & -1 \\ 0 & 0 & 0 \\ 0 & 0 & 0 \end{bmatrix}$.

The general solution of $(3I - A)\mathbf{x} = \mathbf{0}$ is $x_1 = t, x_2 = s, x_3 = t$. In vector form,

$\begin{bmatrix} x_1 \\ x_2 \\ x_3 \end{bmatrix} = \begin{bmatrix} t \\ s \\ t \end{bmatrix} = s\begin{bmatrix} 0 \\ 1 \\ 0 \end{bmatrix} + t\begin{bmatrix} 1 \\ 0 \\ 1 \end{bmatrix}$. A basis for the eigenspace corresponding to $\lambda = 3$ is $\{(0,1,0), (1,0,1)\}$.

13. The matrix $\lambda I - A$ is lower triangular, therefore by Theorem 2.1.2 its determinant is the product of the entries on the main diagonal. Therefore the characteristic equation is $(\lambda-3)(\lambda-7)(\lambda-1) = 0$.

15. $T(x,y) = \begin{bmatrix} x+4y \\ 2x+3y \end{bmatrix} = \begin{bmatrix} 1 & 4 \\ 2 & 3 \end{bmatrix}\begin{bmatrix} x \\ y \end{bmatrix}$; the standard matrix for the operator T is $A = \begin{bmatrix} 1 & 4 \\ 2 & 3 \end{bmatrix}$.

The following results were obtained in the solution of Exercise 5(a). These statements apply to the matrix A, therefore they also apply to the associated operator T:

- the eigenvalues are $\lambda = 5$ and $\lambda = -1$,
- a basis for the eigenspace corresponding to $\lambda = 5$ is $\{(1,1)\}$,
- a basis for the eigenspace corresponding to $\lambda = -1$ is $\{(-2,1)\}$.

17. **(a)** The transformation D^2 maps any function $\mathbf{f} = f(x)$ in $C^\infty(-\infty,\infty)$ into its second derivative, i.e. $D^2(\mathbf{f}) = f''(x)$. From calculus, we have

$D^2(\mathbf{f}+\mathbf{g}) = \frac{d}{dx}\frac{d}{dx}(f(x)+g(x)) = \frac{d}{dx}(f'(x)+g'(x)) = f''(x)+g''(x) = D^2(\mathbf{f}) + D^2(\mathbf{g})$ and

$D^2(k\mathbf{f}) = \frac{d}{dx}\frac{d}{dx}(kf(x)) = \frac{d}{dx}(kf'(x)) = kf''(x) = kD^2(\mathbf{f})$. We conclude that D^2 is linear.

(b) Denote $\mathbf{f} = \sin\sqrt{\omega}x$ and $\mathbf{g} = \cos\sqrt{\omega}x$. We have

$D^2(\mathbf{f}) = \frac{d}{dx}\frac{d}{dx}(\sin\sqrt{\omega}x) = \frac{d}{dx}(\sqrt{\omega}\cos\sqrt{\omega}x) = (\sqrt{\omega})(-\sqrt{\omega}\sin\sqrt{\omega}x) = -\omega\sin\sqrt{\omega}x = -\omega\mathbf{f}$ and

$D^2(\mathbf{g}) = \frac{d}{dx}\frac{d}{dx}(\cos\sqrt{\omega}x) = \frac{d}{dx}(-\sqrt{\omega}\sin\sqrt{\omega}x) = (-\sqrt{\omega})(\sqrt{\omega}\cos\sqrt{\omega}x) = -\omega\cos\sqrt{\omega}x = -\omega\mathbf{g}$.

It follows that $\mathbf{f} = \sin\sqrt{\omega}x$ and $\mathbf{g} = \cos\sqrt{\omega}x$ are eigenvectors of D^2; $\lambda = -\omega$ is the eigenvalue associated with both of these eigenvectors.

19. **(a)** The reflection of any vector on the line $y = x$ is the same vector: an eigenvalue $\lambda = 1$ corresponds to the eigenspace $\text{span}\{(1,1)\}$.

The reflection of any vector perpendicular to the line $y = x$ (i.e., on the line $y = -x$) is the negative of the original vector: an eigenvalue $\lambda = -1$ corresponds to the eigenspace span$\{(-1,1)\}$.

(b) The projection of any vector on the x-axis is the same vector: an eigenvalue $\lambda = 1$ corresponds to the eigenspace span$\{(1,0)\}$.
The projection of any vector perpendicular to the x-axis (i.e., on the y-axis) is the zero vector: an eigenvalue $\lambda = 0$ corresponds to the eigenspace span$\{(0,1)\}$.

(c) The result of the rotation through 90° of a nonzero vector is never a scalar multiple of the original vector. Consequently, this operator has no real eigenvalues.

(d) The result of the contraction of any vector $\mathbf{v}$ is a scalar multiple $k\mathbf{v}$ therefore the only eigenvalue is $\lambda = k$ and the corresponding eigenspace is the entire space R^2.

(e) The result of the shear applied to any vector on the x-axis is the same vector whereas the result of the shear applied to a nonzero vector in any other direction is not a scalar multiple of the original vector. The only eigenvalue is $\lambda = 1$ and the corresponding eigenspace is span$\{(1,0)\}$.

21. (a) The reflection of any vector on the xy-plane is the same vector: an eigenvalue $\lambda = 1$ corresponds to the eigenspace span$\{(1,0,0),(0,1,0)\}$.
The reflection of any vector perpendicular to the xy-plane (i.e., on the z-axis) is the negative of the original vector: an eigenvalue $\lambda = -1$ corresponds to the eigenspace span$\{(0,0,1)\}$.

(b) The projection of any vector on the xz-plane is the same vector: an eigenvalue $\lambda = 1$ corresponds to the eigenspace span$\{(1,0,0),(0,0,1)\}$.
The projection of any vector perpendicular to the xz-plane (i.e., on the y-axis) is the zero vector: an eigenvalue $\lambda = 0$ corresponds to the eigenspace span$\{(0,1,0)\}$.

(c) The result of the rotation applied to any vector on the x-axis is the same vector whereas the result of the rotation applied to a nonzero vector in any other direction is not a scalar multiple of the original vector. The only eigenvalue is $\lambda = 1$ and the corresponding eigenspace is span$\{(1,0,0)\}$.

(d) The result of the contraction of any vector $\mathbf{v}$ is a scalar multiple $k\mathbf{v}$ therefore the only eigenvalue is $\lambda = k$ and the corresponding eigenspace is the entire space R^3.

23. A line through the origin in the direction of $\mathbf{x} \neq \mathbf{0}$ is invariant under A if and only if $\mathbf{x}$ is an eigenvector of A.

(a) $\det(\lambda I - A) = \begin{vmatrix} \lambda - 4 & 1 \\ -2 & \lambda - 1 \end{vmatrix} = (\lambda - 4)(\lambda - 1) - (1)(-2) = \lambda^2 - 5\lambda + 6 = (\lambda - 2)(\lambda - 3)$.
The characteristic equation is $(\lambda - 2)(\lambda - 3) = 0$. The eigenvalues are $\lambda = 2$ and $\lambda = 3$.

The reduced row echelon form of $2I - A = \begin{bmatrix} -2 & 1 \\ -2 & 1 \end{bmatrix}$ is $\begin{bmatrix} 1 & -\frac{1}{2} \\ 0 & 0 \end{bmatrix}$. The general solution of $(2I - A)\mathbf{x} = \mathbf{0}$ is $x = \frac{1}{2}t, y = t$. Therefore $y = 2x$ is an equation of the corresponding invariant line.

The reduced row echelon form of $3I - A = \begin{bmatrix} -1 & 1 \\ -2 & 2 \end{bmatrix}$ is $\begin{bmatrix} 1 & -1 \\ 0 & 0 \end{bmatrix}$. The general solution of $(3I - A)\mathbf{x} = \mathbf{0}$ is $x_1 = t, x_2 = t$. Therefore $y = x$ is an equation of the corresponding invariant line.

(b) $\det(\lambda I - A) = \begin{vmatrix} \lambda & -1 \\ 1 & \lambda \end{vmatrix} = \lambda^2 - (-1)(1) = \lambda^2 + 1$.
There are no real eigenvalues and no invariant lines.

25. **(a)** Since the degree of $p(\lambda)$ is 6, A is a 6×6 matrix (see Exercise 37)

(b) $p(0) \neq 0$, therefore 0 is not an eigenvalue of A. From parts (a) and (t) of Theorem 5.1.6, A is invertible.

(c) A has three eigenspaces since it has three distinct eigenvalues, each corresponding to an eigenspace.

27. Substituting the given eigenvectors $\mathbf{x}$ and the corresponding eigenvalues λ into $A\mathbf{x} = \lambda\mathbf{x}$ yields
$A\begin{bmatrix} 1 \\ -1 \\ 1 \end{bmatrix} = 1\begin{bmatrix} 1 \\ -1 \\ 1 \end{bmatrix} = \begin{bmatrix} 1 \\ -1 \\ 1 \end{bmatrix}$, $A\begin{bmatrix} 1 \\ 1 \\ 0 \end{bmatrix} = -1\begin{bmatrix} 1 \\ 1 \\ 0 \end{bmatrix} = \begin{bmatrix} -1 \\ -1 \\ 0 \end{bmatrix}$, and $A\begin{bmatrix} 1 \\ -1 \\ 0 \end{bmatrix} = 0\begin{bmatrix} 1 \\ -1 \\ 0 \end{bmatrix} = \begin{bmatrix} 0 \\ 0 \\ 0 \end{bmatrix}$.
We can combine these three equations into a single equation $A\left[\begin{array}{c|c|c} 1 & 1 & 1 \\ -1 & 1 & -1 \\ 1 & 0 & 0 \end{array}\right] = \left[\begin{array}{c|c|c} 1 & -1 & 0 \\ -1 & -1 & 0 \\ 1 & 0 & 0 \end{array}\right]$.

Since the matrix $\begin{bmatrix} 1 & 1 & 1 \\ -1 & 1 & -1 \\ 1 & 0 & 0 \end{bmatrix}$ is invertible, we can multiply both sides on the right by its inverse,
$\begin{bmatrix} 0 & 0 & 1 \\ \frac{1}{2} & \frac{1}{2} & 0 \\ \frac{1}{2} & -\frac{1}{2} & -1 \end{bmatrix}$, resulting in $A = \begin{bmatrix} 1 & -1 & 0 \\ -1 & -1 & 0 \\ 1 & 0 & 0 \end{bmatrix}\begin{bmatrix} 0 & 0 & 1 \\ \frac{1}{2} & \frac{1}{2} & 0 \\ \frac{1}{2} & -\frac{1}{2} & -1 \end{bmatrix} = \begin{bmatrix} -\frac{1}{2} & -\frac{1}{2} & 1 \\ -\frac{1}{2} & -\frac{1}{2} & -1 \\ 0 & 0 & 1 \end{bmatrix}$.

(Note that this exercise could also be solved by assigning nine unknown values to the elements of $A = \begin{bmatrix} a & b & c \\ d & e & f \\ g & h & i \end{bmatrix}$, then solving the system of nine equations in nine unknowns resulting from the equation $A\mathbf{x} = \lambda\mathbf{x}$.)

31. It follows from Exercise 28 that if the characteristic polynomial $A = \begin{bmatrix} a_{11} & a_{12} \\ a_{21} & a_{22} \end{bmatrix}$ is $p(\lambda) = \lambda^2 + c_1\lambda + c_2$ then $c_1 = -\text{tr}(A) = -a_{11} - a_{22}$ and $c_2 = \det(A) = a_{11}a_{22} - a_{12}a_{21}$. Therefore
$p(A) = A^2 + c_1 A + c_2 I$

$$= \begin{bmatrix} a_{11} & a_{12} \\ a_{21} & a_{22} \end{bmatrix}\begin{bmatrix} a_{11} & a_{12} \\ a_{21} & a_{22} \end{bmatrix} + (-a_{11} - a_{22})\begin{bmatrix} a_{11} & a_{12} \\ a_{21} & a_{22} \end{bmatrix} + (a_{11}a_{22} - a_{12}a_{21})\begin{bmatrix} 1 & 0 \\ 0 & 1 \end{bmatrix}$$

$$= \begin{bmatrix} a_{11}^2 + a_{12}a_{21} & a_{11}a_{12} + a_{12}a_{22} \\ a_{21}a_{11} + a_{22}a_{21} & a_{21}a_{12} + a_{22}^2 \end{bmatrix} + \begin{bmatrix} -a_{11}^2 - a_{22}a_{11} & -a_{11}a_{12} - a_{22}a_{12} \\ -a_{11}a_{21} - a_{22}a_{21} & -a_{11}a_{22} - a_{22}^2 \end{bmatrix}$$
$$+ \begin{bmatrix} a_{11}a_{22} - a_{12}a_{21} & 0 \\ 0 & a_{11}a_{22} - a_{12}a_{21} \end{bmatrix}$$

$$= \begin{bmatrix} 0 & 0 \\ 0 & 0 \end{bmatrix}.$$

33. By Theorem 5.1.5, it follows from A being invertible that A cannot have a zero eigenvalue. Multiplying both sides of the equation $A\mathbf{x} = \lambda\mathbf{x}$ by A^{-1} on the left and applying Theorem 1.4.1 yields $A^{-1}A\mathbf{x} = \lambda A^{-1}\mathbf{x}$. Since $A^{-1}A = I$, dividing both sides of the equation by λ we obtain $\frac{1}{\lambda}\mathbf{x} = A^{-1}\mathbf{x}$. This shows that $\frac{1}{\lambda}$ is an eigenvalue of A^{-1} associated with an eigenvector $\mathbf{x}$.

35. Multiplying both sides of the equation $A\mathbf{x} = \lambda\mathbf{x}$ by the scalar s yields $s(A\mathbf{x}) = s(\lambda\mathbf{x})$. By Theorem 1.4.1, the equation can be rewritten as $(sA)\mathbf{x} = (s\lambda)\mathbf{x}$. This shows that $s\lambda$ is an eigenvalue of sA associated with the eigenvector $\mathbf{x}$.

True-False Exercises

(a) False. The vector $\mathbf{x}$ must be nonzero (without that requirement, $A\mathbf{x} = \lambda\mathbf{x}$ holds true for <u>all</u> $n \times n$ matrices A and <u>all</u> values λ by taking $\mathbf{x} = \mathbf{0}$).

(b) False. If λ is an eigenvalue of A then $(\lambda I - A)\mathbf{x} = \mathbf{0}$ must have nontrivial solutions.

(c) True. Since $p(0) = 1 \neq 0$, zero is not an eigenvalue of A. By Theorem 5.1.5, we conclude that A is invertible.

(d) False. Every eigenspace must include the zero vector, which is not an eigenvector.

(e) False. E.g., The only eigenvalue of $A = 2I$ is 2. However, the reduced row echelon form of A is I, whose only eigenvalue is 1.

(f) False. By Theorem 5.1.5, the set of columns of A must be linearly dependent.

5.2 Diagonalization

1. $\begin{vmatrix} 1 & 1 \\ 3 & 2 \end{vmatrix} = -1$ does not equal $\begin{vmatrix} 1 & 0 \\ 3 & -2 \end{vmatrix} = -2$ therefore, by Table 1 in Section 5.2, A and B are not similar matrices.

3. $\begin{vmatrix} 1 & 2 & 3 \\ 0 & 1 & 2 \\ 0 & 0 & 1 \end{vmatrix} = (1)(1)(1) = 1$ does not equal $\begin{vmatrix} 1 & 2 & 0 \\ \frac{1}{2} & 1 & 0 \\ 0 & 0 & 1 \end{vmatrix} = 0 - 0 + 1\begin{vmatrix} 1 & 2 \\ \frac{1}{2} & 1 \end{vmatrix} = 0$ therefore, by Table 1 in Section 5.2, A and B are not similar matrices.

5. $\det(\lambda I - A) = \begin{vmatrix} \lambda - 1 & 0 \\ -6 & \lambda + 1 \end{vmatrix} = (\lambda - 1)(\lambda + 1)$ therefore A has eigenvalues 1 and -1.

The reduced row echelon form of $1I - A$ is $\begin{bmatrix} 1 & -\frac{1}{3} \\ 0 & 0 \end{bmatrix}$ so that the eigenspace corresponding to $\lambda_1 = 1$ consists of vectors $\begin{bmatrix} x_1 \\ x_2 \end{bmatrix}$ where $x_1 = \frac{1}{3}t, x_2 = t$. A vector $\mathbf{p}_1 = \begin{bmatrix} 1 \\ 3 \end{bmatrix}$ forms a basis for this eigenspace.

The reduced row echelon form of $-1I - A$ is $\begin{bmatrix} 1 & 0 \\ 0 & 0 \end{bmatrix}$ so that the eigenspace corresponding to $\lambda_2 = -1$ consists of vectors $\begin{bmatrix} x_1 \\ x_2 \end{bmatrix}$ where $x_1 = 0, x_2 = t$. A vector $\mathbf{p}_2 = \begin{bmatrix} 0 \\ 1 \end{bmatrix}$ forms a basis for this eigenspace.

We form a matrix P using the column vectors $\mathbf{p}_1$ and $\mathbf{p}_2$: $P = \begin{bmatrix} 1 & 0 \\ 3 & 1 \end{bmatrix}$. (Note that this answer is not unique. Any nonzero multiples of these columns would also form a valid matrix P. Furthermore, the two columns can be interchanged.)

Calculating $P^{-1} = \frac{1}{(1)(1)-(0)(3)}\begin{bmatrix} 1 & 0 \\ -3 & 1 \end{bmatrix} = \begin{bmatrix} 1 & 0 \\ -3 & 1 \end{bmatrix}$ and performing matrix multiplications we check that $P^{-1}AP = \begin{bmatrix} 1 & 0 \\ -3 & 1 \end{bmatrix}\begin{bmatrix} 1 & 0 \\ 6 & -1 \end{bmatrix}\begin{bmatrix} 1 & 0 \\ 3 & 1 \end{bmatrix} = \begin{bmatrix} 1 & 0 \\ 0 & -1 \end{bmatrix} = \begin{bmatrix} \lambda_1 & 0 \\ 0 & \lambda_2 \end{bmatrix}$.

7. $\det(\lambda I - A) = \begin{vmatrix} \lambda - 2 & 0 & 2 \\ 0 & \lambda - 3 & 0 \\ 0 & 0 & \lambda - 3 \end{vmatrix} = (\lambda - 2)(\lambda - 3)^2$ thus A has eigenvalues 2 and 3 (with algebraic multiplicity 2).

The reduced row echelon form of $2I - A$ is $\begin{bmatrix} 0 & 1 & 0 \\ 0 & 0 & 1 \\ 0 & 0 & 0 \end{bmatrix}$ so that the eigenspace corresponding to $\lambda_1 = 2$ contains vectors $\begin{bmatrix} x_1 \\ x_2 \\ x_3 \end{bmatrix}$ where $x_1 = t, x_2 = 0, x_3 = 0$. A vector $\mathbf{p}_1 = \begin{bmatrix} 1 \\ 0 \\ 0 \end{bmatrix}$ forms a basis for this eigenspace.

The reduced row echelon form of $3I - A$ is $\begin{bmatrix} 1 & 0 & 2 \\ 0 & 0 & 0 \\ 0 & 0 & 0 \end{bmatrix}$ so that the eigenspace corresponding to $\lambda_2 = \lambda_3 = 3$ contains vectors $\begin{bmatrix} x_1 \\ x_2 \\ x_3 \end{bmatrix}$ where $x_1 = -2t, x_2 = s, x_3 = t$. We can write $\begin{bmatrix} x_1 \\ x_2 \\ x_3 \end{bmatrix} = \begin{bmatrix} -2t \\ s \\ t \end{bmatrix} = s\begin{bmatrix} 0 \\ 1 \\ 0 \end{bmatrix} + t\begin{bmatrix} -2 \\ 0 \\ 1 \end{bmatrix}$ therefore vectors $\mathbf{p}_2 = \begin{bmatrix} 0 \\ 1 \\ 0 \end{bmatrix}$ and $\mathbf{p}_3 = \begin{bmatrix} -2 \\ 0 \\ 1 \end{bmatrix}$ form a basis for this eigenspace. Note that the geometric multiplicity of this eigenvalue matches its algebraic multiplicity.

We form a matrix P using the column vectors $\mathbf{p}_1, \mathbf{p}_2$, and $\mathbf{p}_3$: $P = \begin{bmatrix} 1 & 0 & -2 \\ 0 & 1 & 0 \\ 0 & 0 & 1 \end{bmatrix}$. (Note that this answer is not unique. Any nonzero multiples of these columns would also form a valid matrix P. Furthermore, the columns can be interchanged.)

To invert the matrix P, we can employ the procedure introduced in Section 1.5: since the reduced row echelon form of the matrix $\begin{bmatrix} 1 & 0 & -2 & 1 & 0 & 0 \\ 0 & 1 & 0 & 0 & 1 & 0 \\ 0 & 0 & 1 & 0 & 0 & 1 \end{bmatrix}$ is $\begin{bmatrix} 1 & 0 & 0 & 1 & 0 & 2 \\ 0 & 1 & 0 & 0 & 1 & 0 \\ 0 & 0 & 1 & 0 & 0 & 1 \end{bmatrix}$, we have $P^{-1} = \begin{bmatrix} 1 & 0 & 2 \\ 0 & 1 & 0 \\ 0 & 0 & 1 \end{bmatrix}$.

We check that $P^{-1}AP = \begin{bmatrix} 1 & 0 & 2 \\ 0 & 1 & 0 \\ 0 & 0 & 1 \end{bmatrix}\begin{bmatrix} 2 & 0 & -2 \\ 0 & 3 & 0 \\ 0 & 0 & 3 \end{bmatrix}\begin{bmatrix} 1 & 0 & -2 \\ 0 & 1 & 0 \\ 0 & 0 & 1 \end{bmatrix} = \begin{bmatrix} 2 & 0 & 0 \\ 0 & 3 & 0 \\ 0 & 0 & 3 \end{bmatrix} = \begin{bmatrix} \lambda_1 & 0 & 0 \\ 0 & \lambda_2 & 0 \\ 0 & 0 & \lambda_3 \end{bmatrix}$.

9. **(a)** Cofactor expansion along the second column yields $\det(\lambda I - A) = \begin{vmatrix} \lambda-4 & 0 & -1 \\ -2 & \lambda-3 & -2 \\ -1 & 0 & \lambda-4 \end{vmatrix} =$
$(\lambda-3)\begin{vmatrix} \lambda-4 & -1 \\ -1 & \lambda-4 \end{vmatrix} = (\lambda-3)[(\lambda-4)^2-1] = (\lambda-3)^2(\lambda-5)$ therefore A has eigenvalues 3 (with algebraic multiplicity 2) and 5.

(b) The reduced row echelon form of $3I - A$ is $\begin{bmatrix} 1 & 0 & 1 \\ 0 & 0 & 0 \\ 0 & 0 & 0 \end{bmatrix}$, consequently $\operatorname{rank}(3I - A) = 1$.

The reduced row echelon form of $5I - A$ is $\begin{bmatrix} 1 & 0 & -1 \\ 0 & 1 & -2 \\ 0 & 0 & 0 \end{bmatrix}$, consequently $\operatorname{rank}(5I - A) = 2$.

(c) Based on part (b), the geometric multiplicities of the eigenvalues $\lambda = 3$ and $\lambda = 5$ are $3 - 1 = 2$ and $3 - 2 = 1$, respectively. Since these are equal to the corresponding algebraic multiplicities, by Theorem 5.2.4(b) A is diagonalizable.

11. Cofactor expansion along the second row yields

$\det(\lambda I - A) = \begin{vmatrix} \lambda+1 & -4 & 2 \\ 3 & \lambda-4 & 0 \\ 3 & -1 & \lambda-3 \end{vmatrix} = -3\begin{vmatrix} -4 & 2 \\ -1 & \lambda-3 \end{vmatrix} + (\lambda-4)\begin{vmatrix} \lambda+1 & 2 \\ 3 & \lambda-3 \end{vmatrix}$
$= (-3)[(-4)(\lambda-3)-(2)(-1)] + (\lambda-4)[(\lambda+1)(\lambda-3)-(2)(3)] = \lambda^3 - 6\lambda^2 + 11\lambda - 6$.
Following the procedure described in Example 3 of Section 5.1, we determine that the only possibilities for integer solutions of the characteristic equation are $\pm1, \pm2, \pm3$, and ±6.
Since $\det(1I - A) = 0$, $\lambda - 1$ must be a factor of the characteristic polynomial. Dividing $\lambda - 1$ into $\lambda^3 - 6\lambda^2 + 11\lambda - 6$ leads to $\det(\lambda I - A) = (\lambda-1)(\lambda^2 - 5\lambda + 6) = (\lambda-1)(\lambda-2)(\lambda-3)$.
We conclude that the eigenvalues are 1, 2, and 3 - each of them has the algebraic multiplicity 1.

The reduced row echelon form of $1I - A$ is $\begin{bmatrix} 1 & 0 & -1 \\ 0 & 1 & -1 \\ 0 & 0 & 0 \end{bmatrix}$ so that the eigenspace corresponding to $\lambda_1 = 1$ contains vectors $\begin{bmatrix} x_1 \\ x_2 \\ x_3 \end{bmatrix}$ where $x_1 = t, x_2 = t, x_3 = t$. A vector $\mathbf{p}_1 = \begin{bmatrix} 1 \\ 1 \\ 1 \end{bmatrix}$ forms a basis for this eigenspace. This eigenvalue has geometric multiplicity 1.

The reduced row echelon form of $2I - A$ is $\begin{bmatrix} 1 & 0 & -\frac{2}{3} \\ 0 & 1 & -1 \\ 0 & 0 & 0 \end{bmatrix}$ so that the eigenspace corresponding to $\lambda_2 = 2$ contains vectors $\begin{bmatrix} x_1 \\ x_2 \\ x_3 \end{bmatrix}$ where $x_1 = \frac{2}{3}t, x_2 = t, x_3 = t$. A vector $\mathbf{p}_2 = \begin{bmatrix} 2 \\ 3 \\ 3 \end{bmatrix}$ forms a basis for this eigenspace. This eigenvalue has geometric multiplicity 1.

The reduced row echelon form of $3I - A$ is $\begin{bmatrix} 1 & 0 & -\frac{1}{4} \\ 0 & 1 & -\frac{3}{4} \\ 0 & 0 & 0 \end{bmatrix}$ so that the eigenspace corresponding to

$\lambda_3 = 3$ contains vectors $\begin{bmatrix} x_1 \\ x_2 \\ x_3 \end{bmatrix}$ where $x_1 = \frac{1}{4}t, x_2 = \frac{3}{4}t, x_3 = t$. A vector $\mathbf{p}_3 = \begin{bmatrix} 1 \\ 3 \\ 4 \end{bmatrix}$ forms a basis for this eigenspace. This eigenvalue has geometric multiplicity 1.

Since for each eigenvalue the geometric multiplicity matches the algebraic multiplicity, by Theorem 5.2.4(b) A is diagonalizable.

We form a matrix P using the column vectors $\mathbf{p}_1$, $\mathbf{p}_2$, and $\mathbf{p}_3$: $P = \begin{bmatrix} 1 & 2 & 1 \\ 1 & 3 & 3 \\ 1 & 3 & 4 \end{bmatrix}$. (Note that this answer is not unique. Any nonzero multiples of these columns would also form a valid matrix P. Furthermore, the columns can be interchanged.)

$$P^{-1}AP = \begin{bmatrix} \lambda_1 & 0 & 0 \\ 0 & \lambda_2 & 0 \\ 0 & 0 & \lambda_3 \end{bmatrix} = \begin{bmatrix} 1 & 0 & 0 \\ 0 & 2 & 0 \\ 0 & 0 & 3 \end{bmatrix}.$$

13. $\det(\lambda I - A) = \begin{vmatrix} \lambda & 0 & 0 \\ 0 & \lambda & 0 \\ -3 & 0 & \lambda - 1 \end{vmatrix} = \lambda^2(\lambda - 1)$ so the eigenvalues are $\lambda = 0$ with the algebraic multiplicity 2 and $\lambda = 1$ with the algebraic multiplicity 1.

The reduced row echelon form of $0I - A$ is $\begin{bmatrix} 1 & 0 & \frac{1}{3} \\ 0 & 0 & 0 \\ 0 & 0 & 0 \end{bmatrix}$ so that the eigenspace corresponding to $\lambda_1 = \lambda_2 = 0$ contains vectors $\begin{bmatrix} x_1 \\ x_2 \\ x_3 \end{bmatrix}$ where $x_1 = -\frac{1}{3}t, x_2 = s, x_3 = t$. We can write $\begin{bmatrix} x_1 \\ x_2 \\ x_3 \end{bmatrix} = \begin{bmatrix} -\frac{1}{3}t \\ s \\ t \end{bmatrix} = s\begin{bmatrix} 0 \\ 1 \\ 0 \end{bmatrix} + t\begin{bmatrix} -\frac{1}{3} \\ 0 \\ 1 \end{bmatrix}$ therefore vectors $\mathbf{p}_1 = \begin{bmatrix} 0 \\ 1 \\ 0 \end{bmatrix}$ and $\mathbf{p}_2 = \begin{bmatrix} -1 \\ 0 \\ 3 \end{bmatrix}$ form a basis for this eigenspace. This eigenvalue has the geometric multiplicity 2.

The reduced row echelon form of $1I - A$ is $\begin{bmatrix} 1 & 0 & 0 \\ 0 & 1 & 0 \\ 0 & 0 & 0 \end{bmatrix}$ so that the eigenspace corresponding to $\lambda_3 = 1$ contains vectors $\begin{bmatrix} x_1 \\ x_2 \\ x_3 \end{bmatrix}$ where $x_1 = 0, x_2 = 0, x_3 = t$. A vector $\mathbf{p}_3 = \begin{bmatrix} 0 \\ 0 \\ 1 \end{bmatrix}$ forms a basis for this eigenspace. This eigenvalue has the geometric multiplicity 1.

Since for each eigenvalue the geometric multiplicity matches the algebraic multiplicity, by Theorem 5.2.4(b) A is diagonalizable.

We form a matrix P using the column vectors $\mathbf{p}_1$, $\mathbf{p}_2$, and $\mathbf{p}_3$: $P = \begin{bmatrix} 0 & -1 & 0 \\ 1 & 0 & 0 \\ 0 & 3 & 1 \end{bmatrix}$. (Note that this answer is not unique.)

$$P^{-1}AP = \begin{bmatrix} \lambda_1 & 0 & 0 \\ 0 & \lambda_2 & 0 \\ 0 & 0 & \lambda_3 \end{bmatrix} = \begin{bmatrix} 0 & 0 & 0 \\ 0 & 0 & 0 \\ 0 & 0 & 1 \end{bmatrix}.$$

15. **(a)** The degree of the characteristic polynomial of A is 3 therefore A is a 3×3 matrix. All three eigenspaces (for $\lambda = 1, \lambda = -3$, and $\lambda = 5$) must have dimension 1.

(b) The degree of the characteristic polynomial of A is 6 therefore A is a 6×6 matrix.
The possible dimensions of the eigenspace corresponding to $\lambda = 0$ are 1 or 2.
The dimension of the eigenspace corresponding to $\lambda = 1$ must be 1.
The possible dimensions of the eigenspace corresponding to $\lambda = 2$ are 1, 2, or 3.

17. $\det(\lambda I - A) = \begin{vmatrix} \lambda & -3 \\ -2 & \lambda+1 \end{vmatrix} = \lambda(\lambda+1) - (-3)(-2) = \lambda^2 + \lambda - 6 = (\lambda-2)(\lambda+3)$ therefore A has eigenvalues 2 and -3, each with the algebraic multiplicity 1.

The reduced row echelon form of $2I - A$ is $\begin{bmatrix} 1 & -\frac{3}{2} \\ 0 & 0 \end{bmatrix}$ so that the eigenspace corresponding to $\lambda = 2$ consists of vectors $\begin{bmatrix} x_1 \\ x_2 \end{bmatrix}$ where $x_1 = \frac{3}{2}t, x_2 = t$. A vector $\mathbf{p}_1 = \begin{bmatrix} 3 \\ 2 \end{bmatrix}$ forms a basis for this eigenspace.

The reduced row echelon form of $-3I - A$ is $\begin{bmatrix} 1 & 1 \\ 0 & 0 \end{bmatrix}$ so that the eigenspace corresponding to $\lambda = -3$ consists of vectors $\begin{bmatrix} x_1 \\ x_2 \end{bmatrix}$ where $x_1 = -t, x_2 = t$. A vector $\mathbf{p}_2 = \begin{bmatrix} -1 \\ 1 \end{bmatrix}$ forms a basis for this eigenspace.

We form a matrix $P = \begin{bmatrix} 3 & -1 \\ 2 & 1 \end{bmatrix}$ and calculate $P^{-1} = \frac{1}{(3)(1)-(-1)(2)}\begin{bmatrix} 1 & 1 \\ -2 & 3 \end{bmatrix} = \frac{1}{5}\begin{bmatrix} 1 & 1 \\ -2 & 3 \end{bmatrix} = \begin{bmatrix} \frac{1}{5} & \frac{1}{5} \\ -\frac{2}{5} & \frac{3}{5} \end{bmatrix}$ so that $P^{-1}AP = \begin{bmatrix} 2 & 0 \\ 0 & -3 \end{bmatrix} = D$.

Therefore $A^{10} = PD^{10}P^{-1} = \begin{bmatrix} 3 & -1 \\ 2 & 1 \end{bmatrix}\begin{bmatrix} 2^{10} & 0 \\ 0 & (-3)^{10} \end{bmatrix}\begin{bmatrix} \frac{1}{5} & \frac{1}{5} \\ -\frac{2}{5} & \frac{3}{5} \end{bmatrix} = \begin{bmatrix} 3 & -1 \\ 2 & 1 \end{bmatrix}\begin{bmatrix} 1{,}024 & 0 \\ 0 & 59{,}049 \end{bmatrix}\begin{bmatrix} \frac{1}{5} & \frac{1}{5} \\ -\frac{2}{5} & \frac{3}{5} \end{bmatrix}$
$= \begin{bmatrix} 24{,}234 & -34{,}815 \\ -23{,}210 & 35{,}839 \end{bmatrix}$.

19. To invert the matrix P, we can employ the procedure introduced in Section 1.5: since the reduced row echelon form of the matrix $\begin{bmatrix} 1 & 1 & 1 & 1 & 0 & 0 \\ 0 & 0 & 1 & 0 & 1 & 0 \\ 1 & 0 & 5 & 0 & 0 & 1 \end{bmatrix}$ is $\begin{bmatrix} 1 & 0 & 0 & 0 & -5 & 1 \\ 0 & 1 & 0 & 1 & 4 & -1 \\ 0 & 0 & 1 & 0 & 1 & 0 \end{bmatrix}$, we have

$P^{-1} = \begin{bmatrix} 0 & -5 & 1 \\ 1 & 4 & -1 \\ 0 & 1 & 0 \end{bmatrix}$

We verify that $P^{-1}AP = \begin{bmatrix} 0 & -5 & 1 \\ 1 & 4 & -1 \\ 0 & 1 & 0 \end{bmatrix}\begin{bmatrix} -1 & 7 & -1 \\ 0 & 1 & 0 \\ 0 & 15 & -2 \end{bmatrix}\begin{bmatrix} 1 & 1 & 1 \\ 0 & 0 & 1 \\ 1 & 0 & 5 \end{bmatrix} = \begin{bmatrix} -2 & 0 & 0 \\ 0 & -1 & 0 \\ 0 & 0 & 1 \end{bmatrix} = D$ is a diagonal matrix therefore P diagonalizes A.

$$A^{11} = PD^{11}P^{-1} = \begin{bmatrix} 1 & 1 & 1 \\ 0 & 0 & 1 \\ 1 & 0 & 5 \end{bmatrix}\begin{bmatrix} (-2)^{11} & 0 & 0 \\ 0 & (-1)^{11} & 0 \\ 0 & 0 & 1^{11} \end{bmatrix}\begin{bmatrix} 0 & -5 & 1 \\ 1 & 4 & -1 \\ 0 & 1 & 0 \end{bmatrix}$$
$$= \begin{bmatrix} 1 & 1 & 1 \\ 0 & 0 & 1 \\ 1 & 0 & 5 \end{bmatrix}\begin{bmatrix} -2{,}048 & 0 & 0 \\ 0 & -1 & 0 \\ 0 & 0 & 1 \end{bmatrix}\begin{bmatrix} 0 & -5 & 1 \\ 1 & 4 & -1 \\ 0 & 1 & 0 \end{bmatrix} = \begin{bmatrix} -1 & 10{,}237 & -2{,}047 \\ 0 & 1 & 0 \\ 0 & 10{,}245 & -2{,}048 \end{bmatrix}$$

21. Cofactor expansion along the first row yields $\det(\lambda I - A) = \begin{vmatrix} \lambda-3 & 1 & 0 \\ 1 & \lambda-2 & 1 \\ 0 & 1 & \lambda-3 \end{vmatrix}$
$= (\lambda-3)\begin{vmatrix} \lambda-2 & 1 \\ 1 & \lambda-3 \end{vmatrix} - 1\begin{vmatrix} 1 & 1 \\ 0 & \lambda-3 \end{vmatrix} = (\lambda-3)[(\lambda-2)(\lambda-3)-1] - (\lambda-3)$

$= (\lambda - 3)[\lambda^2 - 5\lambda + 6 - 1 - 1] = (\lambda - 3)(\lambda^2 - 5\lambda + 4) = (\lambda - 1)(\lambda - 3)(\lambda - 4)$
therefore A has eigenvalues 1, 3, and 4, each with the algebraic multiplicity 1.

The reduced row echelon form of $1I - A$ is $\begin{bmatrix} 1 & 0 & -1 \\ 0 & 1 & -2 \\ 0 & 0 & 0 \end{bmatrix}$ so that the eigenspace corresponding to $\lambda_1 = 1$ contains vectors $\begin{bmatrix} x_1 \\ x_2 \\ x_3 \end{bmatrix}$ where $x_1 = t, x_2 = 2t, x_3 = t$. A vector $\mathbf{p}_1 = \begin{bmatrix} 1 \\ 2 \\ 1 \end{bmatrix}$ forms a basis for this eigenspace.

The reduced row echelon form of $3I - A$ is $\begin{bmatrix} 1 & 0 & 1 \\ 0 & 1 & 0 \\ 0 & 0 & 0 \end{bmatrix}$ so that the eigenspace corresponding to $\lambda_2 = 3$ contains vectors $\begin{bmatrix} x_1 \\ x_2 \\ x_3 \end{bmatrix}$ where $x_1 = -t, x_2 = 0, x_3 = t$. A vector $\mathbf{p}_2 = \begin{bmatrix} -1 \\ 0 \\ 1 \end{bmatrix}$ forms a basis for this eigenspace.

The reduced row echelon form of $4I - A$ is $\begin{bmatrix} 1 & 0 & -1 \\ 0 & 1 & 1 \\ 0 & 0 & 0 \end{bmatrix}$ so that the eigenspace corresponding to $\lambda_3 = 4$ contains vectors $\begin{bmatrix} x_1 \\ x_2 \\ x_3 \end{bmatrix}$ where $x_1 = t, x_2 = -t, x_3 = t$. A vector $\mathbf{p}_3 = \begin{bmatrix} 1 \\ -1 \\ 1 \end{bmatrix}$ forms a basis for this eigenspace.

We form a matrix $P = \begin{bmatrix} 1 & -1 & 1 \\ 2 & 0 & -1 \\ 1 & 1 & 1 \end{bmatrix}$ and find its inverse using the procedure introduced in Section 1.5. Since the reduced row echelon form of the matrix $\begin{bmatrix} 1 & -1 & 1 & 1 & 0 & 0 \\ 2 & 0 & -1 & 0 & 1 & 0 \\ 1 & 1 & 1 & 0 & 0 & 1 \end{bmatrix}$ is

$\begin{bmatrix} 1 & 0 & 0 & \frac{1}{6} & \frac{1}{3} & \frac{1}{6} \\ 0 & 1 & 0 & -\frac{1}{2} & 0 & \frac{1}{2} \\ 0 & 0 & 1 & \frac{1}{3} & -\frac{1}{3} & \frac{1}{3} \end{bmatrix}$, we have $P^{-1} = \begin{bmatrix} \frac{1}{6} & \frac{1}{3} & \frac{1}{6} \\ -\frac{1}{2} & 0 & \frac{1}{2} \\ \frac{1}{3} & -\frac{1}{3} & \frac{1}{3} \end{bmatrix}$.

We conclude that $A^n = PD^nP^{-1} = \begin{bmatrix} 1 & -1 & 1 \\ 2 & 0 & -1 \\ 1 & 1 & 1 \end{bmatrix} \begin{bmatrix} 1 & 0 & 0 \\ 0 & 3^n & 0 \\ 0 & 0 & 4^n \end{bmatrix} \begin{bmatrix} \frac{1}{6} & \frac{1}{3} & \frac{1}{6} \\ -\frac{1}{2} & 0 & \frac{1}{2} \\ \frac{1}{3} & -\frac{1}{3} & \frac{1}{3} \end{bmatrix}$.

23. By inspection, both A and B have rank 1 (both matrices are in reduced row echelon form).
Let $P = \begin{bmatrix} a & b \\ c & d \end{bmatrix}$. Then $AP = \begin{bmatrix} 1 & 0 \\ 0 & 0 \end{bmatrix}\begin{bmatrix} a & b \\ c & d \end{bmatrix} = \begin{bmatrix} a & b \\ 0 & 0 \end{bmatrix}$ and $PB = \begin{bmatrix} a & b \\ c & d \end{bmatrix}\begin{bmatrix} 0 & 1 \\ 0 & 0 \end{bmatrix} = \begin{bmatrix} 0 & a \\ 0 & c \end{bmatrix}$.
Setting $AP = PB$ requires that $a = 0, b = a$, and $c = 0$.
For any value d, the matrix $P = \begin{bmatrix} 0 & 0 \\ 0 & d \end{bmatrix}$ satisfies the equality $AP = PB$. However, $P = \begin{bmatrix} 0 & 0 \\ 0 & d \end{bmatrix}$ has zero determinant therefore it is not invertible so that the similarity condition $B = P^{-1}AP$ cannot be met.

25. Since there exist matrices P and Q such that $B = P^{-1}AP$ and $C = Q^{-1}BQ$, we can write $C = Q^{-1}(P^{-1}AP)Q = (PQ)^{-1}A(PQ)$. Consequently, A is similar to C.

27. **(a)** The dimension of the eigenspace must be at least 1, but cannot exceed the algebraic multiplicity of the corresponding eigenvalue. Since the algebraic multiplicities of the eigenvalues 1,3, and 4 are 1, 2, and 3, respectively, we conclude that

- The dimension of the eigenspace corresponding to $\lambda = 1$ must be 1.
- The possible dimensions of the eigenspace corresponding to $\lambda = 3$ are 1 or 2.
- The possible dimensions of the eigenspace corresponding to $\lambda = 4$ are 1, 2, or 3.

(b) If A is diagonalizable then by Theorem 5.2.4(b) for each eigenvalue the dimension of the eigenspace must be equal to the algebraic multiplicity. Therefore

- The dimension of the eigenspace corresponding to $\lambda = 1$ must be 1.
- The dimension of the eigenspace corresponding to $\lambda = 3$ must be 2.
- The dimension of the eigenspace corresponding to $\lambda = 4$ must be 3.

(c) If the dimension of the eigenspace were smaller than 3 then by Theorem 4.5.2(a), a set of three vectors from that eigenspace would have to be linearly dependent. Consequently, for the set of the three vectors to be linearly independent, the eigenspace containing the set must be of dimension at least 3. This is only possible for the eigenspace corresponding to the eigenvalue $\lambda = 4$.

29. Using the result obtained in Exercise 30 of Section 5.1, we can take $P = \begin{bmatrix} -b & -b \\ a-\lambda_1 & a-\lambda_2 \end{bmatrix}$ where $\lambda_1 = \frac{1}{2}\left[(a+d) + \sqrt{(a-d)^2 + 4bc}\right]$ and $\lambda_2 = \frac{1}{2}\left[(a+d) - \sqrt{(a-d)^2 + 4bc}\right]$.

31. $T(x_1, x_2) = \begin{bmatrix} -x_2 \\ -x_1 \end{bmatrix} = \begin{bmatrix} 0 & -1 \\ -1 & 0 \end{bmatrix}\begin{bmatrix} x_1 \\ x_2 \end{bmatrix}$; the standard matrix for the operator T is $A = \begin{bmatrix} 0 & -1 \\ -1 & 0 \end{bmatrix}$.

$\det(\lambda I - A) = \begin{vmatrix} \lambda & 1 \\ 1 & \lambda \end{vmatrix} = \lambda^2 - 1 = (\lambda - 1)(\lambda + 1)$ thus A has eigenvalues 1 and -1, both with algebraic multiplicities 1.

The reduced row echelon form of $1I - A$ is $\begin{bmatrix} 1 & 1 \\ 0 & 0 \end{bmatrix}$ so that the eigenspace corresponding to $\lambda_1 = 1$ consists of vectors $\begin{bmatrix} x_1 \\ x_2 \end{bmatrix}$ where $x_1 = -t, x_2 = t$. A vector $\mathbf{p}_1 = \begin{bmatrix} -1 \\ 1 \end{bmatrix}$ forms a basis for this eigenspace.

The reduced row echelon form of $-1I - A$ is $\begin{bmatrix} 1 & -1 \\ 0 & 0 \end{bmatrix}$ so that the eigenspace corresponding to $\lambda_2 = -1$ consists of vectors $\begin{bmatrix} x_1 \\ x_2 \end{bmatrix}$ where $x_1 = t, x_2 = t$. A vector $\mathbf{p}_2 = \begin{bmatrix} 1 \\ 1 \end{bmatrix}$ forms a basis for this eigenspace.

We form a matrix P using the column vectors $\mathbf{p}_1$ and $\mathbf{p}_2$: $P = \begin{bmatrix} -1 & 1 \\ 1 & 1 \end{bmatrix}$. (Note that this answer is not unique. Any nonzero multiples of these columns would also form a valid matrix P. Furthermore, the two columns can be interchanged.)

33. $T(x_1, x_2, x_3) = \begin{bmatrix} 3x_1 \\ x_2 \\ x_1 - x_2 \end{bmatrix} = \begin{bmatrix} 3 & 0 & 0 \\ 0 & 1 & 0 \\ 1 & -1 & 0 \end{bmatrix}\begin{bmatrix} x_1 \\ x_2 \\ x_3 \end{bmatrix}$; the standard matrix for T is $A = \begin{bmatrix} 3 & 0 & 0 \\ 0 & 1 & 0 \\ 1 & -1 & 0 \end{bmatrix}$.

Since $\det(\lambda I - A) = \begin{vmatrix} \lambda - 3 & 0 & 0 \\ 0 & \lambda - 1 & 0 \\ -1 & 1 & \lambda \end{vmatrix} = (\lambda - 3)(\lambda - 1)\lambda$, thus A has eigenvalues 0, 1, and 3, each with algebraic multiplicity 1.

The reduced row echelon form of $0I - A$ is $\begin{bmatrix} 1 & 0 & 0 \\ 0 & 1 & 0 \\ 0 & 0 & 0 \end{bmatrix}$ so that the eigenspace corresponding to $\lambda_1 = 0$ contains vectors $\begin{bmatrix} x_1 \\ x_2 \\ x_3 \end{bmatrix}$ where $x_1 = 0, x_2 = 0, x_3 = t$. A vector $\mathbf{p}_1 = \begin{bmatrix} 0 \\ 0 \\ 1 \end{bmatrix}$ forms a basis for this eigenspace.

The reduced row echelon form of $1I - A$ is $\begin{bmatrix} 1 & 0 & 0 \\ 0 & 1 & 1 \\ 0 & 0 & 0 \end{bmatrix}$ so that the eigenspace corresponding to $\lambda_2 = 1$ contains vectors $\begin{bmatrix} x_1 \\ x_2 \\ x_3 \end{bmatrix}$ where $x_1 = 0, x_2 = -t, x_3 = t$. A vector $\mathbf{p}_2 = \begin{bmatrix} 0 \\ -1 \\ 1 \end{bmatrix}$ forms a basis for this eigenspace.

The reduced row echelon form of $3I - A$ is $\begin{bmatrix} 1 & 0 & -3 \\ 0 & 1 & 0 \\ 0 & 0 & 0 \end{bmatrix}$ so that the eigenspace corresponding to $\lambda_3 = 3$ contains vectors $\begin{bmatrix} x_1 \\ x_2 \\ x_3 \end{bmatrix}$ where $x_1 = 3t, x_2 = 0, x_3 = t$. A vector $\mathbf{p}_3 = \begin{bmatrix} 3 \\ 0 \\ 1 \end{bmatrix}$ forms a basis for this eigenspace.

We form a matrix P using the column vectors $\mathbf{p}_1, \mathbf{p}_2$, and $\mathbf{p}_3$: $P = \begin{bmatrix} 0 & 0 & 3 \\ 0 & -1 & 0 \\ 1 & 1 & 1 \end{bmatrix}$. (Note that this answer is not unique. Any nonzero multiples of these columns would also form a valid matrix P. Furthermore, the columns can be interchanged.)

True-False Exercises

(a) False. E.g., $A = I_2$ has only one eigenvalue $\lambda = 1$, but it is diagonalizable with $P = I_2$.

(b) True. This follows from Theorem 5.2.1.

(c) True. Multiplying $A = P^{-1}BP$ on the left by P yields $PA = BP$.

(d) False. The matrix P is not unique. For instance, interchanging two columns of P results in a different matrix which also diagonalizes A.

(e) True. Since A is invertible, we can take the inverse on both sides of the equality $P^{-1}AP = D = \begin{bmatrix} \lambda_1 & 0 & \cdots & 0 \\ 0 & \lambda_2 & \cdots & 0 \\ \vdots & \vdots & \ddots & \vdots \\ 0 & 0 & \cdots & \lambda_n \end{bmatrix}$ obtaining $P^{-1}A^{-1}P = D^{-1} = \begin{bmatrix} 1/\lambda_1 & 0 & \cdots & 0 \\ 0 & 1/\lambda_2 & \cdots & 0 \\ \vdots & \vdots & \ddots & \vdots \\ 0 & 0 & \cdots & 1/\lambda_n \end{bmatrix}$. Consequently, P diagonalizes both A and A^{-1}.

(f) True. We can transpose both sides of the equality $P^{-1}AP = D$ obtaining $P^TA^T(P^T)^{-1} = D^T = D$, i.e., $((P^T)^{-1})^{-1}A^T(P^T)^{-1} = D$. Consequently, $(P^T)^{-1}$ diagonalizes A^T.

(g) True. A basis for R^n must be a linearly independent set of n vectors, so by Theorem 5.2.1 A is diagonalizable.

(h) True. This follows from Theorem 5.2.2(b).

(i) True. From Theorem 5.1.5 we have $\det(A) = 0$. Since $\det(A^2) = (\det(A))^2 = 0^2 = 0$, it follows from the same Theorem that A^2 is singular.

5.3 Complex Vector Spaces

1. $\bar{\mathbf{u}} = (\overline{2-i}, \overline{4i}, \overline{1+i}) = (2+i, -4i, 1-i)$; $\operatorname{Re}(\mathbf{u}) = (2,0,1)$; $\operatorname{Im}(\mathbf{u}) = (-1,4,1)$;
$\|\mathbf{u}\| = \sqrt{|2-i|^2 + |4i|^2 + |1+i|^2} = \sqrt{(2^2+(-1)^2) + (0^2+4^2) + (1^2+1^2)} = \sqrt{5+16+2} = \sqrt{23}$

3. **(a)** $\bar{\bar{\mathbf{u}}} = \overline{\overline{(3-4i, 2+i, -6i)}} = \overline{(3+4i, 2-i, 6i)} = (3-4i, 2+i, -6i) = \mathbf{u}$

(b) $\overline{k\mathbf{u}} = \overline{i(3-4i, 2+i, -6i)} = \overline{(4+3i, -1+2i, +6)} = (4-3i, -1-2i, 6)$
$\bar{k}\bar{\mathbf{u}} = \bar{i}\,\overline{(3-4i, 2+i, -6i)} = -i(3+4i, 2-i, 6i) = (4-3i, -1-2i, 6)$

(c) $\overline{\mathbf{u}+\mathbf{v}} = \overline{(4-3i, 4, 4-6i)} = (4+3i, 4, 4+6i)$
$\bar{\mathbf{u}} + \bar{\mathbf{v}} = (3+4i, 2-i, 6i) + (1-i, 2+i, 4) = (4+3i, 4, 4+6i)$

(d) $\overline{\mathbf{u}-\mathbf{v}} = \overline{(2-5i, 2i, -4-6i)} = (2+5i, -2i, -4+6i)$
$\bar{\mathbf{u}} - \bar{\mathbf{v}} = (3+4i, 2-i, 6i) - (1-i, 2+i, 4) = (2+5i, -2i, -4+6i)$

5. $i\mathbf{x} - 3\mathbf{v} = \bar{\mathbf{u}}$ can be rewritten as $i\mathbf{x} = 3\mathbf{v} + \bar{\mathbf{u}}$; multiplying both sides by $-i$ and using the fact that $(-i)(i) = 1$, we obtain $\mathbf{x} = (-i)(3\mathbf{v} + \bar{\mathbf{u}}) = (-i)[(3+3i, 6-3i, 12) + (3+4i, 2-i, 6i)]$
$= (-i)(6+7i, 8-4i, 12+6i) = (7-6i, -4-8i, 6-12i)$

7. $\bar{A} = \begin{bmatrix} \overline{-5i} & \overline{4} \\ \overline{2-i} & \overline{1+5i} \end{bmatrix} = \begin{bmatrix} 5i & 4 \\ 2+i & 1-5i \end{bmatrix}$; $\operatorname{Re}(A) = \begin{bmatrix} 0 & 4 \\ 2 & 1 \end{bmatrix}$; $\operatorname{Im}(A) = \begin{bmatrix} -5 & 0 \\ -1 & 5 \end{bmatrix}$;
$\det(A) = (-5i)(1+5i) - (4)(2-i) = -5i + 25 - 8 + 4i = 17 - i$; $\operatorname{tr}(A) = -5i + (1+5i) = 1$

9. **(a)** $\bar{\bar{A}} = \begin{bmatrix} \overline{5i} & \overline{4} \\ \overline{2+i} & \overline{1-5i} \end{bmatrix} = \begin{bmatrix} -5i & 4 \\ 2-i & 1+5i \end{bmatrix} = A$

(b) $\overline{(A^T)} = \begin{bmatrix} \overline{-5i} & \overline{2-i} \\ \overline{4} & \overline{1+5i} \end{bmatrix} = \begin{bmatrix} 5i & 2+i \\ 4 & 1-5i \end{bmatrix}$; $(\bar{A})^T = \begin{bmatrix} 5i & 4 \\ 2+i & 1-5i \end{bmatrix}^T = \begin{bmatrix} 5i & 2+i \\ 4 & 1-5i \end{bmatrix}$

(c) From $AB = \begin{bmatrix} -5i & 4 \\ 2-i & 1+5i \end{bmatrix}\begin{bmatrix} 1-i \\ 2i \end{bmatrix} = \begin{bmatrix} (-5i)(1-i)+(4)(2i) \\ (2-i)(1-i)+(1+5i)(2i) \end{bmatrix}$

$= \begin{bmatrix} -5i-5+8i \\ 2-2i-i-1+2i-10 \end{bmatrix} = \begin{bmatrix} -5+3i \\ -9-i \end{bmatrix}$ we obtain $\overline{AB} = \begin{bmatrix} -5-3i \\ -9+i \end{bmatrix}$

$\overline{A}\,\overline{B} = \begin{bmatrix} 5i & 4 \\ 2+i & 1-5i \end{bmatrix}\begin{bmatrix} 1+i \\ 2i \end{bmatrix} = \begin{bmatrix} (5i)(1+i)+(4)(2i) \\ (2+i)(1+i)+(1-5i)(2i) \end{bmatrix}$

$= \begin{bmatrix} 5i-5-8i \\ 2+2i+i-1-2i-10 \end{bmatrix} = \begin{bmatrix} -5-3i \\ -9+i \end{bmatrix}$

11. $\mathbf{u}\cdot\mathbf{v} = (i)(\overline{4}) + (2i)(\overline{-2i}) + (3)(\overline{1+i}) = (i)(4)+(2i)(2i)+(3)(1-i) = 4i-4+3-3i = -1+i$

$\mathbf{u}\cdot\mathbf{w} = (i)(\overline{2-i}) + (2i)(\overline{2i}) + (3)(\overline{5+3i}) = (i)(2+i)+(2i)(-2i)+(3)(5-3i)$

$= 2i-1+4+15-9i = 18-7i$

$\mathbf{v}\cdot\mathbf{w} = (4)(\overline{2-i}) + (-2i)(\overline{2i}) + (1+i)(\overline{5+3i}) = (4)(2+i)+(-2i)(-2i)+(1+i)(5-3i)$

$= 8+4i-4+5-3i+5i+3 = 12+6i$

Since both $\mathbf{u}^T\overline{\mathbf{v}} = [i \quad 2i \quad 3]\begin{bmatrix} 4 \\ 2i \\ 1-i \end{bmatrix} = [-1+i]$ and $\overline{\mathbf{v}}^T\mathbf{u} = [4 \quad 2i \quad 1-i]\begin{bmatrix} i \\ 2i \\ 3 \end{bmatrix} = [-1+i]$ are equal to $\mathbf{u}\cdot\mathbf{v} = -1+i$, Formula (5) holds.

(a) $\overline{\mathbf{v}\cdot\mathbf{u}} = \overline{(4)(\overline{i}) + (-2i)(\overline{2i}) + (1+i)(\overline{3})} = \overline{(4)(-i)+(-2i)(-2i)+(1+i)(3)}$

$= \overline{-4i-4+3+3i} = \overline{-1-i} = -1+i = \mathbf{u}\cdot\mathbf{v}$

(b) $\mathbf{u}\cdot(\mathbf{v}+\mathbf{w}) = (i)(\overline{4+2-i}) + (2i)(\overline{-2i+2i}) + (3)(\overline{1+i+5+3i})$

$= (i)(6+i)+(2i)(0)+(3)(6-4i) = 6i-1+18-12i = 17-6i$ equals

$\mathbf{u}\cdot\mathbf{v}+\mathbf{u}\cdot\mathbf{w} = -1+i+18-7i = 17-6i$

(c) $k(\mathbf{u}\cdot\mathbf{v}) = (2i)(-1+i) = -2-2i$ equals

$(k\mathbf{u})\cdot\mathbf{v} = (-2)(\overline{4}) + (-4)(\overline{-2i}) + (6i)(\overline{1+i})$

$=(-2)(4)+(-4)(2i)+(6i)(1-i) = -8-8i+6i+6 = -2-2i$

13. $\mathbf{u}\cdot\overline{\mathbf{v}} = (i)(4)+(2i)(-2i)+(3)(1+i) = 4i+4+3+3i = 7+7i$

$\overline{\mathbf{w}\cdot\mathbf{u}} = \overline{(2-i)(\overline{i}) + (2i)(\overline{2i}) + (5+3i)(\overline{3})} = \overline{(2-i)(-i)+(2i)(-2i)+(5+3i)(3)}$

$= \overline{-2i-1+4+15+9i} = \overline{18+7i} = 18-7i$

$\overline{(\mathbf{u}\cdot\overline{\mathbf{v}}) - \overline{\mathbf{w}\cdot\mathbf{u}}} = \overline{7+7i-18+7i} = \overline{-11+14i} = -11-14i$

15. $\det(\lambda I - A) = \begin{vmatrix} \lambda-4 & 5 \\ -1 & \lambda \end{vmatrix} = (\lambda-4)\lambda - (5)(-1) = \lambda^2-4\lambda+5$

Solving the characteristic equation $\lambda^2-4\lambda+5=0$ using the quadratic formula yields $\lambda = \frac{4\pm\sqrt{4^2-4(5)}}{2} = \frac{4\pm\sqrt{-4}}{2} = 2\pm i$ therefore A has eigenvalues $\lambda = 2+i$ and $\lambda = 2-i$.

For the eigenvalue $\lambda = 2+i$, the augmented matrix of the homogeneous system $\big((2+i)I-A\big)\mathbf{x} = \mathbf{0}$ is $\begin{bmatrix} -2+i & 5 & 0 \\ -1 & 2+i & 0 \end{bmatrix}$. The rows of this matrix must be scalar multiples of each other (see Example 3 in Section 5.3) therefore it suffices to solve the equation corresponding to the second row, which yields $-x_1+(2+i)x_2 = 0$. The general solution of this equation (and, consequently, of the entire system) is $x_1 = (2+i)t, x_2 = t$. The vector $\begin{bmatrix} 2+i \\ 1 \end{bmatrix}$ forms a basis for the eigenspace corresponding to $\lambda = 2+i$.

According to Theorem 5.3.4, the vector $\overline{\begin{bmatrix} 2+i \\ 1 \end{bmatrix}} = \begin{bmatrix} 2-i \\ 1 \end{bmatrix}$ forms a basis for the eigenspace corresponding to $\lambda = 2-i$.

17. $\det(\lambda I - A) = \begin{vmatrix} \lambda-5 & 2 \\ -1 & \lambda-3 \end{vmatrix} = (\lambda-5)(\lambda-3) - (2)(-1) = \lambda^2 - 8\lambda + 17$
Solving the characteristic equation $\lambda^2 - 8\lambda + 17 = 0$ using the quadratic formula yields $\lambda = \frac{8\pm\sqrt{8^2-4(17)}}{2} = \frac{8\pm\sqrt{-4}}{2} = 4 \pm i$ therefore A has eigenvalues $\lambda = 4+i$ and $\lambda = 4-i$.

For the eigenvalue $\lambda = 4+i$, the augmented matrix of the homogeneous system $\big((4+i)I - A\big)\mathbf{x} = \mathbf{0}$ is $\begin{bmatrix} -1+i & 2 & 0 \\ -1 & 1+i & 0 \end{bmatrix}$. The rows of this matrix must be scalar multiples of each other (see Example 3 in Section 5.3) therefore it suffices to solve the equation corresponding to the second row, which yields $-x_1 + (1+i)x_2 = 0$. The general solution of this equation (and, consequently, of the entire system) is $x_1 = (1+i)t, x_2 = t$. The vector $\begin{bmatrix} 1+i \\ 1 \end{bmatrix}$ forms a basis for the eigenspace corresponding to $\lambda = 4+i$.

According to Theorem 5.3.4, the vector $\overline{\begin{bmatrix} 1+i \\ 1 \end{bmatrix}} = \begin{bmatrix} 1-i \\ 1 \end{bmatrix}$ forms a basis for the eigenspace corresponding to $\lambda = 4-i$.

19. $\begin{bmatrix} a & -b \\ b & a \end{bmatrix} = \begin{bmatrix} 1 & -1 \\ 1 & 1 \end{bmatrix}$ implies $a = b = 1$. We have $|\lambda| = |1+i| = \sqrt{1+1} = \sqrt{2}$.
The angle inside the interval $(-\pi, \pi]$ from the positive x-axis to the ray that joins the origin to the point $(1,1)$ is $\phi = \frac{\pi}{4}$.

21. $\begin{bmatrix} a & -b \\ b & a \end{bmatrix} = \begin{bmatrix} 1 & \sqrt{3} \\ -\sqrt{3} & 1 \end{bmatrix}$ implies $a = 1$ and $b = -\sqrt{3}$. We have $|\lambda| = |1-\sqrt{3}i| = \sqrt{1+3} = 2$.
The angle inside the interval $(-\pi, \pi]$ from the positive x-axis to the ray that joins the origin to the point $(1, -\sqrt{3})$ is $\phi = -\frac{\pi}{3}$.

23. $\det(\lambda I - A) = \begin{vmatrix} \lambda+1 & 5 \\ -4 & \lambda-7 \end{vmatrix} = (\lambda+1)(\lambda-7) - (5)(-4) = \lambda^2 - 6\lambda + 13$
Solving the characteristic equation $\lambda^2 - 6\lambda + 13 = 0$ using the quadratic formula yields $\lambda = \frac{6\pm\sqrt{6^2-4(13)}}{2} = \frac{6\pm\sqrt{-16}}{2} = 3 \pm 2i$ therefore A has eigenvalues $\lambda = 3+2i$ and $\lambda = 3-2i$.
For the eigenvalue $\lambda = 3-2i$, the augmented matrix of the homogeneous system $\big((3-2i)I - A\big)\mathbf{x} = \mathbf{0}$ is $\begin{bmatrix} 4-2i & 5 & 0 \\ -4 & -4-2i & 0 \end{bmatrix}$. The rows of this matrix must be scalar multiples of each other (see Example 3 in Section 5.3) therefore it suffices to solve the equation corresponding to the second row, which yields $x_1 + \left(1+\frac{1}{2}i\right)x_2 = 0$. The general solution of this equation (and, consequently, of the entire system) is $x_1 = \left(-1-\frac{1}{2}i\right)t, x_2 = t$. Since $\begin{bmatrix} -2-i \\ 2 \end{bmatrix}$ is an eigenvector corresponding to

$\lambda = 3 - 2i$, it follows from Theorem 5.3.8 that the matrices $P = \begin{bmatrix} -2 & -1 \\ 2 & 0 \end{bmatrix}$ and $C = \begin{bmatrix} 3 & -2 \\ 2 & 3 \end{bmatrix}$ satisfy $A = PCP^{-1}$.

25. $\det(\lambda I - A) = \begin{vmatrix} \lambda - 8 & -6 \\ 3 & \lambda - 2 \end{vmatrix} = (\lambda - 8)(\lambda - 2) - (-6)(3) = \lambda^2 - 10\lambda + 34$

Solving the characteristic equation $\lambda^2 - 10\lambda + 34 = 0$ using the quadratic formula yields $\lambda = \frac{10 \pm \sqrt{10^2 - 4(34)}}{2} = \frac{10 \pm \sqrt{-36}}{2} = 5 \pm 3i$ therefore A has eigenvalues $\lambda = 5 + 3i$ and $\lambda = 5 - 3i$.

For the eigenvalue $\lambda = 5 - 3i$, the augmented matrix of the homogeneous system $\left((5 - 3i)I - A\right)\mathbf{x} = \mathbf{0}$ is $\begin{bmatrix} -3 - 3i & -6 & 0 \\ 3 & 3 - 3i & 0 \end{bmatrix}$. The rows of this matrix must be scalar multiples of each other (see Example 3 in Section 5.3) therefore it suffices to solve the equation corresponding to the second row, which yields $x_1 + (1 - i)x_2 = 0$. The general solution of this equation (and, consequently, of the entire system) is $x_1 = (-1 + i)t$, $x_2 = t$. Since $\begin{bmatrix} -1 + i \\ 1 \end{bmatrix}$ is an eigenvector corresponding to $\lambda = 5 - 3i$, it follows from Theorem 5.3.8 that the matrices $P = \begin{bmatrix} -1 & 1 \\ 1 & 0 \end{bmatrix}$ and $C = \begin{bmatrix} 5 & -3 \\ 3 & 5 \end{bmatrix}$ satisfy $A = PCP^{-1}$.

27. **(a)** Letting $k = a + bi$ we have $\mathbf{u} \cdot \mathbf{v} = (2i)(\overline{i}) + (i)(\overline{6i}) + (3i)(\overline{a + bi})$
$= (2i)(-i) + (i)(-6i) + (3i)(a - bi) = 2 + 6 + 3ai + 3b = (8 + 3b) + (3a)i$. Setting this equal to zero yields $a = 0$ and $b = -\frac{8}{3}$ therefore the only complex scalar which satisfies our requirements is $k = -\frac{8}{3}i$.

(b) $\mathbf{u} \cdot \mathbf{v} = (k)(\overline{1}) + (k)(\overline{-1}) + (1 + i)(\overline{1 - i}) = (k)(1) + (k)(-1) + (1 + i)(1 + i) = 2i \neq 0$ therefore no complex scalar k satisfies our requirements.

True-False Exercises

(a) False. By Theorem 5.3.4, complex eigenvalues of a real matrix occur in conjugate pairs, so the total number of complex eigenvalues must be even. Consequently, in a 5×5 matrix at least one eigenvalue must be real.

(b) True. $\lambda^2 - \text{tr}(A)\lambda + \det(A) = 0$ is the characteristic equation of a 2×2 complex matrix A.

(c) False. By Theorem 5.3.5, A has two complex conjugate eigenvalues if $\text{tr}(A)^2 < 4\det(A)$.

(d) True. This follows from Theorem 5.3.4.

(e) False. E.g., $\begin{bmatrix} i & 0 \\ 0 & i \end{bmatrix}$ is symmetric, but its eigenvalue $\lambda = i$ is not real.

(f) False. (This would be true if we assumed $|\lambda| = 1$.)

5.4 Differential Equations

1. **(a)** We begin by diagonalizing the coefficient matrix of the system $A = \begin{bmatrix} 1 & 4 \\ 2 & 3 \end{bmatrix}$.

The characteristic polynomial of A is

$\det(\lambda I - A) = \begin{vmatrix} \lambda-1 & -4 \\ -2 & \lambda-3 \end{vmatrix} = (\lambda-1)(\lambda-3) - (-4)(-2) = \lambda^2 - 4\lambda - 5 = (\lambda-5)(\lambda+1)$
thus the eigenvalues of A are $\lambda = 5$ and $\lambda = -1$.

The reduced row echelon form of $5I - A$ is $\begin{bmatrix} 1 & -1 \\ 0 & 0 \end{bmatrix}$ so that the eigenspace corresponding to $\lambda = 5$ consists of vectors $\begin{bmatrix} x_1 \\ x_2 \end{bmatrix}$ where $x_1 = t, x_2 = t$. A vector $\mathbf{p}_1 = \begin{bmatrix} 1 \\ 1 \end{bmatrix}$ forms a basis for this eigenspace.

The reduced row echelon form of $-1I - A$ is $\begin{bmatrix} 1 & 2 \\ 0 & 0 \end{bmatrix}$ so that the eigenspace corresponding to $\lambda = -1$ consists of vectors $\begin{bmatrix} x_1 \\ x_2 \end{bmatrix}$ where $x_1 = -2t, x_2 = t$. A vector $\mathbf{p}_2 = \begin{bmatrix} -2 \\ 1 \end{bmatrix}$ forms a basis for this eigenspace.

Therefore $P = \begin{bmatrix} 1 & -2 \\ 1 & 1 \end{bmatrix}$ diagonalizes A and $P^{-1}AP = D = \begin{bmatrix} 5 & 0 \\ 0 & -1 \end{bmatrix}$.

The substitution $\mathbf{y} = P\mathbf{u}$ yields the "diagonal system" $\mathbf{u}' = \begin{bmatrix} 5 & 0 \\ 0 & -1 \end{bmatrix} \mathbf{u}$ consisting of equations $u_1' = 5u_1$ and $u_2' = -1u_2$. From Formula (2) in Section 5.4, these equations have the solutions $u_1 = c_1e^{5x}, u_2 = c_2e^{-x}$, i.e., $\mathbf{u} = \begin{bmatrix} c_1e^{5x} \\ c_2e^{-x} \end{bmatrix}$. From $\mathbf{y} = P\mathbf{u}$ we obtain the solution

$\mathbf{y} = \begin{bmatrix} 1 & -2 \\ 1 & 1 \end{bmatrix} \begin{bmatrix} c_1e^{5x} \\ c_2e^{-x} \end{bmatrix} = \begin{bmatrix} c_1e^{5x} - 2c_2e^{-x} \\ c_1e^{5x} + c_2e^{-x} \end{bmatrix}$ thus $y_1 = c_1e^{5x} - 2c_2e^{-x}$ and $y_2 = c_1e^{5x} + c_2e^{-x}$.

(b) Substituting the initial conditions into the general solution obtained in part (a) yields a system

$$\begin{aligned} c_1e^{5(0)} - 2c_2e^{-0} &= 0 \\ c_1e^{5(0)} + c_2e^{-0} &= 0 \end{aligned}$$

which can be rewritten as

$$\begin{aligned} c_1 \; - \; 2c_2 &= 0 \\ c_1 \; + \; c_2 &= 0 \end{aligned}$$

The reduced row echelon form of this system's augmented matrix $\begin{bmatrix} 1 & -2 & 0 \\ 1 & 1 & 0 \end{bmatrix}$ is $\begin{bmatrix} 1 & 0 & 0 \\ 0 & 1 & 0 \end{bmatrix}$ therefore $c_1 = 0$ and $c_2 = 0$.

The solution satisfying the given initial conditions can be expressed as $y_1 = 0$ and $y_2 = 0$.

3. **(a)** We begin by diagonalizing the coefficient matrix of the system $A = \begin{bmatrix} 4 & 0 & 1 \\ -2 & 1 & 0 \\ -2 & 0 & 1 \end{bmatrix}$.

Cofactor expansion along the second column yields

$$\det(\lambda I - A) = \begin{vmatrix} \lambda-4 & 0 & -1 \\ 2 & \lambda-1 & 0 \\ 2 & 0 & \lambda-1 \end{vmatrix} = (\lambda-1)\begin{vmatrix} \lambda-4 & -1 \\ 2 & \lambda-1 \end{vmatrix}$$
$$= (\lambda-1)[(\lambda-4)(\lambda-1) - (-1)(2)]$$
$$= (\lambda-1)(\lambda^2 - 5\lambda + 6) = (\lambda-1)(\lambda-2)(\lambda-3)$$

The characteristic equation of A is $(\lambda-1)(\lambda-2)(\lambda-3) = 0$ thus the eigenvalues of A are 1, 2, and 3 (each with the algebraic multiplicity 1).

The reduced row echelon form of $1I - A$ is $\begin{bmatrix} 1 & 0 & 0 \\ 0 & 0 & 1 \\ 0 & 0 & 0 \end{bmatrix}$ so that the eigenspace corresponding to $\lambda = 1$ consists of vectors $\begin{bmatrix} x_1 \\ x_2 \\ x_3 \end{bmatrix}$ where $x_1 = 0, x_2 = t, x_3 = 0$. A vector $\mathbf{p}_1 = \begin{bmatrix} 0 \\ 1 \\ 0 \end{bmatrix}$ forms a basis for this eigenspace.

The reduced row echelon form of $2I - A$ is $\begin{bmatrix} 1 & 0 & \frac{1}{2} \\ 0 & 1 & -1 \\ 0 & 0 & 0 \end{bmatrix}$ so that the eigenspace corresponding to $\lambda = 2$ consists of vectors $\begin{bmatrix} x_1 \\ x_2 \\ x_3 \end{bmatrix}$ where $x_1 = -\frac{1}{2}t, x_2 = t, x_3 = t$. A vector $\mathbf{p}_2 = \begin{bmatrix} -1 \\ 2 \\ 2 \end{bmatrix}$ forms a basis for this eigenspace.

The reduced row echelon form of $3I - A$ is $\begin{bmatrix} 1 & 0 & 1 \\ 0 & 1 & -1 \\ 0 & 0 & 0 \end{bmatrix}$ so that the eigenspace corresponding to $\lambda = 3$ consists of vectors $\begin{bmatrix} x_1 \\ x_2 \\ x_3 \end{bmatrix}$ where $x_1 = -t, x_2 = t, x_3 = t$. A vector $\mathbf{p}_3 = \begin{bmatrix} -1 \\ 1 \\ 1 \end{bmatrix}$ forms a basis for this eigenspace.

Therefore $P = \begin{bmatrix} 0 & -1 & -1 \\ 1 & 2 & 1 \\ 0 & 2 & 1 \end{bmatrix}$ diagonalizes A and $P^{-1}AP = D = \begin{bmatrix} 1 & 0 & 0 \\ 0 & 2 & 0 \\ 0 & 0 & 3 \end{bmatrix}$.

The substitution $\mathbf{y} = P\mathbf{u}$ yields the "diagonal system" $\mathbf{u}' = \begin{bmatrix} 1 & 0 & 0 \\ 0 & 2 & 0 \\ 0 & 0 & 3 \end{bmatrix}\mathbf{u}$ consisting of equations $u_1' = u_1, u_2' = 2u_2$, and $u_3' = 3u_3$. From Formula (2) in Section 5.4, these equations have the solutions $u_1 = c_1e^x, u_2 = c_2e^{2x}, u_3 = c_3e^{3x}$ i.e., $\mathbf{u} = \begin{bmatrix} c_1e^x \\ c_2e^{2x} \\ c_3e^{3x} \end{bmatrix}$. From $\mathbf{y} = P\mathbf{u}$ we obtain the solution $\mathbf{y} = \begin{bmatrix} 0 & -1 & -1 \\ 1 & 2 & 1 \\ 0 & 2 & 1 \end{bmatrix}\begin{bmatrix} c_1e^x \\ c_2e^{2x} \\ c_3e^{3x} \end{bmatrix} = \begin{bmatrix} -c_2e^{2x} - c_3e^{3x} \\ c_1e^x + 2c_2e^{2x} + c_3e^{3x} \\ 2c_2e^{2x} + c_3e^{3x} \end{bmatrix}$ thus $y_1 = -c_2e^{2x} - c_3e^{3x},\ y_2 = c_1e^x + 2c_2e^{2x} + c_3e^{3x}$, and $y_3 = 2c_2e^{2x} + c_3e^{3x}$.

(b) Substituting the initial conditions into the general solution obtained in part (a) yields a system

$$\begin{aligned} -c_2e^{2(0)} - c_3e^{3(0)} &= -1 \\ c_1e^0 + 2c_2e^{2(0)} + c_3e^{3(0)} &= 1 \\ 2c_2e^{2(0)} + c_3e^{3(0)} &= 0 \end{aligned}$$

which can be rewritten as

$$\begin{array}{rcrcrcr} & - & c_2 & - & c_3 & = & -1 \\ c_1 & + & 2c_2 & + & c_3 & = & 1 \\ & & 2c_2 & + & c_3 & = & 0 \end{array}$$

The reduced row echelon form of this system's augmented matrix $\begin{bmatrix} 0 & -1 & -1 & -1 \\ 1 & 2 & 1 & 1 \\ 0 & 2 & 1 & 0 \end{bmatrix}$ is

$\begin{bmatrix} 1 & 0 & 0 & 1 \\ 0 & 1 & 0 & -1 \\ 0 & 0 & 1 & 2 \end{bmatrix}$ therefore $c_1 = 1, c_2 = -1,$ and $c_3 = 2.$

The solution satisfying the given initial conditions can be expressed as

$y_1 = e^{2x} - 2e^{3x}$, $y_2 = e^{x} - 2e^{2x} + 2e^{3x}$, and $y_3 = -2e^{2x} + 2e^{3x}$.

5. Assume $y = f(x)$ is a solution of $y' = ay$ so that $f'(x) = af(x)$.

We have $\frac{d}{dx}(f(x)e^{-ax}) = f'(x)e^{-ax} + f(x)(-a)e^{-ax} = af(x)e^{-ax} - af(x)e^{-ax} = 0$ for all x

therefore there exists a constant c for which $f(x)e^{-ax} = c$, i.e., $f(x) = \frac{c}{e^{-ax}} = ce^{ax}$. We conclude that every solution of $y' = ay$ must have the form $f(x) = ce^{ax}$.

7. Substituting $y_1 = y$ and $y_2 = y'$ allows us to rewrite the equation $y'' - y' - 6y = 0$ as $y_2' - y_2 - 6y_1 = 0$. Also, $y_2 = y' = y_1'$ so we obtain the system

$$y_1' = y_2$$
$$y_2' = 6y_1 + y_2$$

The coefficient matrix of this system is $A = \begin{bmatrix} 0 & 1 \\ 6 & 1 \end{bmatrix}$. The characteristic polynomial of A is

$\det(\lambda I - A) = \begin{vmatrix} \lambda & -1 \\ -6 & \lambda - 1 \end{vmatrix} = \lambda(\lambda - 1) - (-1)(-6) = \lambda^2 - \lambda - 6 = (\lambda - 3)(\lambda + 2)$ thus the eigenvalues of A are $\lambda = 3$ and $\lambda = -2$.

The reduced row echelon form of $3I - A$ is $\begin{bmatrix} 1 & -\frac{1}{3} \\ 0 & 0 \end{bmatrix}$ so that the eigenspace corresponding to $\lambda = 3$ consists of vectors $\begin{bmatrix} x_1 \\ x_2 \end{bmatrix}$ where $x_1 = \frac{1}{3}t, x_2 = t$. A vector $\mathbf{p}_1 = \begin{bmatrix} 1 \\ 3 \end{bmatrix}$ forms a basis for this eigenspace.

The reduced row echelon form of $-2I - A$ is $\begin{bmatrix} 1 & \frac{1}{2} \\ 0 & 0 \end{bmatrix}$ so that the eigenspace corresponding to $\lambda = -2$ consists of vectors $\begin{bmatrix} x_1 \\ x_2 \end{bmatrix}$ where $x_1 = -\frac{1}{2}t, x_2 = t$. A vector $\mathbf{p}_2 = \begin{bmatrix} -1 \\ 2 \end{bmatrix}$ forms a basis for this eigenspace.

Therefore $P = \begin{bmatrix} 1 & -1 \\ 3 & 2 \end{bmatrix}$ diagonalizes A and $P^{-1}AP = D = \begin{bmatrix} 3 & 0 \\ 0 & -2 \end{bmatrix}$.

The substitution $\mathbf{y} = P\mathbf{u}$ yields the "diagonal system" $\mathbf{u}' = \begin{bmatrix} 3 & 0 \\ 0 & -2 \end{bmatrix}\mathbf{u}$ consisting of equations $u_1' = 3u_1$ and $u_2' = -2u_2$. From Formula (2) in Section 5.4, these equations have the solutions $u_1 = c_1e^{3x}, u_2 = c_2e^{-2x}$, i.e., $\mathbf{u} = \begin{bmatrix} c_1e^{3x} \\ c_2e^{-2x} \end{bmatrix}$. From $\mathbf{y} = P\mathbf{u}$ we obtain the solution

$\mathbf{y} = \begin{bmatrix} 1 & -1 \\ 3 & 2 \end{bmatrix}\begin{bmatrix} c_1e^{3x} \\ c_2e^{-2x} \end{bmatrix} = \begin{bmatrix} c_1e^{3x} - c_2e^{-2x} \\ 3c_1e^{3x} + 2c_2e^{-2x} \end{bmatrix}$ thus $y_1 = c_1e^{3x} - c_2e^{-2x}$ and $y_2 = 3c_1e^{3x} + 2c_2e^{-2x}$.

We conclude that the original equation $y'' - y' - 6y = 0$ has the solution $y = c_1e^{3x} - c_2e^{-2x}$.

9. Substituting $y_1 = y, y_2 = y'$, and $y_3 = y''$ allows us to rewrite the equation $y''' - 6y'' + 11y' - 6y = 0$ as $y_3' - 6y_3 + 11y_2 - 6y_1 = 0$. With $y_2 = y' = y_1'$ and $y_3 = y'' = y_2'$ we obtain the system

$$y_1' = y_2$$

$$y_2' = y_3$$
$$y_3' = 6y_1 - 11y_2 + 6y_3$$

The coefficient matrix of this system is $A = \begin{bmatrix} 0 & 1 & 0 \\ 0 & 0 & 1 \\ 6 & -11 & 6 \end{bmatrix}$.

The characteristic polynomial of A is $\det(\lambda I - A) = \begin{vmatrix} \lambda & -1 & 0 \\ 0 & \lambda & -1 \\ -6 & 11 & \lambda - 6 \end{vmatrix}$

$= \lambda \begin{vmatrix} \lambda & -1 \\ 11 & \lambda - 6 \end{vmatrix} - (-1) \begin{vmatrix} 0 & -1 \\ -6 & \lambda - 6 \end{vmatrix} = \lambda[\lambda(\lambda - 6) + 11] - 6 = \lambda^3 - 6\lambda^2 + 11\lambda - 6.$

Following the procedure described in Example 3 of Section 5.1, we determine that the only possibilities for integer solutions of the characteristic equation are $\pm 1, \pm 2, \pm 3$, and ± 6.

Since $\det(1I - A) = 0$, $\lambda - 1$ must be a factor of the characteristic polynomial. Dividing $\lambda - 1$ into $\lambda^3 - 6\lambda^2 + 11\lambda - 6$ leads to $\det(\lambda I - A) = (\lambda - 1)(\lambda^2 - 5\lambda + 6) = (\lambda - 1)(\lambda - 2)(\lambda - 3)$.

We conclude that the eigenvalues are 1, 2, and 3 - each of them has the algebraic multiplicity 1.

The reduced row echelon form of $1I - A$ is $\begin{bmatrix} 1 & 0 & -1 \\ 0 & 1 & -1 \\ 0 & 0 & 0 \end{bmatrix}$ so that the eigenspace corresponding to $\lambda_1 = 1$ contains vectors $\begin{bmatrix} x_1 \\ x_2 \\ x_3 \end{bmatrix}$ where $x_1 = t, x_2 = t, x_3 = t$. A vector $\mathbf{p}_1 = \begin{bmatrix} 1 \\ 1 \\ 1 \end{bmatrix}$ forms a basis for this eigenspace.

The reduced row echelon form of $2I - A$ is $\begin{bmatrix} 1 & 0 & -\frac{1}{4} \\ 0 & 1 & -\frac{1}{2} \\ 0 & 0 & 0 \end{bmatrix}$ so that the eigenspace corresponding to $\lambda_2 = 2$ contains vectors $\begin{bmatrix} x_1 \\ x_2 \\ x_3 \end{bmatrix}$ where $x_1 = \frac{1}{4}t, x_2 = \frac{1}{2}t, x_3 = t$. A vector $\mathbf{p}_2 = \begin{bmatrix} 1 \\ 2 \\ 4 \end{bmatrix}$ forms a basis for this eigenspace.

The reduced row echelon form of $3I - A$ is $\begin{bmatrix} 1 & 0 & -\frac{1}{9} \\ 0 & 1 & -\frac{1}{3} \\ 0 & 0 & 0 \end{bmatrix}$ so that the eigenspace corresponding to $\lambda_3 = 3$ contains vectors $\begin{bmatrix} x_1 \\ x_2 \\ x_3 \end{bmatrix}$ where $x_1 = \frac{1}{9}t, x_2 = \frac{1}{3}t, x_3 = t$. A vector $\mathbf{p}_3 = \begin{bmatrix} 1 \\ 3 \\ 9 \end{bmatrix}$ forms a basis for this eigenspace.

Therefore $P = \begin{bmatrix} 1 & 1 & 1 \\ 1 & 2 & 3 \\ 1 & 4 & 9 \end{bmatrix}$ diagonalizes A and $P^{-1}AP = D = \begin{bmatrix} 1 & 0 & 0 \\ 0 & 2 & 0 \\ 0 & 0 & 3 \end{bmatrix}$.

The substitution $\mathbf{y} = P\mathbf{u}$ yields the "diagonal system" $\mathbf{u}' = \begin{bmatrix} 1 & 0 & 0 \\ 0 & 2 & 0 \\ 0 & 0 & 3 \end{bmatrix} \mathbf{u}$ consisting of equations $u_1' = u_1, u_2' = 2u_2$, and $u_3' = 3u_3$. From Formula (2) in Section 5.4, these equations have the solutions $u_1 = c_1 e^x, u_2 = c_2 e^{2x}$, and $u_3 = c_3 e^{3x}$, i.e., $\mathbf{u} = \begin{bmatrix} c_1 e^x \\ c_2 e^{2x} \\ c_3 e^{3x} \end{bmatrix}$. From $\mathbf{y} = P\mathbf{u}$ we obtain the solution

$\mathbf{y} = \begin{bmatrix} 1 & 1 & 1 \\ 1 & 2 & 3 \\ 1 & 4 & 9 \end{bmatrix} \begin{bmatrix} c_1 e^x \\ c_2 e^{2x} \\ c_3 e^{3x} \end{bmatrix} = \begin{bmatrix} c_1 e^x + c_2 e^{2x} + c_3 e^{3x} \\ c_1 e^x + 2c_2 e^{2x} + 3c_3 e^{3x} \\ c_1 e^x + 4c_2 e^{2x} + 9c_3 e^{3x} \end{bmatrix}$ thus $y_1 = c_1 e^x + c_2 e^{2x} + c_3 e^{3x}$, $y_2 = c_1 e^x + 2c_2 e^{2x} + 3c_3 e^{3x}$, and $y_3 = c_1 e^x + 4c_2 e^{2x} + 9c_3 e^{3x}$.

We conclude that the original equation $y''' - 6y'' + 11y' - 6y = 0$ has the solution

$$y = c_1 e^x + c_2 e^{2x} + c_3 e^{3x}$$

15. **(a)** Let y and z be functions in $C^\infty(-\infty, \infty)$ and let k be a real number. From calculus, we have
$L(y+z) = \frac{d^3}{dx^3}(y+z) - 2\frac{d^2}{dx^2}(y+z) - \frac{d}{dx}(y+z) + 2(y+z)$
$= y''' + z''' - 2y'' - 2z'' - y' - z' + 2y + 2z = L(y) + L(z)$ and
$L(ky) = \frac{d^3}{dx^3}(ky) - 2\frac{d^2}{dx^2}(ky) - \frac{d}{dx}(ky) + 2(ky) = ky''' - 2ky'' - ky' + 2ky = kL(y)$
therefore L is a linear operator.

(b) Substituting $y_1 = y$, $y_2 = y'$, and $y_3 = y''$ we can rewrite $y''' - 2y'' - y' + 2y = 0$ as the system

$$\begin{aligned} y_1' &= y_2 \\ y_2' &= y_3 \\ y_3' &= -2y_1 + y_2 + 2y_3 \end{aligned}$$

This system can be expressed in the form $\mathbf{y}' = A\mathbf{y}$ where $\mathbf{y} = \begin{bmatrix} y_1 \\ y_2 \\ y_3 \end{bmatrix}$ and $A = \begin{bmatrix} 0 & 1 & 0 \\ 0 & 0 & 1 \\ -2 & 1 & 2 \end{bmatrix}$.

Cofactor expansion along the third column yields $\det(\lambda I - A) = \begin{vmatrix} \lambda & -1 & 0 \\ 0 & \lambda & -1 \\ 2 & -1 & \lambda - 2 \end{vmatrix}$

$= \begin{vmatrix} \lambda & -1 \\ 2 & -1 \end{vmatrix} + (\lambda - 2)\begin{vmatrix} \lambda & -1 \\ 0 & \lambda \end{vmatrix} = (-\lambda + 2) + (\lambda - 2)\lambda^2 = (\lambda - 2)(\lambda^2 - 1) = (\lambda - 2)(\lambda - 1)(\lambda + 1)$.

The characteristic equation is $(\lambda - 2)(\lambda - 1)(\lambda + 1) = 0$ therefore the eigenvalues are 2, 1, and -1 – each of them has the algebraic multiplicity 1.

The reduced row echelon form of $2I - A$ is $\begin{bmatrix} 1 & 0 & -\frac{1}{4} \\ 0 & 1 & -\frac{1}{2} \\ 0 & 0 & 0 \end{bmatrix}$ so that the eigenspace corresponding to $\lambda_1 = 2$ contains vectors $\begin{bmatrix} x_1 \\ x_2 \\ x_3 \end{bmatrix}$ where $x_1 = \frac{1}{4}t$, $x_2 = \frac{1}{2}t$, $x_3 = t$. A vector $\mathbf{p}_1 = \begin{bmatrix} 1 \\ 2 \\ 4 \end{bmatrix}$ forms a basis for this eigenspace.

The reduced row echelon form of $1I - A$ is $\begin{bmatrix} 1 & 0 & -1 \\ 0 & 1 & -1 \\ 0 & 0 & 0 \end{bmatrix}$ so that the eigenspace corresponding to $\lambda_2 = 1$ contains vectors $\begin{bmatrix} x_1 \\ x_2 \\ x_3 \end{bmatrix}$ where $x_1 = t$, $x_2 = t$, $x_3 = t$. A vector $\mathbf{p}_2 = \begin{bmatrix} 1 \\ 1 \\ 1 \end{bmatrix}$ forms a basis for this eigenspace.

The reduced row echelon form of $-1I - A$ is $\begin{bmatrix}1 & 0 & -1\\0 & 1 & 1\\0 & 0 & 0\end{bmatrix}$ so that the eigenspace corresponding to $\lambda_3 = -1$ contains vectors $\begin{bmatrix}x_1\\x_2\\x_3\end{bmatrix}$ where $x_1 = t, x_2 = -t, x_3 = t$. A vector $\mathbf{p}_3 = \begin{bmatrix}1\\-1\\1\end{bmatrix}$ forms a basis for this eigenspace.

Therefore $P = \begin{bmatrix}1 & 1 & 1\\2 & 1 & -1\\4 & 1 & 1\end{bmatrix}$ diagonalizes A and $P^{-1}AP = D = \begin{bmatrix}2 & 0 & 0\\0 & 1 & 0\\0 & 0 & -1\end{bmatrix}$.

The substitution $\mathbf{y} = P\mathbf{u}$ yields the "diagonal system" $\mathbf{u}' = \begin{bmatrix}2 & 0 & 0\\0 & 1 & 0\\0 & 0 & -1\end{bmatrix}\mathbf{u}$ consisting of equations $u_1' = 2u_1, u_2' = u_2$, and $u_3' = -u_3$. From Formula (2) in Section 5.4, these equations have the solutions $u_1 = c_1e^{2x}, u_2 = c_2e^{x}$, and $u_3 = c_3e^{-x}$, i.e., $\mathbf{u} = \begin{bmatrix}c_1e^{2x}\\c_2e^{x}\\c_3e^{-x}\end{bmatrix}$. From $\mathbf{y} = P\mathbf{u}$ we obtain the solution $\mathbf{y} = \begin{bmatrix}1 & 1 & 1\\2 & 1 & -1\\4 & 1 & 1\end{bmatrix}\begin{bmatrix}c_1e^{2x}\\c_2e^{x}\\c_3e^{-x}\end{bmatrix} = \begin{bmatrix}c_1e^{2x} + c_2e^{x} + c_3e^{-x}\\2c_1e^{2x} + c_2e^{x} - c_3e^{-x}\\4c_1e^{2x} + c_2e^{x} + c_3e^{-x}\end{bmatrix}$ thus $y_1 = c_1e^{2x} + c_2e^{x} + c_3e^{-x}$, $y_2 = 2c_1e^{2x} + c_2e^{x} - c_3e^{-x}$, and $y_3 = 4c_1e^{2x} + c_2e^{x} + c_3e^{-x}$.

We conclude that the differential equation $L(y) = 0$ has the solution $y = c_1e^{2x} + c_2e^{x} + c_3e^{-x}$.

True-False Exercises

(a) True. $\mathbf{y} = \mathbf{0}$ is always a solution (called the trivial solution).

(b) False. If a system has a solution $\mathbf{x} \neq \mathbf{0}$ then any for any real number k, $\mathbf{y} = k\mathbf{x}$ is also a solution.

(c) True. $(c\mathbf{x} + d\mathbf{y})' = c\mathbf{x}' + d\mathbf{y}' = c(A\mathbf{x}) + d(A\mathbf{y}) = A(c\mathbf{x}) + A(d\mathbf{y}) = A(c\mathbf{x} + d\mathbf{y})$

(d) True. The solution can be obtained by following the four-step procedure preceding Example 2.

(e) False. If $P = Q^{-1}AQ$ then $\mathbf{u}' = Q^{-1}AQ\mathbf{u}$ implies $(Q\mathbf{u})' = A(Q\mathbf{u})$. Generally, $\mathbf{u}$ and $\mathbf{y} = Q\mathbf{u}$ are not the same.

5.5 Dynamical Systems and Markov Chains

1. **(a)** A is a stochastic matrix: each column vector has nonnegative entries that add up to 1

(b) A is not a stochastic matrix since entries in its columns do not add up to 1

(c) A is a stochastic matrix: each column vector has nonnegative entries that add up to 1

(d) A is not a stochastic matrix since $(A)_{23} = -\frac{1}{2}$ fails to be nonnegative

3. $\mathbf{x}_1 = P\mathbf{x}_0 = \begin{bmatrix}0.5 & 0.6\\0.5 & 0.4\end{bmatrix}\begin{bmatrix}0.5\\0.5\end{bmatrix} = \begin{bmatrix}0.55\\0.45\end{bmatrix}$ $\quad$ $\mathbf{x}_2 = P\mathbf{x}_1 = \begin{bmatrix}0.5 & 0.6\\0.5 & 0.4\end{bmatrix}\begin{bmatrix}0.55\\0.45\end{bmatrix} = \begin{bmatrix}0.545\\0.455\end{bmatrix}$

$\mathbf{x}_3 = P\mathbf{x}_2 = \begin{bmatrix}0.5 & 0.6\\0.5 & 0.4\end{bmatrix}\begin{bmatrix}0.545\\0.455\end{bmatrix} = \begin{bmatrix}0.5455\\0.4545\end{bmatrix}$ $\quad$ $\mathbf{x}_4 = P\mathbf{x}_3 = \begin{bmatrix}0.5 & 0.6\\0.5 & 0.4\end{bmatrix}\begin{bmatrix}0.5455\\0.4545\end{bmatrix} = \begin{bmatrix}0.54545\\0.45455\end{bmatrix}$

An alternate approach is to determine $P^4 = \begin{bmatrix} 0.5455 & 0.5454 \\ 0.4545 & 0.4546 \end{bmatrix}$ then calculate $\mathbf{x}_4 = P^4\mathbf{x}_0 = \begin{bmatrix} 0.54545 \\ 0.45455 \end{bmatrix}$.

5. **(a)** P is a stochastic matrix: each column vector has nonnegative entries that add up to 1; since P has all positive entries, it is also a regular matrix.

(b) By Theorem 1.7.1(b), the product of lower triangular matrices is also lower triangular. Consequently, for all positive integers k, the matrix P^k will have 0 in the first row second column entry. Therefore P is not a regular matrix.

(c) P is a stochastic matrix: each column vector has nonnegative entries that add up to 1; since $P^2 = \begin{bmatrix} \frac{21}{25} & \frac{1}{5} \\ \frac{4}{25} & \frac{4}{5} \end{bmatrix}$ has all positive entries, we conclude that P is a regular matrix.

7. P is a stochastic matrix: each column vector has nonnegative entries that add up to 1; since P has all positive entries, it is also a regular matrix.

To find the steady-state vector, we solve the system $(I - P)\mathbf{q} = \mathbf{0}$, i.e., $\begin{bmatrix} \frac{3}{4} & -\frac{2}{3} \\ -\frac{3}{4} & \frac{2}{3} \end{bmatrix}\begin{bmatrix} q_1 \\ q_2 \end{bmatrix} = \begin{bmatrix} 0 \\ 0 \end{bmatrix}$. The reduced row echelon form of the coefficient matrix of this system is $\begin{bmatrix} 1 & -\frac{8}{9} \\ 0 & 0 \end{bmatrix}$ thus the general solution is $q_1 = \frac{8}{9}t, q_2 = t$. For $\mathbf{q}$ to be a probability vector, its components must add up to 1: $q_1 + q_2 = 1$. Solving the resulting equation $\frac{8}{9}t + t = 1$ for t results in $t = \frac{9}{17}$, consequently the steady-state vector is $\mathbf{q} = \begin{bmatrix} \frac{8}{17} \\ \frac{9}{17} \end{bmatrix}$.

9. P is a stochastic matrix: each column vector has nonnegative entries that add up to 1; since $P^2 = \begin{bmatrix} \frac{3}{8} & \frac{1}{2} & \frac{1}{6} \\ \frac{1}{3} & \frac{3}{8} & \frac{7}{18} \\ \frac{7}{24} & \frac{1}{8} & \frac{4}{9} \end{bmatrix}$ has all positive entries, we conclude that P is a regular matrix.

To find the steady-state vector, we solve the system $(I - P)\mathbf{q} = \mathbf{0}$, i.e., $\begin{bmatrix} \frac{1}{2} & -\frac{1}{2} & 0 \\ -\frac{1}{4} & \frac{1}{2} & -\frac{1}{3} \\ -\frac{1}{4} & 0 & \frac{1}{3} \end{bmatrix}\begin{bmatrix} q_1 \\ q_2 \\ q_3 \end{bmatrix} = \begin{bmatrix} 0 \\ 0 \\ 0 \end{bmatrix}$.

The reduced row echelon form of the coefficient matrix of this system is $\begin{bmatrix} 1 & 0 & -\frac{4}{3} \\ 0 & 1 & -\frac{4}{3} \\ 0 & 0 & 0 \end{bmatrix}$ thus the general solution is $q_1 = \frac{4}{3}t, q_2 = \frac{4}{3}t, q_3 = t$. For $\mathbf{q}$ to be a probability vector, we must have $q_1 + q_2 + q_3 = 1$.

Solving the resulting equation $\frac{4}{3}t + \frac{4}{3}t + t = 1$ for t results in $t = \frac{3}{11}$, consequently the steady-state vector is $\mathbf{q} = \begin{bmatrix} \frac{4}{11} \\ \frac{4}{11} \\ \frac{3}{11} \end{bmatrix}$.

11. **(a)** The entry 0.2 represents the probability that the system will stay in state 1 when it is in state 1.

(b) The entry 0.1 represents the probability that the system will move to state 1 when it is in state 2.

(c) $\begin{bmatrix} 0.2 & 0.1 \\ 0.8 & 0.9 \end{bmatrix}\begin{bmatrix} 1 \\ 0 \end{bmatrix} = \begin{bmatrix} 0.2 \\ 0.8 \end{bmatrix}$ therefore if the system is in state 1 initially, there is 0.8 probability that it will be in state 2 at the next observation.

(d) $\begin{bmatrix} 0.2 & 0.1 \\ 0.8 & 0.9 \end{bmatrix}\begin{bmatrix} 0.5 \\ 0.5 \end{bmatrix} = \begin{bmatrix} 0.15 \\ 0.85 \end{bmatrix}$ therefore if the system has a 50% chance of being in state 1 initially, it will be in state 2 at the next observation with probability 0.85.

13. **(a)** The transition matrix is

$$\begin{array}{c} \\ \textbf{good} \\ \textbf{bad} \end{array} \begin{array}{c} \begin{array}{cc} \textbf{good} & \textbf{bad} \end{array} \\ \begin{bmatrix} 0.95 & 0.55 \\ 0.05 & 0.45 \end{bmatrix} \end{array}$$

(b) $\begin{bmatrix} 0.95 & 0.55 \\ 0.05 & 0.45 \end{bmatrix}\begin{bmatrix} 0.95 & 0.55 \\ 0.05 & 0.45 \end{bmatrix}\begin{bmatrix} 1 \\ 0 \end{bmatrix} = \begin{bmatrix} 0.93 \\ 0.07 \end{bmatrix}$ therefore if the air quality is good today, it will also be good two days from now with probability 0.93 .

(c) $\begin{bmatrix} 0.95 & 0.55 \\ 0.05 & 0.45 \end{bmatrix}\begin{bmatrix} 0.95 & 0.55 \\ 0.05 & 0.45 \end{bmatrix}\begin{bmatrix} 0.95 & 0.55 \\ 0.05 & 0.45 \end{bmatrix}\begin{bmatrix} 0 \\ 1 \end{bmatrix} = \begin{bmatrix} 0.858 \\ 0.142 \end{bmatrix}$ therefore if the air quality is bad today, it will also be bad three days from now with probability 0.142 .

(d) $\begin{bmatrix} 0.95 & 0.55 \\ 0.05 & 0.45 \end{bmatrix}\begin{bmatrix} 0.2 \\ 0.8 \end{bmatrix} = \begin{bmatrix} 0.63 \\ 0.37 \end{bmatrix}$ therefore if there is a 20% chance that air quality will be good today, it will be good tomorrow with probability 0.63 .

15. **(a)** The transition matrix is

$$\begin{array}{c} \\ \textbf{city} \\ \textbf{suburbs} \end{array} \begin{array}{c} \begin{array}{cc} \textbf{city} & \textbf{suburbs} \end{array} \\ \begin{bmatrix} 0.95 & 0.03 \\ 0.05 & 0.97 \end{bmatrix} \end{array} = P$$

The initial state vector $\mathbf{x}_0 = \begin{bmatrix} \frac{100{,}000}{125{,}000} \\ \frac{25{,}000}{125{,}000} \end{bmatrix} = \begin{bmatrix} 0.8 \\ 0.2 \end{bmatrix}$ represents the fractions of the total population (125,000) living in the city and in the suburbs, respectively.

After one year, the corresponding fractions are contained in the state vector $\mathbf{x}_1 = P\mathbf{x}_0 = \begin{bmatrix} 0.95 & 0.03 \\ 0.05 & 0.97 \end{bmatrix}\begin{bmatrix} 0.8 \\ 0.2 \end{bmatrix} = \begin{bmatrix} 0.766 \\ 0.234 \end{bmatrix}$. To determine the populations living in the city and in the suburbs at that time, we can calculate the scalar multiple of the state vector: $125{,}000\ \mathbf{x}_1 = \begin{bmatrix} 95{,}750 \\ 29{,}250 \end{bmatrix}$.

After the second year, the state vector becomes $\mathbf{x}_2 = P\mathbf{x}_1 = \begin{bmatrix} 0.73472 \\ 0.26528 \end{bmatrix}$, and the corresponding population counts are $125{,}000\,\mathbf{x}_2 = \begin{bmatrix}91{,}840\\33{,}160\end{bmatrix}$.

Repeating this process three more times results in

	initial state $k=0$	after 1 year $k=1$	after 2 years $k=2$	after 3 years $k=3$	after 4 years $k=4$	after 5 years $k=5$
state vector $\mathbf{x}_k \approx$	$\begin{bmatrix}0.8\\0.2\end{bmatrix}$	$\begin{bmatrix}0.766\\0.234\end{bmatrix}$	$\begin{bmatrix}0.73472\\0.26528\end{bmatrix}$	$\begin{bmatrix}0.705942\\0.294058\end{bmatrix}$	$\begin{bmatrix}0.679467\\0.320533\end{bmatrix}$	$\begin{bmatrix}0.655110\\0.344890\end{bmatrix}$
city population suburb population	100,000 25,000	95,750 29,250	91,840 33,160	88,243 36,757	84,933 40,067	81,889 43,111

(b) Since P is a regular stochastic matrix, there exists a unique steady-state probability vector. To find the steady-state vector, we solve the system $(I-P)\mathbf{q} = \mathbf{0}$, i.e., $\begin{bmatrix}0.05 & -0.03\\-0.05 & 0.03\end{bmatrix}\begin{bmatrix}q_1\\q_2\end{bmatrix} = \begin{bmatrix}0\\0\end{bmatrix}$. The reduced row echelon form of the coefficient matrix of this system is $\begin{bmatrix}1 & -\frac{3}{5}\\0 & 0\end{bmatrix}$ thus the general solution is $q_1 = \frac{3}{5}t$, $q_2 = t$. The components of the vector $\mathbf{q}$ must add up to 1: $q_1 + q_2 = 1$. Solving the resulting equation $\frac{3}{5}t + t = 1$ for t results in $t = \frac{5}{8}$, consequently over the long term the fractions of the total population living in the city and in the suburbs will approach $\frac{3}{5}\cdot\frac{5}{8} = \frac{3}{8}$ and $\frac{5}{8}$, respectively.

We conclude that the city population will approach $\frac{3}{8}\cdot 125{,}000 = 46{,}875$ and the suburbs population will approach $\frac{5}{8}\cdot 125{,}000 = 78{,}125$.

17. **(a)** The transition matrix is $P = \begin{bmatrix}\frac{1}{10} & \frac{1}{5} & \frac{3}{5}\\ \frac{4}{5} & \frac{3}{10} & \frac{1}{5}\\ \frac{1}{10} & \frac{1}{2} & \frac{1}{5}\end{bmatrix}$. Multiplying this matrix by the initial state vector $\mathbf{x}_0 = \begin{bmatrix}1\\0\\0\end{bmatrix}$ results in $\mathbf{x}_1 = P\mathbf{x}_0 = \begin{bmatrix}\frac{1}{10}\\ \frac{4}{5}\\ \frac{1}{10}\end{bmatrix}$. After the second rental, the state vector becomes

$$\mathbf{x}_2 = P\mathbf{x}_1 = \begin{bmatrix}\frac{23}{100}\\ \frac{17}{50}\\ \frac{43}{100}\end{bmatrix}.$$

The conclude that the car originally rented from location 1 will after two rentals be at location 1 with probability 0.23 .

(b) Since P is a regular stochastic matrix, there exists a unique steady-state probability vector. To find the steady-state vector, we solve the system $(I - P)\mathbf{q} = \mathbf{0}$, i.e., $\begin{bmatrix} \frac{9}{10} & -\frac{1}{5} & -\frac{3}{5} \\ -\frac{4}{5} & \frac{7}{10} & -\frac{1}{5} \\ -\frac{1}{10} & -\frac{1}{2} & \frac{4}{5} \end{bmatrix}\begin{bmatrix} q_1 \\ q_2 \\ q_3 \end{bmatrix} = \begin{bmatrix} 0 \\ 0 \\ 0 \end{bmatrix}$. The reduced row echelon form of the coefficient matrix of this system is $\begin{bmatrix} 1 & 0 & -\frac{46}{47} \\ 0 & 1 & -\frac{66}{47} \\ 0 & 0 & 0 \end{bmatrix}$ thus the general solution is $q_1 = \frac{46}{47}t$, $q_2 = \frac{66}{47}t$, $q_3 = t$. For $\mathbf{q}$ to be a probability vector, we must have $q_1 + q_2 + q_3 = 1$. Solving the resulting equation $\frac{46}{47}t + \frac{66}{47}t + t = 1$ for t results in $t = \frac{47}{159}$, consequently the steady-state vector is $\mathbf{q} = \begin{bmatrix} \frac{46}{159} \\ \frac{66}{159} \\ \frac{47}{159} \end{bmatrix} = \begin{bmatrix} \frac{46}{159} \\ \frac{22}{53} \\ \frac{47}{159} \end{bmatrix} \approx \begin{bmatrix} 0.289 \\ 0.415 \\ 0.296 \end{bmatrix}$.

(c) Over the long term, the number of cars will approach $120\begin{bmatrix} \frac{46}{159} \\ \frac{22}{53} \\ \frac{47}{159} \end{bmatrix} \approx \begin{bmatrix} 35 \\ 50 \\ 35 \end{bmatrix}$ so the rental agency should allocate no fewer than 35, 50, and 35 parking spaces at locations 1, 2, and 3, respectively.

19. For the matrix $P = \begin{bmatrix} \frac{7}{10} & p_{12} & \frac{1}{5} \\ p_{21} & \frac{3}{10} & p_{23} \\ \frac{1}{10} & \frac{3}{5} & \frac{3}{10} \end{bmatrix}$ to be stochastic, each column vector must be a probability vector: a vector with nonnegative entries that add up to one. Applying the latter condition to each column results in three equations, which can be used to solve for the missing entries:

$$\begin{array}{llll} \text{column 1:} & \frac{7}{10} + p_{21} + \frac{1}{10} = 1 & \text{yields} & p_{21} = 1 - \frac{7}{10} - \frac{1}{10} = \frac{2}{10} = \frac{1}{5} \\ \text{column 2:} & p_{12} + \frac{3}{10} + \frac{3}{5} = 1 & \text{yields} & p_{12} = 1 - \frac{3}{10} - \frac{3}{5} = \frac{1}{10} \\ \text{column 3:} & \frac{1}{5} + p_{23} + \frac{3}{10} = 1 & \text{yields} & p_{23} = 1 - \frac{1}{5} - \frac{3}{10} = \frac{5}{10} = \frac{1}{2} \end{array}$$

The resulting transition matrix is $P = \begin{bmatrix} \frac{7}{10} & \frac{1}{10} & \frac{1}{5} \\ \frac{1}{5} & \frac{3}{10} & \frac{1}{2} \\ \frac{1}{10} & \frac{3}{5} & \frac{3}{10} \end{bmatrix}$. Since P is a regular stochastic matrix, there exists a unique steady-state probability vector. To find the steady-state vector, we solve the system

$(I-P)\mathbf{q}=\mathbf{0}$, i.e., $\begin{bmatrix} \frac{3}{10} & -\frac{1}{10} & -\frac{1}{5} \\ -\frac{1}{5} & \frac{7}{10} & -\frac{1}{2} \\ -\frac{1}{10} & -\frac{3}{5} & \frac{7}{10} \end{bmatrix}\begin{bmatrix} q_1 \\ q_2 \\ q_3 \end{bmatrix}=\begin{bmatrix} 0 \\ 0 \\ 0 \end{bmatrix}$. The reduced row echelon form of the coefficient matrix of this system is $\begin{bmatrix} 1 & 0 & -1 \\ 0 & 1 & -1 \\ 0 & 0 & 0 \end{bmatrix}$ thus the general solution is $q_1=t, q_2=t, q_3=t$. For $\mathbf{q}$ to be a probability vector, we must have $q_1+q_2+q_3=1$. Solving the resulting equation $t+t+t=1$ for t results in $t=\frac{1}{3}$, consequently the steady-state vector is $\mathbf{q}=\begin{bmatrix} \frac{1}{3} \\ \frac{1}{3} \\ \frac{1}{3} \end{bmatrix}$.

21. From Theorem 5.5.1(a), we have $P\mathbf{q}=\mathbf{q}$. Therefore for any positive integer k,

$$P^k\mathbf{q}=P^{k-1}(P\mathbf{q})=P^{k-1}\mathbf{q}=P^{k-2}(P\mathbf{q})=P^{k-2}\mathbf{q}=\cdots=\mathbf{q}$$

23. Let A and B be two $n\times n$ stochastic matrices, and let B be partitioned into columns: $B=[\,\mathbf{b}_1 \mid \mathbf{b}_2 \mid \dots \mid \mathbf{b}_n\,]$. Using Formula (6) in Section 1.3, we can now see that the product

$$AB=A[\,\mathbf{b}_1 \mid \mathbf{b}_2 \mid \dots \mid \mathbf{b}_n\,]=[A\,\mathbf{b}_1 \mid A\,\mathbf{b}_2 \mid \dots \mid A\,\mathbf{b}_n\,]$$

has columns that are probability vectors (since each of them is a product of a stochastic matrix and a probability vector). We conclude that AB is stochastic.

True-False Exercises

(a) True. All entries are nonnegative and their sum is 1.

(b) True. This is a stochastic matrix since its columns are probability vectors.

Furthermore, $\begin{bmatrix} 0.2 & 1 \\ 0.8 & 0 \end{bmatrix}^2=\begin{bmatrix} 0.84 & 0.2 \\ 0.16 & 0.8 \end{bmatrix}$ has all positive entries.

(c) True. By definition, a transition matrix is a stochastic matrix.

(d) False. For $\mathbf{q}$ to be a steady-state vector of a regular Markov chain, it must also be a probability vector.

(e) True. (See Exercise 23.)

(f) False. The entries must be nonnegative.

(g) True. This follows from Theorem 5.5.1(a).

Chapter 5 Supplementary Exercises

1. **(a)** The characteristic polynomial is

$$\det(\lambda I-A)=\begin{vmatrix} \lambda-\cos\theta & \sin\theta \\ -\sin\theta & \lambda-\cos\theta \end{vmatrix}=(\lambda-\cos\theta)^2+(\sin\theta)^2.$$

For a real eigenvalue λ to exist, we must have $\lambda=\cos\theta$ and $\sin\theta=0$. However, the latter equation has no solutions on the given interval $0<\theta<\pi$, therefore A has no real eigenvalues, and consequently no real eigenvectors.

(b) According to Table 1 in Section 4.11, A is the standard matrix of the rotation in the plane about the origin through a positive angle θ. Unless the angle is an integer multiple of π, no vector resulting from such a rotation is a scalar multiple of the original nonzero vector.

3. **(a)** If $D = \begin{bmatrix} d_{11} & 0 & \cdots & 0 \\ 0 & d_{22} & \cdots & 0 \\ \vdots & \vdots & \ddots & \vdots \\ 0 & 0 & \cdots & d_{nn} \end{bmatrix}$ with $d_{ii} \geq 0$ for all i then we can take

$S = \begin{bmatrix} \sqrt{d_{11}} & 0 & \cdots & 0 \\ 0 & \sqrt{d_{22}} & \cdots & 0 \\ \vdots & \vdots & \ddots & \vdots \\ 0 & 0 & \cdots & \sqrt{d_{nn}} \end{bmatrix}$ so that $S^2 = D$ holds true. (Note that the answer is not unique: the main diagonal entries of S could be negative square roots instead.)

(b) From our assumptions it follows that there exists a matrix P such that $A = P^{-1}DA$ where

$D = \begin{bmatrix} \lambda_1 & 0 & \cdots & 0 \\ 0 & \lambda_2 & \cdots & 0 \\ \vdots & \vdots & \ddots & \vdots \\ 0 & 0 & \cdots & \lambda_n \end{bmatrix}$ with $\lambda_i \geq 0$ for all i. Taking $R = \begin{bmatrix} \sqrt{\lambda_1} & 0 & \cdots & 0 \\ 0 & \sqrt{\lambda_2} & \cdots & 0 \\ \vdots & \vdots & \ddots & \vdots \\ 0 & 0 & \cdots & \sqrt{\lambda_n} \end{bmatrix}$ so that

$R^2 = D$ (see part (a)), we can form the matrix $S = PRP^{-1}$ so that $S^2 = PRP^{-1}PRP^{-1} = PR^2P^{-1} = PDP^{-1} = A$.

(c) By Theorem 5.1.2, A has eigenvalues $\lambda_1 = 1, \lambda_2 = 4$, and $\lambda_3 = 9$.

The reduced row echelon form of $1I - A$ is $\begin{bmatrix} 0 & 1 & 0 \\ 0 & 0 & 1 \\ 0 & 0 & 0 \end{bmatrix}$ so that the eigenspace corresponding to $\lambda_1 = 1$ contains vectors $\begin{bmatrix} x_1 \\ x_2 \\ x_3 \end{bmatrix}$ where $x_1 = t, x_2 = 0, x_3 = 0$. A vector $\mathbf{p}_1 = \begin{bmatrix} 1 \\ 0 \\ 0 \end{bmatrix}$ forms a basis for this eigenspace.

The reduced row echelon form of $4I - A$ is $\begin{bmatrix} 1 & -1 & 0 \\ 0 & 0 & 1 \\ 0 & 0 & 0 \end{bmatrix}$ so that the eigenspace corresponding to $\lambda_2 = 4$ contains vectors $\begin{bmatrix} x_1 \\ x_2 \\ x_3 \end{bmatrix}$ where $x_1 = t, x_2 = t, x_3 = 0$. A vector $\mathbf{p}_2 = \begin{bmatrix} 1 \\ 1 \\ 0 \end{bmatrix}$ forms a basis for this eigenspace.

The reduced row echelon form of $9I - A$ is $\begin{bmatrix} 1 & 0 & -\frac{1}{2} \\ 0 & 1 & -1 \\ 0 & 0 & 0 \end{bmatrix}$ so that the eigenspace corresponding to $\lambda_3 = 9$ contains vectors $\begin{bmatrix} x_1 \\ x_2 \\ x_3 \end{bmatrix}$ where $x_1 = \frac{1}{2}t, x_2 = t, x_3 = t$. A vector $\mathbf{p}_3 = \begin{bmatrix} 1 \\ 2 \\ 2 \end{bmatrix}$ forms a basis for this eigenspace.

Therefore $P = \begin{bmatrix} 1 & 1 & 1 \\ 0 & 1 & 2 \\ 0 & 0 & 2 \end{bmatrix}$ diagonalizes A and $P^{-1}AP = D = \begin{bmatrix} 1 & 0 & 0 \\ 0 & 4 & 0 \\ 0 & 0 & 9 \end{bmatrix}$.

Since the reduced row echelon form of $\begin{bmatrix}1&1&1&1&0&0\\0&1&2&0&1&0\\0&0&2&0&0&1\end{bmatrix}$ is $\begin{bmatrix}1&0&0&1&-1&\frac{1}{2}\\0&1&0&0&1&-1\\0&0&1&0&0&\frac{1}{2}\end{bmatrix}$, we have $P^{-1}=\begin{bmatrix}1&-1&\frac{1}{2}\\0&1&-1\\0&0&\frac{1}{2}\end{bmatrix}$. As described in the solution of part (b) we can let

$R=\begin{bmatrix}\sqrt{1}&0&0\\0&\sqrt{4}&0\\0&0&\sqrt{9}\end{bmatrix}=\begin{bmatrix}1&0&0\\0&2&0\\0&0&3\end{bmatrix}$ and form

$S=PRP^{-1}=\begin{bmatrix}1&1&1\\0&1&2\\0&0&2\end{bmatrix}\begin{bmatrix}1&0&0\\0&2&0\\0&0&3\end{bmatrix}\begin{bmatrix}1&-1&\frac{1}{2}\\0&1&-1\\0&0&\frac{1}{2}\end{bmatrix}=\begin{bmatrix}1&1&0\\0&2&1\\0&0&3\end{bmatrix}$. This matrix satisfies $S^2=A$.

7. **(a)** The characteristic polynomial is $\det(\lambda I-A)=\begin{vmatrix}\lambda-3&-6\\-1&\lambda-2\end{vmatrix}=-5\lambda+\lambda^2$.

We verify that $-5A+A^2=-5\begin{bmatrix}3&6\\1&2\end{bmatrix}+\begin{bmatrix}3&6\\1&2\end{bmatrix}^2=-\begin{bmatrix}15&30\\5&10\end{bmatrix}+\begin{bmatrix}15&30\\5&10\end{bmatrix}=\begin{bmatrix}0&0\\0&0\end{bmatrix}$.

(b) The characteristic polynomial is $\det(\lambda I-A)=\begin{vmatrix}\lambda&-1&0\\0&\lambda&-1\\-1&3&\lambda-3\end{vmatrix}=-1+3\lambda-3\lambda^2+\lambda^3$.

We verify that

$$-I_3+3A-3A^2+A^3=-\begin{bmatrix}1&0&0\\0&1&0\\0&0&1\end{bmatrix}+3\begin{bmatrix}0&1&0\\0&0&1\\1&-3&3\end{bmatrix}-3\begin{bmatrix}0&1&0\\0&0&1\\1&-3&3\end{bmatrix}^2+\begin{bmatrix}0&1&0\\0&0&1\\1&-3&3\end{bmatrix}^3$$
$$=-\begin{bmatrix}1&0&0\\0&1&0\\0&0&1\end{bmatrix}+\begin{bmatrix}0&3&0\\0&0&3\\3&-9&9\end{bmatrix}-\begin{bmatrix}0&0&3\\3&-9&9\\9&-24&18\end{bmatrix}+\begin{bmatrix}1&-3&3\\3&-8&6\\6&-15&10\end{bmatrix}$$
$$=\begin{bmatrix}0&0&0\\0&0&0\\0&0&0\end{bmatrix}$$

9. Since $\det(\lambda I-A)=\begin{vmatrix}\lambda-3&-6\\-1&\lambda-2\end{vmatrix}=\lambda^2-5\lambda$, it follows from the Cayley-Hamilton Theorem that $A^2-5A=0$. This yields $A^2=5A=5\begin{bmatrix}3&6\\1&2\end{bmatrix}=\begin{bmatrix}15&30\\5&10\end{bmatrix}$, $A^3=5A^2=5\begin{bmatrix}15&30\\5&10\end{bmatrix}=\begin{bmatrix}75&150\\25&50\end{bmatrix}$, $A^4=5A^3=5\begin{bmatrix}75&150\\25&50\end{bmatrix}=\begin{bmatrix}375&750\\125&250\end{bmatrix}$, and $A^5=5A^4=5\begin{bmatrix}375&750\\125&250\end{bmatrix}=\begin{bmatrix}1875&3750\\625&1250\end{bmatrix}$.

11. Method I

For λ to be an eigenvalue of A associated with a nonzero eigenvector $\mathbf{x}=\begin{bmatrix}x_1\\x_2\\\vdots\\x_n\end{bmatrix}$, we must have

$A\mathbf{x}=\lambda\mathbf{x}$ i.e.

$$\begin{bmatrix} c_1x_1 + c_2x_2 + \cdots + c_nx_n \\ c_1x_1 + c_2x_2 + \cdots + c_nx_n \\ \vdots \\ c_1x_1 + c_2x_2 + \cdots + c_nx_n \end{bmatrix} = \lambda \begin{bmatrix} x_1 \\ x_2 \\ \vdots \\ x_n \end{bmatrix}$$

There are two possibilities:

- If $\lambda \neq 0$ then $x_1 = x_2 = \cdots = x_n$. This implies $\lambda = c_1 + \cdots + c_n = \text{tr}(A)$.
- If $\lambda = 0$ then $A\mathbf{x} = \lambda\mathbf{x}$ becomes a homogeneous system $A\mathbf{x} = \mathbf{0}$; its coefficient matrix A can be reduced to $\begin{bmatrix} c_1 & c_2 & \cdots & c_n \\ 0 & 0 & \cdots & 0 \\ \vdots & \vdots & \ddots & \vdots \\ 0 & 0 & \cdots & 0 \end{bmatrix}$. The solution space has dimension of at least $n-1$ therefore $\lambda = 0$ is an eigenvalue whose geometric multiplicity is at least $n-1$.

We conclude that the only eigenvalues of A are 0 and $\text{tr}(A)$.

Method II

$$\det(\lambda I - A) = \begin{vmatrix} \lambda - c_1 & -c_2 & -c_3 & \cdots & -c_{n-1} & -c_n \\ -c_1 & \lambda - c_2 & -c_3 & \cdots & -c_{n-1} & -c_n \\ -c_1 & -c_2 & \lambda - c_3 & \cdots & -c_{n-1} & -c_n \\ \vdots & \vdots & \vdots & \ddots & \vdots & \vdots \\ -c_1 & -c_2 & -c_3 & \cdots & \lambda - c_{n-1} & -c_n \\ -c_1 & -c_2 & -c_3 & \cdots & -c_{n-1} & \lambda - c_n \end{vmatrix}$$

$$= \begin{vmatrix} \lambda - c_1 & -c_2 & -c_3 & \cdots & -c_{n-1} & -c_n \\ -\lambda & \lambda & 0 & \cdots & 0 & 0 \\ -\lambda & 0 & \lambda & \cdots & 0 & 0 \\ \vdots & \vdots & \vdots & \ddots & \vdots & \vdots \\ -\lambda & 0 & 0 & \cdots & \lambda & 0 \\ -\lambda & 0 & 0 & \cdots & 0 & \lambda \end{vmatrix}$$

← -1 times the first row was added to each of the remaining rows.

$$= \begin{vmatrix} \lambda - c_1 - c_2 - c_3 - \cdots - c_{n-1} - c_n & -c_2 & -c_3 & \cdots & -c_{n-1} & -c_n \\ 0 & \lambda & 0 & \cdots & 0 & 0 \\ 0 & 0 & \lambda & \cdots & 0 & 0 \\ \vdots & \vdots & \vdots & \ddots & \vdots & \vdots \\ 0 & 0 & 0 & \cdots & \lambda & 0 \\ 0 & 0 & 0 & \cdots & 0 & \lambda \end{vmatrix}$$

← Each of the columns from the second to the last was added to the first column.

$$= (\lambda - c_1 - c_2 - c_3 - \cdots - c_{n-1} - c_n)\lambda^{n-1}$$

We conclude that the only eigenvalues of A are 0 and $\text{tr}(A) = c_1 + \cdots + c_n$.

13. By Theorem 5.1.2, all eigenvalues of $A^n = 0$ are 0. By Theorem 5.2.3, if A had any eigenvalue $\lambda \neq 0$ then λ^n would be an eigenvalue of A^n. We reached a contradiction, therefore all eigenvalues of A must be 0.

15. The three given eigenvectors can be used as columns of a matrix $P = \begin{bmatrix} 0 & 1 & 0 \\ 1 & -1 & 1 \\ -1 & 1 & 1 \end{bmatrix}$ which diagonalizes A, i.e. $P^{-1}AP = D = \begin{bmatrix} 0 & 0 & 0 \\ 0 & 1 & 0 \\ 0 & 0 & -1 \end{bmatrix}$. The latter equation is equivalent to $A = PDP^{-1}$. The matrix $\begin{bmatrix} 0 & 1 & 0 & 1 & 0 & 0 \\ 1 & -1 & 1 & 0 & 1 & 0 \\ -1 & 1 & 1 & 0 & 0 & 1 \end{bmatrix}$ has the reduced row echelon form $\begin{bmatrix} 1 & 0 & 0 & 1 & \frac{1}{2} & -\frac{1}{2} \\ 0 & 1 & 0 & 1 & 0 & 0 \\ 0 & 0 & 1 & 0 & \frac{1}{2} & \frac{1}{2} \end{bmatrix}$ therefore

$P^{-1} = \begin{bmatrix} 1 & \frac{1}{2} & -\frac{1}{2} \\ 1 & 0 & 0 \\ 0 & \frac{1}{2} & \frac{1}{2} \end{bmatrix}$. We conclude that a matrix A satisfying the given conditions is

$$A = \begin{bmatrix} 0 & 1 & 0 \\ 1 & -1 & 1 \\ -1 & 1 & 1 \end{bmatrix} \begin{bmatrix} 0 & 0 & 0 \\ 0 & 1 & 0 \\ 0 & 0 & -1 \end{bmatrix} \begin{bmatrix} 1 & \frac{1}{2} & -\frac{1}{2} \\ 1 & 0 & 0 \\ 0 & \frac{1}{2} & \frac{1}{2} \end{bmatrix} = \begin{bmatrix} 1 & 0 & 0 \\ -1 & -\frac{1}{2} & -\frac{1}{2} \\ 1 & -\frac{1}{2} & -\frac{1}{2} \end{bmatrix}.$$

17. By Theorem 5.2.3, if A had any eigenvalue λ then λ^3 is an eigenvalue of A^3 corresponding to the same eigenvector. From $A^3 = A$ it follows that $\lambda^3 = \lambda$, so the only possible eigenvalues are -1, 0, and 1.

19. Let a and b denote the two unknown eigenvalues. We solve the system $a + b + 1 = 6$ and $a \cdot b \cdot 1 = 6$. Rewriting the first equation as $b = 5 - a$ and substituting into the second equation yields $a(5 - a) = 6$, therefore $(a - 2)(a - 3) = 0$. Either $a = 2$ (and $b = 3$) or $a = 3$ (and $b = 2$). We conclude that the unknown eigenvalues are 2 and 3.

CHAPTER 6: INNER PRODUCT SPACES

6.1 Inner Products

1. **(a)** $\langle \mathbf{u}, \mathbf{v} \rangle = 2(1)(3) + 3(1)(2) = 12$

(b) $\langle k\mathbf{v}, \mathbf{w} \rangle = 2((3)(3))(0) + 3((3)(2))(-1) = -18$

(c) $\langle \mathbf{u} + \mathbf{v}, \mathbf{w} \rangle = 2(1+3)(0) + 3(1+2)(-1) = -9$

(d) $\|\mathbf{v}\| = \langle \mathbf{v}, \mathbf{v} \rangle^{1/2} = [2(3)(3) + 3(2)(2)]^{1/2} = \sqrt{30}$

(e) $d(\mathbf{u}, \mathbf{v}) = \|\mathbf{u} - \mathbf{v}\| = \langle(-2,-1),(-2,-1)\rangle^{1/2} = [2(-2)(-2) + 3(-1)(-1)]^{1/2} = \sqrt{11}$

(f) $\|\mathbf{u} - k\mathbf{v}\| = \langle(-8,-5),(-8,-5)\rangle^{1/2} = [2(-8)(-8) + 3(-5)(-5)]^{1/2} = \sqrt{203}$

3. **(a)** $\langle \mathbf{u}, \mathbf{v} \rangle = \left(\begin{bmatrix} 2 & 1 \\ 1 & 1 \end{bmatrix}\begin{bmatrix} 1 \\ 1 \end{bmatrix}\right) \cdot \left(\begin{bmatrix} 2 & 1 \\ 1 & 1 \end{bmatrix}\begin{bmatrix} 3 \\ 2 \end{bmatrix}\right) = \begin{bmatrix} 3 \\ 2 \end{bmatrix} \cdot \begin{bmatrix} 8 \\ 5 \end{bmatrix} = 34$

(b) $\langle k\mathbf{v}, \mathbf{w} \rangle = \left(\begin{bmatrix} 2 & 1 \\ 1 & 1 \end{bmatrix}\begin{bmatrix} 9 \\ 6 \end{bmatrix}\right) \cdot \left(\begin{bmatrix} 2 & 1 \\ 1 & 1 \end{bmatrix}\begin{bmatrix} 0 \\ -1 \end{bmatrix}\right) = \begin{bmatrix} 24 \\ 15 \end{bmatrix} \cdot \begin{bmatrix} -1 \\ -1 \end{bmatrix} = -39$

(c) $\langle \mathbf{u} + \mathbf{v}, \mathbf{w} \rangle = \left(\begin{bmatrix} 2 & 1 \\ 1 & 1 \end{bmatrix}\begin{bmatrix} 4 \\ 3 \end{bmatrix}\right) \cdot \left(\begin{bmatrix} 2 & 1 \\ 1 & 1 \end{bmatrix}\begin{bmatrix} 0 \\ -1 \end{bmatrix}\right) = \begin{bmatrix} 11 \\ 7 \end{bmatrix} \cdot \begin{bmatrix} -1 \\ -1 \end{bmatrix} = -18$

(d) $\|\mathbf{v}\| = \langle \mathbf{v}, \mathbf{v} \rangle^{1/2} = \left[\left(\begin{bmatrix} 2 & 1 \\ 1 & 1 \end{bmatrix}\begin{bmatrix} 3 \\ 2 \end{bmatrix}\right) \cdot \left(\begin{bmatrix} 2 & 1 \\ 1 & 1 \end{bmatrix}\begin{bmatrix} 3 \\ 2 \end{bmatrix}\right)\right]^{1/2} = \left(\begin{bmatrix} 8 \\ 5 \end{bmatrix} \cdot \begin{bmatrix} 8 \\ 5 \end{bmatrix}\right)^{1/2} = \sqrt{89}$

(e) $d(\mathbf{u}, \mathbf{v}) = \|\mathbf{u} - \mathbf{v}\| = \left[\left(\begin{bmatrix} 2 & 1 \\ 1 & 1 \end{bmatrix}\begin{bmatrix} -2 \\ -1 \end{bmatrix}\right) \cdot \left(\begin{bmatrix} 2 & 1 \\ 1 & 1 \end{bmatrix}\begin{bmatrix} -2 \\ -1 \end{bmatrix}\right)\right]^{1/2} = \left(\begin{bmatrix} -5 \\ -3 \end{bmatrix} \cdot \begin{bmatrix} -5 \\ -3 \end{bmatrix}\right)^{1/2} = \sqrt{34}$

(f) $\|\mathbf{u} - k\mathbf{v}\| = \left[\left(\begin{bmatrix} 2 & 1 \\ 1 & 1 \end{bmatrix}\begin{bmatrix} -8 \\ -5 \end{bmatrix}\right) \cdot \left(\begin{bmatrix} 2 & 1 \\ 1 & 1 \end{bmatrix}\begin{bmatrix} -8 \\ -5 \end{bmatrix}\right)\right]^{1/2} = \left(\begin{bmatrix} -21 \\ -13 \end{bmatrix} \cdot \begin{bmatrix} -21 \\ -13 \end{bmatrix}\right)^{1/2} = \sqrt{610}$

5. $\begin{bmatrix} \sqrt{2} & 0 \\ 0 & \sqrt{3} \end{bmatrix}$

7. $\langle \mathbf{u}, \mathbf{v} \rangle = \left(\begin{bmatrix} 4 & 1 \\ 2 & -3 \end{bmatrix}\begin{bmatrix} 0 \\ -3 \end{bmatrix}\right) \cdot \left(\begin{bmatrix} 4 & 1 \\ 2 & -3 \end{bmatrix}\begin{bmatrix} 6 \\ 2 \end{bmatrix}\right) = \begin{bmatrix} -3 \\ 9 \end{bmatrix} \cdot \begin{bmatrix} 26 \\ 6 \end{bmatrix} = -24$

9. If $\mathbf{u} = U$ and $\mathbf{v} = V$ then $\langle \mathbf{u}, \mathbf{v} \rangle = \text{tr}(U^T V) = \text{tr}\left(\begin{bmatrix} 1 & 13 \\ 10 & 2 \end{bmatrix}\right) = 3.$

11. $\langle \mathbf{p}, \mathbf{q} \rangle = (-2)(4) + (1)(0) + (3)(-7) = -29$

13. $\begin{bmatrix} \sqrt{3} & 0 \\ 0 & \sqrt{5} \end{bmatrix}$

15. $\langle \mathbf{p}, \mathbf{q} \rangle = p(-2)q(-2) + p(-1)q(-1) + p(0)q(0) + p(1)q(1)$
$= (-10)(5) + (-2)(2) + (0)(1) + (2)(2) = -50$

17. $\|\mathbf{u}\| = \langle \mathbf{u}, \mathbf{u} \rangle^{1/2} = [2(-3)(-3) + 3(2)(2)]^{1/2} = \sqrt{30}$
$d(\mathbf{u}, \mathbf{v}) = \|\mathbf{u} - \mathbf{v}\| = \langle(-4,-5),(-4,-5)\rangle^{1/2} = [2(-4)(-4) + 3(-5)(-5)]^{1/2} = \sqrt{107}$

19. $\|\mathbf{p}\| = \langle \mathbf{p}, \mathbf{p} \rangle^{1/2} = \sqrt{(-2)^2 + 1^2 + 3^2} = \sqrt{14}$; $d(\mathbf{p}, \mathbf{q}) = \|\mathbf{p} - \mathbf{q}\| = \sqrt{(-6)^2 + 1^2 + 10^2} = \sqrt{137}$

21. If $\mathbf{u} = U$ and $\mathbf{v} = V$ then $\|U\| = \langle \mathbf{u}, \mathbf{u} \rangle^{1/2} = \sqrt{\operatorname{tr}(U^T U)} = \sqrt{\operatorname{tr}\left(\begin{bmatrix} 25 & 26 \\ 26 & 68 \end{bmatrix}\right)} = \sqrt{93}$ and

$$d(U, V) = \|U - V\| = \langle \mathbf{u} - \mathbf{v}, \mathbf{u} - \mathbf{v} \rangle^{1/2} = \sqrt{\operatorname{tr}((U - V)^T (U - V))} = \sqrt{\operatorname{tr}\left(\begin{bmatrix} 25 & 1 \\ 1 & 74 \end{bmatrix}\right)} = \sqrt{99} = 3\sqrt{11}$$

23. $\|\mathbf{p}\| = \langle \mathbf{p}, \mathbf{p} \rangle^{1/2} = \sqrt{[p(-2)]^2 + [p(-1)]^2 + [p(0)]^2 + [p(1)]^2} = \sqrt{(-10)^2 + (-2)^2 + 0^2 + 2^2} = 6\sqrt{3}$

$d(\mathbf{p}, \mathbf{q}) = \|\mathbf{p} - \mathbf{q}\| = \sqrt{[p(-2) - q(-2)]^2 + [p(-1) - q(-1)]^2 + [p(0) - q(0)]^2 + [p(1) - q(1)]^2}$

$= \sqrt{(-15)^2 + (-4)^2 + (-1)^2 + 0^2} = 11\sqrt{2}$

25. $\|\mathbf{u}\| = \langle \mathbf{u}, \mathbf{u} \rangle^{1/2} = \left[\left(\begin{bmatrix} 4 & 0 \\ 3 & 5 \end{bmatrix}\begin{bmatrix} -1 \\ 2 \end{bmatrix}\right) \cdot \left(\begin{bmatrix} 4 & 0 \\ 3 & 5 \end{bmatrix}\begin{bmatrix} -1 \\ 2 \end{bmatrix}\right)\right]^{1/2} = \left(\begin{bmatrix} -4 \\ 7 \end{bmatrix} \cdot \begin{bmatrix} -4 \\ 7 \end{bmatrix}\right)^{1/2} = \sqrt{65}$

$d(\mathbf{u}, \mathbf{v}) = \|\mathbf{u} - \mathbf{v}\| = \left[\left(\begin{bmatrix} 4 & 0 \\ 3 & 5 \end{bmatrix}\begin{bmatrix} -3 \\ -3 \end{bmatrix}\right) \cdot \left(\begin{bmatrix} 4 & 0 \\ 3 & 5 \end{bmatrix}\begin{bmatrix} -3 \\ -3 \end{bmatrix}\right)\right]^{1/2} = \left(\begin{bmatrix} -12 \\ -24 \end{bmatrix} \cdot \begin{bmatrix} -12 \\ -24 \end{bmatrix}\right)^{1/2} = 12\sqrt{5}$

27. (a) $\langle 2\mathbf{v} - \mathbf{w}, 3\mathbf{u} + 2\mathbf{w} \rangle = \langle 2\mathbf{v}, 3\mathbf{u} + 2\mathbf{w} \rangle - \langle \mathbf{w}, 3\mathbf{u} + 2\mathbf{w} \rangle = \langle 2\mathbf{v}, 3\mathbf{u} \rangle + \langle 2\mathbf{v}, 2\mathbf{w} \rangle - \langle \mathbf{w}, 3\mathbf{u} \rangle - \langle \mathbf{w}, 2\mathbf{w} \rangle$

$= 6\langle \mathbf{v}, \mathbf{u} \rangle + 4\langle \mathbf{v}, \mathbf{w} \rangle - 3\langle \mathbf{w}, \mathbf{u} \rangle - 2\langle \mathbf{w}, \mathbf{w} \rangle = 6\langle \mathbf{u}, \mathbf{v} \rangle + 4\langle \mathbf{v}, \mathbf{w} \rangle - 3\langle \mathbf{u}, \mathbf{w} \rangle - 2\|\mathbf{w}\|^2$

$= 6(2) + 4(-6) - 3(-3) - 2(49) = -101$

(b) $\|\mathbf{u} + \mathbf{v}\| = \sqrt{\langle \mathbf{u} + \mathbf{v}, \mathbf{u} + \mathbf{v} \rangle} = \sqrt{\langle \mathbf{u}, \mathbf{u} + \mathbf{v} \rangle + \langle \mathbf{v}, \mathbf{u} + \mathbf{v} \rangle} = \sqrt{\langle \mathbf{u}, \mathbf{u} \rangle + \langle \mathbf{u}, \mathbf{v} \rangle + \langle \mathbf{v}, \mathbf{u} \rangle + \langle \mathbf{v}, \mathbf{v} \rangle}$

$= \sqrt{\|\mathbf{u}\|^2 + 2\langle \mathbf{u}, \mathbf{v} \rangle + \|\mathbf{v}\|^2} = \sqrt{1 + 2(2) + 4} = 3$

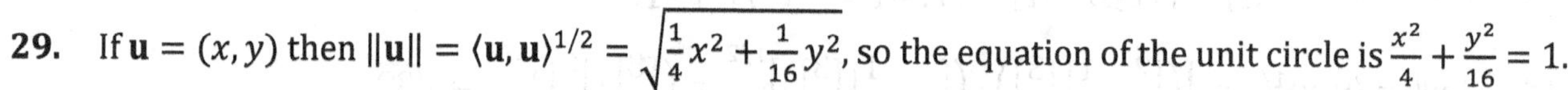

29. If $\mathbf{u} = (x, y)$ then $\|\mathbf{u}\| = \langle \mathbf{u}, \mathbf{u} \rangle^{1/2} = \sqrt{\frac{1}{4}x^2 + \frac{1}{16}y^2}$, so the equation of the unit circle is $\frac{x^2}{4} + \frac{y^2}{16} = 1$.

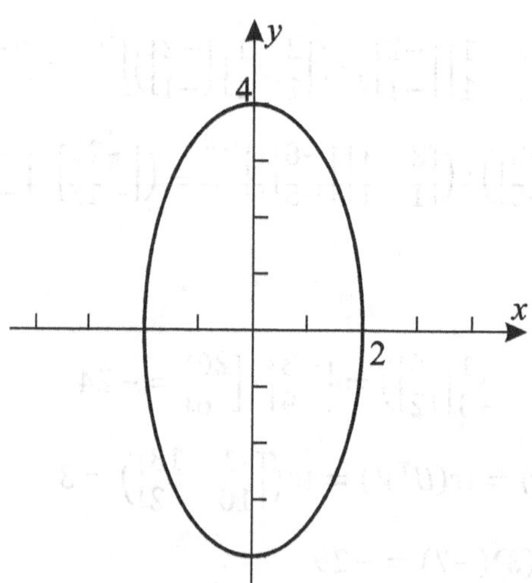

31. $\langle \mathbf{u}, \mathbf{v} \rangle = \frac{1}{9}u_1 v_1 + u_2 v_2$ (see Example 3)

33. Axiom 2 does not hold, e.g., with $\mathbf{u} = \mathbf{v} = \mathbf{w} = (1,0,0)$ we have $\langle \mathbf{u} + \mathbf{v}, \mathbf{w} \rangle = 4$ but $\langle \mathbf{u}, \mathbf{w} \rangle + \langle \mathbf{v}, \mathbf{w} \rangle = 1 + 1 = 2$;

Axiom 3 does not hold either, e.g., with $\mathbf{u} = \mathbf{v} = (1,0,0)$ and $k = 2$, $\langle k\mathbf{u}, \mathbf{v} \rangle = 4$ does not equal

$k\langle \mathbf{u}, \mathbf{v} \rangle = 2;$

This is not an inner product on R^3.

35. By Definition 1, Definition 2, and Theorem 6.1.2, we have

$$\begin{aligned}\langle 2\mathbf{v} - 4\mathbf{u}, \mathbf{u} - 3\mathbf{v} \rangle &= \langle 2\mathbf{v} - 4\mathbf{u}, \mathbf{u} \rangle - \langle 2\mathbf{v} - 4\mathbf{u}, 3\mathbf{v} \rangle \\ &= \langle 2\mathbf{v}, \mathbf{u} \rangle - \langle 4\mathbf{u}, \mathbf{u} \rangle - \langle 2\mathbf{v}, 3\mathbf{v} \rangle + \langle 4\mathbf{u}, 3\mathbf{v} \rangle \\ &= 2\langle \mathbf{v}, \mathbf{u} \rangle - 4\langle \mathbf{u}, \mathbf{u} \rangle - 6\langle \mathbf{v}, \mathbf{v} \rangle + 12\langle \mathbf{u}, \mathbf{v} \rangle \\ &= 2\langle \mathbf{u}, \mathbf{v} \rangle - 4\langle \mathbf{u}, \mathbf{u} \rangle - 6\langle \mathbf{v}, \mathbf{v} \rangle + 12\langle \mathbf{u}, \mathbf{v} \rangle \\ &= 14\langle \mathbf{u}, \mathbf{v} \rangle - 4\|\mathbf{u}\|^2 - 6\|\mathbf{v}\|^2\end{aligned}$$

37. **(a)** $\langle \mathbf{p}, \mathbf{q} \rangle = \left(\int_{-1}^{1} x^2\, dx\right) = \left(\left.\frac{x^3}{3}\right]_{-1}^{1}\right) = \frac{2}{3}$

(b) $d(\mathbf{p}, \mathbf{q}) = \|\mathbf{p} - \mathbf{q}\| = \left(\int_{-1}^{1} (1 - x^2)^2\, dx\right)^{1/2} = \left(\left(x - \frac{2x^3}{3} + \frac{x^5}{5}\right)\Big]_{-1}^{1}\right)^{1/2} = \sqrt{\frac{16}{15}} = \frac{4}{\sqrt{15}}$

(c) $\|\mathbf{p}\| = \langle \mathbf{p}, \mathbf{p} \rangle^{1/2} = \left(\int_{-1}^{1} 1\, dx\right)^{1/2} = (x]_{-1}^{1})^{1/2} = \sqrt{2}$

(d) $\|\mathbf{q}\| = \langle \mathbf{q}, \mathbf{q} \rangle^{1/2} = \left(\int_{-1}^{1} x^4\, dx\right)^{1/2} = \left(\left.\frac{x^5}{5}\right]_{-1}^{1}\right)^{1/2} = \sqrt{\frac{2}{5}}$

39. $\langle \mathbf{f}, \mathbf{g} \rangle = \int_0^1 \cos 2\pi x \sin 2\pi x\, dx = \frac{1}{2\pi}\left.\frac{(\sin 2\pi x)^2}{2}\right]_0^1 = 0$ (substituted $u = \sin 2\pi x$)

41. Part (a) follows directly from Definition 2 and Axiom 4 of Definition 1.

To prove part (b), write

$$\|k\mathbf{v}\| \underset{\text{Def.2}}{=} \sqrt{\langle k\mathbf{v}, k\mathbf{v} \rangle} \underset{\text{Axiom 3}}{=} \sqrt{k\langle \mathbf{v}, k\mathbf{v} \rangle} \underset{\text{Axiom 1}}{=} \sqrt{k\langle k\mathbf{v}, \mathbf{v} \rangle} \underset{\text{Axiom 3}}{=} \sqrt{k^2\langle \mathbf{v}, \mathbf{v} \rangle} \underset{\text{Def.2}}{=} |k|\|\mathbf{v}\|$$

43. **(b)** k_1 and k_2 must both be positive in order for $\langle \mathbf{u}, \mathbf{v} \rangle$ to satisfy the positivity axiom. (Refer to the discussion following Theorem 6.1.1.)

45. By using Definition 2 and Axioms 1, 2, and 3 of Definition 1, we have

$$\begin{aligned}\|\mathbf{u} + \mathbf{v}\|^2 + \|\mathbf{u} - \mathbf{v}\|^2 &= \langle \mathbf{u} + \mathbf{v}, \mathbf{u} + \mathbf{v} \rangle + \langle \mathbf{u} - \mathbf{v}, \mathbf{u} - \mathbf{v} \rangle \\ &= \langle \mathbf{u}, \mathbf{u} \rangle + \langle \mathbf{u}, \mathbf{v} \rangle + \langle \mathbf{v}, \mathbf{u} \rangle + \langle \mathbf{v}, \mathbf{v} \rangle + \langle \mathbf{u}, \mathbf{u} \rangle - \langle \mathbf{u}, \mathbf{v} \rangle - \langle \mathbf{v}, \mathbf{u} \rangle + \langle \mathbf{v}, \mathbf{v} \rangle \\ &= 2\langle \mathbf{u}, \mathbf{u} \rangle + 2\langle \mathbf{v}, \mathbf{v} \rangle \\ &= 2\|\mathbf{u}\|^2 + 2\|\mathbf{v}\|^2\end{aligned}$$

True-False Exercises

(a) True. The dot product is the special case of the weighted inner product with all the weights equal to 1.

(b) False. For example, if $\langle \mathbf{u}, \mathbf{v} \rangle = \mathbf{u} \cdot \mathbf{v}$, $\mathbf{u} = (1,1)$, and $\mathbf{v} = (-2,1)$ then $\langle \mathbf{u}, \mathbf{v} \rangle = -1$.

(c) True. This follows from Axioms 1 and 2 of Definition 1 since $\langle \mathbf{u}, \mathbf{v} + \mathbf{w} \rangle = \langle \mathbf{v} + \mathbf{w}, \mathbf{u} \rangle = \langle \mathbf{v}, \mathbf{u} \rangle + \langle \mathbf{w}, \mathbf{u} \rangle$.

(d) True. This follows from Axiom 3 of Definition 1 as well as part (e) of Theorem 6.1.2.

(e) False. For example, if $\langle \mathbf{u}, \mathbf{v} \rangle = \mathbf{u} \cdot \mathbf{v}$, $\mathbf{u} = (1,1)$, and $\mathbf{v} = (-1,1)$ then $\langle \mathbf{u}, \mathbf{v} \rangle = 0$ even though both vectors are nonzero.

(f) True. By Definition 2, $\|\mathbf{v}\|^2 = \langle \mathbf{v}, \mathbf{v}\rangle$ so by Axiom 4 of Definition 1, $\|\mathbf{v}\|^2 = 0$ implies $\mathbf{v} = \mathbf{0}$.

(g) False. A must be invertible; otherwise $A\mathbf{v} = \mathbf{0}$ has nontrivial solutions $\mathbf{v} \neq \mathbf{0}$ even though $\langle \mathbf{v}, \mathbf{v}\rangle = A\mathbf{v} \cdot A\mathbf{v} = \mathbf{0}$ which would violate Axiom 4 of Definition 1.

6.2 Angle and Orthogonality in Inner Product Spaces

1. **(a)** $\cos\theta = \frac{\langle \mathbf{u},\mathbf{v}\rangle}{\|\mathbf{u}\|\|\mathbf{v}\|} = \frac{(1)(2)+(-3)(4)}{\sqrt{1^2+(-3)^2}\sqrt{2^2+4^2}} = -\frac{10}{\sqrt{10}\sqrt{20}} = -\frac{10}{\sqrt{200}} = -\frac{10}{10\sqrt{2}} = -\frac{1}{\sqrt{2}}$

(b) $\cos\theta = \frac{\langle \mathbf{u},\mathbf{v}\rangle}{\|\mathbf{u}\|\|\mathbf{v}\|} = \frac{(-1)(2)+(5)(4)+(2)(-9)}{\sqrt{(-1)^2+5^2+2^2}\sqrt{2^2+4^2+(-9)^2}} = 0$

(c) $\cos\theta = \frac{\langle \mathbf{u},\mathbf{v}\rangle}{\|\mathbf{u}\|\|\mathbf{v}\|} = \frac{(1)(-3)+(0)(-3)+(1)(-3)+(0)(-3)}{\sqrt{1^2+0^2+1^2+0^2}\sqrt{(-3)^2+(-3)^2+(-3)^2+(-3)^2}} = -\frac{6}{\sqrt{2}\sqrt{36}} = -\frac{1}{\sqrt{2}}$

3. $\cos\theta = \frac{\langle \mathbf{p},\mathbf{q}\rangle}{\|\mathbf{p}\|\|\mathbf{q}\|} = \frac{(-1)(2)+(5)(4)+(2)(-9)}{\sqrt{(-1)^2+5^2+2^2}\sqrt{2^2+4^2+(-9)^2}} = 0$

5. $\cos\theta = \frac{\langle U,V\rangle}{\|U\|\|V\|} = \frac{\operatorname{tr}(U^TV)}{\sqrt{\operatorname{tr}(U^TU)}\sqrt{\operatorname{tr}(V^TV)}} = \frac{(2)(3)+(6)(2)+(1)(1)+(-3)(0)}{\sqrt{2^2+6^2+1^2+(-3)^2}\sqrt{3^2+2^2+1^2+0^2}} = \frac{19}{\sqrt{50}\sqrt{14}} = \frac{19}{10\sqrt{7}}$

7. **(a)** orthogonal: $\langle \mathbf{u}, \mathbf{v}\rangle = -4 + 6 - 2 = 0$

(b) not orthogonal: $\langle \mathbf{u}, \mathbf{v}\rangle = -2 - 2 - 2 = -6 \neq 0$

(c) orthogonal: $\langle \mathbf{u}, \mathbf{v}\rangle = (a)(-b) + (b)(a) = 0$

9. $\langle \mathbf{p}, \mathbf{q}\rangle = (-1)(0) + (-1)(2) + (2)(1) = 0$

11. $\langle U, V\rangle = (2)(-3) + (1)(0) + (-1)(0) + (3)(2) = 0$

13. The vectors are not orthogonal with respect to the Euclidean inner product since $\langle \mathbf{u}, \mathbf{v}\rangle = (1)(2) + (3)(-1) = -1 \neq 0$. Using the weighted inner product instead yields $\langle \mathbf{u}, \mathbf{v}\rangle = 2(1)(2) + k(3)(-1) = 4 - 3k$, so the vectors are orthogonal with respect to this inner product if $k = \frac{4}{3}$.

15. The orthogonality of the two vectors implies $(w_1)(1)(2) + (w_2)(2)(-4) = 0$. The weights must be positive numbers such that $w_1 = 4w_2$.

17. Orthogonality of $\mathbf{p}_1$ and $\mathbf{p}_3$ implies $\langle \mathbf{p}_1, \mathbf{p}_3\rangle = (2)(1) + (k)(2) + (6)(3) = 2k + 20 = 0$ so $k = -10$. Likewise, orthogonality of $\mathbf{p}_2$ and $\mathbf{p}_3$ implies $\langle \mathbf{p}_2, \mathbf{p}_3\rangle = (l)(1) + (5)(2) + (3)(3) = l + 19$ so $l = -19$. Substituting the values of k and l obtained above yields the polynomials $\mathbf{p}_1 = 2 - 10x + 6x^2$ and $\mathbf{p}_2 = -19 + 5x + 3x^2$ which are not orthogonal since $\langle \mathbf{p}_1, \mathbf{p}_2\rangle = (2)(-19) + (-10)(5) + (6)(3) = -70 \neq 0$. We conclude that no scalars k and l exist that make the three vectors mutually orthogonal.

19. $\langle \mathbf{p}, \mathbf{q}\rangle = p(-2)q(-2) + p(0)q(0) + p(2)q(2) = (-2)(4) + (0)(0) + (2)(4) = 0$

21. $|\langle \mathbf{u}, \mathbf{v}\rangle| = |2(1)(2) + 3(0)(1) + (3)(-1)| = |1| = 1$;
$\|\mathbf{u}\| = \sqrt{\langle \mathbf{u}, \mathbf{u}\rangle} = \sqrt{2(1)(1) + 3(0)(0) + (3)(3)} = \sqrt{11}$;
$\|\mathbf{v}\| = \sqrt{\langle \mathbf{v}, \mathbf{v}\rangle} = \sqrt{2(2)(2) + 3(1)(1) + (-1)(-1)} = \sqrt{12}$;
since $\|\mathbf{u}\|\|\mathbf{v}\| = \sqrt{132} \geq 1 = |\langle \mathbf{u}, \mathbf{v}\rangle|$, we conclude that the Cauchy-Schwarz inequality holds

23. $|\langle\mathbf{p},\mathbf{q}\rangle| = |(-1)(2)+(2)(0)+(1)(-4)| = |-6| = 6$;
$\|\mathbf{p}\| = \sqrt{\langle\mathbf{p},\mathbf{p}\rangle} = \sqrt{(-1)(-1)+(2)(2)+(1)(1)} = \sqrt{6}$;
$\|\mathbf{q}\| = \sqrt{\langle\mathbf{q},\mathbf{q}\rangle} = \sqrt{(2)(2)+(0)(0)+(-4)(-4)} = \sqrt{20}$;
since $\|\mathbf{p}\|\|\mathbf{q}\| = \sqrt{120} \ge \sqrt{36} = 6 = |\langle\mathbf{p},\mathbf{q}\rangle|$, we conclude that the Cauchy-Schwarz inequality holds

25. By inspection, $\langle\mathbf{u},\mathbf{w}_1\rangle = -2 \neq 0$. Since $\mathbf{u}$ is not orthogonal to $\mathbf{w}_1$, it is not orthogonal to the subspace.

27. Begin by forming a matrix A whose rows are the given vectors:
$A = \begin{bmatrix} 1 & 4 & 5 & 2 \\ 2 & 1 & 3 & 0 \\ -1 & 3 & 2 & 2 \end{bmatrix}$ has the reduced row echelon form $\begin{bmatrix} 1 & 0 & 1 & -\frac{2}{7} \\ 0 & 1 & 1 & \frac{4}{7} \\ 0 & 0 & 0 & 0 \end{bmatrix}$. The general solution of the homogeneous system $A\mathbf{x} = \mathbf{0}$ is $x_1 = -s + \frac{2}{7}t$, $x_2 = -s - \frac{4}{7}t$, $x_3 = s$, $x_4 = t$ therefore $\mathbf{x} = s(-1,-1,1,0) + t(\frac{2}{7},-\frac{4}{7},0,1)$.
A basis for the orthogonal complement is formed by vectors $(-1,-1,1,0)$ and $(\frac{2}{7},-\frac{4}{7},0,1)$.

29. **(a)** Every vector in W has a form $(x,y) = (x,2x)$, i.e., $W = \text{span}\{(1,2)\}$. By inspection, all vectors in R^2 orthogonal to $(1,2)$ are scalar multiples of the vector $(2,-1)$. Eliminating t from $(x,y) = t(2,-1) = (2t,-t)$ we obtain $x = 2(-y)$, i.e. $W^\perp$ can be represented using an equation $y = -\frac{1}{2}x$.

(An alternate method of solving this exercise is to follow the procedure of Example 6: letting $A = [1 \quad 2]$, the general solution of $A\begin{bmatrix} x \\ y \end{bmatrix} = 0$ is $x = -2t$, $y = t$. Eliminating t yields $y = -\frac{1}{2}x$.)

(b) $W^\perp$ will have dimension 1. A normal to the plane is $\mathbf{u} = (1,-2,-3)$, so $W^\perp$ will consist of all scalar multiples of $\mathbf{u}$ or $t\mathbf{u} = (t,-2t,-3t)$ so parametric equations for $W^\perp$ are $x = t$, $y = -2t$, $z = -3t$.

31. **(a)** $\langle\mathbf{p},\mathbf{q}\rangle = \int_0^1 x^3\,dx = \frac{x^4}{4}\Big]_0^1 = \frac{1}{4}$

(b) $\|\mathbf{p}\| = \langle\mathbf{p},\mathbf{p}\rangle^{\frac{1}{2}} = \left(\int_0^1 x^2\,dx\right)^{\frac{1}{2}} = \left(\frac{x^3}{3}\Big]_0^1\right)^{\frac{1}{2}} = \frac{1}{\sqrt{3}}$

$\|\mathbf{q}\| = \langle\mathbf{q},\mathbf{q}\rangle^{1/2} = \left(\int_0^1 x^4\,dx\right)^{1/2} = \left(\frac{x^5}{5}\Big]_0^1\right)^{1/2} = \frac{1}{\sqrt{5}}$

33. **(a)** $\langle\mathbf{p},\mathbf{q}\rangle = \int_{-1}^1 (x^2-x)(x+1)\,dx = \int_{-1}^1 (x^3-x)\,dx = \left(\frac{x^4}{4}-\frac{x^2}{2}\right)\Big]_{-1}^1 = 0$

(b) $\|\mathbf{p}\| = \langle\mathbf{p},\mathbf{p}\rangle^{\frac{1}{2}} = \left(\int_{-1}^1 (x^2-x)^2\,dx\right)^{\frac{1}{2}} = \left(\left(\frac{x^5}{5}-\frac{x^4}{2}+\frac{x^3}{3}\right)\Big]_{-1}^1\right)^{\frac{1}{2}} = \frac{4}{\sqrt{15}}$

$\|\mathbf{q}\| = \langle\mathbf{q},\mathbf{q}\rangle^{1/2} = \left(\int_{-1}^1 (x+1)^2\,dx\right)^{1/2} = \left(\left(\frac{x^3}{3}+x^2+x\right)\Big]_{-1}^1\right)^{1/2} = \sqrt{\frac{8}{3}} = 2\sqrt{\frac{2}{3}}$

35. **(a)** $\langle\mathbf{p},\mathbf{q}\rangle = \int_0^1 \left(\frac{1}{2}-x\right)dx = \left(\frac{1}{2}x-\frac{x^2}{2}\right)\Big]_0^1 = 0$

(b) $\|\mathbf{p}+\mathbf{q}\|^2 = \int_0^1 \left(\frac{3}{2}-x\right)^2 dx = \left(\frac{9}{4}x - \frac{3x^2}{2} + \frac{x^3}{3}\right)\Big]_0^1 = \frac{13}{12}$; $\|\mathbf{p}\|^2 = \int_0^1 (1)^2\, dx = x]_0^1 = 1$;

$\|\mathbf{q}\|^2 = \int_0^1 \left(\frac{1}{2}-x\right)^2 dx = \left(\frac{x}{4} - \frac{x^2}{2} + \frac{x^3}{3}\right)\Big]_0^1 = \frac{1}{12}$; we conclude that $\|\mathbf{p}+\mathbf{q}\|^2 = \frac{13}{12} = \|\mathbf{p}\|^2 + \|\mathbf{q}\|^2$

37. $\|\mathbf{u}-\mathbf{v}\|^2 = \langle \mathbf{u}-\mathbf{v}, \mathbf{u}-\mathbf{v}\rangle = \langle \mathbf{u}, \mathbf{u}-\mathbf{v}\rangle - \langle \mathbf{v}, \mathbf{u}-\mathbf{v}\rangle = \langle \mathbf{u},\mathbf{u}\rangle - \langle \mathbf{u},\mathbf{v}\rangle - \langle \mathbf{v},\mathbf{u}\rangle + \langle \mathbf{v},\mathbf{v}\rangle$
$= 1 - 0 - 0 + 1 = 2$ therefore $\|\mathbf{u}-\mathbf{v}\| = \sqrt{2}$.

39. Using the trigonometric identity $\cos\alpha\cos\beta = \frac{1}{2}\cos(\alpha-\beta) + \frac{1}{2}\cos(\alpha+\beta)$ we obtain
$\langle \mathbf{f}_k, \mathbf{f}_l\rangle = \frac{1}{2}\int_0^\pi \cos((k-l)x)\, dx + \frac{1}{2}\int_0^\pi \cos((k+l)x)\, dx$ where both $k-l$ and $k+l$ are nonzero integers.
Substituting $u = (k-l)x$ in the first integral, and $t = (k+l)x$ in the second integral yields
$\langle \mathbf{f}_k, \mathbf{f}_l\rangle = \frac{1}{2}\frac{\sin((k-l)x)}{k-l}\Big]_0^\pi + \frac{1}{2}\frac{\sin((k+l)x)}{k+l}\Big]_0^\pi = 0 + 0 = 0$ since $\sin(m\pi) = 0$ for any integer m.

41. $\text{span}\{\mathbf{u}_1, \mathbf{u}_2, \dots, \mathbf{u}_r\}$ contains all linear combinations $k_1\mathbf{u}_1 + k_2\mathbf{u}_2 + \cdots + k_r\mathbf{u}_r$ where $k_1, k_2, \dots, k_r$ are arbitrary scalars. Let $\mathbf{v} \in \text{span}\{\mathbf{u}_1, \mathbf{u}_2, \dots, \mathbf{u}_r\}$.

$\langle \mathbf{w}, \mathbf{v}\rangle = \langle \mathbf{w}, k_1\mathbf{u}_1 + k_2\mathbf{u}_2 + \cdots + k_r\mathbf{u}_r\rangle = k_1\langle \mathbf{w},\mathbf{u}_1\rangle + k_2\langle \mathbf{w},\mathbf{u}_2\rangle + \cdots + k_r\langle \mathbf{w},\mathbf{u}_r\rangle = 0 + 0 + \cdots + 0 = 0$

Thus if $\mathbf{w}$ is orthogonal to each vector $\mathbf{u}_1, \mathbf{u}_2, \dots, \mathbf{u}_r$, then $\mathbf{w}$ must be orthogonal to every vector in $\text{span}\{\mathbf{u}_1, \mathbf{u}_2, \dots, \mathbf{u}_r\}$.

43. Suppose that $\mathbf{v}$ is orthogonal to every basis vector. Then, as in Exercise 41, $\mathbf{v}$ is orthogonal to the span of the set of basis vectors, which is all of W, hence $\mathbf{v}$ is in $W^\perp$. If $\mathbf{v}$ is not orthogonal to every basis vector, then $\mathbf{v}$ clearly cannot be in $W^\perp$. Thus $W^\perp$ consists of all vectors orthogonal to every basis vector.

47. Using the weighted Euclidean inner product of Formula (2) in Section 6.1, the desired inequality follows from the Cauchy-Schwarz inequality.

49. Using the inner product $\langle \mathbf{f}, \mathbf{g}\rangle = \int_0^1 f(x)g(x)dx$, part (a) follows from the Cauchy-Schwarz inequality and part (b) follows from the triangle inequality (part (a) of Theorem 6.2.2).

51. **(a)** We are looking for all vectors $\mathbf{v} = (a, b)$ such that $\langle \mathbf{x}, \mathbf{v}\rangle = a + b$ is equal to $\langle T_A(\mathbf{x}), T_A(\mathbf{v})\rangle = \begin{bmatrix}2\\0\end{bmatrix} \cdot \begin{bmatrix}a+b\\-a+b\end{bmatrix} = 2a + 2b$. The equation $a + b = 2a + 2b$ yields $a + b = 0$, i.e. $b = -a$. Vectors that satisfy $\langle \mathbf{x}, \mathbf{v}\rangle = \langle T_A(\mathbf{x}), T_A(\mathbf{v})\rangle$ must have a form $a(1, -1)$ where a is an arbitrary scalar.

(b) We are looking for all vectors $\mathrm{v} = (a, b)$ such that $\langle \mathbf{x}, \mathbf{v}\rangle = 2a + 3b$ is equal to $\langle T_A(\mathbf{x}), T_A(\mathbf{v})\rangle = \left\langle \begin{bmatrix}2\\0\end{bmatrix}, \begin{bmatrix}a+b\\-a+b\end{bmatrix}\right\rangle = 4a + 4b$. The equation $2a + 3b = 4a + 4b$ yields $2a + b = 0$, i.e. $b = -2a$. Vectors that satisfy $\langle \mathbf{x}, \mathbf{v}\rangle = \langle T_A(\mathbf{x}), T_A(\mathbf{v})\rangle$ must have a form $a(1, -2)$ where a is an arbitrary scalar.

True-False Exercises

(a) False. If $\mathbf{u}$ is orthogonal to every vector of a subspace W, then $\mathbf{u}$ is in $W^\perp$.

(b) True. $W \cap W^\perp = \{\mathbf{0}\}$.

(c) True. For any vector $\mathbf{w}$ in W, $\langle \mathbf{u}+\mathbf{v},\ \mathbf{w}\rangle = \langle \mathbf{u},\ \mathbf{w}\rangle + \langle \mathbf{v},\ \mathbf{w}\rangle = 0$, so $\mathbf{u}+\mathbf{v}$ is in $W^\perp$.

(d) True. For any vector $\mathbf{w}$ in W, $\langle k\mathbf{u},\ \mathbf{w}\rangle = k\langle \mathbf{u},\ \mathbf{w}\rangle = k(0) = 0$, so $k\mathbf{u}$ is in $W^\perp$.

(e) False. If $\mathbf{u}$ and $\mathbf{v}$ are orthogonal $|\langle \mathbf{u},\ \mathbf{v}\rangle| = |0| = 0$.

(f) False. If $\mathbf{u}$ and $\mathbf{v}$ are orthogonal, $\|\mathbf{u}+\mathbf{v}\|^2 = \|\mathbf{u}\|^2 + \|\mathbf{v}\|^2$ thus $\|\mathbf{u}+\mathbf{v}\| = \sqrt{\|\mathbf{u}\|^2 + \|\mathbf{v}\|^2} \neq \|\mathbf{u}\| + \|\mathbf{v}\|$

6.3 Gram-Schmidt Process; QR-Decomposition

1. **(a)** $\langle (0,1),\ (2,0)\rangle = 0+0 = 0$;
$\|(0,1)\| = 1$; $\|(2,0)\| = 2 \neq 1$;
The set is orthogonal, but is not orthonormal.

(b) $\left\langle \left(-\frac{1}{\sqrt{2}},\frac{1}{\sqrt{2}}\right),\ \left(\frac{1}{\sqrt{2}},\frac{1}{\sqrt{2}}\right)\right\rangle = -\frac{1}{2}+\frac{1}{2} = 0$;
$\left\|\left(-\frac{1}{\sqrt{2}},\ \frac{1}{\sqrt{2}}\right)\right\| = \sqrt{\frac{1}{2}+\frac{1}{2}} = 1$; $\left\|\left(\frac{1}{\sqrt{2}},\ \frac{1}{\sqrt{2}}\right)\right\| = \sqrt{\frac{1}{2}+\frac{1}{2}} = 1$
The set is orthogonal and orthonormal.

(c) $\left\langle \left(-\frac{1}{\sqrt{2}},-\frac{1}{\sqrt{2}}\right),\ \left(\frac{1}{\sqrt{2}},\frac{1}{\sqrt{2}}\right)\right\rangle = -\frac{1}{2}-\frac{1}{2} = -1 \neq 0$;
The set is not orthogonal (therefore, it is not orthonormal either).

(d) $\langle (0,0),\ (0,1)\rangle = 0+0 = 0$;
$\|(0,0)\| = 0 \neq 1$; $\|(0,1)\| = 1$;
The set is orthogonal, but is not orthonormal.

3. **(a)** $\langle p_1(x),\ p_2(x)\rangle = \frac{2}{3}\left(\frac{2}{3}\right) - \frac{2}{3}\left(\frac{1}{3}\right) + \frac{1}{3}\left(-\frac{2}{3}\right) = \frac{4}{9}-\frac{2}{9}-\frac{2}{9} = 0$;
$\langle p_1(x),\ p_3(x)\rangle = \frac{2}{3}\left(\frac{1}{3}\right) - \frac{2}{3}\left(\frac{2}{3}\right) + \frac{1}{3}\left(\frac{2}{3}\right) = \frac{2}{9}-\frac{4}{9}+\frac{2}{9} = 0$;
$\langle p_2(x),\ p_3(x)\rangle = \frac{2}{3}\left(\frac{1}{3}\right) + \frac{1}{3}\left(\frac{2}{3}\right) - \frac{2}{3}\left(\frac{2}{3}\right) = \frac{2}{9}+\frac{2}{9}-\frac{4}{9} = 0$;
The set is orthogonal.

(b) $\langle p_1(x),\ p_2(x)\rangle = 1(0) + 0\left(\frac{1}{\sqrt{2}}\right) + 0\left(\frac{1}{\sqrt{2}}\right) = 0$; $\langle p_1(x),\ p_3(x)\rangle = 1(0)+0(0)+0(1) = 0$;
$\langle p_2(x),\ p_3(x)\rangle = 0(0) + \frac{1}{\sqrt{2}}(0) + \frac{1}{\sqrt{2}}(1) = \frac{1}{\sqrt{2}} \neq 0$;
The set is not orthogonal.

5. Let us denote the column vectors $\mathbf{u}_1 = (1,0,-1)$, $\mathbf{u}_2 = (2,0,2)$, and $\mathbf{u}_3 = (0,5,0)$. These vectors are orthogonal since $\langle \mathbf{u}_1,\mathbf{u}_2\rangle = 2+0-2 = 0$, $\langle \mathbf{u}_1,\mathbf{u}_3\rangle = 0+0+0 = 0$, and $\langle \mathbf{u}_2,\mathbf{u}_3\rangle = 0+0+0 = 0$. It follows from Theorem 6.3.1 that the column vectors are linearly independent, therefore they form an orthogonal basis for the column space of A. We proceed to normalize each column vector: $\frac{\mathbf{u}_1}{\|\mathbf{u}_1\|} = \frac{1}{\sqrt{1+0+1}}(1,0,-1) = \left(\frac{1}{\sqrt{2}},0,-\frac{1}{\sqrt{2}}\right)$; $\frac{\mathbf{u}_2}{\|\mathbf{u}_2\|} = \frac{1}{\sqrt{4+0+4}}(2,0,2) = \left(\frac{2}{2\sqrt{2}},0,\ \frac{2}{2\sqrt{2}}\right) = \left(\frac{1}{\sqrt{2}},0,\frac{1}{\sqrt{2}}\right)$; $\frac{\mathbf{u}_3}{\|\mathbf{u}_3\|} = \frac{1}{\sqrt{0+25+0}}(0,5,0) = (0,\ 1,\ 0)$. A resulting orthonormal basis for the column space is $\left\{\left(\frac{1}{\sqrt{2}},0,-\frac{1}{\sqrt{2}}\right),\left(\frac{1}{\sqrt{2}},0,\frac{1}{\sqrt{2}}\right),(0,\ 1,\ 0)\right\}$.

7. $\langle \mathbf{v}_1, \mathbf{v}_2 \rangle = -\frac{12}{25} + \frac{12}{25} + 0 = 0$; $\langle \mathbf{v}_1, \mathbf{v}_3 \rangle = 0 + 0 + 0 = 0$; $\langle \mathbf{v}_2, \mathbf{v}_3 \rangle = 0 + 0 + 0 = 0$;

$\|\mathbf{v}_1\| = \sqrt{\frac{9}{25} + \frac{16}{25} + 0} = 1$; $\|\mathbf{v}_2\| = \sqrt{\frac{16}{25} + \frac{9}{25} + 0} = 1$; $\|\mathbf{v}_3\| = \sqrt{0 + 0 + 1} = 1$;

Since this is an orthogonal set of nonzero vectors, it follows from Theorem 6.3.1 that the set is linearly independent. Because the number of vectors in the set matches $\dim(R^3) = 3$, this set forms a basis for R^3 by Theorem 4.5.4. This basis is orthonormal, so by Theorem 6.3.2(b),

$$\begin{aligned}\mathbf{u} &= \langle \mathbf{u}, \mathbf{v}_1 \rangle \mathbf{v}_1 + \langle \mathbf{u}, \mathbf{v}_2 \rangle \mathbf{v}_2 + \langle \mathbf{u}, \mathbf{v}_3 \rangle \mathbf{v}_3 \\ &= \left(-\frac{3}{5} - \frac{8}{5} + 0\right)\mathbf{v}_1 + \left(\frac{4}{5} - \frac{6}{5} + 0\right)\mathbf{v}_2 + (0 + 0 + 2)\mathbf{v}_3 \\ &= -\frac{11}{5}\mathbf{v}_1 - \frac{2}{5}\mathbf{v}_2 + 2\mathbf{v}_3\end{aligned}$$

9. $\langle \mathbf{v}_1, \mathbf{v}_2 \rangle = 4 - 2 - 2 = 0$; $\langle \mathbf{v}_1, \mathbf{v}_3 \rangle = 2 - 4 + 2 = 0$; $\langle \mathbf{v}_2, \mathbf{v}_3 \rangle = 2 + 2 - 4 = 0$;

Since this is an orthogonal set of nonzero vectors, it follows from Theorem 6.3.1 that the set is linearly independent. Because the number of vectors in the set matches $\dim(R^3) = 3$, this set forms a basis for R^3 by Theorem 4.5.4. By Theorem 6.3.2(a),

$$\begin{aligned}\mathbf{u} &= \frac{\langle \mathbf{u}, \mathbf{v}_1 \rangle}{\|\mathbf{v}_1\|^2}\mathbf{v}_1 + \frac{\langle \mathbf{u}, \mathbf{v}_2 \rangle}{\|\mathbf{v}_2\|^2}\mathbf{v}_2 + \frac{\langle \mathbf{u}, \mathbf{v}_3 \rangle}{\|\mathbf{v}_3\|^2}\mathbf{v}_3 \\ &= \frac{-2 + 0 + 2}{4 + 4 + 1}\mathbf{v}_1 + \frac{-2 + 0 - 4}{4 + 1 + 4}\mathbf{v}_2 + \frac{-1 + 0 + 4}{1 + 4 + 4}\mathbf{v}_3 \\ &= 0\mathbf{v}_1 - \frac{2}{3}\mathbf{v}_2 + \frac{1}{3}\mathbf{v}_3\end{aligned}$$

11. $(\mathbf{u})_S = \left(-\frac{11}{5}, -\frac{2}{5}, 2\right)$

13. $(\mathbf{u})_S = \left(0, -\frac{2}{3}, \frac{1}{3}\right)$

15. **(a)** $\|\mathbf{v}\| = \sqrt{\frac{9}{25} + \frac{16}{25}} = 1$, so $\mathbf{v}$ forms an orthonormal basis for the line $W = \text{span}\{\mathbf{v}\}$.

$\mathbf{w}_1 = \text{proj}_W \mathbf{u} = \langle \mathbf{u}, \mathbf{v} \rangle \mathbf{v} = \left(-\frac{3}{5} + \frac{24}{5}\right)\left(\frac{3}{5}, \frac{4}{5}\right) = \frac{21}{5}\left(\frac{3}{5}, \frac{4}{5}\right) = \left(\frac{63}{25}, \frac{84}{25}\right)$

(b) $\mathbf{w}_2 = \mathbf{u} - \mathbf{w}_1 = (-1, 6) - \left(\frac{63}{25}, \frac{84}{25}\right) = \left(-\frac{88}{25}, \frac{66}{25}\right)$;

$\mathbf{w}_2$ is orthogonal to the line since $\langle \mathbf{w}_2, \mathbf{v} \rangle = -\frac{264}{125} + \frac{264}{125} = 0$

17. **(a)** $\mathbf{w}_1 = \text{proj}_W \mathbf{u} = \frac{\langle \mathbf{u}, \mathbf{v} \rangle}{\|\mathbf{v}\|^2}\mathbf{v} = \frac{2+3}{1+1}(1, 1) = \frac{5}{2}(1, 1) = \left(\frac{5}{2}, \frac{5}{2}\right)$

(b) $\mathbf{w}_2 = \mathbf{u} - \mathbf{w}_1 = (2, 3) - \left(\frac{5}{2}, \frac{5}{2}\right) = \left(-\frac{1}{2}, \frac{1}{2}\right)$;

$\mathbf{w}_2$ is orthogonal to the line since $\langle \mathbf{w}_2, \mathbf{v} \rangle = -\frac{1}{2} + \frac{1}{2} = 0$

19. **(a)** $\langle \mathbf{v}_1, \mathbf{v}_2 \rangle = 0$ and $\|\mathbf{v}_1\| = \|\mathbf{v}_2\| = 1$, so $\{\mathbf{v}_1, \mathbf{v}_2\}$ is an orthonormal basis for the plane $W = \text{span}\{\mathbf{v}_1, \mathbf{v}_2\}$.

$\mathbf{w}_1 = \text{proj}_W \mathbf{u} = \langle \mathbf{u}, \mathbf{v}_1 \rangle \mathbf{v}_1 + \langle \mathbf{u}, \mathbf{v}_2 \rangle \mathbf{v}_2 = 2\left(\frac{1}{3}, \frac{2}{3}, -\frac{2}{3}\right) + 4\left(\frac{2}{3}, \frac{1}{3}, \frac{2}{3}\right) = \left(\frac{10}{3}, \frac{8}{3}, \frac{4}{3}\right)$

(b) $\mathbf{w}_2 = \mathbf{u} - \mathbf{w}_1 = (4,2,1) - \left(\frac{10}{3}, \frac{8}{3}, \frac{4}{3}\right) = \left(\frac{2}{3}, -\frac{2}{3}, -\frac{1}{3}\right)$;

$\mathbf{w}_2$ is orthogonal to the plane since $\langle \mathbf{w}_2, \mathbf{v}_1 \rangle = \frac{2}{9} - \frac{4}{9} + \frac{2}{9} = 0$ and $\langle \mathbf{w}_2, \mathbf{v}_2 \rangle = \frac{4}{9} - \frac{2}{9} - \frac{2}{9} = 0$.

21. **(a)** $\langle \mathbf{v}_1, \mathbf{v}_2 \rangle = 0$, so $\{\mathbf{v}_1, \mathbf{v}_2\}$ is an orthogonal basis for the plane $W = \text{span}\{\mathbf{v}_1, \mathbf{v}_2\}$.

$\mathbf{w}_1 = \text{proj}_W \mathbf{u} = \frac{\langle \mathbf{u}, \mathbf{v}_1 \rangle}{\|\mathbf{v}_1\|^2} \mathbf{v}_1 + \frac{\langle \mathbf{u}, \mathbf{v}_2 \rangle}{\|\mathbf{v}_2\|^2} \mathbf{v}_2 = \frac{4}{6}(1,-2,1) + \frac{2}{5}(2,1,0)$

$= \left(\frac{2}{3}, -\frac{4}{3}, \frac{2}{3}\right) + \left(\frac{4}{5}, \frac{2}{5}, 0\right) = \left(\frac{22}{15}, -\frac{14}{15}, \frac{2}{3}\right)$

(b) $\mathbf{w}_2 = \mathbf{u} - \mathbf{w}_1 = (1,0,3) - \left(\frac{22}{15}, -\frac{14}{15}, \frac{2}{3}\right) = \left(-\frac{7}{15}, \frac{14}{15}, \frac{7}{3}\right)$;

$\mathbf{w}_2$ is orthogonal to the plane since $\langle \mathbf{w}_2, \mathbf{v}_1 \rangle = -\frac{7}{15} - \frac{28}{15} + \frac{7}{3} = \frac{-7-28+35}{15} = 0$ and

$\langle \mathbf{w}_2, \mathbf{v}_2 \rangle = -\frac{14}{15} + \frac{14}{15} + 0 = 0$.

23. $\text{proj}_W \mathbf{b} = \frac{\langle \mathbf{b}, \mathbf{v}_1 \rangle}{\|\mathbf{v}_1\|^2} \mathbf{v}_1 + \frac{\langle \mathbf{b}, \mathbf{v}_2 \rangle}{\|\mathbf{v}_2\|^2} \mathbf{v}_2 = \frac{1+2+0-2}{1+1+1+1}(1,1,1,1) + \frac{1+2+0+2}{1+1+1+1}(1,1,-1,-1)$

$= \frac{1}{4}(1,1,1,1) + \frac{5}{4}(1,1,-1,-1) = \left(\frac{3}{2}, \frac{3}{2}, -1, -1\right)$

25. $\text{proj}_W \mathbf{b} = \langle \mathbf{b}, \mathbf{v}_1 \rangle \mathbf{v}_1 + \langle \mathbf{b}, \mathbf{v}_2 \rangle \mathbf{v}_2 + \langle \mathbf{b}, \mathbf{v}_3 \rangle \mathbf{v}_3$

$= \left(0 + \frac{2}{\sqrt{18}} + 0 + \frac{1}{\sqrt{18}}\right) \mathbf{v}_1 + \left(\frac{1}{2} + \frac{10}{6} + 0 - \frac{1}{6}\right) \mathbf{v}_2 + \left(\frac{1}{\sqrt{18}} + 0 + 0 + \frac{4}{\sqrt{18}}\right) \mathbf{v}_3 = \frac{3}{\sqrt{18}} \mathbf{v}_1 + 2\mathbf{v}_2 + \frac{5}{\sqrt{18}} \mathbf{v}_3$

$= \left(0, \frac{3}{18}, -\frac{12}{18}, -\frac{3}{18}\right) + \left(1, \frac{5}{3}, \frac{1}{3}, \frac{1}{3}\right) + \left(\frac{5}{18}, 0, \frac{5}{18}, -\frac{20}{18}\right) = \left(\frac{23}{18}, \frac{11}{6}, -\frac{1}{18}, -\frac{17}{18}\right)$

27. $\mathbf{v}_1 = \mathbf{u}_1 = (1,-3)$

$\mathbf{v}_2 = \mathbf{u}_2 - \frac{\langle \mathbf{u}_2, \mathbf{v}_1 \rangle}{\|\mathbf{v}_1\|^2} \mathbf{v}_1 = (2,2) - \frac{2-6}{10}(1,-3) = (2,2) - \left(-\frac{2}{5}, \frac{6}{5}\right) = \left(\frac{12}{5}, \frac{4}{5}\right)$

An orthonormal basis is formed by the vectors $\mathbf{q}_1 = \frac{\mathbf{v}_1}{\|\mathbf{v}_1\|} = \frac{1}{\sqrt{10}}(1,-3) = \left(\frac{1}{\sqrt{10}}, -\frac{3}{\sqrt{10}}\right)$ and

$\mathbf{q}_2 = \frac{\mathbf{v}_2}{\|\mathbf{v}_2\|} = \frac{1}{\sqrt{\frac{144}{25} + \frac{16}{25}}}\left(\frac{12}{5}, \frac{4}{5}\right) = \frac{1}{\sqrt{\frac{160}{25}}}\left(\frac{12}{5}, \frac{4}{5}\right) = \frac{5}{4\sqrt{10}}\left(\frac{12}{5}, \frac{4}{5}\right) = \left(\frac{3}{\sqrt{10}}, \frac{1}{\sqrt{10}}\right)$.

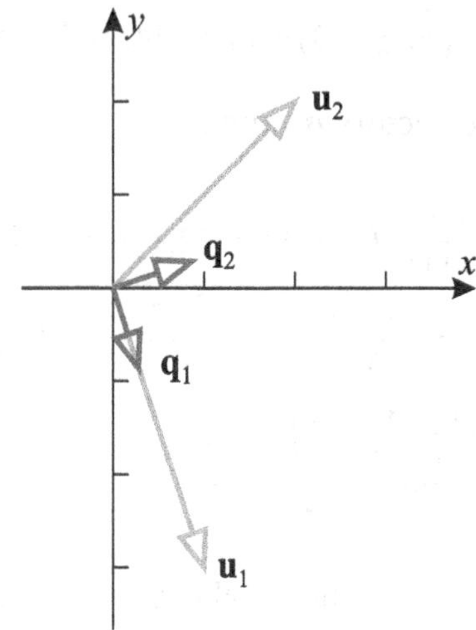

29. $\mathbf{v}_1 = \mathbf{u}_1 = (1,1,1)$

$\mathbf{v}_2 = \mathbf{u}_2 - \frac{\langle \mathbf{u}_2, \mathbf{v}_1 \rangle}{\|\mathbf{v}_1\|^2} \mathbf{v}_1 = (-1,1,0) - \frac{-1+1+0}{1+1+1}(1,1,1) = (-1,1,0) - 0(1,1,1) = (-1,1,0)$

$\mathbf{v}_3 = \mathbf{u}_3 - \frac{\langle \mathbf{u}_3,\mathbf{v}_1\rangle}{\|\mathbf{v}_1\|^2}\mathbf{v}_1 - \frac{\langle \mathbf{u}_3,\mathbf{v}_2\rangle}{\|\mathbf{v}_2\|^2}\mathbf{v}_2 = (1,2,1) - \frac{1+2+1}{1+1+1}(1,1,1) - \frac{-1+2+0}{1+1+0}(-1,1,0)$
$= (1,2,1) - \frac{4}{3}(1,1,1) - \frac{1}{2}(-1,1,0) = \left(\frac{1}{6},\frac{1}{6},-\frac{1}{3}\right)$
An orthonormal basis is formed by the vectors $\mathbf{q}_1 = \frac{\mathbf{v}_1}{\|\mathbf{v}_1\|} = \frac{1}{\sqrt{3}}(1,1,1) = \left(\frac{1}{\sqrt{3}},\frac{1}{\sqrt{3}},\frac{1}{\sqrt{3}}\right)$,
$\mathbf{q}_2 = \frac{\mathbf{v}_2}{\|\mathbf{v}_2\|} = \frac{1}{\sqrt{2}}(-1,1,0) = \left(-\frac{1}{\sqrt{2}},\frac{1}{\sqrt{2}},0\right)$, and $\mathbf{q}_3 = \frac{\mathbf{v}_3}{\|\mathbf{v}_3\|} = \frac{1}{1/\sqrt{6}}\left(\frac{1}{6},\frac{1}{6},-\frac{1}{3}\right) = \left(\frac{1}{\sqrt{6}},\frac{1}{\sqrt{6}},-\frac{2}{\sqrt{6}}\right)$.

31. First, transform the given basis into an orthogonal basis $\{\mathbf{v}_1, \mathbf{v}_2, \mathbf{v}_3, \mathbf{v}_4\}$.
$\mathbf{v}_1 = \mathbf{u}_1 = (0,2,1,0)$
$\mathbf{v}_2 = \mathbf{u}_2 - \frac{\langle \mathbf{u}_2,\mathbf{v}_1\rangle}{\|\mathbf{v}_1\|^2}\mathbf{v}_1 = (1,\ -1,\ 0,\ 0) - \frac{0-2+0+0}{5}(0,\ 2,\ 1,0) = (1,\ -1,\ 0,\ 0) + \left(0,\ \frac{4}{5},\ \frac{2}{5},\ 0\right)$
$=\left(1,\ -\frac{1}{5},\ \frac{2}{5},\ 0\right)$
$\mathbf{v}_3 = \mathbf{u}_3 - \frac{\langle \mathbf{u}_3,\mathbf{v}_1\rangle}{\|\mathbf{v}_1\|^2}\mathbf{v}_1 - \frac{\langle \mathbf{u}_3,\mathbf{v}_2\rangle}{\|\mathbf{v}_2\|^2}\mathbf{v}_2 = (1,\ 2,\ 0,\ -1) - \frac{0+4+0+0}{5}(0,\ 2,\ 1,\ 0) - \frac{1-\frac{2}{5}+0+0}{\frac{6}{5}}\left(1,\ -\frac{1}{5},\ \frac{2}{5},\ 0\right)$
$= (1,\ 2,\ 0,\ -1) - \left(0,\ \frac{8}{5},\ \frac{4}{5},\ 0\right) - \left(\frac{1}{2},\ -\frac{1}{10},\ \frac{1}{5},\ 0\right) = \left(\frac{1}{2},\ \frac{1}{2},\ -1,\ -1\right)$

$$\mathbf{v}_4 = \mathbf{u}_4 - \frac{\langle \mathbf{u}_4,\mathbf{v}_1\rangle}{\|\mathbf{v}_1\|^2}\mathbf{v}_1 - \frac{\langle \mathbf{u}_4,\mathbf{v}_2\rangle}{\|\mathbf{v}_2\|^2}\mathbf{v}_2 - \frac{\langle \mathbf{u}_4,\mathbf{v}_3\rangle}{\|\mathbf{v}_3\|^2}\mathbf{v}_3$$

$= (1,\ 0,\ 0,\ 1) - \frac{0+0+0+0}{5}\mathbf{v}_1 - \frac{1+0+0+0}{\frac{6}{5}}\left(1,\ -\frac{1}{5},\ \frac{2}{5},\ 0\right) - \frac{\frac{1}{2}+0+0-1}{\frac{5}{2}}\left(\frac{1}{2},\ \frac{1}{2},\ -1,\ -1\right)$
$=(1,\ 0,\ 0,\ 1) - \frac{0+0+0+0}{5}\mathbf{v}_1 - \frac{1+0+0+0}{\frac{6}{5}}\left(1,\ -\frac{1}{5},\ \frac{2}{5},\ 0\right) - \frac{\frac{1}{2}+0+0-1}{\frac{5}{2}}\left(\frac{1}{2},\ \frac{1}{2},\ -1,\ -1\right) = \left(\frac{4}{15},\ \frac{4}{15},\ -\frac{8}{15},\ \frac{4}{5}\right)$
An orthonormal basis is formed by the vectors $\mathbf{q}_1 = \frac{\mathbf{v}_1}{\|\mathbf{v}_1\|} = \frac{(0,2,1,0)}{\sqrt{5}} = \left(0,\ \frac{2}{\sqrt{5}},\ \frac{1}{\sqrt{5}},\ 0\right)$,
$\mathbf{q}_2 = \frac{\mathbf{v}_2}{\|\mathbf{v}_2\|} = \frac{\left(1,-\frac{1}{5},\frac{2}{5},0\right)}{\frac{\sqrt{30}}{5}} = \left(\frac{5}{\sqrt{30}},\ -\frac{1}{\sqrt{30}},\ \frac{2}{\sqrt{30}},\ 0\right)$, $\mathbf{q}_3 = \frac{\mathbf{v}_3}{\|\mathbf{v}_3\|} = \frac{\left(\frac{1}{2},\frac{1}{2},-1,-1\right)}{\frac{\sqrt{10}}{2}} = \left(\frac{1}{\sqrt{10}},\ \frac{1}{\sqrt{10}},\ -\frac{2}{\sqrt{10}},\ -\frac{2}{\sqrt{10}}\right)$,
$\mathbf{q}_4 = \frac{\mathbf{v}_4}{\|\mathbf{v}_4\|} = \frac{\left(\frac{4}{15},\frac{4}{15},-\frac{8}{15},\frac{4}{5}\right)}{\frac{4}{\sqrt{15}}} = \left(\frac{1}{\sqrt{15}},\ \frac{1}{\sqrt{15}},\ -\frac{2}{\sqrt{15}},\ \frac{3}{\sqrt{15}}\right)$.

33. From Exercise 23, $\mathbf{w}_1 = \text{proj}_W\mathbf{b} = \left(\frac{3}{2},\ \frac{3}{2},-1,-1\right)$, so $\mathbf{w}_2 = \mathbf{b} - \text{proj}_W\mathbf{b} = \left(-\frac{1}{2},\ \frac{1}{2},1,-1\right)$.

35. Let W be the plane spanned by the vectors $\mathbf{u}_1$ and $\mathbf{u}_2$.
$\mathbf{v}_1 = \mathbf{u}_1 = (1,1,1)$
$\mathbf{v}_2 = \mathbf{u}_2 - \frac{\langle \mathbf{u}_2,\mathbf{v}_1\rangle}{\|\mathbf{v}_1\|^2}\mathbf{v}_1 = (2,0,-1) - \frac{2+0-1}{1+1+1}(1,1,1) = (2,0,-1) - \left(\frac{1}{3},\frac{1}{3},\frac{1}{3}\right) = \left(\frac{5}{3},-\frac{1}{3},-\frac{4}{3}\right)$
$\mathbf{w}_1 = \text{proj}_W\mathbf{w} = \frac{\langle \mathbf{w},\mathbf{v}_1\rangle}{\|\mathbf{v}_1\|^2}\mathbf{v}_1 + \frac{\langle \mathbf{w},\mathbf{v}_2\rangle}{\|\mathbf{v}_2\|^2}\mathbf{v}_2 = \frac{1+2+3}{3}\mathbf{v}_1 + \frac{\frac{5}{3}-\frac{2}{3}-\frac{12}{3}}{\frac{42}{9}}\mathbf{v}_2 = 2(1,\ 1,\ 1) - \frac{9}{14}\left(\frac{5}{3},-\frac{1}{3},-\frac{4}{3}\right)$
$= (2,\ 2,\ 2) - \left(\frac{15}{14},\ -\frac{3}{14},\ -\frac{6}{7}\right) = \left(\frac{13}{14},\ \frac{31}{14},\ \frac{20}{7}\right)$
$\mathbf{w}_2 = \mathbf{w} - \mathbf{w}_1 = (1,2,3) - \left(\frac{13}{14},\ \frac{31}{14},\ \frac{20}{7}\right) = \left(\frac{1}{14},\ -\frac{3}{14},\ \frac{1}{7}\right)$

37. First, transform the given basis into an orthogonal basis $\{\mathbf{v}_1,\ \mathbf{v}_2,\ \mathbf{v}_3\}$.
$\mathbf{v}_1 = \mathbf{u}_1 = (1,\ 1,\ 1)$
$\|\mathbf{v}_1\| = \sqrt{\langle \mathbf{v}_1,\mathbf{v}_1\rangle} = \sqrt{1^2 + 2(1)^2 + 3(1)^2} = \sqrt{1+2+3} = \sqrt{6}$
$\mathbf{v}_2 = \mathbf{u}_2 - \frac{\langle \mathbf{u}_2,\mathbf{v}_1\rangle}{\|\mathbf{v}_1\|^2}\mathbf{v}_1 = (1,\ 1,\ 0) - \frac{1(1)+2(1)(1)+3(0)(1)}{6}(1,\ 1,\ 1) = (1,\ 1,\ 0) - \frac{1}{2}(1,\ 1,\ 1) = \left(\frac{1}{2},\ \frac{1}{2},\ -\frac{1}{2}\right)$

$\|\mathbf{v}_2\| = \sqrt{\langle \mathbf{v}_2, \mathbf{v}_2\rangle} = \sqrt{\left(\frac{1}{2}\right)^2 + 2\left(\frac{1}{2}\right)^2 + 3\left(-\frac{1}{2}\right)^2} = \sqrt{6\left(\frac{1}{4}\right)} = \frac{\sqrt{6}}{2}$

$\mathbf{v}_3 = \mathbf{u}_3 - \frac{\langle \mathbf{u}_3, \mathbf{v}_1\rangle}{\|\mathbf{v}_1\|^2}\mathbf{v}_1 - \frac{\langle \mathbf{u}_3, \mathbf{v}_2\rangle}{\|\mathbf{v}_2\|^2}\mathbf{v}_2 = (1,\ 0,\ 0) - \frac{1+0+0}{6}(1,\ 1,\ 1) - \frac{\frac{1}{2}+0+0}{\frac{6}{4}}\left(\frac{1}{2},\ \frac{1}{2},\ -\frac{1}{2}\right)$

$= (1,\ 0,\ 0) - \left(\frac{1}{6},\ \frac{1}{6},\ \frac{1}{6}\right) - \left(\frac{1}{6},\ \frac{1}{6},\ -\frac{1}{6}\right) = \left(\frac{2}{3},\ -\frac{1}{3},\ 0\right)$

$\|\mathbf{v}_3\| = \sqrt{\langle \mathbf{v}_3, \mathbf{v}_3\rangle} = \sqrt{\left(\frac{2}{3}\right)^2 + 2\left(-\frac{1}{3}\right)^2 + 3(0)^2} = \sqrt{\frac{4}{9} + \frac{2}{9}} = \frac{\sqrt{6}}{3}$

The orthonormal basis is $\mathbf{q}_1 = \frac{\mathbf{v}_1}{\|\mathbf{v}_1\|} = \frac{(1,1,1)}{\sqrt{6}} = \left(\frac{1}{\sqrt{6}},\ \frac{1}{\sqrt{6}},\ \frac{1}{\sqrt{6}}\right)$, $\mathbf{q}_2 = \frac{\mathbf{v}_2}{\|\mathbf{v}_2\|} = \frac{\left(\frac{1}{2},\frac{1}{2},-\frac{1}{2}\right)}{\frac{\sqrt{6}}{2}} = \left(\frac{1}{\sqrt{6}},\ \frac{1}{\sqrt{6}},\ -\frac{1}{\sqrt{6}}\right)$,

and $\mathbf{q}_3 = \frac{\mathbf{v}_3}{\|\mathbf{v}_3\|} = \frac{\left(\frac{2}{3},-\frac{1}{3},0\right)}{\frac{\sqrt{6}}{3}} = \left(\frac{2}{\sqrt{6}},\ -\frac{1}{\sqrt{6}},\ 0\right)$.

39. For example, $\mathbf{x} = \left(\frac{1}{\sqrt{3}}, 0\right)$ and $\mathbf{y} = \left(0, \frac{1}{\sqrt{2}}\right)$.

41. (a) By inspection, $\mathbf{v}_1' = \mathbf{v}_1 + \mathbf{v}_2$ and $\mathbf{v}_2' = \mathbf{v}_1 - 2\mathbf{v}_2$ so $\mathbf{v}_1'$ and $\mathbf{v}_2'$ are in W. The dimension of W is 2 since $\{\mathbf{v}_1, \mathbf{v}_2\}$ is an orthogonal basis for W. By Theorem 6.3.1, $\{\mathbf{v}_1', \mathbf{v}_2'\}$ is linearly independent, so by Theorem 4.5.4 it is a basis for W, hence it spans W.

(b) Calculating $\text{proj}_W\mathbf{u}$ using $\{\mathbf{v}_1, \mathbf{v}_2\}$ we obtain

$\frac{\langle \mathbf{u}, \mathbf{v}_1\rangle}{\|\mathbf{v}_1\|^2}\mathbf{v}_1 + \frac{\langle \mathbf{u}, \mathbf{v}_2\rangle}{\|\mathbf{v}_2\|^2}\mathbf{v}_2 = \frac{-3+0+7}{1+0+1}(1,0,1) + \frac{0+1+0}{0+1+0}(0,1,0) = (2,0,2) + (0,1,0) = (2,1,2)$.

Calculating $\text{proj}_W\mathbf{u}$ using $\{\mathbf{v}_1', \mathbf{v}_2'\}$ instead yields the same vector:

$\frac{\langle \mathbf{u}, \mathbf{v}_1'\rangle}{\|\mathbf{v}_1'\|^2}\mathbf{v}_1' + \frac{\langle \mathbf{u}, \mathbf{v}_2'\rangle}{\|\mathbf{v}_2'\|^2}\mathbf{v}_2' = \frac{-3+1+7}{1+1+1}(1,1,1) + \frac{-3-2+7}{1+4+1}(1,-2,1) = \left(\frac{5}{3},\frac{5}{3},\frac{5}{3}\right) + \left(\frac{1}{3},-\frac{2}{3},\frac{1}{3}\right) = (2,1,2)$.

43. First transform the basis $S = \{\mathbf{p}_1, \mathbf{p}_2, \mathbf{p}_3\} = \{1, x, x^2\}$ into an orthogonal basis $\{\mathbf{v}_1, \mathbf{v}_2, \mathbf{v}_3\}$.

$\mathbf{v}_1 = \mathbf{p}_1 = 1$

$\|\mathbf{v}_1\| = \sqrt{\langle \mathbf{v}_1, \mathbf{v}_1\rangle} = \sqrt{\int_0^1 1^2 dx} = \sqrt{x|_0^1} = 1$

$\langle \mathbf{p}_2,\ \mathbf{v}_1\rangle = \int_0^1 x \cdot 1\, dx = \frac{1}{2}x^2\Big|_0^1 = \frac{1}{2}$

$\mathbf{v}_2 = \mathbf{p}_2 - \frac{\langle \mathbf{p}_2, \mathbf{v}_1\rangle}{\|\mathbf{v}_1\|^2}\mathbf{v}_1 = x - \frac{\frac{1}{2}}{1}(1) = x - \frac{\frac{1}{2}}{1}(1) = -\frac{1}{2} + x$

$\|\mathbf{v}_2\|^2 = \langle \mathbf{v}_2,\ \mathbf{v}_2\rangle = \int_0^1 \left(-\frac{1}{2} + x\right)^2 dx = \int_0^1 \left(\frac{1}{4} - x + x^2\right) dx = \left(\frac{1}{4}x - \frac{1}{2}x^2 + \frac{1}{3}x^3\right)\Big|_0^1 = \frac{1}{12}$

$\|\mathbf{v}_2\| = \sqrt{\frac{1}{12}} = \frac{1}{2\sqrt{3}}$

$\langle \mathbf{p}_3,\ \mathbf{v}_1\rangle = \int_0^1 x^2 \cdot 1\, dx = \frac{1}{3}x^3\Big|_0^1 = \frac{1}{3}$

$\langle \mathbf{p}_3,\ \mathbf{v}_2\rangle = \int_0^1 x^2\left(-\frac{1}{2} + x\right) dx = \int_0^1 \left(-\frac{1}{2}x^2 + x^3\right) dx = \left(-\frac{1}{6}x^3 + \frac{1}{4}x^4\right)\Big|_0^1 = \frac{1}{12}$

$\mathbf{v}_3 = \mathbf{p}_3 - \frac{\langle \mathbf{p}_3, \mathbf{v}_1\rangle}{\|\mathbf{v}_1\|^2}\mathbf{v}_1 - \frac{\langle \mathbf{p}_3, \mathbf{v}_2\rangle}{\|\mathbf{v}_2\|^2}\mathbf{v}_2 = x^2 - \frac{\frac{1}{3}}{1}(1) - \frac{\frac{1}{12}}{\frac{1}{12}}\left(-\frac{1}{2} + x\right) = x^2 - \frac{1}{3} + \frac{1}{2} - x = \frac{1}{6} - x + x^2$

$\|\mathbf{v}_3\|^2 = \langle \mathbf{v}_3,\ \mathbf{v}_3\rangle = \int_0^1 \left(\frac{1}{6} - x + x^2\right)^2 dx = \left(\frac{1}{36}x - \frac{1}{6}x^2 + \frac{4}{9}x^3 - \frac{1}{2}x^4 + \frac{1}{5}x^5\right)\Big|_0^1 = \frac{1}{180}$

$\|\mathbf{v}_3\| = \sqrt{\frac{1}{180}} = \frac{1}{6\sqrt{5}}$

The orthonormal basis is

$\mathbf{q}_1 = \frac{\mathbf{v}_1}{\|\mathbf{v}_1\|} = \frac{1}{1} = 1$

$\mathbf{q}_2 = \frac{\mathbf{v}_2}{\|\mathbf{v}_2\|} = \frac{-\frac{1}{2}+x}{\frac{1}{2\sqrt{3}}} = 2\sqrt{3}\left(-\frac{1}{2}+x\right) = \sqrt{3}(-1+2x)$

$\mathbf{q}_3 = \frac{\mathbf{v}_3}{\|\mathbf{v}_3\|} = \frac{\frac{1}{6}-x+x^2}{\frac{1}{6\sqrt{5}}} = 6\sqrt{5}\left(\frac{1}{6}-x+x^2\right) = \sqrt{5}(1-6x+6x^2)$

45. Let $\mathbf{u}_1 = (1,2)$, $\mathbf{u}_2 = (-1,3)$, $\mathbf{q}_1 = \left(\frac{1}{\sqrt{5}}, \frac{2}{\sqrt{5}}\right)$, and $\mathbf{q}_2 = \left(-\frac{2}{\sqrt{5}}, \frac{1}{\sqrt{5}}\right)$. A QR-decomposition of the matrix A is formed by the given matrix Q and the matrix

$$R = \begin{bmatrix} \langle \mathbf{u}_1, \mathbf{q}_1 \rangle & \langle \mathbf{u}_2, \mathbf{q}_1 \rangle \\ 0 & \langle \mathbf{u}_2, \mathbf{q}_2 \rangle \end{bmatrix} = \begin{bmatrix} \frac{1}{\sqrt{5}} + \frac{4}{\sqrt{5}} & -\frac{1}{\sqrt{5}} + \frac{6}{\sqrt{5}} \\ 0 & \frac{2}{\sqrt{5}} + \frac{3}{\sqrt{5}} \end{bmatrix} = \begin{bmatrix} \sqrt{5} & \sqrt{5} \\ 0 & \sqrt{5} \end{bmatrix}.$$

47. Let $\mathbf{u}_1 = (1,0,1)$, $\mathbf{u}_2 = (0,1,2)$, $\mathbf{u}_3 = (2,1,0)$, $\mathbf{q}_1 = \left(\frac{1}{\sqrt{2}}, 0, \frac{1}{\sqrt{2}}\right)$, $\mathbf{q}_2 = \left(-\frac{1}{\sqrt{3}}, \frac{1}{\sqrt{3}}, \frac{1}{\sqrt{3}}\right)$, and $\mathbf{q}_3 = \left(\frac{1}{\sqrt{6}}, \frac{2}{\sqrt{6}}, -\frac{1}{\sqrt{6}}\right)$. A QR-decomposition of the matrix A is formed by the given matrix Q and the matrix

$$R = \begin{bmatrix} \langle \mathbf{u}_1, \mathbf{q}_1 \rangle & \langle \mathbf{u}_2, \mathbf{q}_1 \rangle & \langle \mathbf{u}_3, \mathbf{q}_1 \rangle \\ 0 & \langle \mathbf{u}_2, \mathbf{q}_2 \rangle & \langle \mathbf{u}_3, \mathbf{q}_2 \rangle \\ 0 & 0 & \langle \mathbf{u}_3, \mathbf{q}_3 \rangle \end{bmatrix} = \begin{bmatrix} \frac{1}{\sqrt{2}} + 0 + \frac{1}{\sqrt{2}} & 0 + 0 + \frac{2}{\sqrt{2}} & \frac{2}{\sqrt{2}} + 0 + 0 \\ 0 & 0 + \frac{1}{\sqrt{3}} + \frac{2}{\sqrt{3}} & -\frac{2}{\sqrt{3}} + \frac{1}{\sqrt{3}} + 0 \\ 0 & 0 & \frac{2}{\sqrt{6}} + \frac{2}{\sqrt{6}} + 0 \end{bmatrix} = \begin{bmatrix} \sqrt{2} & \sqrt{2} & \sqrt{2} \\ 0 & \sqrt{3} & -\frac{1}{\sqrt{3}} \\ 0 & 0 & \frac{4}{\sqrt{6}} \end{bmatrix}.$$

49. In partitioned form, $A = \begin{bmatrix} 1 & 0 & 1 \\ -1 & 1 & 1 \\ 1 & 0 & 1 \\ -1 & 1 & 1 \end{bmatrix} = [\mathbf{u}_1 \quad \mathbf{u}_2 \quad \mathbf{u}_3]$. By inspection, $\mathbf{u}_3 = \mathbf{u}_1 + 2\mathbf{u}_2$, so the column vectors of A are not linearly independent and A does not have a QR-decomposition.

51. The proof of part (a) mirrors the proof of part (b) in the book. By Theorem 6.3.1, an orthogonal set of nonzero vectors in W is linearly independent. It follows from part (b) of Theorem 4.5.5 that this set can be enlarged to form a basis for W. Applying the Gram-Schmidt process (without the normalization step) will yield an enlarged orthogonal set (the original orthogonal set will not be affected).

53. The diagonal entries of R are $\langle \mathbf{u}_i, \mathbf{q}_i \rangle$ for $i = 1, 2, \ldots, n$, where $\mathbf{q}_i = \frac{\mathbf{v}_i}{\|\mathbf{v}_i\|}$ is the normalization of a vector $\mathbf{v}_i$ that is the result of applying the Gram-Schmidt process to $\{\mathbf{u}_1, \mathbf{u}_2, \ldots, \mathbf{u}_n\}$. Thus, $\mathbf{v}_i$ is $\mathbf{u}_i$ minus a linear combination of the vectors $\mathbf{v}_1, \mathbf{v}_2, \ldots, \mathbf{v}_{i-1}$, so $\mathbf{u}_i = \mathbf{v}_i + k_1\mathbf{v}_1 + k_2\mathbf{v}_2 + \cdots + k_{i-1}\mathbf{v}_{i-1}$. Thus, $\langle \mathbf{u}_i, \mathbf{v}_i \rangle = \langle \mathbf{v}_i, \mathbf{v}_i \rangle$ and $\langle \mathbf{u}_i, \mathbf{q}_i \rangle = \left\langle \mathbf{u}_i, \frac{\mathbf{v}_i}{\|\mathbf{v}_i\|} \right\rangle = \frac{1}{\|\mathbf{v}_i\|}\langle \mathbf{v}_i, \mathbf{v}_i \rangle = \|\mathbf{v}_i\|$. Since each vector $\mathbf{v}_i$ is nonzero, each diagonal entry of R is nonzero.

55. **(b)** The range of T is W; the kernel of T is $W^{\perp}$.

True-False Exercises

(a) False. For example, the vectors (1, 0) and (1, 1) in R^2 are linearly independent but not orthogonal.

(b) False. The vectors must be nonzero for this to be true.

(c) True. A nontrivial subspace of R^3 will have a basis, which can be transformed into an orthonormal basis with respect to the Euclidean inner product.

(d) True. A nonzero finite-dimensional inner product space will have finite basis which can be transformed into an orthonormal basis with respect to the inner product via the Gram-Schmidt process with normalization.

(e) False. $\text{proj}_W\mathbf{x}$ is a vector in W.

(f) True. Every invertible $n \times n$ matrix has a QR-decomposition.

6.4 Best Approximation; Least Squares

1. $A = \begin{bmatrix} 1 & -1 \\ 2 & 3 \\ 4 & 5 \end{bmatrix}$; $A^TA = \begin{bmatrix} 1 & 2 & 4 \\ -1 & 3 & 5 \end{bmatrix}\begin{bmatrix} 1 & -1 \\ 2 & 3 \\ 4 & 5 \end{bmatrix} = \begin{bmatrix} 21 & 25 \\ 25 & 35 \end{bmatrix}$; $A^T\mathbf{b} = \begin{bmatrix} 1 & 2 & 4 \\ -1 & 3 & 5 \end{bmatrix}\begin{bmatrix} 2 \\ -1 \\ 5 \end{bmatrix} = \begin{bmatrix} 20 \\ 20 \end{bmatrix}$;

The associated normal equation is $\begin{bmatrix} 21 & 25 \\ 25 & 35 \end{bmatrix}\begin{bmatrix} x_1 \\ x_2 \end{bmatrix} = \begin{bmatrix} 20 \\ 20 \end{bmatrix}$.

3. $A^TA = \begin{bmatrix} 1 & 2 & 4 \\ -1 & 3 & 5 \end{bmatrix}\begin{bmatrix} 1 & -1 \\ 2 & 3 \\ 4 & 5 \end{bmatrix} = \begin{bmatrix} 21 & 25 \\ 25 & 35 \end{bmatrix}$; $A^T\mathbf{b} = \begin{bmatrix} 1 & 2 & 4 \\ -1 & 3 & 5 \end{bmatrix}\begin{bmatrix} 2 \\ -1 \\ 5 \end{bmatrix} = \begin{bmatrix} 20 \\ 20 \end{bmatrix}$;

The normal system $A^TA\mathbf{x} = A^T\mathbf{b}$ is $\begin{bmatrix} 21 & 25 \\ 25 & 35 \end{bmatrix}\begin{bmatrix} x_1 \\ x_2 \end{bmatrix} = \begin{bmatrix} 20 \\ 20 \end{bmatrix}$.

The reduced row echelon form of the augmented matrix of the normal system is $\begin{bmatrix} 1 & 0 & \frac{20}{11} \\ 0 & 1 & -\frac{8}{11} \end{bmatrix}$.

The solution of this system $x_1 = \frac{20}{11}$, $x_2 = -\frac{8}{11}$ is the unique least squares solution of $A\mathbf{x} = \mathbf{b}$.

5. $A^TA = \begin{bmatrix} 1 & 2 & 1 & 1 \\ 0 & 1 & 1 & 1 \\ -1 & -2 & 0 & -1 \end{bmatrix}\begin{bmatrix} 1 & 0 & -1 \\ 2 & 1 & -2 \\ 1 & 1 & 0 \\ 1 & 1 & -1 \end{bmatrix} = \begin{bmatrix} 7 & 4 & -6 \\ 4 & 3 & -3 \\ -6 & -3 & 6 \end{bmatrix}$;

$A^T\mathbf{b} = \begin{bmatrix} 1 & 2 & 1 & 1 \\ 0 & 1 & 1 & 1 \\ -1 & -2 & 0 & -1 \end{bmatrix}\begin{bmatrix} 6 \\ 0 \\ 9 \\ 3 \end{bmatrix} = \begin{bmatrix} 18 \\ 12 \\ -9 \end{bmatrix}$

The normal system $A^TA\mathbf{x} = A^T\mathbf{b}$ is $\begin{bmatrix} 7 & 4 & -6 \\ 4 & 3 & -3 \\ -6 & -3 & 6 \end{bmatrix}\begin{bmatrix} x_1 \\ x_2 \\ x_3 \end{bmatrix} = \begin{bmatrix} 18 \\ 12 \\ -9 \end{bmatrix}$.

The reduced row echelon form of the augmented matrix of the normal system is $\begin{bmatrix} 1 & 0 & 0 & 12 \\ 0 & 1 & 0 & -3 \\ 0 & 0 & 1 & 9 \end{bmatrix}$.

The solution of this system $x_1 = 12$, $x_2 = -3$, $x_3 = 9$ is the unique least squares solution of $A\mathbf{x} = \mathbf{b}$.

7. Least squares error vector: $\mathbf{e} = \mathbf{b} - A\mathbf{x} = \begin{bmatrix} 2 \\ -1 \\ 5 \end{bmatrix} - \begin{bmatrix} 1 & -1 \\ 2 & 3 \\ 4 & 5 \end{bmatrix} \begin{bmatrix} \frac{20}{11} \\ -\frac{8}{11} \end{bmatrix} = \begin{bmatrix} 2 \\ -1 \\ 5 \end{bmatrix} - \begin{bmatrix} \frac{28}{11} \\ \frac{16}{11} \\ \frac{40}{11} \end{bmatrix} = \begin{bmatrix} -\frac{6}{11} \\ -\frac{27}{11} \\ \frac{15}{11} \end{bmatrix}$;

$A^T\mathbf{e} = \begin{bmatrix} 1 & 2 & 4 \\ -1 & 3 & 5 \end{bmatrix} \begin{bmatrix} -\frac{6}{11} \\ -\frac{27}{11} \\ \frac{15}{11} \end{bmatrix} = \begin{bmatrix} 0 \\ 0 \end{bmatrix}$; therefore the least squares error vector is orthogonal to every vector in the column space of A.

Least squares error: $\|\mathbf{b} - A\mathbf{x}\| = \sqrt{\left(-\frac{6}{11}\right)^2 + \left(-\frac{27}{11}\right)^2 + \left(\frac{15}{11}\right)^2} = \frac{3}{11}\sqrt{110} \approx 2.86$.

9. Least squares error vector: $\mathbf{e} = \mathbf{b} - A\mathbf{x} = \begin{bmatrix} 6 \\ 0 \\ 9 \\ 3 \end{bmatrix} - \begin{bmatrix} 1 & 0 & -1 \\ 2 & 1 & -2 \\ 1 & 1 & 0 \\ 1 & 1 & -1 \end{bmatrix} \begin{bmatrix} 12 \\ -3 \\ 9 \end{bmatrix} = \begin{bmatrix} 6 \\ 0 \\ 9 \\ 3 \end{bmatrix} - \begin{bmatrix} 3 \\ 3 \\ 9 \\ 0 \end{bmatrix} = \begin{bmatrix} 3 \\ -3 \\ 0 \\ 3 \end{bmatrix}$; $A^T\mathbf{e} =$

$\begin{bmatrix} 1 & 2 & 1 & 1 \\ 0 & 1 & 1 & 1 \\ -1 & -2 & 0 & -1 \end{bmatrix} \begin{bmatrix} 3 \\ -3 \\ 0 \\ 3 \end{bmatrix} = \begin{bmatrix} 0 \\ 0 \\ 0 \end{bmatrix}$; therefore the least squares error vector is orthogonal to every vector in the column space of A.

Least squares error: $\|\mathbf{b} - A\mathbf{x}\| = \sqrt{3^2 + (-3)^2 + 0^2 + 3^2} = 3\sqrt{3} \approx 5.196$.

11. $A^TA = \begin{bmatrix} 2 & 4 & -2 \\ 1 & 2 & -1 \end{bmatrix} \begin{bmatrix} 2 & 1 \\ 4 & 2 \\ -2 & -1 \end{bmatrix} = \begin{bmatrix} 24 & 12 \\ 12 & 6 \end{bmatrix}$; $A^T\mathbf{b} = \begin{bmatrix} 2 & 4 & -2 \\ 1 & 2 & -1 \end{bmatrix} \begin{bmatrix} 3 \\ 2 \\ 1 \end{bmatrix} = \begin{bmatrix} 12 \\ 6 \end{bmatrix}$;

The normal system $A^TA\mathbf{x} = A^T\mathbf{b}$ is $\begin{bmatrix} 24 & 12 \\ 12 & 6 \end{bmatrix} \begin{bmatrix} x_1 \\ x_2 \end{bmatrix} = \begin{bmatrix} 12 \\ 6 \end{bmatrix}$.

The reduced row echelon form of the augmented matrix of the normal system is $\begin{bmatrix} 1 & \frac{1}{2} & \frac{1}{2} \\ 0 & 0 & 0 \end{bmatrix}$.

The general solution of the normal system is $x_1 = \frac{1}{2} - \frac{1}{2}t, x_2 = t$. All of these are least squares solutions of $A\mathbf{x} = \mathbf{b}$. The error vector is the same for all solutions:

$\mathbf{e} = \mathbf{b} - A\mathbf{x} = \begin{bmatrix} 3 \\ 2 \\ 1 \end{bmatrix} - \begin{bmatrix} 2 & 1 \\ 4 & 2 \\ -2 & -1 \end{bmatrix} \begin{bmatrix} \frac{1}{2} - \frac{1}{2}t \\ t \end{bmatrix} = \begin{bmatrix} 3 \\ 2 \\ 1 \end{bmatrix} - \begin{bmatrix} 1 \\ 2 \\ -1 \end{bmatrix} = \begin{bmatrix} 2 \\ 0 \\ 2 \end{bmatrix}$.

13. $A^TA = \begin{bmatrix} -1 & 2 & 0 \\ 3 & 1 & 1 \\ 2 & 3 & 1 \end{bmatrix} \begin{bmatrix} -1 & 3 & 2 \\ 2 & 1 & 3 \\ 0 & 1 & 1 \end{bmatrix} = \begin{bmatrix} 5 & -1 & 4 \\ -1 & 11 & 10 \\ 4 & 10 & 14 \end{bmatrix}$; $A^T\mathbf{b} = \begin{bmatrix} -1 & 2 & 0 \\ 3 & 1 & 1 \\ 2 & 3 & 1 \end{bmatrix} \begin{bmatrix} 7 \\ 0 \\ -7 \end{bmatrix} = \begin{bmatrix} -7 \\ 14 \\ 7 \end{bmatrix}$;

The normal system $A^TA\mathbf{x} = A^T\mathbf{b}$ is $\begin{bmatrix} 5 & -1 & 4 \\ -1 & 11 & 10 \\ 4 & 10 & 14 \end{bmatrix} \begin{bmatrix} x_1 \\ x_2 \\ x_3 \end{bmatrix} = \begin{bmatrix} -7 \\ 14 \\ 7 \end{bmatrix}$.

The reduced row echelon form of the augmented matrix of the normal system is $\begin{bmatrix} 1 & 0 & 1 & -\frac{7}{6} \\ 0 & 1 & 1 & \frac{7}{6} \\ 0 & 0 & 0 & 0 \end{bmatrix}$.

The general solution of the normal system is $x_1 = -\frac{7}{6} - t$, $x_2 = \frac{7}{6} - t$, $x_3 = t$. All of these are least squares solutions of $A\mathbf{x} = \mathbf{b}$. The error vector is the same for all solutions:

$$\mathbf{e} = \mathbf{b} - A\mathbf{x} = \begin{bmatrix} 7 \\ 0 \\ -7 \end{bmatrix} - \begin{bmatrix} -1 & 3 & 2 \\ 2 & 1 & 3 \\ 0 & 1 & 1 \end{bmatrix} \begin{bmatrix} -\frac{7}{6} - t \\ \frac{7}{6} - t \\ t \end{bmatrix} = \begin{bmatrix} 7 \\ 0 \\ -7 \end{bmatrix} - \begin{bmatrix} \frac{14}{3} \\ -\frac{7}{6} \\ \frac{7}{6} \end{bmatrix} = \begin{bmatrix} \frac{7}{3} \\ \frac{7}{6} \\ -\frac{49}{6} \end{bmatrix}.$$

15. $A^TA = \begin{bmatrix} 1 & 3 & -2 \\ -1 & 2 & 4 \end{bmatrix} \begin{bmatrix} 1 & -1 \\ 3 & 2 \\ -2 & 4 \end{bmatrix} = \begin{bmatrix} 14 & -3 \\ -3 & 21 \end{bmatrix}$; $A^T\mathbf{b} = \begin{bmatrix} 1 & 3 & -2 \\ -1 & 2 & 4 \end{bmatrix} \begin{bmatrix} 4 \\ 1 \\ 3 \end{bmatrix} = \begin{bmatrix} 1 \\ 10 \end{bmatrix}$;

The normal system $A^TA\mathbf{x} = A^T\mathbf{b}$ is $\begin{bmatrix} 14 & -3 \\ -3 & 21 \end{bmatrix} \begin{bmatrix} x_1 \\ x_2 \end{bmatrix} = \begin{bmatrix} 1 \\ 10 \end{bmatrix}$.

The reduced row echelon form of the augmented matrix of the normal system is $\begin{bmatrix} 1 & 0 & \frac{17}{95} \\ 0 & 1 & \frac{143}{285} \end{bmatrix}$.

The solution of this system $x_1 = \frac{17}{95}$, $x_2 = \frac{143}{285}$ is the unique least squares solution of $A\mathbf{x} = \mathbf{b}$.

By Theorem 6.4.2, $\text{proj}_W\mathbf{b} = A\mathbf{x} = \begin{bmatrix} 1 & -1 \\ 3 & 2 \\ -2 & 4 \end{bmatrix} \begin{bmatrix} \frac{17}{95} \\ \frac{143}{285} \end{bmatrix} = \begin{bmatrix} -\frac{92}{285} \\ \frac{439}{285} \\ \frac{94}{57} \end{bmatrix}$.

This matches the result obtained using Theorem 6.4.4:

$$\begin{aligned}
\text{proj}_W\mathbf{b} &= A(A^TA)^{-1}A^T\mathbf{b} \\
&= \begin{bmatrix} 1 & -1 \\ 3 & 2 \\ -2 & 4 \end{bmatrix} \left(\begin{bmatrix} 14 & -3 \\ -3 & 21 \end{bmatrix} \right)^{-1} \begin{bmatrix} 1 & 3 & -2 \\ -1 & 2 & 4 \end{bmatrix} \begin{bmatrix} 4 \\ 1 \\ 3 \end{bmatrix} \\
&= \begin{bmatrix} 1 & -1 \\ 3 & 2 \\ -2 & 4 \end{bmatrix} \left(\frac{1}{(14)(21) - (-3)(-3)} \begin{bmatrix} 21 & 3 \\ 3 & 14 \end{bmatrix} \right) \begin{bmatrix} 1 & 3 & -2 \\ -1 & 2 & 4 \end{bmatrix} \begin{bmatrix} 4 \\ 1 \\ 3 \end{bmatrix} \\
&= \frac{1}{285} \begin{bmatrix} 1 & -1 \\ 3 & 2 \\ -2 & 4 \end{bmatrix} \begin{bmatrix} 21 & 3 \\ 3 & 14 \end{bmatrix} \begin{bmatrix} 1 & 3 & -2 \\ -1 & 2 & 4 \end{bmatrix} \begin{bmatrix} 4 \\ 1 \\ 3 \end{bmatrix} \\
&= \frac{1}{285} \begin{bmatrix} -92 \\ 439 \\ 470 \end{bmatrix} \\
&= \begin{bmatrix} -\frac{92}{285} \\ \frac{439}{285} \\ \frac{94}{57} \end{bmatrix}
\end{aligned}$$

17. We follow the procedure of Example 2. For $A = [\mathbf{v}_1 \mid \mathbf{v}_2] = \begin{bmatrix} -1 & 2 \\ 2 & 2 \\ 1 & 4 \end{bmatrix}$, we have

$$A^TA = \begin{bmatrix} -1 & 2 & 1 \\ 2 & 2 & 4 \end{bmatrix}\begin{bmatrix} -1 & 2 \\ 2 & 2 \\ 1 & 4 \end{bmatrix} = \begin{bmatrix} 6 & 6 \\ 6 & 24 \end{bmatrix} \text{ and } A^T\mathbf{u} = \begin{bmatrix} -1 & 2 & 1 \\ 2 & 2 & 4 \end{bmatrix}\begin{bmatrix} 1 \\ -6 \\ 1 \end{bmatrix} = \begin{bmatrix} -12 \\ -6 \end{bmatrix};$$

The normal system $A^TA\mathbf{x} = A^T\mathbf{u}$ is $\begin{bmatrix} 6 & 6 \\ 6 & 24 \end{bmatrix}\begin{bmatrix} x_1 \\ x_2 \end{bmatrix} = \begin{bmatrix} -12 \\ -6 \end{bmatrix}$. The reduced row echelon form of the augmented matrix of the normal system is $\begin{bmatrix} 1 & 0 & -\frac{7}{3} \\ 0 & 1 & \frac{1}{3} \end{bmatrix}$ so that the least squares solution of $A\mathbf{x} = \mathbf{u}$ is $\mathbf{x} = \begin{bmatrix} -\frac{7}{3} \\ \frac{1}{3} \end{bmatrix}$. Denoting $W = \text{span}\{\mathbf{v}_1, \mathbf{v}_2\}$ we obtain $\text{proj}_W\mathbf{u} = A\mathbf{x} = \begin{bmatrix} -1 & 2 \\ 2 & 2 \\ 1 & 4 \end{bmatrix}\begin{bmatrix} -\frac{7}{3} \\ \frac{1}{3} \end{bmatrix} = \begin{bmatrix} 3 \\ -4 \\ -1 \end{bmatrix}$.

19. Letting $A = \begin{bmatrix} 1 \\ 0 \end{bmatrix}$, we have $P = A(A^TA)^{-1}A^T = \begin{bmatrix} 1 \\ 0 \end{bmatrix}\left(\begin{bmatrix} 1 & 0 \end{bmatrix}\begin{bmatrix} 1 \\ 0 \end{bmatrix}\right)^{-1}\begin{bmatrix} 1 & 0 \end{bmatrix} = \begin{bmatrix} 1 \\ 0 \end{bmatrix}([1])^{-1}\begin{bmatrix} 1 & 0 \end{bmatrix} = \begin{bmatrix} 1 \\ 0 \end{bmatrix}[1]\begin{bmatrix} 1 & 0 \end{bmatrix} = \begin{bmatrix} 1 \\ 0 \end{bmatrix}\begin{bmatrix} 1 & 0 \end{bmatrix} = \begin{bmatrix} 1 & 0 \\ 0 & 0 \end{bmatrix}$. This matches the matrix in Table 3 of Section 4.9.

21. Letting $A = \begin{bmatrix} 1 & 0 \\ 0 & 0 \\ 0 & 1 \end{bmatrix}$, we have $A^TA = \begin{bmatrix} 1 & 0 & 0 \\ 0 & 0 & 1 \end{bmatrix}\begin{bmatrix} 1 & 0 \\ 0 & 0 \\ 0 & 1 \end{bmatrix} = \begin{bmatrix} 1 & 0 \\ 0 & 1 \end{bmatrix}$

$$P = A(A^TA)^{-1}A^T = \begin{bmatrix} 1 & 0 \\ 0 & 0 \\ 0 & 1 \end{bmatrix}\left(\begin{bmatrix} 1 & 0 \\ 0 & 1 \end{bmatrix}\right)^{-1}\begin{bmatrix} 1 & 0 & 0 \\ 0 & 0 & 1 \end{bmatrix} = \begin{bmatrix} 1 & 0 \\ 0 & 0 \\ 0 & 1 \end{bmatrix}\begin{bmatrix} 1 & 0 & 0 \\ 0 & 0 & 1 \end{bmatrix} = \begin{bmatrix} 1 & 0 & 0 \\ 0 & 0 & 0 \\ 0 & 0 & 1 \end{bmatrix}.$$

This matches the matrix in Table 4 of Section 4.9.

23. We use Theorem 6.4.6: $\mathbf{x} = R^{-1}Q^T\mathbf{b} = \frac{1}{(5)\left(\frac{7}{5}\right)}\begin{bmatrix} \frac{7}{5} & \frac{1}{5} \\ 0 & 5 \end{bmatrix}\begin{bmatrix} \frac{3}{5} & -\frac{4}{5} \\ \frac{4}{5} & \frac{3}{5} \end{bmatrix}\begin{bmatrix} 3 \\ 2 \end{bmatrix} = \begin{bmatrix} \frac{1}{5} & \frac{1}{35} \\ 0 & \frac{5}{7} \end{bmatrix}\begin{bmatrix} \frac{1}{5} \\ \frac{18}{5} \end{bmatrix} = \begin{bmatrix} \frac{1}{7} \\ \frac{18}{7} \end{bmatrix}$.

25. **(a)** If $x = s$ and $y = t$, then a point on the plane is $(s, t, -5s + 3t) = s(1, 0, -5) + t(0, 1, 3)$. $\mathbf{w}_1 = (1, 0, -5)$ and $\mathbf{w}_2 = (0, 1, 3)$ form a basis for W (they are linearly independent since neither of them is a scalar multiple of the other).

(b) Letting $A = \begin{bmatrix} 1 & 0 \\ 0 & 1 \\ -5 & 3 \end{bmatrix}$, Formula (11) yields

$$P = A(A^TA)^{-1}A^T = \begin{bmatrix} 1 & 0 \\ 0 & 1 \\ -5 & 3 \end{bmatrix}\left(\begin{bmatrix} 1 & 0 & -5 \\ 0 & 1 & 3 \end{bmatrix}\begin{bmatrix} 1 & 0 \\ 0 & 1 \\ -5 & 3 \end{bmatrix}\right)^{-1}\begin{bmatrix} 1 & 0 & -5 \\ 0 & 1 & 3 \end{bmatrix}$$

$$= \begin{bmatrix} 1 & 0 \\ 0 & 1 \\ -5 & 3 \end{bmatrix}\left(\begin{bmatrix} 26 & -15 \\ -15 & 10 \end{bmatrix}\right)^{-1}\begin{bmatrix} 1 & 0 & -5 \\ 0 & 1 & 3 \end{bmatrix} = \begin{bmatrix} 1 & 0 \\ 0 & 1 \\ -5 & 3 \end{bmatrix}\left(\frac{1}{(26)(10)-(-15)(-15)}\begin{bmatrix} 10 & 15 \\ 15 & 26 \end{bmatrix}\right)\begin{bmatrix} 1 & 0 & -5 \\ 0 & 1 & 3 \end{bmatrix}$$

$$= \frac{1}{35}\begin{bmatrix} 1 & 0 \\ 0 & 1 \\ -5 & 3 \end{bmatrix}\begin{bmatrix} 10 & 15 \\ 15 & 26 \end{bmatrix}\begin{bmatrix} 1 & 0 & -5 \\ 0 & 1 & 3 \end{bmatrix} = \frac{1}{35}\begin{bmatrix} 10 & 15 & -5 \\ 15 & 26 & 3 \\ -5 & 3 & 34 \end{bmatrix}.$$

27. The reduced row echelon form of the augmented matrix of the given homogeneous system is

$\begin{bmatrix} 1 & 0 & \frac{1}{2} & -\frac{1}{2} & 0 \\ 0 & 1 & \frac{1}{2} & \frac{1}{2} & 0 \end{bmatrix}$ so that the general solution is $x_1 = -\frac{1}{2}s + \frac{1}{2}t, x_2 = -\frac{1}{2}s - \frac{1}{2}t, x_3 = s, x_4 = t$.

The solution space W is spanned by vectors $(-1,-1,2,0)$ and $(1,-1,0,2)$.

We construct the matrix with these vectors as its columns $A = \begin{bmatrix} -1 & 1 \\ -1 & -1 \\ 2 & 0 \\ 0 & 2 \end{bmatrix}$ then follow the procedure of Example 2 in Section 6.4.

$$A^T A = \begin{bmatrix} -1 & -1 & 2 & 0 \\ 1 & -1 & 0 & 2 \end{bmatrix} \begin{bmatrix} -1 & 1 \\ -1 & -1 \\ 2 & 0 \\ 0 & 2 \end{bmatrix} = \begin{bmatrix} 6 & 0 \\ 0 & 6 \end{bmatrix}; \; A^T\mathbf{u} = \begin{bmatrix} -1 & -1 & 2 & 0 \\ 1 & -1 & 0 & 2 \end{bmatrix} \begin{bmatrix} 5 \\ 6 \\ 7 \\ 2 \end{bmatrix} = \begin{bmatrix} 3 \\ 3 \end{bmatrix};$$

The normal system $A^T A\mathbf{x} = A^T\mathbf{u}$ is $\begin{bmatrix} 6 & 0 \\ 0 & 6 \end{bmatrix} \begin{bmatrix} x_1 \\ x_2 \end{bmatrix} = \begin{bmatrix} 3 \\ 3 \end{bmatrix}$. The reduced row echelon form of the augmented matrix of the normal system is $\begin{bmatrix} 1 & 0 & \frac{1}{2} \\ 0 & 1 & \frac{1}{2} \end{bmatrix}$ so that the least squares solution of $A\mathbf{x} = \mathbf{u}$ is

$\mathbf{x} = \begin{bmatrix} \frac{1}{2} \\ \frac{1}{2} \end{bmatrix}$. We obtained $\text{proj}_W \mathbf{u} = A\mathbf{x} = \begin{bmatrix} -1 & 1 \\ -1 & -1 \\ 2 & 0 \\ 0 & 2 \end{bmatrix} \begin{bmatrix} \frac{1}{2} \\ \frac{1}{2} \end{bmatrix} = \begin{bmatrix} 0 \\ -1 \\ 1 \\ 1 \end{bmatrix}$.

29. Let W be the row space of A. Since W is also the column space of A^T, by Formula (11) we have $P = A^T((A^T)^T A^T)^{-1}(A^T)^T = A^T(AA^T)^{-1}A$.

31. Since $\mathbf{b}$ is orthogonal to the column space of A, it follows that $A^T\mathbf{b} = \mathbf{0}$. By Theorem 6.4.4, the least squares solution is $\mathbf{x} = (A^T A)^{-1}A^T\mathbf{b} = (A^T A)^{-1}\mathbf{0} = \mathbf{0}$.

True-False Exercises

(a) True. $A^T A$ is an $n \times n$ matrix.

(b) False. Only square matrices have inverses, but $A^T A$ can be invertible when A is not a square matrix.

(c) True. If A is invertible, so is A^T, so the product $A^T A$ is also invertible.

(d) True. Multiplying both sides of $A\mathbf{x} = \mathbf{b}$ on the left by A^T yields $A^T A\mathbf{x} = A^T\mathbf{b}$.

(e) False. By Theorem 6.4.2, the normal system $A^T A\mathbf{x} = A^T\mathbf{b}$ is always consistent.

(f) True. This follows from Theorem 6.4.2.

(g) False. There may be more than one least squares solution as shown in Example 2.

(h) True. This follows from Theorem 6.4.4.

6.5 Mathematical Modeling Using Least Squares

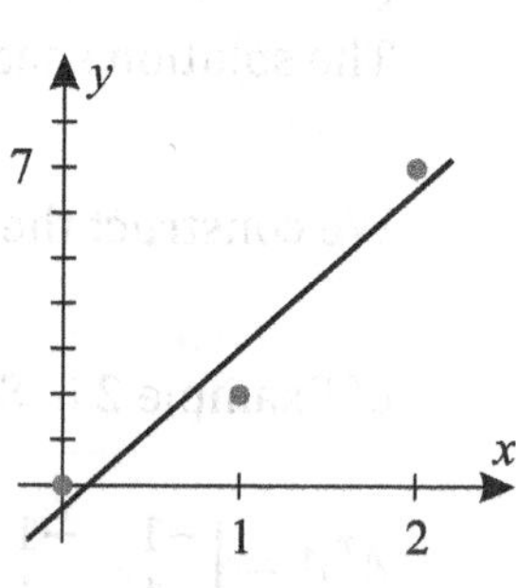

1. We have $M = \begin{bmatrix} 1 & 0 \\ 1 & 1 \\ 1 & 2 \end{bmatrix}$, $M^{\mathrm{T}} = \begin{bmatrix} 1 & 1 & 1 \\ 0 & 1 & 2 \end{bmatrix}$, $M^T M = \begin{bmatrix} 3 & 3 \\ 3 & 5 \end{bmatrix}$,

$(M^T M)^{-1} = \frac{1}{15-9}\begin{bmatrix} 5 & -3 \\ -3 & 3 \end{bmatrix} = \frac{1}{6}\begin{bmatrix} 5 & -3 \\ -3 & 3 \end{bmatrix}$, and

$\mathbf{v}^* = (M^T M)^{-1} M^T \mathbf{y} = \frac{1}{6}\begin{bmatrix} 5 & -3 \\ -3 & 3 \end{bmatrix}\begin{bmatrix} 1 & 1 & 1 \\ 0 & 1 & 2 \end{bmatrix}\begin{bmatrix} 0 \\ 2 \\ 7 \end{bmatrix} = \begin{bmatrix} -\frac{1}{2} \\ \frac{7}{2} \end{bmatrix}$

so the least squares straight line fit to the given data points is $y = -\frac{1}{2} + \frac{7}{2}x$.

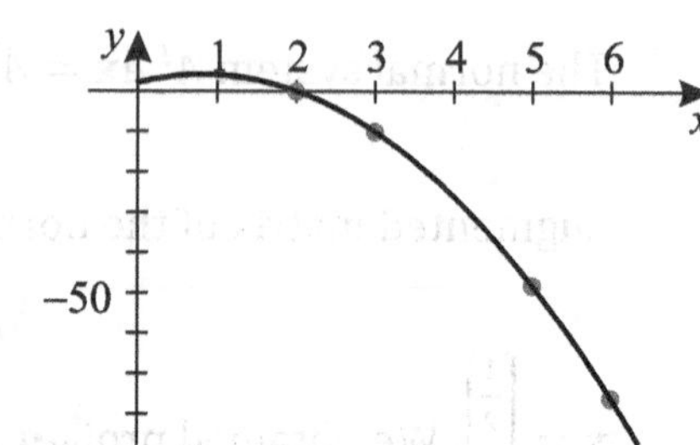

3. We have $M = \begin{bmatrix} 1 & 2 & 2^2 \\ 1 & 3 & 3^2 \\ 1 & 5 & 5^2 \\ 1 & 6 & 6^2 \end{bmatrix} = \begin{bmatrix} 1 & 2 & 4 \\ 1 & 3 & 9 \\ 1 & 5 & 25 \\ 1 & 6 & 36 \end{bmatrix}$,

$M^T M = \begin{bmatrix} 4 & 16 & 74 \\ 16 & 74 & 376 \\ 74 & 376 & 2018 \end{bmatrix}$,

$(M^T M)^{-1} = \frac{1}{90}\begin{bmatrix} 1989 & -1116 & 135 \\ -1116 & 649 & -80 \\ 135 & -80 & 10 \end{bmatrix}$, and

$\mathbf{v}^* = (M^T M)^{-1} M^T \mathbf{y} = \frac{1}{90}\begin{bmatrix} 1989 & -1116 & 135 \\ -1116 & 649 & -80 \\ 135 & -80 & 10 \end{bmatrix}\begin{bmatrix} 1 & 1 & 1 & 1 \\ 2 & 3 & 5 & 6 \\ 4 & 9 & 25 & 36 \end{bmatrix}\begin{bmatrix} 0 \\ -10 \\ -48 \\ -76 \end{bmatrix} = \begin{bmatrix} 2 \\ 5 \\ -3 \end{bmatrix}$

so the least squares quadratic fit to the given data points is $y = 2 + 5x - 3x^2$.

5. With the substitution $X = \frac{1}{x}$, the problem becomes to find a line of the form $y = a + b \cdot X$ that best fits the data points $(1, 7)$, $\left(\frac{1}{3}, 3\right)$, $\left(\frac{1}{6}, 1\right)$.

We have $M = \begin{bmatrix} 1 & 1 \\ 1 & \frac{1}{3} \\ 1 & \frac{1}{6} \end{bmatrix}$, $M^T M = \begin{bmatrix} 3 & \frac{3}{2} \\ \frac{3}{2} & \frac{41}{36} \end{bmatrix}$, $(M^T M)^{-1} = \frac{1}{42}\begin{bmatrix} 41 & -54 \\ -54 & 108 \end{bmatrix}$, and

$\mathbf{v}^* = (M^T M)^{-1} M^T \mathbf{y} = \frac{1}{42}\begin{bmatrix} 41 & -54 \\ -54 & 108 \end{bmatrix}\begin{bmatrix} 1 & 1 & 1 \\ 1 & \frac{1}{3} & \frac{1}{6} \end{bmatrix}\begin{bmatrix} 7 \\ 3 \\ 1 \end{bmatrix} = \begin{bmatrix} \frac{5}{21} \\ \frac{48}{7} \end{bmatrix}$. The line in terms of X is $y = \frac{5}{21} + \frac{48}{7}X$, so the required curve is $y = \frac{5}{21} + \frac{48}{7x}$.

7. The two column vectors of M are linearly independent if and only neither is a multiple of the other. Since all the entries in the first column are equal, the columns are linearly independent if and only if the second column has at least two different entries, i.e., if and only if at least two of the numbers $x_1, x_2, \ldots, x_n$ are distinct.

True-False Exercises

(a) False. There is only a unique least squares straight line fit if the data points do not all lie on a vertical line.

(b) True. If the points are not collinear, there is no solution to the system.

(c) True.

(d) False. The line minimizes the sum of the *squares* of the data errors.

6.6 Function Approximation; Fourier Series

1. $a_0 = \frac{1}{\pi}\int_0^{2\pi}(1+x)\,dx = \frac{1}{\pi}\left(x+\frac{x^2}{2}\right)\Big]_0^{2\pi} = 2+2\pi$

Using integration by parts to integrate both $x\cos(kx)$ and $x\sin(kx)$ we obtain

$a_k = \frac{1}{\pi}\int_0^{2\pi}(1+x)\cos(kx)\,dx = \left(\frac{1+x}{k\pi}\sin(kx)+\frac{1}{k^2\pi}\cos(kx)\right)\Big]_0^{2\pi} = 0$ and

$b_k = \frac{1}{\pi}\int_0^{2\pi}(1+x)\sin(kx)\,dx = \left(-\frac{1+x}{k\pi}\cos(kx)+\frac{1}{k^2\pi}\sin(kx)\right)\Big]_0^{2\pi} = -\frac{2}{k}$

(a) $1+x \approx \frac{a_0}{2} + a_1\cos x + a_2\cos(2x) + a_3\cos(3x) + b_1\sin x + b_2\sin(2x) + b_3\sin(3x)$ yields

$1+x \approx 1+\pi + 0\cos x + 0\cos(2x) - \frac{2}{1}\sin x - \frac{2}{2}\sin(2x) = 1+\pi-2\sin x - \sin 2x$

(b) $1+x \approx \frac{a_0}{2} + a_1\cos x + a_2\cos(2x) + \cdots + a_n\cos(nx) + b_1\sin x + b_2\sin(2x) + \cdots + b_n\sin(nx)$

yields $1+x \approx 1+\pi - \frac{2}{1}\sin x - \frac{2}{2}\sin(2x) - \cdots - \frac{2}{n}\sin(nx)$

3. **(a)** Let us denote $W = \text{span}\{1, e^x\}$. Applying the Gram-Schmidt process to the basis $\mathbf{u}_1 = 1$ and $\mathbf{u}_2 = e^x$ we obtain an orthogonal basis

$\mathbf{v}_1 = 1, \mathbf{v}_2 = \mathbf{u}_2 - \frac{\langle \mathbf{u}_2, \mathbf{v}_1\rangle}{\|\mathbf{v}_1\|^2}\mathbf{v}_1 = e^x - \frac{\int_0^1 e^x dx}{\int_0^1 1dx}1 = e^x - \frac{e^x]_0^1}{x]_0^1}1 = e^x - (e-1)1 = e^x - e + 1.$

Since $\int_0^1 (e^x - e + 1)^2\,dx = \int_0^1 (1 - 2e + e^2 + 2e^x - 2ee^x + e^{2x})dx$

$= \left(x - 2ex + e^2x + 2e^x - 2e^{x+1} + \frac{1}{2}e^{2x}\right)\Big|_0^1 = -\frac{3}{2} + 2e - \frac{1}{2}e^2 = \frac{1}{2}(e-1)(3-e)$, an

orthonormal basis is $\mathbf{q}_1 = \frac{\mathbf{v}_1}{\|\mathbf{v}_1\|} = \frac{1}{\sqrt{\int_0^1 1dx}} = \frac{1}{\sqrt{x]_0^1}} = 1$, $\mathbf{q}_2 = \frac{\mathbf{v}_2}{\|\mathbf{v}_2\|} = \frac{e^x-e+1}{\sqrt{\frac{1}{2}(e-1)(3-e)}}$.

The least squares approximation to $f(x) = x$ from W is

$\text{proj}_W\mathbf{f} = \langle \mathbf{f}, \mathbf{q}_1\rangle\mathbf{q}_1 + \langle \mathbf{f}, \mathbf{q}_2\rangle\mathbf{q}_2 = \int_0^1 x\,dx + \frac{2}{(e-1)(3-e)}\left(\int_0^1 x(e^x - e + 1)\,dx\right)(e^x - e + 1)$

$= \frac{1}{2} + \frac{2}{(e-1)(3-e)}\left(xe^x - e^x - \frac{x^2e}{2} + \frac{x^2}{2}\right)\Big]_0^1 (e^x - e + 1) = \frac{1}{2} + \frac{2}{(e-1)(3-e)}\left(-\frac{e}{2}+\frac{3}{2}\right)(e^x - e + 1)$

$= \frac{1}{2} + \frac{e^x-e+1}{e-1} = \frac{1}{2} + \frac{e^x}{e-1} - 1 = \frac{e^x}{e-1} - \frac{1}{2}$

(b) The mean square error is $\int_0^1\left(x - \left(\frac{e^x}{e-1} - \frac{1}{2}\right)\right)^2 dx = \frac{7e-19}{12e-12} \approx 0.00136$

5. **(a)** Let us denote $W = \text{span}\{1, x, x^2\}$.

Applying the Gram-Schmidt process to the basis $\mathbf{u}_1 = 1$, $\mathbf{u}_2 = x$, and $\mathbf{u}_3 = x^2$ we obtain an orthogonal basis $\mathbf{v}_1 = 1$, $\mathbf{v}_2 = \mathbf{u}_2 - \frac{\langle \mathbf{u}_2, \mathbf{v}_1 \rangle}{\|\mathbf{v}_1\|^2}\mathbf{v}_1 = x - \frac{\int_{-1}^{1} x\,dx}{\int_{-1}^{1} 1\,dx}1 = x - \frac{\left.\frac{x^2}{2}\right]_{-1}^{1}}{\left.x\right]_{-1}^{1}} = x - 0 = x$,

$$\mathbf{v}_3 = \mathbf{u}_3 - \frac{\langle \mathbf{u}_3, \mathbf{v}_1 \rangle}{\|\mathbf{v}_1\|^2}\mathbf{v}_1 - \frac{\langle \mathbf{u}_3, \mathbf{v}_2 \rangle}{\|\mathbf{v}_2\|^2}\mathbf{v}_2 = x^2 - \frac{\int_{-1}^{1} x^2 dx}{\int_{-1}^{1} 1\,dx}1 - \frac{\int_{-1}^{1} x^3 dx}{\int_{-1}^{1} x^2 dx}x = x^2 - \frac{\left.\frac{x^3}{3}\right]_{-1}^{1}}{\left.x\right]_{-1}^{1}} - \frac{\left.\frac{x^4}{4}\right]_{-1}^{1}}{\left.\frac{x^3}{3}\right]_{-1}^{1}}x = x^2 - \frac{1}{3}$$

and an orthonormal basis $\mathbf{q}_1 = \frac{\mathbf{v}_1}{\|\mathbf{v}_1\|} = \frac{1}{\sqrt{\int_{-1}^{1} 1\,dx}} = \frac{1}{\sqrt{\left.x\right]_{-1}^{1}}} = \frac{1}{\sqrt{2}}$,

$$\mathbf{q}_2 = \frac{\mathbf{v}_2}{\|\mathbf{v}_2\|} = \frac{x}{\sqrt{\int_{-1}^{1} x^2 dx}} = \frac{x}{\sqrt{\left.\frac{x^3}{3}\right]_{-1}^{1}}} = \sqrt{\frac{3}{2}}x, \quad \mathbf{q}_3 = \frac{\mathbf{v}_3}{\|\mathbf{v}_3\|} = \frac{x^2 - \frac{1}{3}}{\sqrt{\int_{-1}^{1}\left(x^2 - \frac{1}{3}\right)^2 dx}} = \frac{x^2 - \frac{1}{3}}{\frac{2}{3}\sqrt{\frac{2}{5}}}.$$

The least squares approximation to $f(x) = \sin \pi x$ from W is

$\text{proj}_W \mathbf{f} = \langle \mathbf{f}, \mathbf{q}_1 \rangle \mathbf{q}_1 + \langle \mathbf{f}, \mathbf{q}_2 \rangle \mathbf{q}_2 + \langle \mathbf{f}, \mathbf{q}_3 \rangle \mathbf{q}_3$

$$= \frac{1}{4}\int_{-1}^{1} \sin \pi x\; dx + \frac{3}{2}\left(\int_{-1}^{1} x \sin \pi x\, dx\right)x + \frac{45}{8}\left(\int_{-1}^{1}\left(x^2 - \frac{1}{3}\right)\sin \pi x\, dx\right)\left(x^2 - \frac{1}{3}\right)$$

$$= 0 + \frac{3}{2}\left(-\frac{x \cos \pi x}{\pi} + \frac{\sin \pi x}{\pi^2}\right)\Big|_{-1}^{1} x + 0 = \frac{3}{2}\cdot\frac{2}{\pi}x = \frac{3x}{\pi}$$

(b) The mean square error is $\int_{-1}^{1}\left(\sin \pi x - \frac{3x}{\pi}\right)^2 dx = 1 - \frac{6}{\pi^2} \approx 0.392$

9. Let $f(x) = \begin{cases} 1, & 0 < x < \pi \\ 0, & \pi \le x \le 2\pi \end{cases}$.

$a_0 = \frac{1}{\pi}\int_0^{2\pi} f(x)\, dx = \frac{1}{\pi}\int_0^{\pi} dx = 1$

$a_k = \frac{1}{\pi}\int_0^{2\pi} f(x) \cos kx\, dx = \frac{1}{\pi}\int_0^{\pi} \cos kx\, dx = 0$

$b_k = \frac{1}{\pi}\int_0^{2\pi} f(x) \sin kx\, dx = \frac{1}{\pi}\int_0^{\pi} \sin kx\; dx = \frac{1}{k\pi}(1 - (-1)^k)$

So the Fourier series is $\frac{1}{2} + \sum_{k=1}^{\infty} \frac{1}{k\pi}(1 - (-1)^k) \sin kx$.

True-False Exercises

(a) False. The area between the graphs is the error, not the mean square error.

(b) True.

(c) True.

(d) False. $\|1\| = \langle 1,\ 1 \rangle = \int_0^{2\pi} 1^2 dx = 2\pi \neq 1$.

(e) True.

Chapter 6 Supplementary Exercises

1. **(a)** Let $\mathbf{v} = (v_1,\ v_2,\ v_3,\ v_4)$.

$\langle \mathbf{v},\ \mathbf{u}_1 \rangle = v_1, \langle \mathbf{v},\ \mathbf{u}_2 \rangle = v_2, \langle \mathbf{v},\ \mathbf{u}_3 \rangle = v_3, \langle \mathbf{v},\ \mathbf{u}_4 \rangle = v_4$

If $\langle \mathbf{v}, \mathbf{u}_1 \rangle = \langle \mathbf{v}, \mathbf{u}_4 \rangle = 0$, then $v_1 = v_4 = 0$ and $\mathbf{v} = (0, v_2, v_3, 0)$. Since the angle θ between $\mathbf{u}$ and $\mathbf{v}$ satisfies $\cos\theta = \frac{\langle \mathbf{u},\mathbf{v} \rangle}{\|\mathbf{u}\|\|\mathbf{v}\|}$, $\mathbf{v}$ making equal angles with $\mathbf{u}_2$ and $\mathbf{u}_3$ means that $v_2 = v_3$. In order for the angle between $\mathbf{v}$ and $\mathbf{u}_3$ to be defined $\|\mathbf{v}\| \neq 0$. Thus, $\mathbf{v} = (0, a, a, 0)$ with $a \neq 0$.

(b) As in part (a), since $\langle \mathbf{x}, \mathbf{u}_1 \rangle = \langle \mathbf{x}, \mathbf{u}_4 \rangle = 0$, $x_1 = x_4 = 0$.
Since $\|\mathbf{u}_2\| = \|\mathbf{u}_3\| = 1$ and we want $\|\mathbf{x}\| = 1$, the cosine of the angle between $\mathbf{x}$ and $\mathbf{u}_2$ is $\cos\theta_2 = \langle \mathbf{x}, \mathbf{u}_2 \rangle = x_2$ and, similarly, $\cos\theta_3 = \langle \mathbf{x}, \mathbf{u}_3 \rangle = x_3$, so we want $x_2 = 2x_3$, and $\mathbf{x} = \langle 0, x_2, 2x_2, 0 \rangle$.

$$\|\mathbf{x}\| = \sqrt{x_2^2 + 4x_2^2} = \sqrt{5x_2^2} = |x_2|\sqrt{5}.$$

If $\|\mathbf{x}\| = 1$, then $x_2 = \pm\frac{1}{\sqrt{5}}$, so $\mathbf{x} = \pm\left(0, \frac{1}{\sqrt{5}}, \frac{2}{\sqrt{5}}, 0\right)$.

3. Recall that if $U = \begin{bmatrix} u_1 & u_2 \\ u_3 & u_4 \end{bmatrix}$ and $V = \begin{bmatrix} v_1 & v_2 \\ v_3 & v_4 \end{bmatrix}$, then $\langle U, V \rangle = u_1v_1 + u_2v_2 + u_3v_3 + u_4v_4$.

(a) If U is a diagonal matrix, then $u_2 = u_3 = 0$ and $\langle U, V \rangle = u_1v_1 + u_4v_4$.
For V to be in the orthogonal complement of the subspace of all diagonal matrices, then it must be the case that $v_1 = v_4 = 0$ and V must have zeros on the main diagonal.

(b) If U is a symmetric matrix, then $u_2 = u_3$ and $\langle U, V \rangle = u_1v_1 + u_2(v_2 + v_3) + u_4v_4$.
Since u_1 and u_4 can take on any values, for V to be in the orthogonal complement of the subspace of all symmetric matrices, it must be the case that $v_1 = v_4 = 0$ and $v_2 = -v_3$, thus V must be skew-symmetric.

5. Let $\mathbf{u} = \left(\sqrt{a_1}, \ldots, \sqrt{a_n}\right)$ and $\mathbf{v} = \left(\frac{1}{\sqrt{a_1}}, \ldots, \frac{1}{\sqrt{a_n}}\right)$. By the Cauchy-Schwarz Inequality,

$$\langle \mathbf{u} \cdot \mathbf{v} \rangle^2 = (\underbrace{1 + \cdots + 1}_{n \text{ terms}})^2 \leq \|\mathbf{u}\|^2\|\mathbf{v}\|^2 \text{ or } n^2 \leq (a_1 + \cdots + a_n)\left(\frac{1}{a_1} + \cdots + \frac{1}{a_n}\right).$$

7. Let $\mathbf{x} = (x_1, x_2, x_3)$.

$$\begin{aligned} \langle \mathbf{x}, \mathbf{u}_1 \rangle &= x_1 + x_2 - x_3 \\ \langle \mathbf{x}, \mathbf{u}_2 \rangle &= -2x_1 - x_2 + 2x_3 \\ \langle \mathbf{x}, \mathbf{u}_3 \rangle &= -x_1 + x_3 \end{aligned}$$

$\langle \mathbf{x}, \mathbf{u}_3 \rangle = 0 \Rightarrow -x_1 + x_3 = 0$, so $x_1 = x_3$. Then $\langle \mathbf{x}, \mathbf{u}_1 \rangle = x_2$ and $\langle \mathbf{x}, \mathbf{u}_2 \rangle = -x_2$, so $x_2 = 0$ and $\mathbf{x} = (x_1, 0, x_1)$. Then $\|\mathbf{x}\| = \sqrt{x_1^2 + x_1^2} = \sqrt{2x_1^2} = |x_1|\sqrt{2}$.
If $\|\mathbf{x}\| = 1$ then $x_1 = \pm\frac{1}{\sqrt{2}}$ and the vectors are $\pm\left(\frac{1}{\sqrt{2}}, 0, \frac{1}{\sqrt{2}}\right)$.

9. For $\mathbf{u} = (u_1, u_2)$, $\mathbf{v} = (v_1, v_2)$ in R^2, let $\langle \mathbf{u}, \mathbf{v} \rangle = au_1v_1 + bu_2v_2$ be a weighted inner product.
If $\mathbf{u} = (1, 2)$ and $\mathbf{v} = (3, -1)$ form an orthonormal set, then $\|\mathbf{u}\|^2 = a(1)^2 + b(2)^2 = a + 4b = 1$, $\|\mathbf{v}\|^2 = a(3)^2 + b(-1)^2 = 9a + b = 1$, and $\langle \mathbf{u}, \mathbf{v} \rangle = a(1)(3) + b(2)(-1) = 3a - 2b = 0$.

This leads to the system $\begin{bmatrix} 1 & 4 \\ 9 & 1 \\ 3 & -2 \end{bmatrix}\begin{bmatrix} a \\ b \end{bmatrix} = \begin{bmatrix} 1 \\ 1 \\ 0 \end{bmatrix}$.

Since $\begin{bmatrix} 1 & 4 & 1 \\ 9 & 1 & 1 \\ 3 & -2 & 0 \end{bmatrix}$ reduces to $\begin{bmatrix} 1 & 0 & 0 \\ 0 & 1 & 0 \\ 0 & 0 & 1 \end{bmatrix}$, the system is inconsistent and there is no such weighted inner product.

11. **(a)** Let $\mathbf{u}_1 = (k, 0, 0, \ldots, 0), \mathbf{u}_2 = (0, k, 0, \ldots, 0), \ldots, \mathbf{u}_n = (0, 0, 0, \ldots, k)$ be the edges of the 'cube' in R^n and $\mathbf{u} = (k, k, k, \ldots, k)$ be the diagonal.

Then $\|\mathbf{u}_i\| = k$, $\|\mathbf{u}\| = k\sqrt{n}$, and $\langle \mathbf{u}_i, \mathbf{u} \rangle = k^2$, so $\cos\theta = \frac{\langle \mathbf{u}_i, \mathbf{u} \rangle}{\|\mathbf{u}_i\|\|\mathbf{u}\|} = \frac{k^2}{k(k\sqrt{n})} = \frac{1}{\sqrt{n}}$.

(b) As n approaches ∞, $\frac{1}{\sqrt{n}}$ approaches 0, so θ approaches $\frac{\pi}{2}$.

13. Recall that $\mathbf{u}$ can be expressed as the linear combination $\mathbf{u} = a_1\mathbf{v}_1 + \cdots + a_n\mathbf{v}_n$ where $a_i = \langle \mathbf{u}, \mathbf{v}_i \rangle$ for $i = 1, \ldots, n$. Since $\|\mathbf{v}_i\| = 1$, we have $\cos^2\alpha_i = \left(\frac{\langle \mathbf{u}, \mathbf{v}_i \rangle}{\|\mathbf{u}\|\|\mathbf{v}_i\|}\right)^2 = \left(\frac{a_i}{\|\mathbf{u}\|}\right)^2 = \frac{a_i^2}{a_1^2 + a_2^2 + \cdots + a_n^2}$.

Therefore $\cos^2\alpha_1 + \cdots + \cos^2\alpha_n = \frac{a_1^2 + a_2^2 + \cdots + a_n^2}{a_1^2 + a_2^2 + \cdots + a_n^2} = 1$.

15. To show that $(W^\perp)^\perp = W$, we first show that $W \subseteq (W^\perp)^\perp$. If $\mathbf{w}$ is in W, then $\mathbf{w}$ is orthogonal to every vector in $W^\perp$, so that $\mathbf{w}$ is in $(W^\perp)^\perp$. Thus $W \subseteq (W^\perp)^\perp$.
To show that $(W^\perp)^\perp \subseteq W$, let $\mathbf{v}$ be in $(W^\perp)^\perp$. Since $\mathbf{v}$ is in V, we have, by the Projection Theorem, that $\mathbf{v} = \mathbf{w}_1 + \mathbf{w}_2$ where $\mathbf{w}_1$ is in W and $\mathbf{w}_2$ is in $W^\perp$. By definition, $\langle \mathbf{v}, \mathbf{w}_2 \rangle = \langle \mathbf{w}_1, \mathbf{w}_2 \rangle = 0$. But
$\langle \mathbf{v}, \mathbf{w}_2 \rangle = \langle \mathbf{w}_1 + \mathbf{w}_2, \mathbf{w}_2 \rangle = \langle \mathbf{w}_1, \mathbf{w}_2 \rangle + \langle \mathbf{w}_2, \mathbf{w}_2 \rangle = \langle \mathbf{w}_2, \mathbf{w}_2 \rangle$
so that $\langle \mathbf{w}_2, \mathbf{w}_2 \rangle = 0$. Hence $\mathbf{w}_2 = \mathbf{0}$ and therefore $\mathbf{v} = \mathbf{w}_1$, so that $\mathbf{v}$ is in W. Thus $(W^\perp)^\perp \subseteq W$.

17. $A = \begin{bmatrix} 1 & -1 \\ 2 & 3 \\ 4 & 5 \end{bmatrix}$, $A^T = \begin{bmatrix} 1 & 2 & 4 \\ -1 & 3 & 5 \end{bmatrix}$, $A^TA = \begin{bmatrix} 21 & 25 \\ 25 & 35 \end{bmatrix}$, $A^T\mathbf{b} = \begin{bmatrix} 1 & 2 & 4 \\ -1 & 3 & 5 \end{bmatrix}\begin{bmatrix} 1 \\ 1 \\ s \end{bmatrix} = \begin{bmatrix} 4s + 3 \\ 5s + 2 \end{bmatrix}$

The associated normal system is $\begin{bmatrix} 21 & 25 \\ 25 & 35 \end{bmatrix}\begin{bmatrix} x_1 \\ x_2 \end{bmatrix} = \begin{bmatrix} 4s + 3 \\ 5s + 2 \end{bmatrix}$.

If the least squares solution is $x_1 = 1$ and $x_2 = 2$, then $\begin{bmatrix} 21 & 25 \\ 25 & 35 \end{bmatrix}\begin{bmatrix} 1 \\ 2 \end{bmatrix} = \begin{bmatrix} 71 \\ 95 \end{bmatrix} = \begin{bmatrix} 4s + 3 \\ 5s + 2 \end{bmatrix}$.

The resulting equations have solutions $s = 17$ and $s = 18.6$, respectively, so no such value of s exists.

CHAPTER 7: DIAGONALIZATION AND QUADRATIC FORMS

7.1 Orthogonal Matrices

1. (a) $AA^T = \begin{bmatrix} 1 & 0 \\ 0 & -1 \end{bmatrix}\begin{bmatrix} 1 & 0 \\ 0 & -1 \end{bmatrix} = I$ and $A^TA = \begin{bmatrix} 1 & 0 \\ 0 & -1 \end{bmatrix}\begin{bmatrix} 1 & 0 \\ 0 & -1 \end{bmatrix} = I$ therefore A is an orthogonal matrix; $A^{-1} = A^T = \begin{bmatrix} 1 & 0 \\ 0 & -1 \end{bmatrix}$

(b) $AA^T = \begin{bmatrix} \frac{1}{\sqrt{2}} & -\frac{1}{\sqrt{2}} \\ \frac{1}{\sqrt{2}} & \frac{1}{\sqrt{2}} \end{bmatrix}\begin{bmatrix} \frac{1}{\sqrt{2}} & \frac{1}{\sqrt{2}} \\ -\frac{1}{\sqrt{2}} & \frac{1}{\sqrt{2}} \end{bmatrix} = I$ and $A^TA = \begin{bmatrix} \frac{1}{\sqrt{2}} & \frac{1}{\sqrt{2}} \\ -\frac{1}{\sqrt{2}} & \frac{1}{\sqrt{2}} \end{bmatrix}\begin{bmatrix} \frac{1}{\sqrt{2}} & -\frac{1}{\sqrt{2}} \\ \frac{1}{\sqrt{2}} & \frac{1}{\sqrt{2}} \end{bmatrix} = I$ therefore A is an orthogonal matrix; $A^{-1} = A^T = \begin{bmatrix} \frac{1}{\sqrt{2}} & \frac{1}{\sqrt{2}} \\ -\frac{1}{\sqrt{2}} & \frac{1}{\sqrt{2}} \end{bmatrix}$

3. (a) $\|\mathbf{r}_1\| = \sqrt{0^2 + 1^2 + \left(\frac{1}{\sqrt{2}}\right)^2} = \sqrt{\frac{3}{2}} \neq 1$ so the matrix is not orthogonal.

(b) $AA^T = \begin{bmatrix} -\frac{1}{\sqrt{2}} & \frac{1}{\sqrt{6}} & \frac{1}{\sqrt{3}} \\ 0 & -\frac{2}{\sqrt{6}} & \frac{1}{\sqrt{3}} \\ \frac{1}{\sqrt{2}} & \frac{1}{\sqrt{6}} & \frac{1}{\sqrt{3}} \end{bmatrix}\begin{bmatrix} -\frac{1}{\sqrt{2}} & 0 & \frac{1}{\sqrt{2}} \\ \frac{1}{\sqrt{6}} & -\frac{2}{\sqrt{6}} & \frac{1}{\sqrt{6}} \\ \frac{1}{\sqrt{3}} & \frac{1}{\sqrt{3}} & \frac{1}{\sqrt{3}} \end{bmatrix} = I$ and $A^TA = \begin{bmatrix} -\frac{1}{\sqrt{2}} & 0 & \frac{1}{\sqrt{2}} \\ \frac{1}{\sqrt{6}} & -\frac{2}{\sqrt{6}} & \frac{1}{\sqrt{6}} \\ \frac{1}{\sqrt{3}} & \frac{1}{\sqrt{3}} & \frac{1}{\sqrt{3}} \end{bmatrix}\begin{bmatrix} -\frac{1}{\sqrt{2}} & \frac{1}{\sqrt{6}} & \frac{1}{\sqrt{3}} \\ 0 & -\frac{2}{\sqrt{6}} & \frac{1}{\sqrt{3}} \\ \frac{1}{\sqrt{2}} & \frac{1}{\sqrt{6}} & \frac{1}{\sqrt{3}} \end{bmatrix} = I$

therefore A is an orthogonal matrix; $A^{-1} = A^T = \begin{bmatrix} -\frac{1}{\sqrt{2}} & 0 & \frac{1}{\sqrt{2}} \\ \frac{1}{\sqrt{6}} & -\frac{2}{\sqrt{6}} & \frac{1}{\sqrt{6}} \\ \frac{1}{\sqrt{3}} & \frac{1}{\sqrt{3}} & \frac{1}{\sqrt{3}} \end{bmatrix}$

5. $A^TA = \begin{bmatrix} \frac{4}{5} & -\frac{9}{25} & \frac{12}{25} \\ 0 & \frac{4}{5} & \frac{3}{5} \\ -\frac{3}{5} & -\frac{12}{25} & \frac{16}{25} \end{bmatrix}\begin{bmatrix} \frac{4}{5} & 0 & -\frac{3}{5} \\ -\frac{9}{25} & \frac{4}{5} & -\frac{12}{25} \\ \frac{12}{25} & \frac{3}{5} & \frac{16}{25} \end{bmatrix} = I;$

row vectors of A, $\mathbf{r}_1 = \begin{bmatrix} \frac{4}{5} & 0 & -\frac{3}{5} \end{bmatrix}$, $\mathbf{r}_2 = \begin{bmatrix} -\frac{9}{25} & \frac{4}{5} & -\frac{12}{25} \end{bmatrix}$, $\mathbf{r}_3 = \begin{bmatrix} \frac{12}{25} & \frac{3}{5} & \frac{16}{25} \end{bmatrix}$, form an orthonormal set since $\mathbf{r}_1 \cdot \mathbf{r}_2 = \mathbf{r}_1 \cdot \mathbf{r}_3 = \mathbf{r}_2 \cdot \mathbf{r}_3 = 0$ and $\|\mathbf{r}_1\| = \|\mathbf{r}_2\| = \|\mathbf{r}_3\| = 1$;

column vectors of A, $\mathbf{c}_1 = \begin{bmatrix} \frac{4}{5} \\ -\frac{9}{25} \\ \frac{12}{25} \end{bmatrix}$, $\mathbf{c}_2 = \begin{bmatrix} 0 \\ \frac{4}{5} \\ \frac{3}{5} \end{bmatrix}$, $\mathbf{c}_3 = \begin{bmatrix} -\frac{3}{5} \\ -\frac{12}{25} \\ \frac{16}{25} \end{bmatrix}$, form an orthonormal set since $\mathbf{c}_1 \cdot \mathbf{c}_2 = \mathbf{c}_1 \cdot \mathbf{c}_3 = \mathbf{c}_2 \cdot \mathbf{c}_3 = 0$ and $\|\mathbf{c}_1\| = \|\mathbf{c}_2\| = \|\mathbf{c}_3\| = 1$.

7. $T_A(\mathbf{x}) = \begin{bmatrix} \frac{4}{5} & 0 & -\frac{3}{5} \\ -\frac{9}{25} & \frac{4}{5} & -\frac{12}{25} \\ \frac{12}{25} & \frac{3}{5} & \frac{16}{25} \end{bmatrix}\begin{bmatrix} -2 \\ 3 \\ 5 \end{bmatrix} = \begin{bmatrix} -\frac{23}{5} \\ \frac{18}{25} \\ \frac{101}{25} \end{bmatrix}$; $\|T_A(\mathbf{x})\| = \sqrt{\frac{529}{25} + \frac{324}{625} + \frac{10201}{625}} = \sqrt{38}$ equals $\|\mathbf{x}\| = \sqrt{4+9+25} = \sqrt{38}$

9. Yes, by inspection, the column vectors in each of these matrices form orthonormal sets. By Theorem 7.1.1, these matrices are orthogonal.

11. Let $A = \begin{bmatrix} a+b & b-a \\ a-b & b+a \end{bmatrix}$. Then $A^TA = \begin{bmatrix} 2(a^2+b^2) & 0 \\ 0 & 2(a^2+b^2) \end{bmatrix}$, so a and b must satisfy $a^2+b^2 = \frac{1}{2}$.

13. **(a)** Formula (3) in Section 7.1 yields the transition matrix $P = \begin{bmatrix} \cos\frac{\pi}{3} & -\sin\frac{\pi}{3} \\ \sin\frac{\pi}{3} & \cos\frac{\pi}{3} \end{bmatrix} = \begin{bmatrix} \frac{1}{2} & -\frac{\sqrt{3}}{2} \\ \frac{\sqrt{3}}{2} & \frac{1}{2} \end{bmatrix}$; since

P is orthogonal, $P^{-1} = P^T$ therefore $\begin{bmatrix} x' \\ y' \end{bmatrix} = P^{-1}\begin{bmatrix} x \\ y \end{bmatrix} = \begin{bmatrix} \frac{1}{2} & \frac{\sqrt{3}}{2} \\ -\frac{\sqrt{3}}{2} & \frac{1}{2} \end{bmatrix}\begin{bmatrix} -2 \\ 6 \end{bmatrix} = \begin{bmatrix} -1+3\sqrt{3} \\ 3+\sqrt{3} \end{bmatrix}$

(b) Using the matrix P we obtained in part (a), $\begin{bmatrix} x \\ y \end{bmatrix} = P\begin{bmatrix} x' \\ y' \end{bmatrix} = \begin{bmatrix} \frac{1}{2} & -\frac{\sqrt{3}}{2} \\ \frac{\sqrt{3}}{2} & \frac{1}{2} \end{bmatrix}\begin{bmatrix} 5 \\ 2 \end{bmatrix} = \begin{bmatrix} \frac{5}{2} - \sqrt{3} \\ 1 + \frac{5}{2}\sqrt{3} \end{bmatrix}$

15. **(a)** Following the method of Example 6 in Section 7.1 (also see Table 6 in Section 4.9), we use the orthogonal matrix $P = \begin{bmatrix} \cos\frac{\pi}{4} & -\sin\frac{\pi}{4} & 0 \\ \sin\frac{\pi}{4} & \cos\frac{\pi}{4} & 0 \\ 0 & 0 & 1 \end{bmatrix} = \begin{bmatrix} \frac{1}{\sqrt{2}} & -\frac{1}{\sqrt{2}} & 0 \\ \frac{1}{\sqrt{2}} & \frac{1}{\sqrt{2}} & 0 \\ 0 & 0 & 1 \end{bmatrix}$ to obtain

$$\begin{bmatrix} x' \\ y' \\ z' \end{bmatrix} = P^{-1}\begin{bmatrix} x \\ y \\ z \end{bmatrix} = \begin{bmatrix} \frac{1}{\sqrt{2}} & \frac{1}{\sqrt{2}} & 0 \\ -\frac{1}{\sqrt{2}} & \frac{1}{\sqrt{2}} & 0 \\ 0 & 0 & 1 \end{bmatrix}\begin{bmatrix} -1 \\ 2 \\ 5 \end{bmatrix} = \begin{bmatrix} \frac{1}{\sqrt{2}} \\ \frac{3}{\sqrt{2}} \\ 5 \end{bmatrix}$$

(b) Using the matrix P we obtained in part (a), $\begin{bmatrix} x \\ y \\ z \end{bmatrix} = P\begin{bmatrix} x' \\ y' \\ z' \end{bmatrix} = \begin{bmatrix} \frac{1}{\sqrt{2}} & -\frac{1}{\sqrt{2}} & 0 \\ \frac{1}{\sqrt{2}} & \frac{1}{\sqrt{2}} & 0 \\ 0 & 0 & 1 \end{bmatrix}\begin{bmatrix} 1 \\ 6 \\ -3 \end{bmatrix} = \begin{bmatrix} -\frac{5}{\sqrt{2}} \\ \frac{7}{\sqrt{2}} \\ -3 \end{bmatrix}$

17. **(a)** We follow the method of Example 6 in Section 7.1, with the appropriate orthogonal matrix obtained from Table 6 in Section 4.9: $P = \begin{bmatrix} \cos\frac{\pi}{3} & 0 & \sin\frac{\pi}{3} \\ 0 & 1 & 0 \\ -\sin\frac{\pi}{3} & 0 & \cos\frac{\pi}{3} \end{bmatrix} = \begin{bmatrix} \frac{1}{2} & 0 & \frac{\sqrt{3}}{2} \\ 0 & 1 & 0 \\ -\frac{\sqrt{3}}{2} & 0 & \frac{1}{2} \end{bmatrix}$

$$\begin{bmatrix} x' \\ y' \\ z' \end{bmatrix} = P^{-1}\begin{bmatrix} x \\ y \\ z \end{bmatrix} = \begin{bmatrix} \frac{1}{2} & 0 & -\frac{\sqrt{3}}{2} \\ 0 & 1 & 0 \\ \frac{\sqrt{3}}{2} & 0 & \frac{1}{2} \end{bmatrix}\begin{bmatrix} -1 \\ 2 \\ 5 \end{bmatrix} = \begin{bmatrix} -\frac{1}{2} - \frac{5\sqrt{3}}{2} \\ 2 \\ -\frac{\sqrt{3}}{2} + \frac{5}{2} \end{bmatrix}$$

(b) Using the matrix P we obtained in part (a), $\begin{bmatrix} x \\ y \\ z \end{bmatrix} = P\begin{bmatrix} x' \\ y' \\ z' \end{bmatrix} = \begin{bmatrix} \frac{1}{2} & 0 & \frac{\sqrt{3}}{2} \\ 0 & 1 & 0 \\ -\frac{\sqrt{3}}{2} & 0 & \frac{1}{2} \end{bmatrix}\begin{bmatrix} 1 \\ 6 \\ -3 \end{bmatrix} = \begin{bmatrix} \frac{1}{2} - \frac{3\sqrt{3}}{2} \\ 6 \\ -\frac{\sqrt{3}}{2} - \frac{3}{2} \end{bmatrix}$

19. If $B = \{\mathbf{u}_1, \mathbf{u}_2, \mathbf{u}_3\}$ is the standard basis for R^3 and $B' = \{\mathbf{u'}_1, \mathbf{u'}_2, \mathbf{u'}_3\}$, then $[\mathbf{u}_1']_B = \begin{bmatrix} 1 \\ 0 \\ 0 \end{bmatrix}$,

$[\mathbf{u}_2']_B = \begin{bmatrix} 0 \\ \cos\theta \\ \sin\theta \end{bmatrix}$, and $[\mathbf{u}_3']_B = \begin{bmatrix} 0 \\ -\sin\theta \\ \cos\theta \end{bmatrix}$, so the transition matrix from B' to B is $P = \begin{bmatrix} 1 & 0 & 0 \\ 0 & \cos\theta & -\sin\theta \\ 0 & \sin\theta & \cos\theta \end{bmatrix}$

and $A = \begin{bmatrix} 1 & 0 & 0 \\ 0 & \cos\theta & \sin\theta \\ 0 & -\sin\theta & \cos\theta \end{bmatrix}$.

21. **(a)** Rotations about the origin, reflections about any line through the origin, and any combination of these are rigid operators.

(b) Rotations about the origin, dilations, contractions, reflections about lines through the origin, and combinations of these are angle preserving.

(c) All rigid operators on R^2 are angle preserving. Dilations and contractions are angle preserving operators that are not rigid.

23. **(a)** Denoting $\mathbf{p}_1 = p_1(x) = \frac{1}{\sqrt{3}}$, $\mathbf{p}_2 = p_2(x) = \frac{1}{\sqrt{2}}x$, and $\mathbf{p}_3 = p_3(x) = \sqrt{\frac{3}{2}}x^2 - \sqrt{\frac{2}{3}}$ we have

$$\langle \mathbf{p}, \mathbf{p}_1 \rangle = p(-1)p_1(-1) + p(0)p_1(0) + p(1)p_1(1) = (1)\left(\tfrac{1}{\sqrt{3}}\right) + (1)\left(\tfrac{1}{\sqrt{3}}\right) + (3)\left(\tfrac{1}{\sqrt{3}}\right) = \tfrac{5}{\sqrt{3}}$$
$$\langle \mathbf{p}, \mathbf{p}_2 \rangle = p(-1)p_2(-1) + p(0)p_2(0) + p(1)p_2(1) = (1)\left(\tfrac{-1}{\sqrt{2}}\right) + (1)(0) + (3)\left(\tfrac{1}{\sqrt{2}}\right) = \sqrt{2}$$
$$\langle \mathbf{p}, \mathbf{p}_3 \rangle = p(-1)p_3(-1) + p(0)p_3(0) + p(1)p_3(1) = (1)\left(\tfrac{1}{\sqrt{6}}\right) + (1)\left(-\tfrac{2}{\sqrt{6}}\right) + (3)\left(\tfrac{1}{\sqrt{6}}\right) = \tfrac{\sqrt{2}}{\sqrt{3}}$$
$$\langle \mathbf{q}, \mathbf{p}_1 \rangle = q(-1)p_1(-1) + q(0)p_1(0) + q(1)p_1(1) = (-3)\left(\tfrac{1}{\sqrt{3}}\right) + (0)\left(\tfrac{1}{\sqrt{3}}\right) + (1)\left(\tfrac{1}{\sqrt{3}}\right) = -\tfrac{2}{\sqrt{3}}$$
$$\langle \mathbf{q}, \mathbf{p}_2 \rangle = q(-1)p_2(-1) + q(0)p_2(0) + q(1)p_2(1) = (-3)\left(\tfrac{-1}{\sqrt{2}}\right) + (0)(0) + (1)\left(\tfrac{1}{\sqrt{2}}\right) = 2\sqrt{2}$$
$$\langle \mathbf{q}, \mathbf{p}_3 \rangle = q(-1)p_3(-1) + q(0)p_3(0) + q(1)p_3(1) = (-3)\left(\tfrac{1}{\sqrt{6}}\right) + (0)\left(-\tfrac{2}{\sqrt{6}}\right) + (1)\left(\tfrac{1}{\sqrt{6}}\right) = -\tfrac{\sqrt{2}}{\sqrt{3}}$$

$$(\mathbf{p})_S = (\langle \mathbf{p}, \mathbf{p}_1 \rangle, \langle \mathbf{p}, \mathbf{p}_2 \rangle, \langle \mathbf{p}, \mathbf{p}_3 \rangle) = \left(\tfrac{5}{\sqrt{3}}, \sqrt{2}, \tfrac{\sqrt{2}}{\sqrt{3}}\right)$$
$$(\mathbf{q})_S = (\langle \mathbf{q}, \mathbf{p}_1 \rangle, \langle \mathbf{q}, \mathbf{p}_2 \rangle, \langle \mathbf{q}, \mathbf{p}_3 \rangle) = \left(-\tfrac{2}{\sqrt{3}}, 2\sqrt{2}, -\tfrac{\sqrt{2}}{\sqrt{3}}\right)$$

(b) $\|\mathbf{p}\| = \sqrt{\left(\frac{5}{\sqrt{3}}\right)^2 + \left(\sqrt{2}\right)^2 + \left(\frac{\sqrt{2}}{\sqrt{3}}\right)^2} = \sqrt{\frac{25}{3} + 2 + \frac{2}{3}} = \sqrt{11}$

$$d(\mathbf{p}, \mathbf{q}) = \sqrt{\left(\tfrac{5}{\sqrt{3}} + \tfrac{2}{\sqrt{3}}\right)^2 + \left(\sqrt{2} - 2\sqrt{2}\right)^2 + \left(\tfrac{\sqrt{2}}{\sqrt{3}} + \tfrac{\sqrt{2}}{\sqrt{3}}\right)^2} = \sqrt{\tfrac{49}{3} + 2 + \tfrac{8}{3}} = \sqrt{21}$$
$$\langle \mathbf{p}, \mathbf{q} \rangle = \left(\tfrac{5}{\sqrt{3}}\right)\left(-\tfrac{2}{\sqrt{3}}\right) + (\sqrt{2})(2\sqrt{2}) + \left(\tfrac{\sqrt{2}}{\sqrt{3}}\right)\left(-\tfrac{\sqrt{2}}{\sqrt{3}}\right) = -\tfrac{10}{3} + 4 - \tfrac{2}{3} = 0$$

25. We have $A^T = \left(I_n - \frac{2}{\mathbf{x}^T\mathbf{x}}\mathbf{x}\mathbf{x}^T\right)^T = I_n^T - \frac{2}{\mathbf{x}^T\mathbf{x}}(\mathbf{x}\mathbf{x}^T)^T = I_n^T - \frac{2}{\mathbf{x}^T\mathbf{x}}(\mathbf{x}^T)^T\mathbf{x}^T = I_n - \frac{2}{\mathbf{x}^T\mathbf{x}}\mathbf{x}\mathbf{x}^T = A$ therefore
$A^TA = AA^T = \left(I_n - \frac{2}{\mathbf{x}^T\mathbf{x}}\mathbf{x}\mathbf{x}^T\right)\left(I_n - \frac{2}{\mathbf{x}^T\mathbf{x}}\mathbf{x}\mathbf{x}^T\right) = I_n - \frac{2}{\mathbf{x}^T\mathbf{x}}\mathbf{x}\mathbf{x}^T - \frac{2}{\mathbf{x}^T\mathbf{x}}\mathbf{x}\mathbf{x}^T + \frac{4}{(\mathbf{x}^T\mathbf{x})^2}\mathbf{x}\mathbf{x}^T\mathbf{x}\mathbf{x}^T$
$= I_n - \frac{4}{\mathbf{x}^T\mathbf{x}}\mathbf{x}\mathbf{x}^T + \frac{4(\mathbf{x}^T\mathbf{x})}{(\mathbf{x}^T\mathbf{x})^2}\mathbf{x}\mathbf{x}^T = I_n - \frac{4}{\mathbf{x}^T\mathbf{x}}\mathbf{x}\mathbf{x}^T + \frac{4}{\mathbf{x}^T\mathbf{x}}\mathbf{x}\mathbf{x}^T = I_n$

27. **(a)** Multiplication by $A = \begin{bmatrix}\cos\theta & -\sin\theta\\ \sin\theta & \cos\theta\end{bmatrix}$ is a rotation through θ.
In this case, $\det(A) = \cos^2\theta + \sin^2\theta = 1$.
The determinant of $A = \begin{bmatrix}\cos\theta & \sin\theta\\ \sin\theta & -\cos\theta\end{bmatrix}$ is $\det(A) = -\cos^2\theta + \sin^2\theta = -1$.
We can express this matrix as a product $\begin{bmatrix}\cos\theta & \sin\theta\\ \sin\theta & -\cos\theta\end{bmatrix} = \begin{bmatrix}\cos\theta & -\sin\theta\\ \sin\theta & \cos\theta\end{bmatrix}\begin{bmatrix}1 & 0\\ 0 & -1\end{bmatrix}$.
Multiplying by $\begin{bmatrix}\cos\theta & \sin\theta\\ \sin\theta & -\cos\theta\end{bmatrix}$ is a reflection about the x-axis followed by a rotation through θ.

(b) By Formula (6) of Section 4.9, multiplication by $\begin{bmatrix}\cos\theta & \sin\theta\\ \sin\theta & -\cos\theta\end{bmatrix}$ is a reflection about the line through the origin that makes the angle $\frac{\theta}{2}$ with the positive x-axis.

29. Let A and B be 3×3 standard matrices of two rotations in R^3: T_A and T_B, respectively.
The result stated in this Exercise implies that A and B are both orthogonal and $\det(A) = \det(B) = 1$.
The product AB is a standard matrix of the composition of these rotations $T_A \circ T_B$.
By part (c) of Theorem 7.1.2, AB is an orthogonal matrix.
Furthermore, by Theorem 2.3.4, $\det(AB) = \det(A)\det(B) = 1$.
We conclude that $T_A \circ T_B$ is a rotation in R^3.
(One can show by induction that a composition of more than two rotations in R^3 is also a rotation.)

31. It follows directly from Definition 1 that the transpose of an orthogonal matrix is orthogonal as well (this is also stated as part (a) of Theorem 7.1.2). Since rows of A are columns of A^T, the equivalence of statements (a) and (c) follows from the equivalence of statements (a) and (b) which is shown in the book.

True-False Exercises

(a) False. Only square matrices can be orthogonal.

(b) False. The row and column vectors are not unit vectors.

(c) False. Only square matrices can be orthogonal. (The statement would be true if $m = n$.)

(d) False. The column vectors must form an orthonormal set.

(e) True. Since $A^TA = I$ for an orthogonal matrix A, A must be invertible (and $A^{-1} = A^T$).

(f) True. A product of orthogonal matrices is orthogonal, so A^2 is orthogonal; furthermore, $\det(A^2) = (\det A)^2 = (\pm 1)^2 = 1$.

(g) True. Since $\|A\mathbf{x}\| = \|\mathbf{x}\|$ for an orthogonal matrix.

(h) True. This follows from Theorem 7.1.3.

7.2 Orthogonal Diagonalization

1. $\begin{vmatrix} \lambda-1 & -2 \\ -2 & \lambda-4 \end{vmatrix} = \lambda^2 - 5\lambda$
The characteristic equation is $\lambda^2 - 5\lambda = 0$ and the eigenvalues are $\lambda = 0$ and $\lambda = 5$.
Both eigenspaces are one-dimensional.

3. $\begin{vmatrix} \lambda-1 & -1 & -1 \\ -1 & \lambda-1 & -1 \\ -1 & -1 & \lambda-1 \end{vmatrix} = \lambda^3 - 3\lambda^2 = \lambda^2(\lambda-3)$
The characteristic equation is $\lambda^3 - 3\lambda^2 = 0$ and the eigenvalues are $\lambda = 3$ and $\lambda = 0$.
The eigenspace for $\lambda = 3$ is one-dimensional; the eigenspace for $\lambda = 0$ is two-dimensional.

5. $\begin{vmatrix} \lambda-4 & -4 & 0 & 0 \\ -4 & \lambda-4 & 0 & 0 \\ 0 & 0 & \lambda & 0 \\ 0 & 0 & 0 & \lambda \end{vmatrix} = \lambda^4 - 8\lambda^3 = \lambda^3(\lambda-8)$
The characteristic equation is $\lambda^4 - 8\lambda^3 = 0$ and the eigenvalues are $\lambda = 0$ and $\lambda = 8$.
The eigenspace for $\lambda = 0$ is three-dimensional; the eigenspace for $\lambda = 8$ is one-dimensional.

7. $\det(\lambda I - A) = \begin{vmatrix} \lambda-6 & -2\sqrt{3} \\ -2\sqrt{3} & \lambda-7 \end{vmatrix} = \lambda^2 - 13\lambda + 30 = (\lambda-3)(\lambda-10)$ therefore A has eigenvalues 3 and 10.

The reduced row echelon form of $3I - A$ is $\begin{bmatrix} 1 & \frac{2}{\sqrt{3}} \\ 0 & 0 \end{bmatrix}$ so that the eigenspace corresponding to $\lambda_1 = 3$ consists of vectors $\begin{bmatrix} x_1 \\ x_2 \end{bmatrix}$ where $x_1 = -\frac{2}{\sqrt{3}}t, x_2 = t$. A vector $\mathbf{p}_1 = \begin{bmatrix} -2 \\ \sqrt{3} \end{bmatrix}$ forms a basis for this eigenspace.

The reduced row echelon form of $10I - A$ is $\begin{bmatrix} 1 & -\frac{\sqrt{3}}{2} \\ 0 & 0 \end{bmatrix}$ so that the eigenspace corresponding to $\lambda_2 = 10$ consists of vectors $\begin{bmatrix} x_1 \\ x_2 \end{bmatrix}$ where $x_1 = \frac{\sqrt{3}}{2}t, x_2 = t$. A vector $\mathbf{p}_2 = \begin{bmatrix} \sqrt{3} \\ 2 \end{bmatrix}$ forms a basis for this eigenspace.
Applying the Gram-Schmidt process to both bases $\{\mathbf{p}_1\}$ and $\{\mathbf{p}_2\}$ amounts to simply normalizing the vectors. This yields the columns of a matrix P that orthogonally diagonalizes A:

$P = \begin{bmatrix} -\frac{2}{\sqrt{7}} & \frac{\sqrt{3}}{\sqrt{7}} \\ \frac{\sqrt{3}}{\sqrt{7}} & \frac{2}{\sqrt{7}} \end{bmatrix}$. We have $P^{-1}AP = P^TAP = \begin{bmatrix} \lambda_1 & 0 \\ 0 & \lambda_2 \end{bmatrix} = \begin{bmatrix} 3 & 0 \\ 0 & 10 \end{bmatrix}$.

9. Cofactor expansion along the second row yields $\det(\lambda I - A) = \begin{vmatrix} \lambda+2 & 0 & 36 \\ 0 & \lambda+3 & 0 \\ 36 & 0 & \lambda+23 \end{vmatrix}$

$= (\lambda+3)\begin{vmatrix} \lambda+2 & 36 \\ 36 & \lambda+23 \end{vmatrix} = (\lambda-25)(\lambda+3)(\lambda+50)$ therefore A has eigenvalues 25, −3, and −50.

The reduced row echelon form of $25I - A$ is $\begin{bmatrix} 1 & 0 & \frac{4}{3} \\ 0 & 1 & 0 \\ 0 & 0 & 0 \end{bmatrix}$ so that the eigenspace corresponding to $\lambda_1 = 25$ contains vectors $\begin{bmatrix} x_1 \\ x_2 \\ x_3 \end{bmatrix}$ where $x_1 = -\frac{4}{3}t, x_2 = 0, x_3 = t$. A vector $\mathbf{p}_1 = \begin{bmatrix} -4 \\ 0 \\ 3 \end{bmatrix}$ forms a basis for this eigenspace.

The reduced row echelon form of $-3I - A$ is $\begin{bmatrix} 1 & 0 & 0 \\ 0 & 0 & 1 \\ 0 & 0 & 0 \end{bmatrix}$ so that the eigenspace corresponding to $\lambda_2 = -3$ contains vectors $\begin{bmatrix} x_1 \\ x_2 \\ x_3 \end{bmatrix}$ where $x_1 = 0, x_2 = t, x_3 = 0$. A vector $\mathbf{p}_2 = \begin{bmatrix} 0 \\ 1 \\ 0 \end{bmatrix}$ forms a basis for this eigenspace.

The reduced row echelon form of $-50I - A$ is $\begin{bmatrix} 1 & 0 & -\frac{3}{4} \\ 0 & 1 & 0 \\ 0 & 0 & 0 \end{bmatrix}$ so that the eigenspace corresponding to $\lambda_3 = -50$ contains vectors $\begin{bmatrix} x_1 \\ x_2 \\ x_3 \end{bmatrix}$ where $x_1 = \frac{3}{4}t, x_2 = 0, x_3 = t$. A vector $\mathbf{p}_3 = \begin{bmatrix} 3 \\ 0 \\ 4 \end{bmatrix}$ forms a basis for this eigenspace.

Applying the Gram-Schmidt process to the bases $\{\mathbf{p}_1\}$ and $\{\mathbf{p}_3\}$ amounts to simply normalizing the vectors; the basis $\{\mathbf{p}_2\}$ is already orthonormal. This yields the columns of a matrix P that orthogonally diagonalizes A: $P = \begin{bmatrix} -\frac{4}{5} & 0 & \frac{3}{5} \\ 0 & 1 & 0 \\ \frac{3}{5} & 0 & \frac{4}{5} \end{bmatrix}$.

We have $P^{-1}AP = P^TAP = \begin{bmatrix} \lambda_1 & 0 & 0 \\ 0 & \lambda_2 & 0 \\ 0 & 0 & \lambda_3 \end{bmatrix} = \begin{bmatrix} 25 & 0 & 0 \\ 0 & -3 & 0 \\ 0 & 0 & -50 \end{bmatrix}$.

11. $\det(\lambda I - A) = \begin{vmatrix} \lambda - 2 & 1 & 1 \\ 1 & \lambda - 2 & 1 \\ 1 & 1 & \lambda - 2 \end{vmatrix} = \lambda^3 - 6\lambda^2 + 9\lambda = \lambda(\lambda - 3)^2$ therefore A has eigenvalues 3 and 0.

The reduced row echelon form of $3I - A$ is $\begin{bmatrix} 1 & 1 & 1 \\ 0 & 0 & 0 \\ 0 & 0 & 0 \end{bmatrix}$ so that the eigenspace corresponding to $\lambda_1 = \lambda_2 = 3$ contains vectors $\begin{bmatrix} x_1 \\ x_2 \\ x_3 \end{bmatrix}$ where $x_1 = -s - t, x_2 = s, x_3 = t$. Vectors $\mathbf{p}_1 = \begin{bmatrix} -1 \\ 1 \\ 0 \end{bmatrix}$ and $\mathbf{p}_2 = \begin{bmatrix} -1 \\ 0 \\ 1 \end{bmatrix}$ form a basis for this eigenspace. We apply the Gram-Schmidt process to find an orthogonal basis for this eigenspace: $\mathbf{v}_1 = \mathbf{p}_1 = \begin{bmatrix} -1 \\ 1 \\ 0 \end{bmatrix}$ and $\mathbf{v}_2 = \mathbf{p}_2 - \frac{\langle \mathbf{p}_2, \mathbf{v}_1 \rangle}{\|\mathbf{v}_1\|^2}\mathbf{v}_1 = \begin{bmatrix} -1 \\ 0 \\ 1 \end{bmatrix} - \frac{1}{2}\begin{bmatrix} -1 \\ 1 \\ 0 \end{bmatrix} = \begin{bmatrix} -\frac{1}{2} \\ -\frac{1}{2} \\ 1 \end{bmatrix}$, then proceed to normalize the two vectors to yield an orthonormal basis: $\mathbf{q}_1 = \frac{\mathbf{v}_1}{\|\mathbf{v}_1\|} = \begin{bmatrix} -\frac{1}{\sqrt{2}} \\ \frac{1}{\sqrt{2}} \\ 0 \end{bmatrix}$ and

$\mathbf{q}_2 = \frac{\mathbf{v}_2}{\|\mathbf{v}_2\|} = \begin{bmatrix} -\frac{1}{\sqrt{6}} \\ -\frac{1}{\sqrt{6}} \\ \frac{2}{\sqrt{6}} \end{bmatrix}$.

The reduced row echelon form of $0I - A$ is $\begin{bmatrix} 1 & 0 & -1 \\ 0 & 1 & -1 \\ 0 & 0 & 0 \end{bmatrix}$ so that the eigenspace corresponding to $\lambda_3 = 0$ contains vectors $\begin{bmatrix} x_1 \\ x_2 \\ x_3 \end{bmatrix}$ where $x_1 = t, x_2 = t, x_3 = t$. A vector $\mathbf{p}_3 = \begin{bmatrix} 1 \\ 1 \\ 1 \end{bmatrix}$ forms a basis for this eigenspace.

Applying the Gram-Schmidt process to $\{\mathbf{p}_3\}$ amounts to simply normalizing this vector.

A matrix $P = \begin{bmatrix} -\frac{1}{\sqrt{2}} & -\frac{1}{\sqrt{6}} & \frac{1}{\sqrt{3}} \\ \frac{1}{\sqrt{2}} & -\frac{1}{\sqrt{6}} & \frac{1}{\sqrt{3}} \\ 0 & \frac{2}{\sqrt{6}} & \frac{1}{\sqrt{3}} \end{bmatrix}$ orthogonally diagonalizes A resulting in

$P^{-1}AP = P^TAP = \begin{bmatrix} \lambda_1 & 0 & 0 \\ 0 & \lambda_2 & 0 \\ 0 & 0 & \lambda_3 \end{bmatrix} = \begin{bmatrix} 3 & 0 & 0 \\ 0 & 3 & 0 \\ 0 & 0 & 0 \end{bmatrix}$.

13. $\det(\lambda I - A) = \begin{vmatrix} \lambda+7 & -24 & 0 & 0 \\ -24 & \lambda-7 & 0 & 0 \\ 0 & 0 & \lambda+7 & -24 \\ 0 & 0 & -24 & \lambda-7 \end{vmatrix} = (\lambda+25)^2(\lambda-25)^2$ therefore A has eigenvalues -25 and 25.

The reduced row echelon form of $-25I - A$ is $\begin{bmatrix} 1 & \frac{4}{3} & 0 & 0 \\ 0 & 0 & 1 & \frac{4}{3} \\ 0 & 0 & 0 & 0 \\ 0 & 0 & 0 & 0 \end{bmatrix}$ so that the eigenspace corresponding to $\lambda_1 = \lambda_2 = -25$ contains vectors $\begin{bmatrix} x_1 \\ x_2 \\ x_3 \\ x_4 \end{bmatrix}$ where $x_1 = -\frac{4}{3}s, x_2 = s, x_3 = -\frac{4}{3}t, x_4 = t$. Vectors $\mathbf{p}_1 = \begin{bmatrix} -4 \\ 3 \\ 0 \\ 0 \end{bmatrix}$ and $\mathbf{p}_2 = \begin{bmatrix} 0 \\ 0 \\ -4 \\ 3 \end{bmatrix}$ form a basis for this eigenspace.

The reduced row echelon form of $2I - A$ is $\begin{bmatrix} 1 & -\frac{3}{4} & 0 & 0 \\ 0 & 0 & 1 & -\frac{3}{4} \\ 0 & 0 & 0 & 0 \\ 0 & 0 & 0 & 0 \end{bmatrix}$ so that the eigenspace corresponding to

$\lambda_3 = \lambda_4 = 25$ contains vectors $\begin{bmatrix} x_1 \\ x_2 \\ x_3 \\ x_4 \end{bmatrix}$ where $x_1 = \frac{3}{4}s, x_2 = s, x_3 = \frac{3}{4}t, x_4 = t$. Vectors $\mathbf{p}_3 = \begin{bmatrix} 3 \\ 4 \\ 0 \\ 0 \end{bmatrix}$ and $\mathbf{p}_4 = \begin{bmatrix} 0 \\ 0 \\ 3 \\ 4 \end{bmatrix}$ form a basis for this eigenspace.

Applying the Gram-Schmidt process to the two bases $\{\mathbf{p}_1, \mathbf{p}_2\}, \{\mathbf{p}_3, \mathbf{p}_4\}$ amounts to simply normalizing the vectors since the four vectors are already orthogonal. This yields the columns of a matrix P that orthogonally diagonalizes A: $P = \begin{bmatrix} -\frac{4}{5} & 0 & \frac{3}{5} & 0 \\ \frac{3}{5} & 0 & \frac{4}{5} & 0 \\ 0 & -\frac{4}{5} & 0 & \frac{3}{5} \\ 0 & \frac{3}{5} & 0 & \frac{4}{5} \end{bmatrix}$.

We have $P^{-1}AP = P^TAP = \begin{bmatrix} \lambda_1 & 0 & 0 & 0 \\ 0 & \lambda_2 & 0 & 0 \\ 0 & 0 & \lambda_3 & 0 \\ 0 & 0 & 0 & \lambda_4 \end{bmatrix} = \begin{bmatrix} -25 & 0 & 0 & 0 \\ 0 & -25 & 0 & 0 \\ 0 & 0 & 25 & 0 \\ 0 & 0 & 0 & 25 \end{bmatrix}$.

15. $\det(\lambda I - A) = \begin{vmatrix} \lambda - 3 & -1 \\ -1 & \lambda - 3 \end{vmatrix} = (\lambda - 2)(\lambda - 4)$ therefore A has eigenvalues 2 and 4.

The reduced row echelon form of $2I - A$ is $\begin{bmatrix} 1 & 1 \\ 0 & 0 \end{bmatrix}$ so that the eigenspace corresponding to $\lambda_1 = 2$ consists of vectors $\begin{bmatrix} x_1 \\ x_2 \end{bmatrix}$ where $x_1 = -t, x_2 = t$. A vector $\mathbf{p}_1 = \begin{bmatrix} -1 \\ 1 \end{bmatrix}$ forms a basis for this eigenspace.

The reduced row echelon form of $4I - A$ is $\begin{bmatrix} 1 & -1 \\ 0 & 0 \end{bmatrix}$ so that the eigenspace corresponding to $\lambda_2 = 4$ consists of vectors $\begin{bmatrix} x_1 \\ x_2 \end{bmatrix}$ where $x_1 = t, x_2 = t$. A vector $\mathbf{p}_2 = \begin{bmatrix} 1 \\ 1 \end{bmatrix}$ forms a basis for this eigenspace.

Applying the Gram-Schmidt process to both bases $\{\mathbf{p}_1\}$ and $\{\mathbf{p}_2\}$ amounts to simply normalizing the vectors. This yields the columns of a matrix P that orthogonally diagonalizes A:

$P = \begin{bmatrix} -\frac{1}{\sqrt{2}} & \frac{1}{\sqrt{2}} \\ \frac{1}{\sqrt{2}} & \frac{1}{\sqrt{2}} \end{bmatrix}$. We have $P^{-1}AP = P^TAP = \begin{bmatrix} \lambda_1 & 0 \\ 0 & \lambda_2 \end{bmatrix} = \begin{bmatrix} 2 & 0 \\ 0 & 4 \end{bmatrix}$.

Formula (7) of Section 7.2 yields the spectral decomposition of A:

$$\begin{bmatrix} 3 & 1 \\ 1 & 3 \end{bmatrix} = (2)\begin{bmatrix} -\frac{1}{\sqrt{2}} \\ \frac{1}{\sqrt{2}} \end{bmatrix}\begin{bmatrix} -\frac{1}{\sqrt{2}} & \frac{1}{\sqrt{2}} \end{bmatrix} + (4)\begin{bmatrix} \frac{1}{\sqrt{2}} \\ \frac{1}{\sqrt{2}} \end{bmatrix}\begin{bmatrix} \frac{1}{\sqrt{2}} & \frac{1}{\sqrt{2}} \end{bmatrix} = (2)\begin{bmatrix} \frac{1}{2} & -\frac{1}{2} \\ -\frac{1}{2} & \frac{1}{2} \end{bmatrix} + (4)\begin{bmatrix} \frac{1}{2} & \frac{1}{2} \\ \frac{1}{2} & \frac{1}{2} \end{bmatrix}.$$

17. $\det(\lambda I - A) = \begin{vmatrix} \lambda + 3 & -1 & -2 \\ -1 & \lambda + 3 & -2 \\ -2 & -2 & \lambda \end{vmatrix} = (\lambda + 4)^2(\lambda - 2)$ therefore A has eigenvalues -4 and 2.

The reduced row echelon form of $-4I - A$ is $\begin{bmatrix} 1 & 1 & 2 \\ 0 & 0 & 0 \\ 0 & 0 & 0 \end{bmatrix}$ so that the eigenspace corresponding to

$\lambda = -4$ contains vectors $\begin{bmatrix} x_1 \\ x_2 \\ x_3 \end{bmatrix}$ where $x_1 = -s - 2t, x_2 = s, x_3 = t$. Vectors $\mathbf{p}_1 = \begin{bmatrix} -1 \\ 1 \\ 0 \end{bmatrix}$ and $\mathbf{p}_2 = \begin{bmatrix} -2 \\ 0 \\ 1 \end{bmatrix}$ form a basis for this eigenspace. We apply the Gram-Schmidt process to find an orthogonal basis for this eigenspace: $\mathbf{v}_1 = \mathbf{p}_1 = \begin{bmatrix} -1 \\ 1 \\ 0 \end{bmatrix}$ and $\mathbf{v}_2 = \mathbf{p}_2 - \frac{\langle \mathbf{p}_2, \mathbf{v}_1 \rangle}{\|\mathbf{v}_1\|^2}\mathbf{v}_1 = \begin{bmatrix} -2 \\ 0 \\ 1 \end{bmatrix} - \frac{2}{2}\begin{bmatrix} -1 \\ 1 \\ 0 \end{bmatrix} = \begin{bmatrix} -1 \\ -1 \\ 1 \end{bmatrix}$, then proceed to normalize the two vectors to yield an orthonormal basis: $\mathbf{q}_1 = \frac{\mathbf{v}_1}{\|\mathbf{v}_1\|} = \begin{bmatrix} -\frac{1}{\sqrt{2}} \\ \frac{1}{\sqrt{2}} \\ 0 \end{bmatrix}$ and $\mathbf{q}_2 = \frac{\mathbf{v}_2}{\|\mathbf{v}_2\|} = \begin{bmatrix} -\frac{1}{\sqrt{3}} \\ -\frac{1}{\sqrt{3}} \\ \frac{1}{\sqrt{3}} \end{bmatrix}$.

The reduced row echelon form of $2I - A$ is $\begin{bmatrix} 1 & 0 & -\frac{1}{2} \\ 0 & 1 & -\frac{1}{2} \\ 0 & 0 & 0 \end{bmatrix}$ so that the eigenspace corresponding to $\lambda = 2$ contains vectors $\begin{bmatrix} x_1 \\ x_2 \\ x_3 \end{bmatrix}$ where $x_1 = \frac{1}{2}t, x_2 = \frac{1}{2}t, x_3 = t$. A vector $\mathbf{p}_3 = \begin{bmatrix} 1 \\ 1 \\ 2 \end{bmatrix}$ forms a basis for this eigenspace.

Applying the Gram-Schmidt process to $\{\mathbf{p}_3\}$ amounts to simply normalizing this vector.

A matrix $P = \begin{bmatrix} -\frac{1}{\sqrt{2}} & -\frac{1}{\sqrt{3}} & \frac{1}{\sqrt{6}} \\ \frac{1}{\sqrt{2}} & -\frac{1}{\sqrt{3}} & \frac{1}{\sqrt{6}} \\ 0 & \frac{1}{\sqrt{3}} & \frac{2}{\sqrt{6}} \end{bmatrix}$ orthogonally diagonalizes A resulting in $P^TAP = D = \begin{bmatrix} -4 & 0 & 0 \\ 0 & -4 & 0 \\ 0 & 0 & 2 \end{bmatrix}$.

Formula (7) of Section 7.2 yields the spectral decomposition of A:

$$\begin{bmatrix} -3 & 1 & 2 \\ 1 & -3 & 2 \\ 2 & 2 & 0 \end{bmatrix} = (-4)\begin{bmatrix} -\frac{1}{\sqrt{2}} \\ \frac{1}{\sqrt{2}} \\ 0 \end{bmatrix}\begin{bmatrix} -\frac{1}{\sqrt{2}} & \frac{1}{\sqrt{2}} & 0 \end{bmatrix} + (-4)\begin{bmatrix} -\frac{1}{\sqrt{3}} \\ -\frac{1}{\sqrt{3}} \\ \frac{1}{\sqrt{3}} \end{bmatrix}\begin{bmatrix} -\frac{1}{\sqrt{3}} & -\frac{1}{\sqrt{3}} & \frac{1}{\sqrt{3}} \end{bmatrix} + (2)\begin{bmatrix} \frac{1}{\sqrt{6}} \\ \frac{1}{\sqrt{6}} \\ \frac{2}{\sqrt{6}} \end{bmatrix}\begin{bmatrix} \frac{1}{\sqrt{6}} & \frac{1}{\sqrt{6}} & \frac{2}{\sqrt{6}} \end{bmatrix}$$

$$= (-4)\begin{bmatrix} \frac{1}{2} & -\frac{1}{2} & 0 \\ -\frac{1}{2} & \frac{1}{2} & 0 \\ 0 & 0 & 0 \end{bmatrix} + (-4)\begin{bmatrix} \frac{1}{3} & \frac{1}{3} & -\frac{1}{3} \\ \frac{1}{3} & \frac{1}{3} & -\frac{1}{3} \\ -\frac{1}{3} & -\frac{1}{3} & \frac{1}{3} \end{bmatrix} + (2)\begin{bmatrix} \frac{1}{6} & \frac{1}{6} & \frac{1}{3} \\ \frac{1}{6} & \frac{1}{6} & \frac{1}{3} \\ \frac{1}{3} & \frac{1}{3} & \frac{2}{3} \end{bmatrix}.$$

19. The three vectors are orthogonal, and they can be made into orthonormal vectors by a simple normalization. Forming the columns of a matrix P in this way we obtain an orthogonal matrix $P = \begin{bmatrix} 0 & 1 & 0 \\ \frac{1}{\sqrt{2}} & 0 & \frac{1}{\sqrt{2}} \\ -\frac{1}{\sqrt{2}} & 0 & \frac{1}{\sqrt{2}} \end{bmatrix}$. When the diagonal matrix D contains the corresponding eigenvalues on its main diagonal, $D = \begin{bmatrix} -1 & 0 & 0 \\ 0 & 3 & 0 \\ 0 & 0 & 7 \end{bmatrix}$, then Formula (2) in Section 7.2 yields $PDP^T = A = \begin{bmatrix} 3 & 0 & 0 \\ 0 & 3 & 4 \\ 0 & 4 & 3 \end{bmatrix}$.

21. Yes. The Gram-Schmidt process will ensure that columns of P corresponding to the same eigenvalue are an orthonormal set. Since eigenvectors from distinct eigenvalues are orthogonal, this means that P will be an orthogonal matrix. Then since A is orthogonally diagonalizable, it must be symmetric.

23. **(a)** $\det(\lambda I - A) = \begin{vmatrix} \lambda+1 & -1 \\ -1 & \lambda-1 \end{vmatrix} = \lambda^2 - 2 = (\lambda - \sqrt{2})(\lambda + \sqrt{2})$ therefore A has eigenvalues $\pm\sqrt{2}$.

A is symmetric, so by Theorem 7.2.2(b), eigenvectors from different eigenspaces are orthogonal.

The reduced row echelon form of $\sqrt{2}I - A$ is $\begin{bmatrix} 1 & 1-\sqrt{2} \\ 0 & 0 \end{bmatrix}$ so that the eigenspace corresponding to $\lambda = \sqrt{2}$ consists of vectors $\begin{bmatrix} x_1 \\ x_2 \end{bmatrix}$ where $x_1 = (\sqrt{2}-1)t$, $x_2 = t$. A vector $\begin{bmatrix} \sqrt{2}-1 \\ 1 \end{bmatrix}$ forms a basis for this eigenspace.

The reduced row echelon form of $-\sqrt{2}I - A$ is $\begin{bmatrix} 1 & 1+\sqrt{2} \\ 0 & 0 \end{bmatrix}$ so that the eigenspace corresponding to $\lambda = -\sqrt{2}$ consists of vectors $\begin{bmatrix} x_1 \\ x_2 \end{bmatrix}$ where $x_1 = (-\sqrt{2}-1)t$, $x_2 = t$. A vector $\begin{bmatrix} -\sqrt{2}-1 \\ 1 \end{bmatrix}$ forms a basis for this eigenspace.

Unit eigenvectors chosen from two different eigenspaces will meet our desired condition. For instance, let $\mathbf{u}_1 = \begin{bmatrix} \frac{\sqrt{2}-1}{\sqrt{4-2\sqrt{2}}} \\ \frac{1}{\sqrt{4-2\sqrt{2}}} \end{bmatrix}$ and $\mathbf{u}_2 = \begin{bmatrix} \frac{-\sqrt{2}-1}{\sqrt{4+2\sqrt{2}}} \\ \frac{1}{\sqrt{4+2\sqrt{2}}} \end{bmatrix}$.

(b) $\det(\lambda I - A) = \begin{vmatrix} \lambda-1 & -2 \\ -2 & \lambda-1 \end{vmatrix} = (\lambda+1)(\lambda-3)$ therefore A has eigenvalues -1 and 3.

A is symmetric, so by Theorem 7.2.2(b), eigenvectors from different eigenspaces are orthogonal.

The reduced row echelon form of $-1I - A$ is $\begin{bmatrix} 1 & 1 \\ 0 & 0 \end{bmatrix}$ so that the eigenspace corresponding to $\lambda = -1$ consists of vectors $\begin{bmatrix} x_1 \\ x_2 \end{bmatrix}$ where $x_1 = -t$, $x_2 = t$. A vector $\begin{bmatrix} -1 \\ 1 \end{bmatrix}$ forms a basis for this eigenspace.

The reduced row echelon form of $3I - A$ is $\begin{bmatrix} 1 & -1 \\ 0 & 0 \end{bmatrix}$ so that the eigenspace corresponding to $\lambda = 3$ consists of vectors $\begin{bmatrix} x_1 \\ x_2 \end{bmatrix}$ where $x_1 = t$, $x_2 = t$. A vector $\begin{bmatrix} 1 \\ 1 \end{bmatrix}$ forms a basis for this eigenspace.

Unit eigenvectors chosen from two different eigenspaces will meet our desired condition. For instance, let $\mathbf{u}_1 = \begin{bmatrix} \frac{-1}{\sqrt{2}} \\ \frac{1}{\sqrt{2}} \end{bmatrix}$ and $\mathbf{u}_2 = \begin{bmatrix} \frac{1}{\sqrt{2}} \\ \frac{1}{\sqrt{2}} \end{bmatrix}$.

25. A^TA is a symmetric $n \times n$ matrix since $(A^TA)^T = A^T(A^T)^T = A^TA$. By Theorem 7.2.1 it has an orthonormal set of n eigenvectors.

29. By Theorem 7.1.3(b), if A is an orthogonal $n \times n$ matrix then $\|A\mathbf{x}\| = \|\mathbf{x}\|$ for all $\mathbf{x}$ in R^n. Since the eigenvalues of a symmetric matrix must be real numbers, for every such eigenvalue λ and a corresponding eigenvector $\mathbf{x}$ we have $\|\mathbf{x}\| = \|A\mathbf{x}\| = \|\lambda\mathbf{x}\| = |\lambda|\|\mathbf{x}\|$ hence the only possible eigenvalues for an orthogonal symmetric matrix are 1 and -1.

True-False Exercises

(a) True. For any square matrix A, both AA^T and A^TA are symmetric, hence orthogonally diagonalizable.

(b) True. Since $\mathbf{v}_1$ and $\mathbf{v}_2$ are from distinct eigenspaces of a symmetric matrix, they are orthogonal, so $\|\mathbf{v}_1+\mathbf{v}_2\|^2 = \langle \mathbf{v}_1+\mathbf{v}_2,\ \mathbf{v}_1+\mathbf{v}_2\rangle = \langle \mathbf{v}_1,\ \mathbf{v}_1\rangle + 2\langle \mathbf{v}_1,\ \mathbf{v}_2\rangle + \langle \mathbf{v}_2,\ \mathbf{v}_2\rangle = \|\mathbf{v}_1\|^2 + 0 + \|\mathbf{v}_2\|^2$.

(c) False. An orthogonal matrix is not necessarily symmetric.

(d) True. By Theorem 1.7.4, if A is an invertible symmetric matrix then A^{-1} is also symmetric.

(e) True. By Theorem 7.1.3(b), if A is an orthogonal $n \times n$ matrix then $\|A\mathbf{x}\| = \|\mathbf{x}\|$ for all $\mathbf{x}$ in R^n. For every eigenvalue λ and a corresponding eigenvector $\mathbf{x}$ we have $\|\mathbf{x}\| = \|A\mathbf{x}\| = \|\lambda\mathbf{x}\| = |\lambda|\|\mathbf{x}\|$ hence $|\lambda| = 1$.

(f) True. If A is an $n \times n$ orthogonally diagonalizable matrix, then A has an orthonormal set of n eigenvectors, which form a basis for R^n.

(g) True. This follows from part (a) of Theorem 7.2.2.

7.3 Quadratic Forms

1. **(a)** $3x_1^2 + 7x_2^2 = \begin{bmatrix} x_1 & x_2 \end{bmatrix}\begin{bmatrix} 3 & 0 \\ 0 & 7 \end{bmatrix}\begin{bmatrix} x_1 \\ x_2 \end{bmatrix}$

(b) $4x_1^2 - 9x_2^2 - 6x_1x_2 = \begin{bmatrix} x_1 & x_2 \end{bmatrix}\begin{bmatrix} 4 & -3 \\ -3 & -9 \end{bmatrix}\begin{bmatrix} x_1 \\ x_2 \end{bmatrix}$

(c) $9x_1^2 - x_2^2 + 4x_3^2 + 6x_1x_2 - 8x_1x_3 + x_2x_3 = \begin{bmatrix} x_1 & x_2 & x_3 \end{bmatrix}\begin{bmatrix} 9 & 3 & -4 \\ 3 & -1 & \frac{1}{2} \\ -4 & \frac{1}{2} & 4 \end{bmatrix}\begin{bmatrix} x_1 \\ x_2 \\ x_3 \end{bmatrix}$

3. $\begin{bmatrix} x & y \end{bmatrix}\begin{bmatrix} 2 & -3 \\ -3 & 5 \end{bmatrix}\begin{bmatrix} x \\ y \end{bmatrix} = 2x^2 + 5y^2 - 6xy$

5. $Q = \mathbf{x}^TA\mathbf{x} = \begin{bmatrix} x_1 & x_2 \end{bmatrix}\begin{bmatrix} 2 & -1 \\ -1 & 2 \end{bmatrix}\begin{bmatrix} x_1 \\ x_2 \end{bmatrix}$; the characteristic polynomial of the matrix A is $\lambda^2 - 4\lambda + 3 = (\lambda - 3)(\lambda - 1)$, so the eigenvalues of A are $\lambda = 3,\ 1$.

The reduced row echelon form of $3I - A$ is $\begin{bmatrix} 1 & 1 \\ 0 & 0 \end{bmatrix}$ so that the eigenspace corresponding to $\lambda = 3$ consists of vectors $\begin{bmatrix} x_1 \\ x_2 \end{bmatrix}$ where $x_1 = -t, x_2 = t$. A vector $\mathbf{p}_1 = \begin{bmatrix} -1 \\ 1 \end{bmatrix}$ forms a basis for this eigenspace.

The reduced row echelon form of $1I - A$ is $\begin{bmatrix} 1 & -1 \\ 0 & 0 \end{bmatrix}$ so that the eigenspace corresponding to $\lambda = 1$ consists of vectors $\begin{bmatrix} x_1 \\ x_2 \end{bmatrix}$ where $x_1 = t, x_2 = t$. A vector $\mathbf{p}_2 = \begin{bmatrix} 1 \\ 1 \end{bmatrix}$ forms a basis for this eigenspace.

Applying the Gram-Schmidt process to the bases $\{\mathbf{p}_1\}$ and $\{\mathbf{p}_2\}$ amounts to simply normalizing the vectors. Therefore an orthogonal change of variables $\mathbf{x} = P\mathbf{y}$ that eliminates the cross product terms in Q is $\begin{bmatrix} x_1 \\ x_2 \end{bmatrix} = \begin{bmatrix} -\frac{1}{\sqrt{2}} & \frac{1}{\sqrt{2}} \\ \frac{1}{\sqrt{2}} & \frac{1}{\sqrt{2}} \end{bmatrix} \begin{bmatrix} y_1 \\ y_2 \end{bmatrix}$. In terms of the new variables, we have

$Q = \mathbf{x}^T A \mathbf{x} = \mathbf{y}^T (P^T A P) \mathbf{y} = [y_1 \quad y_2] \begin{bmatrix} 3 & 0 \\ 0 & 1 \end{bmatrix} \begin{bmatrix} y_1 \\ y_2 \end{bmatrix} = 3y_1^2 + y_2^2$.

7. $Q = \mathbf{x}^T A \mathbf{x} = [x_1 \quad x_2 \quad x_3] \begin{bmatrix} 3 & 2 & 0 \\ 2 & 4 & -2 \\ 0 & -2 & 5 \end{bmatrix} \begin{bmatrix} x_1 \\ x_2 \\ x_3 \end{bmatrix}$; the characteristic polynomial of the matrix A is

$\det(\lambda I - A) = \begin{vmatrix} \lambda - 3 & -2 & 0 \\ -2 & \lambda - 4 & 2 \\ 0 & 2 & \lambda - 5 \end{vmatrix} = \lambda^3 - 12\lambda^2 + 39\lambda - 28 = (\lambda - 1)(\lambda - 4)(\lambda - 7)$
so the eigenvalues of A are 1, 4, and 7.

The reduced row echelon form of $1I - A$ is $\begin{bmatrix} 1 & 0 & 2 \\ 0 & 1 & -2 \\ 0 & 0 & 0 \end{bmatrix}$ so that the eigenspace corresponding to

$\lambda = 1$ consists of vectors $\begin{bmatrix} x_1 \\ x_2 \\ x_3 \end{bmatrix}$ where $x_1 = -2t, x_2 = 2t, x_3 = t$. A vector $\mathbf{p}_1 = \begin{bmatrix} -2 \\ 2 \\ 1 \end{bmatrix}$ forms a basis for

this eigenspace.

The reduced row echelon form of $4I - A$ is $\begin{bmatrix} 1 & 0 & -1 \\ 0 & 1 & -\frac{1}{2} \\ 0 & 0 & 0 \end{bmatrix}$ so that the eigenspace corresponding to

$\lambda = 4$ consists of vectors $\begin{bmatrix} x_1 \\ x_2 \\ x_3 \end{bmatrix}$ where $x_1 = t, x_2 = \frac{1}{2}t, x_3 = t$. A vector $\mathbf{p}_2 = \begin{bmatrix} 2 \\ 1 \\ 2 \end{bmatrix}$ forms a basis for this

eigenspace.

The reduced row echelon form of $7I - A$ is $\begin{bmatrix} 1 & 0 & \frac{1}{2} \\ 0 & 1 & 1 \\ 0 & 0 & 0 \end{bmatrix}$ so that the eigenspace corresponding to $\lambda = 7$

consists of vectors $\begin{bmatrix} x_1 \\ x_2 \\ x_3 \end{bmatrix}$ where $x_1 = -\frac{1}{2}t, x_2 = -t, x_3 = t$. A vector $\mathbf{p}_3 = \begin{bmatrix} -1 \\ -2 \\ 2 \end{bmatrix}$ forms a basis for this

eigenspace.
Applying the Gram-Schmidt process to the three bases amounts to simply normalizing the vectors.
Therefore an orthogonal change of variables $\mathbf{x} = P\mathbf{y}$ that eliminates the cross product terms in Q is

$\begin{bmatrix} x_1 \\ x_2 \\ x_3 \end{bmatrix} = \begin{bmatrix} -\frac{2}{3} & \frac{2}{3} & -\frac{1}{3} \\ \frac{2}{3} & \frac{1}{3} & -\frac{2}{3} \\ \frac{1}{3} & \frac{2}{3} & \frac{2}{3} \end{bmatrix} \begin{bmatrix} y_1 \\ y_2 \\ y_3 \end{bmatrix}$. In terms of the new variables, we have

$Q = \mathbf{x}^T A \mathbf{x} = \mathbf{y}^T (P^T A P) \mathbf{y} = [y_1 \quad y_2 \quad y_3] \begin{bmatrix} 1 & 0 & 0 \\ 0 & 4 & 0 \\ 0 & 0 & 7 \end{bmatrix} \begin{bmatrix} y_1 \\ y_2 \\ y_3 \end{bmatrix} = y_1^2 + 4y_2^2 + 7y_3^2$.

9. **(a)** $2x^2 + xy + x - 6y + 2 = 0$ can be expressed as $\begin{bmatrix} x & y \end{bmatrix} \underbrace{\begin{bmatrix} 2 & \frac{1}{2} \\ \frac{1}{2} & 0 \end{bmatrix}}_{A} \begin{bmatrix} x \\ y \end{bmatrix} + \underbrace{\begin{bmatrix} 1 & -6 \end{bmatrix}}_{K} \begin{bmatrix} x \\ y \end{bmatrix} + \underbrace{(2)}_{f} = 0$

(b) $y^2 + 7x - 8y - 5 = 0$ can be expressed as $\begin{bmatrix} x & y \end{bmatrix} \underbrace{\begin{bmatrix} 0 & 0 \\ 0 & 1 \end{bmatrix}}_{A} \begin{bmatrix} x \\ y \end{bmatrix} + \underbrace{\begin{bmatrix} 7 & -8 \end{bmatrix}}_{K} \begin{bmatrix} x \\ y \end{bmatrix} + \underbrace{(-5)}_{f} = 0$

11. **(a)** $2x^2 + 5y^2 = 20$ is $\frac{x^2}{10} + \frac{y^2}{4} = 1$ which is an equation of an ellipse.

(b) $x^2 - y^2 - 8 = 0$ is $x^2 - y^2 = 8$ or $\frac{x^2}{8} - \frac{y^2}{8} = 1$ which is an equation of a hyperbola.

(c) $7y^2 - 2x = 0$ is $x = \frac{7}{2}y^2$ which is an equation of a parabola.

(d) $x^2 + y^2 - 25 = 0$ is $x^2 + y^2 = 25$ which is an equation of a circle.

13. We can rewrite the given equation in the matrix form $\mathbf{x}^T A\mathbf{x} = -8$ with $A = \begin{bmatrix} 2 & -2 \\ -2 & -1 \end{bmatrix}$.

The characteristic polynomial of A is $\det(\lambda I - A) = \begin{vmatrix} \lambda - 2 & 2 \\ 2 & \lambda + 1 \end{vmatrix} = (\lambda - 3)(\lambda + 2)$ so A has eigenvalues 3 and -2.

The reduced row echelon form of $3I - A$ is $\begin{bmatrix} 1 & 2 \\ 0 & 0 \end{bmatrix}$ so that the eigenspace corresponding to $\lambda = 3$ consists of vectors $\begin{bmatrix} x_1 \\ x_2 \end{bmatrix}$ where $x_1 = -2t, x_2 = t$. A vector $\mathbf{p}_1 = \begin{bmatrix} -2 \\ 1 \end{bmatrix}$ forms a basis for this eigenspace.

The reduced row echelon form of $-2I - A$ is $\begin{bmatrix} 1 & -\frac{1}{2} \\ 0 & 0 \end{bmatrix}$ so that the eigenspace corresponding to $\lambda = -2$ consists of vectors $\begin{bmatrix} x_1 \\ x_2 \end{bmatrix}$ where $x_1 = \frac{1}{2}t, x_2 = t$. A vector $\mathbf{p}_2 = \begin{bmatrix} 1 \\ 2 \end{bmatrix}$ forms a basis for this eigenspace.

Applying the Gram-Schmidt process to both bases $\{\mathbf{p}_1\}$ and $\{\mathbf{p}_2\}$ amounts to simply normalizing the vectors.

This yields the columns of a matrix P that orthogonally diagonalizes A - of the two possibilities, $\begin{bmatrix} -\frac{2}{\sqrt{5}} & \frac{1}{\sqrt{5}} \\ \frac{1}{\sqrt{5}} & \frac{2}{\sqrt{5}} \end{bmatrix}$ and $\begin{bmatrix} \frac{1}{\sqrt{5}} & -\frac{2}{\sqrt{5}} \\ \frac{2}{\sqrt{5}} & \frac{1}{\sqrt{5}} \end{bmatrix}$ we choose the latter, i.e., $P = \begin{bmatrix} \frac{1}{\sqrt{5}} & -\frac{2}{\sqrt{5}} \\ \frac{2}{\sqrt{5}} & \frac{1}{\sqrt{5}} \end{bmatrix}$, since its determinant is 1 so that the substitution $\mathbf{x} = P\mathbf{x}'$ performs a rotation of axes. In the rotated coordinates, the equation of the conic becomes $\begin{bmatrix} x' & y' \end{bmatrix} \begin{bmatrix} -2 & 0 \\ 0 & 3 \end{bmatrix} \begin{bmatrix} x' \\ y' \end{bmatrix} = -8$, i.e., $3y'^2 - 2x'^2 = 8$; this equation represents a hyperbola.

Solving $P = \begin{bmatrix} \frac{1}{\sqrt{5}} & -\frac{2}{\sqrt{5}} \\ \frac{2}{\sqrt{5}} & \frac{1}{\sqrt{5}} \end{bmatrix} = \begin{bmatrix} \cos\theta & -\sin\theta \\ \sin\theta & \cos\theta \end{bmatrix}$ we conclude that the angle of rotation is $\theta = \sin^{-1}\left(\frac{2}{\sqrt{5}}\right) \approx 63.4°$.

15. We can rewrite the given equation in the matrix form $\mathbf{x}^T A\mathbf{x} = 15$ with $A = \begin{bmatrix} 11 & 12 \\ 12 & 4 \end{bmatrix}$.

The characteristic polynomial of A is $\det(\lambda I - A) = \begin{vmatrix} \lambda - 11 & -12 \\ -12 & \lambda - 4 \end{vmatrix} = (\lambda - 20)(\lambda + 5)$ so A has eigenvalues 20 and -5.

The reduced row echelon form of $20I - A$ is $\begin{bmatrix} 1 & -\frac{4}{3} \\ 0 & 0 \end{bmatrix}$ so that the eigenspace corresponding to $\lambda = 20$ consists of vectors $\begin{bmatrix} x_1 \\ x_2 \end{bmatrix}$ where $x_1 = \frac{4}{3}t, x_2 = t$. A vector $\mathbf{p}_1 = \begin{bmatrix} 4 \\ 3 \end{bmatrix}$ forms a basis for this eigenspace.

The reduced row echelon form of $-5I - A$ is $\begin{bmatrix} 1 & \frac{3}{4} \\ 0 & 0 \end{bmatrix}$ so that the eigenspace corresponding to $\lambda = -5$ consists of vectors $\begin{bmatrix} x_1 \\ x_2 \end{bmatrix}$ where $x_1 = -\frac{3}{4}t, x_2 = t$. A vector $\mathbf{p}_2 = \begin{bmatrix} -3 \\ 4 \end{bmatrix}$ forms a basis for this eigenspace.

Applying the Gram-Schmidt process to both bases $\{\mathbf{p}_1\}$ and $\{\mathbf{p}_2\}$ amounts to simply normalizing the vectors.

This yields the columns of a matrix P that orthogonally diagonalizes A - of the two possibilities, $\begin{bmatrix} \frac{4}{5} & -\frac{3}{5} \\ \frac{3}{5} & \frac{4}{5} \end{bmatrix}$ and $\begin{bmatrix} -\frac{3}{5} & \frac{4}{5} \\ \frac{4}{5} & \frac{3}{5} \end{bmatrix}$ we choose the former, i.e., $P = \begin{bmatrix} \frac{4}{5} & -\frac{3}{5} \\ \frac{3}{5} & \frac{4}{5} \end{bmatrix}$, since its determinant is 1 so that the substitution $\mathbf{x} = P\mathbf{x}'$ performs a rotation of axes. In the rotated coordinates, the equation of the conic becomes $[x' \quad y'] \begin{bmatrix} 20 & 0 \\ 0 & -5 \end{bmatrix} \begin{bmatrix} x' \\ y' \end{bmatrix} = 15$, i.e., $4x'^2 - y'^2 = 3$; this equation represents a hyperbola.

Solving $P = \begin{bmatrix} \frac{4}{5} & -\frac{3}{5} \\ \frac{3}{5} & \frac{4}{5} \end{bmatrix} = \begin{bmatrix} \cos\theta & -\sin\theta \\ \sin\theta & \cos\theta \end{bmatrix}$ we conclude that the angle of rotation is $\theta = \sin^{-1}\left(\frac{3}{5}\right) \approx 36.9°$.

17. All matrices in this exercise are diagonal, therefore by Theorem 5.1.2, their eigenvalues are the entries on the main diagonal. We use Theorem 7.3.2 (including the remark below it).

(a) positive definite **(b)** negative definite **(c)** indefinite

(d) positive semidefinite **(e)** negative semidefinite

19. For all $(x_1, x_2) \neq (0,0)$, we clearly have $x_1^2 + x_2^2 > 0$ therefore the form is positive definite (an alternate justification would be to calculate eigenvalues of the associated matrix $\begin{bmatrix} 1 & 0 \\ 0 & 1 \end{bmatrix}$ which are $\lambda_1 = \lambda_2 = 1$ then use Theorem 7.3.2).

21. For all $(x_1, x_2) \neq (0,0)$, we clearly have $(x_1 - x_2)^2 \geq 0$, but cannot claim $(x_1 - x_2)^2 > 0$ when $x_1 = x_2$ therefore the form is positive semidefinite

(an alternate justification would be to calculate eigenvalues of the associated matrix $\begin{bmatrix} 1 & -1 \\ -1 & 1 \end{bmatrix}$ which are $\lambda = 2$ and $\lambda = 0$ then use the remark under Theorem 7.3.2).

23. Clearly, the form $x_1^2 - x_2^2$ has both positive and negative values (e.g., $3^2 - 1^2 > 0$ and $2^2 - 4^2 < 0$) therefore this quadratic form is indefinite

(an alternate justification would be to calculate eigenvalues of the associated matrix $\begin{bmatrix} 1 & 0 \\ 0 & -1 \end{bmatrix}$ which are $\lambda = -1$ and $\lambda = 1$ then use Theorem 7.3.2).

25. **(a)** $\det(\lambda I - A) = \begin{vmatrix} \lambda - 5 & 2 \\ 2 & \lambda - 5 \end{vmatrix} = (\lambda - 3)(\lambda - 7)$; since both eigenvalues $\lambda = 3$ and $\lambda = 7$ are positive, by Theorem 7.3.2, A is positive definite.

Determinant of the first principal submatrix of A is $\det([5]) = 5 > 0$.
Determinant of the second principal submatrix of A is $\det(A) = 21 > 0$.
By Theorem 7.3.4, A is positive definite.

(b) $\det(\lambda I - A) = \begin{vmatrix} \lambda - 2 & 1 & 0 \\ 1 & \lambda - 2 & 0 \\ 0 & 0 & \lambda - 5 \end{vmatrix} = (\lambda - 1)(\lambda - 3)(\lambda - 5)$; since all three eigenvalues $\lambda = 1, \lambda = 3$, and $\lambda = 5$ are positive, by Theorem 7.3.2, A is positive definite.

Determinant of the first principal submatrix of A is $\det([2]) = 2 > 0$.
Determinant of the second principal submatrix of A is $\det\left(\begin{bmatrix} 2 & -1 \\ -1 & 2 \end{bmatrix}\right) = 3 > 0$.
Determinant of the third principal submatrix of A is $\det(A) = 15 > 0$.
By Theorem 7.3.4, A is positive definite.

27. **(a)** Determinant of the first principal submatrix of A is $\det([3]) = 3 > 0$.
Determinant of the second principal submatrix of A is $\det\left(\begin{bmatrix} 3 & 1 \\ 1 & -1 \end{bmatrix}\right) = -4 < 0$.
Determinant of the third principal submatrix of A is $\det(A) = -19 < 0$.
By Theorem 7.3.4(c), A is indefinite.

(b) Determinant of the first principal submatrix of A is $\det([-3]) = -3 < 0$.
Determinant of the second principal submatrix of A is $\det\left(\begin{bmatrix} -3 & 2 \\ 2 & -3 \end{bmatrix}\right) = 5 > 0$.
Determinant of the third principal submatrix of A is $\det(A) = -25 < 0$.
By Theorem 7.3.4(b), A is negative definite.

29. The quadratic form $Q = 5x_1^2 + x_2^2 + kx_3^2 + 4x_1x_2 - 2x_1x_3 - 2x_2x_3$ can be expressed in matrix notation as $Q = \mathbf{x}^T A\mathbf{x}$ where $A = \begin{bmatrix} 5 & 2 & -1 \\ 2 & 1 & -1 \\ -1 & -1 & k \end{bmatrix}$. The determinants of the principal submatrices of A are $\det([5]) = 5$, $\det\begin{bmatrix} 5 & 2 \\ 2 & 1 \end{bmatrix} = 1$, and $\det A = k - 2$. Thus Q is positive definite if and only if $k > 2$.

31. **(a)** We assume A is symmetric so that $\mathbf{x}^T A\mathbf{y} = (\mathbf{x}^T A\mathbf{y})^T = \mathbf{y}^T A^T\mathbf{x} = \mathbf{y}^T A\mathbf{x}$. Therefore
$T(\mathbf{x} + \mathbf{y}) = (\mathbf{x} + \mathbf{y})^T A(\mathbf{x} + \mathbf{y}) = \mathbf{x}^T A\mathbf{x} + \mathbf{y}^T A\mathbf{x} + \mathbf{x}^T A\mathbf{y} + \mathbf{y}^T A\mathbf{y} = T(\mathbf{x}) + 2\mathbf{x}^T A\mathbf{y} + T(\mathbf{y})$.

(b) $T(c\mathbf{x}) = (c\mathbf{x})^T A(c\mathbf{x}) = c^2(\mathbf{x}^T A\mathbf{x}) = c^2 T(\mathbf{x})$

33. **(a)** For each $i = 1, \ldots, n$ we have

$$\begin{aligned}(x_i - \bar{x})^2 &= x_i^2 - 2x_i\bar{x} + \bar{x}^2 \\ &= x_i^2 - 2x_i\frac{1}{n}\sum_{j=1}^{n} x_j + \frac{1}{n^2}\left(\sum_{j=1}^{n} x_j\right)^2 \\ &= x_i^2 - \frac{2}{n}\sum_{j=1}^{n} x_i x_j + \frac{1}{n^2}\left(\sum_{j=1}^{n} x_j^2 + 2\sum_{j=1}^{n-1}\sum_{k=j+1}^{n} x_j x_k\right)\end{aligned}$$

Thus in the quadratic form $s_x^2 = \frac{1}{n-1}[(x_1 - \bar{x})^2 + (x_2 - \bar{x})^2 + \cdots + (x_n - \bar{x})^2]$ the coefficient of x_i^2 is $\frac{1}{n-1}\left[1 - \frac{2}{n} + \frac{1}{n^2}n\right] = \frac{1}{n}$, and the coefficient of $x_i x_j$ for $i \neq j$ is $\frac{1}{n-1}\left[-\frac{2}{n} - \frac{2}{n} + \frac{2}{n^2}n\right] = -\frac{2}{n(n-1)}$.

It follows that $s_x^2 = \mathbf{x}^T A\mathbf{x}$ where $A = \begin{bmatrix} \frac{1}{n} & -\frac{1}{n(n-1)} & \cdots & -\frac{1}{n(n-1)} \\ -\frac{1}{n(n-1)} & \frac{1}{n} & \cdots & -\frac{1}{n(n-1)} \\ \vdots & \vdots & \ddots & \vdots \\ -\frac{1}{n(n-1)} & -\frac{1}{n(n-1)} & \cdots & \frac{1}{n} \end{bmatrix}$.

(b) We have $s_x^2 = \frac{1}{n-1}[(x_1 - \bar{x})^2 + (x_2 - \bar{x})^2 + \cdots + (x_n - \bar{x})^2] \geq 0$, and $s_x^2 = 0$ if and only if $x_1 = \bar{x}$, $x_2 = \bar{x}, \ldots, x_n = \bar{x}$, i.e., if and only if $x_1 = x_2 = \cdots = x_n$. Thus s_x^2 is a positive semidefinite form.

35. The eigenvalues of A must be positive and equal to each other. That is, A must have a positive eigenvalue of multiplicity 2.

37. If A is an $n \times n$ symmetric matrix such that its eigenvalues $\lambda_1, \ldots, \lambda_n$ are all nonnegative, then by Theorem 7.3.1 there exists a change of variable $\mathbf{y} = P\mathbf{x}$ for which $\mathbf{x}^T A\mathbf{x} = \lambda_1 y_1^2 + \cdots + \lambda_n y_n^2$. The right hand side is always nonnegative, consequently $\mathbf{x}^T A\mathbf{x} \geq 0$ for all $\mathbf{x}$ in R^n.

True-False Exercises

(a) True. This follows from part (a) of Theorem 7.3.2 and from the margin note next to Definition 1.

(b) False. The term $4x_1x_2x_3$ cannot be included.

(c) True. One can rewrite $(x_1 - 3x_2)^2 = x_1^2 - 6x_1x_2 + 9x_2^2$.

(d) True. None of the eigenvalues will be 0.

(e) False. A symmetric matrix can also be positive semidefinite or negative semidefinite.

(f) True.

(g) True. $\mathbf{x} \cdot \mathbf{x} = x_1^2 + x_2^2 + \cdots + x_n^2$

(h) True. Eigenvalues of A^{-1} are reciprocals of eigenvalues of A. Therefore if all eigenvalues of A are positive, the same is true for all eigenvalues of A^{-1}.

(i) True.

(j) True. This follows from part (a) of Theorem 7.3.4.

(k) True.

(l) False. If $c < 0$, $\mathbf{x}^T A\mathbf{x} = c$ has no graph.

7.4 Optimization Using Quadratic Forms

1. We express the quadratic form in the matrix notation $z = 5x^2 - y^2 = \mathbf{x}^T A\mathbf{x} = [x \quad y]\begin{bmatrix} 5 & 0 \\ 0 & -1 \end{bmatrix}\begin{bmatrix} x \\ y \end{bmatrix}$.

$\det(\lambda I - A) = \begin{vmatrix} \lambda - 5 & 0 \\ 0 & \lambda + 1 \end{vmatrix} = (\lambda - 5)(\lambda + 1)$ therefore the eigenvalues of A are $\lambda = 5$ and $\lambda = -1$.

The reduced row echelon form of $5I - A$ is $\begin{bmatrix} 0 & 1 \\ 0 & 0 \end{bmatrix}$ so that the eigenspace corresponding to $\lambda = 5$ consists of vectors $\begin{bmatrix} x \\ y \end{bmatrix}$ where $x = t, y = 0$. A vector $\begin{bmatrix} 1 \\ 0 \end{bmatrix}$ forms a basis for this eigenspace - this vector is already normalized.

The reduced row echelon form of $-1I - A$ is $\begin{bmatrix} 1 & 0 \\ 0 & 0 \end{bmatrix}$ so that the eigenspace corresponding to $\lambda = -1$ consists of vectors $\begin{bmatrix} x \\ y \end{bmatrix}$ where $x = 0, y = t$. A vector $\begin{bmatrix} 0 \\ 1 \end{bmatrix}$ forms a basis for this eigenspace - this vector is already normalized.

We conclude that the constrained extrema are

- constrained maximum: $z = 5$ at $(x, y) = (\pm 1, 0)$;
- constrained minimum: $z = -1$ at $(x, y) = (0, \pm 1)$.

3. We express the quadratic form in the matrix notation $z = 3x^2 + 7y^2 = \mathbf{x}^T A \mathbf{x} = \begin{bmatrix} x & y \end{bmatrix} \begin{bmatrix} 3 & 0 \\ 0 & 7 \end{bmatrix} \begin{bmatrix} x \\ y \end{bmatrix}$.

$\det(\lambda I - A) = \begin{vmatrix} \lambda - 3 & 0 \\ 0 & \lambda - 7 \end{vmatrix} = (\lambda - 3)(\lambda - 7)$ therefore the eigenvalues of A are $\lambda = 3$ and $\lambda = 7$.

The reduced row echelon form of $3I - A$ is $\begin{bmatrix} 0 & 1 \\ 0 & 0 \end{bmatrix}$ so that the eigenspace corresponding to $\lambda = 3$ consists of vectors $\begin{bmatrix} x \\ y \end{bmatrix}$ where $x = t, y = 0$. A vector $\begin{bmatrix} 1 \\ 0 \end{bmatrix}$ forms a basis for this eigenspace - this vector is already normalized.

The reduced row echelon form of $7I - A$ is $\begin{bmatrix} 1 & 0 \\ 0 & 0 \end{bmatrix}$ so that the eigenspace corresponding to $\lambda = 7$ consists of vectors $\begin{bmatrix} x \\ y \end{bmatrix}$ where $x = 0, y = t$. A vector $\begin{bmatrix} 0 \\ 1 \end{bmatrix}$ forms a basis for this eigenspace - this vector is already normalized.

We conclude that the constrained extrema are

- constrained maximum: $z = 7$ at $(x, y) = (0, \pm 1)$;
- constrained minimum: $z = 3$ at $(x, y) = (\pm 1, 0)$.

5. We express the quadratic form in the matrix notation

$w = 9x^2 + 4y^2 + 3z^2 = \mathbf{x}^T A \mathbf{x} = \begin{bmatrix} x & y & z \end{bmatrix} \begin{bmatrix} 9 & 0 & 0 \\ 0 & 4 & 0 \\ 0 & 0 & 3 \end{bmatrix} \begin{bmatrix} x \\ y \\ z \end{bmatrix}$.

$\det(\lambda I - A) = \begin{vmatrix} \lambda - 9 & 0 & 0 \\ 0 & \lambda - 4 & 0 \\ 0 & 0 & \lambda - 3 \end{vmatrix} = (\lambda - 3)(\lambda - 4)(\lambda - 9)$ therefore the eigenvalues of A are $\lambda = 3, \lambda = 4,$ and $\lambda = 9$.

The reduced row echelon form of $3I - A$ is $\begin{bmatrix} 1 & 0 & 0 \\ 0 & 1 & 0 \\ 0 & 0 & 0 \end{bmatrix}$ so that the eigenspace corresponding to $\lambda = 3$ consists of vectors $\begin{bmatrix} x \\ y \\ z \end{bmatrix}$ where $x = 0, y = 0, z = t$. A vector $\begin{bmatrix} 0 \\ 0 \\ 1 \end{bmatrix}$ forms a basis for this eigenspace - this vector is already normalized.

The reduced row echelon form of $9I - A$ is $\begin{bmatrix} 0 & 1 & 0 \\ 0 & 0 & 1 \\ 0 & 0 & 0 \end{bmatrix}$ so that the eigenspace corresponding to $\lambda = 9$

consists of vectors $\begin{bmatrix} x \\ y \\ z \end{bmatrix}$ where $x = t, y = 0, z = 0$. A vector $\begin{bmatrix} 1 \\ 0 \\ 0 \end{bmatrix}$ forms a basis for this eigenspace eigenspace - this vector is already normalized.

We conclude that the constrained extrema are

- constrained maximum: $w = 9$ at $(x, y, z) = (\pm 1, 0, 0)$;
- constrained minimum: $w = 3$ at $(x, y, z) = (0, 0, \pm 1)$.

7. The constraint $4x^2 + 8y^2 = 16$ can be rewritten as $\left(\frac{x}{2}\right)^2 + \left(\frac{y}{\sqrt{2}}\right)^2 = 1$. We define new variables x_1 and y_1 by $x = 2x_1$ and $y = \sqrt{2}y_1$. Our problem can now be reformulated to find maximum and minimum value of $2\sqrt{2}x_1y_1 = [x_1 \quad y_1] \underbrace{\begin{bmatrix} 0 & \sqrt{2} \\ \sqrt{2} & 0 \end{bmatrix}}_{A} \begin{bmatrix} x_1 \\ y_1 \end{bmatrix}$ subject to the constraint $x_1^2 + y_1^2 = 1$. We have

$\det(\lambda I - A) = \begin{vmatrix} \lambda & -\sqrt{2} \\ -\sqrt{2} & \lambda \end{vmatrix} = \lambda^2 - 2 = (\lambda - \sqrt{2})(\lambda + \sqrt{2})$ thus A has eigenvalues $\pm\sqrt{2}$.

The reduced row echelon form of $\sqrt{2}I - A$ is $\begin{bmatrix} 1 & -1 \\ 0 & 0 \end{bmatrix}$ so that the eigenspace corresponding to $\lambda = \sqrt{2}$ consists of vectors $\begin{bmatrix} x_1 \\ y_1 \end{bmatrix}$ where $x_1 = t, y_1 = t$. A vector $\begin{bmatrix} 1 \\ 1 \end{bmatrix}$ forms a basis for this eigenspace. A normalized eigenvector in this eigenspace is $\begin{bmatrix} \frac{1}{\sqrt{2}} \\ \frac{1}{\sqrt{2}} \end{bmatrix}$. In terms of the original variables, this corresponds to $x = 2x_1 = \sqrt{2}$ and $y = \sqrt{2}y_1 = 1$.

The reduced row echelon form of $-\sqrt{2}I - A$ is $\begin{bmatrix} 1 & 1 \\ 0 & 0 \end{bmatrix}$ so that the eigenspace corresponding to $\lambda = -\sqrt{2}$ consists of vectors $\begin{bmatrix} x_1 \\ y_1 \end{bmatrix}$ where $x_1 = -t, y_1 = t$. A vector $\begin{bmatrix} -1 \\ 1 \end{bmatrix}$ forms a basis for this eigenspace. A normalized eigenvector in this eigenspace is $\begin{bmatrix} -\frac{1}{\sqrt{2}} \\ \frac{1}{\sqrt{2}} \end{bmatrix}$. In terms of the original variables, this corresponds to $x = 2x_1 = -\sqrt{2}$ and $y = \sqrt{2}y_1 = 1$.

We conclude that the constrained extrema are

- constrained maximum value: $\sqrt{2}$ at $(x, y) = (\sqrt{2}, 1)$ and $(x, y) = (-\sqrt{2}, -1)$;
- constrained minimum value: $-\sqrt{2}$ at $(x, y) = (-\sqrt{2}, 1)$ and $(x, y) = (\sqrt{2}, -1)$.

9. The following illustration indicates positions of constrained extrema consistent with the solution that was obtained for Exercise 1.

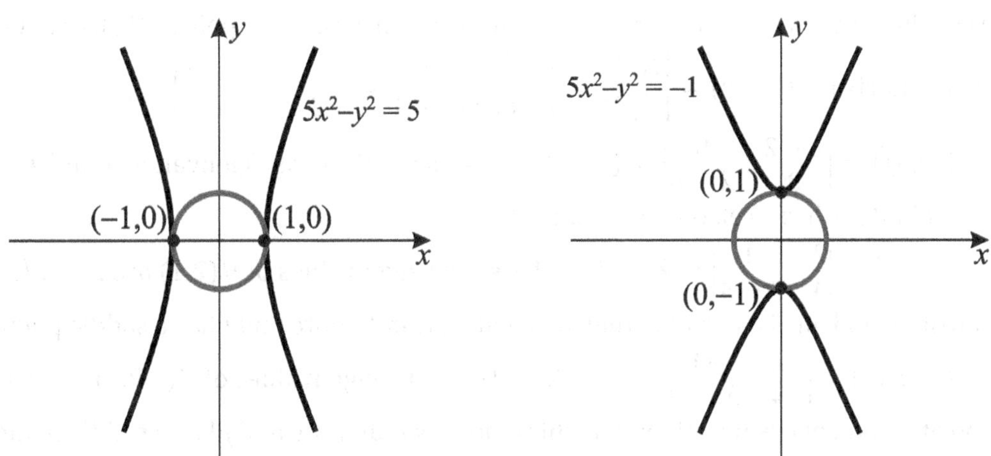

11. **(a)** The first partial derivatives of $f(x,y)$ are $f_x(x,y) = 4y - 4x^3$ and $f_y(x,y) = 4x - 4y^3$. Since $f_x(0,0) = f_y(0,0) = 0$, $f_x(1,1) = f_y(1,1) = 0$, and $f_x(-1,-1) = f_y(-1,-1) = 0$, f has critical points at (0,0), (1,1), and $(-1,-1)$.

(b) The second partial derivatives of $f(x,y)$ are $f_{xx}(x,y) = -12x^2$, $f_{xy}(x,y) = 4$, and $f_{yy}(x,y) = -12y^2$ therefore the Hessian matrix of f is $H(x,y) = \begin{bmatrix} -12x^2 & 4 \\ 4 & -12y^2 \end{bmatrix}$.

$\det(\lambda I - H(0,0)) = \begin{vmatrix} \lambda & -4 \\ -4 & \lambda \end{vmatrix} = (\lambda - 4)(\lambda + 4)$ so $H(0,0)$ has eigenvalues -4 and 4; since $H(0,0)$ is indefinite, f has a saddle point at (0,0);

$\det(\lambda I - H(1,1)) = \begin{vmatrix} \lambda + 12 & -4 \\ -4 & \lambda + 12 \end{vmatrix} = (\lambda + 8)(\lambda + 16)$ so $H(1,1)$ has eigenvalues -8 and -16; since $H(1,1)$ is negative definite, f has a relative maximum at (1,1);

$\det(\lambda I - H(-1,-1)) = \begin{vmatrix} \lambda + 12 & -4 \\ -4 & \lambda + 12 \end{vmatrix} = (\lambda + 8)(\lambda + 16)$ so $H(-1,-1)$ has eigenvalues -8 and -16; since $H(-1,-1)$ is negative definite, f has a relative maximum at $(-1,-1)$

13. The first partial derivatives of f are $f_x(x,\ y) = 3x^2 - 3y$ and $f_y(x,\ y) = -3x - 3y^2$. To find the critical points we set f_x and f_y equal to zero. This yields the equations $y = x^2$ and $x = -y^2$. From this we conclude that $y = y^4$ and so $y = 0$ or $y = 1$. The corresponding values of x are $x = 0$ and $x = -1$ respectively. Thus there are two critical points: (0, 0) and (−1, 1).

The Hessian matrix is $H(x,\ y) = \begin{bmatrix} f_{xx}(x,\ y) & f_{xy}(x,\ y) \\ f_{yx}(x,\ y) & f_{yy}(x,\ y) \end{bmatrix} = \begin{bmatrix} 6x & -3 \\ -3 & -6y \end{bmatrix}$.

$\det(\lambda I - H(0,0)) = \begin{vmatrix} \lambda & 3 \\ 3 & \lambda \end{vmatrix} = (\lambda - 3)(\lambda + 3)$ so $H(0,0)$ has eigenvalues -3 and 3; since $H(0,0)$ is indefinite, f has a saddle point at (0,0);

$\det(\lambda I - H(-1,1)) = \begin{vmatrix} \lambda + 6 & 3 \\ 3 & \lambda + 6 \end{vmatrix} = (\lambda + 3)(\lambda + 9)$ so $H(-1,1)$ has eigenvalues -3 and -9; since $H(-1,1)$ is negative definite, f has a relative maximum at $(-1,1)$.

15. The first partial derivatives of f are $f_x(x, y) = 2x - 2xy$ and $f_y(x, y) = 4y - x^2$. To find the critical points we set f_x and f_y equal to zero. This yields the equations $2x(1-y) = 0$ and $y = \frac{1}{4}x^2$. From the first, we conclude that $x = 0$ or $y = 1$. Thus there are three critical points: (0, 0), (2, 1), and (−2, 1).

The Hessian matrix is $H(x, y) = \begin{bmatrix} f_{xx}(x, y) & f_{xy}(x, y) \\ f_{yx}(x, y) & f_{yy}(x, y) \end{bmatrix} = \begin{bmatrix} 2-2y & -2x \\ -2x & 4 \end{bmatrix}$.

$\det(\lambda I - H(0,0)) = \begin{vmatrix} \lambda - 2 & 0 \\ 0 & \lambda - 4 \end{vmatrix} = (\lambda - 2)(\lambda - 4)$ so $H(0,0)$ has eigenvalues 2 and 4; since $H(0,0)$ is positive definite, f has a relative minimum at (0,0).

$\det(\lambda I - H(2,1)) = \begin{vmatrix} \lambda & 4 \\ 4 & \lambda - 4 \end{vmatrix} = \lambda^2 - 4\lambda - 16$ so the eigenvalues of $H(2,1)$ are $2 \pm 2\sqrt{5}$. One of these is positive and one is negative; thus this matrix is indefinite and f has a saddle point at (2, 1).

$\det(\lambda I - H(-2,1)) = \begin{vmatrix} \lambda & -4 \\ -4 & \lambda - 4 \end{vmatrix} = \lambda^2 - 4\lambda - 16$ so the eigenvalues of $H(-2,1)$ are $2 \pm 2\sqrt{5}$. One of these is positive and one is negative; thus this matrix is indefinite and f has a saddle point at (−2, 1).

17. The problem is to maximize $z = 4xy$ subject to $x^2 + 25y^2 = 25$, or $\left(\frac{x}{5}\right)^2 + \left(\frac{y}{1}\right)^2 = 1$.

Let $x = 5x_1$ and $y = y_1$, so that the problem is to maximize $z = 20x_1y_1$ subject to $\|(x_1, y_1)\| = 1$.

Write $z = \mathbf{x}^T A\mathbf{x} = [x_1 \quad y_1]\begin{bmatrix} 0 & 10 \\ 10 & 0 \end{bmatrix}\begin{bmatrix} x_1 \\ y_1 \end{bmatrix}$.

$$\begin{vmatrix} \lambda & -10 \\ -10 & \lambda \end{vmatrix} = \lambda^2 - 100 = (\lambda + 10)(\lambda - 10).$$

The largest eigenvalue of A is $\lambda = 10$ which has positive unit eigenvector $\begin{bmatrix} \frac{1}{\sqrt{2}} \\ \frac{1}{\sqrt{2}} \end{bmatrix}$. Thus the maximum value of $z = 20\left(\frac{1}{\sqrt{2}}\right)\left(\frac{1}{\sqrt{2}}\right) = 10$ which occurs when $x = 5x_1 = \frac{5}{\sqrt{2}}$ and $y = y_1 = \frac{1}{\sqrt{2}}$, which are the coordinates of one of the corner points of the rectangle.

19. **(a)** The first partial derivatives of $f(x,y)$ are $f_x(x,y) = 4x^3$ and $f_y(x,y) = 4y^3$.
Since $f_x(0,0) = f_y(0,0) = 0$, f has a critical point at (0,0).
The second partial derivatives of $f(x,y)$ are $f_{xx}(x,y) = 12x^2$, $f_{xy}(x,y) = 0$, and $f_{yy}(x,y) = 12y^2$. We have $f_{xx}(0,0)f_{yy}(0,0) - f_{xy}^2(0,0) = 0$ therefore the second derivative test is inconclusive.

The first partial derivatives of $g(x,y)$ are $g_x(x,y) = 4x^3$ and $g_y(x,y) = -4y^3$.
Since $g_x(0,0) = g_y(0,0) = 0$, g has a critical point at (0,0).
The second partial derivatives of $g(x,y)$ are $g_{xx}(x,y) = 12x^2$, $g_{xy}(x,y) = 0$, and $g_{yy}(x,y) = -12y^2$. We have $g_{xx}(0,0)g_{yy}(0,0) - g_{xy}^2(0,0) = 0$ therefore the second derivative test is inconclusive.

(b) Clearly, for all $(x,y) \neq (0,0)$, $f(x,y) > f(0,0) = 0$ therefore f has a relative minimum at (0,0).

For all $x \neq 0$, $g(x,0) > g(0,0) = 0$; however, for all $y \neq 0$, $g(0,y) < g(0,0) = 0$ - consequently, g has a saddle point at (0,0).

21. $\mathbf{x}$ is a unit eigenvector corresponding to λ, then $q(\mathbf{x}) = \mathbf{x}^T A\mathbf{x} = \mathbf{x}^T(\lambda\mathbf{x}) = \lambda(\mathbf{x}^T\mathbf{x}) = \lambda(1) = \lambda$.

True-False Exercises

(a) False. If the only critical point of the quadratic form is a saddle point, then it will have neither a maximum nor a minimum value.

(b) True. This follows from part (b) of Theorem 7.4.1.

(c) True.

(d) False. The second derivative test is inconclusive in this case.

(e) True. If $\det(A) < 0$, then A will have a negative eigenvalue.

7.5 Hermitian, Unitary, and Normal Matrices

1. $\overline{A} = \begin{bmatrix} -2i & 1+i \\ 4 & 3-i \\ 5-i & 0 \end{bmatrix}$ therefore $A^* = \overline{A}^T = \begin{bmatrix} -2i & 4 & 5-i \\ 1+i & 3-i & 0 \end{bmatrix}$

3. $A = \begin{bmatrix} 1 & i & 2-3i \\ -i & -3 & 1 \\ 2+3i & 1 & 2 \end{bmatrix}$

5. **(a)** $(A)_{13} = 2-3i$ does not equal $(A^*)_{13} = 2+3i$

(b) $(A)_{22} = i$ does not equal $(A^*)_{22} = -i$

7. $\det(\lambda I - A) = \begin{vmatrix} \lambda-3 & -2+3i \\ -2-3i & \lambda+1 \end{vmatrix} = \lambda^2 - 2\lambda - 16 = \left(\lambda - (1+\sqrt{17})\right)\left(\lambda - (1-\sqrt{17})\right)$ so A has real eigenvalues $1+\sqrt{17}$ and $1-\sqrt{17}$.

For the eigenvalue $\lambda = 1+\sqrt{17}$, the augmented matrix of the homogeneous system $\left((1+\sqrt{17})I - A\right)\mathbf{x} = \mathbf{0}$ is $\begin{bmatrix} -2+\sqrt{17} & -2+3i & 0 \\ -2-3i & 2+\sqrt{17} & 0 \end{bmatrix}$. The rows of this matrix must be scalar multiples of each other (see Example 3 in Section 5.3) therefore it suffices to solve the equation corresponding to the second row, which yields $x_1 + \frac{2+\sqrt{17}}{13}(-2+3i)x_2 = 0$. The general solution of this equation (and, consequently, of the entire system) is $x_1 = \frac{2+\sqrt{17}}{13}(2-3i)t$, $x_2 = t$. The vector $\mathbf{v}_1 = \begin{bmatrix} \frac{2+\sqrt{17}}{13}(2-3i) \\ 1 \end{bmatrix}$ forms a basis for the eigenspace corresponding to $\lambda = 1+\sqrt{17}$.

For the eigenvalue $\lambda = 1-\sqrt{17}$, the augmented matrix of the homogeneous system $\left((1-\sqrt{17})I - A\right)\mathbf{x} = \mathbf{0}$ is $\begin{bmatrix} -2-\sqrt{17} & -2+3i & 0 \\ -2-3i & 2-\sqrt{17} & 0 \end{bmatrix}$.

As before, this yields $x_1 + \frac{2-\sqrt{17}}{13}(-2+3i)x_2 = 0$. The general solution of this equation (and, consequently, of the entire system) is $x_1 = \frac{2-\sqrt{17}}{13}(2-3i)t$, $x_2 = t$. The vector $\mathbf{v}_2 = \begin{bmatrix} \frac{2-\sqrt{17}}{13}(2-3i) \\ 1 \end{bmatrix}$ forms a basis for the eigenspace corresponding to $\lambda = 1-\sqrt{17}$.

We have

$\mathbf{v}_1 \cdot \mathbf{v}_2 = \left(\frac{2+\sqrt{17}}{13}(2-3i)\right)\left(\overline{\frac{2-\sqrt{17}}{13}(2-3i)}\right) + (1)(\overline{1}) = \left(\frac{2+\sqrt{17}}{13}(2-3i)\right)\left(\frac{2-\sqrt{17}}{13}(2+3i)\right) + (1)(1)$
$= \frac{(2+\sqrt{17})(2-\sqrt{17})}{13^2}(2-3i)(2+3i) + 1 = \frac{4-17}{13^2}(4+9) + 1 = -1 + 1 = 0$ therefore the eigenvectors from different eigenspaces are orthogonal.

9. The following computations show that the row vectors of A are orthonormal:

$\|\mathbf{r}_1\| = \sqrt{\left|\frac{3}{5}\right|^2 + \left|\frac{4}{5}i\right|^2} = \sqrt{\frac{9}{25} + \frac{16}{25}} = 1$; $\|\mathbf{r}_2\| = \sqrt{\left|-\frac{4}{5}\right|^2 + \left|\frac{3}{5}i\right|^2} = \sqrt{\frac{16}{25} + \frac{9}{25}} = 1$;

$\mathbf{r}_1 \cdot \mathbf{r}_2 = \left(\frac{3}{5}\right)\left(-\frac{4}{5}\right) + \left(\frac{4}{5}i\right)\left(-\frac{3}{5}i\right) = -\frac{12}{5} + \frac{12}{5} = 0$

By Theorem 7.5.3, A is unitary, and $A^{-1} = A^* = \begin{bmatrix} \frac{3}{5} & -\frac{4}{5} \\ -\frac{4}{5}i & -\frac{3}{5}i \end{bmatrix}$.

11. The following computations show that the column vectors of A are orthonormal:

$\|\mathbf{c}_1\| = \sqrt{\left|\frac{1}{2\sqrt{2}}(\sqrt{3}+i)\right|^2 + \left|\frac{1}{2\sqrt{2}}(1+i\sqrt{3})\right|^2} = \sqrt{\frac{4}{8} + \frac{4}{8}} = 1$;

$\|\mathbf{c}_2\| = \sqrt{\left|\frac{1}{2\sqrt{2}}(1-i\sqrt{3})\right|^2 + \left|\frac{1}{2\sqrt{2}}(i-\sqrt{3})\right|^2} = \sqrt{\frac{4}{8} + \frac{4}{8}} = 1$;

$\mathbf{c}_1 \cdot \mathbf{c}_2 = \frac{1}{2\sqrt{2}}(\sqrt{3}+i)\frac{1}{2\sqrt{2}}(1+i\sqrt{3}) + \frac{1}{2\sqrt{2}}(1+i\sqrt{3})\frac{1}{2\sqrt{2}}(-i-\sqrt{3}) = 0$

By Theorem 7.5.3, A is unitary, therefore $A^{-1} = A^* = \begin{bmatrix} \frac{1}{2\sqrt{2}}(\sqrt{3}-i) & \frac{1}{2\sqrt{2}}(1-i\sqrt{3}) \\ \frac{1}{2\sqrt{2}}(1+i\sqrt{3}) & \frac{1}{2\sqrt{2}}(-i-\sqrt{3}) \end{bmatrix}$.

13. $\det(\lambda I - A) = \begin{vmatrix} \lambda-4 & -1+i \\ -1-i & \lambda-5 \end{vmatrix} = (\lambda-3)(\lambda-6)$ thus A has eigenvalues $\lambda = 3$ and $\lambda = 6$.

The reduced row echelon form of $3I - A$ is $\begin{bmatrix} 1 & 1-i \\ 0 & 0 \end{bmatrix}$ so that the eigenspace corresponding to $\lambda = 3$ consists of vectors $\begin{bmatrix} x \\ y \end{bmatrix}$ where $x = (-1+i)t$, $y = t$. A vector $\begin{bmatrix} -1+i \\ 1 \end{bmatrix}$ forms a basis for this eigenspace.

The reduced row echelon form of $6I - A$ is $\begin{bmatrix} 1 & -\frac{1}{2}+\frac{1}{2}i \\ 0 & 0 \end{bmatrix}$ so that the eigenspace corresponding to $\lambda = 6$ consists of vectors $\begin{bmatrix} x \\ y \end{bmatrix}$ where $x = \left(\frac{1}{2} - \frac{1}{2}i\right)t$, $y = t$. A vector $\begin{bmatrix} 1-i \\ 2 \end{bmatrix}$ forms a basis for this eigenspace.

Applying the Gram-Schmidt process to both bases amounts to simply normalizing the respective vectors.

Therefore A is unitarily diagonalized by $P = \begin{bmatrix} \frac{-1+i}{\sqrt{3}} & \frac{1-i}{\sqrt{6}} \\ \frac{1}{\sqrt{3}} & \frac{2}{\sqrt{6}} \end{bmatrix}$. Since P is unitary, $P^{-1} = P^* = \begin{bmatrix} \frac{-1-i}{\sqrt{3}} & \frac{1}{\sqrt{3}} \\ \frac{1+i}{\sqrt{6}} & \frac{2}{\sqrt{6}} \end{bmatrix}$. It follows that $P^{-1}AP = \begin{bmatrix} \frac{-1-i}{\sqrt{3}} & \frac{1}{\sqrt{3}} \\ \frac{1+i}{\sqrt{6}} & \frac{2}{\sqrt{6}} \end{bmatrix} \begin{bmatrix} 4 & 1-i \\ 1+i & 5 \end{bmatrix} \begin{bmatrix} \frac{-1+i}{\sqrt{3}} & \frac{1-i}{\sqrt{6}} \\ \frac{1}{\sqrt{3}} & \frac{2}{\sqrt{6}} \end{bmatrix} = \begin{bmatrix} 3 & 0 \\ 0 & 6 \end{bmatrix}$.

15. $\det(\lambda I - A) = \begin{vmatrix} \lambda-6 & -2-2i \\ -2+2i & \lambda-4 \end{vmatrix} = (\lambda-2)(\lambda-8)$ thus A has eigenvalues $\lambda = 2$ and $\lambda = 8$.

The reduced row echelon form of $2I - A$ is $\begin{bmatrix} 1 & \frac{1}{2}+\frac{1}{2}i \\ 0 & 0 \end{bmatrix}$ so that the eigenspace corresponding to $\lambda = 2$ consists of vectors $\begin{bmatrix} x \\ y \end{bmatrix}$ where $x = (-\frac{1}{2}-\frac{1}{2}i)t, y = t$. A vector $\begin{bmatrix} -1-i \\ 2 \end{bmatrix}$ forms a basis for this eigenspace.

The reduced row echelon form of $8I - A$ is $\begin{bmatrix} 1 & -1-i \\ 0 & 0 \end{bmatrix}$ so that the eigenspace corresponding to $\lambda = 8$ consists of vectors $\begin{bmatrix} x \\ y \end{bmatrix}$ where $x = (1+i)t, y = t$. A vector $\begin{bmatrix} 1+i \\ 1 \end{bmatrix}$ forms a basis for this eigenspace.

Applying the Gram-Schmidt process to both bases amounts to simply normalizing the respective vectors.

Therefore A is unitarily diagonalized by $P = \begin{bmatrix} \frac{-1-i}{\sqrt{6}} & \frac{1+i}{\sqrt{3}} \\ \frac{2}{\sqrt{6}} & \frac{1}{\sqrt{3}} \end{bmatrix}$. Since P is unitary, $P^{-1} = P^* = \begin{bmatrix} \frac{-1+i}{\sqrt{6}} & \frac{2}{\sqrt{6}} \\ \frac{1-i}{\sqrt{3}} & \frac{1}{\sqrt{3}} \end{bmatrix}$. It follows that $P^{-1}AP = \begin{bmatrix} \frac{-1+i}{\sqrt{6}} & \frac{2}{\sqrt{6}} \\ \frac{1-i}{\sqrt{3}} & \frac{1}{\sqrt{3}} \end{bmatrix} \begin{bmatrix} 6 & 2+2i \\ 2-1i & 4 \end{bmatrix} \begin{bmatrix} \frac{-1-i}{\sqrt{6}} & \frac{1+i}{\sqrt{3}} \\ \frac{2}{\sqrt{6}} & \frac{1}{\sqrt{3}} \end{bmatrix} = \begin{bmatrix} 2 & 0 \\ 0 & 8 \end{bmatrix}$.

17. The characteristic polynomial of A is $(\lambda-5)(\lambda^2+\lambda-2) = (\lambda+2)(\lambda-1)(\lambda-5)$; thus the eigenvalues of A are $\lambda_1 = -2$, $\lambda_2 = 1$, and $\lambda_3 = 5$. The augmented matrix of the system $(-2I - A)\mathbf{x} = \mathbf{0}$ is $\begin{bmatrix} -7 & 0 & 0 & 0 \\ 0 & -1 & 1-i & 0 \\ 0 & 1+i & -2 & 0 \end{bmatrix}$, which can be reduced to $\begin{bmatrix} 1 & 0 & 0 & 0 \\ 0 & 1 & -1+i & 0 \\ 0 & 0 & 0 & 0 \end{bmatrix}$. Thus $\mathbf{v}_1 = \begin{bmatrix} 0 \\ 1-i \\ 1 \end{bmatrix}$ is a basis for the eigenspace corresponding to $\lambda_1 = -2$, and $\mathbf{p}_1 = \begin{bmatrix} 0 \\ \frac{1-i}{\sqrt{3}} \\ \frac{1}{\sqrt{3}} \end{bmatrix}$ is a unit eigenvector. Similar computations show that $\mathbf{p}_2 = \begin{bmatrix} 0 \\ \frac{-1+i}{\sqrt{6}} \\ \frac{2}{\sqrt{6}} \end{bmatrix}$ is a unit eigenvector corresponding to $\lambda_2 = 1$, and $\mathbf{p}_3 = \begin{bmatrix} 1 \\ 0 \\ 0 \end{bmatrix}$ is a unit eigenvector corresponding to $\lambda_3 = 5$. The vectors $\{\mathbf{p}_1, \mathbf{p}_2, \mathbf{p}_3\}$ form an orthogonal set, and the unitary matrix $P = [\mathbf{p}_1 \quad \mathbf{p}_2 \quad \mathbf{p}_3]$ diagonalizes the matrix A:

$$P^*AP = \begin{bmatrix} 0 & \frac{1+i}{\sqrt{3}} & \frac{1}{\sqrt{3}} \\ 0 & \frac{-1-i}{\sqrt{6}} & \frac{2}{\sqrt{6}} \\ 1 & 0 & 0 \end{bmatrix} \begin{bmatrix} 5 & 0 & 0 \\ 0 & -1 & -1+i \\ 0 & -1-i & 0 \end{bmatrix} \begin{bmatrix} 0 & 0 & 1 \\ \frac{1-i}{\sqrt{3}} & \frac{-1+i}{\sqrt{6}} & 0 \\ \frac{1}{\sqrt{3}} & \frac{2}{\sqrt{6}} & 0 \end{bmatrix} = \begin{bmatrix} -2 & 0 & 0 \\ 0 & 1 & 0 \\ 0 & 0 & 5 \end{bmatrix}.$$

19. $A = \begin{bmatrix} 0 & i & 2-3i \\ i & 0 & 1 \\ -2-3i & -1 & 4i \end{bmatrix}$

21. **(a)** $(-A)_{12} = -i$ does not equal $(A^*)_{12} = i$;
also, $(-A)_{13} = -2+3i$ does not equal $(A^*)_{13} = 2-3i$

(b) $(-A)_{11} = -1$ does not equal $(A^*)_{11} = 1$;
also, $(-A)_{13} = -3 + 5i$ does not equal $(A^*)_{13} = -3 - 5i$ and
$(-A)_{23} = i$ does not equal $(A^*)_{23} = -i$.

23. $\det(\lambda I - A) - \begin{bmatrix} \lambda & 1-i \\ -1+i & \lambda - i \end{bmatrix} = \lambda^2 - i\lambda + 2 = (\lambda - 2i)(\lambda + i)$; thus the eigenvalues of A, $\lambda = 2i$ and $\lambda = -i$, are pure imaginary numbers.

25. $A^* = \begin{bmatrix} 1-2i & 2-i & -2+i \\ 2-i & 1-i & i \\ -2+i & i & 1-i \end{bmatrix}$; we have $AA^* = A^*A = \begin{bmatrix} 15 & 8 & -8 \\ 8 & 8 & -7 \\ -8 & -7 & 8 \end{bmatrix}$

27. **(a)** If $B = \frac{1}{2}(A + A^*)$, then $B^* = \frac{1}{2}(A + A^*)^* = \frac{1}{2}(A^* + A^{**}) = \frac{1}{2}(A^* + A) = B$. Similarly, $C^* = C$.

(b) We have $B + iC = \frac{1}{2}(A + A^*) + \frac{1}{2}(A - A^*) = A$ and $B - iC = \frac{1}{2}(A + A^*) - \frac{1}{2}(A - A^*) = A^*$.

(c) $AA^* = (B + iC)(B - iC) = B^2 - iBC + iCB + C^2$ and $A^*A = B^2 + iBC - iCB + C^2$.
Thus $AA^* = A^*A$ if and only if $-iBC + iCB = iBC - iCB$, or $2iCB = 2iBC$.
Thus A is normal if and only if B and C commute i.e., $CB = BC$.

31. $A\mathbf{x} = \begin{bmatrix} \frac{7}{5} + \frac{11}{5}i \\ -\frac{1}{5} + \frac{2}{5}i \end{bmatrix}$; $\|A\mathbf{x}\| = \sqrt{\left|\frac{7}{5} + \frac{11}{5}i\right|^2 + \left|-\frac{1}{5} + \frac{2}{5}i\right|^2} = \sqrt{\frac{49}{25} + \frac{121}{25} + \frac{1}{25} + \frac{4}{25}} = \sqrt{7}$ equals
$\|\mathbf{x}\| = \sqrt{|1+i|^2 + |2-i|^2} = \sqrt{1+1+4+1} = \sqrt{7}$ which verifies part (b);
$A\mathbf{y} = \begin{bmatrix} \frac{7}{5} + \frac{4}{5}i \\ -\frac{1}{5} + \frac{3}{5}i \end{bmatrix}$; $A\mathbf{x} \cdot A\mathbf{y} = \left(\frac{7}{5} + \frac{11}{5}i\right)\overline{\left(\frac{7}{5} + \frac{4}{5}i\right)} + \left(-\frac{1}{5} + \frac{2}{5}i\right)\overline{\left(-\frac{1}{5} + \frac{3}{5}i\right)}$
$= \left(\frac{7}{5} + \frac{11}{5}i\right)\left(\frac{7}{5} - \frac{4}{5}i\right) + \left(-\frac{1}{5} + \frac{2}{5}i\right)\left(-\frac{1}{5} - \frac{3}{5}i\right) = \left(\frac{93}{25} + \frac{49}{25}i\right) + \left(\frac{7}{25} + \frac{1}{25}i\right) = 4 + 2i$ equals
$\mathbf{x} \cdot \mathbf{y} = (1+i)\overline{(1)} + (2-i)\overline{(1-i)} = (1+i)(1) + (2-i)(1+i) = (1+i) + (3+i) = 4 + 2i$ which verifies part (c).

33. $A^* = \begin{bmatrix} \bar{a} & 0 & 0 \\ 0 & 0 & \bar{b} \\ 0 & \bar{c} & 0 \end{bmatrix}$; $AA^* = \begin{bmatrix} a\bar{a} & 0 & 0 \\ 0 & c\bar{c} & 0 \\ 0 & 0 & b\bar{b} \end{bmatrix} = \begin{bmatrix} |a|^2 & 0 & 0 \\ 0 & |c|^2 & 0 \\ 0 & 0 & |b|^2 \end{bmatrix}$;
$A^*A = \begin{bmatrix} a\bar{a} & 0 & 0 \\ 0 & b\bar{b} & 0 \\ 0 & 0 & c\bar{c} \end{bmatrix} = \begin{bmatrix} |a|^2 & 0 & 0 \\ 0 & |b|^2 & 0 \\ 0 & 0 & |c|^2 \end{bmatrix}$
A is normal if and only if $|b| = |c|$.

35. $A = \begin{bmatrix} \frac{1}{\sqrt{2}} & -\frac{i}{\sqrt{2}} \\ \frac{i}{\sqrt{2}} & -\frac{1}{\sqrt{2}} \end{bmatrix}$ is both Hermitian and unitary.

37. Part (a): $(A^*)^* \underset{\text{Def.1}}{=} \left(\overline{\overline{A}^T}\right)^T \underset{\text{Th. 5.3.2(b)}}{=} \left(\overline{\overline{A^T}}\right)^T \underset{\text{Th. 5.3.2(a)}}{=} (A^T)^T = A$

Part (e): $(AB)^* \underset{\text{Def.1}}{=} \left(\overline{AB}\right)^T \underset{\text{Th. 5.3.2(c)}}{=} \left(\overline{A}\,\overline{B}\right)^T = \left(\overline{B}\right)^T\left(\overline{A}\right)^T \underset{\text{Def.1}}{=} B^*A^*$

39. If A is unitary, then $A^{-1} = A^*$ and so $(A^*)^{-1} = (A^{-1})^* = (A^*)^*$; thus A^* is also unitary.

41. A unitary matrix A has the property that $\|A\mathbf{x}\| = \|\mathbf{x}\|$ for all x in C^n. Thus if A is unitary and $A\mathbf{x} = \lambda\mathbf{x}$ where $\mathbf{x} \neq \mathbf{0}$, we must have $|\lambda|\|\mathbf{x}\| = \|A\mathbf{x}\| = \|\mathbf{x}\|$ and so $|\lambda| = 1$.

43. If $H = I - 2\mathbf{u}\mathbf{u}^*$, then $H^* = (I - 2\mathbf{u}\mathbf{u}^*)^* = I^* - 2\mathbf{u}^{**}\mathbf{u}^* = I - 2\mathbf{u}\mathbf{u}^* = H$; thus H is Hermitian.
$HH^* = (I - 2\mathbf{u}\mathbf{u}^*)(I - 2\mathbf{u}\mathbf{u}^*) = I - 2\mathbf{u}\mathbf{u}^* - 2\mathbf{u}\mathbf{u}^* + 4\mathbf{u}\mathbf{u}^*\mathbf{u}\mathbf{u}^* = I - 4\mathbf{u}\mathbf{u}^* + 4\mathbf{u}\|\mathbf{u}\|^2\mathbf{u}^* = I$
so H is unitary.

45. **(a)** This result can be obtained by mathematical induction.

(b) $\det(A^*) = \det\left(\left(\overline{A}\right)^T\right) = \det(\overline{A}) = \overline{\det(A)}$.

True-False Exercises

(a) False. Denoting $A = \begin{bmatrix} 0 & i \\ i & 2 \end{bmatrix}$, we observe that $(A)_{12} = i$ does not equal $(A^*)_{12} = -i$.

(b) False. For $\mathbf{r}_1 = \begin{bmatrix} -\frac{i}{\sqrt{2}} & \frac{i}{\sqrt{6}} & \frac{i}{\sqrt{3}} \end{bmatrix}$ and $\mathbf{r}_2 = \begin{bmatrix} 0 & -\frac{i}{\sqrt{6}} & \frac{i}{\sqrt{3}} \end{bmatrix}$,
$\mathbf{r}_1 \cdot \mathbf{r}_2 = -\frac{i}{\sqrt{2}}(\overline{0}) + \frac{i}{\sqrt{6}}\left(\overline{-\frac{i}{\sqrt{6}}}\right) + \frac{i}{\sqrt{3}}\left(\overline{\frac{i}{\sqrt{3}}}\right) = 0 + \left(\frac{i}{\sqrt{6}}\right)^2 - \left(\frac{i}{\sqrt{3}}\right)^2 = -\frac{1}{6} + \frac{1}{3} = \frac{1}{6} \neq 0$
thus the row vectors do not form an orthonormal set and the matrix is not unitary by Theorem 7.5.3.

(c) True. If A is unitary, so $A^{-1} = A^*$, then $(A^*)^{-1} = A = (A^*)^*$.

(d) False. Normal matrices that are not Hermitian are also unitarily diagonalizable.

(e) False. If A is skew-Hermitian, then $(A^2)^* = (A^*)(A^*) = (-A)(-A) = A^2 \neq -A^2$.

Chapter 7 Supplementary Exercises

1. **(a)** For $A = \begin{bmatrix} \frac{3}{5} & -\frac{4}{5} \\ \frac{4}{5} & \frac{3}{5} \end{bmatrix}$, $A^TA = \begin{bmatrix} 1 & 0 \\ 0 & 1 \end{bmatrix}$, so $A^{-1} = A^T = \begin{bmatrix} \frac{3}{5} & \frac{4}{5} \\ -\frac{4}{5} & \frac{3}{5} \end{bmatrix}$.

(b) For $A = \begin{bmatrix} \frac{4}{5} & 0 & -\frac{3}{5} \\ -\frac{9}{25} & \frac{4}{5} & -\frac{12}{25} \\ \frac{12}{25} & \frac{3}{5} & \frac{16}{25} \end{bmatrix}$, $A^TA = \begin{bmatrix} 1 & 0 & 0 \\ 0 & 1 & 0 \\ 0 & 0 & 1 \end{bmatrix}$, so $A^{-1} = A^T = \begin{bmatrix} \frac{4}{5} & -\frac{9}{25} & \frac{12}{25} \\ 0 & \frac{4}{5} & \frac{3}{5} \\ -\frac{3}{5} & -\frac{12}{25} & \frac{16}{25} \end{bmatrix}$.

3. Since A is symmetric, there exists an orthogonal matrix P such that $P^TAP = D = \begin{bmatrix} \lambda_1 & 0 & \cdots & 0 \\ 0 & \lambda_2 & \cdots & 0 \\ \vdots & \vdots & \ddots & \vdots \\ 0 & 0 & \cdots & \lambda_n \end{bmatrix}$.

Since A is positive definite, all λ's must be positive. Let us form a diagonal matrix

$C = \begin{bmatrix} \sqrt{\lambda_1} & 0 & \cdots & 0 \\ 0 & \sqrt{\lambda_2} & \cdots & 0 \\ \vdots & \vdots & \ddots & \vdots \\ 0 & 0 & \cdots & \sqrt{\lambda_n} \end{bmatrix}$. Then $A = PDP^T = PCC^TP^T = (PC)(PC)^T$. The matrix $(PC)^T$ is nonsingular (it is a transpose of a product of two nonsingular matrices), therefore it generates an inner product on R^n:

$$\langle \mathbf{u}, \mathbf{v} \rangle = (PC)^T\mathbf{u} \cdot (PC)^T\mathbf{v} = \mathbf{u}^T(PCC^TP^T)\mathbf{v} = \mathbf{u}^TA\mathbf{v}$$

5. The characteristic equation of A is $\lambda^3 - 3\lambda^2 + 2\lambda = \lambda(\lambda - 2)(\lambda - 1)$, so the eigenvalues are $\lambda = 0, 2, 1$.

Orthogonal bases for the eigenspaces are $\lambda = 0$: $\begin{bmatrix} -\frac{1}{\sqrt{2}} \\ 0 \\ \frac{1}{\sqrt{2}} \end{bmatrix}$; $\lambda = 2$: $\begin{bmatrix} \frac{1}{\sqrt{2}} \\ 0 \\ \frac{1}{\sqrt{2}} \end{bmatrix}$; $\lambda = 1$: $\begin{bmatrix} 0 \\ 1 \\ 0 \end{bmatrix}$.

Thus $P = \begin{bmatrix} -\frac{1}{\sqrt{2}} & \frac{1}{\sqrt{2}} & 0 \\ 0 & 0 & 1 \\ \frac{1}{\sqrt{2}} & \frac{1}{\sqrt{2}} & 0 \end{bmatrix}$ orthogonally diagonalizes A, and $P^TAP = \begin{bmatrix} 0 & 0 & 0 \\ 0 & 2 & 0 \\ 0 & 0 & 1 \end{bmatrix}$.

7. In matrix form, the quadratic form is $\mathbf{x}^TA\mathbf{x} = [x_1 \quad x_2] \begin{bmatrix} 1 & -\frac{3}{2} \\ -\frac{3}{2} & 4 \end{bmatrix} \begin{bmatrix} x_1 \\ x_2 \end{bmatrix}$. The characteristic equation of A is $\lambda^2 - 5\lambda + \frac{7}{4} = 0$ which has solutions $\lambda = \frac{5 \pm 3\sqrt{2}}{2}$ or $\lambda \approx 4.62, 0.38$. Since both eigenvalues of A are positive, the quadratic form is positive definite.

9. **(a)** $y - x^2 = 0$ or $y = x^2$ represents a parabola.

(b) $3x - 11y^2 = 0$ or $x = \frac{11}{3}y^2$ represents a parabola.

11. Partitioning U into columns we can write $U = [\mathbf{u}_1 | \mathbf{u}_2 | \dots | \mathbf{u}_n]$. The given product can be rewritten in partitioned form as well:

$$A = U \begin{bmatrix} z_1 & 0 & \cdots & 0 \\ 0 & z_2 & \cdots & 0 \\ \vdots & \vdots & \ddots & \vdots \\ 0 & 0 & \cdots & z_n \end{bmatrix} = [\mathbf{u}_1 | \mathbf{u}_2 | \dots | \mathbf{u}_n] \begin{bmatrix} z_1 & 0 & \cdots & 0 \\ 0 & z_2 & \cdots & 0 \\ \vdots & \vdots & \ddots & \vdots \\ 0 & 0 & \cdots & z_n \end{bmatrix} = [z_1\mathbf{u}_1 | z_2\mathbf{u}_2 | \dots | z_n\mathbf{u}_n]$$

By Theorem 7.5.3, the columns of U form an orthonormal set. Therefore, columns of A must also be orthonormal: $(z_i\mathbf{u}_i) \cdot (z_j\mathbf{u}_j) = (z_i\overline{z_j})(\mathbf{u}_i \cdot \mathbf{u}_j) = 0$ for all $i \neq j$ and $\|z_i\mathbf{u}_i\| = |z_i|\|\mathbf{u}_i\| = 1$ for all i. By Theorem 7.5.3, A is a unitary matrix.

13. Partitioning the given matrix into columns $A = [\mathbf{u}_1 | \mathbf{u}_2 | \mathbf{u}_3]$, we must find $\mathbf{u}_1 = \begin{bmatrix} a \\ b \\ c \end{bmatrix}$ such that

$\mathbf{u}_1 \cdot \mathbf{u}_2 = \frac{a}{\sqrt{2}} + \frac{b}{\sqrt{6}} + \frac{c}{\sqrt{3}} = 0$, $\mathbf{u}_1 \cdot \mathbf{u}_3 = -\frac{a}{\sqrt{2}} + \frac{b}{\sqrt{6}} + \frac{c}{\sqrt{3}} = 0$, and $\|\mathbf{u}_1\|^2 = a^2 + b^2 + c^2 = 1$.

Subtracting the second equation from the first one yields $a = 0$. Therefore $c = -\frac{\sqrt{3}}{\sqrt{6}}b = -\frac{b}{\sqrt{2}}$.

Substituting into $\|\mathbf{u}_1\|^2 = 1$ we obtain $b^2 + \frac{b^2}{2} = 1$ so that $b^2 = \frac{2}{3}$.

There are two possible solutions:

- $a = 0,\ b = \sqrt{\frac{2}{3}},\ c = -\frac{1}{\sqrt{3}}$ and
- $a = 0,\ b = -\sqrt{\frac{2}{3}},\ c = \frac{1}{\sqrt{3}}$.

CHAPTER 8: LINEAR TRANSFORMATIONS

8.1 General Linear Transformations

1. **(a)** $T\left(2\begin{bmatrix}1 & 0\\0 & 1\end{bmatrix}\right) = T\left(\begin{bmatrix}2 & 0\\0 & 2\end{bmatrix}\right) = \begin{bmatrix}2 & 0\\0 & 2\end{bmatrix}^2 = \begin{bmatrix}4 & 0\\0 & 4\end{bmatrix}$ does not equal

$2T\left(\begin{bmatrix}1 & 0\\0 & 1\end{bmatrix}\right) = 2\begin{bmatrix}1 & 0\\0 & 1\end{bmatrix}^2 = 2\begin{bmatrix}1 & 0\\0 & 1\end{bmatrix} = \begin{bmatrix}2 & 0\\0 & 2\end{bmatrix}$ so T does not satisfy the homogeneity property.

Consequently, T is not a linear transformation.

(b) Let A and B be any 2×2 matrices and let k be any real number. We have

$T(kA) = \text{tr}(kA) = ka_{11} + ka_{22} = k(a_{11} + a_{22}) = k\,\text{tr}(A) = kT(A)$ and

$T(A + B) = \text{tr}(A + B) = (a_{11} + b_{11}) + (a_{22} + b_{22}) = (a_{11} + a_{22}) + (b_{11} + b_{22})$

$= \text{tr}(A) + \text{tr}(B) = T(A) + T(B)$

therefore T is a linear transformation.

The kernel of T consists of all matrices $\begin{bmatrix}a & b\\c & d\end{bmatrix}$ such that $T\left(\begin{bmatrix}a & b\\c & d\end{bmatrix}\right) = a + d = 0$, i.e., $d = -a$.

We conclude that the kernel of T consists of all matrices of the form $\begin{bmatrix}a & b\\c & -a\end{bmatrix}$.

(c) Let A and B be any 2×2 matrices and let k be any real number. We have

$T(kA) = kA + (kA)^T = kA + kA^T = k(A + A^T) = kT(A)$ and

$T(A + B) = A + B + (A + B)^T = A + B + A^T + B^T = A + A^T + B + B^T = T(A) + T(B)$

therefore T is a linear transformation.

The kernel of T consists of all matrices $\begin{bmatrix}a & b\\c & d\end{bmatrix}$ such that

$T\left(\begin{bmatrix}a & b\\c & d\end{bmatrix}\right) = \begin{bmatrix}a & b\\c & d\end{bmatrix} + \begin{bmatrix}a & b\\c & d\end{bmatrix}^T = \begin{bmatrix}a & b\\c & d\end{bmatrix} + \begin{bmatrix}a & c\\b & d\end{bmatrix} = \begin{bmatrix}2a & b+c\\b+c & 2d\end{bmatrix} = \begin{bmatrix}0 & 0\\0 & 0\end{bmatrix}$ therefore

$a = d = 0$ and $c = -b$.

We conclude that the kernel of T consists of all matrices of the form $\begin{bmatrix}0 & b\\-b & 0\end{bmatrix}$.

3. For $\mathbf{u} \neq \mathbf{0}$, $T(-1\mathbf{u}) = \|-\mathbf{u}\| = \|\mathbf{u}\| = T(\mathbf{u}) \neq -1T(\mathbf{u})$, so the mapping is not a linear transformation.

5. Let A_1 and A_2 be any 2×2 matrices and let k be any real number. We have

$T(kA_1) = (kA_1)B = k(A_1B) = kT(A_1)$ and

$T(A_1 + A_2) = (A_1 + A_2)B = A_1B + A_2B = T(A_1) + T(A_2)$ therefore T is a linear transformation.

The kernel of T consists of all 2×2 matrices whose rows are orthogonal to all columns of B.

7. Let $p(x) = a_0 + a_1x + a_2x^2$ and $q(x) = b_0 + b_1x + b_2x^2$.

(a) $T(kp(x)) = ka_0 + ka_1(x + 1) + ka_2(x + 1)^2 = kT(p(x))$

$T(p(x) + q(x)) = a_0 + b_0 + (a_1 + b_1)(x + 1) + (a_2 + b_2)(x + 1)^2$

$= a_0 + a_1(x + 1) + a_2(x + 1)^2 + b_0 + b_1(x + 1) + b_2(x + 1)^2 = T(p(x)) + T(q(x))$

Thus T is a linear transformation.

The kernel of T consists of all polynomials $a_0 + a_1x + a_2x^2$ such that

$T(a_0 + a_1x + a_2x^2) = a_0 + a_1(x+1) + a_2(x+1)^2 = 0$. This equality requires that $a_0 = a_1 = a_2 = 0$ therefore $\ker(T) = \{0\}$.

(b) $T(kp(x)) = T(ka_0 + ka_1x + ka_2x^2) = (ka_0 + 1) + (ka_1 + 1)x + (ka_2 + 1)x^2 \neq kT(p(x))$ so T is not a linear transformation.

9. $T(k(a_0, a_1, a_2, \ldots, a_n, \ldots)) = T(ka_0, ka_1, ka_2, \ldots, ka_n, \ldots) = (0, ka_0, ka_1, \ldots, ka_n, \ldots)$
$= k(0, a_0, a_1, \ldots, a_n, \ldots) = kT(a_0, a_1, a_2, \ldots, a_n, \ldots)$
$T((a_0, a_1, a_2, \ldots, a_n, \ldots) + (b_0, b_1, b_2, \ldots, b_n \ldots)) = T(a_0 + b_0, a_1 + b_1, a_2 + b_2, \ldots, a_n + b_n, \ldots)$
$= (0, a_0 + b_0, a_1 + b_1, \ldots, a_n + b_n, \ldots) = (0, a_0, a_1, \ldots, a_n, 0) + (0, b_0, b_1, \ldots, b_n, 0)$
$= T(a_0, a_1, a_2, \ldots, a_n, \ldots) + T(b_0, b_1, b_2, \ldots, b_n, \ldots)$ therefore T is a linear transformation.
The kernel of T contains only $(0,0,0,\ldots)$.

11. **(a)** Since $x + x^2 = x(1+x)$, $x + x^2$ is in $R(T)$.

(b) $1 + x$ cannot be expressed in the form $xp(x)$ for any polynomial $p(x)$ therefore $1 + x$ is not in $R(T)$.

(c) $3 - x^2$ cannot be expressed in the form $xp(x)$for any polynomial $p(x)$ therefore $3 - x^2$ is not in $R(T)$.

(d) $-x = x(-1)$ therefore $-x$ is in $R(T)$.

13. **(a)** nullity(T) = 5 – rank(T) = 2

(b) $\dim(P_4) = 5$, so nullity(T) = 5 – rank(T) = 4

(c) Since $R(T) = R^3$, T has rank 3; $\dim(M_{mn}) = mn$ so nullity(T) $= mn -$rank(T) $= mn - 3$

(d) nullity(T) = 4 – rank(T) = 1

15. **(a)** $T\left(\begin{bmatrix} 1 & 2 \\ -4 & 3 \end{bmatrix}\right) = 3\begin{bmatrix} 1 & 2 \\ -4 & 3 \end{bmatrix} = \begin{bmatrix} 3 & 6 \\ -12 & 9 \end{bmatrix}$

(b) The only 2×2 matrix A such that $3A = 0$ is the zero matrix. Consequently, $\ker(T) = \{0\}$ so the nullity of T is 0.
By Theorem 8.1.4, $\text{rank}(T) = \dim(M_{22}) - \text{nullity}(T) = 4 - 0 = 4$.

17. **(a)** $T(x^2) = ((-1)^2, 0^2, 1^2) = (1,0,1)$

(b) The kernel of T consists of all polynomials $p(x) = a_0 + a_1x + a_2x^2$ such that $(p(-1), p(0), p(1)) = (a_0 - a_1 + a_2,\ a_0,\ a_0 + a_1 + a_2) = (0,0,0)$.
Equating the corresponding components we obtain a linear system

$$\begin{array}{rcrcrcl} a_0 & - & a_1 & + & a_2 & = & 0 \\ a_0 & & & & & = & 0 \\ a_0 & + & a_1 & + & a_2 & = & 0 \end{array}$$

The reduced row echelon form of the coefficient matrix of this system is $\begin{bmatrix} 1 & 0 & 0 \\ 0 & 1 & 0 \\ 0 & 0 & 1 \end{bmatrix}$ hence the system has a unique solution $a_0 = a_1 = a_2 = 0$. We conclude that $\ker(T) = \{0\}$.

(c) It follows from the solution of part (b) that nullity$(T) = 0$.
By Theorem 8.1.4, rank$(T) = \dim(P_2) - \text{nullity}(T) = 3 - 0 = 3$. Consequently, $R(T) = R^3$.

19. For $\mathbf{x} = (x_1,\ x_2) = c_1\mathbf{v}_1 + c_2\mathbf{v}_2$, we have $(x_1,\ x_2) = c_1(1,\ 1) + c_2(1,\ 0) = (c_1 + c_2,\ c_1)$ or

$$\begin{array}{rcrcl} c_1 & + & c_2 & = & x_1 \\ c_1 & & & = & x_2 \end{array}$$

which has the solution $c_1 = x_2, c_2 = x_1 - x_2$.
$(x_1,\ x_2) = x_2(1,\ 1) + (x_1 - x_2)(1,\ 0) = x_2\mathbf{v}_1 + (x_1 - x_2)\mathbf{v}_2$ and
$T(x_1,\ x_2) = x_2T(\mathbf{v}_1) + (x_1 - x_2)T(\mathbf{v}_2) = x_2(1,\ -2) + (x_1 - x_2)(-4,\ 1) = (-4x_1 + 5x_2,\ x_1 - 3x_2)$
$T(5, -3) = (-20 - 15, 5 + 9) = (-35, 14)$.

21. For $\mathbf{x} = (x_1,\ x_2,\ x_3) = c_1\mathbf{v}_1 + c_2\mathbf{v}_2 + c_3\mathbf{v}_3$, we have
$(x_1,\ x_2,\ x_3) = c_1(1,\ 1,\ 1) + c_2(1,\ 1,\ 0) + c_3(1,\ 0,\ 0) = (c_1 + c_2 + c_3,\ c_1 + c_2,\ c_1)$ or

$$\begin{array}{rcrcrcl} c_1 & + & c_2 & + & c_3 & = & x_1 \\ c_1 & + & c_2 & & & = & x_2 \\ c_1 & & & & & = & x_3 \end{array}$$

which has the solution $c_1 = x_3,\ c_2 = x_2 - x_3,\ c_3 = x_1 - (x_2 - x_3) - x_3 = x_1 - x_2$.
$(x_1,\ x_2,\ x_3) = x_3\mathbf{v}_1 + (x_2 - x_3)\mathbf{v}_2 + (x_1 - x_2)\mathbf{v}_3$
$T(x_1,\ x_2,\ x_3) = x_3T(\mathbf{v}_1) + (x_2 - x_3)T(\mathbf{v}_2) + (x_1 - x_2)T(\mathbf{v}_3)$
$= x_3(2,\ -1,\ 4) + (x_2 - x_3)(3,\ 0,\ 1) + (x_1 - x_2)(-1,\ 5,\ 1)$
$= (-x_1 + 4x_2 - x_3,\ 5x_1 - 5x_2 - x_3,\ x_1 + 3x_3)$
$T(2,\ 4,\ -1) = (-2 + 16 + 1,\ 10 - 20 + 1,\ 2 - 3) = (15,\ -9,\ -1)$

23. **(a)** $T\big(k(a_0 + a_1x + a_2x^2 + a_3x^3)\big) = T(ka_0 + ka_1x + ka_2x^2 + ka_3x^3) = 5ka_0 + ka_3x^2$
$= k(5a_0 + a_3x^2) = kT(a_0 + a_1x + a_2x^2 + a_3x^3)$
$T(a_0 + a_1x + a_2x^2 + a_3x^3 + b_0 + b_1x + b_2x^2 + b_3x^3)$
$= T\big((a_0 + b_0) + (a_1 + b_1)x + (a_2 + b_2)x^2 + (a_3 + b_3)x^3\big) = 5(a_0 + b_0) + (a_3 + b_3)x^2$
$= (5a_0 + a_3x^2) + (5b_0 + b_3x^2) = T(a_0 + a_1x + a_2x^2 + a_3x^3) + T(b_0 + b_1x + b_2x^2 + b_3x^3)$
therefore T is linear.

(b) The kernel of T consists of all polynomials $p(x) = a_0 + a_1x + a_2x^2 + a_3x^3$ such that $T(a_0 + a_1x + a_2x^2 + a_3x^3) = 5a_0 + a_3x^2 = 0$, which requires that $a_0 = a_3 = 0$.
Therefore every vector in ker (T) can be written in the form $a_1x + a_2x^2$, i.e., $\ker(T) = \text{span}\{x, x^2\}$. The set $\{x, x^2\}$ is linearly independent since neither polynomial is a scalar multiple of the other one. We conclude that $\{x, x^2\}$ is a basis for $\ker(T)$.

(c) $T(a_0 + a_1x + a_2x^2 + a_3x^3) = 5a_0 + a_3x^2$ so $R(T) = \text{span}\{5, x^2\}$. The set $\{5, x^2\}$ is linearly independent since neither polynomial is a scalar multiple of the other one.
We conclude that $\{5, x^2\}$ is a basis for $R(T)$.

25. **(a)** If $p'(x) = 0$, then $p(x)$ is a constant, so $\ker(D)$ consists of all constant polynomials.

(b) The kernel of J contains all polynomials $a_0 + a_1x$ such that $\int_{-1}^{1}(a_0 + a_1x)dx = 0$. By integration, this condition yields $\left(a_0x + \frac{a_1x^2}{2}\right)\Big]_{-1}^{1} = 0$, i.e., $a_0 + \frac{a_1}{2} + a_0 - \frac{a_1}{2} = 0$, or equivalently, $a_0 = 0$.
The kernel consists of all polynomials of the form a_1x.

27. (a) If $f^4(x) = 0$, then $f'''(x) = a$ for some constant a. Applying Fundamental Theorem of Calculus, we obtain $f''(x) = ax + b$, then $f'(x) = \frac{a}{2}x^2 + bx + c$, and $f(x) = \frac{a}{6}x^3 + \frac{b}{2}x^2 + cx + d$ for constant b, c, and d. We conclude that P_3 is the kernel of $T(f(x)) = f^{(4)}(x)$.

   (b) By similar reasoning, $T(f(x)) = f^{(n+1)}(x)$ has $\ker(T) = P_n$.

29. (a) $R(T)$ must be a subspace of R^3, thus the possibilities are a line through the origin, a plane through the origin, the origin only, or all of R^3.

   (b) The origin, a line through the origin, a plane through the origin, or the entire space R^3.

31. $T(2\mathbf{v}_1 - 3\mathbf{v}_2 + 4\mathbf{v}_3) = 2T(\mathbf{v}_1) - 3T(\mathbf{v}_2) + 4T(\mathbf{v}_3) = (2, -2, 4) - (0, 9, 6) + (-12, 4, 8)$
$= (-10, -7, 6)$

33. Let $\mathbf{v} = c_1\mathbf{v}_1 + c_2\mathbf{v}_2 + \cdots + c_n\mathbf{v}_n$ be any vector in V. Then
$T(\mathbf{v}) = c_1T(\mathbf{v}_1) + c_2T(\mathbf{v}_2) + \cdots + c_nT(\mathbf{v}_n) = c_1\mathbf{v}_1 + c_2\mathbf{v}_2 + \cdots + c_n\mathbf{v}_n = \mathbf{v}$
Since $\mathbf{v}$ was an arbitrary vector in V, T must be the identity operator.

True-False Exercises

(a) True. $c_1 = k$, $c_2 = 0$ gives the homogeneity property and $c_1 = c_2 = 1$ gives the additivity property.

(b) False. Every linear transformation will have $T(-\mathbf{v}) = -T(\mathbf{v})$.

(c) True. Only the zero transformation has this property.

(d) False. $T(\mathbf{0}) = \mathbf{v}_0 + \mathbf{0} = \mathbf{v}_0 \neq \mathbf{0}$, so T is not a linear transformation.

(e) True. This follows from part (a) of Theorem 8.1.3.

(f) True. This follows from part (b) of Theorem 8.1.3.

(g) False. T does not necessarily have rank 4.

(h) False. $\det(A + B) \neq \det(A) + \det(B)$ in general.

(i) False. $\text{nullity}(T) = \text{rank}(T) = 2$

8.2 Compositions and Inverse Transformations

1. (a) By inspection, $\ker(T) = \{\mathbf{0}\}$, so T is one-to-one.

   (b) By inspection, $\ker(T) = \{\mathbf{0}\}$, so T is one-to-one.

   (c) (x, y, z) is in $\ker(T)$ if both $x + y + z = 0$ and $x - y - z = 0$, which is $x = 0$ and $y + z = 0$. Thus, $\ker(T) = \text{span}\{(0, 1, -1)\}$ and T is not one-to-one.

3. (a) The reduced row echelon form of A is $\begin{bmatrix} 1 & -2 \\ 0 & 0 \\ 0 & 0 \end{bmatrix}$, so $\text{nullity}(A) = 1$.

   Multiplication by A is not one-to-one.

(b) The reduced row echelon form of A is $\begin{bmatrix} 1 & 0 & 0 & 30 \\ 0 & 1 & 0 & -10 \\ 0 & 0 & 1 & 7 \end{bmatrix}$, so nullity($A$)= 1.

Multiplication by A is not one-to-one.

5. **(a)** Since nullity(T)= 0, T is one-to-one.

(b) nullity(T) = dim(V) − rank(T) = 0 therefore T is one-to-one.

(c) Since rank(T) ≤ dim(W) < dim (V), we have nullity(T)= dim(V) − rank(T) > 0. We conclude that T is not one-to-one.

7. For example, $T(1-x^2) = (1-(-1)^2, 1-1^2) = (0,0)$.
The transformation is onto since for any real numbers a and b, a polynomial $p(x)$ in P_2 can be found such that $p(-1) = a$ and $p(1) = b$.

9. No; T is not one-to-one because $\ker(T) \neq \{\mathbf{0}\}$ as $T(\mathbf{a}) = \mathbf{a} \times \mathbf{a} = \mathbf{0}$.

11. $(T_2 \circ T_1)(x,\ y) = T_2(2x,\ 3y) = (2x - 3y,\ 2x + 3y)$

13. $(T_3 \circ T_2 \circ T_1)(x,y) = T_3\left(T_2(T_1(x,y))\right) = T_3(T_2(-2y, 3x, x-2y)) = T_3(3x, x-2y, -2y)$
$= \left(3x - 2y, x - 2y - (-2y)\right) = (3x - 2y, x)$

15. **(a)** $(T_1 \circ T_2)(A) = T_1(A^T) = \operatorname{tr}\left(\begin{bmatrix} a & c \\ b & d \end{bmatrix}\right) = a + d$

(b) $(T_2 \circ T_1)(A)$ does not exist because $T_1(A)$ is not a 2×2 matrix.

17. $(T_2 \circ T_1)(a_0 + a_1x + a_2x^2) = T_2(T_1(a_0 + a_1x + a_2x^2)) = T_2(a_0 + a_1(x+1) + a_2(x+1)^2)$
$= x(a_0 + a_1(x+1) + a_2(x+1)^2) = a_0x + a_1x(x+1) + a_2x(x+1)^2$

19. **(a)** $T(1 - 2x) = (1 - 2(0), 1 - 2(1)) = (1, -1)$

(b) $T(kp(x)) = (kp(0), kp(1)) = k(p(0), p(1)) = kT(p(x))$;
$T(p(x) + q(x)) = (p(0) + q(0), p(1) + q(1)) = (p(0), p(1)) + (q(0), q(1))$
$= T(p(x)) + T(q(x))$

(c) Let $p(x) = a_0 + a_1x$, then $T(p(x)) = (a_0,\ a_0 + a_1)$ so if
$T(p(x)) = (0, 0)$, then $a_0 = a_1 = 0$ and p is the zero polynomial, so $\ker(T) = \{\mathbf{0}\}$.

(d) Since $T(p(x)) = (a_0,\ a_0 + a_1)$, then $T^{-1}(2,\ 3)$ has $a_0 = 2$ and $a_0 + a_1 = 3$ or $a_1 = 1$. Thus,
$T^{-1}(2,\ 3) = 2 + x$.

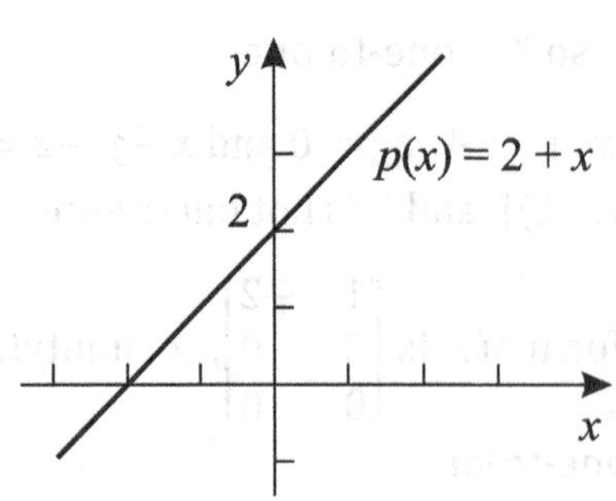

21. **(a)** For T to have an inverse, all the a_i's must be nonzero since otherwise T would have a nonzero kernel.

(b) $T^{-1}(x_1, x_2, \ldots, x_n) = \left(\frac{1}{a_1}x_1, \frac{1}{a_2}x_2, \ldots, \frac{1}{a_n}x_n\right)$

23. **(a)** Since $T_1(p(x)) = xp(x)$, $T_1^{-1}(p(x)) = \frac{1}{x}p(x)$.
Since $T_2(p(x)) = p(x+1)$, $T_2^{-1}(p(x)) = p(x-1)$.
$(T_1^{-1} \circ T_2^{-1})(p(x)) = T_1^{-1}(p(x-1)) = \frac{1}{x}p(x-1)$

(b) Since $(T_2 \circ T_1)(p(x)) = T_2\left(T_1(p(x))\right) = T_2(xp(x)) = (x+1)p(x+1)$, we have
$(T_2 \circ T_1)\left((T_1^{-1} \circ T_2^{-1})(p(x))\right) = (T_2 \circ T_1)\left(\frac{1}{x}p(x-1)\right) = (x+1)\left(\frac{1}{x+1}\right)p(x-1+1) = p(x)$

25. $T_2(\mathbf{v}) = \frac{1}{4}\mathbf{v}$, then $(T_1 \circ T_2)(\mathbf{v}) = T_1\left(\frac{1}{4}\mathbf{v}\right) = 4\left(\frac{1}{4}\mathbf{v}\right) = \mathbf{v}$ and $(T_2 \circ T_1)(\mathbf{v}) = T_2(4\mathbf{v}) = \frac{1}{4}(4\mathbf{v}) = \mathbf{v}$.

27. By inspection, $T(x, y, z) = (x, y, 0)$. Then $T(T(x, y, z)) = T(x, y, 0) = (x, y, 0) = T(x, y, z)$ or $T \circ T = T$.

29. **(a)** $D(kp(x)) = \frac{d}{dx}(kp(x)) = kp'(x) = kD(p(x))$
$D(p(x) + q(x)) = \frac{d}{dx}(p(x) + q(x)) = p'(x) + q'(x) = D(p(x)) + D(q(x))$
$J(kp(x)) = \int_0^x kp(t)\,dt = k\int_0^x p(t)\,dt = kJ(p(x))$
$J(p(x) + q(x)) = \int_0^x (p(t) + q(t))\,dt = \int_0^x p(t)\,dt + \int_0^x q(t)\,dt = J(p(x)) + J(q(x))$

(b) D is not one-to-one (e.g., $D(x^2 + 3) = D(x^2) = 2x$) so it does not have an inverse.

(c) Yes, this can be accomplished by taking $D : V \to P_{n-1}$ and $J : P_{n-1} \to V$ where V is the set of all polynomials $p(x)$ in P_n such that $p(0) = 0$.

31. The kernel of J contains all polynomials $a_0 + a_1x$ such that $\int_{-1}^{1}(a_0 + a_1x)dx = 0$. By integration, this condition yields $\left(a_0x + \frac{a_1x^2}{2}\right)\Big]_{-1}^{1} = 0$, i.e., $a_0 + \frac{a_1}{2} + a_0 - \frac{a_1}{2} = 0$, or equivalently, $a_0 = 0$.
The kernel consists of all polynomials of the form a_1x.
Since $\ker(J) \neq \{\mathbf{0}\}$, by Theorem 8.2.1, J is not one-to-one.

35. $T(kp(x)) = kp(q_0(x)) = kT(p(x))$
$T(p_1(x) + p_2(x)) = p_1(q_0(x)) + p_2(q_0(x)) = T(p_1(x)) + T(p_2(x))$

37. If there were such a transformation T, then it would have nullity 0 (because it would be one-to-one), and have rank no greater than the dimension of W (because its range is a subspace of W). Then by the dimension Theorem, $\dim(V) = \text{rank}(T) + \text{nullity}(T) \leq \dim(W)$ which contradicts the assumption $\dim(W) < \dim(V)$. Thus there is no such transformation.

True-False Exercises

(a) True. This is the statement of Theorem 8.2.3.

(b) False. For example, with $T_1(x, y) = (y, x)$ and $T_2(x, y) = (x, 0)$ we have
$(T_1 \circ T_2)(x, y) = T_1(x, 0) = (0, x)$ which does not equal $(T_2 \circ T_1)(x, y) = T_2(y, x) = (y, 0)$.

(c) True.

(d) True. For T to have an inverse, it must be one-to-one.

(e) False. T^{-1} does not exist.

(f) True. If T_1 is not one-to-one then there is some nonzero vector $\mathbf{v}_1$ with $T_1(\mathbf{v}_1) = \mathbf{0}$. Thus $(T_2 \circ T_1)(\mathbf{v}_1) = T_2(\mathbf{0}) = \mathbf{0}$ and $\ker(T_2 \circ T_1) \neq \{\mathbf{0}\}$.

8.3 Isomorphism

1. The transformation is an isomorphism.

3. The transformation is an isomorphism.

5. The transformation is not an isomorphism (not a linear transformation).

7. The transformation is an isomorphism.

9. **(a)** $T\left(\begin{bmatrix} a & b & c \\ b & d & e \\ c & e & f \end{bmatrix}\right) = \begin{bmatrix} a \\ b \\ c \\ d \\ e \\ f \end{bmatrix}$

(b) $T_1\left(\begin{bmatrix} a & b \\ c & d \end{bmatrix}\right) = \begin{bmatrix} a \\ b \\ c \\ d \end{bmatrix}$ and $T_2\left(\begin{bmatrix} a & b \\ c & d \end{bmatrix}\right) = \begin{bmatrix} a \\ c \\ b \\ d \end{bmatrix}$.

11. $\det(A) = -3 \neq 0$, so by Theorem 4.10.2 T_A is one-to-one and onto. Consequently, T_A is an isomorphism.

13. The reduced row echelon form of A is $\begin{bmatrix} 1 & 1 & 1 & 1 \\ 0 & 0 & 0 & 0 \\ 0 & 0 & 0 & 0 \end{bmatrix}$. The solution space W contains vectors (x_1, x_2, x_3, x_4) such that $x_1 = -r - s - t,\ x_2 = r, x_3 = s, x_4 = t$ so $(x_1, x_2, x_3, x_4) = (-r - s - t, r, s, t) = r(-1,1,0,0) + s(-1,0,1,0) + t(-1,0,0,1)$ hence $\dim(W) = 3$. $(-r - s - t, r, s, t) \to (r, s, t)$ is an isomorphism between W and R^3.

15. Let us denote the given transformation by T. T is a linear transformation, since for any $A = \begin{bmatrix} a & b \\ c & d \end{bmatrix}$ and $B = \begin{bmatrix} a' & b' \\ c' & d' \end{bmatrix}$ in M_{22} and for any scalar k, we have

$$T(kA) = T\left(\begin{bmatrix} ka & kb \\ kc & kd \end{bmatrix}\right) = \begin{bmatrix} ka \\ ka + kb \\ ka + kb + kc \\ ka + kb + kc + kd \end{bmatrix} = k\begin{bmatrix} a \\ a + b \\ a + b + c \\ a + b + c + d \end{bmatrix} = kT(A) \text{ and}$$

$$T(A + B) = T\left(\begin{bmatrix} a + a' & b + b' \\ c + c' & d + d' \end{bmatrix}\right) = \begin{bmatrix} a + a' \\ a + a' + b + b' \\ a + a' + b + b' + c + c' \\ a + a' + b + b' + c + c' + d + d' \end{bmatrix}$$

$$= \begin{bmatrix} a \\ a+b \\ a+b+c \\ a+b+c+d \end{bmatrix} + \begin{bmatrix} a' \\ a'+b' \\ a'+b'+c' \\ a'+b'+c'+d' \end{bmatrix} = T(A) + T(B)$$

By inspection, $\begin{bmatrix} a \\ a+b \\ a+b+c \\ a+b+c+d \end{bmatrix} = \begin{bmatrix} 0 \\ 0 \\ 0 \\ 0 \end{bmatrix}$ implies $a = b = c = d = 0$, therefore $\ker(T) = \left\{\begin{bmatrix} 0 & 0 \\ 0 & 0 \end{bmatrix}\right\}$ so T is one-to-one.

$\text{rank}(T) = \dim(M_{22}) - \text{nullity}(T) = 4 - 0 = 4$ hence T is onto.

We conclude that T is an isomorphism.

17. Yes, $(a, b) \to (0, a, b)$ is an isomorphism between R^2 and the yz-plane in R^3.

19. No. By inspection, $T(x^2 - x) = T(0) = \begin{bmatrix} 0 & 0 \\ 0 & 0 \end{bmatrix}$ so T is not one-to-one.

True-False Exercises

(a) False. $\dim(R^2) = 2$ while $\dim(P_2) = 3$.

(b) True. If $\ker(T) = \{\mathbf{0}\}$ then $\text{rank}(T) = 4$ so T is one-to-one and onto.

(c) False. $\dim(M_{33}) = 9$ while $\dim(P_9) = 10$.

(d) True. For instance, if V consists of all matrices of the form $\begin{bmatrix} a & b & 0 \\ c & d & 0 \end{bmatrix}$, then $T\left(\begin{bmatrix} a & b & 0 \\ c & d & 0 \end{bmatrix}\right) = \begin{bmatrix} a \\ b \\ c \\ d \end{bmatrix}$ is an isomorphism $T: V \to R^4$.

(e) True.

(f) True. For instance if V consists of all vectors in R^{n+1} of the form $(x_1, \dots, x_n, 0)$, then $T(x_1, \dots, x_n) = (x_1, \dots, x_n, 0)$ is an isomorphism $T: R^n \to V$.

8.4 Matrices for General Linear Transformations

1. **(a)** $T(\mathbf{u}_1) = T(1) = x = \mathbf{v}_2$, $T(\mathbf{u}_2) = T(x) = x^2 = \mathbf{v}_3$, and $T(\mathbf{u}_3) = T(x^2) = x^3 = \mathbf{v}_4$ therefore

$[T(\mathbf{u}_1)]_{B'} = \begin{bmatrix} 0 \\ 1 \\ 0 \\ 0 \end{bmatrix}$, $[T(\mathbf{u}_2)]_{B'} = \begin{bmatrix} 0 \\ 0 \\ 1 \\ 0 \end{bmatrix}$, and $[T(\mathbf{u}_3)]_{B'} = \begin{bmatrix} 0 \\ 0 \\ 0 \\ 1 \end{bmatrix}$, thus $[T]_{B',B} = \begin{bmatrix} 0 & 0 & 0 \\ 1 & 0 & 0 \\ 0 & 1 & 0 \\ 0 & 0 & 1 \end{bmatrix}$.

(b) By inspection, $[\mathbf{x}]_B = \begin{bmatrix} c_0 \\ c_1 \\ c_2 \end{bmatrix}$ so $[T]_{B',B}[\mathbf{x}]_B = \begin{bmatrix} 0 & 0 & 0 \\ 1 & 0 & 0 \\ 0 & 1 & 0 \\ 0 & 0 & 0 \end{bmatrix}\begin{bmatrix} c_0 \\ c_1 \\ c_2 \end{bmatrix} = \begin{bmatrix} 0 \\ c_0 \\ c_1 \\ c_2 \end{bmatrix}$.

On the other hand, $[T(\mathbf{x})]_{B'} = [c_0x + c_1x^2 + c_2x^3]_{B'} = \begin{bmatrix} 0 \\ c_0 \\ c_1 \\ c_2 \end{bmatrix}$. Formula (5) is satisfied.

3. **(a)** $T(1) = 1;\ T(x) = x - 1 = -1 + x;\ T(x^2) = (x-1)^2 = 1 - 2x + x^2$

Thus the matrix for T relative to B is $\begin{bmatrix} 1 & -1 & 1 \\ 0 & 1 & -2 \\ 0 & 0 & 1 \end{bmatrix}$.

(b) $[T]_B[\mathbf{x}]_B = \begin{bmatrix} 1 & -1 & 1 \\ 0 & 1 & -2 \\ 0 & 0 & 1 \end{bmatrix}\begin{bmatrix} a_0 \\ a_1 \\ a_2 \end{bmatrix} = \begin{bmatrix} a_0 - a_1 + a_2 \\ a_1 - 2a_2 \\ a_2 \end{bmatrix}$. For $\mathbf{x} = a_0 + a_1x + a_2x^2$,

$T(\mathbf{x}) = a_0 + a_1(x-1) + a_2(x-1)^2 = a_0 - a_1 + a_2 + (a_1 - 2a_2)x + a_2x^2$,

so $[T(\mathbf{x})]_B = \begin{bmatrix} a_0 - a_1 + a_2 \\ a_1 - 2a_2 \\ a_2 \end{bmatrix}$.

5. **(a)** $T(\mathbf{u}_1) = T\left(\begin{bmatrix} 1 \\ 3 \end{bmatrix}\right) = \begin{bmatrix} 7 \\ -1 \\ 0 \end{bmatrix} = 0\mathbf{v}_1 - \frac{1}{2}\mathbf{v}_2 + \frac{8}{3}\mathbf{v}_3;\ T(\mathbf{u}_2) = T\left(\begin{bmatrix} -2 \\ 4 \end{bmatrix}\right) = \begin{bmatrix} 6 \\ 2 \\ 0 \end{bmatrix} = 0\mathbf{v}_1 + \mathbf{v}_2 + \frac{4}{3}\mathbf{v}_3$

$[T]_{B',B} = \begin{bmatrix} 0 & 0 \\ -\frac{1}{2} & 1 \\ \frac{8}{3} & \frac{4}{3} \end{bmatrix}$

(b) For an arbitrary $\mathbf{x} = \begin{bmatrix} x \\ y \end{bmatrix}$ in R^2, solving the system

$$\begin{aligned} c_1 - 2c_2 &= x \\ 3c_1 + 4c_2 &= y \end{aligned}$$

yields $[\mathbf{x}]_B = \begin{bmatrix} \frac{2x+y}{5} \\ \frac{-3x+y}{10} \end{bmatrix}$ therefore $[T]_{B',B}[\mathbf{x}]_B = \begin{bmatrix} 0 & 0 \\ -\frac{1}{2} & 1 \\ \frac{8}{3} & \frac{4}{3} \end{bmatrix}\begin{bmatrix} \frac{2x+y}{5} \\ \frac{-3x+y}{10} \end{bmatrix} = \begin{bmatrix} 0 \\ -\frac{x}{2} \\ \frac{2}{3}x + \frac{2}{3}y \end{bmatrix}$.

On the other hand, $T(\mathbf{x}) = \begin{bmatrix} x + 2y \\ -x \\ 0 \end{bmatrix}$. Solving the system

$$\begin{aligned} c_1 + 2c_2 + 3c_3 &= x + 2y \\ c_1 + 2c_2 &= -x \\ c_1 &= 0 \end{aligned}$$

yields $[T(\mathbf{x})]_{B'} = \begin{bmatrix} 0 \\ -\frac{x}{2} \\ \frac{2}{3}x + \frac{2}{3}y \end{bmatrix}$ showing that Formula (5) holds for every $\mathbf{x}$ in R^2.

7. **(a)** We have $T(1) = 1$, $T(x) = 2x + 1 = 1 + 2x$, and $T(x^2) = (2x+1)^2 = 1 + 4x + 4x^2$.

Therefore, $[T]_B = \begin{bmatrix} 1 & 1 & 1 \\ 0 & 2 & 4 \\ 0 & 0 & 4 \end{bmatrix}$.

(b) Step 1. The coordinate vector of $\mathbf{x} = 2 - 3x + 4x^2$ with respect to the basis B is

$$[\mathbf{x}]_B = \begin{bmatrix} 2 \\ -3 \\ 4 \end{bmatrix}.$$

Step 2. $$[T]_B[\mathbf{x}]_B = \begin{bmatrix} 1 & 1 & 1 \\ 0 & 2 & 4 \\ 0 & 0 & 4 \end{bmatrix}\begin{bmatrix} 2 \\ -3 \\ 4 \end{bmatrix} = \begin{bmatrix} 3 \\ 10 \\ 16 \end{bmatrix} = [T(\mathbf{x})]_B.$$

Step 3. Reconstructing $T(\mathbf{x})$ from the coordinate vector obtained in Step 2, we obtain $T(\mathbf{x}) = 3(1) + 10(x) + 16(x^2) = 3 + 10x + 16x^2$.

(c) $T(2 - 3x + 4x^2) = 2 - 3(2x + 1) + 4(2x + 1)^2 = 2 - 6x - 3 + 16x^2 + 16x + 4$
$= 3 + 10x + 16x^2$

9. **(a)** Since A is the matrix for T relative to B, $A = [[T(\mathbf{v}_1)]_B \quad [T(\mathbf{v}_2)]_B]$.
That is, $[T(\mathbf{v}_1)]_B = \begin{bmatrix} 1 \\ -2 \end{bmatrix}$ and $[T(\mathbf{v}_2)]_B = \begin{bmatrix} 3 \\ 5 \end{bmatrix}$.

(b) Since $[T(\mathbf{v}_1)]_B = \begin{bmatrix} 1 \\ -2 \end{bmatrix}$, $T(\mathbf{v}_1) = 1\mathbf{v}_1 - 2\mathbf{v}_2 = \begin{bmatrix} 1 \\ 3 \end{bmatrix} - \begin{bmatrix} -2 \\ 8 \end{bmatrix} = \begin{bmatrix} 3 \\ -5 \end{bmatrix}$.
Similarly, $T(\mathbf{v}_2) = 3\mathbf{v}_1 + 5\mathbf{v}_2 = \begin{bmatrix} 3 \\ 9 \end{bmatrix} + \begin{bmatrix} -5 \\ 20 \end{bmatrix} = \begin{bmatrix} -2 \\ 29 \end{bmatrix}$.

(c) If $\begin{bmatrix} x_1 \\ x_2 \end{bmatrix} = c_1\mathbf{v}_1 + c_2\mathbf{v}_2 = c_1\begin{bmatrix} 1 \\ 3 \end{bmatrix} + c_2\begin{bmatrix} -1 \\ 4 \end{bmatrix}$, then

$$\begin{aligned} x_1 &= c_1 - c_2 \\ x_2 &= 3c_1 + 4c_2 \end{aligned}$$

Solving for c_1 and c_2 gives $c_1 = \frac{4}{7}x_1 + \frac{1}{7}x_2$, $c_2 = -\frac{3}{7}x_1 + \frac{1}{7}x_2$, so

$$T\left(\begin{bmatrix} x_1 \\ x_2 \end{bmatrix}\right) = c_1T(\mathbf{v}_1) + c_2T(\mathbf{v}_2) = \left(\tfrac{4}{7}x_1 + \tfrac{1}{7}x_2\right)\begin{bmatrix} 3 \\ -5 \end{bmatrix} + \left(-\tfrac{3}{7}x_1 + \tfrac{1}{7}x_2\right)\begin{bmatrix} -2 \\ 29 \end{bmatrix} = \begin{bmatrix} \frac{18}{7}x_1 + \frac{1}{7}x_2 \\ -\frac{107}{7}x_1 + \frac{24}{7}x_2 \end{bmatrix}$$

$$= \begin{bmatrix} \frac{18}{7} & \frac{1}{7} \\ -\frac{107}{7} & \frac{24}{7} \end{bmatrix}\begin{bmatrix} x_1 \\ x_2 \end{bmatrix}.$$

(d) $$T\left(\begin{bmatrix} 1 \\ 1 \end{bmatrix}\right) = \begin{bmatrix} \frac{18}{7} & \frac{1}{7} \\ -\frac{107}{7} & \frac{24}{7} \end{bmatrix}\begin{bmatrix} 1 \\ 1 \end{bmatrix} = \begin{bmatrix} \frac{19}{7} \\ -\frac{83}{7} \end{bmatrix}$$

11. **(a)** Since A is the matrix for T relative to B, the columns of A are $[T(\mathbf{v}_1)]_B$, $[T(\mathbf{v}_2)]_B$, and $[T(\mathbf{v}_3)]_B$, respectively. That is, $[T(\mathbf{v}_1)]_B = \begin{bmatrix} 1 \\ 2 \\ 6 \end{bmatrix}$, $[T(\mathbf{v}_2)]_B = \begin{bmatrix} 3 \\ 0 \\ -2 \end{bmatrix}$, and $[T(\mathbf{v}_3)]_B = \begin{bmatrix} -1 \\ 5 \\ 4 \end{bmatrix}$.

(b) Since $[T(\mathbf{v}_1)]_B = \begin{bmatrix} 1 \\ 2 \\ 6 \end{bmatrix}$,
$T(\mathbf{v}_1) = \mathbf{v}_1 + 2\mathbf{v}_2 + 6\mathbf{v}_3 = 3x + 3x^2 - 2 + 6x + 4x^2 + 18 + 42x + 12x^2 = 16 + 51x + 19x^2$.
Similarly, $T(\mathbf{v}_2) = 3\mathbf{v}_1 - 2\mathbf{v}_3 = 9x + 9x^2 - 6 - 14x - 4x^2 = -6 - 5x + 5x^2$, and

$T(\mathbf{v}_3) = -\mathbf{v}_1 + 5\mathbf{v}_2 + 4\mathbf{v}_3 = -3x - 3x^2 - 5 + 15x + 10x^2 + 12 + 28x + 8x^2$
$= 7 + 40x + 15x^2$.

(c) If $a_0 + a_1x + a_2x^2 = c_1\mathbf{v}_1 + c_2\mathbf{v}_2 + c_3\mathbf{v}_3$
$= c_1(3x + 3x^2) + c_2(-1 + 3x + 2x^2) + c_3(3 + 7x + 2x^2)$ then
$a_0 = -c_2 + 3c_3,\ a_1 = 3c_1 + 3c_2 + 7c_3,\ a_2 = 3c_1 + 2c_2 + 2c_3$.
Solving for c_1, c_2, and c_3 gives
$c_1 = \frac{1}{3}(a_0 - a_1 + 2a_2),\ c_2 = \frac{1}{8}(-5a_0 + 3a_1 - 3a_2),\ c_3 = \frac{1}{8}(a_0 + a_1 - a_2)$, so
$T(a_0 + a_1x + a_2x^2) = c_1T(\mathbf{v}_1) + c_2T(\mathbf{v}_2) + c_3T(\mathbf{v}_3)$
$= \frac{1}{3}(a_0 - a_1 + 2a_2)(16 + 51x + 19x^2) + \frac{1}{8}(-5a_0 + 3a_1 - 3a_2)(-6 - 5x + 5x^2)$
$+ \frac{1}{8}(a_0 + a_1 - a_2)(7 + 40x + 15x^2)$
$= \frac{239a_0 - 161a_1 + 289a_2}{24} + \frac{201a_0 - 111a_1 + 247a_2}{8}x + \frac{61a_0 - 31a_1 + 107a_2}{12}x^2$

(d) In $1 + x^2$, $a_0 = 1, a_1 = 0, a_2 = 1$.
$T(1 + x^2) = \frac{239+289}{24} + \frac{201+247}{8}x + \frac{61+107}{12}x^2 = 22 + 56x + 14x^2$

13. (a) $(T_2 \circ T_1)(1) = T_2(2) = 6x$ and $(T_2 \circ T_1)(x) = T_2(-3x) = -9x^2$ so $[T_2 \circ T_1]_{B',B} = \begin{bmatrix} 0 & 0 \\ 6 & 0 \\ 0 & -9 \\ 0 & 0 \end{bmatrix}$;

$T_1(1) = 2$ and $T_1(x) = -3x$ so $[T_1]_{B'',B} = \begin{bmatrix} 2 & 0 \\ 0 & -3 \\ 0 & 0 \end{bmatrix}$;

$T_2(1) = 3x$, $T_2(x) = 3x^2$, and $T_2(x^2) = 3x^3$ so $[T_2]_{B',B''} = \begin{bmatrix} 0 & 0 & 0 \\ 3 & 0 & 0 \\ 0 & 3 & 0 \\ 0 & 0 & 3 \end{bmatrix}$.

(b) $[T_2 \circ T_1]_{B',B} = [T_2]_{B',B''}[T_1]_{B'',B}$

(c) $\begin{bmatrix} 0 & 0 & 0 \\ 3 & 0 & 0 \\ 0 & 3 & 0 \\ 0 & 0 & 3 \end{bmatrix} \begin{bmatrix} 2 & 0 \\ 0 & -3 \\ 0 & 0 \end{bmatrix} = \begin{bmatrix} 0 & 0 \\ 6 & 0 \\ 0 & -9 \\ 0 & 0 \end{bmatrix}$

15. (a) Since $T(1) = \begin{bmatrix} 1 & 1 \\ 1 & 1 \end{bmatrix}$, $T(x) = \begin{bmatrix} 0 & 1 \\ -1 & 0 \end{bmatrix}$, $T(x^2) = \begin{bmatrix} 0 & 1 \\ 1 & 0 \end{bmatrix}$, we have $[T(1)]_B = \begin{bmatrix} 1 \\ 1 \\ 1 \\ 1 \end{bmatrix}$,

$[T(x)]_B = \begin{bmatrix} 0 \\ 1 \\ -1 \\ 0 \end{bmatrix}$, and $[T(x^2)]_B = \begin{bmatrix} 0 \\ 1 \\ 1 \\ 0 \end{bmatrix}$. Consequently, $[T]_{B,B'} = \begin{bmatrix} 1 & 0 & 0 \\ 1 & 1 & 1 \\ 1 & -1 & 1 \\ 1 & 0 & 0 \end{bmatrix}$.

Since $T(1) = \begin{bmatrix} 1 & 1 \\ 1 & 1 \end{bmatrix}$, $T(1+x) = \begin{bmatrix} 1 & 2 \\ 0 & 1 \end{bmatrix}$, and $T(1+x^2) = \begin{bmatrix} 1 & 2 \\ 2 & 1 \end{bmatrix}$, we have $[T(1)]_B = \begin{bmatrix} 1 \\ 1 \\ 1 \\ 1 \end{bmatrix}$,

$[T(1+x)]_B = \begin{bmatrix} 1 \\ 2 \\ 0 \\ 1 \end{bmatrix}$, and $[T(1+x^2)]_B = \begin{bmatrix} 1 \\ 2 \\ 2 \\ 1 \end{bmatrix}$. Consequently, $[T]_{B,B''} = \begin{bmatrix} 1 & 1 & 1 \\ 1 & 2 & 2 \\ 1 & 0 & 2 \\ 1 & 1 & 1 \end{bmatrix}$.

(b) Applying the three-step procedure to the bases B' and B we obtain:

Step 1. The coordinate vector of $\mathbf{x} = 2 + 2x + x^2$ with respect to the basis B' is $[\mathbf{x}]_{B'} = \begin{bmatrix} 2 \\ 2 \\ 1 \end{bmatrix}$.

Step 2. $[T]_{B,B'}[\mathbf{x}]_{B'} = \begin{bmatrix} 1 & 0 & 0 \\ 1 & 1 & 1 \\ 1 & -1 & 1 \\ 1 & 0 & 0 \end{bmatrix} \begin{bmatrix} 2 \\ 2 \\ 1 \end{bmatrix} = \begin{bmatrix} 2 \\ 5 \\ 1 \\ 2 \end{bmatrix} = [T(\mathbf{x})]_B$.

Step 3. Reconstructing $T(\mathbf{x})$ from the coordinate vector obtained in Step 2, we obtain $T(\mathbf{x}) = 2\begin{bmatrix} 1 & 0 \\ 0 & 0 \end{bmatrix} + 5\begin{bmatrix} 0 & 1 \\ 0 & 0 \end{bmatrix} + 1\begin{bmatrix} 0 & 0 \\ 1 & 0 \end{bmatrix} + 2\begin{bmatrix} 0 & 0 \\ 0 & 1 \end{bmatrix} = \begin{bmatrix} 2 & 5 \\ 1 & 2 \end{bmatrix}$.

Applying the three-step procedure to the bases B'' and B we obtain:

Step 1. The coordinate vector of $\mathbf{x} = 2 + 2x + x^2$ with respect to the basis B'' is $[\mathbf{x}]_{B''} = \begin{bmatrix} 2 \\ 2 \\ 1 \end{bmatrix}$.

To find the coordinate vector of $\mathbf{x} = 2 + 2x + x^2$ with respect to the basis B' we solve the system

$$\begin{aligned} c_1 + c_2 + c_3 &= 2 \\ c_2 &= 2 \\ + c_3 &= 1 \end{aligned}$$

Back-substitution yields $[\mathbf{x}]_{B''} = \begin{bmatrix} -1 \\ 2 \\ 1 \end{bmatrix}$.

Step 2. $[T]_{B,B''}[\mathbf{x}]_{B''} = \begin{bmatrix} 1 & 1 & 1 \\ 1 & 2 & 2 \\ 1 & 0 & 2 \\ 1 & 1 & 1 \end{bmatrix} \begin{bmatrix} -1 \\ 2 \\ 1 \end{bmatrix} = \begin{bmatrix} 2 \\ 5 \\ 1 \\ 2 \end{bmatrix} = [T(\mathbf{x})]_B$.

Step 3. Reconstructing $T(\mathbf{x})$ from the coordinate vector obtained in Step 2, we obtain $T(\mathbf{x}) = 2\begin{bmatrix} 1 & 0 \\ 0 & 0 \end{bmatrix} + 5\begin{bmatrix} 0 & 1 \\ 0 & 0 \end{bmatrix} + 1\begin{bmatrix} 0 & 0 \\ 1 & 0 \end{bmatrix} + 2\begin{bmatrix} 0 & 0 \\ 0 & 1 \end{bmatrix} = \begin{bmatrix} 2 & 5 \\ 1 & 2 \end{bmatrix}$.

(c) $T(2 + 2x + x^2) = \begin{bmatrix} 2 & 5 \\ 1 & 2 \end{bmatrix}$

17. **(a)** $D(\mathbf{p}_1) = 0$; $D(\mathbf{p}_2) = 1$; $D(\mathbf{p}_3) = 2x$; $[D(\mathbf{p}_1)]_B = \begin{bmatrix}0\\0\\0\end{bmatrix}$, $[D(\mathbf{p}_2)]_B = \begin{bmatrix}1\\0\\0\end{bmatrix}$, and $[D(\mathbf{p}_3)]_B = \begin{bmatrix}0\\2\\0\end{bmatrix}$

therefore $[D]_B = \begin{bmatrix}0 & 1 & 0\\0 & 0 & 2\\0 & 0 & 0\end{bmatrix}$.

(b) Denoting $\mathbf{p} = 6 - 6x + 24x^2$, we have $[\mathbf{p}]_B = \begin{bmatrix}6\\-6\\24\end{bmatrix}$;

$[D]_B[\mathbf{p}]_B = \begin{bmatrix}0 & 1 & 0\\0 & 0 & 2\\0 & 0 & 0\end{bmatrix}\begin{bmatrix}6\\-6\\24\end{bmatrix} = \begin{bmatrix}-6\\48\\0\end{bmatrix} = [D(\mathbf{p})]_B$ thus $D(\mathbf{p}) = -6\mathbf{p}_1 + 48\mathbf{p}_2 + 0\mathbf{p}_3 = -6 + 48x$.

19. **(a)** $D(\mathbf{f}_1) = D(1) = 0$; $D(\mathbf{f}_2) = D(\text{sin}x) = \text{cos}x$; $D(\mathbf{f}_3) = D(\text{cos}x) = -\text{sin}x$

The matrix for D relative to this basis is $\begin{bmatrix}0 & 0 & 0\\0 & 0 & -1\\0 & 1 & 0\end{bmatrix}$.

(b) Since $[2 + 3\sin x - 4\cos x]_B = \begin{bmatrix}2\\3\\-4\end{bmatrix}$, $[D(2 + 3\sin x - 4\cos x)]_B = \begin{bmatrix}0 & 0 & 0\\0 & 0 & -1\\0 & 1 & 0\end{bmatrix}\begin{bmatrix}2\\3\\-4\end{bmatrix} = \begin{bmatrix}0\\4\\3\end{bmatrix}$.

Consequently, $D(2 + 3\sin x - 4\cos x) = (0)(1) + 4\sin x + 3\cos x = 4\sin x + 3\cos x$.

21. **(a)** $[T_2 \circ T_1]_{B',B} = [T_2]_{B',B''}[T_1]_{B'',B}$

(b) $[T_3 \circ T_2 \circ T_1]_{B',B} = [T_3]_{B',B'''}[T_2]_{B''',B''}[T_1]_{B'',B}$

23. The matrix for T relative to B is the matrix whose columns are the transforms of the basis vectors in B in terms of the standard basis. Since B is the standard basis for R^n, this matrix is the standard matrix for T. Also, since B' is the standard basis for R^m, the resulting transformation will give vector components relative to the standard basis.

True-False Exercises

(a) False. The conclusion would only be true if $T\colon V \to V$ were a linear operator, i.e., if $V = W$.

(b) False.

(c) True. Since the matrix for T is invertible, by Theorem 8.4.2 $\ker(T) = \{\mathbf{0}\}$.

(d) False. It follows from Theorem 8.4.1 that the matrix of $S \circ T$ relative to B is $[S]_B[T]_B$.

(e) True. This follows from Theorem 8.4.2.

8.5 Similarity

1. **(a)** $\det(A) = -2$ does not equal $\det(B) = -1$

(b) $\text{tr}(A) = 3$ does not equal $\text{tr}(B) = -2$

3. Since $\begin{bmatrix}3 & 2\\1 & 1\end{bmatrix}^{-1} = \frac{1}{(3)(1)-(2)(1)}\begin{bmatrix}1 & -2\\-1 & 3\end{bmatrix} = \begin{bmatrix}1 & -2\\-1 & 3\end{bmatrix}$,

$[T]_{B'} = P_{B\to B'}[T]_B P_{B'\to B} = \begin{bmatrix}3 & 2\\1 & 1\end{bmatrix}\begin{bmatrix}2 & 0\\1 & 1\end{bmatrix}\begin{bmatrix}1 & -2\\-1 & 3\end{bmatrix} = \begin{bmatrix}6 & -10\\2 & -3\end{bmatrix}$

5. Since $\begin{bmatrix}3 & 2\\1 & 1\end{bmatrix}^{-1} = \frac{1}{(3)(1)-(2)(1)}\begin{bmatrix}1 & -2\\-1 & 3\end{bmatrix} = \begin{bmatrix}1 & -2\\-1 & 3\end{bmatrix}$,

$[T]_B = P_{B'\to B}[T]_{B'} P_{B\to B'} = \begin{bmatrix}1 & -2\\-1 & 3\end{bmatrix}\begin{bmatrix}2 & 0\\1 & 1\end{bmatrix}\begin{bmatrix}3 & 2\\1 & 1\end{bmatrix} = \begin{bmatrix}-2 & -2\\6 & 5\end{bmatrix}$

7. From the definition of T we have $T(\mathbf{u}_1) = \begin{bmatrix}1\\0\end{bmatrix} = [T(\mathbf{u}_1)]_B$ and $T(\mathbf{u}_2) = \begin{bmatrix}-2\\-1\end{bmatrix} = [T(\mathbf{u}_2)]_B$ therefore $[T]_B = \begin{bmatrix}1 & -2\\0 & -1\end{bmatrix}$. Since $[\mathbf{v}_1]_B = \begin{bmatrix}4\\1\end{bmatrix}$ and $[\mathbf{v}_2]_B = \begin{bmatrix}7\\2\end{bmatrix}$, we have $P_{B'\to B} = \begin{bmatrix}4 & 7\\1 & 2\end{bmatrix}$.
The inverse of $P_{B'\to B}$ is $P_{B'\to B}^{-1} = \begin{bmatrix}2 & -7\\-1 & 4\end{bmatrix}$. Using Theorem 8.5.2, we obtain

$[T]_{B'} = P_{B'\to B}^{-1}[T]_B P_{B'\to B} = \begin{bmatrix}2 & -7\\-1 & 4\end{bmatrix}\begin{bmatrix}1 & -2\\0 & -1\end{bmatrix}\begin{bmatrix}4 & 7\\1 & 2\end{bmatrix} = \begin{bmatrix}11 & 20\\-6 & -11\end{bmatrix}$.

9. Denoting $\mathbf{e}_1 = (1,0,0)$, $\mathbf{e}_2 = (0,1,0)$, and $\mathbf{e}_3 = (0,0,1)$, we have $T(\mathbf{e}_1) = (-2,1,0) = (T(\mathbf{e}_1))_B$, $T(\mathbf{e}_2) = (-1,0,1) = (T(\mathbf{e}_2))_B$, and $T(\mathbf{e}_3) = (0,1,0) = (T(\mathbf{e}_3))_B$ therefore $[T]_B = \begin{bmatrix}-2 & -1 & 0\\1 & 0 & 1\\0 & 1 & 0\end{bmatrix}$.

Since $(\mathbf{v}_1)_B = (-2,1,0)$, $(\mathbf{v}_2)_B = (-1,0,1)$, and $(\mathbf{v}_3)_B = (0,1,0)$, we have $P_{B'\to B} = \begin{bmatrix}-2 & -1 & 0\\1 & 0 & 1\\0 & 1 & 0\end{bmatrix}$.

Using Theorem 8.5.2, we obtain

$$[T]_{B'} = P_{B'\to B}^{-1}[T]_B P_{B'\to B} = \begin{bmatrix}-2 & -1 & 0\\1 & 0 & 1\\0 & 1 & 0\end{bmatrix}^{-1}\begin{bmatrix}-2 & -1 & 0\\1 & 0 & 1\\0 & 1 & 0\end{bmatrix}\begin{bmatrix}-2 & -1 & 0\\1 & 0 & 1\\0 & 1 & 0\end{bmatrix} = \begin{bmatrix}-2 & -1 & 0\\1 & 0 & 1\\0 & 1 & 0\end{bmatrix}.$$

11. Denoting $\mathbf{e}_1 = (1,0)$ and $\mathbf{e}_2 = (0,1)$ we have $T(\mathbf{e}_1) = (\cos 45°, \sin 45°) = \left(\frac{1}{\sqrt{2}}, \frac{1}{\sqrt{2}}\right) = (T(\mathbf{e}_1))_B$ and $T(\mathbf{e}_2) = (-\sin 45°, \cos 45°) = \left(-\frac{1}{\sqrt{2}}, \frac{1}{\sqrt{2}}\right) = (T(\mathbf{e}_2))_B$ therefore $[T]_B = \begin{bmatrix}\frac{1}{\sqrt{2}} & -\frac{1}{\sqrt{2}}\\ \frac{1}{\sqrt{2}} & \frac{1}{\sqrt{2}}\end{bmatrix}$.

Since $(\mathbf{v}_1)_B = \left(\frac{1}{\sqrt{2}}, \frac{1}{\sqrt{2}}\right)$ and $(\mathbf{v}_2)_B = \left(-\frac{1}{\sqrt{2}}, \frac{1}{\sqrt{2}}\right)$, we have $P_{B'\to B} = \begin{bmatrix}\frac{1}{\sqrt{2}} & -\frac{1}{\sqrt{2}}\\ \frac{1}{\sqrt{2}} & \frac{1}{\sqrt{2}}\end{bmatrix}$. Using Theorem 8.5.2,

we obtain $[T]_{B'} = P_{B'\to B}^{-1}[T]_B P_{B'\to B} = \begin{bmatrix}\frac{1}{\sqrt{2}} & -\frac{1}{\sqrt{2}}\\ \frac{1}{\sqrt{2}} & \frac{1}{\sqrt{2}}\end{bmatrix}^{-1}\begin{bmatrix}\frac{1}{\sqrt{2}} & -\frac{1}{\sqrt{2}}\\ \frac{1}{\sqrt{2}} & \frac{1}{\sqrt{2}}\end{bmatrix}\begin{bmatrix}\frac{1}{\sqrt{2}} & -\frac{1}{\sqrt{2}}\\ \frac{1}{\sqrt{2}} & \frac{1}{\sqrt{2}}\end{bmatrix} = \begin{bmatrix}\frac{1}{\sqrt{2}} & -\frac{1}{\sqrt{2}}\\ \frac{1}{\sqrt{2}} & \frac{1}{\sqrt{2}}\end{bmatrix}$.

13. Denoting $\mathbf{p}_1 = 1$ and $\mathbf{p}_2 = x$ we have $T(\mathbf{p}_1) = -1 + x$ and $T(\mathbf{p}_2) = x$. Thus $(T(\mathbf{p}_1))_B = (-1,1)$ and $(T(\mathbf{p}_2))_B = (0,1)$ so $[T]_B = \begin{bmatrix}-1 & 0\\1 & 1\end{bmatrix}$. Since $(\mathbf{q}_1)_B = (1,1)$ and $(\mathbf{q}_2)_B = (-1,1)$, we have

$P_{B'\to B} = \begin{bmatrix} 1 & -1 \\ 1 & 1 \end{bmatrix}$. The inverse of $P_{B'\to B}$ is $P_{B'\to B}^{-1} = \begin{bmatrix} \frac{1}{2} & \frac{1}{2} \\ -\frac{1}{2} & \frac{1}{2} \end{bmatrix}$. Using Theorem 8.5.2, we obtain

$$[T]_{B'} = P_{B'\to B}^{-1}[T]_B P_{B'\to B} = \begin{bmatrix} \frac{1}{2} & \frac{1}{2} \\ -\frac{1}{2} & \frac{1}{2} \end{bmatrix}\begin{bmatrix} -1 & 0 \\ 1 & 1 \end{bmatrix}\begin{bmatrix} 1 & -1 \\ 1 & 1 \end{bmatrix} = \begin{bmatrix} \frac{1}{2} & \frac{1}{2} \\ \frac{3}{2} & -\frac{1}{2} \end{bmatrix}.$$

15. (a) We have $T(1) = 5 + x^2$, $T(x) = 6 - x$, and $T(x^2) = 2 - 8x - 2x^2$ so the matrix for T relative to the standard basis B is $[T]_B = \begin{bmatrix} 5 & 6 & 2 \\ 0 & -1 & -8 \\ 1 & 0 & -2 \end{bmatrix}$. The characteristic polynomial for $[T]_B$ is $\lambda^3 - 2\lambda^2 - 15\lambda + 36 = (\lambda + 4)(\lambda - 3)^2$ so the eigenvalues of T are $\lambda = -4$ and $\lambda = 3$.

(b) A basis for the eigenspace of $[T]_B$ corresponding to $\lambda = -4$ is $\begin{bmatrix} -2 \\ \frac{8}{3} \\ 1 \end{bmatrix}$, so the polynomial $-2 + \frac{8}{3}x + x^2$ is a basis in P^2 for the corresponding eigenspace of T.

A basis for the eigenspace of $[T]_B$ corresponding to $\lambda = 3$ is $\begin{bmatrix} 5 \\ -2 \\ 1 \end{bmatrix}$, so the polynomial $5 - 2x + x^2$ is a basis in P^2 for the corresponding eigenspace of T.

19. The matrix of T with respect to the standard basis for R^2, $B = \{(1,0), (0,1)\}$, is $[T]_B = \begin{bmatrix} 3 & -4 \\ -1 & 7 \end{bmatrix}$.
By Formula (12), $\det(T) = \det([T]_B) = 17$.
The characteristic polynomial of $[T]_B$ is $\lambda^2 - 10\lambda + 17$. The eigenvalues of T are $5 \pm 2\sqrt{2}$.

21. $T(a_0 + a_1x + a_2x^2) = a_0 + a_1(x-1) + a_2(x-1)^2 = (a_0 - a_1 + a_2) + (a_1 - 2a_2)x + a_2x^2$ hence the matrix of T with respect to the standard basis for P_2, $B = \{1, x, x^2\}$, is $[T]_B = \begin{bmatrix} 1 & -1 & 1 \\ 0 & 1 & -2 \\ 0 & 0 & 1 \end{bmatrix}$.
By Formula (12), $\det(T) = \det([T]_B) = 1$. The characteristic polynomial of $[T]_B$ is $(\lambda - 1)^3$.
The only eigenvalue of T is 1.

23. Step (1) follows from the hypothesis (since $B = P^{-1}AP$).
Step (2) follows from Formula (1) in Section 1.4 (since $I = P^{-1}P$).
Step (3) follows from parts (f), (g), and (m) of Theorem 1.4.1.
Step (4) follows from Theorem 2.3.4.
Step (5) follows from commutativity of real number multiplication.
Step (6) follows from Theorem 2.3.4, Formula (1) in Section 1.4, and from Theorem 2.1.2 (since $\det(P^{-1})\det(P) = \det(P^{-1}P) = \det(I) = 1$).

31. Since $C[\mathbf{x}]_B = D[\mathbf{x}]_B$ for all $\mathbf{x}$ in V, then we can let $\mathbf{x} = \mathbf{v}_i$ for each of the basis vectors $\mathbf{v}_1, \ldots, \mathbf{v}_n$ of V. Since $[\mathbf{v}_i]_B = \mathbf{e}_i$ for each i where $\{\mathbf{e}_1, \ldots, \mathbf{e}_n\}$ is the standard basis for R^n, this yields $C\mathbf{e}_i = D\mathbf{e}_i$ for $i = 1, 2, \ldots, n$. But $C\mathbf{e}_i$ and $D\mathbf{e}_i$ are the ith columns of C and D, respectively. Since corresponding columns of C and D are all equal, $C = D$.

True-False Exercises

(a) False. Every matrix is similar to itself since $A = I^{-1}AI$.

(b) True. If $A = P^{-1}BP$ and $B = Q^{-1}CQ$, then $A = P^{-1}(Q^{-1}CQ)P = (QP)^{-1}C(QP)$.

(c) True. Invertibility is a similarity invariant.

(d) True. If $A = P^{-1}BP$, then $A^{-1} = (P^{-1}BP)^{-1} = P^{-1}B^{-1}P$.

(e) True.

(f) False. For example, if T_1 is the zero operator then $[T_1]_B$ with respect to any basis B is a zero matrix.

(g) True. By Theorem 8.5.2, for any basis B' for R^n there exists P such that $[T]_{B'} = P^{-1}[T]_B P = P^{-1}IP = P^{-1}P = I$.

(h) False. If B and B' are different, let $[T]_B$ be given by the matrix $P_{B\to B'}$. Then $[T]_{B',B} = P_{B\to B'}[T]_B = P_{B\to B'}P_{B'\to B} = I_n$.

Chapter 8 Supplementary Exercises

1. No. $T(\mathbf{x}_1 + \mathbf{x}_2) = A(\mathbf{x}_1 + \mathbf{x}_2) + B \neq (A\mathbf{x}_1 + B) + (A\mathbf{x}_2 + B) = T(\mathbf{x}_1) + T(\mathbf{x}_2)$

3. For instance let A and B have all zero entries except for the nonzero entries $(A)_{11} \neq (B)_{11}$ so that their traces are different. E.g., $A = \begin{bmatrix} 1 & 0 & 0 \\ 0 & 0 & 0 \\ 0 & 0 & 0 \end{bmatrix}$ and $B = \begin{bmatrix} 2 & 0 & 0 \\ 0 & 0 & 0 \\ 0 & 0 & 0 \end{bmatrix}$ are not similar.

5. **(a)** The matrix for T relative to the standard basis is $A = \begin{bmatrix} 1 & 0 & 1 & 1 \\ 2 & 1 & 3 & 1 \\ 1 & 0 & 0 & 1 \end{bmatrix}$. A basis for the range of T is a basis for the column space of A. A reduces to $\begin{bmatrix} 1 & 0 & 0 & 1 \\ 0 & 1 & 0 & -1 \\ 0 & 0 & 1 & 0 \end{bmatrix}$.

Since row operations don't change the dependency relations among columns, the reduced form of A indicates that $T(\mathbf{e}_3)$ and any two of $T(\mathbf{e}_1)$, $T(\mathbf{e}_2)$, $T(\mathbf{e}_4)$ form a basis for the range.

The reduced form of A shows that the general solution of $A\mathbf{x} = \mathbf{0}$ is $x = -s$, $x_2 = s$, $x_3 = 0$, $x_4 = s$ so a basis for the null space of A, which is the kernel of T, is $\begin{bmatrix} -1 \\ 1 \\ 0 \\ 1 \end{bmatrix}$.

(b) Since $R(T)$ is three-dimensional and $\ker(T)$ is one-dimensional, $\text{rank}(T) = 3$ and $\text{nullity}(T) = 1$.

7. **(a)** The matrix for T relative to B is $[T]_B = \begin{bmatrix} 1 & 1 & 2 & -2 \\ 1 & -1 & -4 & 6 \\ 1 & 2 & 5 & -6 \\ 3 & 2 & 3 & -2 \end{bmatrix}$.

$[T]_B$ reduces to $\begin{bmatrix} 1 & 0 & -1 & 2 \\ 0 & 1 & 3 & -4 \\ 0 & 0 & 0 & 0 \\ 0 & 0 & 0 & 0 \end{bmatrix}$ which has rank 2 and nullity 2. Thus, rank$(T) = 2$ and nullity$(T) = 2$.

(b) Since nullity$(T) \neq 0$, T is not one-to-one.

9. **(a)** Since $A = P^{-1}BP$, we have
$A^T = (P^{-1}BP)^T = P^T B^T (P^{-1})^T = ((P^T)^{-1})^{-1} B^T (P^{-1})^T = ((P^{-1})^T)^{-1} B^T (P^{-1})^T$.
Thus, A^T and B^T are similar.

(b) $A^{-1} = (P^{-1}BP)^{-1} = P^{-1}B^{-1}P$ thus A^{-1} and B^{-1} are similar.

11. For $X = \begin{bmatrix} a & b \\ c & d \end{bmatrix}$, $T(X) = \begin{bmatrix} a+c & b+d \\ 0 & 0 \end{bmatrix} + \begin{bmatrix} b & b \\ d & d \end{bmatrix} = \begin{bmatrix} a+b+c & 2b+d \\ d & d \end{bmatrix}$.
$T(X) = \mathbf{0}$ gives the equations

$$\begin{aligned} a + b + c &= 0 \\ 2b + d &= 0 \\ d &= 0 \end{aligned}$$

Thus $b = d = 0$ and $c = -a$ hence X is in ker(T) if it has the form $\begin{bmatrix} k & 0 \\ -k & 0 \end{bmatrix}$, so ker$(T) = \text{span}\left\{\begin{bmatrix} 1 & 0 \\ -1 & 0 \end{bmatrix}\right\}$ which is one-dimensional. We conclude that nullity$(T) = 1$ and rank$(T) = \dim(M_{22}) - \text{nullity}(T) = 3$.

13. The standard basis for M_{22} is $\mathbf{u}_1 = \begin{bmatrix} 1 & 0 \\ 0 & 0 \end{bmatrix}$, $\mathbf{u}_2 = \begin{bmatrix} 0 & 1 \\ 0 & 0 \end{bmatrix}$, $\mathbf{u}_3 = \begin{bmatrix} 0 & 0 \\ 1 & 0 \end{bmatrix}$, $\mathbf{u}_4 = \begin{bmatrix} 0 & 0 \\ 0 & 1 \end{bmatrix}$. Since $L(\mathbf{u}_1) = \mathbf{u}_1$, $L(\mathbf{u}_2) = \mathbf{u}_3$, $L(\mathbf{u}_3) = \mathbf{u}_2$, and $L(\mathbf{u}_4) = \mathbf{u}_4$, the matrix of L relative to the standard basis is $\begin{bmatrix} 1 & 0 & 0 & 0 \\ 0 & 0 & 1 & 0 \\ 0 & 1 & 0 & 0 \\ 0 & 0 & 0 & 1 \end{bmatrix}$

15. The transition matrix from B' to B is $P = \begin{bmatrix} 1 & 1 & 1 \\ 0 & 1 & 1 \\ 0 & 0 & 1 \end{bmatrix}$, thus $[T]_{B'} = P^{-1}[T]_B P = \begin{bmatrix} -4 & 0 & 9 \\ 1 & 0 & -2 \\ 0 & 1 & 1 \end{bmatrix}$.

17. $T\left(\begin{bmatrix} 1 \\ 0 \\ 0 \end{bmatrix}\right) = \begin{bmatrix} 1 \\ 0 \\ 1 \end{bmatrix}$, $T\left(\begin{bmatrix} 0 \\ 1 \\ 0 \end{bmatrix}\right) = \begin{bmatrix} -1 \\ 1 \\ 0 \end{bmatrix}$, and $T\left(\begin{bmatrix} 0 \\ 0 \\ 1 \end{bmatrix}\right) = \begin{bmatrix} 1 \\ 0 \\ -1 \end{bmatrix}$, thus $[T]_B = \begin{bmatrix} 1 & -1 & 1 \\ 0 & 1 & 0 \\ 1 & 0 & -1 \end{bmatrix}$.
Note that this can also be read directly from $[T(\mathbf{x})]_B$.

19. **(a)** $D(\mathbf{f} + \mathbf{g}) = (f(x) + g(x))'' = f''(x) + g''(x)$ and $D(k\mathbf{f}) = (kf(x))'' = kf''(x)$.

(b) If $\mathbf{f}$ is in ker(D), then $\mathbf{f}$ has the form $\mathbf{f} = f(x) = a_0 + a_1 x$, so a basis for ker$(D)$ is $f(x) = 1, g(x) = x$.

(c) The equation $D(\mathbf{f}) = \mathbf{f}$ can be rewritten as $y'' = y$. Substituting $y_1 = y$ and $y_2 = y'$ yields the system

$$y_1' = y_2$$
$$y_2' = y_1$$

The coefficient matrix of this system is $A = \begin{bmatrix} 0 & 1 \\ 1 & 0 \end{bmatrix}$. The characteristic polynomial of A is $\det(\lambda I - A) = \begin{vmatrix} \lambda & -1 \\ -1 & \lambda \end{vmatrix} = \lambda^2 - 1 = (\lambda - 1)(\lambda + 1)$ thus the eigenvalues of A are $\lambda = 1$ and $\lambda = -1$.

The reduced row echelon form of $1I - A$ is $\begin{bmatrix} 1 & -1 \\ 0 & 0 \end{bmatrix}$ so that the eigenspace corresponding to $\lambda = 1$ consists of vectors $\begin{bmatrix} x_1 \\ x_2 \end{bmatrix}$ where $x_1 = t, x_2 = t$. A vector $\mathbf{p}_1 = \begin{bmatrix} 1 \\ 1 \end{bmatrix}$ forms a basis for this eigenspace.

The reduced row echelon form of $-1I - A$ is $\begin{bmatrix} 1 & 1 \\ 0 & 0 \end{bmatrix}$ so that the eigenspace corresponding to $\lambda = -1$ consists of vectors $\begin{bmatrix} x_1 \\ x_2 \end{bmatrix}$ where $x_1 = -t, x_2 = t$. A vector $\mathbf{p}_2 = \begin{bmatrix} -1 \\ 1 \end{bmatrix}$ forms a basis for this eigenspace.

Therefore $P = \begin{bmatrix} 1 & -1 \\ 1 & 1 \end{bmatrix}$ diagonalizes A and $P^{-1}AP = D = \begin{bmatrix} 1 & 0 \\ 0 & -1 \end{bmatrix}$.

The substitution $\mathbf{y} = P\mathbf{u}$ yields the "diagonal system" $\mathbf{u}' = \begin{bmatrix} 1 & 0 \\ 0 & -1 \end{bmatrix}\mathbf{u}$ consisting of equations $u_1' = u_1$ and $u_2' = -u_2$. From Formula (2) in Section 5.4, these equations have the solutions $u_1 = c_1e^x, u_2 = c_2e^{-x}$, i.e., $\mathbf{u} = \begin{bmatrix} c_1e^x \\ c_2e^{-x} \end{bmatrix}$. From $\mathbf{y} = P\mathbf{u}$ we obtain the solution

$\mathbf{y} = \begin{bmatrix} 1 & -1 \\ 1 & 1 \end{bmatrix}\begin{bmatrix} c_1e^x \\ c_2e^{-x} \end{bmatrix} = \begin{bmatrix} c_1e^x - c_2e^{-x} \\ c_1e^x + c_2e^{-x} \end{bmatrix}$ thus $y_1 = c_1e^x - c_2e^{-x}$ and $y_2 = c_1e^x + c_2e^{-x}$.

We conclude that the original equation $y'' = y$ has the solution $y = c_1e^x - c_2e^{-x}$.

Thus, $f(x) = e^x$ and $g(x) = e^{-x}$ form a basis for the subspace of $C^2(-\infty, \infty)$ containing the functions satisfying the equation $D(\mathbf{f}) = \mathbf{f}$. (Other bases are possible, e.g. $\{e^x, -e^{-x}\}$.)

21. **(c)** Note that $a_1P_1(x) + a_2P_2(x) + a_3P_3(x)$ evaluated at x_1, x_2, and x_3 gives the values a_1, a_2, and a_3, respectively, since $P_i(x_i) = 1$ and $P_i(x_j) = 0$ for $i \neq j$.

(d) From the computations in part (c), the points lie on the graph.

23. $D(1) = 0$
$D(x) = 1$
$D(x^2) = 2x$
$\vdots$
$D(x^n) = nx^{n-1}$

This gives the matrix shown.

25. The matrix of J with respect to the given bases is $\begin{bmatrix} 0 & 0 & 0 & \cdots & 0 & 0 \\ 1 & 0 & 0 & \cdots & 0 & 0 \\ 0 & \frac{1}{2} & 0 & \cdots & 0 & 0 \\ 0 & 0 & \frac{1}{3} & \cdots & 0 & 0 \\ \vdots & \vdots & \vdots & \ddots & \vdots & \vdots \\ 0 & 0 & 0 & \cdots & \frac{1}{n} & 0 \\ 0 & 0 & 0 & \cdots & 0 & \frac{1}{n+1} \end{bmatrix}$.

CHAPTER 9: NUMERICAL METHODS

9.1 LU-Decompositions

1. Step 1. Rewrite the system as $\underbrace{\begin{bmatrix} 3 & 0 \\ -2 & 1 \end{bmatrix}}_{L}\underbrace{\begin{bmatrix} 1 & -2 \\ 0 & 1 \end{bmatrix}}_{U}\underbrace{\begin{bmatrix} x_1 \\ x_2 \end{bmatrix}}_{\mathbf{x}} = \underbrace{\begin{bmatrix} 0 \\ 1 \end{bmatrix}}_{\mathbf{b}}$

Step 2. Define y_1 and y_2 by $\underbrace{\begin{bmatrix} 1 & -2 \\ 0 & 1 \end{bmatrix}}_{U}\underbrace{\begin{bmatrix} x_1 \\ x_2 \end{bmatrix}}_{\mathbf{x}} = \underbrace{\begin{bmatrix} y_1 \\ y_2 \end{bmatrix}}_{\mathbf{y}}$

Step 3. Solving $\underbrace{\begin{bmatrix} 3 & 0 \\ -2 & 1 \end{bmatrix}}_{L}\underbrace{\begin{bmatrix} y_1 \\ y_2 \end{bmatrix}}_{\mathbf{y}} = \underbrace{\begin{bmatrix} 0 \\ 1 \end{bmatrix}}_{\mathbf{b}}$ by forward substitution yields $y_1 = 0, y_2 = 1$.

Step 4. Solving $\underbrace{\begin{bmatrix} 1 & -2 \\ 0 & 1 \end{bmatrix}}_{U}\underbrace{\begin{bmatrix} x_1 \\ x_2 \end{bmatrix}}_{\mathbf{x}} = \underbrace{\begin{bmatrix} 0 \\ 1 \end{bmatrix}}_{\mathbf{y}}$ by back substitution yields $x_1 = 2, x_2 = 1$.

3. $A = \begin{bmatrix} 2 & 8 \\ -1 & -1 \end{bmatrix}$ $\qquad \begin{bmatrix} \bullet & 0 \\ \bullet & \bullet \end{bmatrix}$ (we follow the procedure of Example 3)

$\begin{bmatrix} \textcircled{1} & 4 \\ -1 & -1 \end{bmatrix} \leftarrow$ multiplier $= \frac{1}{2}$ $\qquad \begin{bmatrix} 2 & 0 \\ \bullet & \bullet \end{bmatrix}$

$\begin{bmatrix} 1 & 4 \\ \textcircled{0} & 3 \end{bmatrix} \leftarrow$ multiplier $= 1$ $\qquad \begin{bmatrix} 2 & 0 \\ -1 & \bullet \end{bmatrix}$

$U = \begin{bmatrix} 1 & 4 \\ 0 & \textcircled{1} \end{bmatrix} \leftarrow$ multiplier $= \frac{1}{3}$ $\qquad L = \begin{bmatrix} 2 & 0 \\ -1 & 3 \end{bmatrix}$

Step 1. Rewrite the system as $\underbrace{\begin{bmatrix} 2 & 0 \\ -1 & 3 \end{bmatrix}}_{L}\underbrace{\begin{bmatrix} 1 & 4 \\ 0 & 1 \end{bmatrix}}_{U}\underbrace{\begin{bmatrix} x_1 \\ x_2 \end{bmatrix}}_{\mathbf{x}} = \underbrace{\begin{bmatrix} -2 \\ -2 \end{bmatrix}}_{\mathbf{b}}$

Step 2. Define y_1 and y_2 by $\underbrace{\begin{bmatrix} 1 & 4 \\ 0 & 1 \end{bmatrix}}_{U}\underbrace{\begin{bmatrix} x_1 \\ x_2 \end{bmatrix}}_{\mathbf{x}} = \underbrace{\begin{bmatrix} y_1 \\ y_2 \end{bmatrix}}_{\mathbf{y}}$

Step 3. Solving $\underbrace{\begin{bmatrix} 2 & 0 \\ -1 & 3 \end{bmatrix}}_{L}\underbrace{\begin{bmatrix} y_1 \\ y_2 \end{bmatrix}}_{\mathbf{y}} = \underbrace{\begin{bmatrix} -2 \\ -2 \end{bmatrix}}_{\mathbf{b}}$ by forward substitution yields $y_1 = -1, y_2 = -1$.

Step 4. Solving $\underbrace{\begin{bmatrix} 1 & 4 \\ 0 & 1 \end{bmatrix}}_{U}\underbrace{\begin{bmatrix} x_1 \\ x_2 \end{bmatrix}}_{\mathbf{x}} = \underbrace{\begin{bmatrix} -1 \\ -1 \end{bmatrix}}_{\mathbf{y}}$ by back substitution yields $x_1 = 3, x_2 = -1$.

5. $A = \begin{bmatrix} 2 & -2 & -2 \\ 0 & -2 & 2 \\ -1 & 5 & 2 \end{bmatrix}$ $\qquad \begin{bmatrix} \bullet & 0 & 0 \\ \bullet & \bullet & 0 \\ \bullet & \bullet & \bullet \end{bmatrix}$

$$\begin{bmatrix} \textcircled{1} & -1 & -1 \\ 0 & -2 & 2 \\ -1 & 5 & 2 \end{bmatrix} \leftarrow \text{multiplier} = \tfrac{1}{2} \qquad \begin{bmatrix} 2 & 0 & 0 \\ \bullet & \bullet & 0 \\ \bullet & \bullet & \bullet \end{bmatrix}$$

$$\begin{bmatrix} 1 & -1 & -1 \\ \textcircled{0} & -2 & 2 \\ \textcircled{0} & 4 & 1 \end{bmatrix} \begin{matrix} \\ \leftarrow \text{multiplier} = 0 \\ \leftarrow \text{multiplier} = 1 \end{matrix} \qquad \begin{bmatrix} 2 & 0 & 0 \\ 0 & \bullet & 0 \\ -1 & \bullet & \bullet \end{bmatrix}$$

$$\begin{bmatrix} 1 & -1 & -1 \\ 0 & \textcircled{1} & -1 \\ 0 & 4 & 1 \end{bmatrix} \leftarrow \text{multiplier} = -\tfrac{1}{2} \qquad \begin{bmatrix} 2 & 0 & 0 \\ 0 & -2 & 0 \\ -1 & \bullet & \bullet \end{bmatrix}$$

$$\begin{bmatrix} 1 & -1 & -1 \\ 0 & 1 & -1 \\ 0 & \textcircled{0} & 5 \end{bmatrix} \leftarrow \text{multiplier} = -4 \qquad \begin{bmatrix} 2 & 0 & 0 \\ 0 & -2 & 0 \\ -1 & 4 & \bullet \end{bmatrix}$$

$$U = \begin{bmatrix} 1 & -1 & -1 \\ 0 & 1 & -1 \\ 0 & 0 & \textcircled{1} \end{bmatrix} \leftarrow \text{multiplier} = \tfrac{1}{5} \qquad L = \begin{bmatrix} 2 & 0 & 0 \\ 0 & -2 & 0 \\ -1 & 4 & 5 \end{bmatrix}$$

Step 1. Rewrite the system as $\underbrace{\begin{bmatrix} 2 & 0 & 0 \\ 0 & -2 & 0 \\ -1 & 4 & 5 \end{bmatrix}}_{L} \underbrace{\begin{bmatrix} 1 & -1 & -1 \\ 0 & 1 & -1 \\ 0 & 0 & 1 \end{bmatrix}}_{U} \underbrace{\begin{bmatrix} x_1 \\ x_2 \\ x_3 \end{bmatrix}}_{\mathbf{x}} = \underbrace{\begin{bmatrix} -4 \\ -2 \\ 6 \end{bmatrix}}_{\mathbf{b}}$

Step 2. Define y_1, y_2, and y_3 by $\underbrace{\begin{bmatrix} 1 & -1 & -1 \\ 0 & 1 & -1 \\ 0 & 0 & 1 \end{bmatrix}}_{U} \underbrace{\begin{bmatrix} x_1 \\ x_2 \\ x_3 \end{bmatrix}}_{\mathbf{x}} = \underbrace{\begin{bmatrix} y_1 \\ y_2 \\ y_3 \end{bmatrix}}_{\mathbf{y}}$

Step 3. Solving $\underbrace{\begin{bmatrix} 2 & 0 & 0 \\ 0 & -2 & 0 \\ -1 & 4 & 5 \end{bmatrix}}_{L} \underbrace{\begin{bmatrix} y_1 \\ y_2 \\ y_3 \end{bmatrix}}_{\mathbf{y}} = \underbrace{\begin{bmatrix} -4 \\ -2 \\ 6 \end{bmatrix}}_{\mathbf{b}}$ by forward substitution yields $y_1 = -2, y_2 = 1, y_3 = 0$.

Step 4. Solving $\underbrace{\begin{bmatrix} 1 & -1 & -1 \\ 0 & 1 & -1 \\ 0 & 0 & 1 \end{bmatrix}}_{U} \underbrace{\begin{bmatrix} x_1 \\ x_2 \\ x_3 \end{bmatrix}}_{\mathbf{x}} = \underbrace{\begin{bmatrix} -2 \\ 1 \\ 0 \end{bmatrix}}_{\mathbf{y}}$ by back substitution yields $x_1 = -1, x_2 = 1, x_3 = 0$.

7. **(a)** $L^{-1} = \begin{bmatrix} 1 & 0 & 0 \\ -2 & 1 & 0 \\ 1 & 1 & 1 \end{bmatrix}$; $U^{-1} = \begin{bmatrix} \frac{1}{2} & \frac{1}{8} & -\frac{7}{48} \\ 0 & \frac{1}{4} & \frac{5}{24} \\ 0 & 0 & \frac{1}{6} \end{bmatrix}$ **(b)** $A^{-1} = U^{-1}L^{-1} = \begin{bmatrix} \frac{5}{48} & -\frac{1}{48} & -\frac{7}{48} \\ -\frac{7}{24} & \frac{11}{24} & \frac{5}{24} \\ \frac{1}{6} & \frac{1}{6} & \frac{1}{6} \end{bmatrix}$

9. **(a)** Reduce A to upper triangular form.

$$\begin{bmatrix} 2 & 1 & -1 \\ -2 & -1 & 2 \\ 2 & 1 & 0 \end{bmatrix} \to \begin{bmatrix} 1 & \frac{1}{2} & -\frac{1}{2} \\ -2 & -1 & 2 \\ 2 & 1 & 0 \end{bmatrix} \to \begin{bmatrix} 1 & \frac{1}{2} & -\frac{1}{2} \\ 0 & 0 & 1 \\ 2 & 1 & 0 \end{bmatrix} \to \begin{bmatrix} 1 & \frac{1}{2} & -\frac{1}{2} \\ 0 & 0 & 1 \\ 0 & 0 & 1 \end{bmatrix} = U$$

The multipliers used were $\frac{1}{2}$, 2, and −2, which leads to $L = \begin{bmatrix} 2 & 0 & 0 \\ -2 & 1 & 0 \\ 2 & 0 & 1 \end{bmatrix}$ where the 1's on the diagonal reflect that no multiplication was required on the 2nd and 3rd diagonal entries.

(b) To change the 2 on the diagonal of L to a 1, the first column of L is divided by 2 and the diagonal matrix has a 2 as the 1, 1 entry.

$A = L_1DU_1 = \begin{bmatrix} 1 & 0 & 0 \\ -1 & 1 & 0 \\ 1 & 0 & 1 \end{bmatrix}\begin{bmatrix} 2 & 0 & 0 \\ 0 & 1 & 0 \\ 0 & 0 & 1 \end{bmatrix}\begin{bmatrix} 1 & \frac{1}{2} & -\frac{1}{2} \\ 0 & 0 & 1 \\ 0 & 0 & 1 \end{bmatrix}$ is the *LDU*-decomposition of A.

(c) Let $U_2 = DU$, and $L_2 = L_1$, then $A = L_2U_2 = \begin{bmatrix} 1 & 0 & 0 \\ -1 & 1 & 0 \\ 1 & 0 & 1 \end{bmatrix}\begin{bmatrix} 2 & 1 & -1 \\ 0 & 0 & 1 \\ 0 & 0 & 1 \end{bmatrix}$.

11. $P^{-1} = \begin{bmatrix} 0 & 1 & 0 \\ 1 & 0 & 0 \\ 0 & 0 & 1 \end{bmatrix}$ and $P^{-1}\mathbf{b} = \begin{bmatrix} 1 \\ 2 \\ 5 \end{bmatrix}$, so the system $P^{-1}A\mathbf{x} = P^{-1}\mathbf{b}$ is $\begin{bmatrix} 1 & 0 & 0 \\ 0 & 1 & 0 \\ 3 & -5 & 1 \end{bmatrix}\begin{bmatrix} 1 & 2 & 2 \\ 0 & 1 & 4 \\ 0 & 0 & 17 \end{bmatrix}\begin{bmatrix} x_1 \\ x_2 \\ x_3 \end{bmatrix} = \begin{bmatrix} 1 \\ 2 \\ 5 \end{bmatrix}$

$\begin{bmatrix} 1 & 0 & 0 \\ 0 & 1 & 0 \\ 3 & -5 & 1 \end{bmatrix}\begin{bmatrix} y_1 \\ y_2 \\ y_3 \end{bmatrix} = \begin{bmatrix} 1 \\ 2 \\ 5 \end{bmatrix}$ is

$$\begin{aligned} y_1 \qquad\quad &= 1 \\ y_2 \quad &= 2 \\ 3y_1 - 5y_2 + y_3 &= 5 \end{aligned}$$

which has the solution $y_1 = 1, y_2 = 2, y_3 = 12$.

$\begin{bmatrix} 1 & 2 & 2 \\ 0 & 1 & 4 \\ 0 & 0 & 17 \end{bmatrix}\begin{bmatrix} x_1 \\ x_2 \\ x_3 \end{bmatrix} = \begin{bmatrix} 1 \\ 2 \\ 12 \end{bmatrix}$ is

$$\begin{aligned} x_1 + 2x_2 + 2x_3 &= 1 \\ x_2 + 4x_3 &= 2 \\ 17x_3 &= 12 \end{aligned}$$

which gives the solution of the original system: $x_1 = \frac{21}{17}, x_2 = -\frac{14}{17}, x_3 = \frac{12}{17}$.

13. $A = \begin{bmatrix} 2 & 2 \\ 4 & 1 \end{bmatrix}$ $\qquad\qquad \begin{bmatrix} \bullet & 0 \\ \bullet & \bullet \end{bmatrix}$

$\begin{bmatrix} \textcircled{1} & 1 \\ 4 & 1 \end{bmatrix} \leftarrow$ multiplier $= \frac{1}{2}$ $\qquad\qquad \begin{bmatrix} 2 & 0 \\ \bullet & \bullet \end{bmatrix}$

$\begin{bmatrix} 1 & 1 \\ \textcircled{0} & -3 \end{bmatrix} \leftarrow$ multiplier $= -4$ $\qquad\qquad \begin{bmatrix} 2 & 0 \\ 4 & \bullet \end{bmatrix}$

$U = \begin{bmatrix} 1 & 1 \\ 0 & \textcircled{1} \end{bmatrix} \leftarrow$ multiplier $= -\frac{1}{3}$ $\qquad\qquad \begin{bmatrix} 2 & 0 \\ 4 & -3 \end{bmatrix}$

A general 2×2 lower triangular matrix with nonzero main diagonal entries can be factored as $\begin{bmatrix} a_{11} & 0 \\ a_{21} & a_{22} \end{bmatrix} = \begin{bmatrix} 1 & 0 \\ a_{21}/a_{11} & 1 \end{bmatrix}\begin{bmatrix} a_{11} & 0 \\ 0 & a_{22} \end{bmatrix}$ therefore $\begin{bmatrix} 2 & 0 \\ 4 & -3 \end{bmatrix} = \begin{bmatrix} 1 & 0 \\ 2 & 1 \end{bmatrix}\begin{bmatrix} 2 & 0 \\ 0 & -3 \end{bmatrix}$. We conclude that an LDU-decomposition of A is $A = \begin{bmatrix} 2 & 2 \\ 4 & 1 \end{bmatrix} = \begin{bmatrix} 1 & 0 \\ 2 & 1 \end{bmatrix}\begin{bmatrix} 2 & 0 \\ 0 & -3 \end{bmatrix}\begin{bmatrix} 1 & 1 \\ 0 & 1 \end{bmatrix} = LDU.$

15. If rows 2 and 3 of A are interchanged, then the resulting matrix has an LU-decomposition.

For $P = \begin{bmatrix} 1 & 0 & 0 \\ 0 & 0 & 1 \\ 0 & 1 & 0 \end{bmatrix}$, $PA = \begin{bmatrix} 3 & -1 & 0 \\ 0 & 2 & 1 \\ 3 & -1 & 1 \end{bmatrix}$. Reduce PA to upper triangular form:

$$\begin{bmatrix} 3 & -1 & 0 \\ 0 & 2 & 1 \\ 3 & -1 & 1 \end{bmatrix} \to \begin{bmatrix} 1 & -\frac{1}{3} & 0 \\ 0 & 2 & 1 \\ 3 & -1 & 1 \end{bmatrix} \to \begin{bmatrix} 1 & -\frac{1}{3} & 0 \\ 0 & 2 & 1 \\ 0 & 0 & 1 \end{bmatrix} \to \begin{bmatrix} 1 & -\frac{1}{3} & 0 \\ 0 & 1 & \frac{1}{2} \\ 0 & 0 & 1 \end{bmatrix} = U$$

The multipliers used were $\frac{1}{3}$, -3, and $\frac{1}{2}$, so $L = \begin{bmatrix} 3 & 0 & 0 \\ 0 & 2 & 0 \\ 3 & 0 & 1 \end{bmatrix}$. Since $P = P^{-1}$, we have $A = PLU =$

$$\begin{bmatrix} 1 & 0 & 0 \\ 0 & 0 & 1 \\ 0 & 1 & 0 \end{bmatrix}\begin{bmatrix} 3 & 0 & 0 \\ 0 & 2 & 0 \\ 3 & 0 & 1 \end{bmatrix}\begin{bmatrix} 1 & -\frac{1}{3} & 0 \\ 0 & 1 & \frac{1}{2} \\ 0 & 0 & 1 \end{bmatrix}.$$

Since $P\mathbf{b} = \begin{bmatrix} -2 \\ 4 \\ 1 \end{bmatrix}$, the system to solve is $\begin{bmatrix} 3 & 0 & 0 \\ 0 & 2 & 0 \\ 3 & 0 & 1 \end{bmatrix}\begin{bmatrix} 1 & -\frac{1}{3} & 0 \\ 0 & 1 & \frac{1}{2} \\ 0 & 0 & 1 \end{bmatrix}\begin{bmatrix} x_1 \\ x_2 \\ x_3 \end{bmatrix} = \begin{bmatrix} -2 \\ 4 \\ 1 \end{bmatrix}$.

$\begin{bmatrix} 3 & 0 & 0 \\ 0 & 2 & 0 \\ 3 & 0 & 1 \end{bmatrix}\begin{bmatrix} y_1 \\ y_2 \\ y_3 \end{bmatrix} = \begin{bmatrix} -2 \\ 4 \\ 1 \end{bmatrix}$ is

$$\begin{aligned} 3y_1 \qquad\qquad &= -2 \\ 2y_2 \qquad &= 4 \\ 3y_1 \qquad + y_3 &= 1 \end{aligned}$$

which has the solution $y_1 = -\frac{2}{3}$, $y_2 = 2$, $y_3 = 3$.

$\begin{bmatrix} 1 & -\frac{1}{3} & 0 \\ 0 & 1 & \frac{1}{2} \\ 0 & 0 & 1 \end{bmatrix}\begin{bmatrix} x_1 \\ x_2 \\ x_3 \end{bmatrix} = \begin{bmatrix} -\frac{2}{3} \\ 2 \\ 3 \end{bmatrix}$ is

$$\begin{aligned} x_1 - \frac{1}{3}x_2 &= -\frac{2}{3} \\ x_2 + \frac{1}{2}x_3 &= 2 \\ x_3 &= 3 \end{aligned}$$

which gives the solution to the original system: $x_1 = -\frac{1}{2}$, $x_2 = \frac{1}{2}$, $x_3 = 3$.

17. Approximately $\frac{2}{3}n^3$ additions and multiplications are required – see Section 9.3.

True-False Exercises

(a) False. If the matrix cannot be reduced to row echelon form without interchanging rows, then it does not have an LU-decomposition.

(b) False. If the row equivalence of A and U requires interchanging rows of A, then A does not have an LU-decomposition.

(c) True. This follows from part (b) of Theorem 1.7.1.

(d) True. (Refer to the subsection "LDU-Decompositions" for the relevant result.)

(e) True. The procedure for obtaining a PLU-decomposition of a matrix A has been described in the subsection "PLU-Decompositions".

9.2 The Power Method

1. **(a)** $\lambda_3 = -8$ is the dominant eigenvalue since $|\lambda_3| = 8$ is greater than the absolute values of all remaining eigenvalues

(b) $|\lambda_1| = |\lambda_4| = 5$; no dominant eigenvalue

3. $A\mathbf{x}_0 = \begin{bmatrix} 5 & -1 \\ -1 & -1 \end{bmatrix}\begin{bmatrix} 1 \\ 0 \end{bmatrix} = \begin{bmatrix} 5 \\ -1 \end{bmatrix}$ $\qquad \mathbf{x}_1 = \frac{A\mathbf{x}_0}{\|A\mathbf{x}_0\|} = \frac{1}{\sqrt{26}}\begin{bmatrix} 5 \\ -1 \end{bmatrix} \approx \begin{bmatrix} 0.98058 \\ -0.19612 \end{bmatrix}$

$A\mathbf{x}_1 \approx \begin{bmatrix} 5 & -1 \\ -1 & -1 \end{bmatrix}\begin{bmatrix} 0.98058 \\ -0.19612 \end{bmatrix} \approx \begin{bmatrix} 5.09902 \\ -0.78446 \end{bmatrix}$ $\qquad \mathbf{x}_2 = \frac{A\mathbf{x}_1}{\|A\mathbf{x}_1\|} \approx \begin{bmatrix} 0.98837 \\ -0.15206 \end{bmatrix}$

$A\mathbf{x}_2 \approx \begin{bmatrix} 5 & -1 \\ -1 & -1 \end{bmatrix}\begin{bmatrix} 0.98837 \\ -0.15206 \end{bmatrix} \approx \begin{bmatrix} 5.09391 \\ -0.83631 \end{bmatrix}$ $\qquad \mathbf{x}_3 = \frac{A\mathbf{x}_2}{\|A\mathbf{x}_2\|} \approx \begin{bmatrix} 0.98679 \\ -0.16201 \end{bmatrix}$

$A\mathbf{x}_3 \approx \begin{bmatrix} 5 & -1 \\ -1 & -1 \end{bmatrix}\begin{bmatrix} 0.98679 \\ -0.16201 \end{bmatrix} \approx \begin{bmatrix} 5.09596 \\ -0.82478 \end{bmatrix}$ $\qquad \mathbf{x}_4 = \frac{A\mathbf{x}_3}{\|A\mathbf{x}_3\|} \approx \begin{bmatrix} 0.98715 \\ -0.15977 \end{bmatrix}$

$$\lambda^{(1)} = A\mathbf{x}_1 \cdot \mathbf{x}_1 = (A\mathbf{x}_1)^T\mathbf{x}_1 \approx 5.15385$$

$$\lambda^{(2)} = A\mathbf{x}_2 \cdot \mathbf{x}_2 = (A\mathbf{x}_2)^T\mathbf{x}_2 \approx 5.16185$$

$$\lambda^{(3)} = A\mathbf{x}_3 \cdot \mathbf{x}_3 = (A\mathbf{x}_3)^T\mathbf{x}_3 \approx 5.16226$$

$$\lambda^{(4)} = A\mathbf{x}_4 \cdot \mathbf{x}_4 = (A\mathbf{x}_4)^T\mathbf{x}_4 \approx 5.16228$$

$\det(\lambda I - A) = \begin{vmatrix} \lambda - 5 & 1 \\ 1 & \lambda + 1 \end{vmatrix} = \lambda^2 - 4\lambda - 6 = (\lambda - 2 - \sqrt{10})(\lambda - 2 + \sqrt{10})$; the dominant eigenvalue is $2 + \sqrt{10} \approx 5.16228$.

The reduced row echelon form of $(2 + \sqrt{10})I - A$ is $\begin{bmatrix} 1 & 3 + \sqrt{10} \\ 0 & 0 \end{bmatrix}$ so that the eigenspace corresponding to $\lambda = 2 + \sqrt{10}$ contains vectors (x_1, x_2) where $x_1 = -(3 + \sqrt{10})t$, $x_2 = t$,. A vector $(-3 - \sqrt{10}, 1)$ forms a basis for this eigenspace. We see that $\mathbf{x}_4$ approximates a unit eigenvector $\frac{1}{\sqrt{20+6\sqrt{10}}}(3\sqrt{10}, -1) \approx (0.98709, -0.16018)$ and $\lambda^{(4)}$ approximates the dominant eigenvalue $2 + \sqrt{10} \approx 5.16228$.

5. $A\mathbf{x}_0 = \begin{bmatrix} 1 & -3 \\ -3 & 5 \end{bmatrix}\begin{bmatrix} 1 \\ 1 \end{bmatrix} = \begin{bmatrix} -2 \\ 2 \end{bmatrix}$ $\qquad \mathbf{x}_1 = \dfrac{A\mathbf{x}_0}{\max(A\mathbf{x}_0)} = \begin{bmatrix} -1 \\ 1 \end{bmatrix}$

$A\mathbf{x}_1 \approx \begin{bmatrix} 1 & -3 \\ -3 & 5 \end{bmatrix}\begin{bmatrix} -1 \\ 1 \end{bmatrix} = \begin{bmatrix} -4 \\ 8 \end{bmatrix}$ $\qquad \mathbf{x}_2 = \dfrac{A\mathbf{x}_1}{\max(A\mathbf{x}_1)} = \begin{bmatrix} -0.5 \\ 1 \end{bmatrix}$

$A\mathbf{x}_2 \approx \begin{bmatrix} 1 & -3 \\ -3 & 5 \end{bmatrix}\begin{bmatrix} -0.5 \\ 1 \end{bmatrix} \approx \begin{bmatrix} -3.5 \\ 6.5 \end{bmatrix}$ $\qquad \mathbf{x}_3 = \dfrac{A\mathbf{x}_2}{\max(A\mathbf{x}_2)} \approx \begin{bmatrix} -0.53846 \\ 1 \end{bmatrix}$

$A\mathbf{x}_3 \approx \begin{bmatrix} 1 & -3 \\ -3 & 5 \end{bmatrix}\begin{bmatrix} -0.53846 \\ 1 \end{bmatrix} \approx \begin{bmatrix} -3.53846 \\ 6.61538 \end{bmatrix}$ $\qquad \mathbf{x}_4 = \dfrac{A\mathbf{x}_3}{\max(A\mathbf{x}_3)} \approx \begin{bmatrix} -0.53488 \\ 1 \end{bmatrix}$

$$\lambda^{(1)} = \frac{A\mathbf{x}_1 \cdot \mathbf{x}_1}{\mathbf{x}_1 \cdot \mathbf{x}_1} = 6$$
$$\lambda^{(2)} = \frac{A\mathbf{x}_2 \cdot \mathbf{x}_2}{\mathbf{x}_2 \cdot \mathbf{x}_2} = 6.6$$
$$\lambda^{(3)} = \frac{A\mathbf{x}_3 \cdot \mathbf{x}_3}{\mathbf{x}_3 \cdot \mathbf{x}_3} \approx 6.60550$$
$$\lambda^{(4)} = \frac{A\mathbf{x}_4 \cdot \mathbf{x}_4}{\mathbf{x}_4 \cdot \mathbf{x}_4} \approx 6.60555$$

$\det(\lambda I - A) = \begin{vmatrix} \lambda - 1 & 3 \\ 3 & \lambda - 5 \end{vmatrix} = \lambda^2 - 6\lambda - 4$, so the eigenvalues of A are $\lambda = 3 \pm \sqrt{13}$. The dominant eigenvalue is $3 + \sqrt{13} \approx 6.60555$ with corresponding scaled eigenvector $\begin{bmatrix} \frac{2-\sqrt{13}}{3} \\ 1 \end{bmatrix} \approx \begin{bmatrix} -0.53518 \\ 1 \end{bmatrix}$.

7. **(a)** $A\mathbf{x}_0 = \begin{bmatrix} 2 & -1 \\ -1 & 2 \end{bmatrix}\begin{bmatrix} 1 \\ 0 \end{bmatrix} = \begin{bmatrix} 2 \\ -1 \end{bmatrix}$ $\qquad \mathbf{x}_1 = \dfrac{A\mathbf{x}_0}{\max(A\mathbf{x}_0)} = \begin{bmatrix} 1 \\ -0.5 \end{bmatrix}$

$A\mathbf{x}_1 = \begin{bmatrix} 2 & -1 \\ -1 & 2 \end{bmatrix}\begin{bmatrix} 1 \\ -0.5 \end{bmatrix} = \begin{bmatrix} 2.5 \\ 2 \end{bmatrix}$ $\qquad \mathbf{x}_2 = \dfrac{A\mathbf{x}_1}{\max(A\mathbf{x}_1)} = \begin{bmatrix} 1 \\ -0.8 \end{bmatrix}$

$A\mathbf{x}_2 = \begin{bmatrix} 2 & -1 \\ -1 & 2 \end{bmatrix}\begin{bmatrix} 1 \\ -0.8 \end{bmatrix} = \begin{bmatrix} 2.8 \\ -2.6 \end{bmatrix}$ $\qquad \mathbf{x}_3 = \dfrac{A\mathbf{x}_2}{\max(A\mathbf{x}_2)} \approx \begin{bmatrix} 1 \\ -0.929 \end{bmatrix}$

(b) $\lambda^{(1)} = \frac{A\mathbf{x}_1 \cdot \mathbf{x}_1}{\mathbf{x}_1 \cdot \mathbf{x}_1} = 2.8$; $\lambda^{(2)} = \frac{A\mathbf{x}_2 \cdot \mathbf{x}_2}{\mathbf{x}_2 \cdot \mathbf{x}_2} \approx 2.976$; $\lambda^{(3)} = \frac{A\mathbf{x}_3 \cdot \mathbf{x}_3}{\mathbf{x}_3 \cdot \mathbf{x}_3} \approx 2.997$

(c) $\det(\lambda I - A) = \begin{vmatrix} \lambda - 2 & 1 \\ 1 & \lambda - 2 \end{vmatrix} = (\lambda - 3)(\lambda - 1)$; the dominant eigenvalue is 3.

The reduced row echelon form of $10I - A$ is $\begin{bmatrix} 1 & 1 \\ 0 & 0 \end{bmatrix}$ so that the eigenspace corresponding to $\lambda = 10$ contains vectors (x_1, x_2) where $x_1 = -t, x_2 = t$. A vector $(-1, 1)$ forms a basis for this eigenspace. We see that $\mathbf{x}_3$ approximates the eigenvector $(1, -1)$ and $\lambda^{(3)}$ approximates the dominant eigenvalue 3.

(d) The percentage error is $\left|\frac{\lambda - \lambda^{(3)}}{\lambda}\right| \approx \left|\frac{3 - 2.997}{3}\right| = 0.001 = 0.1\%$.

9. By Formula (10), $\mathbf{x}_5 = \dfrac{A^5\mathbf{x}_0}{\max(A^5\mathbf{x}_0)} \approx \begin{bmatrix} 0.99180 \\ 1 \end{bmatrix}$. Thus $\lambda^{(5)} = \frac{A\mathbf{x}_5 \cdot \mathbf{x}_5}{\mathbf{x}_5 \cdot \mathbf{x}_5} \approx 2.99993$.

11. By inspection, A is symmetric and has a dominant eigenvalue -1.
Assuming $a \neq 0$, the power sequence is

$$A\mathbf{x}_0 = \begin{bmatrix} -1 & 0 \\ 0 & 0 \end{bmatrix}\begin{bmatrix} a \\ b \end{bmatrix} = \begin{bmatrix} -a \\ 0 \end{bmatrix} \qquad \mathbf{x}_1 = \frac{A\mathbf{x}_0}{\|A\mathbf{x}_0\|} = \frac{1}{|a|}\begin{bmatrix} -a \\ 0 \end{bmatrix} = \begin{bmatrix} -a/|a| \\ 0 \end{bmatrix}$$

$$A\mathbf{x}_1 \approx \begin{bmatrix} -1 & 0 \\ 0 & 0 \end{bmatrix}\begin{bmatrix} -a/|a| \\ 0 \end{bmatrix} = \begin{bmatrix} a/|a| \\ 0 \end{bmatrix} \qquad \mathbf{x}_2 = \frac{A\mathbf{x}_1}{\|A\mathbf{x}_1\|} = \frac{1}{1}\begin{bmatrix} a/|a| \\ 0 \end{bmatrix} = \begin{bmatrix} a/|a| \\ 0 \end{bmatrix}$$

$$A\mathbf{x}_2 \approx \begin{bmatrix} -1 & 0 \\ 0 & 0 \end{bmatrix}\begin{bmatrix} a/|a| \\ 0 \end{bmatrix} = \begin{bmatrix} -a/|a| \\ 0 \end{bmatrix} \qquad \mathbf{x}_3 = \frac{A\mathbf{x}_2}{\|A\mathbf{x}_2\|} = \frac{1}{1}\begin{bmatrix} -a/|a| \\ 0 \end{bmatrix} = \begin{bmatrix} -a/|a| \\ 0 \end{bmatrix}$$

$$A\mathbf{x}_3 \approx \begin{bmatrix} -1 & 0 \\ 0 & 0 \end{bmatrix}\begin{bmatrix} -a/|a| \\ 0 \end{bmatrix} \approx \begin{bmatrix} a/|a| \\ 0 \end{bmatrix} \qquad \mathbf{x}_4 = \frac{A\mathbf{x}_3}{\|A\mathbf{x}_3\|} = \frac{1}{1}\begin{bmatrix} a/|a| \\ 0 \end{bmatrix} = \begin{bmatrix} a/|a| \\ 0 \end{bmatrix}$$

$$\vdots \qquad\qquad \vdots$$

The quantity $a/|a|$ is equal to 1 if $a > 0$ and -1 if $a < 0$. Since the power sequence continues to oscillate between $\begin{bmatrix} -1 \\ 0 \end{bmatrix}$ and $\begin{bmatrix} 1 \\ 0 \end{bmatrix}$, it does not converge.

13. **(a)** Starting with $\mathbf{x}_0 = \begin{bmatrix} 1 \\ 0 \\ 0 \end{bmatrix}$, it takes 8 iterations.

$$\mathbf{x}_1 \approx \begin{bmatrix} 0.229 \\ 0.668 \\ 0.668 \end{bmatrix}, \qquad \lambda^{(1)} \approx 7.632$$

$$\mathbf{x}_2 \approx \begin{bmatrix} 0.507 \\ 0.320 \\ 0.800 \end{bmatrix}, \qquad \lambda^{(2)} \approx 9.968$$

$$\mathbf{x}_3 \approx \begin{bmatrix} 0.380 \\ 0.197 \\ 0.904 \end{bmatrix}, \qquad \lambda^{(3)} \approx 10.622$$

$$\mathbf{x}_4 \approx \begin{bmatrix} 0.344 \\ 0.096 \\ 0.934 \end{bmatrix}, \qquad \lambda^{(4)} \approx 10.827$$

$$\mathbf{x}_5 \approx \begin{bmatrix} 0.317 \\ 0.044 \\ 0.948 \end{bmatrix}, \qquad \lambda^{(5)} \approx 10.886$$

$$\mathbf{x}_6 \approx \begin{bmatrix} 0.302 \\ 0.016 \\ 0.953 \end{bmatrix}, \qquad \lambda^{(6)} \approx 10.903$$

$$\mathbf{x}_7 \approx \begin{bmatrix} 0.294 \\ 0.002 \\ 0.956 \end{bmatrix}, \qquad \lambda^{(7)} \approx 10.908$$

$$\mathbf{x}_8 \approx \begin{bmatrix} 0.290 \\ -0.006 \\ 0.957 \end{bmatrix}, \qquad \lambda^{(8)} \approx 10.909$$

(b) Starting with $\mathbf{x}_0 = \begin{bmatrix} 1 \\ 0 \\ 0 \\ 0 \end{bmatrix}$, it takes 8 iterations.

$$\mathbf{x}_1 \approx \begin{bmatrix} 0.577 \\ 0 \\ 0.577 \\ 0.577 \end{bmatrix}, \qquad \lambda^{(1)} \approx 6.333$$

$$\mathbf{x}_2 \approx \begin{bmatrix} 0.249 \\ 0 \\ 0.498 \\ 0.830 \end{bmatrix}, \quad \lambda^{(2)} \approx 8.062$$

$$\mathbf{x}_3 = \begin{bmatrix} 0.193 \\ 0.041 \\ 0.376 \\ 0.905 \end{bmatrix}, \quad \lambda^{(3)} \approx 8.382$$

$$\mathbf{x}_4 \approx \begin{bmatrix} 0.175 \\ 0.073 \\ 0.305 \\ 0.933 \end{bmatrix}, \quad \lambda^{(4)} \approx 8.476$$

$$\mathbf{x}_5 \approx \begin{bmatrix} 0.167 \\ 0.091 \\ 0.266 \\ 0.945 \end{bmatrix}, \quad \lambda^{(5)} \approx 8.503$$

$$\mathbf{x}_6 \approx \begin{bmatrix} 0.162 \\ 0.101 \\ 0.245 \\ 0.951 \end{bmatrix}, \quad \lambda^{(6)} \approx 8.511$$

$$\mathbf{x}_7 \approx \begin{bmatrix} 0.159 \\ 0.107 \\ 0.234 \\ 0.953 \end{bmatrix}, \quad \lambda^{(7)} \approx 8.513$$

$$\mathbf{x}_8 \approx \begin{bmatrix} 0.158 \\ 0.110 \\ 0.228 \\ 0.954 \end{bmatrix}, \quad \lambda^{(8)} \approx 8.513$$

9.3 Comparison of Procedures for Solving Linear Systems

1. **(a)** For $n = 1000 = 10^3$, the flops for both phases is $\frac{2}{3}(10^3)^3 + \frac{3}{2}(10^3)^2 - \frac{7}{6}(10^3) = 668{,}165{,}500$, which is 0.6681655 gigaflops, so it will take $0.6681655 \times 10^{-1} \approx 0.067$second.

(b) $n = 10{,}000 = 10^4$: $\frac{2}{3}(10^4)^3 + \frac{3}{2}(10^4)^2 - \frac{7}{6}(10^4) = 666{,}816{,}655{,}000$ flops or 666.816655 gigaflops. The time is about 66.68 seconds.

(c) $n = 100{,}000 = 10^5$; $\frac{2}{3}(10^5)^3 + \frac{3}{2}(10^5)^2 - \frac{7}{6}(10^5) \approx 666{,}682 \times 10^9$flops or 666,682 gigaflops. The time is about 66,668 seconds which is about 18.5 hours.

3. $n = 10{,}000 = 10^4$

(a) $\frac{2}{3}n^3 \approx \frac{2}{3}(10^{12}) \approx 666.67 \times 10^9$;
666.67 gigaflops are required, which will take $\frac{666.67}{70} \approx 9.52$ seconds.

(b) $n^2 \approx 10^8 = 0.1 \times 10^9$; 0.1 gigaflop is required, which will take about 0.0014 second.

(c) This is the same as part (a); about 9.52 seconds.

(d) $2n^3 \approx 2 \times 10^{12} = 2000 \times 10^9$;
2000 gigaflops are required, which will take about 28.57 seconds.

5. **(a)** $n = 100{,}000 = 10^5$; $\frac{2}{3}n^3 \approx \frac{2}{3} \times 10^{15} \approx 0.667 \times 10^{15} = 6.67 \times 10^5 \times 10^9$;

Thus, the forward phase would require about 6.67×10^5seconds.
$n^2 = 10^{10} = 10 \times 10^9$; The backward phase would require about 10 seconds.

(b) $n = 10{,}000 = 10^4$; $\frac{2}{3}n^3 \approx \frac{2}{3} \times 10^{12} \approx 0.667 \times 10^{12} \approx 6.67 \times 10^2 \times 10^9$;
About 667 gigaflops are required, so the computer would have to execute 2(667) = 1334 gigaflops per second.

7. Multiplying each of the n^2 entries of A by c requires n^2 flops.

9. Let $C = [c_{ij}] = AB$. Computing c_{ij} requires first multiplying each of the n entries a_{ik} by the corresponding entry b_{kj}, which requires n flops. Then the n terms $a_{ik}b_{kj}$ must be summed, which requires $n - 1$ flops. Thus, each of the n^2 entries in AB requires $2n - 1$ flops, for a total of $n^2(2n - 1) = 2n^3 - n^2$ flops.
Note that adding two numbers requires 1 flop, adding three numbers requires 2 flops, and in general, $n - 1$ flops are required to add n numbers.

9.4 Singular Value Decomposition

1. The characteristic polynomial of $A^TA = \begin{bmatrix}1\\2\\0\end{bmatrix}\begin{bmatrix}1 & 2 & 0\end{bmatrix} = \begin{bmatrix}1 & 2 & 0\\2 & 4 & 0\\0 & 0 & 0\end{bmatrix}$ is $\lambda^2(\lambda - 5)$; thus the eigenvalues of A^TA are $\lambda_1 = 5$ and $\lambda_2 = 0$, and $\sigma_1 = \sqrt{5}$ and $\sigma_2 = 0$ are singular values of A.

3. The eigenvalues of $A^TA = \begin{bmatrix}1 & 2\\-2 & 1\end{bmatrix}\begin{bmatrix}1 & -2\\2 & 1\end{bmatrix} = \begin{bmatrix}5 & 0\\0 & 5\end{bmatrix}$ are $\lambda_1 = 5$ and $\lambda_2 = 5$ (i.e., $\lambda = 5$ is an eigenvalue of multiplicity 2); thus the singular value of A is $\sigma_1 = \sqrt{5}$.

5. The only eigenvalue of $A^TA = \begin{bmatrix}1 & 1\\-1 & 1\end{bmatrix}\begin{bmatrix}1 & -1\\1 & 1\end{bmatrix} = \begin{bmatrix}2 & 0\\0 & 2\end{bmatrix}$ is $\lambda = 2$ (multiplicity 2), and the vectors $\mathbf{v}_1 = \begin{bmatrix}1\\0\end{bmatrix}$ and $\mathbf{v}_2 = \begin{bmatrix}0\\1\end{bmatrix}$ form an orthonormal basis for the eigenspace (which is all of R^2).

The singular values of A are $\sigma_1 = \sqrt{2}$ and $\sigma_2 = \sqrt{2}$. We have $\mathbf{u}_1 = \frac{1}{\sigma_1}A\mathbf{v}_1 = \frac{1}{\sqrt{2}}\begin{bmatrix}1 & -1\\1 & 1\end{bmatrix}\begin{bmatrix}1\\0\end{bmatrix} = \begin{bmatrix}\frac{1}{\sqrt{2}}\\\frac{1}{\sqrt{2}}\end{bmatrix}$, and

$\mathbf{u}_2 = \frac{1}{\sigma_2}A\mathbf{v}_2 = \frac{1}{\sqrt{2}}\begin{bmatrix}1 & -1\\1 & 1\end{bmatrix}\begin{bmatrix}0\\1\end{bmatrix} = \begin{bmatrix}-\frac{1}{\sqrt{2}}\\\frac{1}{\sqrt{2}}\end{bmatrix}$.

This results in the following singular value decomposition of A:

$$A = U\Sigma V^T = \begin{bmatrix} \frac{1}{\sqrt{2}} & -\frac{1}{\sqrt{2}} \\ \frac{1}{\sqrt{2}} & \frac{1}{\sqrt{2}} \end{bmatrix} \begin{bmatrix} \sqrt{2} & 0 \\ 0 & \sqrt{2} \end{bmatrix} \begin{bmatrix} 1 & 0 \\ 0 & 1 \end{bmatrix}$$

7. The eigenvalues of $A^TA = \begin{bmatrix} 4 & 0 \\ 6 & 4 \end{bmatrix}\begin{bmatrix} 4 & 6 \\ 0 & 4 \end{bmatrix} = \begin{bmatrix} 16 & 24 \\ 24 & 52 \end{bmatrix}$ are $\lambda_1 = 64$ and $\lambda_2 = 4$, with corresponding unit eigenvectors $\mathbf{v}_1 = \begin{bmatrix} \frac{1}{\sqrt{5}} \\ \frac{2}{\sqrt{5}} \end{bmatrix}$ and $\mathbf{v}_2 = \begin{bmatrix} -\frac{2}{\sqrt{5}} \\ \frac{1}{\sqrt{5}} \end{bmatrix}$ respectively. The singular values of A are $\sigma_1 = 8$ and $\sigma_2 = 2$.

We have $\mathbf{u}_1 = \frac{1}{\sigma_1}A\mathbf{v}_1 = \frac{1}{8}\begin{bmatrix} 4 & 6 \\ 0 & 4 \end{bmatrix}\begin{bmatrix} \frac{1}{\sqrt{5}} \\ \frac{2}{\sqrt{5}} \end{bmatrix} = \begin{bmatrix} \frac{2}{\sqrt{5}} \\ \frac{1}{\sqrt{5}} \end{bmatrix}$, and $\mathbf{u}_2 = \frac{1}{\sigma_2}A\mathbf{v}_2 = \frac{1}{2}\begin{bmatrix} 4 & 6 \\ 0 & 4 \end{bmatrix}\begin{bmatrix} -\frac{2}{\sqrt{5}} \\ \frac{1}{\sqrt{5}} \end{bmatrix} = \begin{bmatrix} -\frac{1}{\sqrt{5}} \\ \frac{2}{\sqrt{5}} \end{bmatrix}$.

This results in the following singular value decomposition:

$$A = U\Sigma V^T = \begin{bmatrix} \frac{2}{\sqrt{5}} & -\frac{1}{\sqrt{5}} \\ \frac{1}{\sqrt{5}} & \frac{2}{\sqrt{5}} \end{bmatrix} \begin{bmatrix} 8 & 0 \\ 0 & 2 \end{bmatrix} \begin{bmatrix} \frac{1}{\sqrt{5}} & \frac{2}{\sqrt{5}} \\ -\frac{2}{\sqrt{5}} & \frac{1}{\sqrt{5}} \end{bmatrix}$$

9. The eigenvalues of $A^TA = \begin{bmatrix} -2 & -1 & 2 \\ 2 & 1 & -2 \end{bmatrix}\begin{bmatrix} -2 & 2 \\ -1 & 1 \\ 2 & -2 \end{bmatrix} = \begin{bmatrix} 9 & -9 \\ -9 & 9 \end{bmatrix}$ are $\lambda_1 = 18$ and $\lambda_2 = 0$, with corresponding unit eigenvectors $\mathbf{v}_1 = \begin{bmatrix} -\frac{1}{\sqrt{2}} \\ \frac{1}{\sqrt{2}} \end{bmatrix}$ and $\mathbf{v}_2 = \begin{bmatrix} \frac{1}{\sqrt{2}} \\ \frac{1}{\sqrt{2}} \end{bmatrix}$ respectively. The only nonzero singular value of A is $\sigma_1 = \sqrt{18} = 3\sqrt{2}$, and we have $\mathbf{u}_1 = \frac{1}{\sigma_1}A\mathbf{v}_1 = \frac{1}{3\sqrt{2}}\begin{bmatrix} -2 & 2 \\ -1 & 1 \\ 2 & -2 \end{bmatrix}\begin{bmatrix} -\frac{1}{\sqrt{2}} \\ \frac{1}{\sqrt{2}} \end{bmatrix} = \begin{bmatrix} \frac{2}{3} \\ \frac{1}{3} \\ -\frac{2}{3} \end{bmatrix}$. We must choose the vectors $\mathbf{u}_2$ and $\mathbf{u}_3$ so that $\{\mathbf{u}_1, \ \mathbf{u}_2, \ \mathbf{u}_3\}$ is an orthonormal basis R^3.

A possible choice is $\mathbf{u}_2 = \begin{bmatrix} \frac{1}{\sqrt{2}} \\ 0 \\ \frac{1}{\sqrt{2}} \end{bmatrix}$ and $\mathbf{u}_3 = \begin{bmatrix} \frac{\sqrt{2}}{6} \\ -\frac{2\sqrt{2}}{3} \\ -\frac{\sqrt{2}}{6} \end{bmatrix}$. This results in the following singular value decomposition: $A = U\Sigma V^T = \begin{bmatrix} \frac{2}{3} & \frac{1}{\sqrt{2}} & \frac{\sqrt{2}}{6} \\ \frac{1}{3} & 0 & -\frac{2\sqrt{2}}{3} \\ -\frac{2}{3} & \frac{1}{\sqrt{2}} & -\frac{\sqrt{2}}{6} \end{bmatrix}\begin{bmatrix} 3\sqrt{2} & 0 \\ 0 & 0 \\ 0 & 0 \end{bmatrix}\begin{bmatrix} -\frac{1}{\sqrt{2}} & \frac{1}{\sqrt{2}} \\ \frac{1}{\sqrt{2}} & \frac{1}{\sqrt{2}} \end{bmatrix}$.

Note: The singular value decomposition is not unique. It depends on the choice of the (extended) orthonormal basis for R^3. This is just one possibility.

11. The eigenvalues of $A^TA = \begin{bmatrix} 1 & 1 & -1 \\ 0 & 1 & 1 \end{bmatrix}\begin{bmatrix} 1 & 0 \\ 1 & 1 \\ -1 & 1 \end{bmatrix} = \begin{bmatrix} 3 & 0 \\ 0 & 2 \end{bmatrix}$ are $\lambda_1 = 3$ and $\lambda_2 = 2$, with corresponding unit eigenvectors $\mathbf{v}_1 = \begin{bmatrix} 1 \\ 0 \end{bmatrix}$ and $\mathbf{v}_2 = \begin{bmatrix} 0 \\ 1 \end{bmatrix}$ respectively. The singular values of A are $\sigma_1 = \sqrt{3}$ and $\sigma_2 = \sqrt{2}$. We have $\mathbf{u}_1 = \frac{1}{\sigma_1}A\mathbf{v}_1 = \frac{1}{\sqrt{3}}\begin{bmatrix} 1 & 0 \\ 1 & 1 \\ -1 & 1 \end{bmatrix}\begin{bmatrix} 1 \\ 0 \end{bmatrix} = \begin{bmatrix} \frac{1}{\sqrt{3}} \\ \frac{1}{\sqrt{3}} \\ -\frac{1}{\sqrt{3}} \end{bmatrix}$ and $\mathbf{u}_2 = \frac{1}{\sqrt{2}}\begin{bmatrix} 1 & 0 \\ 1 & 1 \\ -1 & 1 \end{bmatrix}\begin{bmatrix} 0 \\ 1 \end{bmatrix} = \begin{bmatrix} 0 \\ \frac{1}{\sqrt{2}} \\ \frac{1}{\sqrt{2}} \end{bmatrix}$. We choose $\mathbf{u}_3 = \begin{bmatrix} \frac{2}{\sqrt{6}} \\ -\frac{1}{\sqrt{6}} \\ \frac{1}{\sqrt{6}} \end{bmatrix}$ so that $\{\mathbf{u}_1,\ \mathbf{u}_2,\ \mathbf{u}_3\}$ is an orthonormal basis for R^3. This results in the following singular value decomposition: $A = U\Sigma V^T = \begin{bmatrix} \frac{1}{\sqrt{3}} & 0 & \frac{2}{\sqrt{6}} \\ \frac{1}{\sqrt{3}} & \frac{1}{\sqrt{2}} & -\frac{1}{\sqrt{6}} \\ -\frac{1}{\sqrt{3}} & \frac{1}{\sqrt{2}} & \frac{1}{\sqrt{6}} \end{bmatrix}\begin{bmatrix} \sqrt{3} & 0 \\ 0 & \sqrt{2} \\ 0 & 0 \end{bmatrix}\begin{bmatrix} 1 & 0 \\ 0 & 1 \end{bmatrix}$.

19. (b) In the solution of Exercise 5, we obtained a singular value decomposition

$$A = U\Sigma V^T = \begin{bmatrix} \frac{1}{\sqrt{2}} & -\frac{1}{\sqrt{2}} \\ \frac{1}{\sqrt{2}} & \frac{1}{\sqrt{2}} \end{bmatrix}\begin{bmatrix} \sqrt{2} & 0 \\ 0 & \sqrt{2} \end{bmatrix}\begin{bmatrix} 1 & 0 \\ 0 & 1 \end{bmatrix}$$

A polar decomposition of A is

$$A = (U\Sigma U^T)(UV^T)$$

$$= \left(\begin{bmatrix} \frac{1}{\sqrt{2}} & -\frac{1}{\sqrt{2}} \\ \frac{1}{\sqrt{2}} & \frac{1}{\sqrt{2}} \end{bmatrix}\begin{bmatrix} \sqrt{2} & 0 \\ 0 & \sqrt{2} \end{bmatrix}\begin{bmatrix} \frac{1}{\sqrt{2}} & \frac{1}{\sqrt{2}} \\ -\frac{1}{\sqrt{2}} & \frac{1}{\sqrt{2}} \end{bmatrix}\right)\left(\begin{bmatrix} \frac{1}{\sqrt{2}} & -\frac{1}{\sqrt{2}} \\ \frac{1}{\sqrt{2}} & \frac{1}{\sqrt{2}} \end{bmatrix}\begin{bmatrix} 1 & 0 \\ 0 & 1 \end{bmatrix}\right)$$

$$= \begin{bmatrix} \sqrt{2} & 0 \\ 0 & \sqrt{2} \end{bmatrix}\begin{bmatrix} \frac{1}{\sqrt{2}} & -\frac{1}{\sqrt{2}} \\ \frac{1}{\sqrt{2}} & \frac{1}{\sqrt{2}} \end{bmatrix}$$

True-False Exercises

(a) False. If A is an $m \times n$ matrix, then A^T is an $n \times m$ matrix, and A^TA is an $n \times n$ matrix.

(b) True. $(A^TA)^T = A^T(A^T)^T = A^TA$.

(c) False. A^TA may have eigenvalues that are 0.

(d) False. A would have to be symmetric to be orthogonally diagonalizable.

(e) True. This follows since A^TA is a symmetric $n \times n$ matrix.

(f) False. The eigenvalues of A^TA are the squares of the singular values of A.

(g) True. This follows from Theorem 9.4.3.

9.5 Data Compression Using Singular Value Decomposition

1. From Exercise 9 in Section 9.4, A has the singular value decomposition

$$A = \begin{bmatrix} \frac{2}{3} & \frac{1}{\sqrt{2}} & \frac{\sqrt{2}}{6} \\ \frac{1}{3} & 0 & -\frac{2\sqrt{2}}{3} \\ -\frac{2}{3} & \frac{1}{\sqrt{2}} & -\frac{\sqrt{2}}{6} \end{bmatrix} \begin{bmatrix} 3\sqrt{2} & 0 \\ 0 & 0 \\ 0 & 0 \end{bmatrix} \begin{bmatrix} -\frac{1}{\sqrt{2}} & \frac{1}{\sqrt{2}} \\ \frac{1}{\sqrt{2}} & \frac{1}{\sqrt{2}} \end{bmatrix}.$$

Thus the reduced singular value decomposition of A is $A = U_1\Sigma_1V_1^T = \begin{bmatrix} \frac{2}{3} \\ \frac{1}{3} \\ -\frac{2}{3} \end{bmatrix} [3\sqrt{2}] \begin{bmatrix} -\frac{1}{\sqrt{2}} & \frac{1}{\sqrt{2}} \end{bmatrix}$.

3. From Exercise 11 in Section 9.4, A has the singular value decomposition

$$A = U\Sigma V^T = \begin{bmatrix} \frac{1}{\sqrt{3}} & 0 & \frac{2}{\sqrt{6}} \\ \frac{1}{\sqrt{3}} & \frac{1}{\sqrt{2}} & -\frac{1}{\sqrt{6}} \\ -\frac{1}{\sqrt{3}} & \frac{1}{\sqrt{2}} & \frac{1}{\sqrt{6}} \end{bmatrix} \begin{bmatrix} \sqrt{3} & 0 \\ 0 & \sqrt{2} \\ 0 & 0 \end{bmatrix} \begin{bmatrix} 1 & 0 \\ 0 & 1 \end{bmatrix}.$$

Thus the reduced singular value decomposition of A is $A = U_1\Sigma_1V_1^T = \begin{bmatrix} \frac{1}{\sqrt{3}} & 0 \\ \frac{1}{\sqrt{3}} & \frac{1}{\sqrt{2}} \\ -\frac{1}{\sqrt{3}} & \frac{1}{\sqrt{2}} \end{bmatrix} \begin{bmatrix} \sqrt{3} & 0 \\ 0 & \sqrt{2} \end{bmatrix} \begin{bmatrix} 1 & 0 \\ 0 & 1 \end{bmatrix}$.

5. The reduced singular value expansion of A is $3\sqrt{2} \begin{bmatrix} \frac{2}{3} \\ \frac{1}{3} \\ -\frac{2}{3} \end{bmatrix} \begin{bmatrix} -\frac{1}{\sqrt{2}} & \frac{1}{\sqrt{2}} \end{bmatrix}$.

7. The reduced singular value decomposition of A is $\sqrt{3} \begin{bmatrix} \frac{1}{\sqrt{3}} \\ \frac{1}{\sqrt{3}} \\ -\frac{1}{\sqrt{3}} \end{bmatrix} [1 \quad 0] + \sqrt{2} \begin{bmatrix} 0 \\ \frac{1}{\sqrt{2}} \\ \frac{1}{\sqrt{2}} \end{bmatrix} [0 \quad 1]$.

9. A rank 100 approximation of A requires storage space for 100(200 + 500 + 1) = 70,100 numbers, while A has 200(500) = 100,000 entries.

True-False Exercises

(a) True. This follows from the definition of a reduced singular value decomposition.

(b) True. This follows from the definition of a reduced singular value decomposition.

(c) False. V_1 has size $n \times k$ so that V_1^T has size $k \times n$.

Chapter 9 Supplementary Exercises

1. Reduce A to upper triangular form:

$$\begin{bmatrix} -6 & 2 \\ 6 & 0 \end{bmatrix} \to \begin{bmatrix} -3 & 1 \\ 6 & 0 \end{bmatrix} \to \begin{bmatrix} -3 & 1 \\ 0 & 2 \end{bmatrix} = U$$

The multipliers used were $\frac{1}{2}$ and 2, so $L = \begin{bmatrix} 2 & 0 \\ -2 & 1 \end{bmatrix}$ and $A = \begin{bmatrix} 2 & 0 \\ -2 & 1 \end{bmatrix}\begin{bmatrix} -3 & 1 \\ 0 & 2 \end{bmatrix}$.

3. Reduce A to upper triangular form.

$$\begin{bmatrix} 2 & 4 & 6 \\ 1 & 4 & 7 \\ 1 & 3 & 7 \end{bmatrix} \to \begin{bmatrix} 1 & 2 & 3 \\ 1 & 4 & 7 \\ 1 & 3 & 7 \end{bmatrix} \to \begin{bmatrix} 1 & 2 & 3 \\ 0 & 2 & 4 \\ 1 & 3 & 7 \end{bmatrix} \to \begin{bmatrix} 1 & 2 & 3 \\ 0 & 2 & 4 \\ 0 & 1 & 4 \end{bmatrix}$$

$$\to \begin{bmatrix} 1 & 2 & 3 \\ 0 & 1 & 2 \\ 0 & 1 & 4 \end{bmatrix} \to \begin{bmatrix} 1 & 2 & 3 \\ 0 & 1 & 2 \\ 0 & 0 & 2 \end{bmatrix} \to \begin{bmatrix} 1 & 2 & 3 \\ 0 & 1 & 2 \\ 0 & 0 & 1 \end{bmatrix} = U$$

The multipliers used were $\frac{1}{2}, -1, -1, \frac{1}{2}, -1$, and $\frac{1}{2}$ so $L = \begin{bmatrix} 2 & 0 & 0 \\ 1 & 2 & 0 \\ 1 & 1 & 2 \end{bmatrix}$ and $A = \begin{bmatrix} 2 & 0 & 0 \\ 1 & 2 & 0 \\ 1 & 1 & 2 \end{bmatrix}\begin{bmatrix} 1 & 2 & 3 \\ 0 & 1 & 2 \\ 0 & 0 & 1 \end{bmatrix}$.

5. **(a)** The characteristic equation of A is $\lambda^2 - 4\lambda + 3 = (\lambda - 3)(\lambda - 1) = 0$ so the dominant eigenvalue of A is $\lambda_1 = 3$, with corresponding positive unit eigenvector

$$\mathbf{v} = \begin{bmatrix} \frac{1}{\sqrt{2}} \\ \frac{1}{\sqrt{2}} \end{bmatrix} \approx \begin{bmatrix} 0.7071 \\ 0.7071 \end{bmatrix}.$$

(b) $A\mathbf{x}_0 = \begin{bmatrix} 2 & 1 \\ 1 & 2 \end{bmatrix}\begin{bmatrix} 1 \\ 0 \end{bmatrix} = \begin{bmatrix} 2 \\ 1 \end{bmatrix}$

$$\mathbf{x}_1 = \frac{A\mathbf{x}_0}{\|A\mathbf{x}_0\|} = \frac{1}{\sqrt{5}}\begin{bmatrix} 2 \\ 1 \end{bmatrix} \approx \begin{bmatrix} 0.8944 \\ 0.4472 \end{bmatrix}$$

$$\mathbf{x}_2 = \frac{A\mathbf{x}_1}{\|A\mathbf{x}_1\|} \approx \begin{bmatrix} 0.7809 \\ 0.6247 \end{bmatrix}$$

$$\mathbf{x}_3 = \frac{A\mathbf{x}_2}{\|A\mathbf{x}_2\|} \approx \begin{bmatrix} 0.7328 \\ 0.6805 \end{bmatrix}$$

$$\mathbf{x}_4 = \frac{A\mathbf{x}_3}{\|A\mathbf{x}_3\|} \approx \begin{bmatrix} 0.7158 \\ 0.6983 \end{bmatrix}$$

$$\mathbf{x}_5 = \frac{A\mathbf{x}_4}{\|A\mathbf{x}_4\|} \approx \begin{bmatrix} 0.7100 \\ 0.7042 \end{bmatrix}$$

$\mathbf{x}_5 \approx \begin{bmatrix} 0.7100 \\ 0.7042 \end{bmatrix}$ as compared to $\mathbf{v} \approx \begin{bmatrix} 0.7071 \\ 0.7071 \end{bmatrix}$.

(c) $A\mathbf{x}_0 = \begin{bmatrix} 2 \\ 1 \end{bmatrix}$

$$\mathbf{x}_1 = \frac{A\mathbf{x}_0}{\max(A\mathbf{x}_0)} = \begin{bmatrix} 1 \\ 0.5 \end{bmatrix}$$

$$\mathbf{x}_2 = \frac{A\mathbf{x}_1}{\max(A\mathbf{x}_1)} = \begin{bmatrix}1\\0.8\end{bmatrix}$$
$$\mathbf{x}_3 = \frac{A\mathbf{x}_2}{\max(A\mathbf{x}_2)} \approx \begin{bmatrix}1\\0.9286\end{bmatrix}$$
$$\mathbf{x}_4 = \frac{A\mathbf{x}_3}{\max(A\mathbf{x}_3)} \approx \begin{bmatrix}1\\0.9756\end{bmatrix}$$
$$\mathbf{x}_5 = \frac{A\mathbf{x}_4}{\max(A\mathbf{x}_4)} \approx \begin{bmatrix}1\\0.9918\end{bmatrix}$$

$\mathbf{x}_5 \approx \begin{bmatrix}1\\0.9918\end{bmatrix}$ as compared to the exact eigenvector $\mathbf{v} = \begin{bmatrix}1\\1\end{bmatrix}$.

7. The Rayleigh quotients will converge to the dominant eigenvalue $\lambda_4 = -8.1$. However, since the ratio $\frac{|\lambda_4|}{|\lambda_1|} = \frac{8.1}{8} = 1.0125$ is very close to 1, the rate of convergence is likely to be quite slow.

9. The eigenvalues of $A^TA = \begin{bmatrix}2 & 2\\2 & 2\end{bmatrix}$ are $\lambda_1 = 4$ and $\lambda_2 = 0$ with corresponding unit eigenvectors $\mathbf{v}_1 = \begin{bmatrix}-\frac{1}{\sqrt{2}}\\-\frac{1}{\sqrt{2}}\end{bmatrix}$ and $\mathbf{v}_2 = \begin{bmatrix}-\frac{1}{\sqrt{2}}\\\frac{1}{\sqrt{2}}\end{bmatrix}$, respectively. The only nonzero singular value A is $\sigma_1 = \sqrt{4} = 2$, and we have $\mathbf{u}_1 = \frac{1}{\sigma_1}A\mathbf{v}_1 = \frac{1}{2}\begin{bmatrix}1 & 1\\0 & 0\\1 & 1\end{bmatrix}\begin{bmatrix}-\frac{1}{\sqrt{2}}\\-\frac{1}{\sqrt{2}}\end{bmatrix} = \begin{bmatrix}-\frac{1}{\sqrt{2}}\\0\\-\frac{1}{\sqrt{2}}\end{bmatrix}$.

We must choose the vectors $\mathbf{u}_2$ and $\mathbf{u}_3$ so that $\{\mathbf{u}_1,\ \mathbf{u}_2,\ \mathbf{u}_3\}$ is an orthonormal basis for R^3. A possible choice is $\mathbf{u}_2 = \begin{bmatrix}0\\1\\0\end{bmatrix}$ and $\mathbf{u}_3 = \begin{bmatrix}\frac{1}{\sqrt{2}}\\0\\-\frac{1}{\sqrt{2}}\end{bmatrix}$. This results in the following singular value decomposition:

$$A = U\Sigma V^T = \begin{bmatrix}-\frac{1}{\sqrt{2}} & 0 & \frac{1}{\sqrt{2}}\\0 & 1 & 0\\-\frac{1}{\sqrt{2}} & 0 & -\frac{1}{\sqrt{2}}\end{bmatrix}\begin{bmatrix}2 & 0\\0 & 0\\0 & 0\end{bmatrix}\begin{bmatrix}-\frac{1}{\sqrt{2}} & -\frac{1}{\sqrt{2}}\\-\frac{1}{\sqrt{2}} & \frac{1}{\sqrt{2}}\end{bmatrix}.$$

11. A has rank 2, thus $U_1 = [\mathbf{u}_1\ \ \mathbf{u}_2]$ and $V_1^T = \begin{bmatrix}\mathbf{v}_1^T\\\mathbf{v}_2^T\end{bmatrix}$ and the reduced singular value decomposition of A is

$$A = U_1\Sigma_1V_1^T = \begin{bmatrix}\frac{1}{2} & \frac{1}{2}\\\frac{1}{2} & -\frac{1}{2}\\\frac{1}{2} & -\frac{1}{2}\\\frac{1}{2} & \frac{1}{2}\end{bmatrix}\begin{bmatrix}24 & 0\\0 & 12\end{bmatrix}\begin{bmatrix}\frac{2}{3} & -\frac{1}{3} & \frac{2}{3}\\\frac{2}{3} & \frac{2}{3} & -\frac{1}{3}\end{bmatrix}.$$

CHAPTER 10: APPLICATIONS OF LINEAR ALGEBRA

10.1 Constructing Curves and Surfaces Through Specified Points

1. **(a)** Substituting the coordinates of the points into Equation (4) yields $\begin{vmatrix} x & y & 1 \\ 1 & -1 & 1 \\ 2 & 2 & 1 \end{vmatrix} = 0$ which, upon cofactor expansion along the first row, yields $-3x + y + 4 = 0$; that is, $y = 3x - 4$.

(b) As in (a), $\begin{vmatrix} x & y & 1 \\ 0 & 1 & 1 \\ 1 & -1 & 1 \end{vmatrix} = 0$ yields $2x + y - 1 = 0$ or $y = -2x + 1$.

2. **(a)** Equation (9) yields $\begin{vmatrix} x^2+y^2 & x & y & 1 \\ 40 & 2 & 6 & 1 \\ 4 & 2 & 0 & 1 \\ 34 & 5 & 3 & 1 \end{vmatrix} = 0$ which, upon first-row cofactor expansion, yields $18(x^2 + y^2) - 72x - 108y + 72 = 0$ or, dividing by 18, $x^2 + y^2 - 4x - 6y + 4 = 0$. Completing the squares in x and y yields the standard form $(x-2)^2 + (y-3)^2 = 9$.

(b) As in (a), $\begin{vmatrix} x^2+y^2 & x & y & 1 \\ 8 & 2 & -2 & 1 \\ 34 & 3 & 5 & 1 \\ 52 & -4 & 6 & 1 \end{vmatrix} = 0$ yields $50(x^2 + y^2) + 100x - 200y - 1000 = 0$; that is, $x^2 + y^2 + 2x - 4y - 20 = 0$. In standard form this is $(x+1)^2 + (y-2)^2 = 25$.

3. Using Equation (10) we obtain $\begin{vmatrix} x^2 & xy & y^2 & x & y & 1 \\ 0 & 0 & 0 & 0 & 0 & 1 \\ 0 & 0 & 1 & 0 & -1 & 1 \\ 4 & 0 & 0 & 2 & 0 & 1 \\ 4 & -10 & 25 & 2 & -5 & 1 \\ 16 & -4 & 1 & 4 & -1 & 1 \end{vmatrix} = 0$ which is the same as

$\begin{vmatrix} x^2 & xy & y^2 & x & y \\ 0 & 0 & 1 & 0 & -1 \\ 4 & 0 & 0 & 2 & 0 \\ 4 & -10 & 25 & 2 & -5 \\ 16 & -4 & 1 & 4 & -1 \end{vmatrix} = 0$ by expansion along the second row (taking advantage of the zeros there). Add column five to column three and take advantage of another row of all but one zero to get $\begin{vmatrix} x^2 & xy & y^2+y & x \\ 4 & 0 & 0 & 2 \\ 4 & -10 & 20 & 2 \\ 16 & -4 & 0 & 4 \end{vmatrix} = 0.$

Now expand along the first row and get $160x^2 + 320xy + 160(y^2 + y) - 320x = 0$; that is, $x^2 + 2xy + y^2 - 2x + y = 0$, which is the equation of a parabola.

4. **(a)** From Equation (11), the equation of the plane is $\begin{vmatrix} x & y & z & 1 \\ 1 & 1 & -3 & 1 \\ 1 & -1 & 1 & 1 \\ 0 & -1 & 2 & 1 \end{vmatrix} = 0.$

Expansion along the first row yields $-2x - 4y - 2z = 0$; that is, $x + 2y + z = 0$.

(b) As in (a), $\begin{vmatrix} x & y & z & 1 \\ 2 & 3 & 1 & 1 \\ 2 & -1 & -1 & 1 \\ 1 & 2 & 1 & 1 \end{vmatrix} = 0$ yields $-2x + 2y - 4z + 2 = 0$; that is $-x + y - 2z + 1 = 0$ for the equation of the plane.

5. **(a)** Equation (11) involves the determinant of the coefficient matrix of the system

$$\begin{bmatrix} x & y & z & 1 \\ x_1 & y_1 & z_1 & 1 \\ x_2 & y_2 & z_2 & 1 \\ x_3 & y_3 & z_3 & 1 \end{bmatrix} \begin{bmatrix} c_1 \\ c_2 \\ c_3 \\ c_4 \end{bmatrix} = \begin{bmatrix} 0 \\ 0 \\ 0 \\ 0 \end{bmatrix}.$$

Rows 2 through 4 show that the plane passes through the three points (x_i, y_i, z_i), $i = 1, 2, 3$, while row 1 gives the equation $c_1x + c_2y + c_3z + c_4 = 0$. For the plane passing through the origin parallel to the plane passing through the three points, the constant term in the final equation will be 0, which is accomplished by using

$$\begin{vmatrix} x & y & z & 0 \\ x_1 & y_1 & z_1 & 1 \\ x_2 & y_2 & z_2 & 1 \\ x_3 & y_3 & z_3 & 1 \end{vmatrix} = 0.$$

(b) The parallel planes passing through the origin are $x + 2y + z = 0$ and $-x + y - 2z = 0$, respectively.

6. **(a)** Using Equation (12), the equation of the sphere is $\begin{vmatrix} x^2+y^2+z^2 & x & y & z & 1 \\ 14 & 1 & 2 & 3 & 1 \\ 6 & -1 & 2 & 1 & 1 \\ 2 & 1 & 0 & 1 & 1 \\ 6 & 1 & 2 & -1 & 1 \end{vmatrix} = 0.$

Expanding by cofactors along the first row yields $16(x^2 + y^2 + z^2) - 32x - 64y - 32z + 32 = 0$; that is, $(x^2 + y^2 + z^2) - 2x - 4y - 2x + 2 = 0$. Completing the squares in each variable yields the standard form $(x - 1)^2 + (y - 2)^2 + (z - 1)^2 = 4$.
Note: When evaluating the cofactors, it is useful to take advantage of the column of ones and elementary row operations; for example, the cofactor of $x^2 + y^2 + z^2$ above can be evaluated as follows: $\begin{vmatrix} 1 & 2 & 3 & 1 \\ -1 & 2 & 1 & 1 \\ 1 & 0 & 1 & 1 \\ 1 & 2 & -1 & 1 \end{vmatrix} = \begin{vmatrix} 1 & 2 & 3 & 1 \\ -2 & 0 & -2 & 0 \\ 0 & -2 & -2 & 0 \\ 0 & 0 & -4 & 0 \end{vmatrix} = 16$ by cofactor expansion of the latter determinant along the last column.

(b) As in (a), $\begin{vmatrix} x^2+y^2+z^2 & x & y & z & 1 \\ 5 & 0 & 1 & -2 & 1 \\ 11 & 1 & 3 & 1 & 1 \\ 5 & 2 & -1 & 0 & 1 \\ 11 & 3 & 1 & -1 & 1 \end{vmatrix} = 0$ yields $-24(x^2+y^2+z^2)+48x+48y+72=0$; that is, $x^2+y^2+z^2-2x-2y-3=0$ or in standard form, $(x-1)^2+(y-1)^2+z^2=5$.

7. Substituting each of the points (x_1, y_1), (x_2, y_2), (x_3, y_3), (x_4, y_4), and (x_5, y_5) into the equation $c_1x^2+c_2xy+c_3y^2+c_4x+c_5y+c_6=0$ yields

$$\begin{aligned} c_1x_1^2+c_2x_1y_1+c_3y_1^2+c_4x_1+c_5y_1+c_6&=0 \\ \vdots \qquad\qquad \vdots \qquad\qquad &\vdots \\ c_1x_5^2+c_2x_5y_5+c_3y_5^2+c_4x_5+c_5y_5+c_6&=0. \end{aligned}$$

These together with the original equation form a homogeneous linear system with a nontrivial solution for $c_1, c_2, \ldots, c_6$. Thus the determinant of the coefficient matrix is zero, which is exactly Equation (10).

8. As in the previous problem, substitute the coordinates (x_i, y_i, z_i) of each of the three points into the equation $c_1x+c_2y+c_3z+c_4=0$ to obtain a homogeneous system with nontrivial solution for $c_1, \ldots, c_4$. Thus the determinant of the coefficient matrix is zero, which is exactly Equation (11).

9. Substituting the coordinates (x_i, y_i, z_i) of the four points into the equation $c_1(x^2+y^2+z^2)+c_2x+c_3y+c_4z+c_5=0$ of the sphere yields four equations, which together with the above sphere equation form a homogeneous linear system for $c_1, \ldots, c_5$ with a nontrivial solution. Thus the determinant of this system is zero, which is Equation (12).

10. Upon substitution of the coordinates of the three points (x_1, y_1), (x_2, y_2) and (x_3, y_3), we obtain the equations:

$$\begin{aligned} c_1y+c_2x^2+c_3x+c_4&=0 \\ c_1y_1+c_2x_1^2+c_3x_1+c_4&=0 \\ c_1y_2+c_2x_2^2+c_3x_2+c_4&=0 \\ c_1y_3+c_2x_3^2+c_3x_3+c_4&=0. \end{aligned}$$

This is a homogeneous system with a nontrivial solution for c_1, c_2, c_3, c_4, so the determinant of the coefficient matrix is zero; that is, $\begin{vmatrix} y & x^2 & x & 1 \\ y_1 & x_1^2 & x_1 & 1 \\ y_2 & x_2^2 & x_2 & 1 \\ y_3 & x_3^2 & x_3 & 1 \end{vmatrix} = 0.$

11. Expanding the determinant in Equation (9) by cofactors of the first row makes it apparent that the coefficient of x^2+y^2 in the final equation is $\begin{vmatrix} x_1 & y_1 & 1 \\ x_2 & y_2 & 1 \\ x_3 & y_3 & 1 \end{vmatrix}$. If the points are collinear, then the columns are linearly dependent ($y_i = mx_i + b$), so the coefficient of $x^2 + y^2 = 0$ and the resulting equation is that of the line through the three points.

12. If the three distinct points are collinear then two of the coordinates can be expressed in terms of the third. Without loss of generality, we can say that y and z can be expressed in terms of x, i.e., x is the parameter. If the line is $(x, ax + b, cx + d)$, then the determinant in Equation (11) is $\begin{vmatrix} x & ax+b & cx+d & 1 \\ x_1 & ax_1+b & cx_1+d & 1 \\ x_2 & ax_2+b & cx_2+d & 1 \\ x_3 & ax_3+b & cx_3+d & 1 \end{vmatrix}$. Performing the following column operations:

- add $-a$ times the first column to the second column,
- add $-b$ times the fourth column to the second column,

yields the determinant $\begin{vmatrix} x & 0 & cx+d & 1 \\ x_1 & 0 & cx_1+d & 1 \\ x_2 & 0 & cx_2+d & 1 \\ x_3 & 0 & cx_3+d & 1 \end{vmatrix}$ equal to the original one. Expanding along the second column, it is clear that the determinant is 0 and Equation (11) becomes $0 = 0$.

13. As in Exercise 11, the coefficient of $x^2 + y^2 + z^2$ will be 0, so Equation (12) gives the equation of the plane in which the four points lie.

10.2 The Earliest Applications of Linear Algebra

1. The number of oxen is 50 per herd, and there are 7 herds, so there are 350 oxen. Hence the total number of oxen and sheep is 350 + 350 = 700.

2. **(a)** The equations are $B = 2A$, $C = 3(A + B)$, $D = 4(A + B + C)$, $300 = A + B + C + D$. Solving this linear system gives $A = 5$ (and $B = 10$, $C = 45$, $D = 240$).

(b) The equations are $B = 2A$, $C = 3B$, $D = 4C$, $132 = A + B + C + D$. Solving this linear system gives $A = 4$ (and $B = 8$, $C = 24$, $D = 96$).

3. Note that this is, effectively, Gaussian elimination applied to the augmented matrix $\begin{bmatrix} 1 & 1 & 10 \\ 1 & \frac{1}{4} & 7 \end{bmatrix}$.

4. **(a)** Let x represent oxen and y represent sheep, then the equations are $5x + 2y = 10$ and $2x + 5y = 8$. The corresponding array is

2	5
5	2
8	10

and the elimination step subtracts twice column 2 from five times column 1, giving

	5
21	2
20	10

and so $y = \frac{20}{21}$ unit for a sheep, and $x = \frac{34}{21}$ units for an ox.

(b) Let x, y, and z represent the number of bundles of each class. Then the equations are

$$\begin{aligned} 2x + y \qquad &= 1 \\ 3y + z &= 1 \\ x \qquad + 4z &= 1 \end{aligned}$$

and the corresponding array is

	2	1
3	1	
1		4
1	1	1

Subtract two times the numbers in the third column from the second column to get

		1
3	1	
1	−8	4
1	−1	1

Now subtract three times the numbers in the second column from the first column to get

		1
	1	
25	−8	4
4	−1	1

This is equivalent to the linear system

$$\begin{aligned} x + 4z &= 1 \\ y - 8z &= -1. \\ 25z &= 4 \end{aligned}$$

From this, the solution is that a bundle of the first class contains $\frac{9}{25}$ measure, second class contains $\frac{7}{25}$ measure, and third class contains $\frac{4}{25}$ measure.

5. **(a)** From equations 2 through n, $x_j = a_j - x_1$ (j = 2, ..., n). Using these equations in equation 1 gives

$$x_1 + (a_2 - x_1) + (a_3 - x_1) + \cdots + (a_n - x_1) = a_1$$

$$x_1 = \frac{a_2 + a_3 + \cdots + a_n - a_1}{n - 2}$$

First find x_1 in terms of the known quantities n and the a_i. Then we can use $x_j = a_j - x_1$ (j = 2, ..., n) to find the other x_i.

(b) Exercise 7(b) may be solved using this technique. x_1 represents gold, x_2 represents brass, x_3 represents tin, and x_4 represents much-wrought iron, so n = 4 and

$$a_1 = 60,$$
$$a_2 = \frac{2}{3}(60) = 40,$$
$$a_3 = \frac{3}{4}(60) = 45,$$
$$a_4 = \frac{3}{5}(60) = 36.$$
$$x_1 = \frac{(a_2 + a_3 + a_4) - a_1}{n - 2} = \frac{40 + 45 + 36 - 60}{4 - 2} = \frac{61}{2}$$
$$x_2 = a_2 - x_1 = 40 - \frac{61}{2} = \frac{19}{2}$$
$$x_3 = a_3 - x_1 = 45 - \frac{61}{2} = \frac{29}{2}$$
$$x_4 = a_4 - x_1 = 36 - \frac{61}{2} = \frac{11}{2}$$

The crown was made with $30\frac{1}{2}$ minae of gold, $9\frac{1}{2}$ minae of brass, $14\frac{1}{2}$ minae of tin, and $5\frac{1}{2}$ minae of iron.

6. **(a)** We can write this as

$$\begin{aligned} 5x + y + z - K &= 0 \\ x + 7y + z - K &= 0 \\ x + y + 8z - K &= 0 \end{aligned}$$

(a 3×4 system). Since the coefficient matrix of equations (5) is invertible (its determinant is 262), there is a unique solution x, y, z for every K; hence, K is an arbitrary parameter.

(b) Gaussian elimination gives $x = \frac{21K}{131}, y = \frac{14K}{131}, z = \frac{12K}{131}$, for any choice of K. Since 131 is prime, we must choose K to be an integer multiple of 131 to get integer solutions. The obvious choice is K = 131, giving x = 21, y = 14, z = 12, and K = 131.

(c) This solution corresponds to $K = 262 = 2 \cdot 131$.

7. **(a)** The system is $x + y = 1000$, $\left(\frac{1}{5}\right)x - \left(\frac{1}{4}\right)y = 10$, with solution $x = 577$ and $\frac{7}{9}$, $y = 422$ and $\frac{2}{9}$. The legitimate son receives $577\frac{7}{9}$ staters, the illegitimate son receives $422\frac{2}{9}$.

(b) The system is $G + B = \left(\frac{2}{3}\right)60$, $G + T = \left(\frac{3}{4}\right)60$, $G + I = \left(\frac{3}{5}\right)60$, $G + B + T + I = 60$, with solution $G = 30.5$, $B = 9.5$, $T = 14.5$, and $I = 5.5$. The crown was made with $30\frac{1}{2}$ minae of gold, $9\frac{1}{2}$ minae of brass, $14\frac{1}{2}$ minae of tin, and $5\frac{1}{2}$ minae of iron.

(c) The system is $A = B + \left(\frac{1}{3}C\right)$, $B = C + \left(\frac{1}{3}\right)A$, $C = \left(\frac{1}{3}\right)B + 10$, with solution $A = 45$, $B = 37.5$, and $C = 22.5$.
The first person has 45 minae, the second has $37\frac{1}{2}$, and the third has $22\frac{1}{2}$.

10.3 Cubic Spline Interpolation

2. **(a)** Set $h = .2$ and

$$\begin{aligned} x_1 &= 0, & y_1 &= .00000 \\ x_2 &= .2, & y_2 &= .19867 \\ x_3 &= .4, & y_3 &= .38942 \\ x_4 &= .6, & y_4 &= .56464 \\ x_5 &= .8, & y_5 &= .71736 \\ x_6 &= 1.0, & y_6 &= .84147 \end{aligned}$$

Then

$$\frac{6(y_1 - 2y_2 + y_3)}{h^2} = -1.1880$$
$$\frac{6(y_2 - 2y_3 + y_4)}{h^2} = -2.3295$$
$$\frac{6(y_3 - 2y_4 + y_5)}{h^2} = -3.3750$$
$$\frac{6(y_4 - 2y_5 + y_6)}{h^2} = -4.2915$$

and the linear system (21) for the parabolic runout spline becomes

$$\begin{bmatrix} 5 & 1 & 0 & 0 \\ 1 & 4 & 1 & 0 \\ 0 & 1 & 4 & 1 \\ 0 & 0 & 1 & 5 \end{bmatrix} \begin{bmatrix} M_2 \\ M_3 \\ M_4 \\ M_5 \end{bmatrix} = \begin{bmatrix} -1.1880 \\ -2.3295 \\ -3.3750 \\ -4.2915 \end{bmatrix}.$$

Solving this system yields $M_2 = -.15676$, $M_3 = -.40421$, $M_4 = -.55592$, $M_5 = -.74712$. From (19) and (20) we have $M_1 = M_2 = -.15676$, $M_6 = M_5 = -.74712$. The specific interval $.4 \le x \le .6$ is the third interval. Using (14) to solve for a_3, b_3, c_3, and d_3 gives

$$a_3 = \frac{(M_4 - M_3)}{6h} = -.12643$$
$$b_3 = \frac{M_3}{2} = -.20211$$
$$c_3 = \frac{(y_4 - y_3)}{h} - \frac{(M_4 + 2M_3)h}{6} = .92158$$
$$d_3 = y_3 = .38942.$$

The interpolating parabolic runout spline for $.4 \le x \le .6$ is thus

$$S(x) = -.12643(x - .4)^3 - .20211(x - .4)^2 + .92158(x - .4) + .38942.$$

(b) $S(.5) = -.12643(.1)^3 - .20211(.1)^2 + .92158(.1) + .38942 = .47943$. Since $\sin(.5) = S(.5) = .47943$ to five decimal places, the percentage error is zero.

3. (a) Given that the points lie on a single cubic curve, the cubic runout spline will agree exactly with the single cubic curve.

(b) Set $h = 1$ and

$$\begin{aligned} x_1 &= 0, & y_1 &= 1 \\ x_2 &= 1, & y_2 &= 7 \\ x_3 &= 2, & y_3 &= 27 \\ x_4 &= 3, & y_4 &= 79 \\ x_5 &= 4, & y_5 &= 181. \end{aligned}$$

Then

$$\frac{6(y_1 - 2y_2 + y_3)}{h^2} = 84$$
$$\frac{6(y_2 - 2y_3 + y_4)}{h^2} = 192$$
$$\frac{6(y_3 - 2y_4 + y_5)}{h^2} = 300$$

and the linear system (24) for the cubic runout spline becomes $\begin{bmatrix} 6 & 0 & 0 \\ 1 & 4 & 1 \\ 0 & 0 & 6 \end{bmatrix}\begin{bmatrix} M_2 \\ M_3 \\ M_4 \end{bmatrix} = \begin{bmatrix} 84 \\ 192 \\ 300 \end{bmatrix}$.

Solving this system yields $M_2 = 14$, $M_3 = 32$, $M_4 = 50$.
From (22) and (23) we have $M_1 = 2M_2 - M_3 = -4$, $M_5 = 2M_4 - M_3 = 68$.
Using (14) to solve for the a_i's, b_i's, c_i's, and d_i's we have

$a_1 = \frac{(M_2 - M_1)}{6h} = 3$, $a_2 = \frac{(M_3 - M_2)}{6h} = 3$, $a_3 = \frac{(M_4 - M_3)}{6h} = 3$, $a_4 = \frac{(M_5 - M_4)}{6h} = 3$,
$b_1 = \frac{M_1}{2} = -2$, $b_2 = \frac{M_2}{2} = 7$, $b_3 = \frac{M_3}{2} = 16$, $b_4 = \frac{M_4}{2} = 25$,
$c_1 = \frac{(y_2 - y_1)}{h} - \frac{(M_2 + 2M_1)h}{6} = 5$, $c_2 = \frac{(y_3 - y_2)}{h} - \frac{(M_3 + 2M_2)h}{6} = 10$,
$c_3 = \frac{(y_4 - y_3)}{h} - \frac{(M_4 + 2M_3)h}{6} = 33$, $c_4 = \frac{(y_5 - y_4)}{h} - \frac{(M_5 + 2M_4)h}{6} = 74$,
$d_1 = y_1 = 1$, $d_2 = y_2 = 7$, $d_3 = y_3 = 27$, $d_4 = y_4 = 79$.

For $0 \le x \le 1$ we have

$$S(x) = S_1(x) = 3x^3 - 2x^2 + 5x + 1.$$

For $1 \le x \le 2$ we have

$$S(x) = S_2(x) = 3(x-1)^3 + 7(x-1)^2 + 10(x-1) + 7 = 3x^3 - 2x^2 + 5x + 1.$$

For $2 \le x \le 3$ we have

$$S(x) = S_3(x) = 3(x-2)^3 + 16(x-2)^2 + 33(x-2) + 27 = 3x^3 - 2x^2 + 5x + 1.$$

For $3 \le x \le 4$ we have

$$S(x) = S_4(x) = 3(x-3)^3 + 25(x-3)^2 + 74(x-3) + 79 = 3x^3 - 2x^2 + 5x + 1.$$

Thus $S_1(x) = S_2(x) = S_3(x) = S_4(x)$, or $S(x) = 3x^3 - 2x^2 + 5x + 1$ for $0 \le x \le 4$.

4. The linear system (16) for the natural spline becomes $\begin{bmatrix} 4 & 1 & 0 \\ 1 & 4 & 1 \\ 0 & 1 & 4 \end{bmatrix} \begin{bmatrix} M_2 \\ M_3 \\ M_4 \end{bmatrix} = \begin{bmatrix} -.0001116 \\ -.0000816 \\ -.0000636 \end{bmatrix}$.

Solving this system yields $M_2 = -.0000252$, $M_3 = -.0000108$, $M_4 = -.0000132$.
From (17) and (18) we have $M_1 = 0$, $M_5 = 0$.
Solving for the a_i's, b_i's, c_i's, and d_i's from Equations (14) we have $a_1 = \frac{(M_2-M_1)}{6h} = -.00000042$,

$a_2 = \frac{(M_3-M_2)}{6h} = .00000024$, $a_3 = \frac{(M_4-M_3)}{6h} = -.00000004$, $a_4 = \frac{(M_5-M_4)}{6h} = -.00000022$,

$b_1 = \frac{M_1}{2} = 0$, $b_2 = \frac{M_2}{2} = -.0000126$, $b_3 = \frac{M_3}{2} = -.0000054$, $b_4 = \frac{M_4}{2} = -.0000066$, $b_5 = \frac{M_5}{2} = 0$.

$c_1 = \frac{(y_2-y_1)}{h} - \frac{(M_2+2M_1)h}{6} = .000214$, $c_2 = \frac{(y_3-y_2)}{h} - \frac{(M_3+2M_2)h}{6} = .000088$,
$c_3 = \frac{(y_4-y_3)}{h} - \frac{(M_4+2M_3)h}{6} = -.000092$, $c_4 = \frac{(y_5-y_4)}{h} - \frac{(M_5+2M_4)h}{6} = -.000212$,
$d_1 = y_1 = .99815$, $d_2 = y_2 = .99987$, $d_3 = y_3 = .99973$, $d_4 = y_4 = .99823$.
The resulting natural spline is

$$S(x) = \begin{cases} -.00000042(x+10)^3 + .000214(x+10) + .99815, & -10 \le x \le 0 \\ .00000024(x)^3 - .0000126(x)^2 + .000088(x) + .99987, & 0 \le x \le 10 \\ -.00000004(x-10)^3 - .0000054(x-10)^2 - .000092(x-10) + .99973, & 10 \le x \le 20 \\ -.00000022(x-20)^3 - .0000066(x-20)^2 - .000212(x-20) + .99823, & 20 \le x \le 30. \end{cases}$$

Assuming the maximum is attained in the interval [0, 10] we set $S'(x)$ equal to zero in this interval: $S'(x) = .00000072x^2 - .0000252x + .000088 = 0$.
To three significant digits the root of this quadratic equation in the interval [0, 10] is $x = 3.93$, and $S(3.93) = 1.00004$.

5. The linear system (24) for the cubic runout spline becomes $\begin{bmatrix} 6 & 0 & 0 \\ 1 & 4 & 1 \\ 0 & 0 & 6 \end{bmatrix} \begin{bmatrix} M_2 \\ M_3 \\ M_4 \end{bmatrix} = \begin{bmatrix} -.0001116 \\ -.0000816 \\ -.0000636 \end{bmatrix}$.

Solving this system yields $M_2 = -.0000186$, $M_3 = -.0000131$, $M_4 = -.0000106$.
From (22) and (23) we have $M_1 = 2M_2 - M_3 = -.0000241$, $M_5 = 2M_4 - M_3 = -.0000081$.
Solving for the a_i's, b_i's, c_i's, and d_i's from Equations (14) we have $a_1 = \frac{(M_2-M_1)}{6h} = .00000009$,

$a_2 = \frac{(M_3-M_2)}{6h} = .00000009$, $a_3 = \frac{(M_4-M_3)}{6h} = .00000004$, $a_4 = \frac{(M_5-M_4)}{6h} = .00000004$.

$b_1 = \frac{M_1}{2} = -.0000121$, $b_2 = \frac{M_2}{2} = -.0000093$, $b_3 = \frac{M_3}{2} = -.0000066$, $b_4 = \frac{M_4}{2} = -.0000053$,

$c_1 = \frac{(y_2-y_1)}{h} - \frac{(M_2+2M_1)h}{6} = .000282, c_2 = \frac{(y_3-y_2)}{h} - \frac{(M_3+2M_2)h}{6} = .000070,$

$c_3 = \frac{(y_4-y_3)}{h} - \frac{(M_4+2M_3)h}{6} = -.000087, c_4 = \frac{(y_5-y_4)}{h} - \frac{(M_5+2M_4)h}{6} = -.000207,$

$d_1 = y_1 = .99815, d_2 = y_2 = .99987, d_3 = y_3 = .99973, d_4 = y_4 = .99823.$

The resulting cubic runout spline is

$S(x) =$

$$\begin{cases} .00000009(x+10)^3 - .0000121(x+10)^2 + .000282(x+10) + .99815, & -10 \le x \le 0 \\ .00000009(x)^3 - .0000093(x)^2 + .000070(x) + .99987, & 0 \le x \le 10 \\ .00000004(x-10)^3 - .0000066(x-10)^2 - .000087(x-10) + .99973, & 10 \le x \le 20 \\ .00000004(x-20)^3 - .0000053(x-20)^2 - .000207(x-20) + .99823, & 20 \le x \le 30. \end{cases}$$

Assuming the maximum is attained in the interval [0, 10], we set $S'(x)$ equal to zero in this interval:

$$S'(x) = .00000027x^2 - .0000186x + .000070 = 0.$$

To three significant digits the root of this quadratic equation in the interval [0, 10] is 4.00 and $S(4.00) = 1.00001$.

6. **(a)** Set $h = .5$ and $x_1 = 0, y_1 = 0,\ x_2 = .5,\ y_2 = 1, x_3 = 1,\ y_3 = 0.$

For a natural spline with $n = 3$, $M_1 = M_3 = 0$ and $\frac{6(y_1-2y_2+y_3)}{h^2} = -48 = 4M_2$. Thus $M_2 = -12$.

$a_1 = \frac{M_2-M_1}{6h} = -4, a_2 = \frac{M_3-M_2}{6h} = 4, b_1 = \frac{M_1}{2} = 0, b_2 = \frac{M_2}{2} = -6, c_1 = \frac{y_2-y_1}{h} - \frac{(M_2+2M_1)h}{6} = 3,$

$c_2 = \frac{y_3-y_2}{h} - \frac{(M_3+2M_2)h}{6} = 0, d_1 = 0, d_2 = 1$

For $0 \le x \le .5$, we have $S(x) = S_1(x) = -4x^3 + 3x$.

For $.5 \le x \le 1$, we have

$$S(x) = S_2(x) = 4(x-.5)^3 - 6(x-.5)^2 + 1 = 4x^3 - 12x^2 + 9x - 1.$$

The resulting natural spline is $S(x) = \begin{cases} -4x^3 + 3x & 0 \le x \le 0.5 \\ 4x^3 - 12x^2 + 9x - 1 & 0.5 \le x \le 1 \end{cases}$.

(b) Again $h = .5$ and

$$x_1 = .5, \quad y_1 = 1$$
$$x_2 = 1, \quad y_2 = 0$$
$$x_3 = 1.5, \quad y_3 = -1$$
$$4M_2 = \frac{6(y_1 - 2y_2 + y_3)}{h^2} = 0$$

Thus $M_1 = M_2 = M_3 = 0$, hence all a_i and b_i are also 0.

$c_1 = \frac{y_2-y_1}{h} = -2, c_2 = \frac{y_3-y_2}{h} = -2,\ d_1 = y_1 = 1,\ d_2 = y_2 = 0$

For $.5 \le x \le 1$, we have $S(x) = S_1(x) = -2(x-.5) + 1 = -2x + 2$.

For $1 \le x \le 1.5$, we have $S(x) = S_2(x) = -2(x-1) + 0 = -2x + 2$.

The resulting natural spline is $S(x) = \begin{cases} -2x + 2 & 0.5 \le x \le 1 \\ -2x + 2 & 1 \le x \le 1.5 \end{cases}$.

(c) The three data points are collinear, so the spline is just the line the points lie on.

7. **(b)** Equations (15) together with the three equations in part (a) of the exercise statement give

$$4M_1 + M_2 + M_{n-1} = \frac{6(y_{n-1} - 2y_1 + y_2)}{h^2}$$
$$M_1 + 4M_2 + M_3 = \frac{6(y_1 - 2y_2 + y_3)}{h^2}$$
$$M_2 + 4M_3 + M_4 = \frac{6(y_2 - 2y_3 + y_4)}{h^2}$$
$$\vdots$$
$$M_{n-3} + 4M_{n-2} + M_{n-1} = \frac{6(y_{n-3} - 2y_{n-2} + y_{n-1})}{h^2}$$
$$M_1 + M_{n-2} + 4M_{n-1} = \frac{6(y_{n-2} - 2y_{n-1} + y_1)}{h^2}.$$

The linear system for $M_1, M_2, \ldots, M_{n-1}$ in matrix form is

$$\begin{bmatrix} 4 & 1 & 0 & 0 & \cdots & 0 & 0 & 0 & 1 \\ 1 & 4 & 1 & 0 & \cdots & 0 & 0 & 0 & 0 \\ 0 & 1 & 4 & 1 & \cdots & 0 & 0 & 0 & 0 \\ \vdots & \vdots & \vdots & \vdots & & \vdots & \vdots & \vdots & \vdots \\ 0 & 0 & 0 & 0 & \cdots & 0 & 1 & 4 & 1 \\ 1 & 0 & 0 & 0 & \cdots & 0 & 0 & 1 & 4 \end{bmatrix} \begin{bmatrix} M_1 \\ M_2 \\ M_3 \\ \vdots \\ M_{n-2} \\ M_{n-1} \end{bmatrix} = \frac{6}{h^2} \begin{bmatrix} y_{n-1} - 2y_1 + y_2 \\ y_1 - 2y_2 + y_3 \\ y_2 - 2y_3 + y_4 \\ \vdots \\ y_{n-3} - 2y_{n-2} + y_{n-1} \\ y_{n-2} - 2y_{n-1} + y_1 \end{bmatrix}.$$

8. **(b)** Equations (15) together with the two equations in part (a) give

$$2M_1 + M_2 = \frac{6(y_2 - y_1 - hy'_1)}{h^2}$$
$$M_1 + 4M_2 + M_3 = \frac{6(y_1 - 2y_2 + y_3)}{h^2}$$
$$M_2 + 4M_3 + M_4 = \frac{6(y_2 - 2y_3 + y_4)}{h^2}$$
$$\vdots$$
$$M_{n-2} + 4M_{n-1} + M_n = \frac{6(y_{n-2} - 2y_{n-1} + y_n)}{h^2}$$
$$M_{n-1} + 2M_n = \frac{6(y_{n-1} - y_n + hy'_n)}{h^2}.$$

This linear system for $M_1, M_2, \ldots, M_n$ in matrix form is

$$\begin{bmatrix} 2 & 1 & 0 & 0 & \cdots & 0 & 0 & 0 & 0 \\ 1 & 4 & 1 & 0 & \cdots & 0 & 0 & 0 & 0 \\ 0 & 1 & 4 & 1 & \cdots & 0 & 0 & 0 & 0 \\ 0 & 0 & 1 & 4 & \cdots & 0 & 0 & 0 & 0 \\ \vdots & \vdots & \vdots & \vdots & & \vdots & \vdots & \vdots & \vdots \\ 0 & 0 & 0 & 0 & \cdots & 1 & 4 & 1 & 0 \\ 0 & 0 & 0 & 0 & \cdots & 0 & 1 & 4 & 1 \\ 0 & 0 & 0 & 0 & \cdots & 0 & 0 & 1 & 2 \end{bmatrix} \begin{bmatrix} M_1 \\ M_2 \\ M_3 \\ \vdots \\ M_{n-2} \\ M_{n-1} \\ M_n \end{bmatrix} = \frac{6}{h^2} \begin{bmatrix} -hy'_1 - y_1 + y_2 \\ y_1 - 2y_2 + y_3 \\ y_2 - 2y_3 + y_4 \\ \vdots \\ y_{n-2} - 2y_{n-1} + y_n \\ y_{n-1} - y_n + hy'_n \end{bmatrix}.$$

10.4 Markov Chains

1. **(a)** $\mathbf{x}^{(1)} = P\mathbf{x}^{(0)} = \begin{bmatrix} .4 \\ .6 \end{bmatrix}, \mathbf{x}^{(2)} = P\mathbf{x}^{(1)} = \begin{bmatrix} .46 \\ .54 \end{bmatrix}.$

Continuing in this manner yields $\mathbf{x}^{(3)} = \begin{bmatrix} .454 \\ .546 \end{bmatrix}$, $\mathbf{x}^{(4)} = \begin{bmatrix} .4546 \\ .5454 \end{bmatrix}$ and $\mathbf{x}^{(5)} = \begin{bmatrix} .45454 \\ .54546 \end{bmatrix}$.

(b) P is regular because all of the entries of P are positive. Its steady-state vector $\mathbf{q}$ solves $(I - P)\mathbf{q} = \mathbf{0}$; that is, $\begin{bmatrix} .6 & -.5 \\ -.6 & .5 \end{bmatrix}\begin{bmatrix} q_1 \\ q_2 \end{bmatrix} = \begin{bmatrix} 0 \\ 0 \end{bmatrix}$.

This yields one independent equation, $.6q_1 - .5q_2 = 0$, or $q_1 = \frac{5}{6}q_2$. Solutions are thus of the form $\mathbf{q} = s\begin{bmatrix} \frac{5}{6} \\ 1 \end{bmatrix}$. Set $s = \frac{1}{\frac{5}{6}+1} = \frac{6}{11}$ to obtain $\mathbf{q} = \begin{bmatrix} \frac{5}{11} \\ \frac{6}{11} \end{bmatrix}$.

2. **(a)** $\mathbf{x}^{(1)} = P\mathbf{x}^{(0)} = \begin{bmatrix} .7 \\ .2 \\ .1 \end{bmatrix}$; likewise $\mathbf{x}^{(2)} = \begin{bmatrix} .23 \\ .52 \\ .25 \end{bmatrix}$ and $\mathbf{x}^{(3)} = \begin{bmatrix} .273 \\ .396 \\ .331 \end{bmatrix}$.

(b) P is regular because all of its entries are positive. To solve $(I - P)\mathbf{q} = \mathbf{0}$, i.e. $\begin{bmatrix} .8 & -.1 & -.7 \\ -.6 & .6 & -.2 \\ -.2 & -.5 & .9 \end{bmatrix}\begin{bmatrix} q_1 \\ q_2 \\ q_3 \end{bmatrix} = \begin{bmatrix} 0 \\ 0 \\ 0 \end{bmatrix}$, reduce the coefficient matrix to row-echelon form:

$$\begin{bmatrix} 8 & -1 & -7 \\ -6 & 6 & -2 \\ -2 & -5 & 9 \end{bmatrix} \rightarrow \begin{bmatrix} 2 & 5 & -9 \\ 0 & -21 & 29 \\ 0 & 0 & 0 \end{bmatrix} \rightarrow \begin{bmatrix} 1 & 0 & -\frac{22}{21} \\ 0 & 1 & -\frac{29}{21} \\ 0 & 0 & 0 \end{bmatrix}.$$

This yields solutions (setting $q_3 = s$) of the form $\begin{bmatrix} \frac{22}{21} \\ \frac{29}{21} \\ 1 \end{bmatrix} s$.

To obtain a probability vector, take $s = \frac{1}{\frac{22}{21}+\frac{29}{21}+1} = \frac{21}{72}$, yielding $\mathbf{q} = \begin{bmatrix} \frac{22}{72} \\ \frac{29}{72} \\ \frac{21}{72} \end{bmatrix}$.

3. **(a)** Solve $(I - P)\mathbf{q} = \mathbf{0}$, i.e., $\begin{bmatrix} \frac{2}{3} & -\frac{3}{4} \\ -\frac{2}{3} & \frac{3}{4} \end{bmatrix}\begin{bmatrix} q_1 \\ q_2 \end{bmatrix} = \begin{bmatrix} 0 \\ 0 \end{bmatrix}$.

The only independent equation is $\frac{2}{3}q_1 = \frac{3}{4}q_2$, yielding $\mathbf{q} = \begin{bmatrix} \frac{9}{8} \\ 1 \end{bmatrix} s$.

Setting $s = \frac{8}{17}$ yields $\mathbf{q} = \begin{bmatrix} \frac{9}{17} \\ \frac{8}{17} \end{bmatrix}$.

(b) As in (a), solve $\begin{bmatrix} .19 & -.26 \\ -.19 & .26 \end{bmatrix}\begin{bmatrix} q_1 \\ q_2 \end{bmatrix} = \begin{bmatrix} 0 \\ 0 \end{bmatrix}$

i.e., $.19q_1 = .26q_2$. Solutions have the form $\mathbf{q} = \begin{bmatrix} \frac{26}{19} \\ 1 \end{bmatrix} s$. Set $s = \frac{19}{45}$ to get $\mathbf{q} = \begin{bmatrix} \frac{26}{45} \\ \frac{19}{45} \end{bmatrix}$.

(c) Again, solve $\begin{bmatrix} \frac{2}{3} & -\frac{1}{2} & 0 \\ -\frac{1}{3} & 1 & -\frac{1}{4} \\ -\frac{1}{3} & -\frac{1}{2} & \frac{1}{4} \end{bmatrix}\begin{bmatrix} q_1 \\ q_2 \\ q_3 \end{bmatrix} = \begin{bmatrix} 0 \\ 0 \\ 0 \end{bmatrix}$ by reducing the coefficient matrix to row-echelon form: $\begin{bmatrix} 1 & 0 & -\frac{1}{4} \\ 0 & 1 & -\frac{1}{3} \\ 0 & 0 & 0 \end{bmatrix}$ yielding solutions of the form $\mathbf{q} = \begin{bmatrix} \frac{1}{4} \\ \frac{1}{3} \\ 1 \end{bmatrix} s$.

Set $s = \frac{12}{19}$ to get $\mathbf{q} = \begin{bmatrix} \frac{3}{19} \\ \frac{4}{19} \\ \frac{12}{19} \end{bmatrix}$.

4. **(a)** Prove by induction that $p_{12}^{(n)} = 0$: Already true for $n = 1$. If true for $n - 1$, we have $P^n = P^{n-1}P$, so $p_{12}^{(n)} = p_{11}^{(n-1)}p_{12} + p_{12}^{(n-1)}p_{22}$. But $p_{12} = p_{12}^{(n-1)} = 0$ so $p_{12}^{(n)} = 0 + 0 = 0$. Thus, no power of P can have all positive entries, so P is not regular.

(b) If $\mathbf{x} = \begin{bmatrix} x_1 \\ x_2 \end{bmatrix}$, $P\mathbf{x} = \begin{bmatrix} \frac{1}{2}x_1 \\ \frac{1}{2}x_1 + x_2 \end{bmatrix}$, $P^2\mathbf{x} = \begin{bmatrix} \frac{1}{4}x_1 \\ \frac{1}{4}x_1 + \frac{1}{2}x_1 + x_2 \end{bmatrix}$ etc. We use induction to show

$$P^n\mathbf{x} = \begin{bmatrix} \left(\frac{1}{2}\right)^n x_1 \\ \left(1 - \left(\frac{1}{2}\right)^n\right)x_1 + x_2 \end{bmatrix}.$$

Already true for $n = 1, 2$. If true for $n - 1$, then

$$P^n = P(P^{n-1}\mathbf{x}) = P\begin{bmatrix} \left(\frac{1}{2}\right)^{n-1} x_1 \\ \left(1 - \left(\frac{1}{2}\right)^{n-1}\right)x_1 + x_2 \end{bmatrix}$$

$$= \begin{bmatrix} \left(\frac{1}{2}\right)\left(\frac{1}{2}\right)^{n-1} x_1 \\ \left(1 - \left(\frac{1}{2}\right)^{n-1} + \left(\frac{1}{2}\right)^n\right)x_1 + x_2 \end{bmatrix} = \begin{bmatrix} \left(\frac{1}{2}\right)^n x_1 \\ \left(1 - \left(\frac{1}{2}\right)^n\right)x_1 + x_2 \end{bmatrix}.$$

Then $\lim_{n\to\infty} P^n\mathbf{x} = \begin{bmatrix} 0 \\ x_1 + x_2 \end{bmatrix} = \begin{bmatrix} 0 \\ 1 \end{bmatrix}$ if $\mathbf{x}$ is a state vector.

(c) The Theorem says that the entries of the steady state vector should be positive; they are not for $\begin{bmatrix} 0 \\ 1 \end{bmatrix}$.

5. Let $\mathbf{q} = \begin{bmatrix} \frac{1}{k} \\ \frac{1}{k} \\ \vdots \\ \frac{1}{k} \end{bmatrix}$. Then $(P\mathbf{q})_i = \sum_{j=1}^{k} p_{ij}q_j = \sum_{j=1}^{k} \frac{1}{k} p_{ij} = \frac{1}{k}\sum_{j=1}^{k} p_{ij} = \frac{1}{k}$, since the row sums of P are 1. Thus $(P\mathbf{q})_i = q_i$ for all *i*. By Theorem 10.4.4, $\mathbf{q} = \begin{bmatrix} \frac{1}{k} \\ \frac{1}{k} \\ \vdots \\ \frac{1}{k} \end{bmatrix}$ is the steady-state vector.

6. Since P has zeros entries, consider $P^2 = \begin{bmatrix} \frac{1}{2} & \frac{1}{4} & \frac{1}{4} \\ \frac{1}{4} & \frac{1}{2} & \frac{1}{4} \\ \frac{1}{4} & \frac{1}{4} & \frac{1}{2} \end{bmatrix}$, so P is regular. Note that all the rows of P sum to 1. Since P is 3×3, Exercise 5 implies $\mathbf{q} = \begin{bmatrix} \frac{1}{3} \\ \frac{1}{3} \\ \frac{1}{3} \end{bmatrix}$.

7. Let $\mathbf{x} = [x_1 \quad x_2]^T$ be the state vector, with x_1 = probability that John is happy and x_2 = probability that John is sad. The transition matrix P will be $P = \begin{bmatrix} \frac{4}{5} & \frac{2}{3} \\ \frac{1}{5} & \frac{1}{3} \end{bmatrix}$ since the columns must sum to one. We find the steady state vector for P by solving $\begin{bmatrix} \frac{1}{5} & -\frac{2}{3} \\ -\frac{1}{5} & \frac{2}{3} \end{bmatrix} \begin{bmatrix} q_1 \\ q_2 \end{bmatrix} = \begin{bmatrix} 0 \\ 0 \end{bmatrix}$, i.e., $\frac{1}{5}q_1 = \frac{2}{3}q_2$, so $\mathbf{q} = \begin{bmatrix} \frac{10}{3} \\ 1 \end{bmatrix} s$. Let $s = \frac{3}{13}$ and get $\mathbf{q} = \begin{bmatrix} \frac{10}{13} \\ \frac{3}{13} \end{bmatrix}$, so $\frac{10}{13}$ is the probability that John will be happy on a given day.

8. The state vector $\mathbf{x} = [x_1 \quad x_2 \quad x_3]^T$ will represent the proportion of the population living in regions 1, 2, and 3, respectively. In the transition matrix, p_{ij} will represent the proportion of the people in region *j* who move to region *i*, yielding $P = \begin{bmatrix} .90 & .15 & .10 \\ .05 & .75 & .05 \\ .05 & .10 & .85 \end{bmatrix}$.

$$\begin{bmatrix} .10 & -.15 & -.10 \\ -.05 & .25 & -.05 \\ -.05 & -.10 & .15 \end{bmatrix} \begin{bmatrix} q_1 \\ q_2 \\ q_3 \end{bmatrix} = \begin{bmatrix} 0 \\ 0 \\ 0 \end{bmatrix}$$

First reduce to row echelon form $\begin{bmatrix} 1 & 0 & -\frac{13}{7} \\ 0 & 1 & -\frac{4}{7} \\ 0 & 0 & 0 \end{bmatrix}$, yielding $\mathbf{q} = \begin{bmatrix} \frac{13}{7} \\ \frac{4}{7} \\ 1 \end{bmatrix} s$. Set $s = \frac{7}{24}$ and get

$\mathbf{q} = \begin{bmatrix} \frac{13}{24} \\ \frac{4}{24} \\ \frac{7}{24} \end{bmatrix}$, i.e., in the long run $\frac{13}{24}$ $\left(\text{or } 54\frac{1}{6}\%\right)$ of the people reside in region 1, $\frac{4}{24}$ $\left(\text{or } 16\frac{2}{3}\%\right)$ in region 2, and $\frac{7}{24}$ $\left(\text{or } 29\frac{1}{6}\%\right)$ in region 3.

10.5 Graph Theory

1. Note that the matrix has the same number of rows and columns as the graph has vertices, and that ones in the matrix correspond to arrows in the graph.

(a) $\begin{bmatrix} 0 & 0 & 0 & 1 \\ 1 & 0 & 1 & 1 \\ 1 & 1 & 0 & 1 \\ 0 & 0 & 0 & 0 \end{bmatrix}$ **(b)** $\begin{bmatrix} 0 & 1 & 1 & 0 & 0 \\ 0 & 0 & 0 & 0 & 1 \\ 1 & 0 & 0 & 1 & 0 \\ 0 & 0 & 1 & 0 & 0 \\ 0 & 0 & 1 & 0 & 0 \end{bmatrix}$ **(c)** $\begin{bmatrix} 0 & 1 & 0 & 1 & 0 & 0 \\ 1 & 0 & 0 & 0 & 0 & 0 \\ 0 & 1 & 0 & 1 & 1 & 1 \\ 0 & 0 & 0 & 0 & 0 & 1 \\ 0 & 0 & 0 & 0 & 0 & 1 \\ 0 & 0 & 1 & 0 & 1 & 0 \end{bmatrix}$

2. See the remark in problem 1; we obtain

(a)

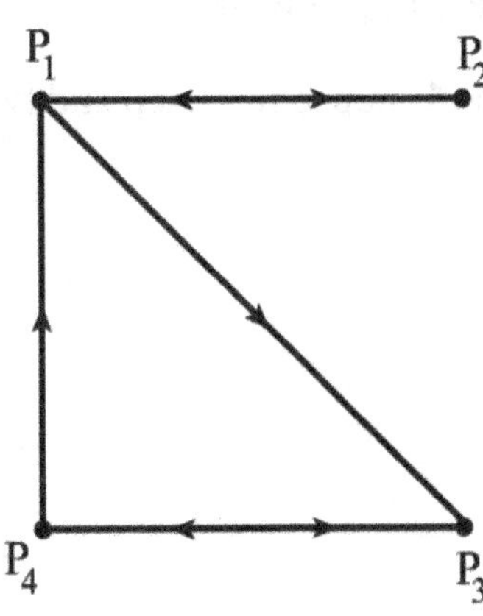

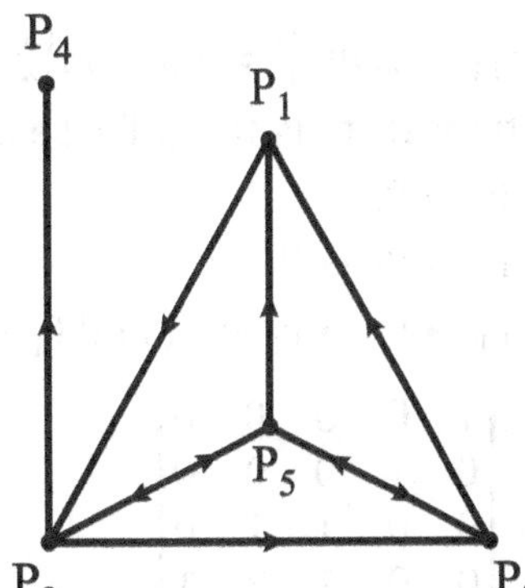

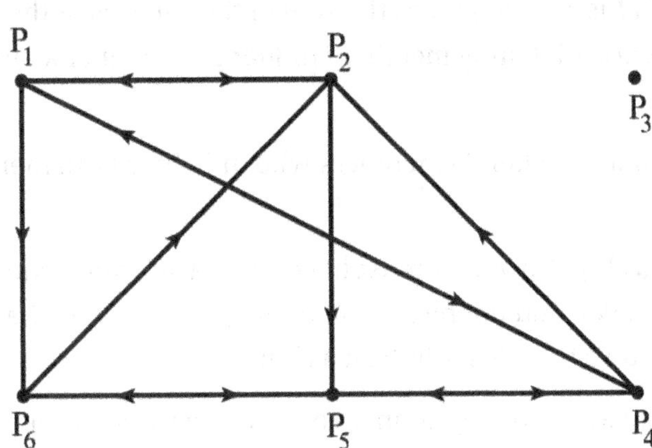

3. **(a)**

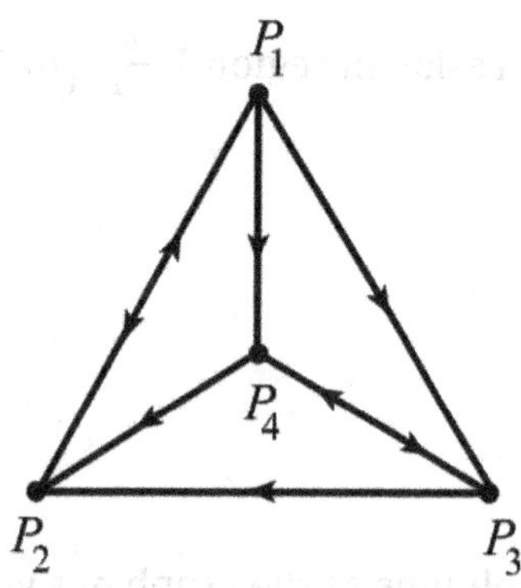

(b) $m_{12} = 1$, so there is one 1-step connection from P_1 to P_2.

$$M^2 = \begin{bmatrix} 1 & 2 & 1 & 1 \\ 0 & 1 & 1 & 1 \\ 1 & 1 & 1 & 0 \\ 1 & 1 & 0 & 1 \end{bmatrix} \text{ and } M^3 = \begin{bmatrix} 2 & 3 & 2 & 2 \\ 1 & 2 & 1 & 1 \\ 1 & 2 & 1 & 2 \\ 1 & 2 & 2 & 1 \end{bmatrix}.$$

So $m_{12}^{(2)} = 2$ and $m_{12}^{(3)} = 3$ meaning there are two 2-step and three 3-step connections from P_1 to P_2 by Theorem 10.6.1. These are:

1-step: $P_1 \to P_2$

2-step: $P_1 \to P_4 \to P_2$ and $P_1 \to P_3 \to P_2$

3-step: $P_1 \to P_2 \to P_1 \to P_2$, $P_1 \to P_3 \to P_4 \to P_2$, and $P_1 \to P_4 \to P_3 \to P_2$.

(c) Since $m_{14} = 1$, $m_{14}^{(2)} = 1$ and $m_{14}^{(3)} = 2$, there are one 1-step, one 2-step and two 3-step connections from P_1 to P_4. These are:

1-step: $P_1 \to P_4$

2-step: $P_1 \to P_3 \to P_4$

3-step: $P_1 \to P_2 \to P_1 \to P_4$ and $P_1 \to P_4 \to P_3 \to P_4$.

4. **(a)** $M^TM = \begin{bmatrix} 1 & 0 & 0 & 0 & 0 \\ 0 & 1 & 0 & 0 & 0 \\ 0 & 0 & 1 & 1 & 0 \\ 0 & 0 & 1 & 2 & 1 \\ 0 & 0 & 0 & 1 & 2 \end{bmatrix}$

(b) The kth diagonal entry of M^TM is $\sum_{i=1}^{5} m_{ik}{}^2$, i.e., the sum of the squares of the entries in column k of M. These entries are 1 if family member i influences member k and 0 otherwise.

(c) The ij entry of M^TM is the number of family members who influence both member i and member j.

5. **(a)** Note that to be contained in a clique, a vertex must have "two-way" connections with at least two other vertices. Thus, P_4 could not be in a clique, so $\{P_1, P_2, P_3\}$ is the only possible clique. Inspection shows that this is indeed a clique.

(b) Not only must a clique vertex have two-way connections to at least two other vertices, but the vertices to which it is connected must share a two-way connection. This consideration eliminates P_1 and P_2, leaving $\{P_3, P_4, P_5\}$ as the only possible clique. Inspection shows that it is indeed a clique.

(c) The above considerations eliminate P_1, P_3 and P_7 from being in a clique. Inspection shows that each of the sets $\{P_2, P_4, P_6\}$, $\{P_4, P_6, P_8\}$, $\{P_2, P_6, P_8\}$, $\{P_2, P_4, P_8\}$ and $\{P_4, P_5, P_6\}$ satisfy conditions (i) and (ii) in the definition of a clique. But note that P_8 can be added to the first set and we still satisfy the conditions. P_5 may not be added, so $\{P_2, P_4, P_6, P_8\}$ is a clique, containing all the other possibilities except $\{P_4, P_5, P_6\}$, which is also a clique.

6. **(a)** With the given M we get $S = \begin{bmatrix} 0 & 1 & 0 & 1 & 0 \\ 1 & 0 & 1 & 0 & 0 \\ 0 & 1 & 0 & 0 & 1 \\ 1 & 0 & 0 & 0 & 1 \\ 0 & 0 & 1 & 1 & 0 \end{bmatrix}$, $S^3 = \begin{bmatrix} 0 & 3 & 1 & 3 & 1 \\ 3 & 0 & 3 & 1 & 1 \\ 1 & 3 & 0 & 1 & 3 \\ 3 & 1 & 1 & 0 & 3 \\ 1 & 1 & 3 & 3 & 0 \end{bmatrix}$.

Since $s_{ii}^{(3)} = 0$ for all i, there are no cliques in the graph represented by M.

(b) Here $S = \begin{bmatrix} 0 & 1 & 0 & 1 & 0 & 0 \\ 1 & 0 & 1 & 0 & 1 & 0 \\ 0 & 1 & 0 & 1 & 0 & 1 \\ 1 & 0 & 1 & 0 & 1 & 1 \\ 0 & 1 & 0 & 1 & 0 & 0 \\ 0 & 0 & 1 & 1 & 0 & 0 \end{bmatrix}$, $S^3 = \begin{bmatrix} 0 & 6 & 1 & 7 & 0 & 2 \\ 6 & 0 & 7 & 1 & 6 & 3 \\ 1 & 7 & 2 & 8 & 1 & 4 \\ 7 & 1 & 8 & 2 & 7 & 5 \\ 0 & 6 & 1 & 7 & 0 & 2 \\ 2 & 3 & 4 & 5 & 2 & 2 \end{bmatrix}$.

The elements along the main diagonal tell us that only P_3, P_4, and P_6 are members of a clique. Since a clique contains at least three vertices, we must have $\{P_3, P_4, P_6\}$ as the only clique.

7. $M = \begin{bmatrix} 0 & 0 & 1 & 1 \\ 1 & 0 & 0 & 0 \\ 0 & 1 & 0 & 1 \\ 0 & 1 & 0 & 0 \end{bmatrix}$. Then $M^2 = \begin{bmatrix} 0 & 2 & 0 & 1 \\ 0 & 0 & 1 & 1 \\ 1 & 1 & 0 & 0 \\ 1 & 0 & 0 & 0 \end{bmatrix}$ and $M + M^2 = \begin{bmatrix} 0 & 2 & 1 & 2 \\ 1 & 0 & 1 & 1 \\ 1 & 2 & 0 & 1 \\ 1 & 1 & 0 & 0 \end{bmatrix}$.

By summing the rows of $M + M^2$, we get that the power of P_1 is 2 + 1 + 2 = 5, the power of P_2 is 3, of P_3 is 4, and of P_4 is 2.

8. Associating vertex P_1 with team A, P_2 with B, ..., P_5 with E, the game results yield the following dominance-directed graph:

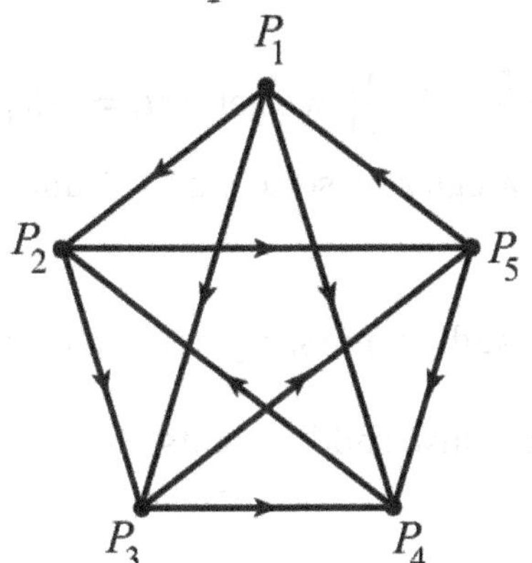

which has vertex matrix $M = \begin{bmatrix} 0 & 1 & 1 & 1 & 0 \\ 0 & 0 & 1 & 0 & 1 \\ 0 & 0 & 0 & 1 & 1 \\ 0 & 1 & 0 & 0 & 0 \\ 1 & 0 & 0 & 1 & 0 \end{bmatrix}$.

Then $M^2 = \begin{bmatrix} 0 & 1 & 1 & 1 & 2 \\ 1 & 0 & 0 & 2 & 1 \\ 1 & 1 & 0 & 1 & 0 \\ 0 & 0 & 1 & 0 & 1 \\ 0 & 2 & 1 & 1 & 0 \end{bmatrix}$, $M + M^2 = \begin{bmatrix} 0 & 2 & 2 & 2 & 2 \\ 1 & 0 & 1 & 2 & 2 \\ 1 & 1 & 0 & 2 & 1 \\ 0 & 1 & 1 & 0 & 1 \\ 1 & 2 & 1 & 2 & 0 \end{bmatrix}$. Summing the rows, we get that the power of A is 8, of B is 6, of C is 5, of D is 3, and of E is 6. Thus ranking in decreasing order we get A in first place, B and E tie for second place, C in fourth place, and D last.

10.6 Games of Strategy

1. **(a)** From Equation (2), the expected payoff of the game is

$$\mathbf{p}A\mathbf{q} = \begin{bmatrix} \frac{1}{2} & 0 & \frac{1}{2} \end{bmatrix} \begin{bmatrix} -4 & 6 & -4 & 1 \\ 5 & -7 & 3 & 8 \\ -8 & 0 & 6 & -2 \end{bmatrix} \begin{bmatrix} \frac{1}{4} \\ \frac{1}{4} \\ \frac{1}{4} \\ \frac{1}{4} \end{bmatrix} = -\frac{5}{8}.$$

(b) If player R uses strategy $[p_1 \quad p_2 \quad p_3]$ against player C's strategy $\begin{bmatrix} \frac{1}{4} \\ \frac{1}{4} \\ \frac{1}{4} \\ \frac{1}{4} \end{bmatrix}$ his payoff will be

$\mathbf{p}A\mathbf{q} = \left(-\frac{1}{4}\right)p_1 + \left(\frac{9}{4}\right)p_2 - p_3$. Since p_1, p_2, and p_3 are nonnegative and add up to 1, this is a weighted average of the numbers $-\frac{1}{4}, \frac{9}{4}$, and -1. Clearly this is the largest if $p_1 = p_3 = 0$ and $p_2 = 1$; that is, $\mathbf{p} = [0 \quad 1 \quad 0]$.

(c) As in (b), if player C uses $[q_1 \quad q_2 \quad q_3 \quad q_4]^T$ against $\begin{bmatrix} \frac{1}{2} & 0 & \frac{1}{2} \end{bmatrix}$, we get $\mathbf{p}A\mathbf{q} = -6q_1 + 3q_2 + q_3 - \frac{1}{2}q_4$. Clearly this is minimized over all strategies by setting $q_1 = 1$ and $q_2 = q_3 = q_4 = 0$. That is $\mathbf{q} = [1 \quad 0 \quad 0 \quad 0]^T$.

2. As per the hint, we will construct a 3×3 matrix with two saddle points, say $a_{11} = a_{33} = 1$. Such a matrix is $A = \begin{bmatrix} 1 & 2 & 1 \\ 0 & 7 & 0 \\ 1 & 2 & 1 \end{bmatrix}$. Note that $a_{13} = a_{31} = 1$ are also saddle points.

3. **(a)** Calling the matrix A, we see a_{22} is a saddle point, so the optimal strategies are pure, namely: $\mathbf{p}^* = [0 \quad 1]$, $\mathbf{q}^* = \begin{bmatrix}0\\1\end{bmatrix}$, the value of the game is $v = a_{22} = 3$.

(b) As in (a), a_{21} is a saddle point, so optimal strategies are $\mathbf{p}^* = [0 \quad 1 \quad 0]$, $\mathbf{q}^* = \begin{bmatrix}1\\0\end{bmatrix}$, the value of the game is $v = a_{21} = 2$.

(c) Here, a_{32} is a saddle point, so optimal strategies are $\mathbf{p}^* = [0 \quad 0 \quad 1]$, $\mathbf{q}^* = \begin{bmatrix}0\\1\\0\end{bmatrix}$ and $v = a_{32} = 2$.

(d) Here, a_{21} is a saddle point, so $\mathbf{p}^* = [0 \quad 1 \quad 0 \quad 0]$, $\mathbf{q}^* = \begin{bmatrix}1\\0\\0\end{bmatrix}$ and $v = a_{21} = -2$.

4. **(a)** Calling the matrix A, the formulas of Theorem 10.6.2 yield $\mathbf{p}^* = \begin{bmatrix}\frac{5}{8} & \frac{3}{8}\end{bmatrix}$, $\mathbf{q}^* = \begin{bmatrix}\frac{1}{8}\\\frac{7}{8}\end{bmatrix}$, $v = \frac{27}{8}$ (A has no saddle points).

(b) As in (a), $\mathbf{p}^* = \begin{bmatrix}\frac{40}{60} & \frac{20}{60}\end{bmatrix} = \begin{bmatrix}\frac{2}{3} & \frac{1}{3}\end{bmatrix}$, $\mathbf{q}^* = \begin{bmatrix}\frac{10}{60}\\\frac{50}{60}\end{bmatrix} = \begin{bmatrix}\frac{1}{6}\\\frac{5}{6}\end{bmatrix}$, $v = \frac{1400}{60} = \frac{70}{3}$ (Again, A has no saddle points).

(c) For this matrix, a_{11} is a saddle point, so $\mathbf{p}^* = [1 \quad 0]$, $\mathbf{q}^* = \begin{bmatrix}1\\0\end{bmatrix}$, and $v = a_{11} = 3$.

(d) This matrix has no saddle points, so, as in (a), $\mathbf{p}^* = \begin{bmatrix}\frac{-3}{-5} & \frac{-2}{-5}\end{bmatrix} = \begin{bmatrix}\frac{3}{5} & \frac{2}{5}\end{bmatrix}$, $\mathbf{q}^* = \begin{bmatrix}\frac{-3}{-5}\\\frac{-2}{-5}\end{bmatrix} = \begin{bmatrix}\frac{3}{5}\\\frac{2}{5}\end{bmatrix}$, and $v = \frac{-19}{-5} = \frac{19}{5}$.

(e) Again, A has no saddle points, so as in (a), $\mathbf{p}^* = \begin{bmatrix}\frac{3}{13} & \frac{10}{13}\end{bmatrix}$, $\mathbf{q}^* = \begin{bmatrix}\frac{1}{13}\\\frac{12}{13}\end{bmatrix}$, and $v = \frac{-29}{13}$.

5. Let

a_{11} = payoff to R if the black ace and black two are played = 3.
a_{12} = payoff to R if the black ace and red three are played = –4.
a_{21} = payoff to R if the red four and black two are played = –6.
a_{22} = payoff to R if the red four and red three are played = 7.
So, the payoff matrix for the game is $A = \begin{bmatrix}3 & -4\\-6 & 7\end{bmatrix}$.

A has no saddle points, so from Theorem 10.7.2, $\mathbf{p}^* = \begin{bmatrix}\frac{13}{20} & \frac{7}{20}\end{bmatrix}$, $\mathbf{q}^* = \begin{bmatrix}\frac{11}{20}\\\frac{9}{20}\end{bmatrix}$; that is, player R should play the black ace 65 percent of the time, and player C should play the black two 55 percent of the time. The value of the game is $-\frac{3}{20}$, that is, player C can expect to collect on the average 15 cents per game.

10.7 Leontief Economic Models

1. **(a)** Calling the given matrix E, we need to solve $(I-E)\mathbf{p} = \begin{bmatrix} \frac{1}{2} & -\frac{1}{3} \\ -\frac{1}{2} & \frac{1}{3} \end{bmatrix}\begin{bmatrix} p_1 \\ p_2 \end{bmatrix} = \begin{bmatrix} 0 \\ 0 \end{bmatrix}$.

This yields $\frac{1}{2}p_1 = \frac{1}{3}p_2$, that is, $\mathbf{p} = s\begin{bmatrix} 1 \\ \frac{3}{2} \end{bmatrix}$. Set $s = 2$ and get $\mathbf{p} = \begin{bmatrix} 2 \\ 3 \end{bmatrix}$.

(b) As in (a), solve $(I-E)\mathbf{p} = \begin{bmatrix} \frac{1}{2} & 0 & -\frac{1}{2} \\ -\frac{1}{3} & 1 & -\frac{1}{2} \\ -\frac{1}{6} & -1 & 1 \end{bmatrix}\begin{bmatrix} p_1 \\ p_2 \\ p_3 \end{bmatrix} = \begin{bmatrix} 0 \\ 0 \\ 0 \end{bmatrix}$.

In row-echelon form, this reduces to $\begin{bmatrix} 1 & 0 & -1 \\ 0 & 1 & -\frac{5}{6} \\ 0 & 0 & 0 \end{bmatrix}\begin{bmatrix} p_1 \\ p_2 \\ p_3 \end{bmatrix} = \begin{bmatrix} 0 \\ 0 \\ 0 \end{bmatrix}$.

Solutions of this system have the form $\mathbf{p} = s\begin{bmatrix} 1 \\ \frac{5}{6} \\ 1 \end{bmatrix}$. Set $s = 6$ and get $\mathbf{p} = \begin{bmatrix} 6 \\ 5 \\ 6 \end{bmatrix}$.

(c) As in (a), solve $(I-E)\mathbf{p} = \begin{bmatrix} .65 & -.50 & -.30 \\ -.25 & .80 & -.30 \\ -.40 & -.30 & .60 \end{bmatrix}\begin{bmatrix} p_1 \\ p_2 \\ p_3 \end{bmatrix} = \begin{bmatrix} 0 \\ 0 \\ 0 \end{bmatrix}$, which reduces to

$\begin{bmatrix} 1 & 0 & -\frac{78}{79} \\ 0 & 1 & -\frac{54}{79} \\ 0 & 0 & 0 \end{bmatrix}\begin{bmatrix} p_1 \\ p_2 \\ p_3 \end{bmatrix} = \begin{bmatrix} 0 \\ 0 \\ 0 \end{bmatrix}$. Solutions are of the form $\mathbf{p} = s\begin{bmatrix} \frac{78}{79} \\ \frac{54}{79} \\ 1 \end{bmatrix}$.

Let $s = 79$ to obtain $\mathbf{p} = \begin{bmatrix} 78 \\ 54 \\ 79 \end{bmatrix}$.

2. **(a)** By Corollary 10.7.4, this matrix is productive, since each of its row sums is .9.

(b) By Corollary 10.7.5, this matrix is productive, since each of its column sums is less than one.

(c) Try $\mathbf{x} = \begin{bmatrix} 2 \\ 1 \\ 1 \end{bmatrix}$. Then $C\mathbf{x} = \begin{bmatrix} 1.9 \\ .9 \\ .9 \end{bmatrix}$, i.e., $\mathbf{x} > C\mathbf{x}$, so this matrix is productive by Theorem 10.7.3.

3. Theorem 10.8.2 says there will be one linearly independent price vector for the matrix E if some positive power of E is positive. Since E is not positive, try E^2.

$$E^2 = \begin{bmatrix} .2 & .34 & .1 \\ .2 & .54 & .6 \\ .6 & .12 & .3 \end{bmatrix} > 0$$

4. The exchange matrix for this arrangement (using A, B, and C in that order) is $\begin{bmatrix} \frac{1}{2} & \frac{1}{3} & \frac{1}{4} \\ \frac{1}{3} & \frac{1}{3} & \frac{1}{4} \\ \frac{1}{6} & \frac{1}{3} & \frac{1}{2} \end{bmatrix}$.

For equilibrium, we must solve $(I - E)\mathbf{p} = 0$. That is $\begin{bmatrix} \frac{1}{2} & -\frac{1}{3} & -\frac{1}{4} \\ -\frac{1}{3} & \frac{2}{3} & -\frac{1}{4} \\ -\frac{1}{6} & -\frac{1}{3} & \frac{1}{2} \end{bmatrix}\begin{bmatrix} p_1 \\ p_2 \\ p_3 \end{bmatrix} = \begin{bmatrix} 0 \\ 0 \\ 0 \end{bmatrix}$.

Row reduction yields solutions of the form $\mathbf{p} = \begin{bmatrix} \frac{18}{16} \\ \frac{15}{16} \\ 1 \end{bmatrix} s$. Set $s = \frac{1600}{15}$ and obtain $\mathbf{p} = \begin{bmatrix} 120 \\ 100 \\ 106.67 \end{bmatrix}$; i.e., the price of tomatoes was \$120, corn was \$100, and lettuce was \$106.67.

5. Taking the CE, EE, and ME in that order, we form the consumption matrix C, where c_{ij} = the amount (per consulting dollar) of the i-th engineer's services purchased by the j-th engineer. Thus, $C = \begin{bmatrix} 0 & .2 & .3 \\ .1 & 0 & .4 \\ .3 & .4 & 0 \end{bmatrix}$.

We want to solve $(I - C)\mathbf{x} = \mathbf{d}$, where $\mathbf{d}$ is the demand vector, i.e.

$$\begin{bmatrix} 1 & -.2 & -.3 \\ -.1 & 1 & -.4 \\ -.3 & -.4 & 1 \end{bmatrix}\begin{bmatrix} x_1 \\ x_2 \\ x_3 \end{bmatrix} = \begin{bmatrix} 500 \\ 700 \\ 600 \end{bmatrix}.$$

In row-echelon form this reduces to $\begin{bmatrix} 1 & -.2 & -.3 \\ 0 & 1 & -.43877 \\ 0 & 0 & 1 \end{bmatrix}\begin{bmatrix} x_1 \\ x_2 \\ x_3 \end{bmatrix} = \begin{bmatrix} 500.00 \\ 765.31 \\ 1556.19 \end{bmatrix}$.

Back-substitution yields the solution $\mathbf{x} = \begin{bmatrix} 1256.48 \\ 1448.12 \\ 1556.19 \end{bmatrix}$.

The CE received \$1256, the EE received \$1448, and the ME received \$1556.

6. **(a)** The solution of the system $(I - C)\mathbf{x} = \mathbf{d}$ is $\mathbf{x} = (I - C)^{-1}\mathbf{d}$. The effect of increasing the demand d_i for the ith industry by one unit is the same as adding $\begin{bmatrix} 0 \\ \vdots \\ 0 \\ 1 \\ 0 \\ \vdots \\ 0 \end{bmatrix}$ to $\mathbf{d}$ where the 1 is in the ith row. The new solution is $(I - C)^{-1}\left(\mathbf{d} + \begin{bmatrix} 0 \\ \vdots \\ 0 \\ 1 \\ 0 \\ \vdots \\ 0 \end{bmatrix}\right) = \mathbf{x} + (I - C)^{-1}\begin{bmatrix} 0 \\ \vdots \\ 0 \\ 1 \\ 0 \\ \vdots \\ 0 \end{bmatrix}$ which has the effect of adding the ith column of $(I - C)^{-1}$ to the original solution.

(b) The increase in value is the second column of $(I - C)^{-1}$, $\frac{1}{503}\begin{bmatrix}542\\690\\170\end{bmatrix}$. Thus the value of the coal-mining operation must increase by $\frac{542}{503}$.

7. The i-th column sum of E is $\sum_{j=1}^{n} e_{ji}$, and the elements of the i-th column of $I - E$ are the negatives of the elements of E, except for the ii-th, which is $1 - e_{ii}$. So, the i-th column sum of $I - E$ is $1 - \sum_{j=1}^{n} e_{ji} = 1 - 1 = 0$. Now, $(I - E)^{\mathrm{T}}$ has zero row sums, so the vector $\mathbf{x} = [1 \;\; 1 \;\; \cdots \;\; 1]^{\mathrm{T}}$ solves $(I - E)^{\mathrm{T}}\mathbf{x} = \mathbf{0}$. This implies $\det(I - E)^{\mathrm{T}} = 0$. But $\det(I - E)^{\mathrm{T}} = \det(I - E)$, so $(I - E)\mathbf{p} = \mathbf{0}$ must have nontrivial (i.e., nonzero) solutions.

8. Let C be a consumption matrix whose column sums are less than one; then the row sums of C^T are less than one. By Corollary 10.7.4, C^T is productive so $(\mathrm{I} - C^T)^{-1} \geq 0$. But

$$(I - C)^{-1} = (((I - C)^{\mathrm{T}}))^{-1})^T = ((I - C^T)^{-1})^T \geq 0.$$

Thus, C is productive.

10.8 Forest Management

1. Using Equation (18), we calculate

$$Yld_2 = \frac{30s}{2} = 15s$$

$$Yld_3 = \frac{50s}{2 + \frac{3}{2}} = \frac{100s}{7}.$$

So all the trees in the second class should be harvested for an optimal yield (since $s = 1000$) of \$15,000.

2. From the solution to Example 1, we see that for the fifth class to be harvested in the optimal case we must have $\frac{p_5 s}{(.28^{-1}+.31^{-1}+.25^{-1}+.23^{-1})} > 14.7s$, yielding $p_5 > \$222.63$.

3. Assume $p_2 = 1$, then $Yld_2 = \frac{s}{(.28)^{-1}} = .28s$. Thus, for all the yields to be the same we must have

$$\frac{p_3 s}{(.28^{-1} + .31^{-1})} = .28s$$

$$\frac{p_4 s}{(.28^{-1} + .31^{-1} + .25^{-1})} = .28s$$

$$\frac{p_5 s}{(.28^{-1} + .31^{-1} + .25^{-1} + .23^{-1})} = .28s$$

$$\frac{p_6 s}{(.28^{-1} + .31^{-1} + .25^{-1} + .23^{-1} + .37^{-1})} = .28s$$

Solving these successively yields $p_3 = 1.90$, $p_4 = 3.02$, $p_5 = 4.24$ and $p_6 = 5.00$. Thus the ratio $p_2{:}p_3{:}p_4{:}p_5{:}p_6 = 1{:}1.90{:}3.02{:}4.24{:}5.00$.

5. Since $\mathbf{y}$ is the harvest vector, $N = \sum_{i=1}^{n} y_i$ is the number of trees removed from the forest. Then Equation (7) and the first of Equations (8) yield $N = g_1 x_1$, and from Equation (17) we obtain $N = \frac{g_1 s}{1+\frac{g_1}{g_2}+\cdots+\frac{g_1}{g_{k-1}}} = \frac{s}{\frac{1}{g_1}+\cdots+\frac{1}{g_{k-1}}}$.

6. Set $g_1 = \cdots = g_{n-1} = g$, and $p_2 = 1$. Then from Equation (18), $Yld_2 = \frac{p_2 s}{\frac{1}{g_1}} = gs$. Since we want all of the Yld_k's to be the same, we need to solve $Yld_k = \frac{p_k s}{(k-1)\frac{1}{g}} = gs$ for p_k for $3 \le k \le n$. Thus $p_k = k-1$. So the ratio $p_2 : p_3 : p_4 : \cdots : p_n = 1 : 2 : 3 : \cdots : (n-1)$.

10.9 Computer Graphics

1. **(a)** Using the coordinates of the points as the columns of a matrix we obtain $\begin{bmatrix} 0 & 1 & 1 & 0 \\ 0 & 0 & 1 & 1 \\ 0 & 0 & 0 & 0 \end{bmatrix}$.

(b) The scaling is accomplished by multiplication of the coordinate matrix on the left by $\begin{bmatrix} \frac{3}{2} & 0 & 0 \\ 0 & \frac{1}{2} & 0 \\ 0 & 0 & 1 \end{bmatrix}$, resulting in the matrix $\begin{bmatrix} 0 & \frac{3}{2} & \frac{3}{2} & 0 \\ 0 & 0 & \frac{1}{2} & \frac{1}{2} \\ 0 & 0 & 0 & 0 \end{bmatrix}$, which represents the vertices $(0, 0, 0)$, $\left(\frac{3}{2}, 0, 0\right)$, $\left(\frac{3}{2}, \frac{1}{2}, 0\right)$ and $\left(0, \frac{1}{2}, 0\right)$ as shown below.

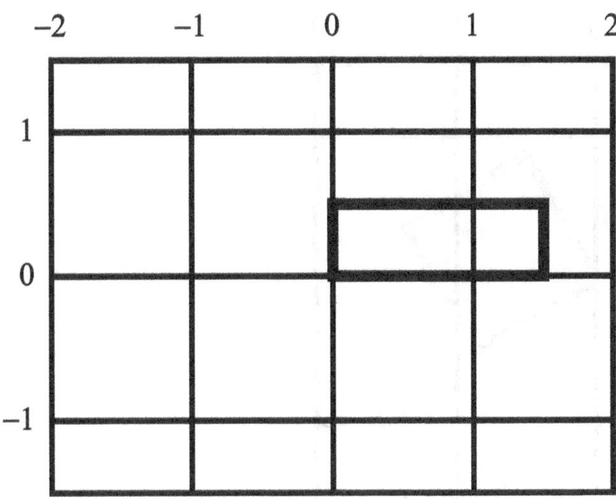

(c) Adding the matrix $\begin{bmatrix} -2 & -2 & -2 & -2 \\ -1 & -1 & -1 & -1 \\ 3 & 3 & 3 & 3 \end{bmatrix}$ to the original matrix yields $\begin{bmatrix} -2 & -1 & -1 & -2 \\ -1 & -1 & 0 & 0 \\ 3 & 3 & 3 & 3 \end{bmatrix}$, which represents the vertices $(-2, -1, 3)$, $(-1, -1, 3)$, $(-1, 0, 3)$, and (–2, 0, 3) as shown below.

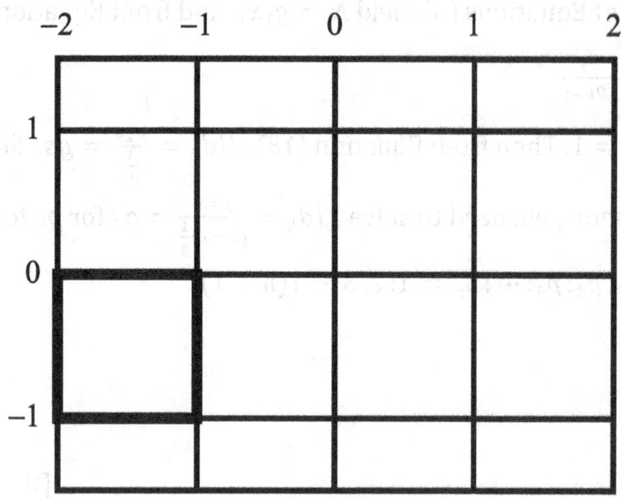

(d) Multiplying by the matrix $\begin{bmatrix} \cos(-30°) & -\sin(-30°) & 0 \\ \sin(-30°) & \cos(-30°) & 0 \\ 0 & 0 & 1 \end{bmatrix}$, we obtain

$$\begin{bmatrix} 0 & \cos(-30°) & \cos(-30°) - \sin(-30°) & -\sin(-30°) \\ 0 & \sin(-30°) & \cos(-30°) + \sin(-30°) & \cos(-30°) \\ 0 & 0 & 0 & 0 \end{bmatrix} = \begin{bmatrix} 0 & .866 & 1.366 & .500 \\ 0 & -.500 & .366 & .866 \\ 0 & 0 & 0 & 0 \end{bmatrix}.$$

The vertices are then (0, 0, 0), (.866, –.500, 0), (1.366, .366, 0), and (.500, .866, 0) as shown:

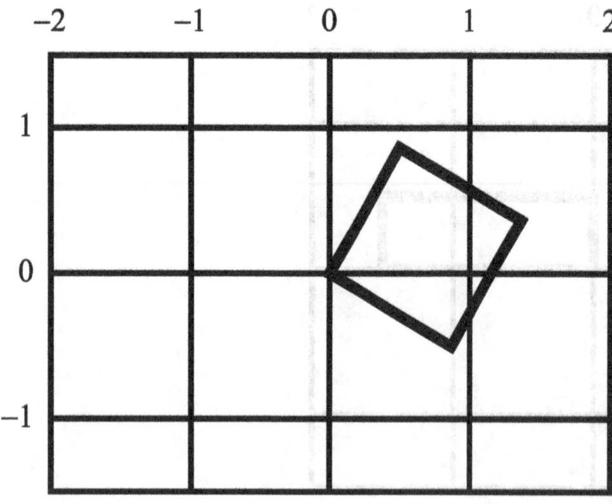

2. **(a)** Simply perform the matrix multiplication $\begin{bmatrix}1 & \frac{1}{2} & 0\\ 0 & 1 & 0\\ 0 & 0 & 1\end{bmatrix}\begin{bmatrix}x_i\\ y_i\\ z_i\end{bmatrix} = \begin{bmatrix}x_i + \frac{1}{2}y_i\\ y_i\\ z_i\end{bmatrix}$.

(b) We multiply $\begin{bmatrix}1 & \frac{1}{2} & 0\\ 0 & 1 & 0\\ 0 & 0 & 1\end{bmatrix}\begin{bmatrix}0 & 1 & 1 & 0\\ 0 & 0 & 1 & 1\\ 0 & 0 & 0 & 0\end{bmatrix} = \begin{bmatrix}0 & 1 & \frac{3}{2} & \frac{1}{2}\\ 0 & 0 & 1 & 1\\ 0 & 0 & 0 & 0\end{bmatrix}$ yielding the vertices (0, 0, 0), (1, 0, 0), $\left(\frac{3}{2}, 1, 0\right)$, and $\left(\frac{1}{2}, 1, 0\right)$.

(c) Obtain the vertices via $\begin{bmatrix}1 & 0 & 0\\ .6 & 1 & 0\\ 0 & 0 & 1\end{bmatrix}\begin{bmatrix}0 & 1 & 1 & 0\\ 0 & 0 & 1 & 1\\ 0 & 0 & 0 & 0\end{bmatrix} = \begin{bmatrix}0 & 1 & 1 & 0\\ 0 & .6 & 1.6 & 1\\ 0 & 0 & 0 & 0\end{bmatrix}$, yielding (0, 0, 0), (1, .6, 0), (1, 1.6, 0), and (0, 1, 0), as shown:

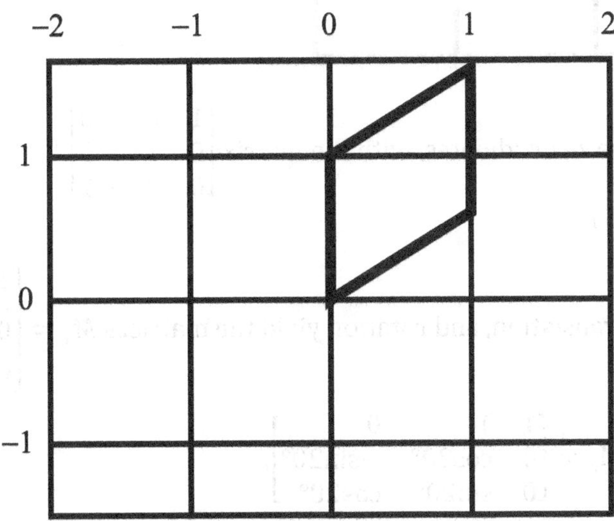

3. **(a)** This transformation looks like scaling by the factors 1, −1, 1, respectively and indeed its matrix is $\begin{bmatrix}1 & 0 & 0\\ 0 & -1 & 0\\ 0 & 0 & 1\end{bmatrix}$.

(b) For this reflection we want to transform (x_i, y_i, z_i) to $(-x_i, y_i, z_i)$ with the matrix $\begin{bmatrix}-1 & 0 & 0\\ 0 & 1 & 0\\ 0 & 0 & 1\end{bmatrix}$. Negating the x-coordinates of the 12 points in view 1 yields the 12 points

(−1.000, −.800, .000), (−.500, −.800, −.866), etc., as shown:

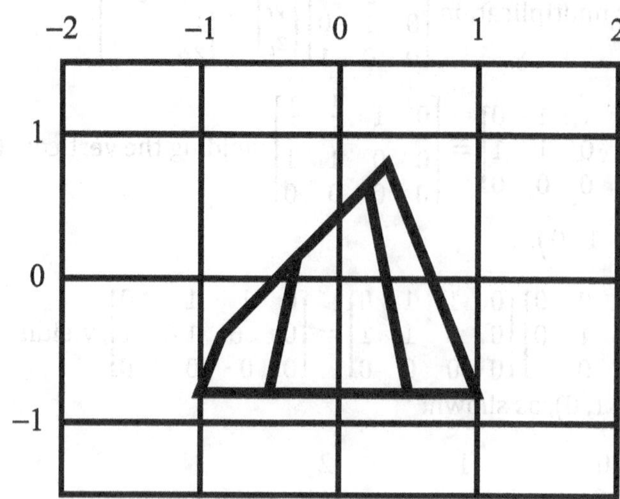

(c) Here we want to negate the z-coordinates, with the matrix $\begin{bmatrix}1 & 0 & 0\\ 0 & 1 & 0\\ 0 & 0 & -1\end{bmatrix}$.

This does not change View 1.

4. **(a)** The formulas for scaling, translation, and rotation yield the matrices $M_1 = \begin{bmatrix}\frac{1}{2} & 0 & 0\\ 0 & 2 & 0\\ 0 & 0 & \frac{1}{3}\end{bmatrix}$,

$M_2 = \begin{bmatrix}\frac{1}{2} & \frac{1}{2} & \frac{1}{2} & \cdots & \frac{1}{2}\\ 0 & 0 & 0 & \cdots & 0\\ 0 & 0 & 0 & \cdots & 0\end{bmatrix}$, $\mathrm{M}_3 = \begin{bmatrix}1 & 1 & 0\\ 0 & \cos 20° & -\sin 20°\\ 0 & \sin 20° & \cos 20°\end{bmatrix}$,

$M_4 = \begin{bmatrix}\cos(-45°) & 0 & \sin(-45)°\\ 0 & 1 & 0\\ -\sin(-45°) & 0 & \cos(-45°)\end{bmatrix}$, and $M_5 = \begin{bmatrix}\cos 90° & -\sin 90° & 0\\ \sin 90° & \cos 90° & 0\\ 0 & 0 & 1\end{bmatrix}$.

(b) Clearly $P' = M_5M_4M_3(M_1P + M_2)$.

5. **(a)** As in 4(a), $M_1 = \begin{bmatrix}.3 & 0 & 0\\ 0 & .5 & 0\\ 0 & 0 & 1\end{bmatrix}$, $M_2 = \begin{bmatrix}1 & 0 & 0\\ 0 & \cos 45° & -\sin 45°\\ 0 & \sin 45° & \cos 45°\end{bmatrix}$, $M_3 = \begin{bmatrix}1 & 1 & 1 & \cdots & 1\\ 0 & 0 & 0 & \cdots & 0\\ 0 & 0 & 0 & \cdots & 0\end{bmatrix}$,

$M_4 = \begin{bmatrix}\cos 35° & 0 & \sin 35°\\ 0 & 1 & 0\\ -\sin 35° & 0 & \cos 35°\end{bmatrix}$, $M_5 = \begin{bmatrix}\cos(-45°) & -\sin(-45°) & 0\\ \sin(-45°) & \cos(-45°) & 0\\ 0 & 0 & 1\end{bmatrix}$,

$M_6 = \begin{bmatrix}0 & 0 & 0 & \cdots & 0\\ 0 & 0 & 0 & \cdots & 0\\ 1 & 1 & 1 & \cdots & 1\end{bmatrix}$, and $M_7 = \begin{bmatrix}2 & 0 & 0\\ 0 & 1 & 0\\ 0 & 0 & 1\end{bmatrix}$.

(b) As in 4(b), $P' = M_7(M_6 + M_5M_4(M_3 + M_2M_1P))$.

6. Using the hint given, we have $R_1 = \begin{bmatrix} \cos\beta & 0 & \sin\beta \\ 0 & 1 & 0 \\ -\sin\beta & 0 & \cos\beta \end{bmatrix}$, $R_2 = \begin{bmatrix} \cos\alpha & -\sin\alpha & 0 \\ \sin\alpha & \cos\alpha & 0 \\ 0 & 0 & 1 \end{bmatrix}$,

$R_3 = \begin{bmatrix} \cos\theta & 0 & \sin\theta \\ 0 & 1 & 0 \\ -\sin\theta & 0 & \cos\theta \end{bmatrix}$, $R_4 = \begin{bmatrix} \cos(-\alpha) & -\sin(-\alpha) & 0 \\ \sin(-\alpha) & \cos(-\alpha) & 0 \\ 0 & 0 & 1 \end{bmatrix}$, and

$R_5 = \begin{bmatrix} \cos(-\beta) & 0 & \sin(-\beta) \\ 0 & 1 & 0 \\ -\sin(-\beta) & 0 & \cos(-\beta) \end{bmatrix}$.

7. **(a)** We rewrite the formula for v'_i as $v'_i = \begin{bmatrix} 1 \cdot x_i + x_0 \cdot 1 \\ 1 \cdot y_i + y_0 \cdot 1 \\ 1 \cdot z_i + z_0 \cdot 1 \\ 1 \cdot 1 \end{bmatrix}$. So $v'_i = \begin{bmatrix} 1 & 0 & 0 & x_0 \\ 0 & 1 & 0 & y_0 \\ 0 & 0 & 1 & z_0 \\ 0 & 0 & 0 & 1 \end{bmatrix} \begin{bmatrix} x_i \\ y_i \\ z_i \\ 1 \end{bmatrix}$.

(b) We want to translate x_i by -5, y_i by $+9$, z_i by -3, so $x_0 = -5, y_0 = 9, z_0 = -3$.

The matrix is $\begin{bmatrix} 1 & 0 & 0 & -5 \\ 0 & 1 & 0 & 9 \\ 0 & 0 & 1 & -3 \\ 0 & 0 & 0 & 1 \end{bmatrix}$.

8. This can be done most easily performing the multiplication RR^T and showing that this is I. For example, for the rotation matrix about the x-axis we obtain

$$RR^T = \begin{bmatrix} 1 & 0 & 0 \\ 0 & \cos\theta & -\sin\theta \\ 0 & \sin\theta & \cos\theta \end{bmatrix} \begin{bmatrix} 1 & 0 & 0 \\ 0 & \cos\theta & \sin\theta \\ 0 & -\sin\theta & \cos\theta \end{bmatrix} = \begin{bmatrix} 1 & 0 & 0 \\ 0 & 1 & 0 \\ 0 & 0 & 1 \end{bmatrix}.$$

10.10 Equilibrium Temperature Distributions

1. **(a)** The discrete mean value property yields the four equations

$$t_1 = \frac{1}{4}(t_2 + t_3)$$
$$t_2 = \frac{1}{4}(t_1 + t_4 + 1 + 1)$$
$$t_3 = \frac{1}{4}(t_1 + t_4)$$
$$t_4 = \frac{1}{4}(t_2 + t_3 + 1 + 1).$$

Translated into matrix notation, this becomes $\begin{bmatrix} t_1 \\ t_2 \\ t_3 \\ t_4 \end{bmatrix} = \begin{bmatrix} 0 & \frac{1}{4} & \frac{1}{4} & 0 \\ \frac{1}{4} & 0 & 0 & \frac{1}{4} \\ \frac{1}{4} & 0 & 0 & \frac{1}{4} \\ 0 & \frac{1}{4} & \frac{1}{4} & 0 \end{bmatrix} \begin{bmatrix} t_1 \\ t_2 \\ t_3 \\ t_4 \end{bmatrix} + \begin{bmatrix} 0 \\ \frac{1}{2} \\ 0 \\ \frac{1}{2} \end{bmatrix}$.

(b) To solve the system in part (a), we solve $(I - M)\mathbf{t} = \mathbf{b}$ for $\mathbf{t}$:

$$\begin{bmatrix} 1 & -\frac{1}{4} & -\frac{1}{4} & 0 \\ -\frac{1}{4} & 1 & 0 & -\frac{1}{4} \\ -\frac{1}{4} & 0 & 1 & -\frac{1}{4} \\ 0 & -\frac{1}{4} & -\frac{1}{4} & 1 \end{bmatrix}\begin{bmatrix} t_1 \\ t_2 \\ t_3 \\ t_4 \end{bmatrix} = \begin{bmatrix} 0 \\ \frac{1}{2} \\ 0 \\ \frac{1}{2} \end{bmatrix}$$

In row-echelon form, this is $\begin{bmatrix} 1 & -\frac{1}{4} & -\frac{1}{4} & 0 \\ 0 & 1 & -15 & 4 \\ 0 & 0 & 1 & -\frac{1}{2} \\ 0 & 0 & 0 & 1 \end{bmatrix}\begin{bmatrix} t_1 \\ t_2 \\ t_3 \\ t_4 \end{bmatrix} = \begin{bmatrix} 0 \\ 0 \\ -\frac{1}{8} \\ \frac{3}{4} \end{bmatrix}$.

Back substitution yields the result $\mathbf{t} = \begin{bmatrix} \frac{1}{4} \\ \frac{3}{4} \\ \frac{1}{4} \\ \frac{3}{4} \end{bmatrix}$.

(c) $\mathbf{t}^{(1)} = M\mathbf{t}^{(0)} + \mathbf{b} = \begin{bmatrix} 0 & \frac{1}{4} & \frac{1}{4} & 0 \\ \frac{1}{4} & 0 & 0 & \frac{1}{4} \\ \frac{1}{4} & 0 & 0 & \frac{1}{4} \\ 0 & \frac{1}{4} & \frac{1}{4} & 0 \end{bmatrix}\begin{bmatrix} 0 \\ 0 \\ 0 \\ 0 \end{bmatrix} + \begin{bmatrix} 0 \\ \frac{1}{2} \\ 0 \\ \frac{1}{2} \end{bmatrix} = \begin{bmatrix} 0 \\ \frac{1}{2} \\ 0 \\ \frac{1}{2} \end{bmatrix}$, $\mathbf{t}^{(2)} = M\mathbf{t}^{(1)} + \mathbf{b} = \begin{bmatrix} \frac{1}{8} \\ \frac{5}{8} \\ \frac{1}{8} \\ \frac{5}{8} \end{bmatrix}$,

$\mathbf{t}^{(3)} = M\mathbf{t}^{(2)} + \mathbf{b} = \begin{bmatrix} \frac{3}{16} \\ \frac{11}{16} \\ \frac{3}{16} \\ \frac{11}{16} \end{bmatrix}$, $\mathbf{t}^{(4)} = M\mathbf{t}^{(3)} + \mathbf{b} = \begin{bmatrix} \frac{7}{32} \\ \frac{23}{32} \\ \frac{7}{32} \\ \frac{23}{32} \end{bmatrix}$, $\mathbf{t}^{(5)} = M\mathbf{t}^{(4)} + \mathbf{b} = \begin{bmatrix} \frac{15}{64} \\ \frac{47}{64} \\ \frac{15}{64} \\ \frac{47}{64} \end{bmatrix}$

$$\mathbf{t}^{(5)} - \mathbf{t} = \begin{bmatrix} \frac{15}{64} \\ \frac{47}{64} \\ \frac{15}{64} \\ \frac{47}{64} \end{bmatrix} - \begin{bmatrix} \frac{1}{4} \\ \frac{3}{4} \\ \frac{1}{4} \\ \frac{3}{4} \end{bmatrix} = \begin{bmatrix} -\frac{1}{64} \\ -\frac{1}{64} \\ -\frac{1}{64} \\ -\frac{1}{64} \end{bmatrix}$$

(d) Using percentage error $= \frac{\text{computed value} - \text{actual value}}{\text{actual value}} \times 100\%$ we have that the percentage error for t_1 and t_3 was $\frac{-.0371}{.2871} \times 100\% = -12.9\%$, and for t_2 and t_4 was $\frac{.0371}{.7129} \times 100\% = 5.2\%$.

2. The average value of the temperature on the circle is $\frac{1}{2\pi r}\int_{-\pi}^{\pi} f(\theta)r\,d\theta$, where r is the radius of the circle and $f(\theta)$ is the temperature at the point of the circumference where the radius to that point makes the angle θ with the horizontal. Clearly $f(\theta) = 1$ for $\frac{-\pi}{2} < \theta < \frac{\pi}{2}$ and is zero otherwise. Consequently, the value of the integral above (which equals the temperature at the center of the circle) is $\frac{1}{2}$.

3. As in 1(c), but using M and $\mathbf{b}$ as in the problem statement, we obtain

$$\mathbf{t}^{(1)} = M\mathbf{t}^{(0)} + \mathbf{b} = \begin{bmatrix} \frac{3}{4} & \frac{5}{4} & \frac{1}{2} & \frac{5}{4} & 1 & \frac{1}{2} & \frac{5}{4} & 1 & \frac{3}{4} \end{bmatrix}^{\mathrm{T}}$$

$$\mathbf{t}^{(2)} = M\mathbf{t}^{(1)} + \mathbf{b} = \begin{bmatrix} \frac{13}{16} & \frac{9}{8} & \frac{9}{16} & \frac{11}{8} & \frac{13}{16} & \frac{7}{16} & \frac{21}{16} & 1 & \frac{5}{8} \end{bmatrix}^{\mathrm{T}}.$$

10.11 Computed Tomography

1. (c) The linear system

$$x_{31}^* = \frac{1}{20}[28 + x_{31}^* - x_{32}^*]$$
$$x_{32}^* = \frac{1}{20}[24 + 3x_{31}^* - 3x_{32}^*]$$

can be rewritten as

$$19x_{31}^* + x_{32}^* = 28$$
$$-3x_{31}^* + 23x_{32}^* = 24,$$

which has the solution $x_{31}^* = \frac{31}{22}$, $x_{32}^* = \frac{27}{22}$.

2. (a) Setting $\mathbf{x}_0^{(1)} = \left(x_{01}^{(1)}, x_{02}^{(1)}\right) = \left(x_{31}^{(0)}, x_{32}^{(0)}\right) = (0,\ 0)$, and using part (b) of Exercise 1, we have

$$x_{31}^{(1)} = \frac{1}{20}[28] = 1.40000$$
$$x_{32}^{(1)} = \frac{1}{20}[24] = 1.20000$$
$$x_{31}^{(2)} = \frac{1}{20}[28 + 1.4 - 1.2] = 1.41000$$
$$x_{32}^{(2)} = \frac{1}{20}[24 + 3(1.4) - 3(1.2)] = 1.23000$$
$$x_{31}^{(3)} = \frac{1}{20}[28 + 1.41 - 1.23] = 1.40900$$
$$x_{32}^{(3)} = \frac{1}{20}[24 + 3(1.41) - 3(1.23)] = 1.22700$$
$$x_{31}^{(4)} = \frac{1}{20}[28 + 1.409 - 1.227] = 1.40910$$
$$x_{32}^{(4)} = \frac{1}{20}[24 + 3(1.409) - 3(1.227) = 1.22730$$

$$x_{31}^{(5)} = \frac{1}{20}[28 + 1.4091 - 1.2273] = 1.40909$$
$$x_{32}^{(5)} = \frac{1}{20}[24 + 3(1.4091) - 3(1.2273)] = 1.22727$$
$$x_{31}^{(6)} = \frac{1}{20}[28 + 1.40909 - 1.22727] = 1.40909$$
$$x_{32}^{(6)} = \frac{1}{20}[24 + 3(1.40909) - 3(1.22727)] = 1.22727.$$

(b) $\mathbf{x}_0^{(1)} = (1,\ 1) = \left(x_{31}^{(0)},\ x_{32}^{(0)}\right)$

$$x_{31}^{(1)} = \frac{1}{20}[28 + 1 - 1] = 1.4$$
$$x_{32}^{(2)} = \frac{1}{20}[24 + 3(1) - 3(1)] = 1.2$$

Since $\mathbf{x}_3^{(1)}$ in this part is the same as $\mathbf{x}_3^{(1)}$ in part (a), we will get $\mathbf{x}_3^{(2)}$ as in part (a) and therefore $\mathbf{x}_3^{(3)}, \ldots, \mathbf{x}_3^{(6)}$ will also be the same as in part (a).

(c) $\mathbf{x}_0^{(1)} = (148,\ -15) = \left(x_{31}^{(0)},\ x_{32}^{(0)}\right)$

$$x_{31}^{(1)} = \frac{1}{20}[28 + 148 - (-15)] = 9.55000$$
$$x_{32}^{(1)} = \frac{1}{20}[24 + 3(148) - 3(-15)] = 25.65000$$
$$x_{31}^{(2)} = \frac{1}{20}[28 + 9.55 - 25.65] = 0.59500$$
$$x_{32}^{(2)} = \frac{1}{20}[24 + 3(9.55) - 3(25.65)] = -1.21500$$
$$x_{31}^{(3)} = \frac{1}{20}[28 + 0.595 + 1.215] = 1.49050$$
$$x_{32}^{(3)} = \frac{1}{20}[24 + 3(0.595) + 3(1.215)] = 1.47150$$
$$x_{31}^{(4)} = \frac{1}{20}[28 + 1.4905 - 1.4715] = 1.40095$$
$$x_{32}^{(4)} = \frac{1}{20}[24 + 3(1.4905) - 3(1.4715)] = 1.20285$$
$$x_{31}^{(5)} = \frac{1}{20}[28 + 1.40095 - 1.20285] = 1.40991$$
$$x_{32}^{(5)} = \frac{1}{20}[24 + 3(1.40095) - 3(1.20285)] = 1.22972$$
$$x_{31}^{(6)} = \frac{1}{20}[28 + 1.40991 - 1.22972] = 1.40901$$
$$x_{32}^{(6)} = \frac{1}{20}[24 + 3(1.40991) - 3(1.22972)] = 1.22703$$

4. Referring to the figure below and starting with $\mathbf{x}_0^{(1)} = (0,\ 0)$:

$\mathbf{x}_0^{(1)}$ is projected to $\mathbf{x}_1^{(1)}$ on L_1, $\mathbf{x}_1^{(1)}$ is projected to $\mathbf{x}_2^{(1)}$ on L_2, $\mathbf{x}_2^{(1)}$ is projected to $\mathbf{x}_3^{(1)}$ on L_3, and so on.

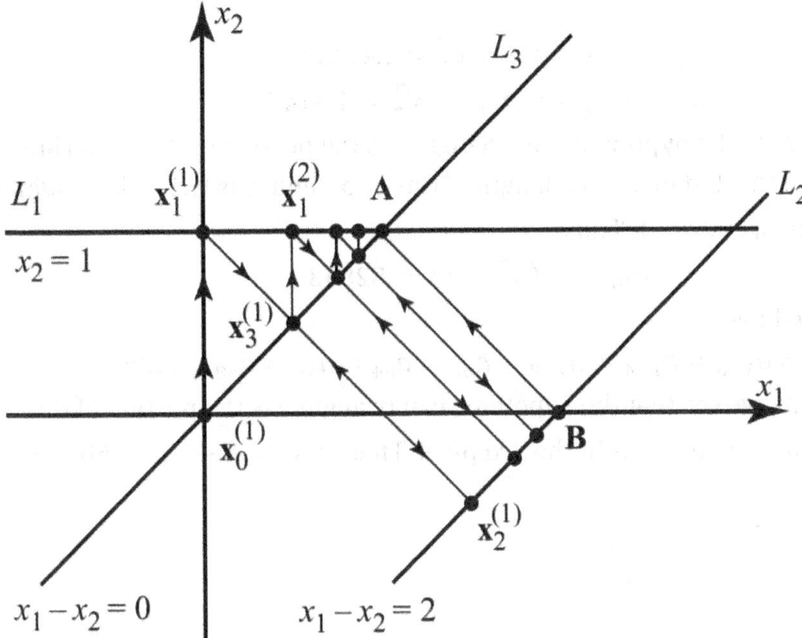

As seen from the graph the points of the limit cycle are $\mathbf{x}_1^* = \mathbf{A}, \mathbf{x}_2^* = \mathbf{B}, \mathbf{x}_3^* = \mathbf{A}$.
Since $\mathbf{x}_1^*$ is the point of intersection of L_1 and L_3 it follows on solving the system

$$x_{12}^* = 1$$
$$x_{11}^* - x_{12}^* = 0$$

that $\mathbf{x}_1^* = (1,\ 1)$. Since $\mathbf{x}_2^* = (x_{21}^*,\ x_{22}^*)$ is on L_2, it follows that $x_{21}^* - x_{22}^* = 2$. Now $\overline{\mathbf{x}_1^*\mathbf{x}_2^*}$ is perpendicular to L_2, therefore $\left(\frac{x_{22}^*-1}{x_{21}^*-1}\right)(1) = -1$ so we have $x_{22}^* - 1 = 1 - x_{21}^*$ or $x_{21}^* + x_{22}^* = 2$.

Solving the system

$$x_{21}^* - x_{22}^* = 2$$
$$x_{21}^* + x_{22}^* = 2$$

gives $x_{21}^* = 2$ and $x_{22}^* = 0$. Thus the points on the limit cycle are $\mathbf{x}_1^* = (1,\ 1), \mathbf{x}_2^* = (2,\ 0)$, $\mathbf{x}_3^* = (1,\ 1)$.

7. Let us choose units so that each pixel is one unit wide. Then a_{ij} =length of the center line of the i-th beam that lies in the j-th pixel.
If the i-th beam crosses the j-th pixel squarely, it follows that $a_{ij} = 1$. From Fig. 10.11.11 in the text, it is then clear that

$$a_{17} = a_{18} = a_{19} = 1$$
$$a_{24} = a_{25} = a_{26} = 1$$
$$a_{31} = a_{32} = a_{33} = 1$$
$$a_{73} = a_{76} = a_{79} = 1$$
$$a_{82} = a_{85} = a_{88} = 1$$
$$a_{91} = a_{94} = a_{97} = 1$$

since beams 1, 2, 3, 7, 8, and 9 cross the pixels squarely. Next, the centerlines of beams 5 and 11 lie along the diagonals of pixels 3, 5, 7 and 1, 5, 9, respectively. Since these diagonals have length $\sqrt{2}$,we have

$$a_{53} = a_{55} = a_{57} = \sqrt{2} = 1.41421$$
$$a_{11,1} = a_{11,5} = a_{11,9} = \sqrt{2} = 1.41421.$$

In the following diagram, the hypotenuse of triangle A is the portion of the centerline of the 10th beam that lies in the 2nd pixel. The length of this hypotenuse is twice the height of triangle A, which in turn is $\sqrt{2} - 1$. Thus,

$$a_{10,2} = 2(\sqrt{2} - 1) = .82843.$$

By symmetry we also have

$$a_{10,2} = a_{10,6} = a_{12,4} = a_{12,8} = a_{62} = a_{64} = a_{46} = a_{48} = .82843.$$

Also from the diagram, we see that the hypotenuse of triangle B is the portion of the centerline of the 10th beam that lies in the 3rd pixel. Thus, $a_{10,3} = 2 - \sqrt{2} = .58579$.

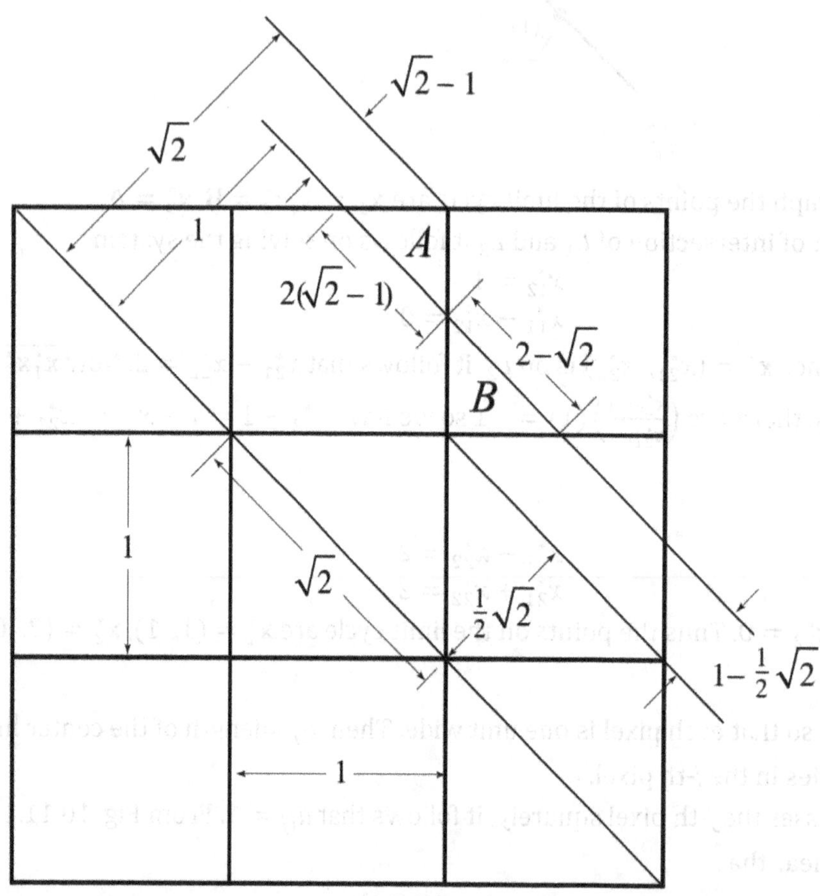

By symmetry we have $a_{10,3} = a_{12,7} = a_{61} = a_{49} = .58579$.

The remaining a_{ij}'s are all zero, and so the 12 beam equations (4) are

$$x_7 + x_8 + x_9 = 13.00$$
$$x_4 + x_5 + x_6 = 15.00$$
$$x_1 + x_2 + x_3 = 8.00$$
$$.82843(x_6 + x_8) + .58579x_9 = 14.79$$
$$1.41421(x_3 + x_5 + x_7) = 14.31$$

$$.82843(x_2 + x_4) + .58579x_1 = 3.81$$
$$x_3 + x_6 + x_9 = 18.00$$
$$x_2 + x_5 + x_8 = 12.00$$
$$x_1 + x_4 + x_7 = 6.00$$
$$.82843(x_2 + x_6) + .58579x_3 = 10.51$$
$$1.41421(x_1 + x_5 + x_9) = 16.13$$
$$.82843(x_4 + x_8) + .58579x_7 = 7.04$$

8. Let us choose units so that each pixel is one unit wide. Then a_{ij} =area of the *i*-th beam that lies in the *j*-th pixel. Since the width of each beam is also one unit it follows that $a_{ij} = 1$ if the *i*-th beam crosses the *j*-th pixel squarely. From Fig. 10.12.11 in the text, it is then clear that

$$a_{17} = a_{18} = a_{19} = 1$$
$$a_{24} = a_{25} = a_{26} = 1$$
$$a_{31} = a_{32} = a_{33} = 1$$
$$a_{73} = a_{76} = a_{79} = 1$$
$$a_{82} = a_{85} = a_{88} = 1$$
$$a_{91} = a_{94} = a_{97} = 1$$

since beams 1, 2, 3, 7, 8, and 9 cross the pixels squarely.
For the remaining a_{ij}'s, first observe from the figure that an isosceles right triangle of height h, as indicated, has area h^2. From the diagram of the nine pixels, we then have

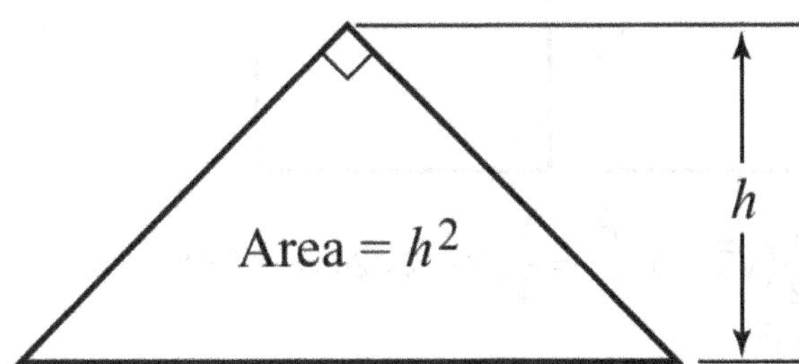

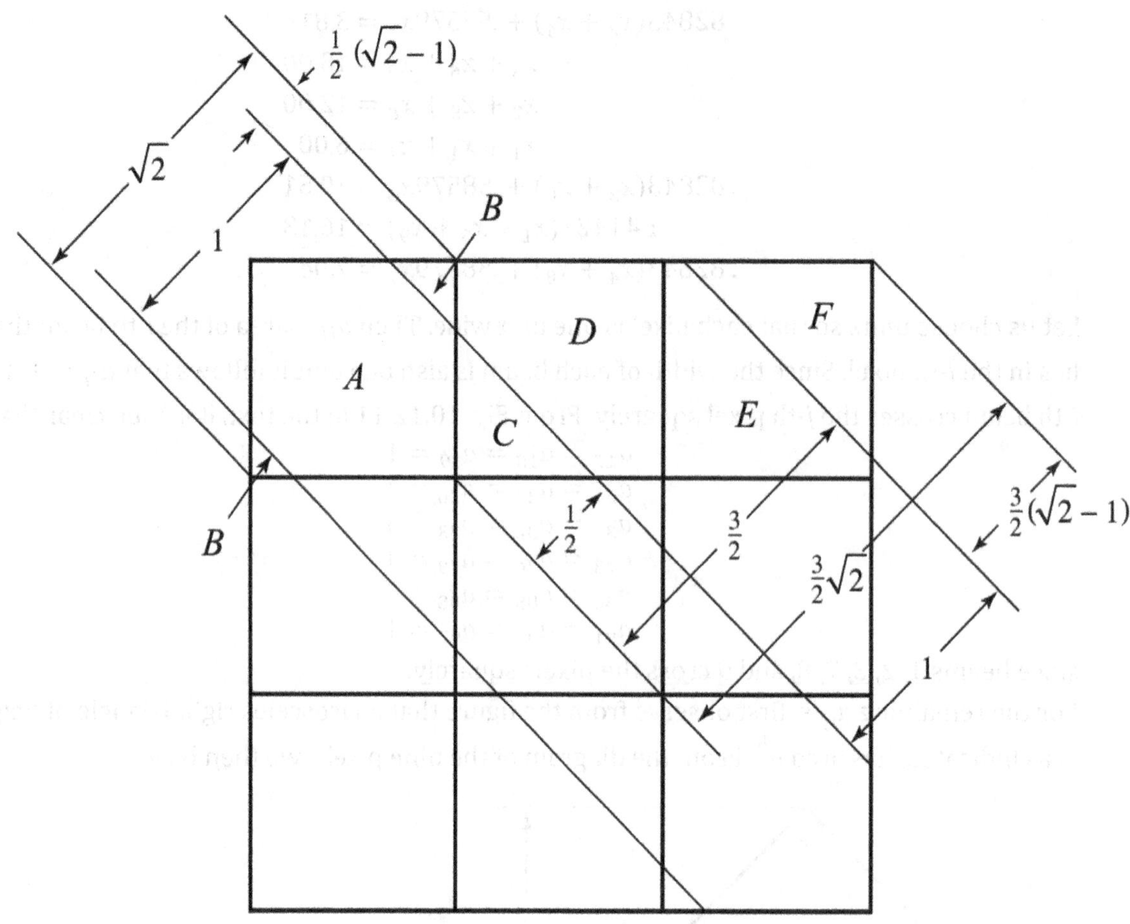

$$\text{Area of triangle } B = \left[\frac{1}{2}(\sqrt{2}-1)\right]^2 = \frac{1}{4}(3-2\sqrt{2}) = 0.4289$$

$$\text{Area of triangle } C = \left[\frac{1}{2}\right]^2 = \frac{1}{4} = .25000$$

$$\text{Area of triangle } F = \left[\frac{3}{2}(\sqrt{2}-1)\right]^2 = \frac{9}{4}(3-2\sqrt{2}) = .38604$$

We also have

$$\text{Area of polygon } A = 1 - 2 \times (\text{Area of triangle } B) = \sqrt{2} - \frac{1}{2} = .91421$$

$$\text{Area of polygon } D = 1 - (\text{Area of triangle } C) = 1 - \frac{1}{4} = \frac{3}{4} = .7500$$

$$\text{Area of polygon } E = 1 - (\text{Area of triangle } F) = \frac{1}{4}(18\sqrt{2} - 23) = .61396$$

Referring back to Fig. 10.11.11, we see that

$$a_{11,1} = \text{Area of polygon } A = .91421$$
$$a_{10,1} = \text{Area of triangle } B = .04289$$
$$a_{11,2} = \text{Area of triangle } C = .25000$$
$$a_{10,2} = \text{Area of polygon } D = .75000$$
$$a_{10,3} = \text{Area of polygon } E = .61396$$

By symmetry we then have

$a_{11,1} = a_{11,5} = a_{11,9} = a_{53} = a_{55} = a_{57} = .91421$

$a_{10,1} = a_{10,5} = a_{10,9} = a_{12,1} = a_{12,5} = a_{12,9} = a_{63} = a_{65} = a_{67} = a_{43} = a_{45} = a_{47} = .04289$

$a_{11,2} = a_{11,4} = a_{11,6} = a_{11,8} = a_{52} = a_{54} = a_{56} = a_{58} = .25000$

$a_{10,2} = a_{10,6} = a_{12,4} = a_{12,8} = a_{62} = a_{64} = a_{46} = a_{48} = .75000$

$a_{10,3} = a_{12,7} = a_{61} = a_{49} = .61396.$

The remaining a_{ij}'s are all zero, and so the 12 beam equations (4) are

$$\begin{aligned}
x_7 + x_8 + x_9 &= 13.00 \\
x_4 + x_5 + x_6 &= 15.00 \\
x_1 + x_2 + x_3 &= 8.00 \\
0.04289(x_3 + x_5 + x_7) + 0.75(x_6 + x_8) + 0.61396x_9 &= 14.79 \\
0.91421(x_3 + x_5 + x_7) + 0.25(x_2 + x_4 + x_6 + x_8) &= 14.31 \\
0.04289(x_3 + x_5 + x_7) + 0.75(x_2 + x_4) + 0.61396x_1 &= 3.81 \\
x_3 + x_6 + x_9 &= 18.00 \\
x_2 + x_5 + x_8 &= 12.00 \\
x_1 + x_4 + x_7 &= 6.00 \\
0.04289(x_1 + x_5 + x_9) + 0.75(x_2 + x_6) + 0.61396x_3 &= 10.51 \\
0.91421(x_1 + x_5 + x_9) + 0.25(x_2 + x_4 + x_6 + x_8) &= 16.13 \\
0.04289(x_1 + x_5 + x_9) + 0.75(x_4 + x_8) + 0.61396x_7 &= 7.04
\end{aligned}$$

10.12 Fractals

1. Each of the subsets S_1, S_2, S_3, S_4 in the figure is congruent to the entire set scaled by a factor of $\frac{12}{25}$. Also, the rotation angles for the four subsets are all 0°. The displacement distances can be determined from the figure to find the four similitudes that map the entire set onto the four subsets S_1, S_2, S_3, S_4. These are, respectively,

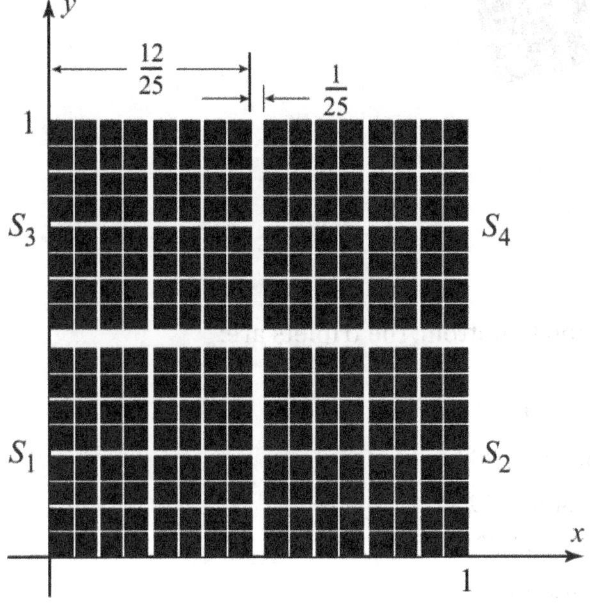

$T_i\left(\begin{bmatrix}x\\y\end{bmatrix}\right) = \frac{12}{25}\begin{bmatrix}1 & 0\\0 & 1\end{bmatrix}\begin{bmatrix}x\\y\end{bmatrix} + \begin{bmatrix}e_i\\f_i\end{bmatrix}$, i = 1, 2, 3, 4, where the four values of $\begin{bmatrix}e_i\\f_i\end{bmatrix}$ are $\begin{bmatrix}0\\0\end{bmatrix}$, $\begin{bmatrix}\frac{13}{25}\\0\end{bmatrix}$, $\begin{bmatrix}0\\\frac{13}{25}\end{bmatrix}$, and $\begin{bmatrix}\frac{13}{25}\\\frac{13}{25}\end{bmatrix}$. Because $s = \frac{12}{25}$ and k = 4 in the definition of a self-similar set, the Hausdorff dimension of the set is $d_H(S) = \frac{\ln(k)}{\ln\left(\frac{1}{s}\right)} = \frac{\ln(4)}{\ln\left(\frac{25}{12}\right)} = 1.888\ldots$ The set is a fractal because its Hausdorff dimension is not an integer.

2. The rough measurements indicated in the figure give an approximate scale factor of $s \approx \frac{\left(\frac{15}{16}\right)}{2} = .47$ to two decimal places. Since k = 4, the Hausdorff dimension of the set is approximately $d_H(S) = \frac{\ln(k)}{\ln\left(\frac{1}{s}\right)} = \frac{\ln(4)}{\ln\left(\frac{1}{.47}\right)} = 1.8$ to two significant digits. Examination of the figure reveals rotation angles of 180°, 180°, 0°, and −90° for the sets S_1, S_2, S_3, and S_4, respectively.

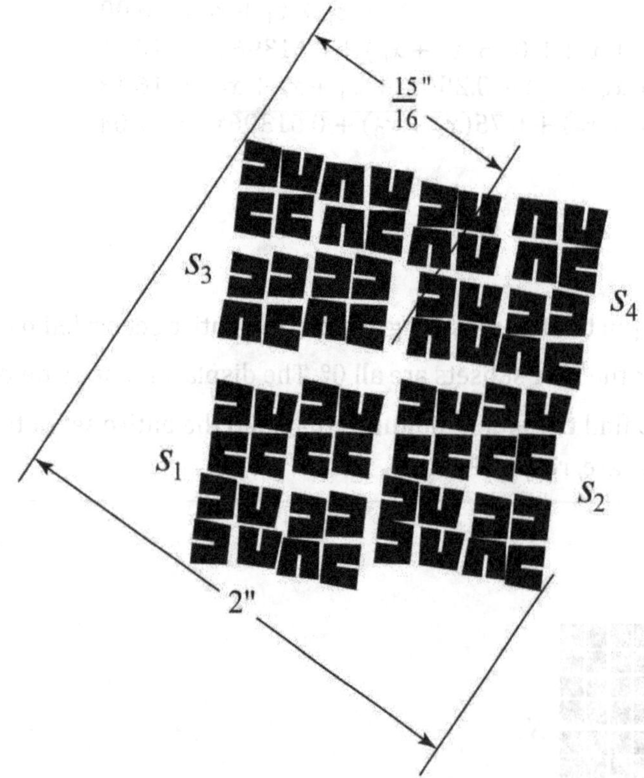

3. By inspection, reading left to right and top to bottom, the triplets are:

(0, 0, 0) none are rotated
(1, 0, 0) the upper right iteration is rotated 90°
(2, 0, 0) the upper right iteration is rotated 180°
(3, 0, 0) the upper right iteration is rotated 270°
(0, 0, 1) the lower right iteration is rotated 90°
(0, 0, 2) the lower right iteration is rotated 180°
(1, 2, 0) the upper right iteration is rotated 90° and the lower left is rotated 180°

(2, 1, 3) the upper right iteration is rotated 180°, the lower left is rotated 90°, and the lower right is rotated 270°

(2, 0, 1) the upper right iteration is rotated 180° and the lower right is rotated 90°

(2, 0, 2) the upper right and lower right iterations are both rotated 180°

(2, 2, 0) the upper right and lower left iterations are both rotated 180°

(0, 3, 3) the lower left and lower right iterations are both rotated 270°

4. **(a)** The figure shows the original self-similar set and a decomposition of the set into seven nonoverlapping congruent subsets, each of which is congruent to the original set scaled by a factor $s = \frac{1}{3}$. By inspection, the rotations angles are 0° for all seven subsets. The Hausdorff dimension of the set is $d_H(S) = \frac{\ln(k)}{\ln\left(\frac{1}{s}\right)} = \frac{\ln(7)}{\ln(3)} = 1.771\ldots$. Because its Hausdorff dimension is not an integer, the set is a fractal.

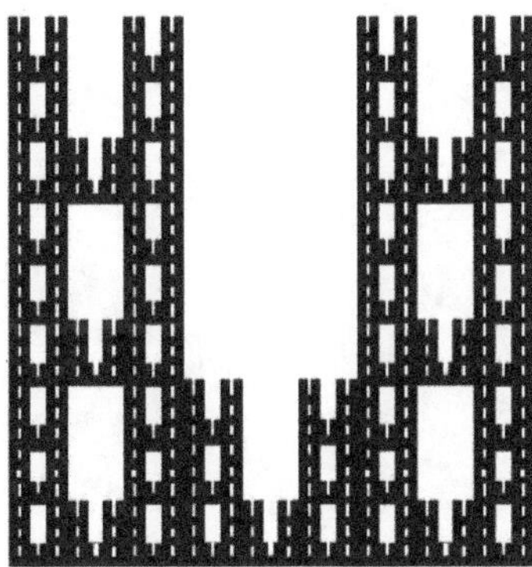

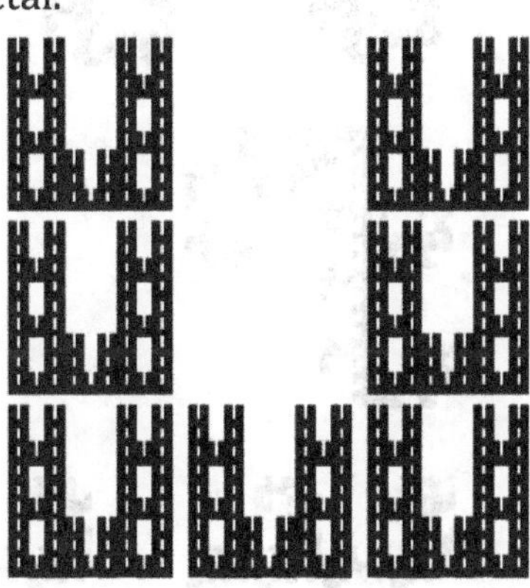

(b) The figure shows the original self-similar set and a decomposition of the set into three nonoverlapping congruent subsets, each of which is congruent to the original set scaled by a factor $s = \frac{1}{2}$. By inspection, the rotation angles are 180° for all three subsets. The Hausdorff dimension of the set is $d_H(S) = \frac{\ln(k)}{\ln\left(\frac{1}{s}\right)} = \frac{\ln(3)}{\ln(2)} = 1.584\ldots$. Because its Hausdorff dimension is not an integer, the set is a fractal.

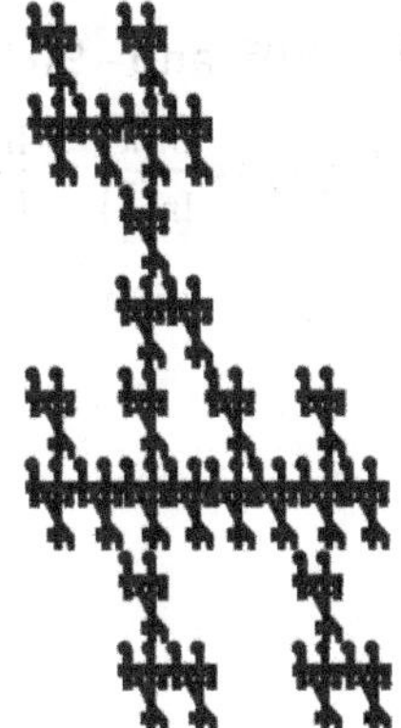

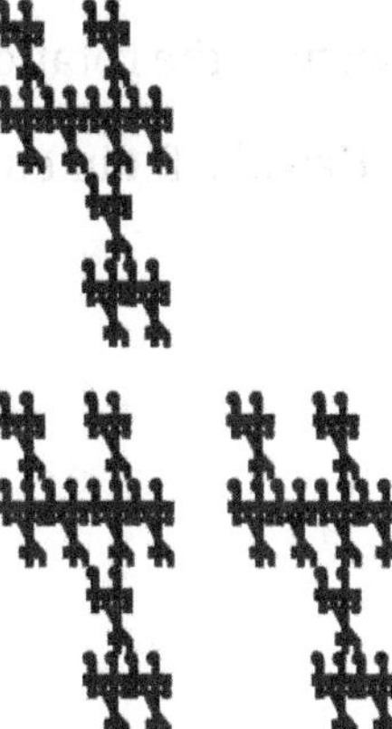

(c) The figure shows the original self-similar set and a decomposition of the set into three nonoverlapping congruent subsets, each of which is congruent to the original set scaled by a factor $s = \frac{1}{2}$. By inspection, the rotation angles are 180°, 180°, and −90° for S_1, S_2, and S_3, respectively. The Hausdorff dimension of the set is $d_H(S) = \frac{\ln(k)}{\ln\left(\frac{1}{s}\right)} = \frac{\ln(3)}{\ln(2)} =$ 1.584…. Because its Hausdorff dimension is not an integer, the set is a fractal.

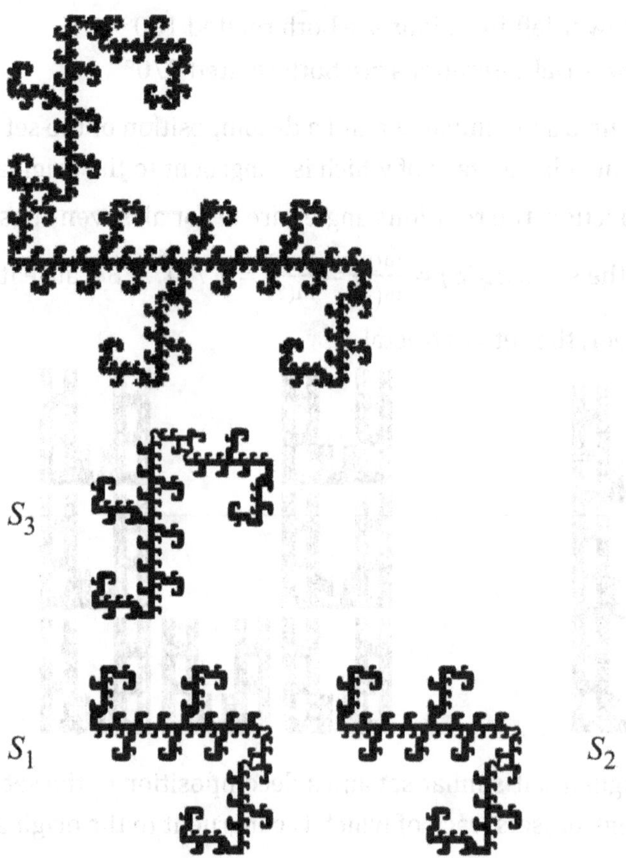

(d) The figure shows the original self-similar set and a decomposition of the set into three nonoverlapping congruent subsets, each of which is congruent to the original set scaled by a factor $s = \frac{1}{2}$. By inspection, the rotation angles are 180°, 180°, and −90° for S_1, S_2, and S_3, respectively. The Hausdorff dimension of the set is $d_H(S) = \frac{\ln(k)}{\ln\left(\frac{1}{s}\right)} = \frac{\ln(3)}{\ln(2)} =$

1.584.... Because its Hausdorff dimension is not an integer, the set is a fractal.

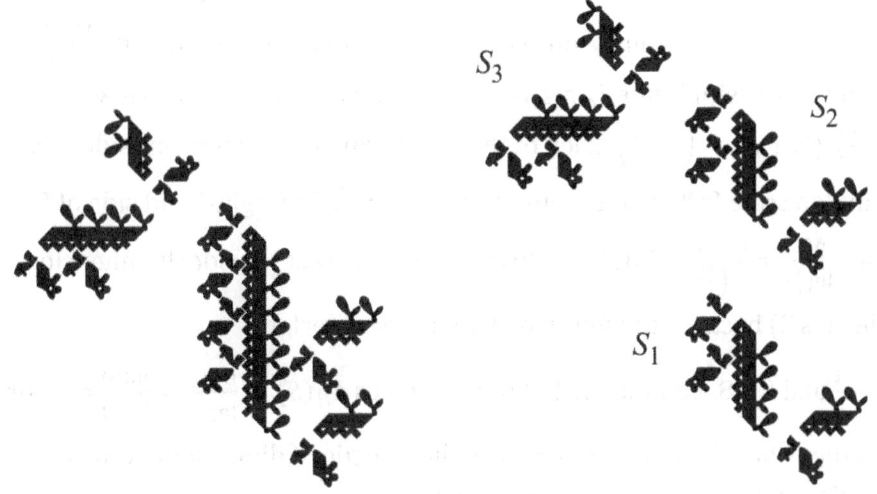

5. The matrix of the affine transformation in question is $\begin{bmatrix} .85 & .04 \\ -.04 & .85 \end{bmatrix}$. The matrix portion of a similitude is of the form $s\begin{bmatrix} \cos\theta & -\sin\theta \\ \sin\theta & \cos\theta \end{bmatrix}$. Consequently, we must have $s\cos\theta = .85$ and $s\sin\theta = -.04$. Solving this pair of equations gives $s = \sqrt{(.85)^2 + (-.04)^2} = .8509\ldots$ and $\theta = \tan^{-1}\left(\frac{-.04}{.85}\right) = -2.69\ldots°$.

6. Letting $\begin{bmatrix} x \\ y \end{bmatrix}$ be the vector to the tip of the fern and using the hint, we have $\begin{bmatrix} x \\ y \end{bmatrix} = T_2\left(\begin{bmatrix} x \\ y \end{bmatrix}\right)$ or $\begin{bmatrix} x \\ y \end{bmatrix} = \begin{bmatrix} .85 & .04 \\ -.04 & .85 \end{bmatrix}\begin{bmatrix} x \\ y \end{bmatrix} + \begin{bmatrix} .075 \\ .180 \end{bmatrix}$. Solving this matrix equation gives $\begin{bmatrix} x \\ y \end{bmatrix} = \begin{bmatrix} .15 & -.04 \\ .04 & .15 \end{bmatrix}^{-1}\begin{bmatrix} .075 \\ .180 \end{bmatrix} = \begin{bmatrix} .766 \\ .996 \end{bmatrix}$ rounded to three decimal places.

7. As the figure indicates, the unit square can be expressed as the union of 16 nonoverlapping congruent squares, each of side length $\frac{1}{4}$. Consequently, the Hausdorff dimension of the unit square as given by Equation (2) of the text is $d_H(S) = \frac{\ln(k)}{\ln\left(\frac{1}{s}\right)} = \frac{\ln(16)}{\ln(4)} = 2$.

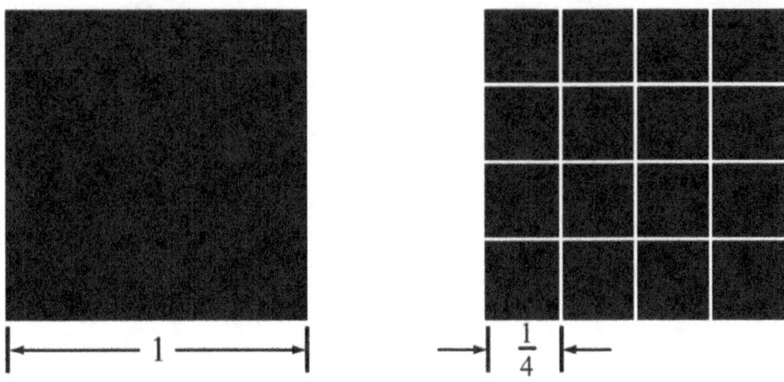

8. The similitude T_1 maps the unit square (whose vertices are (0, 0), (1, 0), (1, 1), and (0, 1)) onto the square whose vertices are (0, 0), $\left(\frac{3}{4}, 0\right)$, $\left(\frac{3}{4}, \frac{3}{4}\right)$, and $\left(0, \frac{3}{4}\right)$. The similitude T_2 maps the unit square onto the square whose vertices are $\left(\frac{1}{4},\ 0\right)$, $(1, 0)$, $\left(1,\ \frac{3}{4}\right)$, and $\left(\frac{1}{4},\ \frac{3}{4}\right)$. The similitude T_3 maps the unit square onto the square whose vertices are $\left(0,\ \frac{1}{4}\right)$, $\left(\frac{3}{4},\ \frac{1}{4}\right)$, $\left(\frac{3}{4},\ 1\right)$, and $(0, 1)$. Finally the similitude T_4 maps the unit square onto the square whose vertices are $\left(\frac{1}{4},\ \frac{1}{4}\right)$, $\left(1,\ \frac{1}{4}\right)$, $(1, 1)$, and $\left(\frac{1}{4},\ 1\right)$. Each of these four smaller squares has side length of $\frac{3}{4}$, so that the common scale factor of the similitudes is $s = \frac{3}{4}$. The right-hand side of Equation (2) of the text gives $\frac{\ln(k)}{\ln\left(\frac{1}{s}\right)} = \frac{\ln(4)}{\ln\left(\frac{4}{3}\right)} = 4.818\ldots$. This is not the correct Hausdorff dimension of the square (which is 2) because the four smaller squares overlap.

9. Because $s = \frac{1}{2}$ and $k = 8$, Equation (2) of the text gives $d_H(S) = \frac{\ln(k)}{\ln\left(\frac{1}{s}\right)} = \frac{\ln(8)}{\ln(2)} = 3$ for the Hausdorff dimension of a unit cube. Because the Hausdorff dimension of the cube is the same as its topological dimension, the cube is not a fractal.

10. A careful examination of Figure Ex-10 in the text shows that the Menger sponge can be expressed as the union of 20 smaller nonoverlapping congruent Menger sponges each of side length $\frac{1}{3}$. Consequently, $k = 20$ and $s = \frac{1}{3}$, and so the Hausdorff dimension of the Menger sponge is $d_H(S) = \frac{\ln(k)}{\ln\left(\frac{1}{s}\right)} = \frac{\ln(20)}{\ln(3)} = 2.726\ldots$. Because its Hausdorff dimension is not an integer, the Menger sponge is a fractal.

11. The figure shows the first four iterates as determined by Algorithm 1 and starting with the unit square as the initial set. Because $k = 2$ and $s = \frac{1}{3}$, the Hausdorff dimension of the Cantor set is $d_H(S) = \frac{\ln(k)}{\ln\left(\frac{1}{s}\right)} = \frac{\ln(2)}{\ln(3)} = 0.6309\ldots$. Notice that the Cantor set is a subset of the unit interval along the x-axis and that its topological dimension must be 0 (since the topological dimension of any set is a nonnegative integer less than or equal to its Hausdorff dimension).

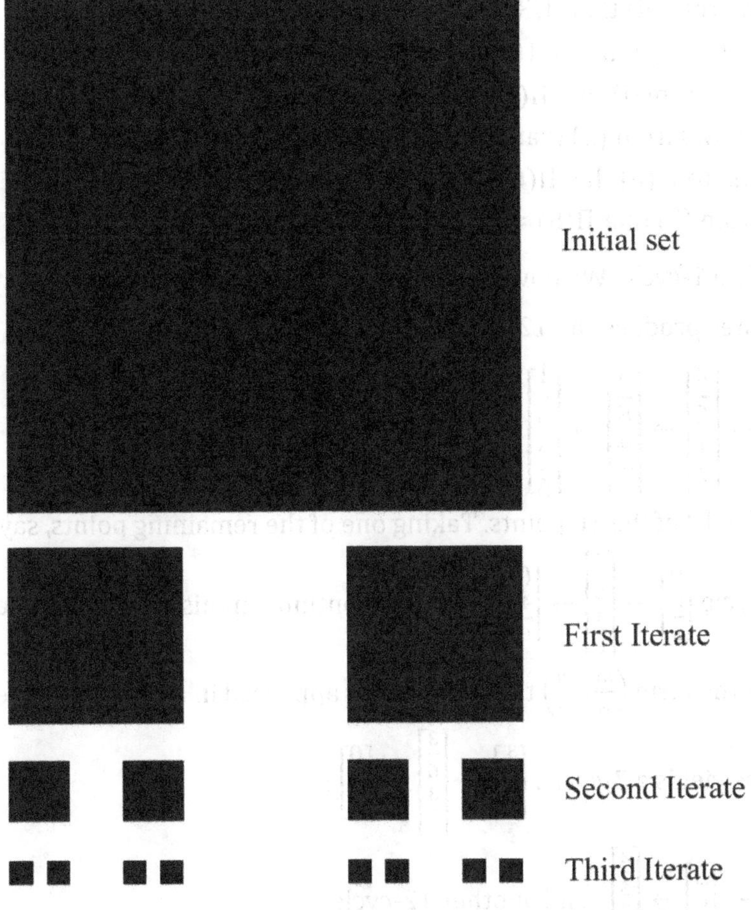

12. The area of the unit square S_0 is, of course, 1. Each of the eight similitudes $T_1, T_2, \ldots, T_8$ given in Equation (8) of the text has scale factor $s = \frac{1}{3}$, and so each maps the unit square onto a smaller square of area $\frac{1}{9}$. Because these eight smaller squares are nonoverlapping, their total area is $\frac{8}{9}$, which is then the area of the set S_1. By a similar argument, the area of the set S_2 is $\frac{8}{9}$-th the area of the set S_1. Continuing the argument further, we find that the areas of $S_0, S_1, S_2, S_3, S_4, \ldots$, form the geometric sequence $1, \frac{8}{9}, \left(\frac{8}{9}\right)^2, \left(\frac{8}{9}\right)^3, \left(\frac{8}{9}\right)^4, \ldots$. (Notice that this implies that the area of the Sierpinski carpet is 0, since the limit of $\left(\frac{8}{9}\right)^n$ as n tends to infinity is 0.)

10.13 Chaos

1. Because $250 = 2 \cdot 5^3$ it follows from (i) that $\Pi(250) = 3 \cdot 250 = 750$.
Because $25 = 5^2$ it follows from (ii) that $\Pi(25) = 2 \cdot 25 = 50$.

Because $125 = 5^3$ it follows from (ii) that $\Pi(125) = 2 \cdot 125 = 250$.
Because $30 = 6 \cdot 5$ it follows from (ii) that $\Pi(30) = 2 \cdot 30 = 60$.
Because $10 = 2 \cdot 5$ it follows from (i) that $\Pi(10) = 3 \cdot 10 = 30$.
Because $50 = 2 \cdot 5^2$ it follows from (i) that $\Pi(50) = 3 \cdot 50 = 150$.
Because $3750 = 6 \cdot 5^4$ it follows from (ii) that $\Pi(3750) = 2 \cdot 3750 = 7500$.
Because $6 = 6 \cdot 5^0$ it follows from (ii) that $\Pi(6) = 2 \cdot 6 = 12$
Because $5 = 5^1$ it follows from (ii) that $\Pi(5) = 2 \cdot 5 = 10$.

2. The point (0, 0) is obviously a 1-cycle. We now choose another of the 36 points of the form $\left(\frac{m}{6}, \frac{n}{6}\right)$, say $\left(0, \frac{1}{6}\right)$. Its iterates produce the 12-cycle

$$\begin{bmatrix} 0 \\ \frac{1}{6} \end{bmatrix} \to \begin{bmatrix} \frac{1}{6} \\ \frac{2}{6} \end{bmatrix} \to \begin{bmatrix} \frac{3}{6} \\ \frac{5}{6} \end{bmatrix} \to \begin{bmatrix} \frac{2}{6} \\ \frac{1}{6} \end{bmatrix} \to \begin{bmatrix} \frac{3}{6} \\ \frac{4}{6} \end{bmatrix} \to \begin{bmatrix} \frac{1}{6} \\ \frac{5}{6} \end{bmatrix} \to \begin{bmatrix} 0 \\ \frac{5}{6} \end{bmatrix} \to \begin{bmatrix} \frac{5}{6} \\ \frac{4}{6} \end{bmatrix} \to \begin{bmatrix} \frac{3}{6} \\ \frac{1}{6} \end{bmatrix} \to \begin{bmatrix} \frac{4}{6} \\ \frac{5}{6} \end{bmatrix} \to \begin{bmatrix} \frac{3}{6} \\ \frac{2}{6} \end{bmatrix} \to \begin{bmatrix} \frac{5}{6} \\ \frac{1}{6} \end{bmatrix}.$$

So far we have accounted for 13 of the 36 points. Taking one of the remaining points, say $\left(0, \frac{2}{6}\right)$, we arrive at the 4-cycle $\begin{bmatrix} 0 \\ \frac{2}{6} \end{bmatrix} \to \begin{bmatrix} \frac{2}{6} \\ \frac{4}{6} \end{bmatrix} \to \begin{bmatrix} 0 \\ \frac{4}{6} \end{bmatrix} \to \begin{bmatrix} \frac{4}{6} \\ \frac{2}{6} \end{bmatrix}$. We continue in this way, each time starting with some point of the form $\left(\frac{m}{6}, \frac{n}{6}\right)$ that has not yet appeared in a cycle, until we exhaust all such points. This yields a 3-cycle: $\begin{bmatrix} \frac{3}{6} \\ 0 \end{bmatrix} \to \begin{bmatrix} \frac{3}{6} \\ \frac{3}{6} \end{bmatrix} \to \begin{bmatrix} 0 \\ \frac{3}{6} \end{bmatrix}$;

another 4-cycle: $\begin{bmatrix} \frac{4}{6} \\ 0 \end{bmatrix} \to \begin{bmatrix} \frac{4}{6} \\ \frac{4}{6} \end{bmatrix} \to \begin{bmatrix} \frac{2}{6} \\ 0 \end{bmatrix} \to \begin{bmatrix} \frac{2}{6} \\ \frac{2}{6} \end{bmatrix}$; and another 12-cycle:

$$\begin{bmatrix} \frac{1}{6} \\ 0 \end{bmatrix} \to \begin{bmatrix} \frac{1}{6} \\ \frac{1}{6} \end{bmatrix} \to \begin{bmatrix} \frac{2}{6} \\ \frac{3}{6} \end{bmatrix} \to \begin{bmatrix} \frac{5}{6} \\ \frac{2}{6} \end{bmatrix} \to \begin{bmatrix} \frac{1}{6} \\ \frac{3}{6} \end{bmatrix} \to \begin{bmatrix} \frac{4}{6} \\ \frac{1}{6} \end{bmatrix} \to \begin{bmatrix} \frac{5}{6} \\ 0 \end{bmatrix} \to \begin{bmatrix} \frac{5}{6} \\ \frac{5}{6} \end{bmatrix} \to \begin{bmatrix} \frac{4}{6} \\ \frac{3}{6} \end{bmatrix} \to \begin{bmatrix} \frac{1}{6} \\ \frac{4}{6} \end{bmatrix} \to \begin{bmatrix} \frac{5}{6} \\ \frac{3}{6} \end{bmatrix} \to \begin{bmatrix} \frac{2}{6} \\ \frac{5}{6} \end{bmatrix}.$$

The possible periods of points for the form $\left(\frac{m}{6}, \frac{n}{6}\right)$ are thus 1, 3, 4 and 12. The least common multiple of these four numbers is 12, and so $\Pi(6) = 12$.

3. **(a)** We are given that $x_0 = 3$ and $x_1 = 7$. With $p = 15$ we have

$$\begin{aligned}
x_2 &= x_1 + x_0 \mod 15 = 7+3 \mod 15 = 10 \mod 15 = 10,\\
x_3 &= x_2 + x_1 \mod 15 = 10+7 \mod 15 = 17 \mod 15 = 2,\\
x_4 &= x_3 + x_2 \mod 15 = 2+10 \mod 15 = 12 \mod 15 = 12\\
x_5 &= x_4 + x_3 \mod 15 = 12+2 \mod 15 = 14 \mod 15 = 14,\\
x_6 &= x_5 + x_4 \mod 15 = 14+12 \mod 15 = 26 \mod 15 = 11,\\
x_7 &= x_6 + x_5 \mod 15 = 11+14 \mod 15 = 25 \mod 15 = 10\\
x_8 &= x_7 + x_6 \mod 15 = 10+11 \mod 15 = 21 \mod 15 = 6,\\
x_9 &= x_8 + x_7 \mod 15 = 6+10 \mod 15 = 16 \mod 15 = 1,\\
x_{10} &= x_9 + x_8 \mod 15 = 1+6 \mod 15 = 7 \mod 15 = 7,\\
x_{11} &= x_{10} + x_9 \mod 15 = 7+1 \mod 15 = 8 \mod 15 = 8\\
x_{12} &= x_{11} + x_{10} \mod 15 = 8+7 \mod 15 = 15 \mod 15 = 0,\\
x_{13} &= x_{12} + x_{11} \mod 15 = 0+8 \mod 15 = 8 \mod 15 = 8,\\
x_{14} &= x_{13} + x_{12} \mod 15 = 8+0 \mod 15 = 8 \mod 15 = 8,\\
x_{15} &= x_{14} + x_{13} \mod 15 = 8+8 \mod 15 = 16 \mod 15 = 1,\\
x_{16} &= x_{15} + x_{14} \mod 15 = 1+8 \mod 15 = 9 \mod 15 = 9,\\
x_{17} &= x_{16} + x_{15} \mod 15 = 9+1 \mod 15 = 10 \mod 15 = 10,\\
x_{18} &= x_{17} + x_{16} \mod 15 = 10+9 \mod 15 = 19 \mod 15 = 4,\\
x_{19} &= x_{18} + x_{17} \mod 15 = 4+10 \mod 15 = 14 \mod 15 = 14,\\
x_{20} &= x_{19} + x_{18} \mod 15 = 14+4 \mod 15 = 18 \mod 15 = 3,\\
x_{21} &= x_{20} + x_{19} \mod 15 = 3+14 \mod 15 = 17 \mod 15 = 2,\\
x_{22} &= x_{21} + x_{20} \mod 15 = 2+3 \mod 15 = 5 \mod 15 = 5,\\
x_{23} &= x_{22} + x_{21} \mod 15 = 5+2 \mod 15 = 7 \mod 15 = 7,\\
x_{24} &= x_{23} + x_{22} \mod 15 = 7+5 \mod 15 = 12 \mod 15 = 12,\\
x_{25} &= x_{24} + x_{23} \mod 15 = 12+7 \mod 15 = 19 \mod 15 = 4,\\
x_{26} &= x_{25} + x_{24} \mod 15 = 4+12 \mod 15 = 16 \mod 15 = 1,\\
x_{27} &= x_{26} + x_{25} \mod 15 = 1+4 \mod 15 = 5 \mod 15 = 5,\\
x_{28} &= x_{27} + x_{26} \mod 15 = 5+1 \mod 15 = 6 \mod 15 = 6,\\
x_{29} &= x_{28} + x_{27} \mod 15 = 6+5 \mod 15 = 11 \mod 15 = 11,\\
x_{30} &= x_{29} + x_{28} \mod 15 = 11+6 \mod 15 = 17 \mod 15 = 2,\\
x_{31} &= x_{30} + x_{29} \mod 15 = 2+11 \mod 15 = 13 \mod 15 = 13,\\
x_{32} &= x_{31} + x_{30} \mod 15 = 13+2 \mod 15 = 15 \mod 15 = 0,\\
x_{33} &= x_{32} + x_{31} \mod 15 = 0+13 \mod 15 = 13 \mod 15 = 13,\\
x_{34} &= x_{33} + x_{32} \mod 15 = 13+0 \mod 15 = 13 \mod 15 = 13,\\
x_{35} &= x_{34} + x_{33} \mod 15 = 13+13 \mod 15 = 26 \mod 15 = 11,\\
x_{36} &= x_{35} + x_{34} \mod 15 = 11+13 \mod 15 = 24 \mod 15 = 9,\\
x_{37} &= x_{36} + x_{35} \mod 15 = 9+11 \mod 15 = 20 \mod 15 = 5,\\
x_{38} &= x_{37} + x_{36} \mod 15 = 5+9 \mod 15 = 14 \mod 15 = 14,\\
x_{39} &= x_{38} + x_{37} \mod 15 = 14+5 \mod 15 = 19 \mod 15 = 4,\\
x_{40} &= x_{39} + x_{38} \mod 15 = 4+14 \mod 15 = 18 \mod 15 = 3,\\
x_{41} &= x_{40} + x_{39} \mod 15 = 3+4 \mod 15 = 7 \mod 15 = 7,
\end{aligned}$$

and finally: $x_{40} = x_0$ and $x_{41} = x_1$. Thus this sequence is periodic with period 40.

(b) Step (ii) of the algorithm is $x_{n+1} = x_n + x_{n-1} \mathrm{mod} p$.
Replacing n in this formula by $n + 1$ gives

$$x_{n+2} = x_{n+1} + x_n \mathrm{mod} p = (x_n + x_{n-1}) + x_n \mathrm{mod} p = 2x_n + x_{n-1} \mathrm{mod} p.$$

These equations can be written as

$$x_{n+1} = x_{n-1} + x_n \bmod p$$
$$x_{n+2} = x_{n-1} + 2x_n \bmod p$$

which in matrix form are $\begin{bmatrix} x_{n+1} \\ x_{n+2} \end{bmatrix} = \begin{bmatrix} 1 & 1 \\ 1 & 2 \end{bmatrix} \begin{bmatrix} x_{n-1} \\ x_n \end{bmatrix} \bmod p$.

(c) Beginning with $\begin{bmatrix} x_0 \\ x_1 \end{bmatrix} = \begin{bmatrix} 5 \\ 5 \end{bmatrix}$, we obtain

$$\begin{bmatrix} x_2 \\ x_3 \end{bmatrix} = \begin{bmatrix} 1 & 1 \\ 1 & 2 \end{bmatrix} \begin{bmatrix} 5 \\ 5 \end{bmatrix} \bmod 21 = \begin{bmatrix} 10 \\ 15 \end{bmatrix} \bmod 21 = \begin{bmatrix} 10 \\ 15 \end{bmatrix}$$
$$\begin{bmatrix} x_4 \\ x_5 \end{bmatrix} = \begin{bmatrix} 1 & 1 \\ 1 & 2 \end{bmatrix} \begin{bmatrix} 10 \\ 15 \end{bmatrix} \bmod 21 = \begin{bmatrix} 25 \\ 40 \end{bmatrix} \bmod 21 = \begin{bmatrix} 4 \\ 19 \end{bmatrix}$$
$$\begin{bmatrix} x_6 \\ x_7 \end{bmatrix} = \begin{bmatrix} 1 & 1 \\ 1 & 2 \end{bmatrix} \begin{bmatrix} 4 \\ 19 \end{bmatrix} \bmod 21 = \begin{bmatrix} 23 \\ 42 \end{bmatrix} \bmod 21 = \begin{bmatrix} 2 \\ 0 \end{bmatrix}$$
$$\begin{bmatrix} x_8 \\ x_9 \end{bmatrix} = \begin{bmatrix} 1 & 1 \\ 1 & 2 \end{bmatrix} \begin{bmatrix} 2 \\ 0 \end{bmatrix} \bmod 21 = \begin{bmatrix} 2 \\ 2 \end{bmatrix} \bmod 21 = \begin{bmatrix} 2 \\ 2 \end{bmatrix}$$
$$\begin{bmatrix} x_{10} \\ x_{11} \end{bmatrix} = \begin{bmatrix} 1 & 1 \\ 1 & 2 \end{bmatrix} \begin{bmatrix} 2 \\ 2 \end{bmatrix} \bmod 21 = \begin{bmatrix} 4 \\ 6 \end{bmatrix} \bmod 21 = \begin{bmatrix} 4 \\ 6 \end{bmatrix}$$
$$\begin{bmatrix} x_{12} \\ x_{13} \end{bmatrix} = \begin{bmatrix} 1 & 1 \\ 1 & 2 \end{bmatrix} \begin{bmatrix} 4 \\ 6 \end{bmatrix} \bmod 21 = \begin{bmatrix} 10 \\ 16 \end{bmatrix} \bmod 21 = \begin{bmatrix} 10 \\ 16 \end{bmatrix}$$
$$\begin{bmatrix} x_{14} \\ x_{15} \end{bmatrix} = \begin{bmatrix} 1 & 1 \\ 1 & 2 \end{bmatrix} \begin{bmatrix} 10 \\ 16 \end{bmatrix} \bmod 21 = \begin{bmatrix} 26 \\ 42 \end{bmatrix} \bmod 21 = \begin{bmatrix} 5 \\ 0 \end{bmatrix}$$
$$\begin{bmatrix} x_{16} \\ x_{17} \end{bmatrix} = \begin{bmatrix} 1 & 1 \\ 1 & 2 \end{bmatrix} \begin{bmatrix} 5 \\ 0 \end{bmatrix} \bmod 21 = \begin{bmatrix} 5 \\ 5 \end{bmatrix} \bmod 21 = \begin{bmatrix} 5 \\ 5 \end{bmatrix}$$

and we see that $\begin{bmatrix} x_{16} \\ x_{17} \end{bmatrix} = \begin{bmatrix} x_0 \\ x_1 \end{bmatrix}$.

4. **(c)** We have that

$$C\left(\begin{bmatrix} \frac{1}{101} \\ 0 \end{bmatrix}\right) = \begin{bmatrix} 1 & 1 \\ 1 & 2 \end{bmatrix} \begin{bmatrix} \frac{1}{101} \\ 0 \end{bmatrix} \bmod 1 = \begin{bmatrix} \frac{1}{101} \\ \frac{1}{101} \end{bmatrix},$$
$$C^2\left(\begin{bmatrix} \frac{1}{101} \\ 0 \end{bmatrix}\right) = \begin{bmatrix} 1 & 1 \\ 1 & 2 \end{bmatrix} \begin{bmatrix} \frac{1}{101} \\ \frac{1}{101} \end{bmatrix} \bmod 1 = \begin{bmatrix} \frac{2}{101} \\ \frac{3}{101} \end{bmatrix},$$
$$C^3\left(\begin{bmatrix} \frac{1}{101} \\ 0 \end{bmatrix}\right) = \begin{bmatrix} 1 & 1 \\ 1 & 2 \end{bmatrix} \begin{bmatrix} \frac{2}{101} \\ \frac{3}{101} \end{bmatrix} \bmod 1 = \begin{bmatrix} \frac{5}{101} \\ \frac{8}{101} \end{bmatrix},$$
$$C^4\left(\begin{bmatrix} \frac{1}{101} \\ 0 \end{bmatrix}\right) = \begin{bmatrix} 1 & 1 \\ 1 & 2 \end{bmatrix} \begin{bmatrix} \frac{5}{101} \\ \frac{8}{101} \end{bmatrix} \bmod 1 = \begin{bmatrix} \frac{13}{101} \\ \frac{21}{101} \end{bmatrix},$$
$$C^5\left(\begin{bmatrix} \frac{1}{101} \\ 0 \end{bmatrix}\right) = \begin{bmatrix} 1 & 1 \\ 1 & 2 \end{bmatrix} \begin{bmatrix} \frac{13}{101} \\ \frac{21}{101} \end{bmatrix} \bmod 1 = \begin{bmatrix} \frac{34}{101} \\ \frac{55}{101} \end{bmatrix}.$$

Because all five iterates are different, the period of the periodic point $\left(\frac{1}{101}, 0\right)$ must be greater than 5.

5. If $0 \le x < 1$ and $0 \le y < 1$, then $T(x, y) = \left(x + \frac{5}{12}, y\right) \text{mod}1$, and so
$T^2(x, y) = \left(x + \frac{10}{12}, y\right) \text{mod}1, T^3(x, y) = \left(x + \frac{15}{12}, y\right) \text{mod}1, \ldots,$
$T^{12}(x, y) = \left(x + \frac{60}{12}, y\right) \text{mod}1 = (x + 5, y)\text{mod}1 = (x, y).$
Thus every point in S returns to its original position after 12 iterations and so every point in S is a periodic point with period at most 12. Because every point is a periodic point, no point can have a dense set of iterates, and so the mapping cannot be chaotic.

6. **(a)** The matrix of Arnold's cat map, $\begin{bmatrix} 1 & 1 \\ 1 & 2 \end{bmatrix}$, is one in which (i) the entries are all integers, (ii) the determinant is 1, and (iii) the eigenvalues, $\frac{3+\sqrt{5}}{2} = 2.6180\ldots$ and $\frac{3-\sqrt{5}}{2} = 0.3819\ldots$, do not have magnitude 1. The three conditions of an Anosov automorphism are thus satisfied.

(b) The eigenvalues of the matrix $\begin{bmatrix} 0 & 1 \\ 1 & 0 \end{bmatrix}$ are ± 1, both of which have magnitude 1. By part (iii) of the definition of an Anosov automorphism, this matrix is not the matrix of an Anosov automorphism.

The entries of the matrix $\begin{bmatrix} 3 & 2 \\ 1 & 1 \end{bmatrix}$ are integers, its determinant is 1; and neither of its eigenvalues, $2 \pm \sqrt{3}$, has magnitude 1. Consequently, this is the matrix of an Anosov automorphism.

The eigenvalues of the matrix $\begin{bmatrix} 1 & 0 \\ 0 & 1 \end{bmatrix}$ are both equal to 1, and so both have magnitude 1. By part (iii) of the definition, this is not the matrix of an Anosov automorphism.

The entries of the matrix $\begin{bmatrix} 5 & 7 \\ 2 & 3 \end{bmatrix}$ are integers; its determinant is 1; and neither of its eigenvalues, $4 \pm \sqrt{15}$, has magnitude 1. Consequently, this is the matrix of an Anosov automorphism.

The determinant of the matrix $\begin{bmatrix} 6 & 2 \\ 5 & 2 \end{bmatrix}$ is 2, and so by part (ii) of the definition, this is not the matrix of an Anosov automorphism.

(c) The eigenvalues of the matrix $\begin{bmatrix} 0 & 1 \\ -1 & 0 \end{bmatrix}$ are $\pm i$; both of which have magnitude 1. By part (iii) of the definition, this cannot be the matrix of an Anosov automorphism.

Starting with an arbitrary point $\begin{bmatrix} x \\ y \end{bmatrix}$ in the interior of S, (that is, with $0 < x < 1$ and $0 < y < 1$) we obtain

$$\begin{bmatrix} 0 & 1 \\ -1 & 0 \end{bmatrix}\begin{bmatrix} x \\ y \end{bmatrix} \text{mod}1 = \begin{bmatrix} y \\ -x \end{bmatrix} \text{mod}1 = \begin{bmatrix} y \\ 1-x \end{bmatrix},$$
$$\begin{bmatrix} 0 & 1 \\ -1 & 0 \end{bmatrix}\begin{bmatrix} y \\ 1-x \end{bmatrix} \text{mod}1 = \begin{bmatrix} 1-x \\ -y \end{bmatrix} \text{mod}1 = \begin{bmatrix} 1-x \\ 1-y \end{bmatrix},$$
$$\begin{bmatrix} 0 & 1 \\ -1 & 0 \end{bmatrix}\begin{bmatrix} 1-x \\ 1-y \end{bmatrix} \text{mod}1 = \begin{bmatrix} 1-y \\ -1+x \end{bmatrix} \text{mod}1 = \begin{bmatrix} 1-y \\ x \end{bmatrix},$$
$$\begin{bmatrix} 0 & 1 \\ -1 & 0 \end{bmatrix}\begin{bmatrix} 1-y \\ x \end{bmatrix} \text{mod}1 = \begin{bmatrix} x \\ -1+y \end{bmatrix} \text{mod}1 = \begin{bmatrix} x \\ y \end{bmatrix}.$$

Thus every point in the interior of S is a periodic point with period at most 4. The geometric effect of this transformation, as seen by the iterates, is to rotate each point in the interior of S clockwise by 90° about the center point $\begin{bmatrix} \frac{1}{2} \\ \frac{1}{2} \end{bmatrix}$ of S. Consequently, each point in the interior of S has period 4 with the exception of the center point $\begin{bmatrix} \frac{1}{2} \\ \frac{1}{2} \end{bmatrix}$ which is a fixed point.

For points not in the interior of S, we first observe that the origin $\begin{bmatrix} 0 \\ 0 \end{bmatrix}$ is a fixed point, which can easily be verified. Starting with a point of the form $\begin{bmatrix} x \\ 0 \end{bmatrix}$ with $0 < x < 1$, we obtain a 4-cycle $\begin{bmatrix} x \\ 0 \end{bmatrix} \to \begin{bmatrix} 0 \\ 1-x \end{bmatrix} \to \begin{bmatrix} 1-x \\ 0 \end{bmatrix} \to \begin{bmatrix} 0 \\ x \end{bmatrix} \to \begin{bmatrix} x \\ 0 \end{bmatrix}$ if $x \neq \frac{1}{2}$, otherwise we obtain the 2-cycle $\begin{bmatrix} \frac{1}{2} \\ 0 \end{bmatrix} \to \begin{bmatrix} 0 \\ \frac{1}{2} \end{bmatrix} \to \begin{bmatrix} \frac{1}{2} \\ 0 \end{bmatrix}$.

Similarly, starting with a point of the form $\begin{bmatrix} 0 \\ y \end{bmatrix}$ with $0 < y < 1$, we obtain a 4-cycle $\begin{bmatrix} 0 \\ y \end{bmatrix} \to \begin{bmatrix} y \\ 0 \end{bmatrix} \to \begin{bmatrix} 0 \\ 1-y \end{bmatrix} \to \begin{bmatrix} 1-y \\ 0 \end{bmatrix} \to \begin{bmatrix} 0 \\ y \end{bmatrix}$ if $y \neq \frac{1}{2}$, otherwise we obtain the 2-cycle $\begin{bmatrix} 0 \\ \frac{1}{2} \end{bmatrix} \to \begin{bmatrix} \frac{1}{2} \\ 0 \end{bmatrix} \to \begin{bmatrix} 0 \\ \frac{1}{2} \end{bmatrix}$. Thus every point not in the interior of S is a periodic point with 1, 2, or 4. Finally because no point in S can have a dense set of iterates, it follows that the mapping cannot be chaotic.

9. As per the hint, we wish to find the regions in S that map onto the four indicated regions in the figure below.

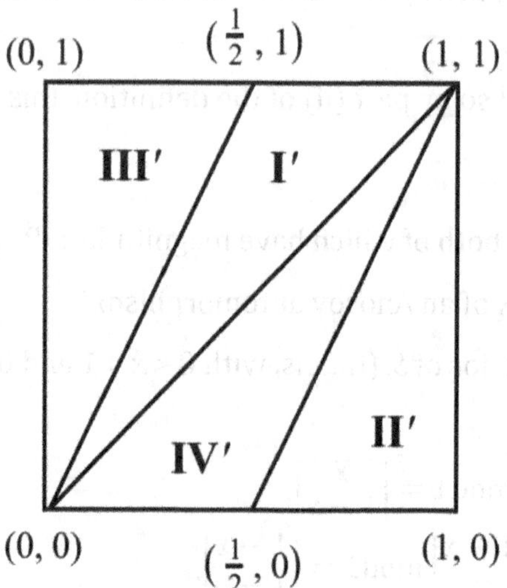

We first consider region **I**′ with vertices (0, 0), $\left(\frac{1}{2}, 1\right)$, and (1, 1). We seek points (x_1, y_1), (x_2, y_2), and (x_3, y_3), with entries that lie in [0, 1], that map onto these three points under the mapping $\begin{bmatrix} x \\ y \end{bmatrix} \to \begin{bmatrix} 1 & 1 \\ 1 & 2 \end{bmatrix} \begin{bmatrix} x \\ y \end{bmatrix} + \begin{bmatrix} a \\ b \end{bmatrix}$ for certain integer values of a and b to be determined. This leads to the three equations

$$\begin{bmatrix} 1 & 1 \\ 1 & 2 \end{bmatrix} \begin{bmatrix} x_1 \\ y_2 \end{bmatrix} + \begin{bmatrix} a \\ b \end{bmatrix} = \begin{bmatrix} 0 \\ 0 \end{bmatrix},$$

$$\begin{bmatrix} 1 & 1 \\ 1 & 2 \end{bmatrix} \begin{bmatrix} x_2 \\ y_2 \end{bmatrix} + \begin{bmatrix} a \\ b \end{bmatrix} = \begin{bmatrix} \frac{1}{2} \\ 1 \end{bmatrix},$$

$$\begin{bmatrix} 1 & 1 \\ 1 & 2 \end{bmatrix} \begin{bmatrix} x_3 \\ y_3 \end{bmatrix} + \begin{bmatrix} a \\ b \end{bmatrix} = \begin{bmatrix} 1 \\ 1 \end{bmatrix}.$$

The inverse of the matrix $\begin{bmatrix} 1 & 1 \\ 1 & 2 \end{bmatrix}$ is $\begin{bmatrix} 2 & -1 \\ -1 & 1 \end{bmatrix}$. We multiply the above three matrix equations by this inverse and set $\begin{bmatrix} c \\ d \end{bmatrix} = \begin{bmatrix} 2 & -1 \\ -1 & 1 \end{bmatrix} \begin{bmatrix} a \\ b \end{bmatrix}$. Notice that c and d must be integers. This leads to

$$\begin{bmatrix} x_1 \\ y_1 \end{bmatrix} = \begin{bmatrix} 2 & -1 \\ -1 & 1 \end{bmatrix} \begin{bmatrix} 0 \\ 0 \end{bmatrix} - \begin{bmatrix} c \\ d \end{bmatrix} = -\begin{bmatrix} c \\ d \end{bmatrix},$$

$$\begin{bmatrix} x_2 \\ y_2 \end{bmatrix} = \begin{bmatrix} 2 & -1 \\ -1 & 1 \end{bmatrix} \begin{bmatrix} \frac{1}{2} \\ 1 \end{bmatrix} - \begin{bmatrix} c \\ d \end{bmatrix} = \begin{bmatrix} 0 \\ \frac{1}{2} \end{bmatrix} - \begin{bmatrix} c \\ d \end{bmatrix},$$

$$\begin{bmatrix} x_3 \\ y_3 \end{bmatrix} = \begin{bmatrix} 2 & -1 \\ -1 & 1 \end{bmatrix} \begin{bmatrix} 1 \\ 1 \end{bmatrix} - \begin{bmatrix} c \\ d \end{bmatrix} = \begin{bmatrix} 1 \\ 0 \end{bmatrix} - \begin{bmatrix} c \\ d \end{bmatrix}.$$

The only integer values of c and d that will give values of x_i and y_i in the interval [0, 1] are $c = d = 0$. This then gives $a = b = 0$ and the mapping $\begin{bmatrix} x \\ y \end{bmatrix} \to \begin{bmatrix} 1 & 1 \\ 1 & 2 \end{bmatrix} \begin{bmatrix} x \\ y \end{bmatrix}$ maps the three points (0, 0), $\left(0, \frac{1}{2}\right)$, and (1, 0) to the three points (0, 0), $\left(\frac{1}{2}, 1\right)$, and (1, 1), respectively. The three points (0, 0), $\left(0, \frac{1}{2}\right)$, and (1, 0) define the triangular region labeled **I** in the diagram below, which then maps onto the region **I**′.

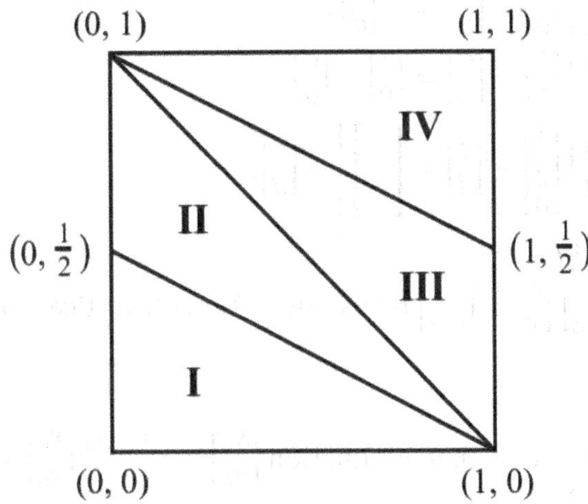

For region **II**′, the calculations are as follows:

$$\begin{bmatrix} 1 & 1 \\ 1 & 2 \end{bmatrix} \begin{bmatrix} x_1 \\ y_1 \end{bmatrix} + \begin{bmatrix} a \\ b \end{bmatrix} = \begin{bmatrix} \frac{1}{2} \\ 0 \end{bmatrix}, \quad \begin{bmatrix} 1 & 1 \\ 1 & 2 \end{bmatrix} \begin{bmatrix} x_2 \\ y_2 \end{bmatrix} + \begin{bmatrix} a \\ b \end{bmatrix} = \begin{bmatrix} 1 \\ 1 \end{bmatrix}, \quad \begin{bmatrix} 1 & 1 \\ 1 & 2 \end{bmatrix} \begin{bmatrix} x_3 \\ y_3 \end{bmatrix} + \begin{bmatrix} a \\ b \end{bmatrix} = \begin{bmatrix} 1 \\ 0 \end{bmatrix};$$

$$\begin{bmatrix} x_1 \\ y_1 \end{bmatrix} = \begin{bmatrix} 2 & -1 \\ -1 & 1 \end{bmatrix} \begin{bmatrix} \frac{1}{2} \\ 0 \end{bmatrix} - \begin{bmatrix} c \\ d \end{bmatrix} = \begin{bmatrix} 1 \\ -\frac{1}{2} \end{bmatrix} - \begin{bmatrix} c \\ d \end{bmatrix},$$

$$\begin{bmatrix}x_2\\y_2\end{bmatrix}=\begin{bmatrix}2&-1\\-1&1\end{bmatrix}\begin{bmatrix}1\\1\end{bmatrix}-\begin{bmatrix}c\\d\end{bmatrix}=\begin{bmatrix}1\\0\end{bmatrix}-\begin{bmatrix}c\\d\end{bmatrix},$$
$$\begin{bmatrix}x_3\\y_3\end{bmatrix}=\begin{bmatrix}2&-1\\-1&1\end{bmatrix}\begin{bmatrix}1\\0\end{bmatrix}-\begin{bmatrix}c\\d\end{bmatrix}=\begin{bmatrix}2\\-1\end{bmatrix}-\begin{bmatrix}c\\d\end{bmatrix}.$$

Only $c = 1$ and $d = -1$ will work. This leads to

$a = 0, b = -1$ and the mapping $\begin{bmatrix}x\\y\end{bmatrix}\to\begin{bmatrix}1&1\\1&2\end{bmatrix}\begin{bmatrix}x\\y\end{bmatrix}+\begin{bmatrix}0\\-1\end{bmatrix}$ maps region **II** with vertices $\left(0, \frac{1}{2}\right)$, (0, 1), and (1, 0) onto region **II′**. For region **III′**, the calculations are as follows:

$$\begin{bmatrix}1&1\\1&2\end{bmatrix}\begin{bmatrix}x_1\\y_1\end{bmatrix}+\begin{bmatrix}a\\b\end{bmatrix}=\begin{bmatrix}0\\0\end{bmatrix},\quad \begin{bmatrix}1&1\\1&2\end{bmatrix}\begin{bmatrix}x_2\\y_2\end{bmatrix}+\begin{bmatrix}a\\b\end{bmatrix}=\begin{bmatrix}\frac{1}{2}\\1\end{bmatrix},\quad \begin{bmatrix}1&1\\1&2\end{bmatrix}\begin{bmatrix}x_3\\y_3\end{bmatrix}+\begin{bmatrix}a\\b\end{bmatrix}=\begin{bmatrix}0\\1\end{bmatrix};$$

$$\begin{bmatrix}x_1\\y_1\end{bmatrix}=\begin{bmatrix}2&-1\\-1&1\end{bmatrix}\begin{bmatrix}0\\0\end{bmatrix}-\begin{bmatrix}c\\d\end{bmatrix}=-\begin{bmatrix}c\\d\end{bmatrix},$$
$$\begin{bmatrix}x_2\\y_2\end{bmatrix}=\begin{bmatrix}2&-1\\-1&1\end{bmatrix}\begin{bmatrix}\frac{1}{2}\\1\end{bmatrix}-\begin{bmatrix}c\\d\end{bmatrix}=\begin{bmatrix}0\\\frac{1}{2}\end{bmatrix}-\begin{bmatrix}c\\d\end{bmatrix},$$
$$\begin{bmatrix}x_3\\y_3\end{bmatrix}=\begin{bmatrix}2&-1\\-1&1\end{bmatrix}\begin{bmatrix}0\\1\end{bmatrix}-\begin{bmatrix}c\\d\end{bmatrix}=\begin{bmatrix}-1\\1\end{bmatrix}-\begin{bmatrix}c\\d\end{bmatrix}.$$

Only $c = -1$ and $d = 0$ will work. This leads to

$a = -1, b = -1$ and the mapping $\begin{bmatrix}x\\y\end{bmatrix}\to\begin{bmatrix}1&1\\1&2\end{bmatrix}\begin{bmatrix}x\\y\end{bmatrix}+\begin{bmatrix}-1\\-1\end{bmatrix}$ maps region **III** with vertices (1, 0), $\left(1, \frac{1}{2}\right)$, and (0, 1) onto region **III′**.

For region **IV′**, the calculations are as follows:

$$\begin{bmatrix}1&1\\1&2\end{bmatrix}\begin{bmatrix}x_1\\y_1\end{bmatrix}+\begin{bmatrix}a\\b\end{bmatrix}=\begin{bmatrix}0\\0\end{bmatrix},\quad \begin{bmatrix}1&1\\1&2\end{bmatrix}\begin{bmatrix}x_2\\y_2\end{bmatrix}+\begin{bmatrix}a\\b\end{bmatrix}=\begin{bmatrix}1\\1\end{bmatrix},\quad \begin{bmatrix}1&1\\1&2\end{bmatrix}\begin{bmatrix}x_3\\y_3\end{bmatrix}+\begin{bmatrix}a\\b\end{bmatrix}=\begin{bmatrix}\frac{1}{2}\\0\end{bmatrix};$$

$$\begin{bmatrix}x_1\\y_1\end{bmatrix}=\begin{bmatrix}2&-1\\-1&1\end{bmatrix}\begin{bmatrix}0\\0\end{bmatrix}-\begin{bmatrix}c\\d\end{bmatrix}=-\begin{bmatrix}c\\d\end{bmatrix},$$
$$\begin{bmatrix}x_2\\y_2\end{bmatrix}=\begin{bmatrix}2&-1\\-1&1\end{bmatrix}\begin{bmatrix}1\\1\end{bmatrix}-\begin{bmatrix}c\\d\end{bmatrix}=\begin{bmatrix}1\\0\end{bmatrix}-\begin{bmatrix}c\\d\end{bmatrix},$$
$$\begin{bmatrix}x_3\\y_3\end{bmatrix}=\begin{bmatrix}2&-1\\-1&1\end{bmatrix}\begin{bmatrix}\frac{1}{2}\\0\end{bmatrix}-\begin{bmatrix}c\\d\end{bmatrix}=\begin{bmatrix}1\\-\frac{1}{2}\end{bmatrix}-\begin{bmatrix}c\\d\end{bmatrix}.$$

Only $c = 0$ and $d = -1$ will work. This leads to

$a = -1, b = -2$ and the mapping $\begin{bmatrix}x\\y\end{bmatrix}\to\begin{bmatrix}1&1\\1&2\end{bmatrix}\begin{bmatrix}x\\y\end{bmatrix}+\begin{bmatrix}-1\\-2\end{bmatrix}$ maps region **IV** with vertices (0, 1), (1, 1), and $\left(1, \frac{1}{2}\right)$ onto region **IV′**.

12. As per the hint, we want to find all solutions of the matrix equation $\begin{bmatrix}x_0\\y_0\end{bmatrix}=\begin{bmatrix}2&3\\3&5\end{bmatrix}\begin{bmatrix}x_0\\y_0\end{bmatrix}-\begin{bmatrix}r\\s\end{bmatrix}$ where $0 \le x_0 < 1, 0 \le y_0 < 1$, and r and s are nonnegative integers. This equation can be rewritten as $\begin{bmatrix}1&3\\3&4\end{bmatrix}\begin{bmatrix}x_0\\y_0\end{bmatrix}=\begin{bmatrix}r\\s\end{bmatrix}$, which has the solution $x_0 = \frac{-4r+3s}{5}$ and $y_0 = \frac{3r-s}{5}$. First trying $r = 0$ and $s = 0, 1, 2, \ldots$, then $r = 1$ and
$s = 0, 1, 2, \ldots$, etc., we find that the only values of r and s that yield values of x_0 and y_0 lying in [0, 1) are:

$r = 1$ and $s = 2$, which give $x_0 = \frac{2}{5}$ and $y_0 = \frac{1}{5}$;

$r = 2$ and $s = 3$, which give $x_0 = \frac{1}{5}$ and $y_0 = \frac{3}{5}$;

$r = 2$ and $s = 4$, which give $x_0 = \frac{4}{5}$ and $y_0 = \frac{2}{5}$;

$r = 3$ and $s = 5$, which give $x_0 = \frac{3}{5}$ and $y_0 = \frac{4}{5}$.

We can then check that $\left(\frac{2}{5}, \frac{1}{5}\right)$ and $\left(\frac{3}{5}, \frac{4}{5}\right)$ form one 2-cycle and $\left(\frac{1}{5}, \frac{3}{5}\right)$ and $\left(\frac{4}{5}, \frac{2}{5}\right)$ form another 2-cycle.

14. Begin with a 101×101 array of white pixels and add the letter 'A' in black pixels to it. Apply the mapping T to this image, which will scatter the black pixels throughout the image. Then superimpose the letter 'B' in black pixels onto this image. Apply the mapping T again and then superimpose the letter 'C' in black pixels onto the resulting image. Repeat this procedure with the letter 'D' and 'E'. The next application of the mapping will return you to the letter 'A' with the pixels for the letters 'B' through 'E' scattered in the background. Four subsequent applications of T to this image will produce the remaining images.

10.14 Cryptography

1. First group the plaintext into pairs, add the dummy letter T, and get the numerical equivalents from Table 1.

DA	*RK*	*NI*	*GH*	*TT*
4 1	18 11	14 9	7 8	20 20

(a) For the enciphering matrix $A = \begin{bmatrix} 1 & 3 \\ 2 & 1 \end{bmatrix}$, reducing everything mod 26, we have

$$\begin{bmatrix} 1 & 3 \\ 2 & 1 \end{bmatrix}\begin{bmatrix} 4 \\ 1 \end{bmatrix} = \begin{bmatrix} 7 \\ 9 \end{bmatrix} \qquad \begin{matrix} G \\ I \end{matrix}$$

$$\begin{bmatrix} 1 & 3 \\ 2 & 1 \end{bmatrix}\begin{bmatrix} 18 \\ 11 \end{bmatrix} = \begin{bmatrix} 51 \\ 47 \end{bmatrix} = \begin{bmatrix} 25 \\ 21 \end{bmatrix} \qquad \begin{matrix} Y \\ U \end{matrix}$$

$$\begin{bmatrix} 1 & 3 \\ 2 & 1 \end{bmatrix}\begin{bmatrix} 14 \\ 9 \end{bmatrix} = \begin{bmatrix} 41 \\ 37 \end{bmatrix} = \begin{bmatrix} 15 \\ 11 \end{bmatrix} \qquad \begin{matrix} O \\ K \end{matrix}$$

$$\begin{bmatrix} 1 & 3 \\ 2 & 1 \end{bmatrix}\begin{bmatrix} 7 \\ 8 \end{bmatrix} = \begin{bmatrix} 31 \\ 22 \end{bmatrix} = \begin{bmatrix} 5 \\ 22 \end{bmatrix} \qquad \begin{matrix} E \\ V \end{matrix}$$

$$\begin{bmatrix} 1 & 3 \\ 2 & 1 \end{bmatrix}\begin{bmatrix} 20 \\ 20 \end{bmatrix} = \begin{bmatrix} 80 \\ 60 \end{bmatrix} = \begin{bmatrix} 2 \\ 8 \end{bmatrix} \qquad \begin{matrix} B \\ H \end{matrix}$$

The Hill cipher is *GIYUOKEVBH*.

(b) For the enciphering matrix $A = \begin{bmatrix}4 & 3\\1 & 2\end{bmatrix}$, reducing everything mod 26, we have

$\begin{bmatrix}4 & 3\\1 & 2\end{bmatrix}\begin{bmatrix}4\\1\end{bmatrix} = \begin{bmatrix}19\\6\end{bmatrix}$ $\quad \begin{matrix}S\\F\end{matrix}$

$\begin{bmatrix}4 & 3\\1 & 2\end{bmatrix}\begin{bmatrix}18\\11\end{bmatrix} = \begin{bmatrix}105\\40\end{bmatrix} = \begin{bmatrix}1\\14\end{bmatrix}$ $\quad \begin{matrix}A\\N\end{matrix}$

$\begin{bmatrix}4 & 3\\1 & 2\end{bmatrix}\begin{bmatrix}14\\9\end{bmatrix} = \begin{bmatrix}83\\32\end{bmatrix} = \begin{bmatrix}5\\6\end{bmatrix}$ $\quad \begin{matrix}E\\F\end{matrix}$

$\begin{bmatrix}4 & 3\\1 & 2\end{bmatrix}\begin{bmatrix}7\\8\end{bmatrix} = \begin{bmatrix}52\\23\end{bmatrix} = \begin{bmatrix}0\\23\end{bmatrix}$ $\quad \begin{matrix}Z\\W\end{matrix}$

$\begin{bmatrix}4 & 3\\1 & 2\end{bmatrix}\begin{bmatrix}20\\20\end{bmatrix} = \begin{bmatrix}140\\60\end{bmatrix} = \begin{bmatrix}10\\8\end{bmatrix}$ $\quad \begin{matrix}J\\H\end{matrix}$

The Hill cipher is *SFANEFZWJH*.

2. **(a)** For $A = \begin{bmatrix}9 & 1\\7 & 2\end{bmatrix}$, $\det(A) = 18 - 7 = 11$, which is not divisible by 2 or 13. Therefore by Corollary 10.14.3, A is invertible. From Equation (2):

$$A^{-1} = (11)^{-1}\begin{bmatrix}2 & -1\\-7 & 9\end{bmatrix} = 19\begin{bmatrix}2 & -1\\-7 & 9\end{bmatrix} = \begin{bmatrix}38 & -19\\-133 & 171\end{bmatrix} = \begin{bmatrix}12 & 7\\23 & 15\end{bmatrix} \text{(mod26)}.$$

Checking:

$$AA^{-1} = \begin{bmatrix}9 & 1\\7 & 2\end{bmatrix}\begin{bmatrix}12 & 7\\23 & 15\end{bmatrix} = \begin{bmatrix}131 & 78\\130 & 79\end{bmatrix} = \begin{bmatrix}1 & 0\\0 & 1\end{bmatrix} \text{(mod26)}$$

$$A^{-1}A = \begin{bmatrix}12 & 7\\23 & 15\end{bmatrix}\begin{bmatrix}9 & 1\\7 & 2\end{bmatrix} = \begin{bmatrix}157 & 26\\312 & 53\end{bmatrix} = \begin{bmatrix}1 & 0\\0 & 1\end{bmatrix} \text{(mod26)}.$$

(b) For $A = \begin{bmatrix}3 & 1\\5 & 3\end{bmatrix}$, $\det(A) = 9 - 5 = 4$, which is divisible by 2. Therefore by Corollary 10.14.3, A is not invertible.

(c) For $A = \begin{bmatrix}8 & 11\\1 & 9\end{bmatrix}$, $\det(A) = 72 - 11 = 61 = 9$ (mod 26), which is not divisible by 2 or 13. Therefore by Corollary 10.14.3, A is invertible. From (2):

$$A^{-1} = (9)^{-1}\begin{bmatrix}9 & -11\\-1 & 8\end{bmatrix} = 3\begin{bmatrix}9 & -11\\-1 & 8\end{bmatrix} = \begin{bmatrix}27 & -33\\-3 & 24\end{bmatrix} = \begin{bmatrix}1 & 19\\23 & 24\end{bmatrix} \text{(mod26)}.$$

Checking:

$$AA^{-1} = \begin{bmatrix}8 & 11\\1 & 9\end{bmatrix}\begin{bmatrix}1 & 19\\23 & 24\end{bmatrix} = \begin{bmatrix}261 & 416\\208 & 235\end{bmatrix} = \begin{bmatrix}1 & 0\\0 & 1\end{bmatrix} \text{(mod26)}$$

$$A^{-1}A = \begin{bmatrix}1 & 19\\23 & 24\end{bmatrix}\begin{bmatrix}8 & 11\\1 & 9\end{bmatrix} = \begin{bmatrix}27 & 182\\208 & 469\end{bmatrix} = \begin{bmatrix}1 & 0\\0 & 1\end{bmatrix} \text{(mod26)}.$$

(d) For $A = \begin{bmatrix}2 & 1\\1 & 7\end{bmatrix}$, $\det(A) = 14 - 1 = 13$, which is divisible by 13. Therefore by Corollary 10.14.4, A is not invertible.

(e) For $A = \begin{bmatrix}3 & 1\\6 & 2\end{bmatrix}$, $\det(A) = 6 - 6 = 0$, so that A is not invertible by Corollary 10.14.4.

(f) For $A = \begin{bmatrix}1 & 8\\1 & 3\end{bmatrix}$, $\det(A) = 3 - 8 = -5 = 21$ (mod 26), which is not divisible by 2 or 13. Therefore by Corollary 10.14.4, A is invertible. From (2):

$$A^{-1} = (21)^{-1}\begin{bmatrix}3 & -8\\-1 & 1\end{bmatrix} = 5\begin{bmatrix}3 & -8\\-1 & 1\end{bmatrix} = \begin{bmatrix}15 & -40\\-5 & 5\end{bmatrix} = \begin{bmatrix}15 & 12\\21 & 5\end{bmatrix} \text{(mod26)}.$$

Checking:

$$AA^{-1} = \begin{bmatrix} 1 & 8 \\ 1 & 3 \end{bmatrix}\begin{bmatrix} 15 & 12 \\ 21 & 5 \end{bmatrix} = \begin{bmatrix} 183 & 52 \\ 78 & 27 \end{bmatrix} = \begin{bmatrix} 1 & 0 \\ 0 & 1 \end{bmatrix} \pmod{26}$$

$$A^{-1}A = \begin{bmatrix} 15 & 12 \\ 21 & 5 \end{bmatrix}\begin{bmatrix} 1 & 8 \\ 1 & 3 \end{bmatrix} = \begin{bmatrix} 27 & 156 \\ 26 & 183 \end{bmatrix} = \begin{bmatrix} 1 & 0 \\ 0 & 1 \end{bmatrix} \pmod{26}.$$

3. From Table 1 the numerical equivalent of this ciphertext is

19 1   11 14   15 24   1 15   10 24

Now we have to find the inverse of $A = \begin{bmatrix} 4 & 1 \\ 3 & 2 \end{bmatrix}$.

Since $\det(A) = 8 - 3 = 5$, we have by (2):

$$A^{-1} = (5)^{-1}\begin{bmatrix} 2 & -1 \\ -3 & 4 \end{bmatrix} = 21\begin{bmatrix} 2 & -1 \\ -3 & 4 \end{bmatrix} = \begin{bmatrix} 42 & -21 \\ -63 & 84 \end{bmatrix} = \begin{bmatrix} 16 & 5 \\ 15 & 6 \end{bmatrix} \pmod{26}.$$

To obtain the plaintext, multiply each ciphertext vector by A^{-1} and reduce the results modulo 26.

$$\begin{bmatrix} 16 & 5 \\ 15 & 6 \end{bmatrix}\begin{bmatrix} 19 \\ 1 \end{bmatrix} = \begin{bmatrix} 309 \\ 291 \end{bmatrix} = \begin{bmatrix} 23 \\ 5 \end{bmatrix} \quad \begin{matrix} W \\ E \end{matrix}$$

$$\begin{bmatrix} 16 & 5 \\ 15 & 6 \end{bmatrix}\begin{bmatrix} 11 \\ 14 \end{bmatrix} = \begin{bmatrix} 246 \\ 249 \end{bmatrix} = \begin{bmatrix} 12 \\ 15 \end{bmatrix} \quad \begin{matrix} L \\ O \end{matrix}$$

$$\begin{bmatrix} 16 & 5 \\ 15 & 6 \end{bmatrix}\begin{bmatrix} 15 \\ 24 \end{bmatrix} = \begin{bmatrix} 360 \\ 369 \end{bmatrix} = \begin{bmatrix} 22 \\ 5 \end{bmatrix} \quad \begin{matrix} V \\ E \end{matrix}$$

$$\begin{bmatrix} 16 & 5 \\ 15 & 6 \end{bmatrix}\begin{bmatrix} 1 \\ 15 \end{bmatrix} = \begin{bmatrix} 91 \\ 105 \end{bmatrix} = \begin{bmatrix} 13 \\ 1 \end{bmatrix} \quad \begin{matrix} M \\ A \end{matrix}$$

$$\begin{bmatrix} 16 & 5 \\ 15 & 6 \end{bmatrix}\begin{bmatrix} 10 \\ 24 \end{bmatrix} = \begin{bmatrix} 280 \\ 294 \end{bmatrix} = \begin{bmatrix} 20 \\ 8 \end{bmatrix} \quad \begin{matrix} T \\ H \end{matrix}$$

The plaintext is thus WE LOVE MATH.

4. From Table 1 the numerical equivalent of the known plaintext is

AR   *MY*

1 18   13 25

and the numerical equivalent of the corresponding ciphertext is

SL   *HK*

19 12   8 11

so the corresponding plaintext and ciphertext vectors are

$$\mathbf{p}_1 = \begin{bmatrix} 1 \\ 18 \end{bmatrix} \leftrightarrow \mathbf{c}_1 = \begin{bmatrix} 19 \\ 12 \end{bmatrix}$$

$$\mathbf{p}_2 = \begin{bmatrix} 13 \\ 25 \end{bmatrix} \leftrightarrow \mathbf{c}_2 = \begin{bmatrix} 8 \\ 11 \end{bmatrix}$$

We want to reduce $C = \begin{bmatrix} \mathbf{c}_1^T \\ \mathbf{c}_2^T \end{bmatrix} = \begin{bmatrix} 19 & 12 \\ 8 & 11 \end{bmatrix}$ to I by elementary row operations and simultaneously apply these operations to $P = \begin{bmatrix} \mathbf{p}_1^T \\ \mathbf{p}_2^T \end{bmatrix} = \begin{bmatrix} 1 & 18 \\ 13 & 25 \end{bmatrix}$.

The calculations are as follows:

$\begin{bmatrix} 19 & 12 & 1 & 18 \\ 8 & 11 & 13 & 25 \end{bmatrix}$ Form the matrix $[C \quad P]$.

$\begin{bmatrix} 1 & 132 & 11 & 198 \\ 8 & 11 & 13 & 25 \end{bmatrix}$ Multiply the first row by $19^{-1} = 11 \pmod{26}$.

$\begin{bmatrix} 1 & 2 & 11 & 16 \\ 8 & 11 & 13 & 25 \end{bmatrix}$ Replace 132 and 198 by their residues modulo 26.

$\begin{bmatrix} 1 & 2 & 11 & 16 \\ 0 & -5 & -75 & -103 \end{bmatrix}$ −8 times the first row to the second.

$\begin{bmatrix} 1 & 2 & 11 & 16 \\ 0 & 21 & 3 & 1 \end{bmatrix}$ Replace the entries in the second row by their residues modulo 26.

$\begin{bmatrix} 1 & 2 & 11 & 16 \\ 0 & 1 & 15 & 5 \end{bmatrix}$ Multiply the second row by $21^{-1} = 5(\text{mod}26)$.

$\begin{bmatrix} 1 & 0 & -19 & 6 \\ 0 & 1 & 15 & 5 \end{bmatrix}$ Add −2 times the second row to the first.

$\begin{bmatrix} 1 & 0 & 7 & 6 \\ 0 & 1 & 15 & 5 \end{bmatrix}$ Replace −19 by its residue modulo 26.

Thus $(A^{-1})^T = \begin{bmatrix} 7 & 6 \\ 15 & 5 \end{bmatrix}$ so the deciphering matrix is $A^{-1} = \begin{bmatrix} 7 & 15 \\ 6 & 5 \end{bmatrix}$ (mod 26).

Since $\det(A^{-1}) = 35 - 90 = -55 = 23(\text{mod}26)$,

$$A = (A^{-1})^{-1} = 23^{-1}\begin{bmatrix} 5 & -15 \\ -6 & 7 \end{bmatrix} = 17\begin{bmatrix} 5 & -15 \\ -6 & 7 \end{bmatrix} = \begin{bmatrix} 85 & -255 \\ -102 & 119 \end{bmatrix} = \begin{bmatrix} 7 & 5 \\ 2 & 15 \end{bmatrix} (\text{mod}26)$$

is the enciphering matrix.

5. From Table 1 the numerical equivalent of the known plaintext is

AT	*OM*
1 20	15 13

and the numerical equivalent of the corresponding ciphertext is

JY	*QO*
10 25	17 15

The corresponding plaintext and ciphertext vectors are:

$$\mathbf{p}_1 = \begin{bmatrix} 1 \\ 20 \end{bmatrix} \leftrightarrow \mathbf{c}_1 = \begin{bmatrix} 10 \\ 25 \end{bmatrix}$$

$$\mathbf{p}_2 = \begin{bmatrix} 15 \\ 13 \end{bmatrix} \leftrightarrow \mathbf{c}_2 = \begin{bmatrix} 17 \\ 15 \end{bmatrix}$$

We want to reduce $C = \begin{bmatrix} 10 & 25 \\ 17 & 15 \end{bmatrix}$ to I by elementary row operations and simultaneously apply these operations to $P = \begin{bmatrix} 1 & 20 \\ 15 & 13 \end{bmatrix}$.

The calculations are as follows:

$\begin{bmatrix} 10 & 25 & 1 & 20 \\ 17 & 15 & 15 & 13 \end{bmatrix}$ Form the matrix $[C \quad P]$.

$\begin{bmatrix} 27 & 40 & 16 & 33 \\ 17 & 15 & 15 & 13 \end{bmatrix}$ Add the second row to the first (since 10^{-1} does not exist mod 26).

$\begin{bmatrix} 1 & 14 & 16 & 7 \\ 17 & 15 & 15 & 13 \end{bmatrix}$ Replace the entries in the first row by their residues modulo 26.

$\begin{bmatrix} 1 & 14 & 16 & 7 \\ 0 & -223 & -257 & -106 \end{bmatrix}$ Add −17 times the first row to the second.

$\begin{bmatrix} 1 & 14 & 16 & 7 \\ 0 & 11 & 3 & 24 \end{bmatrix}$ Replace the entries in the 2nd row by their residues modulo 26.

$\begin{bmatrix} 1 & 14 & 16 & 7 \\ 0 & 1 & 57 & 456 \end{bmatrix}$ Multiply the second row by $11^{-1} = 19$ (mod 26).

$\begin{bmatrix} 1 & 14 & 16 & 7 \\ 0 & 1 & 5 & 14 \end{bmatrix}$ Replace the entries in the 2nd row by their residues modulo 26.

$\begin{bmatrix} 1 & 0 & -54 & -189 \\ 0 & 1 & 5 & 14 \end{bmatrix}$ Add −14 times the second row to the first.

$\begin{bmatrix} 1 & 0 & 24 & 19 \\ 0 & 1 & 5 & 14 \end{bmatrix}$ Replace −54 and −189 by their residues modulo 26.

Thus $(A^{-1})^T = \begin{bmatrix} 24 & 19 \\ 5 & 14 \end{bmatrix}$, and so the deciphering matrix is $A^{-1} = \begin{bmatrix} 24 & 5 \\ 19 & 14 \end{bmatrix}$.

From Table 1 the numerical equivalent of the given ciphertext is

LN GI HG YB VR EN JY QO

12 14 7 9 8 7 25 2 22 18 5 14 10 25 17 15

To obtain the plaintext pairs, we multiply each ciphertext vector by A^{-1}:

$$\begin{aligned}
\begin{bmatrix} 24 & 5 \\ 19 & 14 \end{bmatrix}\begin{bmatrix} 12 \\ 14 \end{bmatrix} &= \begin{bmatrix} 358 \\ 424 \end{bmatrix} = \begin{bmatrix} 20 \\ 8 \end{bmatrix} && \begin{matrix} T \\ H \end{matrix} \\
\begin{bmatrix} 24 & 5 \\ 19 & 14 \end{bmatrix}\begin{bmatrix} 7 \\ 9 \end{bmatrix} &= \begin{bmatrix} 213 \\ 259 \end{bmatrix} = \begin{bmatrix} 5 \\ 25 \end{bmatrix} && \begin{matrix} E \\ Y \end{matrix} \\
\begin{bmatrix} 24 & 5 \\ 19 & 14 \end{bmatrix}\begin{bmatrix} 8 \\ 7 \end{bmatrix} &= \begin{bmatrix} 227 \\ 250 \end{bmatrix} = \begin{bmatrix} 19 \\ 16 \end{bmatrix} && \begin{matrix} S \\ P \end{matrix} \\
\begin{bmatrix} 24 & 5 \\ 19 & 14 \end{bmatrix}\begin{bmatrix} 25 \\ 2 \end{bmatrix} &= \begin{bmatrix} 610 \\ 503 \end{bmatrix} = \begin{bmatrix} 12 \\ 9 \end{bmatrix} && \begin{matrix} L \\ I \end{matrix} \pmod{26} \\
\begin{bmatrix} 24 & 5 \\ 19 & 14 \end{bmatrix}\begin{bmatrix} 22 \\ 18 \end{bmatrix} &= \begin{bmatrix} 618 \\ 670 \end{bmatrix} = \begin{bmatrix} 20 \\ 20 \end{bmatrix} && \begin{matrix} T \\ T \end{matrix} \\
\begin{bmatrix} 24 & 5 \\ 19 & 14 \end{bmatrix}\begin{bmatrix} 5 \\ 14 \end{bmatrix} &= \begin{bmatrix} 190 \\ 291 \end{bmatrix} = \begin{bmatrix} 8 \\ 5 \end{bmatrix} && \begin{matrix} H \\ E \end{matrix} \\
\begin{bmatrix} 24 & 5 \\ 19 & 14 \end{bmatrix}\begin{bmatrix} 10 \\ 25 \end{bmatrix} &= \begin{bmatrix} 365 \\ 540 \end{bmatrix} = \begin{bmatrix} 1 \\ 20 \end{bmatrix} && \begin{matrix} A \\ T \end{matrix} \\
\begin{bmatrix} 24 & 5 \\ 19 & 14 \end{bmatrix}\begin{bmatrix} 17 \\ 15 \end{bmatrix} &= \begin{bmatrix} 483 \\ 533 \end{bmatrix} = \begin{bmatrix} 15 \\ 13 \end{bmatrix} && \begin{matrix} O \\ M \end{matrix}
\end{aligned}$$

which yields the message *THEY SPLIT THE ATOM.*

6. Since we want a Hill 3-cipher, we will group the letters in triples. From Table 1 the numerical equivalents of the known plaintext are

I	H	A	V	E	C	O	M	E
9	8	1	22	5	3	15	13	5

and the numerical equivalent of the corresponding ciphertext are

H	P	A	F	Q	G	G	D	U
8	16	1	6	17	7	7	4	21

The corresponding plaintext and ciphertext vectors are

$$\mathbf{p}_1 = \begin{bmatrix} 9 \\ 8 \\ 1 \end{bmatrix} \leftrightarrow \mathbf{c}_1 = \begin{bmatrix} 8 \\ 16 \\ 1 \end{bmatrix}$$

$$\mathbf{p}_2 = \begin{bmatrix} 22 \\ 5 \\ 3 \end{bmatrix} \leftrightarrow \mathbf{c}_2 = \begin{bmatrix} 6 \\ 17 \\ 7 \end{bmatrix}$$

$$\mathbf{p}_3 = \begin{bmatrix} 15 \\ 13 \\ 5 \end{bmatrix} \leftrightarrow \mathbf{c}_3 = \begin{bmatrix} 7 \\ 4 \\ 21 \end{bmatrix}$$

We want to reduce $C = \begin{bmatrix} 8 & 16 & 1 \\ 6 & 17 & 7 \\ 7 & 4 & 21 \end{bmatrix}$ to I by elementary row operations and simultaneously apply these operations to $P = \begin{bmatrix} 9 & 8 & 1 \\ 22 & 5 & 3 \\ 15 & 13 & 5 \end{bmatrix}$. The calculations are as follows:

$\begin{bmatrix} 8 & 16 & 1 & 9 & 8 & 1 \\ 6 & 17 & 7 & 22 & 5 & 3 \\ 7 & 4 & 21 & 15 & 13 & 5 \end{bmatrix}$ Form the matrix $[C \quad P]$.

$\begin{bmatrix} 15 & 20 & 22 & 24 & 21 & 6 \\ 6 & 17 & 7 & 22 & 5 & 3 \\ 7 & 4 & 21 & 15 & 13 & 5 \end{bmatrix}$ Add the third row to the first since 8^{-1} does not exist modulo 26.

$\begin{bmatrix} 1 & 140 & 154 & 168 & 147 & 42 \\ 6 & 17 & 7 & 22 & 5 & 3 \\ 7 & 4 & 21 & 15 & 13 & 5 \end{bmatrix}$ Multiply the first row by $15^{-1} = 7 \pmod{26}$.

$\begin{bmatrix} 1 & 10 & 24 & 12 & 17 & 16 \\ 6 & 17 & 7 & 22 & 5 & 3 \\ 7 & 4 & 21 & 15 & 13 & 5 \end{bmatrix}$ Replace the entries in the first row by their residues modulo 26.

$\begin{bmatrix} 1 & 10 & 24 & 12 & 17 & 16 \\ 0 & -43 & -137 & -50 & -97 & -93 \\ 0 & -66 & -147 & -69 & -106 & -107 \end{bmatrix}$ Add –6 times the first row to the second and –7 times the first row to the third.

$\begin{bmatrix} 1 & 10 & 24 & 12 & 17 & 16 \\ 0 & 9 & 19 & 2 & 7 & 11 \\ 0 & 12 & 9 & 9 & 24 & 23 \end{bmatrix}$ Replace the entries in the second and third rows by their residues modulo 26.

$\begin{bmatrix} 1 & 10 & 24 & 12 & 17 & 16 \\ 0 & 1 & 57 & 6 & 21 & 33 \\ 0 & 12 & 9 & 9 & 24 & 23 \end{bmatrix}$ Multiply the second row by $9^{-1} = 3 \pmod{26}$.

$\begin{bmatrix} 1 & 10 & 24 & 12 & 17 & 16 \\ 0 & 1 & 5 & 6 & 21 & 7 \\ 0 & 12 & 9 & 9 & 24 & 23 \end{bmatrix}$ Replace the entries in the second row by their residues modulo 26.

$\begin{bmatrix} 1 & 0 & -26 & -48 & -193 & -54 \\ 0 & 1 & 5 & 6 & 21 & 7 \\ 0 & 0 & -51 & -63 & -228 & -61 \end{bmatrix}$ Add –10 times the second row to the first and –12 times the second row to the third.

$\begin{bmatrix} 1 & 0 & 0 & 4 & 15 & 24 \\ 0 & 1 & 5 & 6 & 21 & 7 \\ 0 & 0 & 1 & 15 & 6 & 17 \end{bmatrix}$ Replace the entries in the first and second row by their residues modulo 26.

$\begin{bmatrix} 1 & 0 & 0 & 4 & 15 & 24 \\ 0 & 1 & 0 & -69 & -9 & -78 \\ 0 & 0 & 1 & 15 & 6 & 17 \end{bmatrix}$ Add –5 times the third row to the second.

$$\begin{bmatrix} 1 & 0 & 0 & 4 & 15 & 24 \\ 0 & 1 & 0 & 9 & 17 & 0 \\ 0 & 0 & 1 & 15 & 6 & 17 \end{bmatrix}$$

Replace the entries in the second row by their residues modulo 26.

Thus, $(A^{-1})^T = \begin{bmatrix} 4 & 15 & 24 \\ 9 & 17 & 0 \\ 15 & 6 & 17 \end{bmatrix}$ and so the deciphering matrix is $A^{-1} = \begin{bmatrix} 4 & 9 & 15 \\ 15 & 17 & 6 \\ 24 & 0 & 17 \end{bmatrix}$.

From Table 1 the numerical equivalent of the given ciphertext is

H	P	A	F	Q	G	G	D	U	G	G	D	H	P	G	O	D	Y	N	O	R
8	16	1	6	17	7	7	4	21	7	4	4	8	16	7	15	4	25	14	15	18

To obtain the plaintext triples, we multiply each ciphertext vector by A^{-1}:

$$\begin{bmatrix} 4 & 9 & 15 \\ 15 & 17 & 6 \\ 24 & 0 & 17 \end{bmatrix}\begin{bmatrix} 8 \\ 16 \\ 1 \end{bmatrix} = \begin{bmatrix} 191 \\ 398 \\ 209 \end{bmatrix} = \begin{bmatrix} 9 \\ 8 \\ 1 \end{bmatrix} \quad \begin{matrix} I \\ H \\ A \end{matrix}$$

$$\begin{bmatrix} 4 & 9 & 15 \\ 15 & 17 & 6 \\ 24 & 0 & 17 \end{bmatrix}\begin{bmatrix} 6 \\ 17 \\ 7 \end{bmatrix} = \begin{bmatrix} 282 \\ 421 \\ 263 \end{bmatrix} = \begin{bmatrix} 22 \\ 5 \\ 3 \end{bmatrix} \quad \begin{matrix} V \\ E \\ C \end{matrix}$$

$$\begin{bmatrix} 4 & 9 & 15 \\ 15 & 17 & 6 \\ 24 & 0 & 17 \end{bmatrix}\begin{bmatrix} 7 \\ 4 \\ 21 \end{bmatrix} = \begin{bmatrix} 379 \\ 299 \\ 525 \end{bmatrix} = \begin{bmatrix} 15 \\ 13 \\ 5 \end{bmatrix} \quad \begin{matrix} O \\ M \\ E \end{matrix}$$

$$\begin{bmatrix} 4 & 9 & 15 \\ 15 & 17 & 6 \\ 24 & 0 & 17 \end{bmatrix}\begin{bmatrix} 7 \\ 4 \\ 4 \end{bmatrix} = \begin{bmatrix} 124 \\ 197 \\ 236 \end{bmatrix} = \begin{bmatrix} 20 \\ 15 \\ 2 \end{bmatrix} \quad \begin{matrix} T \\ O \\ B \end{matrix}$$

$$\begin{bmatrix} 4 & 9 & 15 \\ 15 & 17 & 6 \\ 24 & 0 & 17 \end{bmatrix}\begin{bmatrix} 8 \\ 16 \\ 7 \end{bmatrix} = \begin{bmatrix} 281 \\ 434 \\ 311 \end{bmatrix} = \begin{bmatrix} 21 \\ 18 \\ 25 \end{bmatrix} \quad \begin{matrix} U \\ R \\ Y \end{matrix}$$

$$\begin{bmatrix} 4 & 9 & 15 \\ 15 & 17 & 6 \\ 24 & 0 & 17 \end{bmatrix}\begin{bmatrix} 15 \\ 4 \\ 25 \end{bmatrix} = \begin{bmatrix} 471 \\ 443 \\ 785 \end{bmatrix} = \begin{bmatrix} 3 \\ 1 \\ 5 \end{bmatrix} \quad \begin{matrix} C \\ A \\ E \end{matrix}$$

$$\begin{bmatrix} 4 & 9 & 15 \\ 15 & 17 & 6 \\ 24 & 0 & 17 \end{bmatrix}\begin{bmatrix} 14 \\ 15 \\ 18 \end{bmatrix} = \begin{bmatrix} 461 \\ 573 \\ 642 \end{bmatrix} = \begin{bmatrix} 19 \\ 1 \\ 18 \end{bmatrix} \quad \begin{matrix} S \\ A \\ R \end{matrix}$$

Finally, the message is *I HAVE COME TO BURY CAESAR.*

7. **(a)** Multiply each of the triples of the message by $A = \begin{bmatrix} 1 & 1 & 0 \\ 0 & 1 & 1 \\ 1 & 1 & 1 \end{bmatrix}$ and reduce the results modulo 2.

$$\begin{bmatrix} 1 & 1 & 0 \\ 0 & 1 & 1 \\ 1 & 1 & 1 \end{bmatrix}\begin{bmatrix} 1 \\ 1 \\ 0 \end{bmatrix} = \begin{bmatrix} 2 \\ 1 \\ 2 \end{bmatrix} = \begin{bmatrix} 0 \\ 1 \\ 0 \end{bmatrix}$$

$$\begin{bmatrix} 1 & 1 & 0 \\ 0 & 1 & 1 \\ 1 & 1 & 1 \end{bmatrix}\begin{bmatrix} 1 \\ 0 \\ 1 \end{bmatrix} = \begin{bmatrix} 1 \\ 1 \\ 2 \end{bmatrix} = \begin{bmatrix} 1 \\ 1 \\ 0 \end{bmatrix}$$

$$\begin{bmatrix} 1 & 1 & 0 \\ 0 & 1 & 1 \\ 1 & 1 & 1 \end{bmatrix}\begin{bmatrix} 1 \\ 1 \\ 1 \end{bmatrix} = \begin{bmatrix} 2 \\ 2 \\ 3 \end{bmatrix} = \begin{bmatrix} 0 \\ 0 \\ 1 \end{bmatrix}$$

The encoded message is 010110001.

(b) Reduce $[A \quad I]$ to $[I \quad A^{-1}]$ modulo 2.

$$\begin{bmatrix} 1 & 1 & 0 & 1 & 0 & 0 \\ 0 & 1 & 1 & 0 & 1 & 0 \\ 1 & 1 & 1 & 0 & 0 & 1 \end{bmatrix}$$ Form the matrix $[A \quad I]$.

$$\begin{bmatrix} 1 & 1 & 0 & 1 & 0 & 0 \\ 0 & 1 & 1 & 0 & 1 & 0 \\ 2 & 2 & 1 & 1 & 0 & 1 \end{bmatrix}$$ Add the first row to the third row.

$$\begin{bmatrix} 1 & 1 & 0 & 1 & 0 & 0 \\ 0 & 1 & 1 & 0 & 1 & 0 \\ 0 & 0 & 1 & 1 & 0 & 1 \end{bmatrix}$$ Replace 2 by its residue modulo 2.

$$\begin{bmatrix} 1 & 1 & 0 & 1 & 0 & 0 \\ 0 & 1 & 2 & 1 & 1 & 1 \\ 0 & 0 & 1 & 1 & 0 & 1 \end{bmatrix}$$ Add the third row to the second row.

$$\begin{bmatrix} 1 & 1 & 0 & 1 & 0 & 0 \\ 0 & 1 & 0 & 1 & 1 & 1 \\ 0 & 0 & 1 & 1 & 0 & 1 \end{bmatrix}$$ Replace 2 by its residue modulo 2.

$$\begin{bmatrix} 1 & 2 & 0 & 2 & 1 & 1 \\ 0 & 1 & 0 & 1 & 1 & 1 \\ 0 & 0 & 1 & 1 & 0 & 1 \end{bmatrix}$$ Add the second row to the first row.

$$\begin{bmatrix} 1 & 0 & 0 & 0 & 1 & 1 \\ 0 & 1 & 0 & 1 & 1 & 1 \\ 0 & 0 & 1 & 1 & 0 & 1 \end{bmatrix}$$ Replace 2 by its residue modulo 2.

Thus $A^{-1} = \begin{bmatrix} 0 & 1 & 1 \\ 1 & 1 & 1 \\ 1 & 0 & 1 \end{bmatrix}$.

$$\begin{bmatrix} 0 & 1 & 1 \\ 1 & 1 & 1 \\ 1 & 0 & 1 \end{bmatrix}\begin{bmatrix} 0 \\ 1 \\ 0 \end{bmatrix} = \begin{bmatrix} 1 \\ 1 \\ 0 \end{bmatrix}$$

$$\begin{bmatrix} 0 & 1 & 1 \\ 1 & 1 & 1 \\ 1 & 0 & 1 \end{bmatrix}\begin{bmatrix} 1 \\ 1 \\ 0 \end{bmatrix} = \begin{bmatrix} 1 \\ 2 \\ 1 \end{bmatrix} = \begin{bmatrix} 1 \\ 0 \\ 1 \end{bmatrix}$$

$$\begin{bmatrix} 0 & 1 & 1 \\ 1 & 1 & 1 \\ 1 & 0 & 1 \end{bmatrix}\begin{bmatrix} 0 \\ 0 \\ 1 \end{bmatrix} = \begin{bmatrix} 1 \\ 1 \\ 1 \end{bmatrix}$$

The decoded message is 110101111, which is the original message.

8. Since 29 is a prime number, by Corollary 10.15.2 a matrix A with entries in Z_{29} is invertible if and only if $\det(A) \neq 0 \pmod{29}$.

10.15 Genetics

1. Use induction on n, the case $n = 1$ being already given. If the result is true for $n - 1$, then

$$M^n = M^{n-1}M = (PD^{n-1}P^{-1})(PDP^{-1}) = PD^{n-1}(P^{-1}P)DP^{-1} = PD^{n-1}DP^{-1} = PD^nP^{-1},$$

proving the result.

2. Using Table 1 and notations of Example 1, we derive the following equations:

$$a_n = \frac{1}{2}a_{n-1} + \frac{1}{4}b_{n-1}$$
$$b_n = \frac{1}{2}a_{n-1} + \frac{1}{2}b_{n-1} + \frac{1}{2}c_{n-1}$$
$$c_n = \frac{1}{4}b_{n-1} + \frac{1}{2}c_{n-1}.$$

The transition matrix is thus $M = \begin{bmatrix} \frac{1}{2} & \frac{1}{4} & 0 \\ \frac{1}{2} & \frac{1}{2} & \frac{1}{2} \\ 0 & \frac{1}{4} & \frac{1}{2} \end{bmatrix}$. The characteristic polynomial of M is

$$\det(\lambda I - M) = \lambda^3 - \left(\frac{3}{2}\right)\lambda^2 + \left(\frac{1}{2}\right)\lambda = \lambda(\lambda - 1)\left(\lambda - \frac{1}{2}\right),$$

so the eigenvalues of M are $\lambda = 1$, $\lambda_2 = \frac{1}{2}$, and $\lambda_3 = 0$. Corresponding eigenvectors (found by solving $(\lambda I - M)\mathbf{x} = 0$) are $\mathbf{e}_1 = [1 \quad 2 \quad 1]^T$, $\mathbf{e}_2 = [1 \quad 0 \quad -1]^T$, and $\mathbf{e}_3 = [1 \quad -2 \quad 1]^T$. Thus

$$M^n = PD^nP^{-1} = \begin{bmatrix} 1 & 1 & 1 \\ 2 & 0 & -2 \\ 1 & -1 & 1 \end{bmatrix} \begin{bmatrix} 1 & 0 & 0 \\ 0 & \left(\frac{1}{2}\right)^n & 0 \\ 0 & 0 & 0 \end{bmatrix} \begin{bmatrix} \frac{1}{4} & \frac{1}{4} & \frac{1}{4} \\ \frac{1}{2} & 0 & -\frac{1}{2} \\ \frac{1}{4} & -\frac{1}{4} & \frac{1}{4} \end{bmatrix}.$$

This yields

$$\mathbf{x}^{(n)} = \begin{bmatrix} a_n \\ b_n \\ c_n \end{bmatrix} = \begin{bmatrix} \frac{1}{4} + \left(\frac{1}{2}\right)^{n+1} & \frac{1}{4} & \frac{1}{4} - \left(\frac{1}{2}\right)^{n+1} \\ \frac{1}{2} & \frac{1}{2} & \frac{1}{2} \\ \frac{1}{4} - \left(\frac{1}{2}\right)^{n+1} & \frac{1}{4} & \frac{1}{4} + \left(\frac{1}{2}\right)^{n+1} \end{bmatrix} \begin{bmatrix} a_0 \\ b_0 \\ c_0 \end{bmatrix}.$$

Remembering that $a_0 + b_0 + c_0 = 1$, we obtain

$$a_n = \frac{1}{4}a_0 + \frac{1}{4}b_0 + \frac{1}{4}c_0 + \left(\frac{1}{2}\right)^{n+1} a_0 - \left(\frac{1}{2}\right)^{n+1} c_0 = \frac{1}{4} + \left(\frac{1}{2}\right)^{n+1} (a_0 - c_0)$$
$$b_n = \frac{1}{2}a_0 + \frac{1}{2}b_0 + \frac{1}{2}c_0 = \frac{1}{2}$$

$$c_n = \frac{1}{4}a_0 + \frac{1}{4}b_0 + \frac{1}{2}c_0 - \left(\frac{1}{2}\right)^{n+1} a_0 + \left(\frac{1}{2}\right)^{n+1} c_0 = \frac{1}{4} - \left(\frac{1}{2}\right)^{n+1} (a_0 - c_0).$$

Since $\left(\frac{1}{2}\right)^{n+1}$ approaches zero as $n \to \infty$, we obtain $a_n \to \frac{1}{4}$, $b_n \to \frac{1}{2}$, and $c_n \to \frac{1}{4}$ as $n \to \infty$.

3. Call M_1 the matrix of Example 1, and M_2 the matrix of Exercise 2. Then $\mathbf{x}^{(2n)} = (M_2M_1)^n\mathbf{x}^{(0)}$ and $\mathbf{x}^{(2n+1)} = M_1(M_2M_1)^n\mathbf{x}^{(0)}$. We have

$$M_2M_1 = \begin{bmatrix} \frac{1}{2} & \frac{1}{4} & 0 \\ \frac{1}{2} & \frac{1}{2} & \frac{1}{2} \\ 0 & \frac{1}{4} & \frac{1}{2} \end{bmatrix} \begin{bmatrix} 1 & \frac{1}{2} & 0 \\ 0 & \frac{1}{2} & 1 \\ 0 & 0 & 0 \end{bmatrix} = \begin{bmatrix} \frac{1}{2} & \frac{3}{8} & \frac{1}{4} \\ \frac{1}{2} & \frac{1}{2} & \frac{1}{2} \\ 0 & \frac{1}{8} & \frac{1}{4} \end{bmatrix}.$$

The characteristic polynomial of this matrix is $\lambda^3 - \frac{5}{4}\lambda^2 + \frac{1}{4}\lambda$, so the eigenvalues are $\lambda_1 = 1$, $\lambda_2 = \frac{1}{4}$, $\lambda_3 = 0$. Corresponding eigenvectors are $\mathbf{e}_1 = [5 \quad 6 \quad 1]^T$, $\mathbf{e}_2 = [-1 \quad 0 \quad 1]^T$, and $\mathbf{e}_3 = [1 \quad -2 \quad 1]^T$. Thus,

$$(M_2M_1)^n = PD^nP^{-1} = \begin{bmatrix} 5 & -1 & 1 \\ 6 & 0 & -2 \\ 1 & 1 & 1 \end{bmatrix} \begin{bmatrix} 1 & 0 & 0 \\ 0 & \left(\frac{1}{4}\right)^n & 0 \\ 0 & 0 & 0 \end{bmatrix} \begin{bmatrix} \frac{1}{12} & \frac{1}{12} & \frac{1}{12} \\ -\frac{1}{3} & \frac{1}{6} & \frac{2}{3} \\ \frac{1}{4} & -\frac{1}{4} & \frac{1}{4} \end{bmatrix}.$$

Using the notation of Example 1 (recall that $a_0 + b_0 + c_0 = 1$), we obtain

$$\begin{aligned} a_{2n} &= \frac{5}{12} + \frac{1}{6 \cdot 4^n}(2a_0 - b_0 - 4c_0) \\ b_{2n} &= \frac{1}{2} \\ c_{2n} &= \frac{1}{12} - \frac{1}{6 \cdot 4^n}(2a_0 - b_0 - 4c_0) \end{aligned}$$

and

$$\begin{aligned} a_{2n+1} &= \frac{2}{3} + \frac{1}{6 \cdot 4^n}(2a_0 - b_0 - 4c_0) \\ b_{2n+1} &= \frac{1}{3} - \frac{1}{6 \cdot 4^n}(2a_0 - b_0 - 4c_0) \\ c_{2n+1} &= 0. \end{aligned}$$

4. The characteristic polynomial of M is $(\lambda - 1)\left(\lambda - \frac{1}{2}\right)$, so the eigenvalues are $\lambda_1 = 1$ and $\lambda_2 = \frac{1}{2}$. Corresponding eigenvectors are easily found to be $\mathbf{e}_1 = [1 \quad 0]^T$ and $\mathbf{e}_2 = [1 \quad -1]^T$. From this point, the verification of Equation (7) is in the text.

5. From Equation (9), if $b_0 = .25 = \frac{1}{4}$, we get $b_1 = \frac{\frac{1}{4}}{\frac{9}{8}} = \frac{2}{9}$, then $b_2 = \frac{\frac{2}{9}}{\frac{10}{9}} = \frac{1}{5}$, $b_3 = \frac{\frac{1}{5}}{\frac{11}{10}} = \frac{2}{11}$, and, in general, $b_n = \frac{2}{8+n}$. We will reach $\frac{2}{20} = .10$ in 12 generations. According to Equation (8), under the controlled program the percentage would be $\frac{1}{2^{14}}$ in 12 generations, or $\frac{1}{16{,}384} \approx .00006 =$ .006%.

6. $P^{-1}\mathbf{x}^{(0)} = \begin{bmatrix} \frac{1}{2} \\ \frac{1}{2} \\ 0 \\ 0 \\ \frac{1}{12}(1+\sqrt{5}) \\ \frac{1}{12}(1-\sqrt{5}) \end{bmatrix}$, $D^nP^{-1}\mathbf{x}^{(0)} = \begin{bmatrix} \frac{1}{2} \\ \frac{1}{2} \\ 0 \\ 0 \\ \frac{1}{3}\left[\frac{1}{4}(1+\sqrt{5})\right]^{n+1} \\ \frac{1}{3}\left[\frac{1}{4}(1-\sqrt{5})\right]^{n+1} \end{bmatrix}$

$$PD^nP^{-1}\mathbf{x}^{(0)} = \begin{bmatrix} \frac{1}{2} + \frac{1}{3}\cdot\frac{1}{4^{n+2}}\left[(-3-\sqrt{5})(1+\sqrt{5})^{n+1} + (-3+\sqrt{5})(1-\sqrt{5})^{n+1}\right] \\ \frac{1}{3}\cdot\frac{1}{4^{n+1}}\left[(1+\sqrt{5})^{n+1} + (1-\sqrt{5})^{n+1}\right] \\ \frac{1}{3}\cdot\frac{1}{4^{n+1}}\left[(1+\sqrt{5})^{n} + (1-\sqrt{5})^{n}\right] \\ \frac{1}{3}\cdot\frac{1}{4^{n+1}}\left[(1+\sqrt{5})^{n} + (1-\sqrt{5})^{n}\right] \\ \frac{1}{3}\cdot\frac{1}{4^{n+1}}\left[(1+\sqrt{5})^{n+1} + (1-\sqrt{5})^{n+1}\right] \\ \frac{1}{2} + \frac{1}{3}\cdot\frac{1}{4^{n+2}}\left[(-3-\sqrt{5})(1+\sqrt{5})^{n+1} + (-3+\sqrt{5})(1-\sqrt{5})^{n+1}\right] \end{bmatrix}$$

As n tends to infinity, $\frac{1}{4^{n+2}}$ and $\frac{1}{4^{n+1}}$ approach 0, so $\mathbf{x}^{(n)}$ approaches $\begin{bmatrix} \frac{1}{2} \\ 0 \\ 0 \\ 0 \\ 0 \\ \frac{1}{2} \end{bmatrix}$.

7. From (13) we have that the probability that the limiting sibling-pairs will be type (A, AA) is $a_0 + \frac{2}{3}b_0 + \frac{1}{3}c_0 + \frac{2}{3}d_0 + \frac{1}{3}e_0$.
The proportion of A genes in the population at the outset is as follows: all the type (A, AA) genes, $\frac{2}{3}$ of the type (A, Aa) genes, $\frac{1}{3}$ the type (A, aa) genes, etc. ...yielding

$$a_0 + \frac{2}{3}b_0 + \frac{1}{3}c_0 + \frac{2}{3}d_0 + \frac{1}{3}e_0.$$

8. From an (A, AA) pair we get only (A, AA) pairs and similarly for (a, aa). From either (A, aa) or (a, AA) pairs we must get an Aa female, who will not mature. Thus no offspring will come from such pairs. The transition matrix is then $\begin{bmatrix} 1 & 0 & 0 & 0 \\ 0 & 0 & 0 & 0 \\ 0 & 0 & 0 & 0 \\ 0 & 0 & 0 & 1 \end{bmatrix}$.

9. For the first column of M we realize that parents of type (A, AA) can produce offspring only of that type, and similarly for the last column. The fifth column is like the second column, and follows the analysis in the text. For the middle two columns, say the third, note that male offspring from (A, aa) must be of type a, and females are of type Aa, because of the way the genes are inherited.

10.16 Age-Specific Population Growth

1. **(a)** The characteristic polynomial of L is $\lambda^2 - \lambda - \frac{3}{4}$, so the eigenvalues of L are $\lambda = \frac{3}{2}$ and $\lambda = -\frac{1}{2}$, thus $\lambda_1 = \frac{3}{2}$ and $\mathbf{x}_1 = \begin{bmatrix} 1 \\ \frac{1}{\frac{3}{2}} \end{bmatrix} = \begin{bmatrix} 1 \\ \frac{1}{3} \end{bmatrix}$ is the corresponding eigenvector.

(b) $\mathbf{x}^{(1)} = L\mathbf{x}^{(0)} = \begin{bmatrix} 100 \\ 50 \end{bmatrix}$, $\mathbf{x}^{(2)} = L\mathbf{x}^{(1)} = \begin{bmatrix} 175 \\ 50 \end{bmatrix}$, $\mathbf{x}^{(3)} \approx L\mathbf{x}^{(2)} \approx \begin{bmatrix} 250 \\ 88 \end{bmatrix}$, $\mathbf{x}^{(4)} \approx L\mathbf{x}^{(3)} \approx \begin{bmatrix} 382 \\ 125 \end{bmatrix}$, $\mathbf{x}^5 \approx L\mathbf{x}^{(4)} \approx \begin{bmatrix} 570 \\ 191 \end{bmatrix}$

(c) $\mathbf{x}^{(6)} \approx L\mathbf{x}^{(5)} \approx \begin{bmatrix} 857 \\ 285 \end{bmatrix}$; $\mathbf{x}^{(6)} \approx \lambda_1 \mathbf{x}^{(5)} \approx \begin{bmatrix} 855 \\ 287 \end{bmatrix}$

5. a_1 is the average number of offspring produced in the first age period. $a_2 b_1$ is the number of offspring produced in the second period times the probability that the female will live into the second period, i.e., it is the expected number of offspring per female during the second period, and so on for all the periods. Thus, the sum of these, which is the net reproduction rate, is the expected number of offspring produced by a given female during her expected lifetime.

7. $R = 0 + 4\left(\frac{1}{2}\right) + 3\left(\frac{1}{2}\right)\left(\frac{1}{4}\right) = \frac{19}{8} = 2.375$

8. $R = 0 + (.00024)(.99651) + \cdots + (.00240)(.99651)\cdots(.987) = 1.49611.$

10.17 Harvesting of Animal Populations

1. **(a)** The characteristic polynomial of L is $\lambda^3 - 2\lambda - \frac{3}{8} = \left(\lambda - \frac{3}{2}\right)\left[\lambda^2 + \left(\frac{3}{2}\right)\lambda + \frac{1}{4}\right]$, so $\lambda_1 = \frac{3}{2}$. Thus h, the fraction harvested of each age group, is $1 - \frac{2}{3} = \frac{1}{3}$ so the yield is $33\frac{1}{3}\%$ of the population. The eigenvector corresponding to $\lambda = \frac{3}{2}$ is $\begin{bmatrix} 1 \\ \frac{1}{3} \\ \frac{1}{18} \end{bmatrix}$; this is the age distribution vector after each harvest.

(b) From Equation (10), the age distribution vector $\mathbf{x}_1$ is $\begin{bmatrix} 1 & \frac{1}{2} & \frac{1}{8} \end{bmatrix}^T$. Equation (9) tells us that $h_1 = 1 - \frac{1}{\frac{19}{8}} = \frac{11}{19}$, so we harvest $\frac{11}{19}$ or 57.9% of the youngest age class. Since

$L\mathbf{x}_1 = \begin{bmatrix} \frac{19}{8} & \frac{1}{2} & \frac{1}{8} \end{bmatrix}^T$, the youngest class contains 79.2% of the population. Thus the yield is 57.9% of 79.2%, or 45.8% of the population.

2. The Leslie matrix of Example 1 has $b_1 = .845$, $b_2 = .975$, $b_3 = .965$, etc. This, together with the harvesting data from Equations (13) and the formula of Equation (5) yields

$$\mathbf{x}_1 = [1 \quad .845 \quad .824 \quad .795 \quad .755 \quad .699 \quad .626 \quad .5323 \quad 0 \quad 0 \quad 0 \quad 0]^T$$
$$L\mathbf{x}_1 = [2.090 \quad .845 \quad .824 \quad .795 \quad .755 \quad .699 \quad .626 \quad .532 \quad .418 \quad 0 \quad 0 \quad 0]^T.$$

The total of the entries of $L\mathbf{x}_1$ is 7.584. The proportion of sheep harvested is $h_1(L\mathbf{x}_1)_2 + h_9(L\mathbf{x}_1)_9 = 1.51$, or 19.9% of the population.

4. In this situation we have $h_I \neq 0$, and $h_1 = h_2 = \cdots = h_{I-1} = h_{I+1} = \cdots = h_n = 0$. Equation (4) then takes the form $a_1 + a_2b_1 + a_3b_1b_2 + \cdots + a_Ib_1b_2 \cdots b_{I-1}(1 - h_I) + a_{I+1}b_1b_2 \cdots b_I(1 - h_I) + \cdots + a_nb_1b_2 \cdots b_{n-1}(1 - h_I) = 1$.

$$(1 - h_I)[a_Ib_1b_2 \cdots b_{I-1} + a_{I+1}b_1b_2 \cdots b_I + \cdots + a_nb_1b_2 \cdots b_{n-1}]$$
$$= 1 - a_1 - a_2b_1 - \cdots - a_{I-1}b_1b_2 \cdots b_{I-2}$$

So,

$$h_I = \frac{a_1 + a_2b_1 + \cdots + a_{I-1}b_1b_2 \cdots b_{I-2} - 1}{a_Ib_1b_2 \cdots b_{I-1} + \cdots + a_nb_1b_2 \cdots b_{n-1}} + 1$$
$$= \frac{a_1 + a_2b_1 + \cdots + a_{I-1}b_1b_2 \cdots b_{I-2} - 1 + a_Ib_1b_2 \cdots b_{I-1} + \cdots + a_nb_1b_2 \cdots b_{n-1}}{a_Ib_1b_2 \cdots b_{I-1} + \cdots + a_nb_1b_2 \cdots b_{n-1}}$$
$$= \frac{R - 1}{a_Ib_1b_2 \cdots b_{I-1} + \cdots + a_nb_1b_2 \cdots b_{n-1}}$$

5. Here $h_J = 1$, $h_I \neq 0$, and all the other h_k's are zero. Then Equation (4) becomes

$$a_1 + a_2b_1 + \cdots + a_{I-1}b_1b_2 \cdots b_{I-2} + (1 - h_I)[a_Ib_1b_2 \cdots b_{I-1} + \cdots + a_{J-1}b_1b_2 \cdots b_{J-2}] = 1.$$

We solve for h_I to obtain

$$h_I = \frac{a_1 + a_2b_1 + \cdots + a_{I-1}b_1b_2 \cdots b_{I-2} - 1}{a_Ib_1b_2 \cdots b_{I-1} + \cdots + a_{J-1}b_1b_2 \cdots b_{J-2}} + 1 = \frac{a_1 + a_2b_1 + \cdots + a_{J-1}b_1b_2 \cdots b_{J-2} - 1}{a_Ib_1b_2 \cdots b_{I-1} + \cdots + a_{J-1}b_1b_2 \cdots b_{J-2}}.$$

10.18 A Least Squares Model for Human Hearing

1. From Theorem 10.18.1, we compute $a_0 = \frac{1}{\pi}\int_0^{2\pi}(t - \pi)^2 dt = \frac{2}{3}\pi^2$, $a_k = \frac{1}{\pi}\int_0^{2\pi}(t - \pi)^2 \cos kt\, dt = \frac{4}{k^2}$, and $b_k = \frac{1}{\pi}\int_0^{2\pi}(t - \pi)^2 \sin kt\, dt = 0$.

So the least-squares trigonometric polynomial of order 3 is $\frac{\pi^2}{3} + 4\cos t + \cos 2t + \frac{4}{9}\cos 3t$.

2. From Theorem 10.18.2, we compute $a_0 = \frac{2}{T}\int_0^T t^2 dt = \frac{2}{3}T^2$, $a_k = \frac{2}{T}\int_0^T t^2 \cos\frac{2k\pi t}{T} dt = \frac{T^2}{k^2\pi^2}$ and $b_k = \frac{2}{T}\int_0^T t^2 \sin\frac{2k\pi t}{T} dt = -\frac{T^2}{k\pi}$.

So the least-squares trigonometric polynomial of order 4 is

$$\frac{T^2}{3}+\frac{T^2}{\pi^2}\left(\cos\frac{2\pi t}{T}+\frac{1}{4}\cos\frac{4\pi t}{T}+\frac{1}{9}\cos\frac{6\pi t}{T}+\frac{1}{16}\cos\frac{8\pi t}{T}\right)$$
$$-\frac{T^2}{\pi}\left(\sin\frac{2\pi t}{T}+\frac{1}{2}\sin\frac{4\pi t}{T}+\frac{1}{3}\sin\frac{6\pi t}{T}+\frac{1}{4}\sin\frac{8\pi t}{T}\right).$$

3. From Theorem 10.18.2, $a_0=\frac{1}{\pi}\int_0^{2\pi}f(t)dt=\frac{1}{\pi}\int_0^{\pi}\sin t\,dt=\frac{2}{\pi}$. (Note the upper limit on the second integral),

$$a_k=\frac{1}{\pi}\int_0^{\pi}\sin t\cos kt\,dt=\frac{1}{\pi}\left(\frac{1}{k^2-1}[k\sin kt\sin t+\cos kt\cos t]\right)\Big|_0^{\pi}$$
$$=\frac{1}{\pi}\left(\frac{1}{k^2-1}[0+(-1)^{k-1}-1]\right)=\begin{cases}0 & \text{if } k \text{ is odd}\\ -\dfrac{2}{\pi(k^2-1)} & \text{if } k \text{ is even.}\end{cases}$$

$$b_k=\frac{1}{\pi}\int_0^{\pi}\sin kt\sin t\,dt=\begin{cases}\frac{1}{2} & \text{if } k=1\\ 0 & \text{if } k>1\end{cases}.$$

So the least-squares trigonometric polynomial of order 4 is $\frac{1}{\pi}+\frac{1}{2}\sin t-\frac{2}{3\pi}\cos 2t-\frac{2}{15\pi}\cos 4t.$

4. From Theorem 10.18.2, $a_0=\frac{1}{\pi}\int_0^{2\pi}\sin\frac{1}{2}t\,dt=\frac{4}{\pi}$.

$$a_k=\frac{1}{\pi}\int_0^{2\pi}\sin\frac{1}{2}t\cos kt\,dt=-\frac{4}{\pi(4k^2-1)}=-\frac{4}{\pi(2k-1)(2k+1)},$$
$$b_k=\frac{1}{\pi}\int_0^{2\pi}\sin\frac{1}{2}t\sin kt\,dt=0.$$

So the least-square trigonometric polynomial of order n is

$$\frac{2}{\pi}-\frac{4}{\pi}\left(\frac{\cos t}{1\cdot 3}+\frac{\cos 2t}{3\cdot 5}+\frac{\cos 3t}{5\cdot 7}+\cdots+\frac{\cos nt}{(2n-1)(2n+1)}\right).$$

5. From Theorem 10.18.2, $a_0=\frac{2}{T}\int_0^T f(t)dt=\frac{2}{T}\int_0^{\frac{1}{2}T}t\,dt+\frac{2}{T}\int_{\frac{1}{2}T}^{T}(T-t)dt=\frac{T}{2}$.

$$a_k=\frac{2}{T}\int_0^{\frac{1}{2}T}t\cos\frac{2k\pi t}{T}dt+\frac{2}{T}\int_{\frac{1}{2}T}^{T}(T-t)\cos\frac{2k\pi t}{T}dt=\frac{4T}{4k^2\pi^2}((-1)^k-1)=\begin{cases}0 & \text{if } k \text{ is even}\\ \frac{8T}{(2k)^2\pi^2} & \text{if } k \text{ is odd}\end{cases}$$

$$b_k=\frac{2}{T}\int_0^{\frac{1}{2}T}t\sin\frac{2k\pi t}{T}dt+\frac{2}{T}\int_{\frac{1}{2}T}^{T}(T-t)\sin\frac{2k\pi t}{T}dt=\frac{T(-1)^{k+1}}{2k\pi}+\frac{T(-1)^k}{2k\pi}=\frac{T(-1)^k}{2k\pi}(-1+1)=0$$

So the least-squares trigonometric polynomial of order n is:

$$\frac{T}{4}-\frac{8T}{\pi^2}\left(\frac{1}{2^2}\cos\frac{2\pi t}{T}+\frac{1}{6^2}\cos\frac{6\pi t}{T}+\frac{1}{10^2}\cos\frac{10\pi t}{T}+\cdots+\frac{1}{(2n)^2}\cos\frac{2\pi nt}{T}\right)$$

if n is even; the last term involves $n-1$ if n is odd.

6. **(a)** $\|1\|=\sqrt{\int_0^{2\pi}dt}=\sqrt{2\pi}$

(b) $\|\cos kt\| = \sqrt{\int_0^{2\pi} \cos^2 kt\, dt} = \sqrt{\int_0^{2\pi} \left(\frac{1+\cos 2t}{2}\right) dt} = \sqrt{\pi}$

(c) $\|\sin kt\| = \sqrt{\int_0^{2\pi} \sin^2 kt\, dt} = \sqrt{\int_0^{2\pi} \left(\frac{1-\cos 2t}{2}\right) dt} = \sqrt{\pi}$

10.19 Warps and Morphs

1. (a) Equation (2) is $c_1\begin{bmatrix}1\\1\end{bmatrix} + c_2\begin{bmatrix}3\\5\end{bmatrix} + c_3\begin{bmatrix}4\\2\end{bmatrix} = \begin{bmatrix}3\\3\end{bmatrix}$ and Equation (3) is $c_1 + c_2 + c_3 = 1$. These equations can be written in combined matrix form as $\begin{bmatrix}1 & 3 & 4\\1 & 5 & 2\\1 & 1 & 1\end{bmatrix}\begin{bmatrix}c_1\\c_2\\c_3\end{bmatrix} = \begin{bmatrix}3\\3\\1\end{bmatrix}$.

   This system has the unique solution $c_1 = \frac{1}{5}, c_2 = \frac{2}{5}$, and $c_3 = \frac{2}{5}$. Because these coefficients are all nonnegative, it follows that $\mathbf{v}$ is a convex combination of the vectors $\mathbf{v}_1$, $\mathbf{v}_2$, and $\mathbf{v}_3$.

   (b) As in part (a) the system for c_1, c_2, and c_3 is $\begin{bmatrix}1 & 3 & 4\\1 & 5 & 2\\1 & 1 & 1\end{bmatrix}\begin{bmatrix}c_1\\c_2\\c_3\end{bmatrix} = \begin{bmatrix}2\\4\\1\end{bmatrix}$ which has the unique solution $c_1 = \frac{2}{5}, c_2 = \frac{4}{5}$, and $c_3 = -\frac{1}{5}$. Because one of these coefficients is negative, it follows that $\mathbf{v}$ is not a convex combination of the vectors $\mathbf{v}_1$, $\mathbf{v}_2$, and $\mathbf{v}_3$.

   (c) As in part (a) the system for c_1, c_2, and c_3 is $\begin{bmatrix}3 & -2 & 3\\3 & -2 & 0\\1 & 1 & 1\end{bmatrix}\begin{bmatrix}c_1\\c_2\\c_3\end{bmatrix} = \begin{bmatrix}0\\0\\1\end{bmatrix}$ which has the unique solution $c_1 = \frac{2}{5}, c_2 = \frac{3}{5}$, and $c_3 = 0$. Because these coefficients are all nonnegative, it follows that $\mathbf{v}$ is a convex combination of the vectors $\mathbf{v}_1$, $\mathbf{v}_2$, and $\mathbf{v}_3$.

   (d) As in part (a) the system for c_1, c_2, and c_3 is $\begin{bmatrix}3 & -2 & 3\\3 & -2 & 0\\1 & 1 & 1\end{bmatrix}\begin{bmatrix}c_1\\c_2\\c_3\end{bmatrix} = \begin{bmatrix}1\\0\\1\end{bmatrix}$ which has the unique solution $c_1 = \frac{4}{15}, c_2 = \frac{6}{15}$, and $c_3 = \frac{5}{15}$. Because these coefficients are all nonnegative, it follows that $\mathbf{v}$ is a convex combination of the vectors $\mathbf{v}_1$, $\mathbf{v}_2$, and $\mathbf{v}_3$.

2. For both triangulations the number of triangles, m, is equal to 7; the number of vertex points, n, is equal to 7; and the number of boundary vertex points, k, is equal to 5. Equation (7), $m = 2n - 2 - k$, becomes $7 = 2(7) - 2 - 5$, or $7 = 7$.

3. Combining everything that is given in the statement of the problem, we obtain:

$$\begin{aligned}\mathbf{w} = M\mathbf{v} + \mathbf{b} &= M(c_1\mathbf{v}_1 + c_2\mathbf{v}_2 + c_3\mathbf{v}_3) + (c_1 + c_2 + c_3)\mathbf{b}\\ &= c_1(M\mathbf{v}_1 + \mathbf{b}) + c_2(M\mathbf{v}_2 + \mathbf{b}) + c_3(M\mathbf{v}_3 + \mathbf{b}) = c_1\mathbf{w}_1 + c_2\mathbf{w}_2 + c_3\mathbf{w}_3.\end{aligned}$$

4. (a)

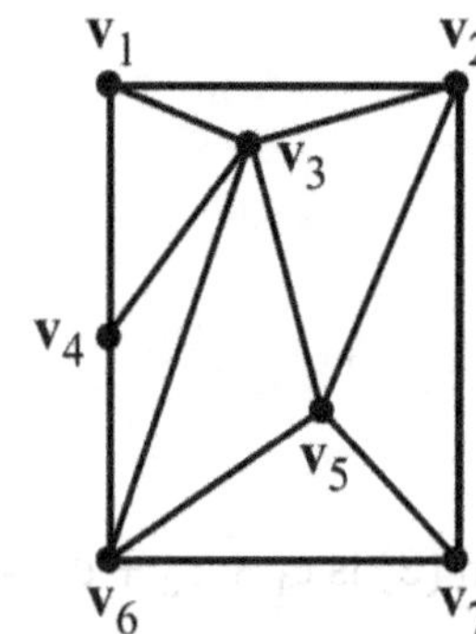

(b)

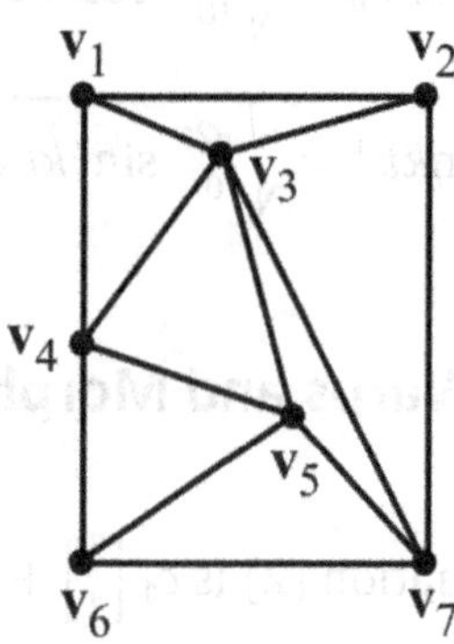

5. (a) Let $M = \begin{bmatrix} m_{11} & m_{12} \\ m_{21} & m_{22} \end{bmatrix}$ and $\mathbf{b} = \begin{bmatrix} b_1 \\ b_2 \end{bmatrix}$. Then the three matrix equations $M\mathbf{v}_i + \mathbf{b} = \mathbf{w}_i$, $i = 1, 2, 3$, can be written as the six scalar equations

$$\begin{aligned} m_{11} + m_{12} + b_1 &= 4 \\ m_{21} + m_{22} + b_2 &= 3 \end{aligned}$$

$$\begin{aligned} 2m_{11} + 3m_{12} + b_1 &= 9 \\ 2m_{21} + 3m_{22} + b_2 &= 5 \end{aligned}$$

$$\begin{aligned} 2m_{11} + m_{12} + b_1 &= 5 \\ 2m_{21} + m_{22} + b_2 &= 3 \end{aligned}$$

The first, third, and fifth equations can be written in matrix form as

$\begin{bmatrix} 1 & 1 & 1 \\ 2 & 3 & 1 \\ 2 & 1 & 1 \end{bmatrix} \begin{bmatrix} m_{11} \\ m_{12} \\ b_1 \end{bmatrix} = \begin{bmatrix} 4 \\ 9 \\ 5 \end{bmatrix}$ and the second, fourth, and sixth equations as

$\begin{bmatrix} 1 & 1 & 1 \\ 2 & 3 & 1 \\ 2 & 1 & 1 \end{bmatrix} \begin{bmatrix} m_{21} \\ m_{22} \\ b_2 \end{bmatrix} = \begin{bmatrix} 3 \\ 5 \\ 3 \end{bmatrix}$.

The first system has the solution $m_{11} = 1, m_{12} = 2, b_1 = 1$ and the second system has the solution $m_{21} = 0, m_{22} = 1, b_2 = 2$. Thus we obtain $M = \begin{bmatrix} 1 & 2 \\ 0 & 1 \end{bmatrix}$ and $\mathbf{b} = \begin{bmatrix} 1 \\ 2 \end{bmatrix}$.

(b) As in part (a), we are led to the following two linear systems:

$\begin{bmatrix} -2 & 2 & 1 \\ 0 & 0 & 1 \\ 2 & 1 & 1 \end{bmatrix} \begin{bmatrix} m_{11} \\ m_{12} \\ b_1 \end{bmatrix} = \begin{bmatrix} -8 \\ 0 \\ 5 \end{bmatrix}$ and $\begin{bmatrix} -2 & 2 & 1 \\ 0 & 0 & 1 \\ 2 & 1 & 1 \end{bmatrix} \begin{bmatrix} m_{21} \\ m_{22} \\ b_2 \end{bmatrix} = \begin{bmatrix} 1 \\ 1 \\ 4 \end{bmatrix}$.

Solving these two linear systems leads to $M = \begin{bmatrix} 3 & -1 \\ 1 & 1 \end{bmatrix}$ and $\mathbf{b} = \begin{bmatrix} 0 \\ 1 \end{bmatrix}$.

(c) As in part (a), we are led to the following two linear systems:

$\begin{bmatrix} -2 & 1 & 1 \\ 3 & 5 & 1 \\ 1 & 0 & 1 \end{bmatrix} \begin{bmatrix} m_{11} \\ m_{12} \\ b_1 \end{bmatrix} = \begin{bmatrix} 0 \\ 5 \\ 3 \end{bmatrix}$ and $\begin{bmatrix} -2 & 1 & 1 \\ 3 & 5 & 1 \\ 1 & 0 & 1 \end{bmatrix} \begin{bmatrix} m_{21} \\ m_{22} \\ b_2 \end{bmatrix} = \begin{bmatrix} -2 \\ 2 \\ -3 \end{bmatrix}$.

Solving these two linear systems leads to $M = \begin{bmatrix} 1 & 0 \\ 0 & 1 \end{bmatrix}$ and $\mathbf{b} = \begin{bmatrix} 2 \\ -3 \end{bmatrix}$.

(d) As in part (a), we are led to the following two linear systems:

$$\begin{bmatrix} 0 & 2 & 1 \\ 2 & 2 & 1 \\ -4 & -2 & 1 \end{bmatrix}\begin{bmatrix} m_{11} \\ m_{12} \\ b_1 \end{bmatrix} = \begin{bmatrix} \frac{5}{2} \\ \frac{7}{2} \\ -\frac{7}{2} \end{bmatrix} \text{ and } \begin{bmatrix} 0 & 2 & 1 \\ 2 & 2 & 1 \\ -4 & -2 & 1 \end{bmatrix}\begin{bmatrix} m_{21} \\ m_{22} \\ b_2 \end{bmatrix} = \begin{bmatrix} -1 \\ 3 \\ -9 \end{bmatrix}.$$

Solving these two linear systems leads to $M = \begin{bmatrix} \frac{1}{2} & 1 \\ 2 & 0 \end{bmatrix}$ and $\mathbf{b} = \begin{bmatrix} \frac{1}{2} \\ -1 \end{bmatrix}$.

7. **(a)** The vertices $\mathbf{v}_1$, $\mathbf{v}_2$, and $\mathbf{v}_3$ of a triangle can be written as the convex combinations $\mathbf{v}_1 = (1)\mathbf{v}_1 + (0)\mathbf{v}_2 + (0)\mathbf{v}_3$, $\mathbf{v}_2 = (0)\mathbf{v}_1 + (1)\mathbf{v}_2 + (0)\mathbf{v}_3$, and $\mathbf{v}_3 = (0)\mathbf{v}_1 + (0)\mathbf{v}_2 + (1)\mathbf{v}_3$. In each of these cases, precisely two of the coefficients are zero and one coefficient is one.

(b) If, for example, $\mathbf{v}$ lies on the side of the triangle determined by the vectors $\mathbf{v}_1$ and $\mathbf{v}_2$ then from Exercise 6(a) we must have that $\mathbf{v} = c_1\mathbf{v}_1 + c_2\mathbf{v}_2 + (0)\mathbf{v}_3$ where $c_1 + c_2 = 1$. Thus at least one of the coefficients, in this example c_3, must equal zero.

(c) From part (b), if at least one of the coefficients in the convex combination is zero, then the vector must lie on one of the sides of the triangle. Consequently, none of the coefficients can be zero if the vector lies in the interior of the triangle.

8. **(a)** Consider the vertex $\mathbf{v}_1$ of the triangle and its opposite side determined by the vectors $\mathbf{v}_2$ and $\mathbf{v}_3$. The midpoint $\mathbf{v}_m$ of this opposite side is $\frac{(\mathbf{v}_2+\mathbf{v}_3)}{2}$ and the point on the line segment from $\mathbf{v}_1$ to $\mathbf{v}_m$ that is two-thirds of the distance to $\mathbf{v}_m$ is given by

$$\frac{1}{3}\mathbf{v}_1 + \frac{2}{3}\mathbf{v}_m = \frac{1}{3}\mathbf{v}_1 + \frac{2}{3}\left(\frac{\mathbf{v}_2 + \mathbf{v}_3}{2}\right) = \frac{1}{3}\mathbf{v}_1 + \frac{1}{3}\mathbf{v}_2 + \frac{1}{3}\mathbf{v}_3.$$

(b) $\frac{1}{3}\begin{bmatrix} 2 \\ 3 \end{bmatrix} + \frac{1}{3}\begin{bmatrix} 5 \\ 2 \end{bmatrix} + \frac{1}{3}\begin{bmatrix} 1 \\ 1 \end{bmatrix} = \frac{1}{3}\begin{bmatrix} 8 \\ 6 \end{bmatrix} = \begin{bmatrix} \frac{8}{3} \\ 2 \end{bmatrix}$

10.20 Internet Search Engines

1. **(a)** The probability transition matrix is $B = \begin{bmatrix} 0 & \frac{1}{2} & 0 \\ 0 & 0 & 1 \\ 1 & \frac{1}{2} & 0 \end{bmatrix}$. Since $I - B = \begin{bmatrix} 1 & -\frac{1}{2} & 0 \\ 0 & 1 & -1 \\ -1 & -\frac{1}{2} & 1 \end{bmatrix}$ has the reduced row echelon form $\begin{bmatrix} 1 & 0 & -\frac{1}{2} \\ 0 & 1 & -1 \\ 0 & 0 & 0 \end{bmatrix}$, the eigenspace associated with $\lambda = 1$ is span{(1,2,2)}. The normalized eigenvector in that eigenspace is $\left(\frac{1}{5}, \frac{2}{5}, \frac{2}{5}\right)$. We obtain:

- page rank $\frac{2}{5}$ for pages 2 and 3 (a tie),

- page rank $\frac{1}{5}$ for page 1.

(b) The probability transition matrix is $B = \begin{bmatrix} 0 & \frac{1}{2} & 1 \\ 0 & 0 & 0 \\ 1 & \frac{1}{2} & 0 \end{bmatrix}$. Since $I - B = \begin{bmatrix} 1 & -\frac{1}{2} & -1 \\ 0 & 1 & 0 \\ -1 & -\frac{1}{2} & 1 \end{bmatrix}$ has the reduced row echelon form $\begin{bmatrix} 1 & 0 & -1 \\ 0 & 1 & 0 \\ 0 & 0 & 0 \end{bmatrix}$, the eigenspace associated with $\lambda = 1$ is span{(1,0,1)}. The normalized eigenvector in that eigenspace is $\left(\frac{1}{2}, 0, \frac{1}{2}\right)$. We obtain:

- page rank $\frac{1}{2}$ for pages 1 and 3 (a tie),
- page rank 0 for page 2.

(Note that the matrix B in this part is not regular and the state vector sequence $\mathbf{x}^{(k)}$ does not converge. However, it could be shown that the fractional page count for this webgraph will approach $\left(\frac{1}{2}, 0, \frac{1}{2}\right)$.)

9. The following directed graph represents the given slide show:

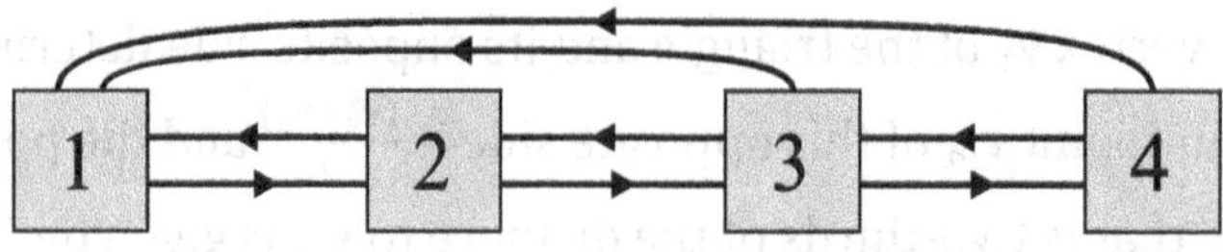

The probability transition matrix is $B = \begin{bmatrix} 0 & \frac{1}{2} & \frac{1}{3} & \frac{1}{2} \\ 1 & 0 & \frac{1}{3} & 0 \\ 0 & \frac{1}{2} & 0 & \frac{1}{2} \\ 0 & 0 & \frac{1}{3} & 0 \end{bmatrix}$. Since $I - B = \begin{bmatrix} 1 & -\frac{1}{2} & -\frac{1}{3} & -\frac{1}{2} \\ -1 & 1 & -\frac{1}{3} & 0 \\ 0 & -\frac{1}{2} & 1 & -\frac{1}{2} \\ 0 & 0 & -\frac{1}{3} & 1 \end{bmatrix}$ has the reduced row echelon form $\begin{bmatrix} 1 & 0 & 0 & -4 \\ 0 & 1 & 0 & -5 \\ 0 & 0 & 1 & -3 \\ 0 & 0 & 0 & 0 \end{bmatrix}$, the eigenspace associated with $\lambda = 1$ is span{(4,5,3,1)}. The normalized eigenvector in that eigenspace is $\left(\frac{4}{13}, \frac{5}{13}, \frac{3}{13}, \frac{1}{13}\right)$. We obtain:

- page rank $\frac{5}{13}$ for slide 2,
- page rank $\frac{4}{13}$ for slide 1,
- page rank $\frac{3}{13}$ for slide 3,
- page rank $\frac{1}{13}$ for slide 4.

10. The following directed graph represents the given slide show:

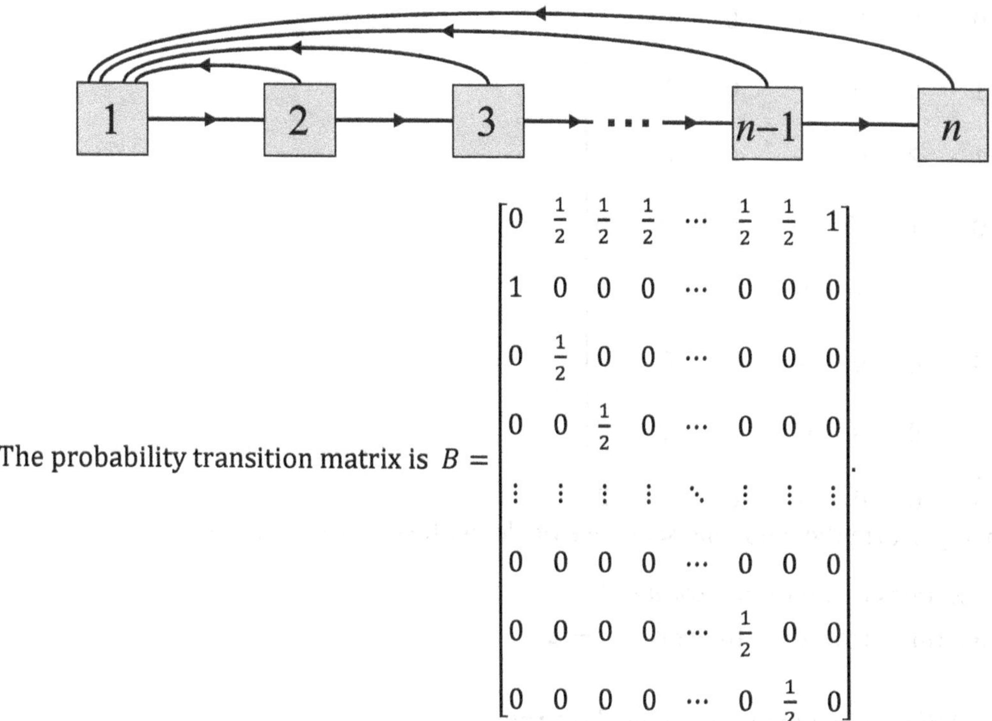

The probability transition matrix is $B = \begin{bmatrix} 0 & \frac{1}{2} & \frac{1}{2} & \frac{1}{2} & \cdots & \frac{1}{2} & \frac{1}{2} & 1 \\ 1 & 0 & 0 & 0 & \cdots & 0 & 0 & 0 \\ 0 & \frac{1}{2} & 0 & 0 & \cdots & 0 & 0 & 0 \\ 0 & 0 & \frac{1}{2} & 0 & \cdots & 0 & 0 & 0 \\ \vdots & \vdots & \vdots & \vdots & \ddots & \vdots & \vdots & \vdots \\ 0 & 0 & 0 & 0 & \cdots & 0 & 0 & 0 \\ 0 & 0 & 0 & 0 & \cdots & \frac{1}{2} & 0 & 0 \\ 0 & 0 & 0 & 0 & \cdots & 0 & \frac{1}{2} & 0 \end{bmatrix}$.

We have

$$I - B = \begin{bmatrix} 1 & -\frac{1}{2} & -\frac{1}{2} & -\frac{1}{2} & \cdots & -\frac{1}{2} & -\frac{1}{2} & -1 \\ -1 & 1 & 0 & 0 & \cdots & 0 & 0 & 0 \\ 0 & -\frac{1}{2} & 1 & 0 & \cdots & 0 & 0 & 0 \\ 0 & 0 & -\frac{1}{2} & 1 & \cdots & 0 & 0 & 0 \\ \vdots & \vdots & \vdots & \vdots & \ddots & \vdots & \vdots & \vdots \\ 0 & 0 & 0 & 0 & \cdots & 1 & 0 & 0 \\ 0 & 0 & 0 & 0 & \cdots & -\frac{1}{2} & 1 & 0 \\ 0 & 0 & 0 & 0 & \cdots & 0 & -\frac{1}{2} & 1 \end{bmatrix}.$$

Adding each of the rows 2 through n to the first row results in the zeros in the entire first row. Multiplying the second row by -1, and each of the rows 3 through n by -2 now yields

$$\begin{bmatrix} 0 & 0 & 0 & 0 & \cdots & 0 & 0 & 0 \\ 1 & -1 & 0 & 0 & \cdots & 0 & 0 & 0 \\ 0 & 1 & -2 & 0 & \cdots & 0 & 0 & 0 \\ 0 & 0 & 1 & -2 & \cdots & 0 & 0 & 0 \\ \vdots & \vdots & \vdots & \vdots & \ddots & \vdots & \vdots & \vdots \\ 0 & 0 & 0 & 0 & \cdots & -2 & 0 & 0 \\ 0 & 0 & 0 & 0 & \cdots & 1 & -2 & 0 \\ 0 & 0 & 0 & 0 & \cdots & 0 & 1 & -2 \end{bmatrix}.$$

We then perform the following sequence of elementary row operations

- add 2 times row n to the row $n-1$;
- add 2 times row $n-1$ to the row $n-2$;
- ...
- add 2 times the fourth row to the third row;
- add the third row to the second row

then interchange rows 1 and 2, 2 and 3, etc., to obtain the reduced row echelon form

$$\begin{bmatrix} 1 & 0 & 0 & 0 & \cdots & 0 & 0 & -2^{n-2} \\ 0 & 1 & 0 & 0 & \cdots & 0 & 0 & -2^{n-2} \\ 0 & 0 & 1 & 0 & \cdots & 0 & 0 & -2^{n-3} \\ 0 & 0 & 0 & 1 & \cdots & 0 & 0 & -2^{n-4} \\ \vdots & \vdots & \vdots & \vdots & \ddots & \vdots & \vdots & \vdots \\ 0 & 0 & 0 & 0 & \cdots & 1 & 0 & -2^{2} \\ 0 & 0 & 0 & 0 & \cdots & 0 & 1 & -2^{1} \\ 0 & 0 & 0 & 0 & \cdots & 0 & 0 & 0 \end{bmatrix}$$. The eigenspace associated with $\lambda = 1$ is $\text{span}\{(2^{n-2}, 2^{n-2}, 2^{n-3}, \dots, 2^2, 2^1, 2^0)\}$. We can express

$$\begin{aligned} 2^{n-2} + 2^{n-3} + \cdots + 2^1 + 2^0 &= (2^{n-2} + 2^{n-3} + \cdots + 2^1 + 2^0)\frac{2-1}{2-1} \\ &= 2^{n-1} - 2^{n-2} + 2^{n-2} - 2^{n-3} + \cdots + 2^2 - 2^2 + 2^1 - 2^0 \\ &= 2^{n-1} - 1 \end{aligned}$$

therefore the normalized eigenvector in that eigenspace is $\frac{1}{2^{n-1}+2^{n-2}-1}(2^{n-2}, 2^{n-2}, 2^{n-3}, \dots, 2^2, 2^1, 2^0)$.

The slides can be arranged in the order of the decreasing rank as follows:

- page rank $\frac{2^{n-2}}{2^{n-1}+2^{n-2}-1}$ for slides 1 and 2 (a tie),
- page rank $\frac{2^{n-3}}{2^{n-1}+2^{n-2}-1}$ for slide 3,
- ...
- page rank $\frac{2^{1}}{2^{n-1}+2^{n-2}-1}$ for slide $n-1$,
- page rank $\frac{2^{0}}{2^{n-1}+2^{n-2}-1}$ for slide n.

11. The following directed graph represents the given slide show:

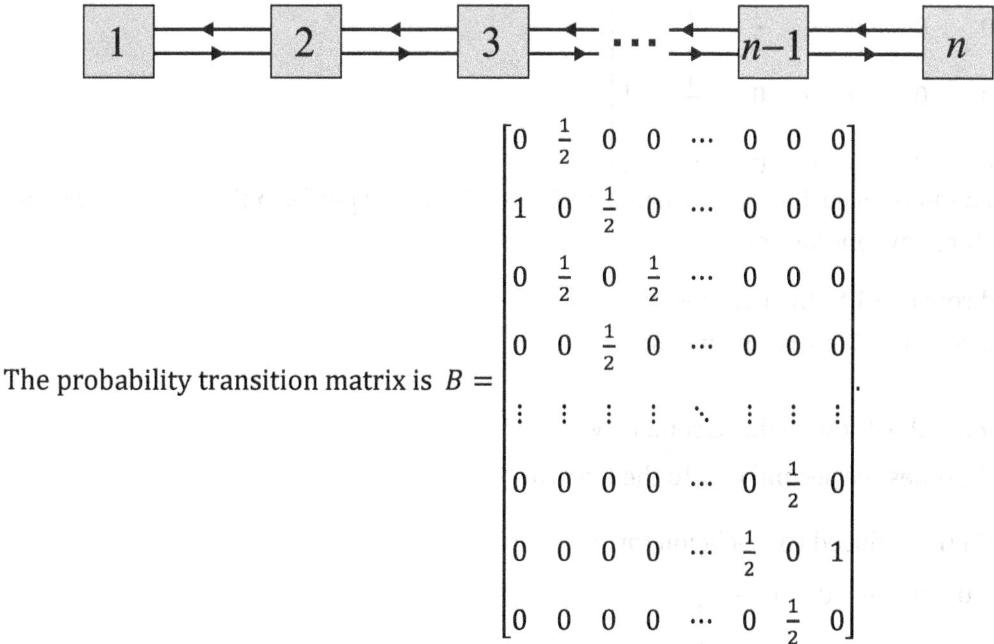

The probability transition matrix is $B = \begin{bmatrix} 0 & \frac{1}{2} & 0 & 0 & \cdots & 0 & 0 & 0 \\ 1 & 0 & \frac{1}{2} & 0 & \cdots & 0 & 0 & 0 \\ 0 & \frac{1}{2} & 0 & \frac{1}{2} & \cdots & 0 & 0 & 0 \\ 0 & 0 & \frac{1}{2} & 0 & \cdots & 0 & 0 & 0 \\ \vdots & \vdots & \vdots & \vdots & \ddots & \vdots & \vdots & \vdots \\ 0 & 0 & 0 & 0 & \cdots & 0 & \frac{1}{2} & 0 \\ 0 & 0 & 0 & 0 & \cdots & \frac{1}{2} & 0 & 1 \\ 0 & 0 & 0 & 0 & \cdots & 0 & \frac{1}{2} & 0 \end{bmatrix}$.

We have

$$I - B = \begin{bmatrix} 1 & -\frac{1}{2} & 0 & 0 & \cdots & 0 & 0 & 0 \\ -1 & 1 & -\frac{1}{2} & 0 & \cdots & 0 & 0 & 0 \\ 0 & -\frac{1}{2} & 1 & -\frac{1}{2} & \cdots & 0 & 0 & 0 \\ 0 & 0 & -\frac{1}{2} & 1 & \cdots & 0 & 0 & 0 \\ \vdots & \vdots & \vdots & \vdots & \ddots & \vdots & \vdots & \vdots \\ 0 & 0 & 0 & 0 & \cdots & 1 & -\frac{1}{2} & 0 \\ 0 & 0 & 0 & 0 & \cdots & -\frac{1}{2} & 1 & -1 \\ 0 & 0 & 0 & 0 & \cdots & 0 & -\frac{1}{2} & 1 \end{bmatrix}.$$

Adding the first row to the second row, then the resulting second row to the third, etc. until row $n-1$ is added to row n, we obtain

$$\begin{bmatrix} 1 & -\frac{1}{2} & 0 & 0 & \cdots & 0 & 0 & 0 \\ 0 & \frac{1}{2} & -\frac{1}{2} & 0 & \cdots & 0 & 0 & 0 \\ 0 & 0 & \frac{1}{2} & -\frac{1}{2} & \cdots & 0 & 0 & 0 \\ 0 & 0 & 0 & \frac{1}{2} & \cdots & 0 & 0 & 0 \\ \vdots & \vdots & \vdots & \vdots & \ddots & \vdots & \vdots & \vdots \\ 0 & 0 & 0 & 0 & \cdots & \frac{1}{2} & -\frac{1}{2} & 0 \\ 0 & 0 & 0 & 0 & \cdots & 0 & \frac{1}{2} & -1 \\ 0 & 0 & 0 & 0 & \cdots & 0 & 0 & 0 \end{bmatrix}.$$

We then multiply each row between 2 and $n-1$ by 2, and perform the following sequence of elementary row operations

- add row $n-1$ to the row $n-2$;
- add row $n-2$ to the row $n-3$;
- ...
- add the third row to the second row;
- add $\frac{1}{2}$ times the second row to the first row

to obtain the reduced row echelon form

$$\begin{bmatrix} 1 & 0 & 0 & 0 & \cdots & 0 & 0 & -1 \\ 0 & 1 & 0 & 0 & \cdots & 0 & 0 & -2 \\ 0 & 0 & 1 & 0 & \cdots & 0 & 0 & -2 \\ 0 & 0 & 0 & 1 & \cdots & 0 & 0 & -2 \\ \vdots & \vdots & \vdots & \vdots & \ddots & \vdots & \vdots & \vdots \\ 0 & 0 & 0 & 0 & \cdots & 1 & 0 & -2 \\ 0 & 0 & 0 & 0 & \cdots & 0 & 1 & -2 \\ 0 & 0 & 0 & 0 & \cdots & 0 & 0 & 0 \end{bmatrix}$$

. The eigenspace associated with $\lambda = 1$ is span$\{(1,2,2,\ldots,2,2,1)\}$ therefore the normalized eigenvector in that eigenspace is $\frac{1}{2n-2}(1,2,2,\ldots,2,2,1)$.

The slides can be arranged in the order of the decreasing rank as follows:

- page rank $\frac{2}{2n-2} = \frac{1}{n-1}$ for slides 2 through $n-1$ (an $n-2$-way tie),
- page rank $\frac{1}{2n-2}$ for slides 1 and n.

(Note that the matrix B in this problem is not regular and the state vector sequence $\mathbf{x}^{(k)}$ does not converge. However, it could be shown that the fractional page count for this webgraph will approach $\frac{1}{2n-2}(1,2,2,\ldots,2,2,1)$.)

12. The following directed graph represents the given slide show:

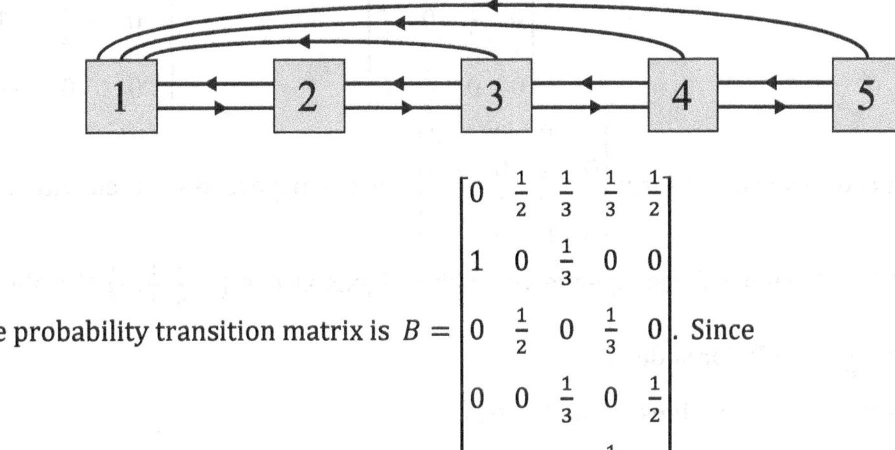

The probability transition matrix is $B = \begin{bmatrix} 0 & \frac{1}{2} & \frac{1}{3} & \frac{1}{3} & \frac{1}{2} \\ 1 & 0 & \frac{1}{3} & 0 & 0 \\ 0 & \frac{1}{2} & 0 & \frac{1}{3} & 0 \\ 0 & 0 & \frac{1}{3} & 0 & \frac{1}{2} \\ 0 & 0 & 0 & \frac{1}{3} & 0 \end{bmatrix}$. Since

$I - B = \begin{bmatrix} 1 & -\frac{1}{2} & -\frac{1}{3} & -\frac{1}{3} & -\frac{1}{2} \\ -1 & 1 & -\frac{1}{3} & 0 & 0 \\ 0 & -\frac{1}{2} & 1 & -\frac{1}{3} & 0 \\ 0 & 0 & -\frac{1}{3} & 1 & -\frac{1}{2} \\ 0 & 0 & 0 & -\frac{1}{3} & 1 \end{bmatrix}$ has the reduced row echelon form $\begin{bmatrix} 1 & 0 & 0 & 0 & -10.5 \\ 0 & 1 & 0 & 0 & -13 \\ 0 & 0 & 1 & 0 & -7.5 \\ 0 & 0 & 0 & 1 & -3 \\ 0 & 0 & 0 & 0 & 0 \end{bmatrix}$, the eigenspace associated with $\lambda = 1$ is span$\{(10.5,13,7.5,3,1)\}$. The normalized eigenvector in that eigenspace is $\left(\frac{3}{10}, \frac{13}{35}, \frac{3}{14}, \frac{3}{35}, \frac{1}{35}\right)$. We obtain:

- page rank $\frac{13}{35} \approx 0.371$ for slide 2,
- page rank $\frac{3}{10} = 0.3$ for slide 1,
- page rank $\frac{3}{14} \approx 0.214$ for slide 3,
- page rank $\frac{3}{35} \approx 0.086$ for slide 4,
- page rank $\frac{1}{35} \approx 0.029$ for slide 5.

13. The following directed graph represents the given slide show:

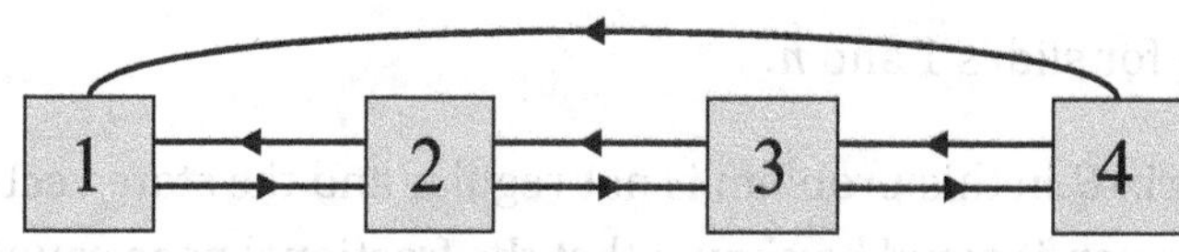

The probability transition matrix is $B = \begin{bmatrix} 0 & \frac{1}{2} & 0 & \frac{1}{2} \\ 1 & 0 & \frac{1}{2} & 0 \\ 0 & \frac{1}{2} & 0 & \frac{1}{2} \\ 0 & 0 & \frac{1}{2} & 0 \end{bmatrix}$. Since $I - B = \begin{bmatrix} 1 & -\frac{1}{2} & 0 & -\frac{1}{2} \\ -1 & 1 & -\frac{1}{2} & 0 \\ 0 & -\frac{1}{2} & 1 & -\frac{1}{2} \\ 0 & 0 & -\frac{1}{2} & 1 \end{bmatrix}$

has the reduced row echelon form $\begin{bmatrix} 1 & 0 & 0 & -2 \\ 0 & 1 & 0 & -3 \\ 0 & 0 & 1 & -2 \\ 0 & 0 & 0 & 0 \end{bmatrix}$, the eigenspace associated with $\lambda = 1$ is span{(2,3,2,1)}. The normalized eigenvector in that eigenspace is $\left(\frac{1}{4}, \frac{3}{8}, \frac{1}{4}, \frac{1}{8}\right)$. We obtain:

- page rank $\frac{3}{8} = 0.375$ for slide 2,
- page rank $\frac{1}{4} = 0.25$ for slides 1 and 3 (a tie),
- page rank $\frac{1}{8} = 0.125$ for slide 4.

(Note that the matrix B in this problem is not regular and the state vector sequence $\mathbf{x}^{(k)}$ does not converge. However, it could be shown that the fractional page count for this webgraph will approach $\left(\frac{1}{4}, \frac{3}{8}, \frac{1}{4}, \frac{1}{8}\right)$.)